Duyong A (DDP-A)

The Petroleum Geology and Resources of Malaysia

PETRONAS

Copyright © 1999, by Petroliam Nasional Berhad (PETRONAS)

All Rights Reserved.

The opinions, views, conclusions and observations expressed in each of the chapters of this book are the personal opinions, views, conclusions and observations of the individual writers. They may not necessarily reflect that of PETRONAS.

Published by
Petroliam Nasional Berhad (PETRONAS)
Exploration & Production Business Unit and
Legal & Corporate Affairs Division
Tower 1, PETRONAS Twin Towers
Kuala Lumpur City Centre
50088 Kuala Lumpur, Malaysia.
http://www.petronas.com.my

Printed by
PERCETAKAN MEGA SDN. BHD.
392-394, Jalan 25/39,
Taman Petaling, Kepong
52100 Kuala Lumpur

Colour Separation by
MEDIALINE COLOR SEPARATION SDN. BHD.
Lot 40 & 40M, Jalan 13/34A,
Kepong Entrepreneur's Park
Bt. 7, Jalan Kepong, Kepong,
52100 Kuala Lumpur

International Standard Book Number: 983-9738-10-0

The Petroleum Geology and Resources of Malaysia

Foreword by

Tan Sri Dato' Mohd Hassan Marican
President & Chief Executive, PETRONAS

I would like to congratulate the project team and all those involved for successfully completing this special publication on the petroleum geology and resources of Malaysia. The publication of this book is indeed timely especially at this important juncture of PETRONAS' journey into the new millennium, marked by the corporation's 25th Anniversary celebration on 17 August, 1999.

Signifying yet another milestone achievement not only for PETRONAS but also the Malaysian petroleum industry as a whole, "The Petroleum Geology and Resources of Malaysia" is no doubt a commendable initiative in making available for the first time in this country such a useful specialised petroleum reference book. Presenting a comprehensive and consolidated account of Malaysia's petroleum geology, this book offers valuable data and information which, I believe, will encourage deeper geoscientific studies and research to help further promote the development of the country's petroleum resources and reserves base.

Indeed, the Malaysian petroleum industry has undergone tremendous transformation since its beginning in 1910 when the first commercial discovery of oil was made in Miri, Sarawak. Although the country's petroleum exploration history dates back more than a century ago, it was only the last two and a half decades that saw the accelerated growth and expansion of the industry. The promulgation of the Petroleum Development Act and the incorporation of PETRONAS as the national oil company in 1974 had provided the required stimulus for the systematic development and continual growth of the industry with rapid intensification of exploration and development activities domestically.

As a result, the country's oil and gas production grew significantly from 81,000 barrels of oil per day (bopd) and 272 million standard cubic feet of gas per day (mmscf/d) in 1974 to 630,000 bopd and 5,261 mmscf/d in 1998. Contributing to this growth is the rapid evolution of PETRONAS from being a mere custodian and manager of the country's petroleum resources to become a fully integrated international petroleum corporation today, continuously providing the catalyst needed for the development of the domestic petroleum industry while actively expanding its global business activities to augment Malaysia's oil and gas reserves.

To complement this effort in ensuring the continual growth of the petroleum industry in Malaysia, "The Petroleum Geology and Resources of Malaysia" will serve as an effective and informative avenue to promote greater interest and investment as well as geoscientific research in our domestic exploration, development and production operations.

Finally, I would like to take this opportunity to commend the project team for their commitment and untiring efforts in making this book project a success. Syabas!

TAN SRI DATO' MOHD HASSAN MARICAN

Foreword by

Dato' (Dr) Mohamad Idris Mansor
Senior Vice President
Exploration And Production Business PETRONAS

"The Petroleum Geology and Resources of Malaysia" is a step forward taken by PETRONAS to put together for the first time, using both published and unpublished data and information, a comprehensive account on the petroleum geology of the country's mainly offshore Tertiary basins and petroleum resources under one cover. Malaysia indeed has a wealth of geoscientific data and information gathered from more than 100 years of petroleum exploration and 90 years of oil production, 40 years of which are in the offshore areas.

Since its formation 25 years ago, PETRONAS has always encouraged the release of offshore geology and petroleum engineering data, including information on Malaysia's oil and gas fields, through the support of and participation in local and international conferences as well as regional bodies such as ASEAN Council on Petroleum (ASCOPE) and Committee for Coordination of Joint Prospecting for Mineral Resources in Asian Offshore Areas (CCOP). The publication of this book represents a further manifestation of PETRONAS' support to the geoscientific community in promoting petroleum studies and research. The ultimate objective is to encourage new exploration and development investments and activities to increase the country's petroleum resources and reserves base.

I hope that "The Petroleum Geology and Resources of Malaysia" will serve as a valuable reference book for the practising petroleum geologists and engineers in the upstream petroleum business contemplating to operate or currently operating in Malaysia. The book is also aimed at the geoscientific community at large, both local and international, and should, thus, find its rightful place in earth science and petroleum engineering libraries of universities as a useful source of reference for academicians and research students.

Finally, I would like to thank all those who have contributed in one way or another towards making this book project a success. I would like to specially thank PETRONAS Research and Scientific Services Sdn Bhd for its provision of full time contributors to write the articles; Legal and Corporate Affairs Division for managing the publication of the book; and staff of the Exploration and Production Business for their contributions. I would also like to take this opportunity to congratulate all the authors of the book for their diligence and patience in completing their respective chapters and for the scientific ideas expressed therein.

DATO' (DR) MOHAMAD IDRIS MANSOR

INTRODUCTION AND ACKNOWLEDGEMENTS

There has always been a need in the geoscientific community for a comprehensive and coherent account of the geology of Malaysia, covering both onshore and offshore areas. "The Petroleum Geology and Resources of Malaysia" is an effort towards meeting that need, although the emphasis is on petroleum resources and the focus on the Tertiary sedimentary basins.

The first comprehensive geological accounts of Malaysia were given separately in Liechti et al. (1960) and in Gobbett and Hutchison (1973) for Sarawak/Sabah and Peninsular Malaysia, respectively. However, these early and later publications, mainly by the Geological Survey of Malaysia and staff of local universities, were on onshore geology, though certain offshore areas were covered (e.g. Haile et al., 1964). The oil industry kicked off with the 1910 Miri-1 oil discovery in eastern Malaysia, and it was not surprising that early published papers on petroleum geology were limited to this region (e.g. Schaub and Jackson, 1958). The next major discoveries were in the early 60s in offshore Sarawak, late 60s in offshore Peninsular Malaysia, and early 70s in offshore Sabah. The offshore geology of Malaysia became progressively better known through intensive seismic data aquisition and exploration drilling activities. However, there was still a dearth of published papers of regional extent on offshore, though a few papers on offshore geology and stratigraphy began to appear in various publications (e.g. Bell and Jessop, 1974; Murphy, 1975; Schaar, 1976; Armitage and Viotti, 1977; Ho, 1978; Whittle and Short, 1978; the United Nations (ESCAP,1978); and CCOP (Committee for Co-ordination of Joint Prospecting for Mineral Resources in Asian Offshore Areas; Chung et al., 1977). The 'jump start' came with the launching of the petroleum geology seminar (now an annual conference) in 1977 organized by the Geological Society of Malaysia (GSM) and supported by contributions of both finance and technical papers from PETRONAS, the oil companies, notably Shell (Sarawak Shell Bhd., Sabah Shell Petroleum Co. Ltd.) and Esso (Esso Production Malaysia Inc.), and the petroleum industry service companies. Many of the presented papers have been published in the bulletins of the GSM from 1980 onwards. The next important milestone was the establishment of ASCOPE (Asean Council on Petroleum), of which PETRONAS is a member, and the publications on several offshore basins of Southeast Asia (e.g. ASCOPE, 1981, 1984, 1985). In 1986, over 3000 km of deepwater (to 2900m depth) multichannel reflection seismic lines, together with magnetic and gravity measurements, were acquired on the continental margin of Sabah, including the Sabah Trough, by PETRONAS-BGR (Germany Federal Institute for Geosciences and Natural Resources). The results of this important far-reaching deepwater survey were published in Hinz et al. (1989). The latest addition to the regional data on the offshore areas, with PETRONAS participation, came from CCOP (1991) publication of the Total Sedimentary Isopach Map, Offshore East Asia on a scale 1:4 million.

This book is an effort to describe the petroleum geology of Malaysia in a comprehensive and consolidated approach, making use of both published and unpublished material. Although it is focused on the offshore Tertiary basins, the pre-Tertiary and Tertiary geological settings covering onshore and offshore Peninsular Malaysia, Sarawak and Sabah are also given. The book may be considered the first attempt in putting together under one publication the geology, albeit with emphasis on the Tertiary and petroleum resources, of the entire country both onshore and offshore. It should fill the need of the petroleum industry and geoscientific community for the latest comprehensive account on the geology of Malaysia. It should also serve as a reference text, not only on the petroleum geology and resources of Malaysia for local and international readers, but also on PETRONAS unique and successful upstream petroleum business arrangement.

It is imperative that exploration and production (E&P) activities continue to contribute to the economic wealth of Malaysia. Although geoscience plays an undisputed role in the search and development of petroleum resources, the economics of E&P ventures are as important. The book consists essentially of 5 parts with an account of the pre-Tertiary petroleum potential (Chapter 26) and the scope of petroleum geoscience research (Chapter 27). Part 1 consists of 3 chapters including a brief introduction to Malaysia. Chapter 2 describes the formation of PETRONAS, the introduction of the Production Sharing Contract, its evolutionary changes, and the ensuing successes in attracting E&P business investments. Chapter 3 gives an overview of the country's petroleum resources, the phenomenal increases in both oil and gas resource/reserves base, and in the oil and gas production levels. Active exploration activities have indeed contributed to increased geoscientific knowledge, and have been sustained through several commercial oil and gas discoveries.

Part 2 also consists of 3 chapters which give comprehensive accounts of the geological evolution of Malaysia and its position in regional tectonics, the various sedimentary basins, types and structural styles and their possible origins, and on the complex Tertiary stratigraphic schemes, and the role and importance of the applications of biostratigraphy and sequence stratigraphy in improving the correlation of key horizons regionally across basins as well as within oil and gas fields. Parts 3 to 5 deal in depth the petroleum geology, with considerable emphasis on petroleum geochemistry, of the Tertiary basins and provinces of the 3 regions in Malaysia - Peninsular Malaysia, Sarawak and Sabah. Each of these parts begin with the geological setting of the region and conclude with an account of the petroleum resources and the potential for new discoveries.

Figure 1 shows the location and chapter numbers of the Tertiary sedimentary basins and provinces of Malaysia that are described in this book. Readers will note the uneven length and varied style of the contributions in the different chapters. There is good reason for this variation. In some of the basins, e.g. Malay Basin, considerable data are available, both published and unpublished, in contrast to others, e.g. Southeast Sabah Basin. In others, like the West Baram Delta, detailed and illustrated descriptions of exposed onshore outcrops are given emphasis as the province contains not only the first oilfield of Malaysia (Miri) but also the latest onshore oilfield (Asam Paya). The book illustrates that there is sufficient variety of basin types, tectonic styles, traps, reservoir facies, oil and gas fields, etc. in Malaysia to wet the appetite of the readers to seek further information, to carry out additional studies or to invest in new E&P business ventures.

Many people and organizations have helped make this book a reality. Dato' (Dr.) Mohamad Idris Mansor, Senior Vice President, E&P Business, initiated and spearheaded the project with a firm belief that PETRONAS staff can do it. In late 1996, PETRONAS Management Committee approved and provided a budget for the work. PETRONAS Research and Scientific Services Sdn. Bhd. (**PRSS**) released staff to write the bulk of the chapters. Petroleum Management Unit (**PMU**) provided access to unpublished reports and data, mostly submitted by the oil companies/PSCs, and also released staff to contribute chapters on petroleum resources. PETRONAS Carigali Sdn. Bhd. (**Carigali**) assigned staff to contribute selected field data and provided almost all the draughting work. PETRONAS Legal and Corporate Affairs Division (**LCAD**) managed the publication of the book and tirelessly prepared the layouts of all the chapters and design works including the cover. The **Geological Survey of Malaysia** allowed the use of their library and granted permission for the publication of onshore gravity data. The authors are thanked for their diligence and patience in completing their chapters. Their brief biodata are given in the back pages. All chapters are authored by PETRONAS staff, except for Chapter 13 (West Baram Delta) in which Shell and University of Malaya staff also participated.

The acknowledgements would not be complete without the mention of the following people who have contributed to the book's success:

PRSS : The support of Mohamed Zohari Mohamed Shaharun, Vice President, Technology Resource Management Division (TRM) and PRSS Chairman, and Dr Zainal Abidin Hj. Kasim; the technical inputs and assistance from Dato' Dr. Khalid Ngah (ex-PETRONAS), Mohd Kassim Kinchu (ex-PETRONAS) and Abdul Jalil Mohamad at the early stage of project; provision of digital data by Ramly Khairuddin and Ku Izhar Ku Baharudin; discussion on heat flow and Low Resistivity and Low Contrast (LRLC) with Wan Ismail Wan Yusoff and Ahmad Sharby Abdul Hamid respectively; administrative assistance was provided by Dr. Malcolm Basil Ralph Pereira and Awalludin Harun, and IT matters by Mohd Khir Abdul Jalil, Hairol Adenan Kasim, Mohd Rozaidee Harun and Sri Rohati Mat; library resource centre staff Ng Yoke Sum also assisted; Ahmad Khairiri Harun assisted with photographic work for Chapter 16.

PMU : Akbar Tajudin Abdul Wahab, Hoh Swee Chee and Abu Bakar Mohamed for support and access to unpublished reports including petroleum resource data; Hoh Swee Chee and Abu Bakar Mohamed also assigned dedicated staff to contribute to chapters 2, 3, 11, 16, 20, 25, with the support of Abdul Manaf Mohamad (in Carigali), Barney Mahendran S. Ganesan and Chua Beng Yap. The following assisted in retrieving and locating data : Mohd Badri Haji Hassan, Andrew Sypkerman, Abdul Rahman Eusoff, Abdul Gani Razak, Sarak Ali Gulam Rasul, Hasana Hussein and Mohd Tahir Ali; Chua Bee Lye for retrieving and printing out maps with well and seismic data; Mohd Jefri Jamaludin for draughting works; Idrus Mohd Shuhud for printing out seismic sections for Chapter 23; and Mohd Badri Haji Hassan and Samberah Shariff for administrative staff support and assigning Rashidah A. Aziz for duration of project.

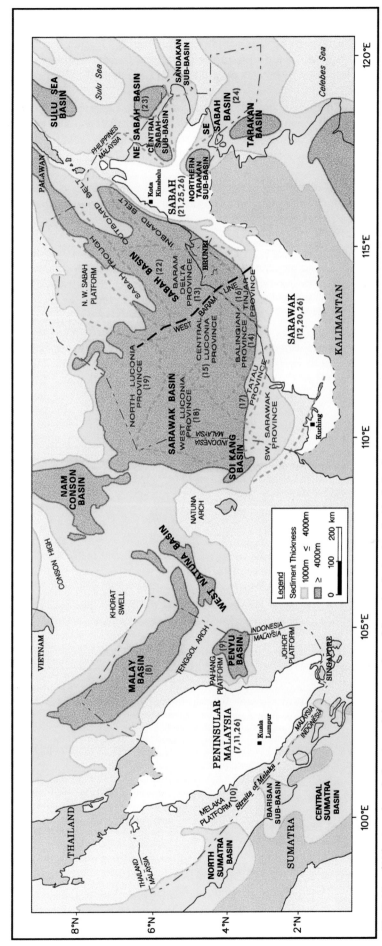

Fig. 1 Map of Malaysia to show the location and chapter numbers of the various basins described in the book.

Carigali: Ahmad Said Fazal Mohamad (now in Group Planning), Hashim Wahir, Effendy Cheng Abdullah, Tai Say Ann and Khairul Anuar Husin for support; Sy Kharil Anuar Sy Othman and Mohd Hashim Abas for assigning Raja Azmir Shah Raja Aznan full time to the project for the draughting work; Noor Hayati Mior Bahari also assisted in draughting; Rosli Mohd Noor and Michael Gattrall provided oil and gas field data.

LCAD: Mohd Azhar Osman Khairuddin and Eileen Chua for support; Kristine Low in managing all aspects of the publication and assigned staff to write the first chapter; Mohd Ishak Mohd Din prepared the layouts and designed the cover and liaised with printers; Lili Suryani Mohd Idris tirelessly prepared the dummy copies of all chapters. Legal advice was given by Noorudin Abdullah, Lai Hoi Kean and Aizan Azrina Rosli.

The Director-General of the **Geological Survey of Malaysia** (now known as Minerals and Geoscience Department Malaysia), Chen Shick Pei, for the use of the library and permission for the publication of onshore gravity data in Chapter 7.

Rusli Jusoh (in Carigali), Jamaludin Obeng (ex-PETRONAS), Baharul Aizal Baharuddin and Maggie Boey of Senior Vice President's Office for administrative assistance; the full time support staff Raja Azmir Shah Raja Aznan for almost all the draughting work and Rashidah A.Aziz for secretarial assistance, including data search, and for both their utmost patience and dedication during the entire duration of the project.

Professor R.C. Selley of Imperial College of Science, Technology and Medicine, University of London, for reviewing the entire manuscript of the book, preparing the Index and his visits to Kuala Lumpur for hands-on advice on the project; to Dr. C.S. Hutchison, formerly of University of Malaya, and Denis N.K. Tan, Sarawak Shell Bhd., for the review of several chapters; and to both Professor Selley and Dr B.K.Tan, formerly of University of Malaya, for editorial consultancy and proof-reading.

Leong Khee Meng

REFERENCES

Armitage, J.H. and Viotti, C., 1977. Stratigraphic nomenclature southern end Malay Basin. Proceeding Indonesian Petroleum Association 6th Annual Convention, Jakarta, May 1977, I, 69-94.

ASCOPE, 1981. Tertiary sedimentary basins of the Gulf of Thailand and South China Sea : stratigraphy, structure and hydrocarbon occurrences. Asean Council on Petroleum (ASCOPE).

ASCOPE, 1984. Tertiary sedimentary basins of the Southwest Sulu Sea, Makassar Strait and Java Sea : stratigraphy, structure and hydrocarbon occurrences. Technical Paper TP/3 Asean Council on Petroleum (ASCOPE).

ASCOPE, 1985. The stratigraphic correlation study of the Andaman Sea-Strait of Malacca. Technical Paper TP/4 Asean Council on Petroleum (ASCOPE).

Bell, R.M. and Jessop, R.G.C., 1974. Exploration and Geology of the West Sulu Basin, Philippines. Australian Petroleum Exploration Association Journal, 1, 21-28.

CCOP, 1991. Total Sedimentary Isopach Map, Offshore East Asia, Scale 1:4,000,000.

Chung, S.K., Gan, A.S., Leong, K.M. and Kho, C.H., 1977. Ten years of petroleum exploration in Malaysia. United Nations ECAFE, CCOP Technical Bulletin, 11, 111-142.

ESCAP, 1978. ESCAP Atlas of Stratigraphy I. Burma, Malaysia, Thailand, Indonesia, Philippines. United Nations, Mineral Resources Development Series, 44.

Gobbett, D.J. and Hutchison, C.S. (Editors), 1973. "Geology of the Malay Peninsula". John Wiley, New York, 438 pp.

Haile, N.S., Keij, A.J. and Pimm, A.C., 1964. Preliminary report on the oceanography cruise of HMS Dampier in the South China Sea. British Borneo Geological Survey Annual Report for 1963, 119-145.

Hinz, K., Fritsch, J., Kempter, E.H.K., Mohammad, A.M., Meyer, J., Mohamed, D., Vosberg, H., Weber, J. and Benavidez, J., 1989. Thrust tectonics along the North-western Continental Margin of Sabah/Borneo. Geologische Rundschau, Band 78, Heft 3, 705-730.

Ho, K.F., 1978. Stratigraphic framework for oil exploration in Sarawak. Bulletin of the Geological Society of Malaysia, 10, 1-14.

Liechti, P., Roe, F.W. and Haile, N.S., 1960. The geology of Sarawak, Brunei, and the western part of North Borneo. British Borneo Geological Survey Department, Bulletin 3,

Murphy, R.W., 1975. Tertiary basins of Southeast Asia. Southeast Asia Petroleum Exploration Society (SEAPEX) Proceedings, 2, 1-36.

Schaar, G., 1976. The occurrence of hydrocarbons in overpressured reservoirs of the Baram Delta (offshore Sarawak, Malaysia). Proceedings Indonesian Petroleum Association 5th Annual Convention, Jakarta, June 1976, II, 163-169.

Schaub, H.P. and Jackson, A., 1958. The northwestern oil basin of Borneo. In : Lewis G. Weeks (Editor), Habitat of Oil: A Symposium. American Association of Petroleum Geologists, 1330-1336.

Whittle, A.P. and Short, G.A., 1978. The petroleum geology of the Tembungo Field, East Malaysia. Southeast Asia Petroleum Exploration (SEAPEX) Conference Paper.

TABLE OF CONTENTS

FOREWORDS	(iv)
INTRODUCTION AND ACKNOWLEDGEMENTS	(vii)

PART 1 MALAYSIA AND THE UPSTREAM PETROLEUM INDUSTRY

Chapter 1 : Malaysia in Brief — 1
MUNIRAH KHAIRUDDIN AND MAZLAN B. HJ. MADON

Chapter 2 : Malaysia E&P Business — 17
HO WANG KIN

Chapter 3 : Overview of Petroleum Resources of Malaysia — 33
ABDUL JALIL BIN ZAINUL, ROSNAN BIN MISMAN AND ABDUL JALIL BIN ALI

PART 2 REGIONAL TECTONICS AND SEDIMENTARY BASINS OF MALAYSIA

Chapter 4 : Plate Tectonic Elements and the Evolution of Southeast Asia — 59
MAZLAN B. HJ. MADON

Chapter 5 : Basin Types, Tectono-Stratigraphic Provinces, Structural Styles — 77
MAZLAN B. HJ. MADON

Chapter 6 : Tertiary Stratigraphy and Correlation Schemes, Sequence Stratigraphy — 113
MAZLAN B. HJ. MADON, RASHIDAH BT ABD. KARIM AND ROBERT WONG HIN FATT

PART 3 PENINSULAR MALAYSIA SEDIMENTARY BASINS AND PETROLEUM RESOURCES

Chapter 7 : Geological Setting of Peninsular Malaysia — 139
H.D. TJIA

Chapter 8 : Malay Basin — 171
MAZLAN B. HJ. MADON, PETER ABOLINS, MOHAMMAD JAMAAL BIN HOESNI AND MANSOR BIN AHMAD

Chapter 9 : **Penyu Basin**	219
MAZLAN B. HJ. MADON AND AZLINA ANUAR	

Chapter 10: **Basins in the Straits of Melaka**	235
MAZLAN B. HJ. MADON AND MANSOR BIN AHMAD	

Chapter 11: **Petroleum Resources, Peninsular Malaysia**	251
ROBERT WONG HIN FATT	

PART 4 SARAWAK SEDIMENTARY BASINS AND PETROLEUM RESOURCES

Chapter 12: **Geological Setting of Sarawak**	273
MAZLAN B. HJ. MADON	

Chapter 13: **West Baram Delta**	291
DENIS N.K. TAN, ABDUL HADI ABD. RAHMAN, AZLINA ANUAR, BONIFACE BAIT AND CHOW KOK THO	

Chapter 14: **Balingian Province**	343
MAZLAN B. HJ. MADON AND PETER ABOLINS	

Chapter 15: **Central Luconia Province**	369
MOHAMMAD YAMIN BIN ALI AND PETER ABOLINS	

Chapter 16: **Tinjar Province**	393
MOHD IDRUS BIN ISMAIL AND REDZUAN BIN ABU HASSAN	

Chapter 17: **Tatau Province**	411
MAZLAN B. HJ. MADON AND REDZUAN BIN ABU HASSAN	

Chapter 18: **West Luconia Province**	427
MAZLAN B. HJ. MADON AND REDZUAN BIN ABU HASSAN	

Chapter 19: **North Luconia Province**	441
MAZLAN B. HJ. MADON	

Chapter 20 : **Petroleum Resources, Sarawak**	455
OTHMAN ALI BIN MAHMUD AND SALAHUDDIN BIN SALEH	

PART 5 SABAH SEDIMENTARY BASINS AND PETROLEUM RESOURCES

Chapter 21: Geological Setting of Sabah — 473
LEONG KHEE MENG

Chapter 22: Sabah Basin — 499
MAZLAN B. HJ. MADON, LEONG KHEE MENG AND AZLINA ANUAR

Chapter 23: Northeast Sabah Basin — 543
LEONG KHEE MENG AND AZLINA ANUAR

Chapter 24: Southeast Sabah Basin — 571
LEONG KHEE MENG AND AZLINA ANUAR

Chapter 25: Petroleum Resources, Sabah — 591
MOHD IDRUS BIN ISMAIL

PART 6 MALAYSIA PRE-TERTIARY

Chapter 26: Malaysia Pre-Tertiary Hydrocarbon Potential — 603
H.D. TJIA

PART 7 GEOSCIENCE RESEARCH

Chapter 27: Scope for Further Geoscience Research — 637
MAZLAN B. HJ. MADON

BIODATA OF AUTHORS — 649

GLOSSARY — 653

INDEX — 657

Part 1

MALAYSIA AND THE UPSTREAM PETROLEUM INDUSTRY

Chapter 1

Malaysia in Brief

*Munirah Khairuddin and
Mazlan B. Hj. Madon*

INTRODUCTION

Malaysia is a developing country that is rapidly transforming itself into a newly-industrialized country. It is located at the centre of Southeast Asia, and is strategically positioned as a major commercial link between the East and the West. Several decades of political stability and rapid economic growth have made Malaysia one of the most buoyant and wealthy countries in the region. The Malaysian government has effectively implemented fiscal and monetary policies that are conducive for economic growth. It is vigorously undertaking core infrastructure projects to improve road, rail, and air links to the business capital, Kuala Lumpur. Since the early 1980s its economy which was commodity-dependent during the 60s and 70s, has since diversified to include light industrial manufacturing and heavy industries. Today, the country boasts state-of-the-art infrastructure and communications network, and the capital city, Kuala Lumpur, has become one of Asia's prime urban attractions for business. The country is now gearing towards a new phase of development based on the multimedia and information technology. This chapter gives a brief introduction to Malaysia, its culture, economy, industries, and infrastructure, and a glimpse of the new visionary Malaysia in the 21st century.

GEOGRAPHY

Malaysia is located near the equator between 1°-7° N and 100°-119° E (Fig. 1.1). It has a total land area of about 330,400 km^2, straddled 650 km across the South China Sea. The country has two distinct parts. At the tip of mainland Southeast Asia is Peninsular Malaysia, which comprises 11 states, while on Borneo there are two states, Sabah and Sarawak. Peninsular Malaysia has a mountainous spine known as the Main Range (Banjaran Titiwangsa). The highest peak is Gunung Tahan (2187 m), while the longest river is Sungai Pahang (475 m). Sarawak has extensive lowland coastal plains with abundant peat swamps. These pass inland into rugged mountain ranges drained by massive river systems. The longest river in Sarawak is the Rajang (563 km). Among the highest peaks in Sarawak is Gunung Mulu (2371 m) which also has one of the largest natural caves in the world. Sabah also has prominent mountains, the Crocker Range, which culminates in Gunung Kinabalu (4101 m), the highest peak in Southeast Asia.

About 80% of Malaysia, particularly the mountainous areas, is covered by equatorial rainforest. Large parts of the forest have been conserved as national parks which are popular eco-tourism sites (Fig. 1.2). Taman Negara, which includes Gunung Tahan, is Malaysia's oldest and largest protected tract of primary rainforest covering 4343 km^2, sprawled over parts of Kelantan, Pahang and Terengganu. Sarawak and Sabah are also famous for their national parks, such as the Mulu National Park, northeastern Sarawak, and the Kinabalu Park of Sabah, which includes Gunung Kinabalu. Most of Malaysia's population is concentrated on the coastal lowlands, which also have large areas that are covered with rice, oil palm, rubber and coconut plantations.

Malaysia's offshore areas total 332,300 km^2. These include parts of the Straits of Melaka, South China Sea, Sulu Sea and Celebes (Sulawesi) Sea (Fig. 1.1). About 60% of these offshore areas have water depths of less than 200m. Petroleum is produced from the sedimentary basins offshore Terengganu, Sarawak and northwestern Sabah, in water depths of between 25 and 180 m. The sedimentary basins are described in Chapter 5. The deeper water areas in the northern parts of offshore Sarawak and Sabah are now being actively explored for petroleum.

Malaysia lies entirely in the equatorial zone with fairly uniform temperatures throughout the year. The average daily temperature is about

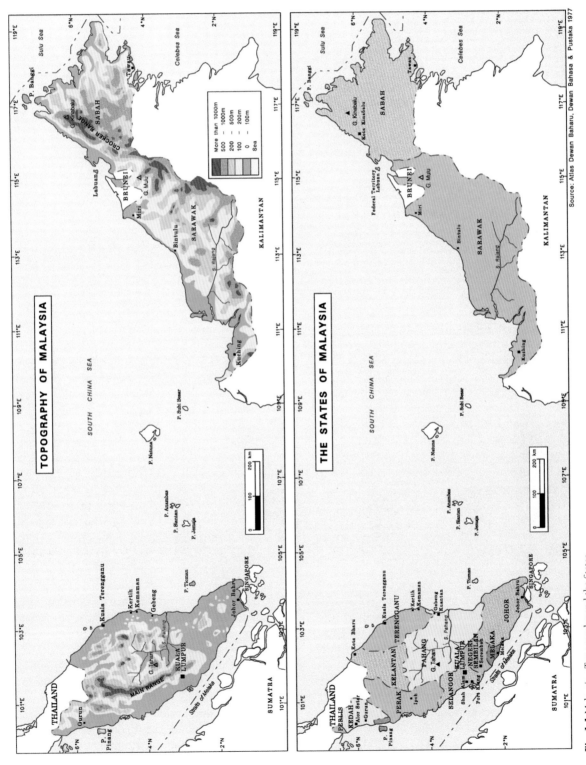

Fig. 1.1 Malaysia : Topography and the States.

Fig. 1.2 Rain forest.

26° C, with a diurnal temperature range of about 7°C. Humidity is high, at about 80%, because of the high temperature and consequently high evaporation rate. Rainfall is relatively heavy throughout the year, averaging 200-250 cm annually (Fig 1.3). The variation in temperature and rainfall is due to topographical relief and the effect of tropical monsoons. During the months of October to March the NE Monsoon brings extra rain to the east coast of Peninsular Malaysia, while the SW Monsoon affects the west coast of Peninsular Malaysia during April to September, but with less rainfall. Malaysia is generally not affected by earthquakes.

Malaysia is a harmonious multi-racial country with a population of about 22 million people (Fig. 1.4). Peninsular Malaysia has an average population density of about 83 persons per km^2, while Sabah and Sarawak have about 10-13 persons per km^2. The three largest ethnic

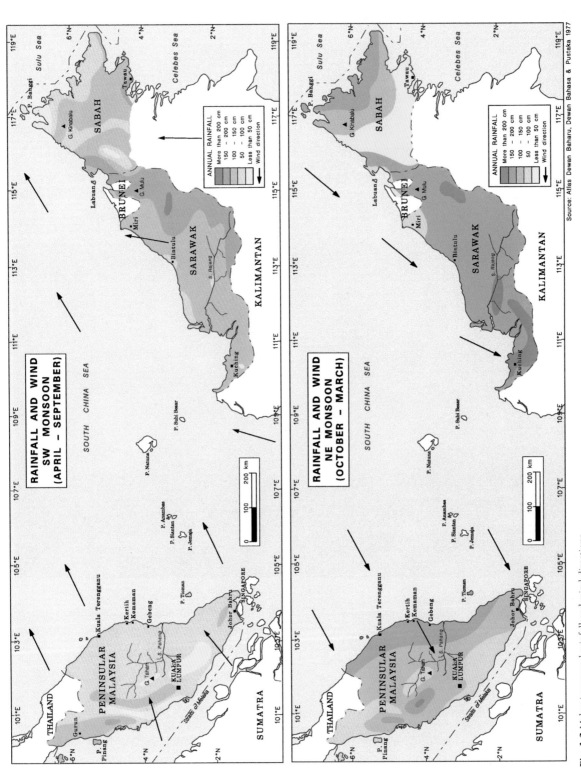

Fig. 1.3 Malaysia: annual rainfall and wind directions.

groups in Peninsular Malaysia are Malay (59%), Chinese (32%), and Indian (8%). The indigenous people of Peninsular Malaysia are known as the Orang Asli (Aborigines). In Sarawak the largest ethnic group is the Iban, while in Sabah the majority is Kadazandusun. There are many other indigenous peoples, among others the Bidayuh, Melanau, Kenyah, Kelabit, Kedayan and Kayan of Sarawak, and the Bajau and Murut of Sabah. Religious affinities are just as diverse. Although Islam is the official religion, the constitution guarantees freedom of worship. Hence, other major religions including Buddhism, Christianity, and Hinduism, are widely practised by the various ethnic groups.

The Malay language is the official language of Malaysia, but the various ethnic groups have their own languages, including Tamil for the Indians and several major dialects for the Chinese. English is widely spoken by most Malaysians as a second language and is also the preferred language for trade and commerce. There are national and regional daily newspapers in all the four main languages, plus the indigenous languages of Sabah and Sarawak. Malaysia has an educated, highly skilled workforce as a result of a very comprehensive education system, from schools to universities. There are now 10 universities, including the International Islamic University, which attracts students of many nationalities. There is also a widespread and expanding system of private schools, colleges, and universities, many of which are affiliated with overseas institutions.

Malaysia's capital is Kuala Lumpur (Fig 1.5). This city of about 1.4 million people is fast becoming an international commercial centre, with many multinational companies locating their regional offices here. The most outstanding landmark in the city is the 88-storey PETRONAS Twin Towers which, at 452 m, are the tallest buildings in the world (Fig. 1.6). The towers are part of the Kuala Lumpur City Centre (KLCC), a 'city-within-a-city', sprawled across 40 hectares of land. It ranks among the largest real-estate developments in the world. Situated strategically in the 'golden triangle' of Kuala Lumpur, KLCC has commercial, residential, entertainment, convention, and exhibition facilities. Besides providing office space for PETRONAS employees, the towers have a world-class 865-seat concert hall. Adjacent to the towers is a popular shopping mall which also houses an interactive petroleum discovery centre and an art gallery.

POLITICS AND GOVERNMENT

The Federation of Malaysia comprises 13 states and two federal territories (Kuala Lumpur, the capital city, and Labuan, an island off the coast of Sabah). Nine of the states have a hereditary ruler (the Sultan). Malaysia practises constitutional monarchy, with the Yang Di-Pertuan Agong (King) as the head of state, who is elected every five years among the Malay rulers of the nine states of the federation. The King acts upon the advice of the Government which is led by the Prime Minister as the executive head of state appointed by the King. Malaysia has a bicameral parliament comprising a 69-member Senate and 177-member House of

Fig. 1.4 Malaysia's diverse racial population.

Fig 1.5 Kuala Lumpur, capital of Malaysia.

Representatives. General elections are held every five years.

Malaysia's political stability has been a major factor in its rapid economic development. For over four decades the country was led by a coalition government headed by the Malay-majority party, the United Malay National Organization (UMNO), together with the Malaysian Chinese Association (MCA) and the Malaysian Indian Congress (MIC). In Sarawak the Parti Bersatu Bumiputera Sarawak (PBB) represents the largest indigenous group, and is part of the ruling national coalition.

INFRASTRUCTURE AND COMMUNICATIONS

Electricity

Malaysia has a well-developed infrastructure and communication network. The excellent infrastructure for electricity supply is one of the key factors drawing foreign investments into Malaysia's manufacturing sector. The industrial sector is the largest user of electricity in Peninsular Malaysia, consuming about 53% of the country's power supply. Malaysia's electricity and other power needs are supplied by privatised state corporations. In Peninsular Malaysia, Tenaga Nasional Berhad (TNB) has been providing electricity for over four decades. Apart from TNB, independent power producers have also started operations since 1995. In Sabah and Sarawak, electricity supply is provided by the Sabah Electricity Board and Sarawak Electricity Supply Corporation, respectively.

Telecommunications

The main provider of telecommunications services is Telekom Malaysia Berhad, which has about 3 million subscribers nationwide. This translates to about 15 telephones per 100 people. Telekom Malaysia also connects the country with the rest of the world by submarine cable systems and satellites. There are five international satellite stations in Kuala Lumpur, and 12 earth stations around the country. To facilitate the overseas and domestic needs of telecommunication services, the Government has erected a 420-m tall communications tower (the KL Tower) in the heart of the capital. This gigantic concrete structure is the third tallest

Fig 1.6 PETRONAS Twin Towers, night view.

concrete structure in the world, and is part of the Government's infrastructure plan costing well over RM250 million. International Direct Dialing (IDD) facilities are available to over 200 countries. Apart from Telekom Malaysia, there are also many other telecommunication companies offering competitive telecommunication services. Telekom Malaysia is also one of the several internet service providers in Malaysia

Malaysia launched its first communication satellite, MEASAT 1, in January 1996 from Kourou in the French Guiana. MEASAT, or Malaysia East Asia Satellite, represents the new-generation satellites specifically designed to meet the increasing demands of satellite broadcasting and communication technology in the region, and is estimated to serve more than 1 billion people.

Transport

Malaysia has a well-developed transportation infrastructure, including land, sea, and air transport systems. Malaysian roads are among the best in the region. Almost 70% of the roads on Peninsular Malaysia are paved. Cross-country journeys are made convenient via the 845-km long North-South Expressway, stretching from the Malaysia-Thai border to Johor Bahru near the border with Singapore. The East-West Highway cuts across the Main Range to link major cities along the east coast with those on the west coast. Buses and coaches provide inter-state services between major cities. Trunk roads link the major cities on the northwestern coast of Sarawak and Sabah, while roads over the Crocker Range link the coastal towns of eastern Sabah.

The major cities on Peninsular Malaysia are also linked by a railway network. The now privatised railway services have been upgraded and modernised through double tracking and electrification, and the introduction of the tilting train, and a new signalling and communication system. The corporatised railway offers efficient intercity passenger transport and freight. During the past several years, the Government has taken steps to mitigate the increasing traffic congestion in Kuala Lumpur and its enclaves in the Klang Valley. A privately run commuter train network, Light Rail Transit (LRT) system, complements the existing bus and taxi services, providing another convenient alternative to public transport within the city.

Malaysia has many ports to cater for the increasing volume of export-import cargo. There are five major ports which come under the purview of the federal Ministry of Transport (Bintulu, Johor, Klang, Kuantan, and Penang), while several other ports in Sabah and Sarawak are operated by the respective state governments. Port Klang, Malaysia's largest port, handles 55% of Peninsular Malaysia's sea traffic, which is about 30% of Malaysia's total traffic.

Malaysia's international sea trade has been handled by Malaysia International Shipping Corporation Bhd (MISC) since 1968. MISC's shipping services cover regions such as the Mediterranean, Europe, South Africa, Australia, New Zealand, East Asia, Middle East, Southeast Asia and eastern Malaysia. PETRONAS is a major shareholder of MISC, which has more than 100 vessels. Besides the major ports, there are three supply bases that provide oil companies with the facilities for storage of oilfield equipment and supplies. These are located at Kemaman (Terengganu), Miri (Sarawak) and Labuan (offshore west Sabah).

Aviation

Malaysia's ideal geographical location in this fast-growing Asia-Pacific region has helped in the development of the aviation industry. Malaysia is now served by more than 45 international airlines, including the national carrier, Malaysia Airlines, which services 75 international routes around the globe. In addition, there are several smaller airline companies servicing mainly domestic tourist destinations. Malaysia has five international airports: Kuala Lumpur, Pulau Pinang, Kota Kinabalu, Johor Bahru and Kuching, and a dozen domestic airports that link the major towns. The RM19 (US$5) billion Kuala Lumpur International Airport (KLIA) (Fig. 1.7), which

Fig 1.7 Kuala Lumpur International Airport (KLIA).

opened on 28th June 1998, replaces the Sultan Azlan Shah International Airport in Subang, and is the new gateway into Malaysia. The airport is built on 25,000 acres of land, 50 km south of Kuala Lumpur, and is one of the largest airports in the world. KLIA also serves as the home of Malaysia's emerging aerospace industry.

ECONOMY AND INDUSTRY

Malaysia is rich in natural resources. This provided the impetus for the rapid economic growth during the 1970s. In the last two decades Malaysia has sustained a high growth rate, averaging 6.7% per annum. Malaysia is one of the world's largest producers of natural rubber, tin, palm oil, timber and pepper. Other main agricultural commodities include cocoa, pineapple and tobacco. The country has now largely diversified from a resource-based economy, dependant on agricultural products and commodities, to manufacturing and heavy industry. Great emphasis has been given to value-added and high-technology industries with strong research and development (R&D) backing. By the early 1990s the export of manufactured goods exceeded 50% of the total exports (Fig. 1.8). The first half of the 1990s marked the emergence of industrial development policies to broaden, deepen and modernise the industrial structure. Increased local and foreign investments in capital-intensive projects and the manufacturing sector contribute towards the GDP and total exports. Manufacturing is now the single largest component of Malaysia's economy, with electrical and electronic products contributing about a fifth of the total production. The rapid growth in the manufacturing sector has made Malaysia the largest exporter of semi-conductor components to the United States.

There are many commercial banks in Malaysia, a total of 35 in July 1997, and about a third of these were incorporated overseas. Together with the Central Bank (Bank Negara) they ensure smooth operation and development of the payment systems in the country. Apart from banks, there are many financial companies and merchant banks providing financial services. Malaysia's capital market has also evolved rapidly, with over 700 companies listed on the Kuala Lumpur Stock Exchange (KLSE), representing a total market capitalisation of over RM375 billion. As a result of the regional economic melt-down during 1997/1998, Malaysia's economy slipped below the average 8% GDP growth achieved in the previous eight consecutive years. The economy had shown signs of recovery by the middle of 1999.

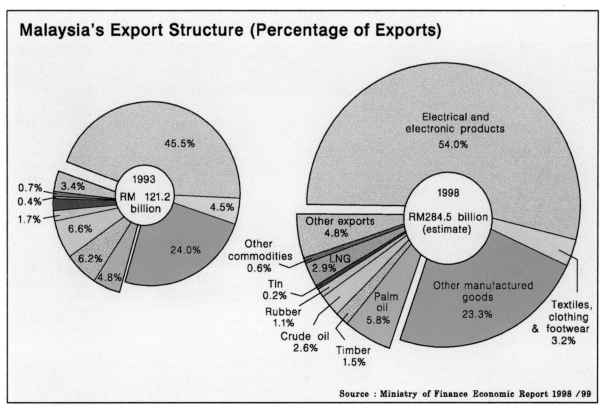

Fig. 1.8 Malaysia's export structure.

Petroleum

Oil was discovered in Malaysia in 1910 with the momentous strike in Miri, Sarawak. A modest production of 83 barrels per day 90 years ago has now become a multi-billion dollar industry. The petroleum industry underwent dramatic expansion in 1974 with the incorporation of the state-owned oil company Petroliam Nasional Berhad (PETRONAS) (see Chapter 2). PETRONAS is now a fully integrated multinational corporation involved in both domestic and international oil and gas exploration and production, oil refining, trading and marketing, as well as in the downstream gas and petrochemical businesses.

In the upstream petroleum sector, Malaysia attracts many multinational companies to explore for and produce petroleum, based on production sharing contracts (PSCs) with PETRONAS. Oil and gas are being produced from offshore Terengganu, Sarawak, and Sabah. PETRONAS operates two oil refineries, one in Kertih, Terengganu and one in Melaka, with a combined capacity of 240,000 barrels per day (Fig. 1.9). This contributes almost 51% of the total capacity of the country's five refineries of 475,000 barrels per day.

Malaysia has abundant gas reserves, more than 80 TSCF (Chapter 3). Various gas development projects are being implemented to maximise the use of this resource, including the two joint venture liquefied natural gas (LNG) plants in Bintulu, Sarawak. A third LNG plant is under construction and is due to be in operation by 2002. Malaysia is now the third largest LNG producer in the world, with a total production capacity of almost 16 million tonnes per year. Most of the LNG is exported to Japan, South Korea and Taiwan.

Another major project is the Peninsular Gas Utilisation (PGU) project, which processes and transmits gas to power, industrial and residential users via a trans-peninsular gas transmission pipeline system. This is linked to PETRONAS' gas processing plants (GPP) and related facilities. Some 2,000 million standard cubic feet of feed gas are being piped daily from fields offshore Terengganu to the GPPs.

PETRONAS has also introduced natural gas for vehicles to promote and encourage the use of gas in the country. In addition, natural gas is also used in the Gas District Cooling systems that supply chilled water for air conditioning and power co-generation in the PETRONAS Twin Towers, Putrajaya, the new Federal government's administrative centre and the Kuala Lumpur International Airport (KLIA).

Petrochemicals

PETRONAS is playing a significant role in turning Malaysia into a regional petrochemical hub. The Integrated Petrochemical Complexes (IPC) in Kertih, Terengganu and Gebeng, Pahang, have been set up to meet the growing

Fig 1.9 PETRONAS' refinery in Melaka.

Fig 1.10 PETRONAS fertilizer plant in Kedah.

demand for petrochemical products in the region. The IPC in Kertih is an ethylene-based complex whilst the Gebeng IPC is propane-based. Among the other petrochemical plants in Malaysia are the methanol plant in Labuan, the ammonia/urea plant in Bintulu, Sarawak, an ethylbenzene/styrene monomer plant in Johor, and a fertilizer plant in Gurun, Kedah (Fig. 1.10). The two fertilizer plants supply urea and ammonia to the Asia-Pacific Region. Other petrochemical projects include the vinyl chloride monomer, aromatics and acrylic monomer plants in Kertih. More plants to manufacture primary and intermediate petrochemical products are being set up along the east coast of Peninsular Malaysia. These are being undertaken by PETRONAS on a joint-venture basis, with foreign multinational corporations that have the technological expertise, experience and market access, and share a common vision with the corporation.

Heavy Industries

Malaysia has been undergoing rapid industrialisation, with manufacturing as a major contributor to the economy. Also aggressively pursued is the heavy industry, spear headed by Heavy Industries Company of Malaysia Bhd (HICOM) which was incorporated in 1980. HICOM is involved in five core industries: transportation, engineering, building materials, realty development and services. Its biggest achievement is the manufacture of Malaysia's first national car by its subsidiary, Perusahaan Otomobil Nasional Berhad (PROTON). By 1996, just over 10 years after its introduction, PROTON cars accounted for over 60% of passenger car sales. By 1997, approximately 20% of PROTON's production volume is exported to about 50 countries. This national venture has resulted in many spin-offs, such as the transfer of technology and the creation of many manufacturers of automobile components. HICOM has also teamed up with manufacturing giants from Japan to produce motorcycle engines. There has since been a second national car, manufactured by Perusahaan Otomobil Kedua Bhd (PERODUA).

21ST CENTURY MALAYSIA

Malaysia aims to become the regional centre for information technology, telecommunications, and broadcasting industries, as inspired by the Prime Minister Dato Seri Dr Mahathir Mohamad's Vision 2020. An advanced telecommunication infrastructure is essential for the country to achieve this vision. The establishment of the National Information Technology Council, under the chairmanship of the Prime Minister, emphasises the role of information technology (IT) in taking Malaysia into the new millennium. The council is formulating a comprehensive national framework for IT and multimedia planning, development and management.

A major project that will transform Malaysia into a high-technology and knowledge-based society by 2020 is the development of the Multimedia Super-Corridor (MSC). This is a 15-by-50 km zone extending from PETRONAS Twin Towers to the Kuala Lumpur International Airport (KLIA), covering some 750 km^2 of land. The zone is wired with the latest fibre-optic technology that will link up various industries, R&D laboratories and universities. The nerve-centre for the MSC is the futuristic, multimedia garden city, Cyberjaya, Malaysia's answer to the Silicon Valley. Cyberjaya is designed to be an intelligent, high-tech, low-density and environmentally friendly city equipped with a world-class infrastructure. High-profile international companies, including some IT giants, have already set up office there to spearhead the development of value-adding IT applications. The MSC will have eight flagship applications: R&D clusters, telemedicine, electronic government, smart schools, borderless marketing, world-wide manufacturing web, and a multipurpose card. A new multimedia university located within the MSC has just been opened in 1999. There will also be electronic shopping and banking, on-line libraries and a central billing system. Tight protection of intellectual property, further development of cyber laws, access to funds at competitive rates, tax exemptions and equity participation are some of the incentives offered to companies to operate in the MSC.

Besides being the technology hub, the MSC is also the home to Malaysia's new administrative capital city, Putrajaya. Putrajaya is the first major intelligent city in Malaysia and serves as a model for electronic government and IT-based administration and management.

CONCLUSION

Prime Minister Dato Seri Dr Mahathir Mohamad's Vision 2020 represents Malaysia's aspiration to become a fully developed and industrialised country by the year 2020. One of the challenges faced by Malaysia in its effort to realise this vision is the creation of a competitive, dynamic, robust and resilient economy. The rapid economic development achieved since independence is an indication that this vision is within reach. Malaysia now has a high domestic savings rate, a relatively well-developed physical, social and institutional infrastructure, political, economic and financial stability, and a relatively developed manufacturing and industrial-based private sector. The excellent infrastructure, the various incentives offered by the Government, and the availability of a highly skilled work force, put Malaysia at an advantage as the preferred choice location for multinational companies to invest and do business, and to help realise this vision.

BIBLIOGRAPHY

Information Malaysia Yearbook 1998. Berita Publishing, Kuala Lumpur.

Selected websites:
- Cyberjaya www.cyberjaya-msc.com
- Government of Malaysia (Prime Minister's Office) www.smpke.jpm.my
- Kuala Lumpur International Airport www.kiat. net/klia
- Malaysia homepage www. mymalaysia.net.my
- Malaysian Technology Development Corporation www.mtdc.com.my
- Multimedia Super Corridor www.mdc.com.my
- PETRONAS www.petronas.com.my

Chapter 2

Malaysia E & P Business

Ho Wang Kin

INTRODUCTION

Malaysia's Exploration and Production (E & P) business has matured from its humble beginning at the turn of the twentieth century to a fully integrated petroleum industry today. The evolution and transformation of Malaysia's petroleum industry is to a very large extent attributable to the efforts of the Malaysian Government in formulating and implementing policies and institutional frameworks pertaining to the petroleum industry.

The formation of PETRONAS, the acronym for Petroliam Nasional Berhad, Malaysia's national oil company, on 17 August 1974 marked the beginning of this transformation process. In managing the petroleum resources of Malaysia, PETRONAS is guided by the national policies directly related to the industry itself which amongst others includes; Petroleum Development Act (PDA) 1974, National Petroleum Policy, National Energy Policy and the National Depletion Policy. On a broader perspective, the direction of the petroleum industry is guided by the New Economic Policy (NEP) and the National Development Policy (NDP) within the purview of the First Outline Perspective Plan (OPP1) over the period 1971 - 90 and the Second Outline Perspective Plan (OPP2) over the period 1991 - 2000 respectively. The NDP falls within the ambit of a grander and larger goal known as "Vision 2020", in which Malaysia attains the status of a "developed" nation economically, politically and socially by the year 2020.

The PDA enacted in 1974 transfers the entire ownership in, and the exclusive rights, powers, liberties and privileges of exploring, exploiting, winning and obtaining petroleum, whether onshore or offshore Malaysia, to PETRONAS. PETRONAS, which is a fully Government-owned company, is designed to run the industry like any other commercial or profit-oriented organisation. It would, above all, ensure that Malaysians enjoy to the fullest the benefits to be obtained from their indigenous petroleum wealth.

The National Petroleum Policy which was enacted in the same year as the PDA has the following objectives :

- to put to good use the petroleum resources of the country, as a first priority to serve national needs by making available supplies at reasonable prices to meet domestic consumption, including the requirements of power generation, industry and agriculture;
- to enhance the favourable investment climate of the country by opening up new opportunities for the establishment of heavy energy-intensive and petrochemical industries both for the domestic and the export markets;
- to take advantage of the option of increasing revenue and export earnings by the export of oil and gas to overseas markets;
- to ensure that Malaysians are adequately represented in terms of ownership, management and control in all phases of petroleum operations ranging from exploration at one end to marketing and distribution at both the local and international levels at the other, and
- to effect an optimal social economic pace of exploration of the nation's endowment of exhaustible oil and natural gas resources, taking into account the need for conservation of these depletable assets and the protection of the environment.

On the other hand, the National Energy Policy was later introduced by the Government to further emphasize the need to provide the nation with adequate and secured energy supplies, towards reducing the dependence on oil, and by developing and utilising alternative sources of energy. A 4-fuel energy strategy based on oil, hydro, natural gas and coal was formulated towards diversifying the country's energy base as well as guaranteeing assured energy supplies for continued economic growth.

Being a proactive and prudent resource manager, the Government introduced the National Depletion Policy in June 1980. The objectives of the National Depletion Policy are two fold i.e. to plan for optimal development of major oil fields in order to prolong the production life of the nation's oil resource and to avoid excessive pre-investment.

With the timely formulation and implementation of policies and institutional frameworks, the Malaysian Government has ensured that wealth derived from the indigenous petroleum resources is exploited for the benefit of the nation and the general well being of its population.

PRE-WORLD WAR II

In 1882, the first official record of oil discovery was documented in Miri, Sarawak in the Baram district by a Mr Claude Champion de Crespigny, the then Resident of Baram. However the exploration and commercial exploitation of petroleum in Malaysia did not begin until the next century. In 1909, the Rajah of Sarawak awarded the Anglo-Saxon Petroleum Company, the forerunner of the present day Sarawak Shell Bhd, the sole right to explore petroleum resources throughout Sarawak. It was the first petroleum licence awarded in Malaysia. Slow and laborious drilling process took place but oil was finally struck on 22 December 1910, marking the beginning of the Malaysian petroleum industry. Daily oil production from the Miri field started at about 80 barrels of oil per day (BOPD) and peaked at about 15,000 BOPD in 1929. During the Japanese occupation of Sarawak, the Miri field was severely damaged and despite massive efforts of reconstruction, post-war production never returned to pre-war levels. The Miri field finally shut down in October 1972 after having produced about 80 million barrels of oil.

PRE-PETRONAS FORMATION (THE CONCESSION ERA)

Prior to the formation of Malaysia in 1963, the Sarawak Oil Mining Ordinance of 1958 controlled the issue of oil exploration licences and leases in the State. Sabah claimed oil rights "in Malaysian waters surrounding her coastline". Following the formation of Malaysia and interest shown by oil companies in exploring offshore Peninsular Malaysia, the Federal Government passed the following Acts pertaining to the development of the country's petroleum resources : Continental Shelf Act, 1966, Petroleum Mining Act, 1966, Petroleum Income Tax Act, 1967, and Petroleum Mining Rules, 1968. These legislations originally covered only the states of Peninsular Malaysia; their applications were extended to Sabah and Sarawak by Emergency (Essential Powers) Ordinance Number 10, 1969. Consequently, all rights and liabilities accruing to Sabah and Sarawak under offshore petroleum licences issued prior to the 1969 Ordinance have been transferred to the Federal Government.

Encouraged by the success in Miri, further onshore exploration attempts were made but without success. By the 1960s, attention turned to offshore drilling made possible by new development in offshore petroleum technology in the west. Marine seismic surveys were introduced and carried out prior to the first offshore drilling campaign. In 1962 significant oil was discovered in two areas offshore Sarawak. Other hydrocarbon finds followed in rapid succession. These discoveries encouraged new players to the industry and between 1965 and 1969 five companies, namely Shell (Sabah Shell), Esso, Elf Aquitaine, Oceanic and Sabah Teiseki Oil Co. signed concession agreements with the Sabah and Sarawak State governments.

In Peninsular Malaysia, Esso and Conoco were awarded the first concession in 1968 followed by Mobil in 1971. The first commercially exploitable oil field - the Tapis field - was discovered by Esso in 1969. By the end of 1973, a total of 19 oil fields had been discovered in Malaysia's offshore areas, of which four had already been brought into production. Consequently, Malaysian crude oil production increased from only 4,000 BOPD in 1968 to 90,000 BOPD in 1973. Significant quantities of natural gas had also been discovered, notably in the Central Luconia Province, offshore Sarawak and the Northern Malay Basin, offshore Peninsular Malaysia.

By the early 1970s the petroleum industry was recognised as the most important hydrocarbon resource in Malaysia and forms the life blood of the nation's economy whereby it paves the way for industrialisation of Malaysia's energy future.

The industry continued to grow as more companies participated in the search for hydrocarbon in Malaysia under the concession system (Fig. 2.1). Under such arrangement, the Government had essentially no control over

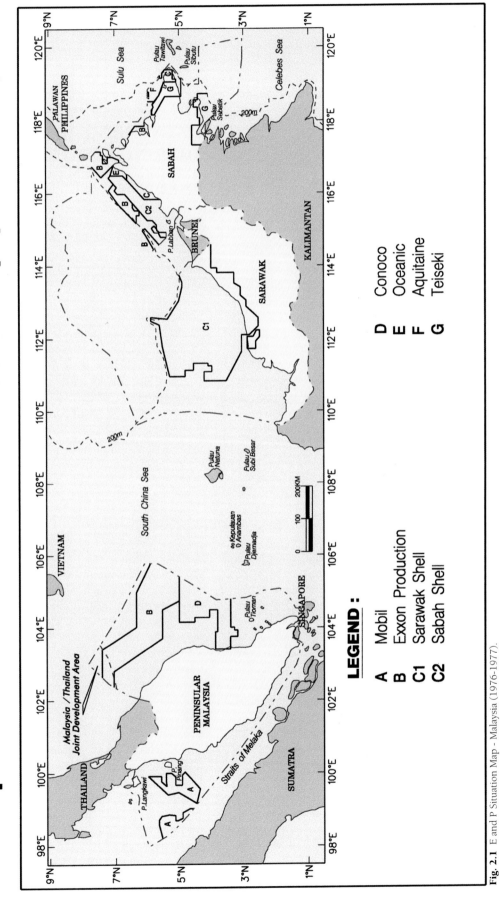

Fig. 2.1 E and P Situation Map - Malaysia (1976-1977).

petroleum activities except collecting taxes/royalties from the operating oil companies. However, the oil crisis in 1973 permanently changed the system of oil exploration in Malaysia, leading to the introduction of the Petroleum Development Act in October 1974 and setting up of PETRONAS earlier in the same year. The immediate task of PETRONAS then was to negotiate with existing concessionaires for the smooth transformation of the prevailing Concession Agreements held by Esso, Conoco, Mobil, Shell, Elf Aquitaine, Oceanic and Teiseki to the Production Sharing Agreements which PETRONAS adopted in discharging its role as the nation's resource manager.

THE PETRONAS ERA

The Government decided to set up a national petroleum corporation; a Government-owned company that would run the industry like any other commercial company. To realise the objective of ensuring that all Malaysians will enjoy the full benefits obtained from their indigenous petroleum wealth, PETRONAS was incorporated on 17 August 1974 under the Companies Act 1965. A major change in the management style of the nation's petroleum resources was introduced with the enactment of the Petroleum Development Act (PDA) on 1 October 1974 under which PETRONAS has been vested with the entire ownership in, and the exclusive rights, powers, liberties and privileges of exploring, exploiting, winning and obtaining petroleum whether onshore or offshore Malaysia. PETRONAS is under the control and direction of the Prime Minister and by the virtues of the PDA, is the sole concessionaire of the petroleum resources in Malaysia.

Initially, PETRONAS started out as a regulator of the upstream sector, collecting revenue in the form of profit-sharing derived from oil production by the upstream contractors. Its function was primarily to ensure the orderly development of the upstream industry by foreign oil companies in Malaysia. This was later expanded when it was allowed to participate as an active player in the domestic upstream industry to speed up the development of our resources as well as to augment the exploration and development activities of foreign oil companies. Thus, PETRONAS Carigali Sdn Bhd (Carigali), a wholly owned subsidiary of PETRONAS, was incorporated in 1978 as an exploration and production arm of PETRONAS. Within two years, it was able to start the development of its first gas field, Duyong, in 1980 and first oil field, Dulang, in 1991. Carigali accounts for about 40% and 31% of Malaysia's total oil and natural gas production respectively, through its equity holdings in 35 oil and 10 gas producing fields.

THE EVOLUTION OF THE MALAYSIAN PRODUCTION SHARING CONTRACT

Background

PETRONAS was incorporated to serve as the Government's instrument to take charge of petroleum matters and to exercise, on behalf of the country, its sovereign rights over its own oil and gas resources. In turn, PETRONAS has opted to adopt the Production Sharing mechanism to manage the exploration, development and production of the nation's petroleum resources. In brief, the production sharing concept is summarised in Figure 2.2.

PETRONAS began to develop strategies to encourage investment in the petroleum industry and to ensure meaningful Malaysian participation in the ownership, management and control in all phases of the petroleum operations; as well as to secure additional hydrocarbon reserves. A conscious effort was made to encourage multinational oil companies to invest in the petroleum industry to support its continued growth.

The challenge for PETRONAS then was to balance the divergent needs of the country for growth and development against the bottomline needs of foreign oil companies. This called for some reconciliation of the differences between the national aspirations and the expectations of the foreign oil companies. To be too nationalistic would discourage investments, but on the other hand, giving too much flexibility to the oil companies may be detrimental to the nation's interests.

The Production Sharing Contracts (PSCs) provide increased shares of revenue to the government which enables the government to implement an accelerated programme for the long term development of the Malaysian petroleum industry, to spur and support the nation's economic growth. For the past 20 years, Malaysia has used the PSC as a granting instrument which, in our view, is best suited in balancing the goals and aspirations of the nation and the oil companies.

RELATIONSHIP BETWEEN GOVERNMENT, PETRONAS AND PSC PARTNERS

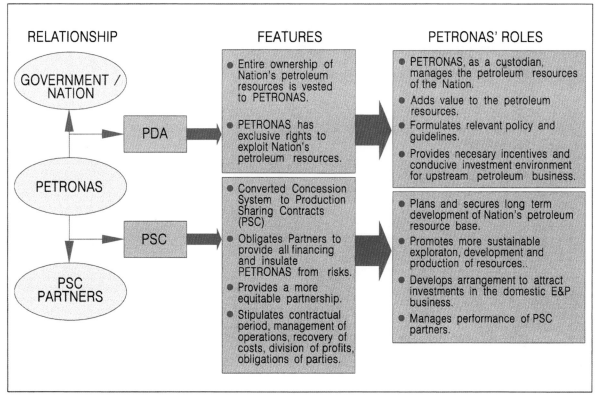

Fig. 2.2 Relationship between Government, PETRONAS and PS Contractors.

1976 PSCs

The Malaysian petroleum industry entered into a completely new era in 1976, when the concession system was converted into a Production Sharing Contract to ensure a more equitable partnership between Malaysia and the multi-national oil companies. PETRONAS entered into the first round of PSCs with oil companies in 1976, i.e. referred to as '1976 PSC'. The key players were the two major multi-national oil companies, Esso and Shell and later joined by Carigali in 1978.

The design of the PSC adopted by PETRONAS not only ensures a more equitable partnership between Malaysia and the MNOCs, but also fulfills its obligation to disseminate the wealth derived from this indigenous petroleum to the people of the nation in line with the New Economic Policy (NEP) and the National Petroleum Policy.

To complement the aims of the NEP i.e. the eradication of poverty and restructuring of the economic balance of society, PETRONAS through its policies had encouraged the growth and development of the local ancillary and supporting industries to achieve greater participation and dissemination of benefits to the local populace.

1985 PSCs

The prevailing business environment surrounding the petroleum industry in Malaysia necessitated the revitalisation of the dwindling exploration interest and activities during the early part of 1980s. Recognising that the fiscal terms play an important role in determining the attractiveness of the PSCs offered for the exploration, development and production of oil and gas resources, PETRONAS decided to introduce a new package of incentives to attract more foreign investment in the exploration for oil and gas in its attempt to revitalise the upstream business. In 1985, PETRONAS introduced revised terms to the 1976 PSC, which was known as the "1985 PSC".

The 1985 PSC offered better terms to the investors, with major improvements being made to the cost oil ceiling which was increased from 20% to 50%, and cost gas from 25% to 60%. A sliding scale on the profit split for both oil and gas was also introduced for the first time. The 1985 PSC drew a tremendous interest from investors (Fig. 2.3) with a total of 28 PSCs signed involving minimum investment commitments of US$ 368 million and drilling of 126 exploratory wells. The '1985 PSC' also stipulates a minimum level of state participation, through Carigali.

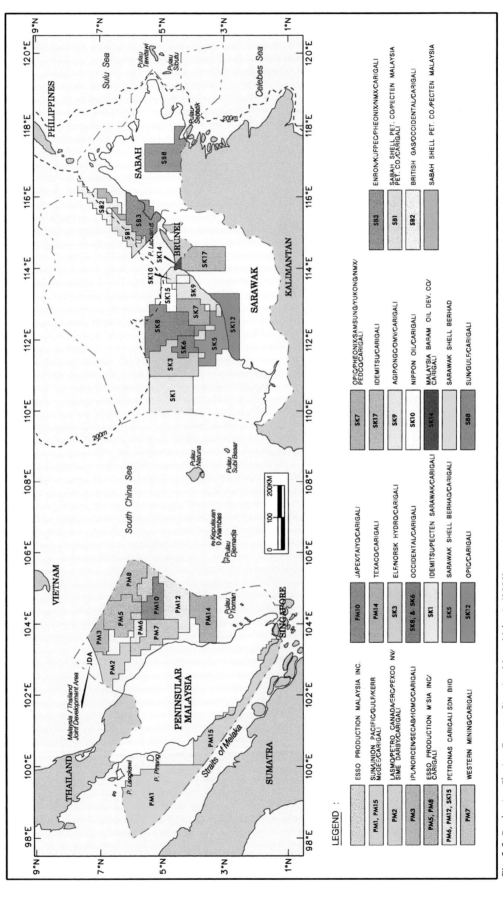

Fig. 2.3 Production Sharing Contract Situation Map - Malaysia (1990).

Deepwater PSCs

Keeping in tandem with the advancement in drilling technology in deeper water, large tracts of unchartered deepwater acreage were opened up to the search for oil and gas worldwide. Such exploration efforts were pioneered by MNOCs having such technology edge and resources in the 1990s. Recognising the hydrocarbon potential of Malaysian deepwater areas, PETRONAS introduced the "Deepwater PSC" in 1993. This was based on the 1985 PSC, but further improvement was made to the cost recovery, profit split and the exploration, development and production periods, to reflect the higher costs, higher risks and state-of-the-art technology required in the deepwater operations. Given the longer lead time required, the exploration period was extended to 7 years, the development period to 6 years and the production period to 25 years as compared to 5 years for exploration, 4 years for development and 15 years for production under the shallow water PSCs i.e. 1976 PSC and 1985 PSC.

The Deepwater PSC allows the PSC partners to recover their investment cost up to 75 per cent on oil and 60 per cent on gas and gives a better profit split for the PSC partners. Two types of deepwater contracts were introduced, one for water depths in the range of 200 to 1,000 metres and the other for water depths of greater than 1,000 metres. Ten deepwater blocks were demarcated to be offered to MNOCs under the Deepwater PSC terms. As at 1 October 1999, 5 such blocks are operated by Esso, 3 by Shell and one by Murphy (Fig. 2.4).

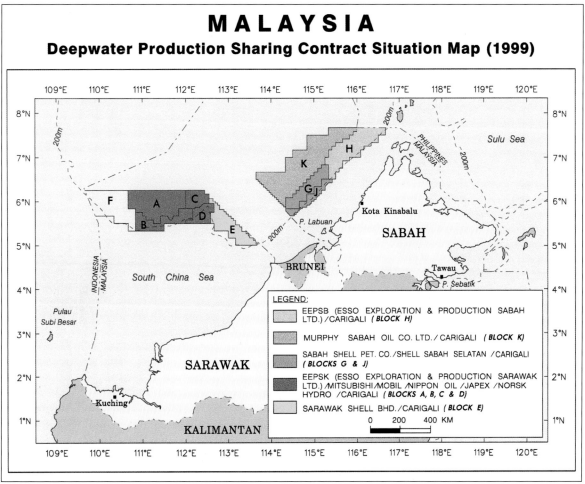

Fig. 2.4 Deepwater Production Sharing Contract Situation Map - Malaysia (1999).

Revenue Over Cost PSCs

In early 1997, PETRONAS introduced a new profitability-based fiscal regime using the "Revenue Over Cost" (R/C) concept to further stimulate and revitalise exploration activities and investments in Malaysia because 1985 PSCs were approaching their expiry dates and prices for Tapis crude oil and domestic petroleum products were softening. On the other hand, world-wide development and operating costs were generally increasing.

The R/C PSC is more progressive as it has a built-in self-adjusting mechanism vis-a-vis the changing price and costs environments. It

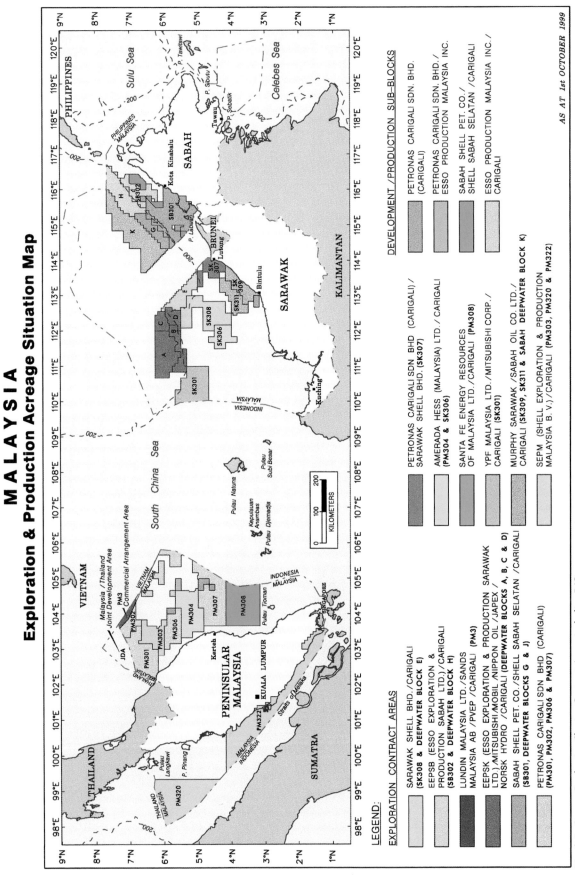

Fig. 2.5 Current Production Sharing Contract Areas, including R/C arrangements.

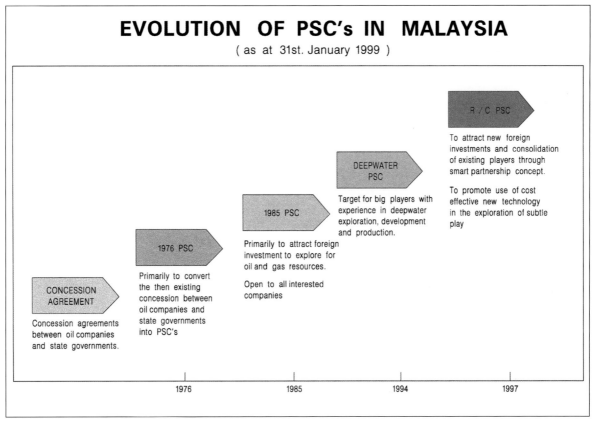

Fig. 2.6 Evolution of Malaysia's PSC.

allows PSCs to have a larger share of revenue from a project when their profitability is low and for PETRONAS to progressively increase its share as PSCs profitability improves. As at 1 October 1999, 13 PSCs under the R/C terms have been awarded (Fig. 2.5). A summary of the Malaysian petroleum arrangement evolution is depicted in Figure 2.6.

Non-Traditional PSCs

In addition to the above mentioned PSCs which cater for exploration investment, different hybrids of PSC and contractual arrangements have also been developed to cater for specific areas and purposes. They are namely, the Seismic Option, BDO PSC, MLNG-Dua PSC, Global Arrangement Gas PSC (GPSC).

The Seismic Option was introduced for very high risk acreages. The other hybrids of PSCs mentioned above cater for continuity of previous arrangements, whereby the contracts of otherwise relinquished fields are renewed for existing oil companies to continue participation with Carigali as a partner and equity holder. In addition to continued participation, specific provisions are formulated in the arrangements to address certain particular concerns of the oil companies to make the projects viable for all involved parties.

Government Incentives

The Government also played its role by reducing taxes to make the petroleum economics and arrangement of Malaysian acreages more competitive. The export duty on oil was lowered from 25% to 20% while the Petroleum Income Tax (PITA) rate was also lowered from 45% to 40% in 1995. Subsequently, the Government in 1998 further reduced PITA and Export Duty on oil to 38% and 10% respectively to stimulate the upstream oil and gas industries and promote re-investment.

THE CHANGING ROLE OF PETRONAS IN THE MALAYSIAN E&P INDUSTRY

The Malaysian experience in attracting continued E&P investments demonstrates the need to constantly adapt to the changing environment. This is a prevalent theme in Malaysia as reflected by the adaptability of the PSC fiscal terms, which have been restructured and reviewed accordingly to meet the nation's needs at different times. Due consideration was given, not only to the economic environment, but also to other factors such as the existing fiscal and legal regimes.

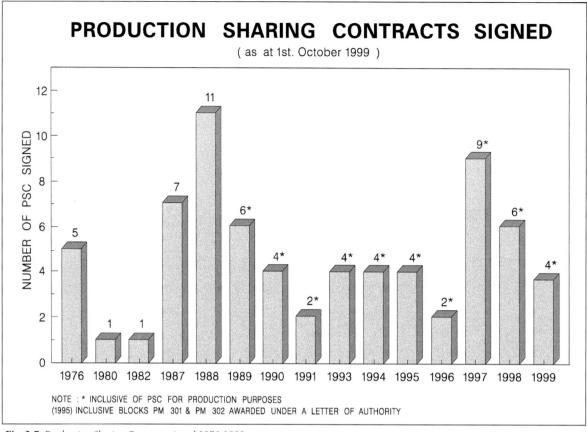

Fig. 2.7 Production Sharing Contracts signed 1976-1999.

PETRONAS initially started out as a regulator of the upstream sector. However, its role has increasingly focused on resource management and attracting E&P investments through competitive and innovative contractual arrangements. Carigali was formed to assume the active role in the search for oil and gas domestically on its own, and in partnership with foreign oil and gas companies. Since its inception, Carigali had established itself as an important domestic E&P company. To date, Carigali has also spread its wings to cover several international upstream ventures.

The involvement of foreign multi-national oil companies is an important element in the successful replenishment of reserves and for the development of our human resources. They have been instrumental in our efforts to provide the necessary foundation and infrastructure to support the development of the petroleum industry in Malaysia.

PETRONAS has adopted a more pragmatic and flexible approach in managing the PSC with the view of achieving a "win-win" and equitable arrangement with its PSC partners. This is reflective of PETRONAS' commitment to promoting a long term relationship with its PSC partners.

The relationship between PETRONAS and these oil and gas companies in Malaysia has evolved from a regulatory or adversarial nature, to long term partnerships founded on trust and mutual respect. The number of PSCs signed throughout the years since 1976 is evidence of the continuous interest of the oil and gas companies in doing E&P business in Malaysia (Fig. 2.7).

CONTRIBUTION TO THE MALAYSIAN ECONOMY

The importance of petroleum is reflected in its twin role as a principal source of energy and a major foreign exchange earner for Malaysia. From its beginning as an insignificant primary commodity produced by Malaysia, petroleum has today emerged as an important contributor to the nation's foreign exchange earnings. The petroleum industry generates income for the government through corporate income taxes payable by oil companies; import duties; export duties and excise duties. Non-tax revenues include cash payments from petroleum production and dividends from PETRONAS.

In the 1970s and 1980s, when Malaysia was on its journey towards industrialisation, the petroleum sector has been the engine of growth

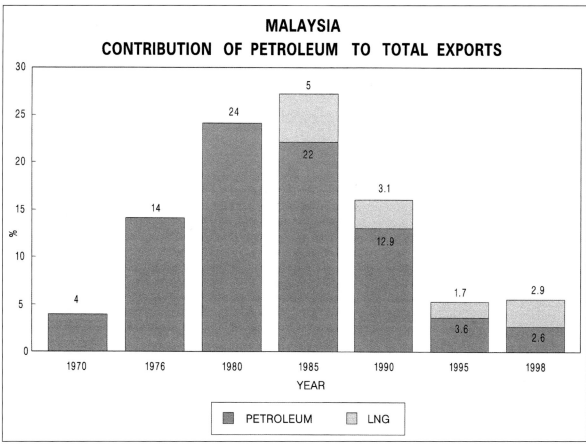

Fig. 2.8 Malaysia - Breakdown of Total Exports.

for transforming its then resource and agricultural based economy, relying mainly on the production and export or primary commodities, to industrial based economy represented by the well-diversified manufacturing sector of today.

In the early 70s and prior to the formation of PETRONAS, production of crude oil increased from about 18,000 BOPD in 1970 to about 80,000 BOPD in 1974. The crude oil production buildup was due to the bringing onstream of some major oilfields discovered in the states of Sarawak and Sabah. Inheriting a headstart of 80,000 BOPD production, the formation of PETRONAS in 1974 had further elevated the growth of the petroleum industry to a new height. Successes were not only recorded in the states of Sarawak and Sabah, but more significantly off the coast of the state of Terengganu. By 1990, the production of crude oil had surpassed the 600,000 BOPD mark and since then has been maintained at this rate to the present time. There was an increase of foreign exchange earnings from this primary commodity from a mere MR200 million in 1970, to more than MR10 billion through the 1990s. This is an increase in the share of total export earnings from 4% in 1970s to about 10% at present after a peak in the 1980s of about 24%.

Another significant contribution of the petroleum sector to Malaysia GDP is from the gas resources both for export as LNG starting in the early 1980s, and for domestic utilisation as fuel in the race for industrialisation in the 1990s. Domestic utilisation of gas as a cheap source of energy, and the foreign exchange earnings from sale of LNG has contributed significantly in building Malaysia's road to Vision 2020.

The relative importance of the petroleum sector as a foreign exchange earner within the context of the Malaysian economy is depicted in Figure 2.8. Prior to the formation of PETRONAS, the petroleum sector constituted a small percentage (about 4% in 1970) of Malaysia's total export earning. This has increased by 10% to about 14% in 1976 and again an additional 10% to 24% in 1980. During the early part of the 1980s, the contribution of this sector to the Malaysian economy had been significant. This sector had nurtured the growth of the manufacturing industries and before the turn of the decade, export earning from the manufacturing sector had established itself as the leading foreign exchange earner with a 58.9% relative contribution to Malaysia's total export in 1990. During the last decade or so, the manufacturing sector has overtaken in terms of relative importance to the overall export earnings of Malaysia.

In summary, Malaysia is blessed with an abundance of primary commodities which have contributed significantly in achieving a developing nation status. Rubber and tin were the prime commodities in the 1950s and 1960s, supplemented by timber and palm oil in the 1970s. Crude oil became dominant in the late 1970s and early 1980s before manufacturing established its ascendancy from the mid-1980s. Today, manufacturing is number one, followed by tourism, palm oil and then petroleum. Although the lead position of the petroleum sector's contribution towards the GDP of the nation has been overtaken by some other sectors it continues to play an important role in fuelling Malaysia's industrialisation programme into the next millennium, where emphasis will be focused on high technology, capital-intensive and skill-oriented industries.

CONCLUSION

The establishment of PETRONAS marked the beginning of the revolutionary change of E&P business in Malaysia. The commitment and support of the Malaysian Government to legislate policies and institutional frameworks for the growth of the petroleum industry, coupled with the adoption of the PSC system by PETRONAS in managing the nation's petroleum resources, has served the industry well over the last few decades. Since the first PSC was signed in 1976, PETRONAS had to date concluded more than 60 PSCs with a multitude of foreign and local oil companies in the search for and development and production of Malaysia's indigenous petroleum.

The growth of the E&P business has spawned a large pool of both local and foreign ancillary and support businesses related to the petroleum industry. Significant achievements in line with the objectives of the New Economic Policy (NEP) and National Development Policy (NDP) were accomplished through the establishment and involvement of local participation in the petroleum industry. Not forgetting the need for foreign expertise, the Government of Malaysia and PETRONAS have always encouraged the introduction of new business skills and technologies in support of the growth of this sector. Incentives and policies were proactively introduced to manage the nation's petroleum resources. PETRONAS' administration and management of the nation's petroleum resources, starting from a regulatory role to today's smart partnership management, is testimony to its commitment to ensuring the survival of this industry for years to come. The experience and skills gained in the domestic E&P business had also been put to good use in maturing PETRONAS' quest for international

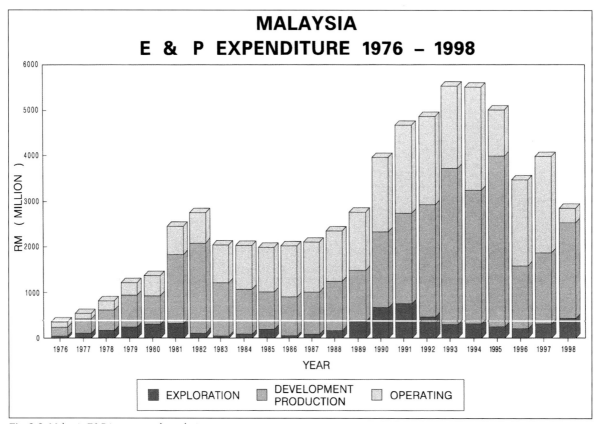

Fig. 2.9 Malaysia E&P investment through time.

E&P business worldwide through Carigali.

The growth of E&P investments through the years in the domestic exploration and exploitation of petroleum are shown in Figure 2.9. The E&P expenditure trend demonstrates PETRONAS' ability and achievement in attracting foreign E&P spending in Malaysia up to the commercialisation of the discoveries made. Domestic E&P investments are complemented by the involvement of Carigali, which is currently producing a third of Malaysia's total production. Many medium to small oil fields have yet to be developed. As such, vast development and production investment opportunities still exist for prospective investors having the right mix of skills and resources to commercialise these remaining oil and gas fields in the future.

The contribution of the petroleum industry to the Malaysian economy has been significant when it represented the engine of growth for the Malaysian economy towards its industralisation plan in the 1980s. Today, this industry still plays a major role in providing the fuel for the growth of the Malaysian economy towards Malaysia's vision to achieve the status of a "fully developed" nation by Year 2020.

Chapter 3

Overview of Petroleum Resources of Malaysia

Abdul Jalil B. Zainul, Rosnan B. Misman and Abdul Jalil B. Ali

INTRODUCTION

This chapter provides an overview of the petroleum resources of Malaysia based on the Annual Review of Petroleum Resources assessment report as at 1.1.1998. The resource inventory is updated annually to provide a basis for the management of domestic petroleum resources.

Malaysia's 8 sedimentary basins have been proven to be petroliferous, and currently commercial oil and gas are being produced in the Malay, Sarawak and Sabah basins (Fig. 3.1). Sedimentary deposition and subsequent structural overprints in these basins have resulted in a variety of structural and stratigraphic configurations which provide conducive conditions for petroleum generation, migration and entrapment. The petroleum exploration activities in Malaysia have shifted to deeper water areas with the first Deepwater PSC signed in 1993.

Following the first Production Sharing Contract in 1976, more than 60 PSCs have been awarded. As at 1.1.1998, 35 Production Sharing Contracts were in operations, 25 in the exploration stage and 10 in the production phase. (Up to 1 October 1999, an additional 10 PSCs have been awarded - see chapter 2)

The petroleum resource definitions, classification systems and terminologies used by PETRONAS are summarized in Abdul Jalil Zainul et al. (1997a, 1997b). The new resource classification system is targeted for full compliance in 1999. The latest 1.1.1998 reserves inventory has adopted the new classification for most of the major fields. The national reserves reported for both oil and gas as at 1.1.1998 reflected the reserves category from PETRONAS commercial perspective.

PETROLEUM RESOURCES OF MALAYSIA

Status of Petroleum Resources

As at 1.1.1998, some 7.7 billion stock tank barrels (BSTB) and 98.7 trillion standard cubic feet (TSCF) of recoverable crude oil and natural gas, respectively, have been discovered in Malaysia. The estimated remaining crude oil and natural gas reserves stand at 3.9 BSTB and 87.0 TSCF, respectively (Figs. 3.2 and 3.3). These reserves include oil and gas fields with estimated ultimate recovery greater than 8 million stock tank barrel (MMSTB) and 50 BSCF respectively.

The total natural gas reserves are equivalent to 17.9 BSTB of oil equivalent (BOE) of which 3.8 BOE (13.9 TSCF) are attributed to associated gas and the remainder to non-associated gas. In total, Malaysia has an equivalent of 21.8 BOE petroleum reserves in the form of crude oil and natural gas.

As at 1.1.1998, gas fields that have been dedicated to existing and future firm gas projects such as PGU, MLNG, MLNG-Dua, MLNG-Tiga, Sabah Gas and SMDS are estimated to contain 1.1 BSTB of condensate.

Exploration Successes

As at 1.1.1998, a total of 122 oil fields and 208 gas fields have been discovered, irrespective of the recovery size. These include 40 oil fields having estimated recovery of less than 8 MMSTB.

The discoveries made are mainly in the shallower water area of less than 200 metres water depth. A total of 64 oil and 95 gas fields are located in Peninsular Malaysia; 39 oil and 86 gas fields in Sarawak; and 19 oil and 27 gas fields in Sabah.

The exploration and development activities in Malaysia have been on the upswing since 1976, when the first PSC was awarded. The levels of exploration well drilling and seismic acquisition are cyclical in nature corresponding to

OIL AND GAS FI

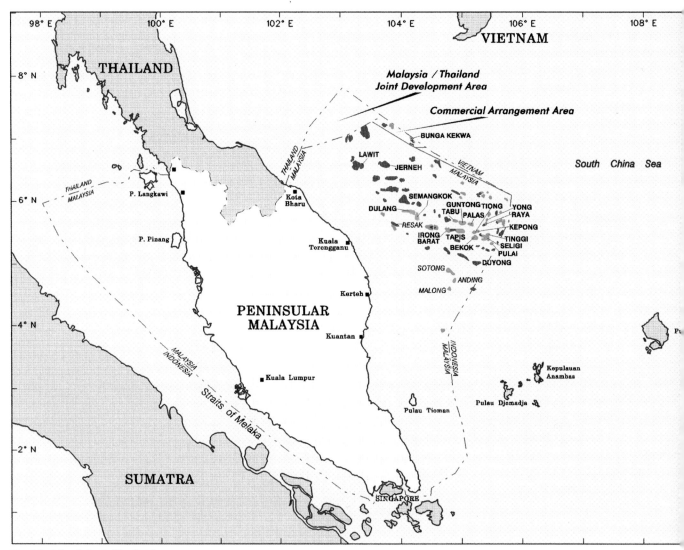

Fig. 3.1 Oil and Gas Fields of Malaysia.

the exploration period cycles of the respective PSC rounds (Fig. 3.4). The last two exploration campaigns have been able to partly sustain the level of Malaysia's petroleum resource portfolio.

Generally, a peak of 60 exploration wells were drilled per year in each of the PSC cycles. 3D seismic data acquisition has played a major role in exploration. Historically, an average of 60 development wells were drilled and 5 new platforms were installed per year (Fig. 3.5).

The success ratio for oil discovery in Malaysia was one oil discovery for every 4.5 exploration wildcat wells drilled during the period from 1990 through end of calendar year 1997. If gas discoveries are included, the success ratio is significantly improved to one in 2.6. Meanwhile, the finding cost for the 5-years moving average for 1997, i.e. for 1992 - 1997 period, is about US$ 0.25/Barrel Oil Equivalent (BOE). These figures indicate that Malaysia still offers tremendous opportunities for finding new discoveries in the future with a relatively low exploration cost.

First Production

Following production from the Miri Field in 1910, Malaysia's first offshore producing oil field was West Lutong off Sarawak which came onstream in June 1968. Tembungo was the first oil field in Sabah with production commencing in 1974. Pulai offshore Peninsular Malaysia started production in 1978. In the search for oil, significant natural gas reserves were also discovered by both Shell (SSB/SSPC) and Esso Malaysia (EPMI). Shell's first gas discovery was made in the Central Luconia Province off Sarawak when it found the F6 Field in July 1969,

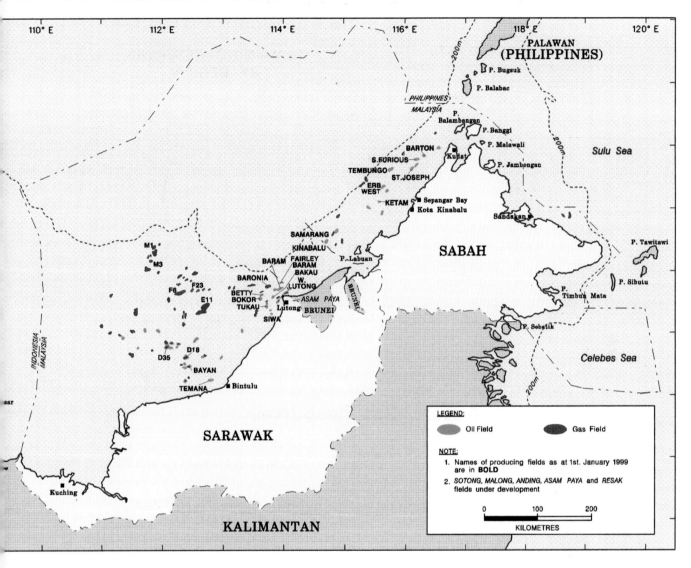

followed by Esso when it found the Jerneh gas field in October the same year in offshore Terengganu. As at 1.1.1998, a total of 3.8 BSTB crude oil and 11.7 TSCF natural gas have been produced, and 35 oil fields (40 excluding Ketam as at 1 October 1999) and 10 gas fields are producing 629,000 BOPD and 4,600 MMSCF/D respectively.

Global Petroleum Resources

Extent of Crude Oil Reserves Globally

Malaysia's crude oil reserves of 3.9 BSTB and natural gas reserves of 87.0 TSCF are small when compared to the global oil and gas reserves.

Despite strong production gains and the absence of major discoveries, world-wide crude oil reserves increased in 1997. Reserves of crude oil rose by 0.7 BSTB during the year to stand at about 1,020 BSTB (Figs 3.6, 3.7), with more than half of the world oil reserves in the Middle East region. This indicates the importance of countries in the region to the international crude oil supply.

The Asia-Pacific region has total crude oil reserves of 42 BSTB, representing only about 4% of world oil reserves. Within the Asia-Pacific region, China has 24 BSTB or 57% of the crude oil reserves in the region. Malaysia represents only about 9% of the region's oil reserves.

Extent of Natural Gas Reserves Globally

Total world natural gas reserves are about 5,086 TSCF with countries in eastern Europe and the CIS having total gas reserves of more than 2,000 TSCF, representing about 40% of the total world

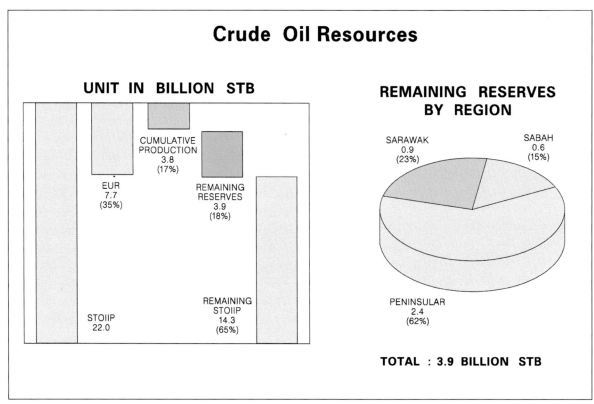

Fig. 3.2 Crude Oil Resources in Malaysia. Status of crude oil resources in Malaysia as at 1.1.1998.

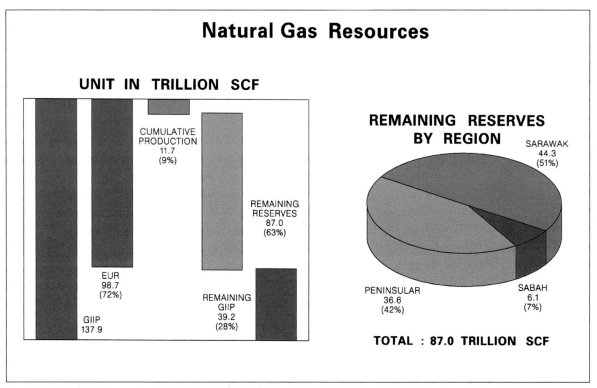

Fig. 3.3 Natural Gas Resources in Malaysia Status of natural gas resources in Malaysia as at 1.1.1998.

gas reserves (Fig. 3.8).

The Asia-Pacific region has a total natural gas reserves of about 321 TSCF, which represents about 6% of world gas reserves. Malaysia only represents about 27% of the region's gas reserves. Figure 3.9 shows countries in the world having more than 20 TSCF of gas reserves.

Global Reserves Trends

International surveys conducted on global reserves trend have shown that the increasing reserves in 1990s have been largely attributed to activities other than new exploration. While such exploration continues to contribute to reserves,

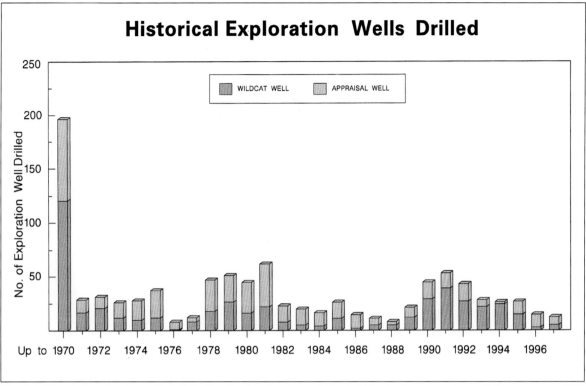

Fig. 3.4 Historical exploration wells drilled. Trends of exploration activities in Malaysia since concession period up to current PSC arrangement. A total of 16 exploration wells were drilled during 1998.

the proportion added by in-fill drilling, new pays in existing fields and adjustments based on new data acquired has been very encouraging. Even as new discoveries have diminished in number and average size, the reserves have been able to grow dramatically in and around fields already in production. This underscores the importance of development activities on the discovered fields for optimum exploitation of the available resources.

CRUDE OIL RESERVES

Malaysian crude oil reserves as at 1.1.1998 are estimated at 3.9 BSTB. About 2.6 BSTB or 67% of the remaining 3.9 BSTB of crude oil reserves are found within the 35 producing fields.

Breakdown of Crude Oil Reserves

As shown in Fig 3.10, the crude oil reserves are increasing. Since 1980, some 5.9 BSTB of recoverable crude oil were added to the resource base (ultimate recovery) of 1.8 BSTB to total 7.7 BSTB. During the same period, a total of 3.8 BSTB (49%) was produced leaving about 3.9 BSTB (51%). Out of the total 3.9 BSTB remaining crude oil reserves, 1.4 BSTB have been developed and 2.5 BSTB are to be developed.

The breakdown of crude oil reserves by field ultimate recovery size is shown in Figure 3.11. The Seligi Field is the biggest discovered in Peninsular Malaysia, with estimated recoverable reserves of 589 MMSTB. 27 oil fields are in the 'giant' category, containing about 90% of the discovered oil ('giant' meaning reserves in excess of 100 million barrels of oil or 1 trillion cu. ft. of gas based on American Association of Petroleum Geologists classifications).

Crude oil reserves in Peninsular Malaysia represent about 62% of total reserves, 23% in Sarawak, and 15% in Sabah. In terms of crude oil quality, most of the crude oil produced is within the "light oil" category (Fig. 3.12). This is based on the cut-off for medium crude oil quality of between 22 and 31 degrees API as defined by World Energy Conference (WEC).

Trends in Crude Oil Reserves Addition

Despite the increase in production in the past, oil reserves were maintained above 3.0 BSTB over the past 9 years, which indicates the success in replacing the crude oil produced. This success is partly attributed to exploration success and also to reserves growth in the discovered fields (Figs. 3.13, 3.14).

The above reserves growth is attributed to better estimation of reserves following the

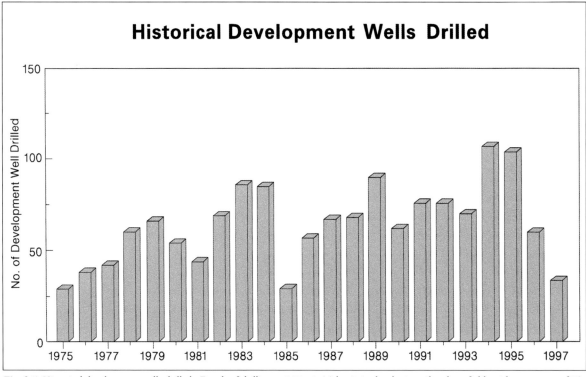

Fig. 3.5 Historical development wells drilled. Trends of drilling activities in Malaysia in developing oil and gas fields with an average of 50-60 development wells drilled per year. A total of 84 development wells were drilled during 1998.

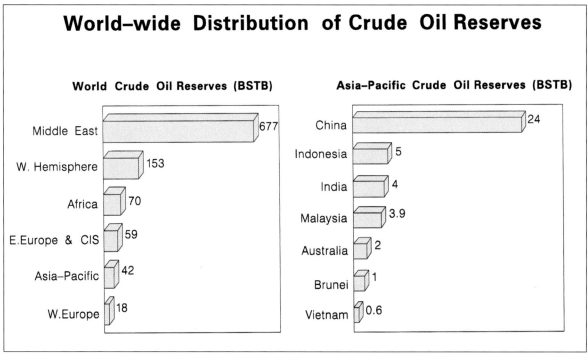

Fig. 3.6 Worldwide Distribution of Crude Oil Reserves as at 1.1.1998. Data from Oil & Gas Journal, Dec 29, 1997.

acquisition of more data from appraisal and development drillings and improved recovery performance of the producing fields. There is always some degree of uncertainty in the reserves estimate of which the extent is dependent upon the latest available information and the field actual performance. However, it is recognised that growth from discovered fields contributes to about half of the total reserves addition.

Crude Oil Development and Production

All discovered resources made to-date, except Asam Paya oil field found onshore Sarawak, are within the Continental Shelf area in water depths of less than 200 metres. Offshore oil field development and production require special consideration during the exploration, appraisal, field development, project implementation and

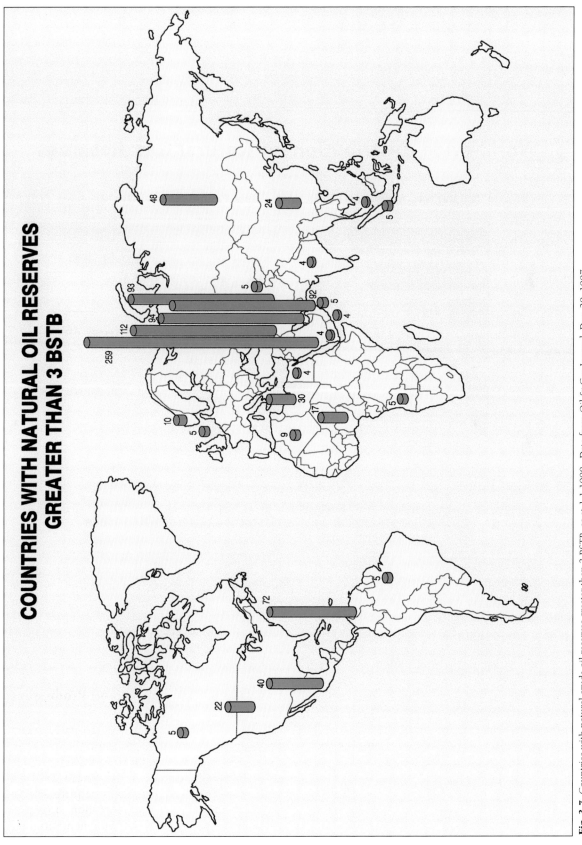

Fig. 3.7 Countries with natural crude oil reserves greater than 3 BSTB, as at 1.1.1998. Data from Oil & Gas Journal, Dec. 29, 1997.

field production phases. The reserves and well productivity need to have certain threshold values to offset the greater expenditure (capital/operating) and long project lead time in offshore environment. However, the Malaysian oil industry underwent a dramatic expansion in the 80s when a total of 18 oil fields were brought into production. The historical oil field development sequence and growth in production, excluding the Miri Field, are shown in Figures 3.15 and 3.16.

Esso also brought onstream its oil discoveries made under the 1985 PSC when the Raya and Yong fields in Block PM-8 were put into production in June 1998. Meanwhile, Carigali is in the advanced stage of developing the oil fields, Malong, Sotong and Anding, known as the MASA fields, in Block PM-12 for production start-up by May 1999. Plans have also been made to bring Sarawak onshore Asam Paya field into production in 1999.

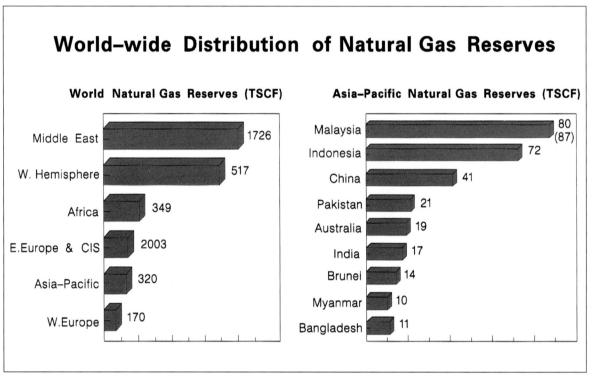

Fig. 3.8 Worldwide distribution of natural gas reserves as at 1.1.1998. Data from Oil & Gas Journal, Dec. 29, 1997.

To sustain the production level at around 600,000 BOPD (Fig.3.16), extensive development projects are being planned for further development in the existing producing fields, and to bring up new fields into production. A number of new development projects have been committed and some are in the preliminary stage of project scoping and techno-commercial evaluation.

The development of oil discoveries made under the 1985 PSC terms began with offshore Peninsular Malaysia Bunga Kekwa oil field in the PM3 Commercial Area in July 1997. The crude is evacuated via Floating Production, Storage and Offloading Vessel (FPSO). This was later followed by the Kinabalu Field in December 1997. Production from the Kinabalu Field is through Samarang facilities. Kinabalu is the first oil field in the region which is remotely operated from the shore in Labuan, about 57 km away.

Historical Crude Oil Prices

The Malaysian crude oil price movement is a benchmark based on Tapis crude price. The historical trend is shown in Figure 3.17. With the exception of price spikes during short-run political and economic crises, crude oil prices since the early 1970s have tended to move closely with the "OPEC Basket". After a short increase during the Gulf War in the early 1990s, the Tapis crude oil price remains below US$20.00 per barrel. It has reached as low as US$14.00 per barrel in August 1998, as most part of the world, including the Asia-Pacific region and Russia, were facing an economic downturn due to devaluation of their currencies. Unless there is a dramatic cut in the world crude oil supply initiated by the OPEC countries, the projected soft oil price is expected to continue for several years.

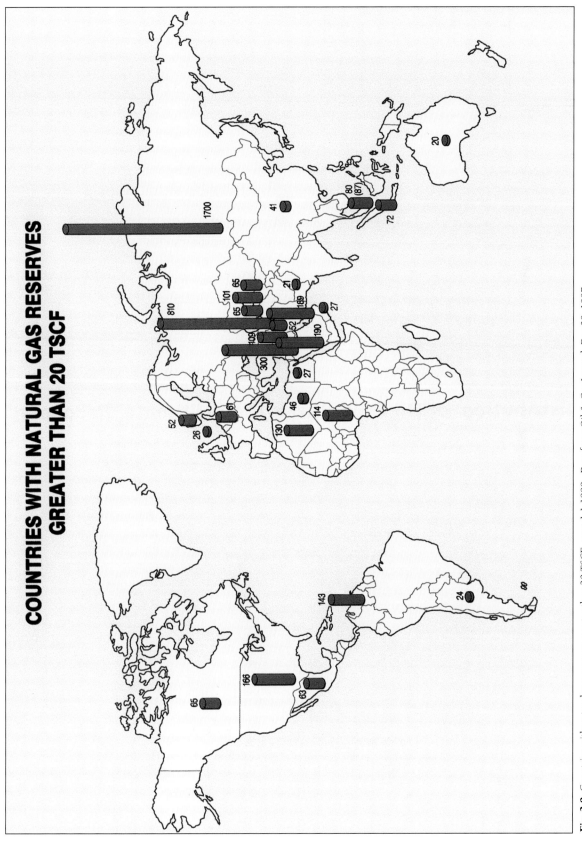

Fig. 3.9 Countries with natural gas reserves greater than 20 TSCF, as at 1.1.1998. Data from Oil & Gas Journal, Dec 29, 1997.

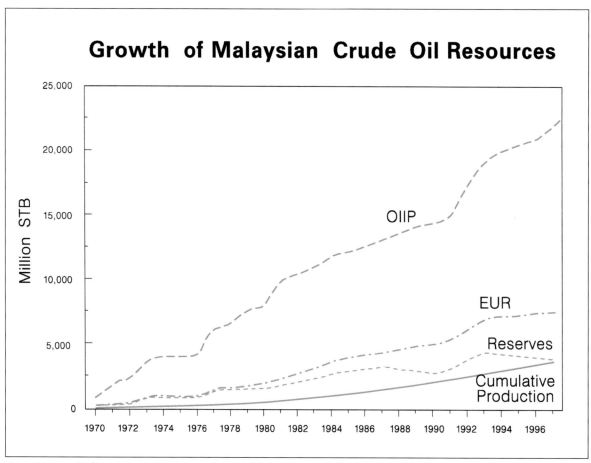

Fig. 3.10 Growth of Malaysian Crude oil resources. As at 1.1.1998, the total EUR stands at 7.7 BSTB.

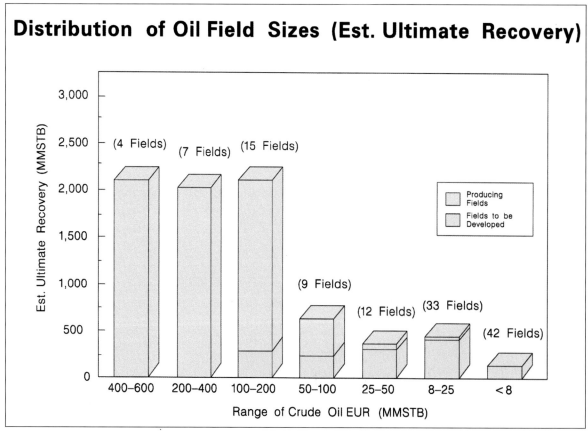

Fig. 3.11 Distribution of oil field sizes (EUR). Majority of major oil fields in Malaysia are currently producing. Efforts are now underway to develop the remaining fields.

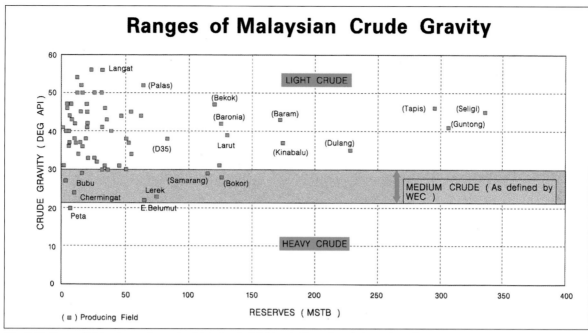

Fig. 3.12 Range of Malaysian crude gravity. Majority of oil fields in Malaysia have light crude quality.

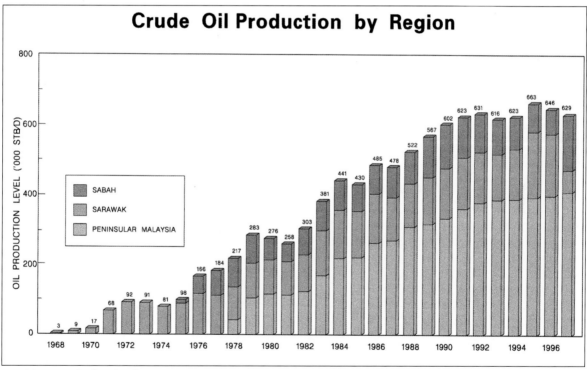

Fig. 3.13 Growth of crude oil reserves in Malaysia since 1970. Reserves addition has been able to sustain Malaysian crude oil reserves above 3 BSTB for the last 10 years. As at 1.1.1998.

NATURAL GAS RESERVES

Malaysia's total natural gas reserves as at 1.1.1998 are estimated at 87.0 TSCF.

Breakdown of Natural Gas Reserves

Figure 3.18 shows that the natural gas reserves additions are increasing. Since 1980 some 56.3 TSCF of recoverable natural gas were added to the resource base (ultimate recovery) to total 98.7 TSCF. This corresponds to an average growth rate of 4.8% per year over the past 18 years. For the same period, a total of 126 gas fields were discovered offshore.

The Gas-initially-in-place (GIIP) and the Estimated Ultimate Recovery (EUR) increased in tandem. Two obvious jumps in GIIP and EUR are noticeable from the plot, in the early 80s and early 90s. These are due to new gas discoveries made by the 1976 and 1985 PSC Contractors during their exploration campaigns. The major

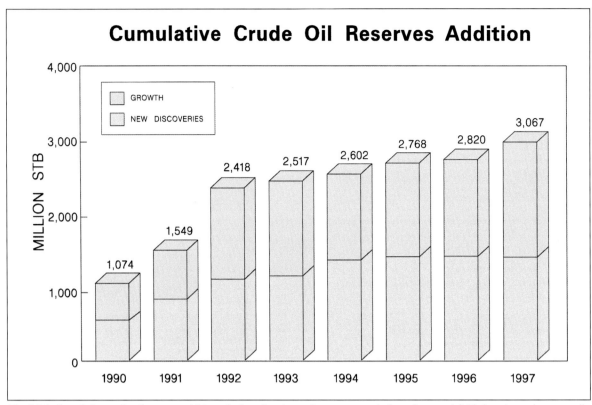

Fig. 3.14 Cumulative crude oil reserves addition. Both exploration and reserves growth from discovered fields have been almost equally significant for crude oil reserves addition in Malaysia.

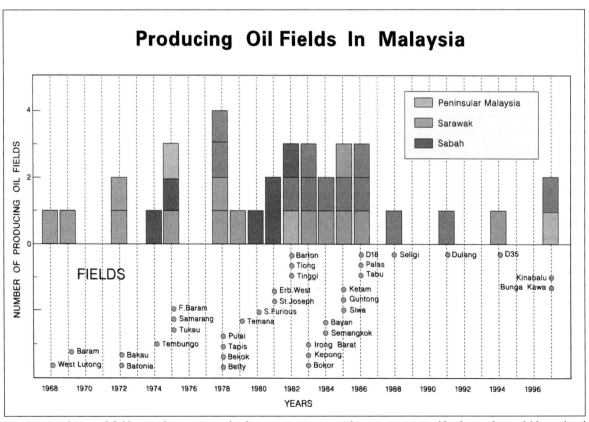

Fig. 3.15 Producing oil fields in Malaysia. Active development activities in Malaysia continue to add value to the available crude oil resources.

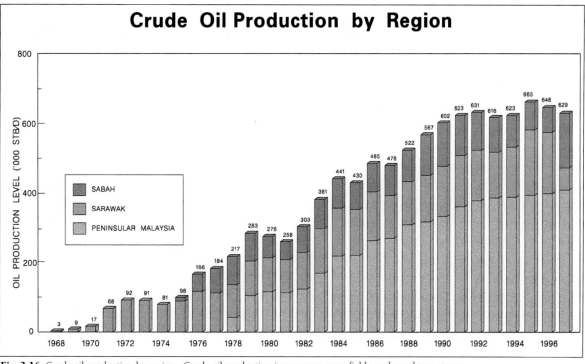

Fig. 3.16 Crude oil production by region. Crude oil production increases as more fields are brought on-stream.

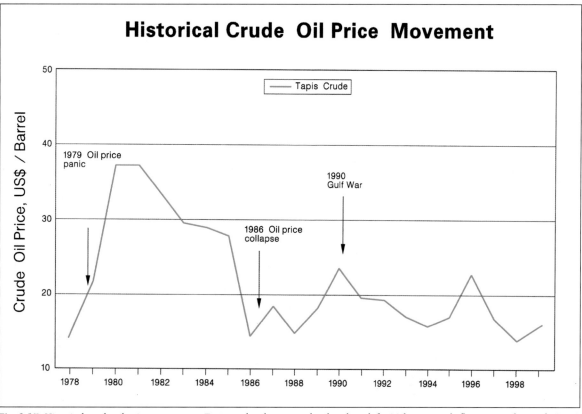

Fig. 3.17 Historical crude oil price movement. Tapis crude oil price used as benchmark for Malaysian crude fluctuates in line with OPEC basket movement.

discoveries made include: Resak and Bergading Deep gas fields in Peninsular Malaysia; F28, Jintan, Helang and Layang gas fields in Sarawak and Kebabangan and considerable gas sands in Kinabalu in Sabah.

As at 1.1.1998, a total of 208 gas fields have been discovered offshore in water depths of less than 200 metres. However, 63 of the total discovered gas fields are located within oil fields. These gas fields, as shown in an example illustrated in Figure 3.19, are stacked gas reservoirs (non-associated gas) within the same trap or structure predominantly containing reservoirs of oil and associated gas, which are

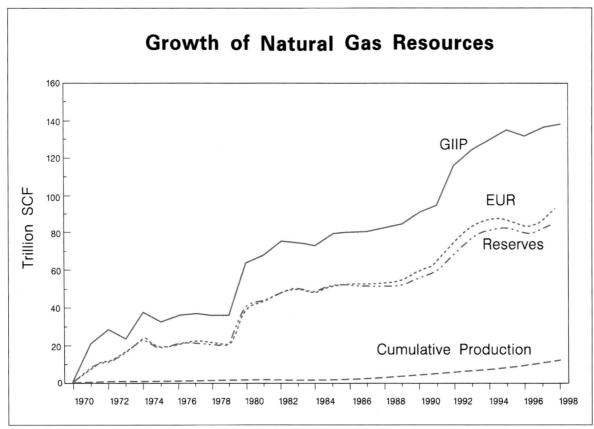

Fig. 3.18 Growth of natural gas resources. As at 1. 1. 1998, total natural gas EUR stands at 98.7 TSCF.

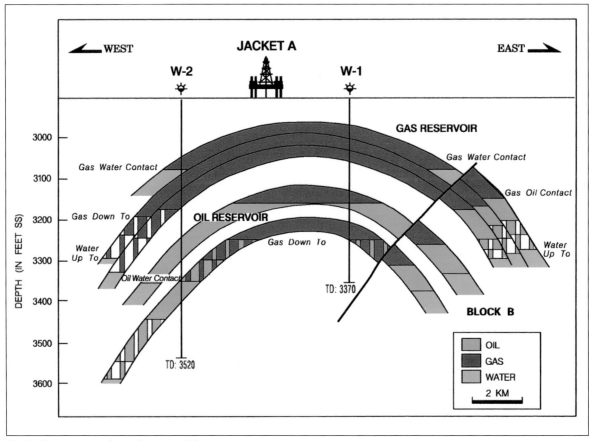

Fig. 3.19 An example of a gas field in an oil field.

considered previously as oil fields. For the purpose of reporting the gas reserves, only gas fields having UR of more than 50 BSCF are included. Of the 208 gas fields, only 122 gas fields met this criterion and have an estimated total GIIP of 138 TSCF. The estimated total UR stands at 98.7 TSCF which correspondingly gives an average gas field's recovery factor of about 72%.

From the 122 gas fields having reserves of more than 50 BSCF, a total of 25 gas fields are considered as 'giant' discoveries with reserves of more than 1.0 TSCF each and form about 65% of the total non-associated gas UR discovered thus far. F6 is the biggest gas field discovered to-date with UR of 6.2 TSCF. Details of field size distribution and total UR discovered are shown in Figure 3.20.

Trends in Natural Gas Reserves Addition

The gas reserves base has continued to increase for the last two decades (Fig. 3.21), indicating the success in adding more reserves than the volume produced annually. The historical trend in the natural gas reserves replacement from growth within the existing discovered fields and additions from new gas discoveries made under the 1985 PSC era over the past 10 years is shown in Figure 3.22.

The remaining reserves addition are attributed to field growth as a result of better field performance, extensive reservoir and geological and geophysical studies, and the incorporation of appraisal and development wells' drilling results.

Geographical Distribution of Non-Associated and Associated Gas Reserves

Geographically, 36.6 TSCF (42%) of the natural gas reserves are in Peninsular Malaysia, 44.3 TSCF (51%) in Sarawak and the remaining 6.1 TSCF (7%) in Sabah (Fig. 3.23). In energy equivalent terms, the total remaining natural gas reserves are estimated to have a heating value equivalent to 17.9 BOE (3.8 BOE for associated gas and 14.1 BOE for non-associated gas) or about 4.6 times larger than the total remaining crude oil reserves.

In summary, 15.9 TSCF or 18% of the total remaining natural gas reserves are developed of which 8.4 TSCF or 53% are in Peninsular Malaysia, 7.2 TSCF or 45% are in Sarawak and only 0.3 TSCF are in Sabah. The majority of the undeveloped natural gas reserves are located in Sarawak as shown in Figure 3.24.

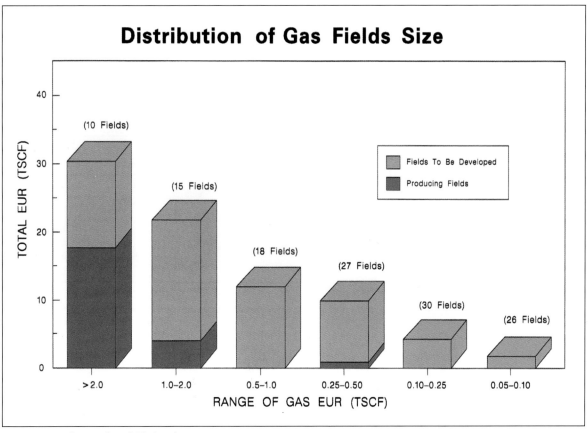

Fig. 3.20 Distribution of gas fields (EUR).

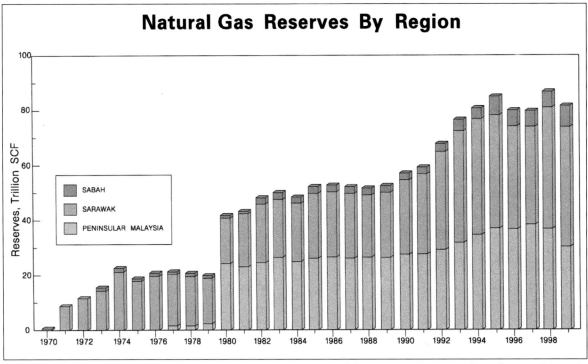

Fig. 3.21 Growth of natural gas reserves in Malaysia since 1970. Majority of the reserves are in Peninsular Malaysia and Sarawak. As at 1.1.1998.

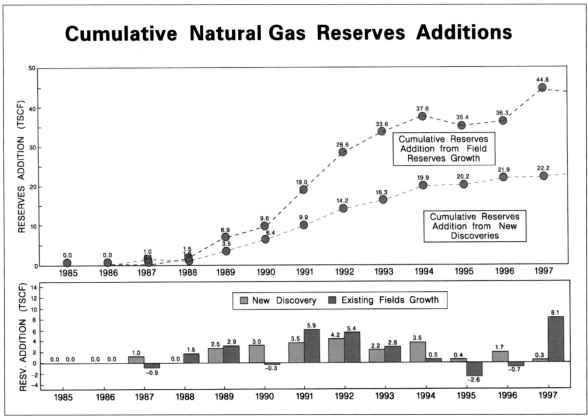

Fig. 3.22 Cumulative natural gas reserves addition.

Non-Associated Gas

Non-associated gas reserves in Peninsular Malaysia account for 38% (27.6 TSCF) of the total non-associated gas reserves and are located in 23 oil fields and 37 gas fields. About 55% (40.6 TSCF) of the reserves are located in 3 oil fields and 47 gas fields in Sarawak and the balance of 7% (4.9 TSCF) in 3 oil fields and 10 gas fields in Sabah (Fig. 3.24).

Out of the total 73.1 TSCF non-associated gas reserves, some 10.2 TSCF (14%) are considered as developed reserves residing in ten producing gas fields including two non-

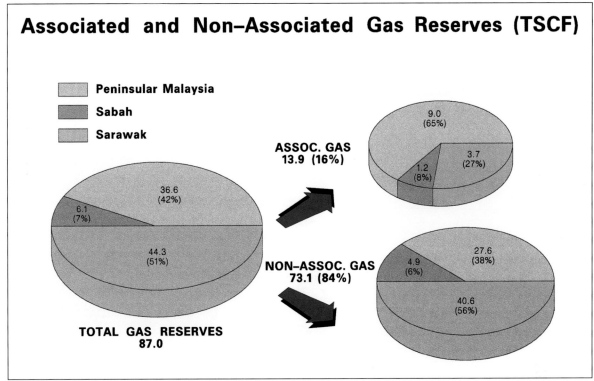

Fig. 3.23 Associated and non-associated gas reserves distribution.

associated gas fields within the Bekok and Tiong oil fields, Duyong, Jerneh and Lawit gas fields in Peninsular Malaysia and E11, F6, F23, M1 and M3 gas fields in Sarawak. Developed non-associated gas reserves in Peninsular Malaysia accounts for 4.2 TSCF and the balance of 6.0 TSCF are in Sarawak.

Associated Gas

Peninsular Malaysia's associated gas reserves account for 65% (9.0 TSCF) of the total associated gas reserves which are contained in 45 oil fields. About 27% (3.7 TSCF) are in 26 oil fields in Sarawak and the balance 8% (1.2 TSCF) in 8 oil fields in Sabah.

Of the total 13.9 TSCF of associated gas reserves, 11.5 TSCF are accumulated in the producing oil fields and 5.7 TSCF (41%) are categorised as developed. The developed reserves are in the 35 producing oil fields i.e. 4.1 TSCF in 14 oil fields in Peninsular Malaysia, 1.3 TSCF in 13 oil fields in Sarawak and 0.3 TSCF in 8 oil fields in Sabah.

Gas Quality, Development and Production

In general, gas quality varies from one area to another. In Malaysia, most of the gas fields have sweet gas, except for a few which have a relatively high CO_2 and H_2S content.

In the 80's, natural gas, which was previously ignored by the industry in the early 1960s, began to gain prominence. As Malaysia is well endowed with gas, with reserves of 4.6 times larger relative to oil in energy terms, PETRONAS began looking at developing this energy source in line with the Government's fuel diversification policy as enacted under the National Energy Policy. This move has enabled Malaysia to gradually reduce its dependence on crude oil from 90% in 1970s to about 51% in 1998. With the available gas infrastructure and pipelines already in place, it has also created spin-off projects such as gas-based petrochemicals, gas-based district cooling, co-generation and the use of natural gas for transportation.

The producing gas fields and historical natural gas production in Malaysia are shown in Figures 3.25, 3.26 and 3.27. Natural gas production in the 1970s was maintained at below 300 MMSCF/D of which mostly were associated gas produced from oil fields in Sabah and Sarawak. Production from Peninsular Malaysia only started in 1978 when the first two EPMI's oil fields, Pulai and Tapis, were brought onstream. The gas production started to gain its momentum in 1983, which exceeded the 300 MMSCF/D level, when the first gas field in Malaysia i.e. E11 in offshore Sarawak started production for supply to Malaysia's first LNG plant in Bintulu. Since then production continued to increase gradually and reached about 1,240 MMSCF/D three years

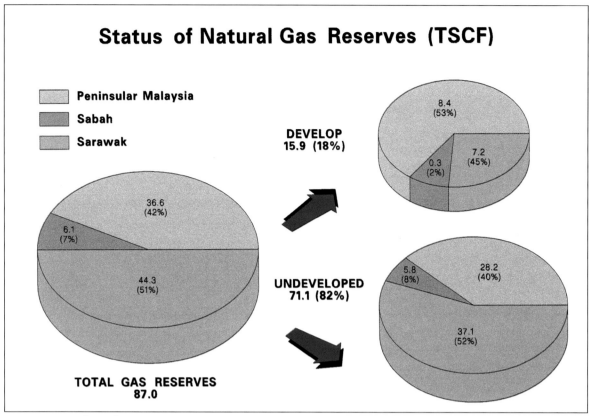

Fig. 3.24 Status of natural gas reserves by region.

later with the start-up of PGU 1 in July 1985. The setting up of gas based industries and the coming onstream of the MLNG-Dua plant in mid 1995 further increased the production two fold to about 3,500 MMSCF/D. In 1997, associated gas from the 35 producing oil fields contributed to about 1,400 MMSCF/D or 30% of Malaysia's total natural gas production of about 4,600 MMSCF/D. Meanwhile, the non-associated gas production in Malaysia has also increased gradually from around 23 MMSCF/D in 1982 to 3,200 MMSCF/D in 1998 from 8 producing gas fields and 2 non-associated gas reservoirs within oil fields.

The 1997 production corresponded to a depletion rate of 1.9% of Malaysia's total gas reserves. The development of gas resources sustains and ensures the security of gas supply to the committed gas projects where demand is expected to peak at 6 BSCF/D dry gas in 2005. Supplying at this rate, the existing gas reserves would last for another 30 years. Based on the current committed gas projects; namely the PGU, Miri-Lutong Sales, SMDS, MLNG/ABF/SESCO/PGB, MLNG-Dua, MLNG-Tiga and Sabah Gas Projects, some 57.5 TSCF (67%) of the natural gas reserves would be required to satisfy demand by end 2020 assuming that all existing gas projects are operateable up to 2020 (Fig. 3.28).

In the next few years, several new gas fields for PGU in Peninsular Malaysia, and MLNG-Dua and MLNG Tiga in Sarawak are planned for development to satisfy the projected near term gas demand.

Based on the projected natural gas demand of 6 BSCF/D in year 2005, and assuming that there is no positive reserves replacement from existing discovered fields and potential future finds, the remaining gas reserves are anticipated to decline to some 33 TSCF by year 2020. This is a reduction of about 56% from 1998's reserves base or at an average depletion rate of 4.3% per year.

Associated Gas Conservation

In the early days, associated gas produced from oil fields was used for petroleum operations including re-injection for petroleum recovery. The remainder was flared or vented. However, its utilisation started to gain momentum when the first gas gathering scheme known as Baram Delta Gas Gathering (BARDEGG) was implemented offshore Sarawak and commissioned in 1993. This was later followed by the Balingian gas collection scheme also in Sarawak.

The BARDEGG scheme collects the associated gas produced from the 5 oil fields which are Baram, Baronia, Bokor, Betty and West Lutong in Baram Delta area operated by Carigali on a 50:50 joint venture with Shell. The collected gas, less the petroleum operations usage, is compressed for re-injection in Baronia oil field and for sales to Miri-Lutong and Bintulu areas. It is currently supplying 60 MMSCF/D of gas to MLNG plant in Bintulu via the E11-Bintulu trunkline and up to 21 MMSCF/D to Miri-Lutong consumers.

Similar schemes installed in the Balingian area, specifically the D35 and Bayan oil fields, are meant to collect associated gas produced in these fields for supply to SMDS and MLNG plant in Bintulu. Currently, the two fields are supplying a total of 50 MMSCF/D of gas to MLNG plant as the SMDS plant is shut down for re-building after an explosion occurred in December 1997.

Historically, the associated gas production has increased tremendously starting from only 28 MMSCF/D in 1970 and later reaching about 1,400 MMSCF/D in 1997 and 1998, an increase of 50-fold or 15% per year as shown in Figure 3.27.

The trend in the associated gas utilisation/conservation has been very encouraging as shown in Figure 3.29. The success in optimising associated gas utilisation resulted from the close surveillance of PS Contractors' production level, securing market for the gas, implementing gas gathering schemes for sales and re-injection for reservoir management and storage, exercising prudent and modern oil practices and through regular discussions with PSC contractors. Despite an increase in the associated gas production, the percentage of total gas flared and vented has been reduced steadily from more than 96% before 1974 to about 86% in 1981 and to about 12% in 1996. However, a slight increase was recorded in 1997 and 1998 simply because some oil fields are cutting higher GOR, and the production from new fields that do not have a gas collection scheme due to economic reason. Nevertheless, the total flaring and venting level of the associated gas is anticipated to decline further to less than 10% by the year 2001.

Associated gas conservation efforts in future oil field development are expected to pose a great challenge to the upstream business in Malaysia as it may further erode projected economics as the fields are smaller. As such initiatives to conserve gas would therefore be pursued only if there are net benefits to all parties involved unless there is a law imposed by the Government.

The start-up of Kinabalu Field production in December 1997 indicates the success of another gas conservation effort, where all associated gas produced from the field, less the petroleum operation usage, are collected and piped to the Labuan demand centre.

To date, a total of 14 oil fields have already had gas re-injection facilities and two more will be implemented in St. Joseph and

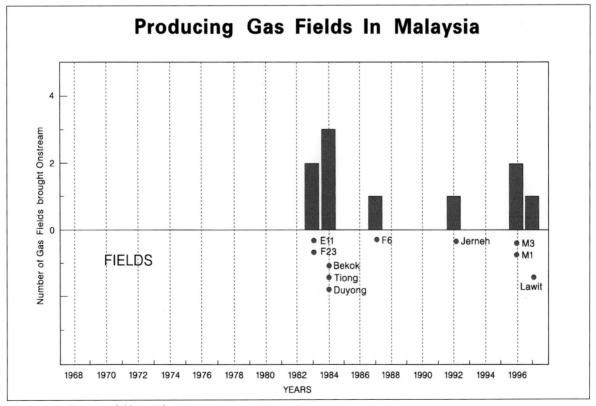

Fig. 3.25 Producing gas fields in Malaysia.

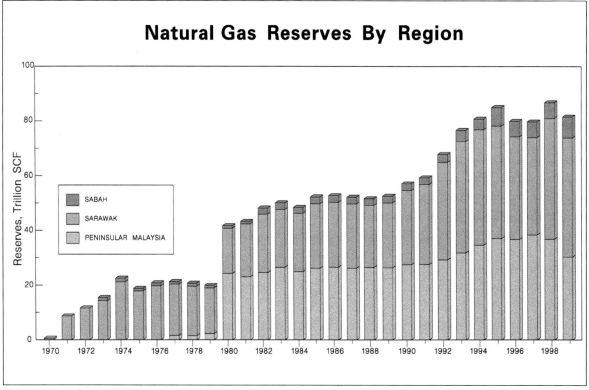

Fig. 3.26 Natural gas production by region. Gas supply increases to meet demand.

Barton fields to enhance oil recovery and to reduce wastage of associated gas produced. If the planned utilisation of high CO_2 gas for improved oil recovery (IOR) activities proves viable, the flaring/venting level is anticipated to be further reduced significantly.

CONDENSATE RESERVES

Condensate reserves refers to the hydrocarbon liquids which are stable at ambient conditions based on the volume to be recovered from the projected natural gas supply to the existing and

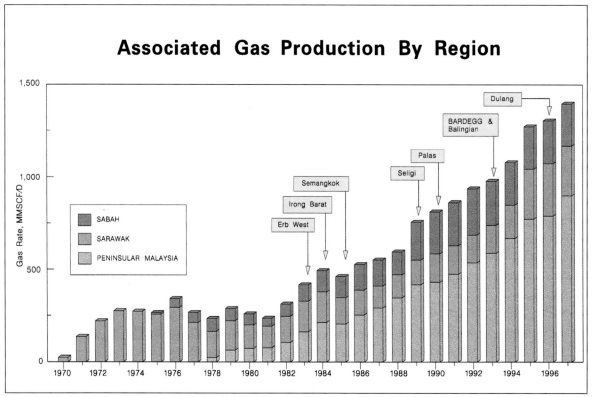

Fig. 3.27 Associated gas production by region, contributing about 30% of total natural production.

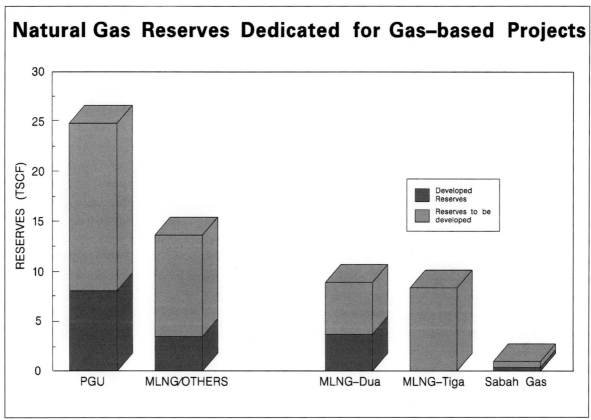

Fig. 3.28 Natural gas reserves dedicated for gas-based projects.

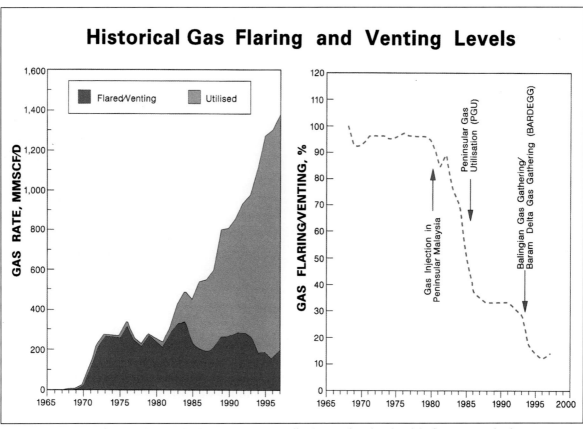

Fig. 3.29 Historical gas flaring and venting levels. Gas conservation effort has significantly reduced the flaring/venting level.

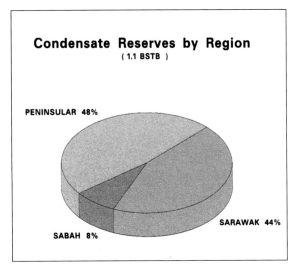

Fig. 3.30 Condensate reserves by region. As at 1.1.1998.

future firm gas projects and from dedicated and potential future source of supply.

The total condensate reserves as at 1.1.1998 is 1.1 BSTB, of which 0.3 BSTB or 27% are currently categorised as developed reserves. The reserves are expected to be recoverable from gas fields which have been dedicated to gas projects such as PGU, MLNG, MLNG-Dua, MLNG-Tiga, Sabah Gas and SMDS. The breakdown of total condensate reserves by region is illustrated in Figure 3.30.

The condensate reserves include the estimated volume of potentially recoverable condensate from only the C_5 and C_5+ components of the total natural gas reserves. The reserves are mainly contributed by the Guntong, Palas, Tabu, Seligi and Jerneh fields in Peninsular Malaysia, and by the M1 and M3 fields in Sarawak and Kinabalu Field in Sabah. Within the next 5 years, condensate production is expected to reach a peak level of 100,000 STB/D as a number of condensate rich gas fields were recently brought on-stream.

CONCLUSIONS

Petroleum resources are valuable assets to Malaysia. The revenues and foreign exchange earnings generated from the production and exports of these finite and depletable resources are recognised as one of the major contributors to Malaysian economic growth in the 1970s and 1980s.

As such, appropriate near and long term planning of these resources are important in providing a more sustainable revenues to the Government as well as ensuring better security of energy supply for future development of the national economy.

To sustain the reserves base, a net positive reserves replacement rate needs to be maintained or even improved. The process involved in replacing the petroleum reserves is illustrated in Figure 3.31.

The major challenge ahead for us is to further add and find new oil and gas resources domestically in order to have a more sustainable oil and gas exploration, development and production businesses and also to prolong our status as a net exporter of oil and gas/LNG. Another challenge is to dynamically position ourselves in this fast changing environment so that values could be continuously created/added domestically. The upstream oil and gas business will continue to explore and appraise for additional reserves in the existing and new unexplored acreages. Since the 1976 PSC era, Production Sharing Contractors operating in Malaysia have invested over US$ 25 billion for the exploration, development and production of oil and gas in Malaysia.

The prevailing environment surrounding world-wide upstream exploration and production activities has been much driven by the relatively soft crude oil prices and escalating costs. To be continuously able to attract foreign investment in the exploration activities in the country, PETRONAS has proactively been introducing new terms and improvements to our Production Sharing Contracts to ensure its competitiveness. This will help to ensure that Malaysia remain a net oil exporter during next decade. The Production Sharing Contracts which were improved in 1985 provided incentives for such ventures. The third generation PSC, after the 1976 PSC and 1985 PSC, adopts the profitability based fiscal regime concept. The "R/C PSC" provide incentives to develop smaller discoveries.

It remains a priority to maximise the petroleum recovery rate in order to achieve the optimum value creation from fields in production and fields approved for development. As production from existing fields is declining and smaller fields are brought on stream, the average costs to develop and produce a barrel of oil will increase. It will be a major challenge to create conditions that permit the establishment of an economic basis for the development of these resources. The economic feasibility for developing smaller fields is more sensitive to oil price, cost, geographical location vis-a-vis distance from existing infra-structure, etc..

The challenge is to further reduce development cost using latest state-of-the-art technology to make development of smaller and marginal fields feasible. In addition, effort to

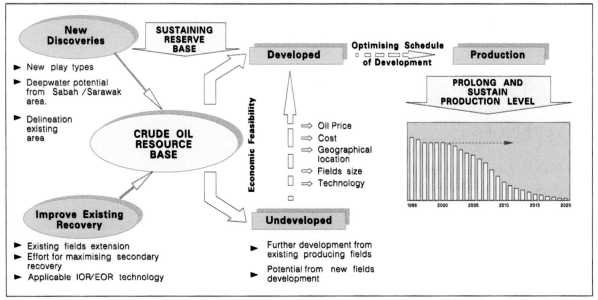

Fig. 3.31 Reserves replenishment process (Example on crude oil). Reserves addition mainly achieved from the intensive and effective exploration activities and efforts in maximising recovery from the discovered fields.

acquire additional reserves from existing fields through improved and/or enhanced oil recovery will continue. There is also a need to look for other sources in new territory and frontier areas.

Since the petroleum resources are finite and depletable assets, PETRONAS must also venture overseas to add to the petroleum resource base and to sustain the business portfolios. This business venture is important to help to provide Malaysia with security of energy supply for a more sustainable socio-economic development.

REFERENCES

Abdul Jalil Zainul, Rosly M. Nor, Teh, Y.H., Egbogah, E.O., Hamdan, M.K., Foo, W.Y., Musbah, M.W., 1997a. 'An Integrated Approach to Petroleum Resources Definitions, Classification and Reporting', paper SPE 38044 presented at the 1997 APOGCE, Kuala Lumpur, April 14-17.

Abdul Jalil Zainul, Rosly M. Nor, Teh, Y.H., 1997b. 'Petroleum Resources Definitions, Classification and Reporting from National Oil Company Perspective', Proceedings 97' ASCOPE Conference, Jakarta, November 26-28, v.1, 99-106.

Part 2

REGIONAL TECTONICS AND SEDIMENTARY BASINS OF MALAYSIA

Chapter 4

Plate Tectonic Elements and Evolution of Southeast Asia

Mazlan B. Hj. Madon

INTRODUCTION

Southeast Asia is a geologically complex region that has evolved through various phases of continental accretion, mountain building and rifting. Malaysia occupies a central position in Southeast Asia such that its geological history is inextricably linked to that of the whole region. The geology of Peninsular Malaysia and western Sarawak is much older than in other parts of Sarawak and Sabah, and is represented by rocks as old as Cambrian that occur in Langkawi, NW Peninsular Malaysia. Much of Sarawak and Sabah and the adjacent continental margins evolved during the Late Cretaceous and Tertiary from an older continental nucleus by accretion and amalgamation of continental fragments welded together with subduction complexes and island-arcs. The contrasting tectonic histories of western and eastern Malaysia have had a significant influence on the structural styles of the major petroleum-bearing basins that had developed in the last 40 million years. It is imperative, therefore, to have a good understanding of the tectonic history of the region as a framework for oil exploration.

This chapter summarises the geological evolution of Southeast Asia including the major tectonic elements, the Palaeozoic and Mesozoic tectonic events leading to the Indosinian Orogeny, and the Cenozoic history. The region has been the focus of many studies by academia, government geological survey departments, and the oil industry, and is probably one of the finest natural laboratories in which to study the processes of plate tectonics. This is an area of active plate movement where, processes such as volcanic activity associated with subduction zones, the formation of back-arc or marginal basins, and the collision of continents can be observed. There are many sources of information on the geology of Southeast Asia for the interested reader. Two earlier publications, *Tectonics of the Indonesian Region* by Hamilton (1979) and *Geological Evolution of Southeast Asia* by Hutchison (1989), contain summaries of Southeast Asian geology. Metcalfe (1988, 1996) discussed the pre-Tertiary tectonics of Southeast Asia. Daly et al. (1987, 1991) and Hall (1996, 1997) described the Cenozoic plate tectonic evolution of Southeast Asia. Rangin et al. (1990) and Jolivet et al. (1989) discussed the evolution of the western Pacific margin, particularly the Philippines. Holloway (1981, 1982) proposed a plate tectonic model for the evolution of the South China Sea Basin. Subsequent authors (e.g. Hutchison, 1986; 1996a) have adopted this general model in their reconstructions. Packham (1990, 1996) reviewed the Tertiary tectonic framework of Southeast Asia and discussed the India-Asia collision, the opening of the South China Sea Basin, the kinematics of the Philippine Sea Plate, and the tectonics of eastern Indonesia. Hall (1996, 1997) presented an animated reconstruction of the plate tectonic evolution of Southeast Asia.

TECTONIC ELEMENTS

Much of present-day Southeast Asia is covered by seas with numerous islands. Continental and transitional crusts generally underlie the shallow shelf seas, which have water depths of 200 m or less. Ocean crust underlies the deeper ocean basins. The megatectonic features of Southeast Asia can be explained generally by plate tectonics, the interaction of torsionally rigid lithospheric entities called plates. Southeast Asia is surrounded by three major lithospheric plates: the Indian-Australian, Pacific, and Philippine Sea plates. These are converging towards Eurasia at a relative velocity of about 6 to 8 cm per yr (Fig. 4.1). The boundaries between these plates are active subduction zones that circle the region almost continuously from the Ryukyu arc in the north-east, through the Philippines, Java, and Sumatra, to Myanmar in the northwest. These active plate margins are in most places associated with shallow to deep seismicity, well-developed Benioff-Wadati zones, trenches, accretionary

Evolution of Southeast Asia

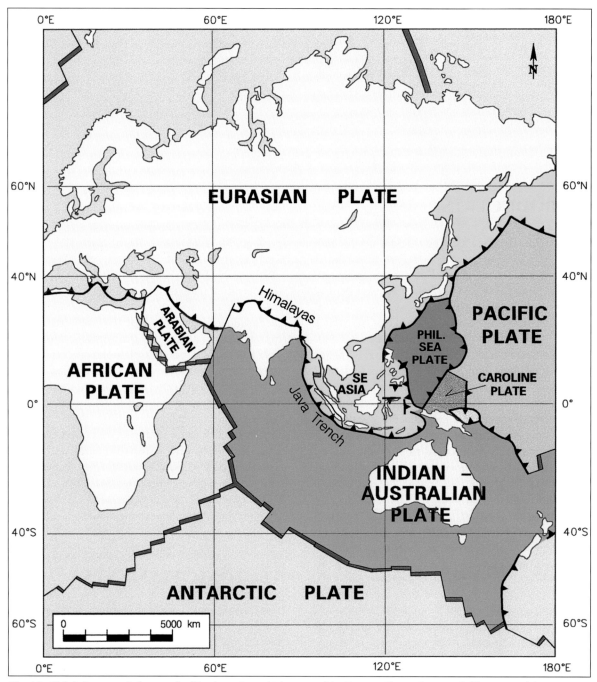

Fig. 4.1. Global plate tectonic framework of Southeast Asia, which is part of the Eurasian Plate and surrounded by active subduction zones (bold lines with barbs) that mark the boundaries with major oceanic plates: India-Australia (which is partly continental), the Philippine Sea, and the Pacific. Red lines represent mid-ocean ridges. After Hamilton (1979).

complexes, and volcanic arcs. The northward motion of the India-Australia Plate results in a highly oblique convergent plate boundary along the Sumatra arc, which results in a strong right-lateral shear component to the subduction-related deformation (McCaffrey, 1996; Malod and Kemal, 1996).

Southeast Asia consists of a pre-Tertiary continental core called Sundaland that underlies most of its western region from Myanmar to western Borneo (Fig. 4.2). The oldest outcropping rocks in Sundaland are the Kontum Massif of eastern Vietnam. Sundaland is effectively a southern appendage of the Eurasian Plate that is welded to, and sandwiched between, India and South China. Peninsular Malaysia and southwestern Borneo form contiguous parts of this continental terrane that once formed a major landmass during the Mesozoic and early Cenozoic. In contrast, eastern Southeast Asia is geologically young, having evolved during the last 50 to 60 million years (Late Cretaceous-Tertiary). The geology here is more oceanic in character, dominated by accreted island-arcs interspersed with numerous microcontinental blocks of diverse origins and obducted ophiolites. The

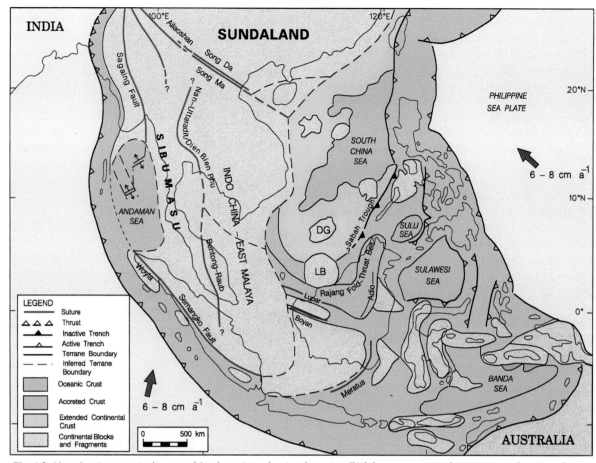

Fig. 4.2. Map of major tectonic elements of Southeast Asia, showing the major allochthonous continental terranes, crustal types, and suture zones. Note the different tectonic settings of Peninsular Malaysia, Sarawak and Sabah. Peninsular Malaysia and adjacent areas are underlain by continental crust whereas Sarawak and Sabah are underlain by accreted crust that may include exotic terranes such as the Luconia Block (LB). DG - Dangerous Grounds. Arrows show present-day direction and rate of relative plate motion. Modified from Metcalfe (1996).

Rajang Fold-Thrust Belt, which extends across almost the entire northern part of Borneo, in Sarawak and Sabah (Fig. 4.2), is the relict of a Late Cretaceous-Early Miocene deep oceanic basin whose sediment fill has been deformed and exhumed into a major accretionary wedge complex (Hutchison, 1996b). The deep marine rocks found in this complex were almost coeval with those in the East Sabah Ophiolite-Melange Zone (Hutchison, 1988), which comprises mainly Lower Cretaceous ophiolitic association of ultrabasic rocks, chert, and spilite, overlain unconformably by Middle Miocene and younger deepwater sediments and olistostromes.

The offshore region between South China/Vietnam and Borneo is underlain by transitional/oceanic crust of the South China Sea basin that opened during the Oligocene (about 32 Ma) (Taylor and Hayes, 1980; Briais et al., 1993). The margins of this basin are underlain by continental blocks produced during the rifting event, including the Luconia and Reed Bank blocks on the southern margin of the basin. The Rajang Fold-Thrust Belt is thought to represent an uplifted accretionary prism produced by the collision of the Luconia and Reed Bank microcontinental fragments with the West Borneo Basement (Hazebroek and Tan, 1993).

PRE-TERTIARY EVOLUTION

The tectonic evolution of Southeast Asia during the Palaeozoic and Mesozoic is closely related to the gradual break-up of the supercontinent Pangaea and the evolution of the eastern Tethys seaway. During Palaeozoic and Mesozoic times the Tethys Ocean occupied a triangular gap between two prongs of the Pangaean continent, Gondwanaland to the south and Laurasia to the north (Fig. 4.3). Stratigraphic, palaeobiological, and palaeomagnetic studies have shown that continental Southeast Asia consists of several major tectonostratigraphic terranes that were formed during the break-up of Pangaea and the opening and closure of successive Tethys oceans (Metcalfe 1988, 1991, 1996). Metcalfe (1996) recognised at least three successive Tethys (Palaeo-, Meso-, and Neo-Tethys) that closed during different collisional events. Sundaland itself is formed of two composite terranes that are separated by a major ophiolitic suture zone that runs N-S from Thailand to southern Peninsular Malaysia and probably into Sumatra. It is well-established now that these two major blocks were once separated by the Palaeo-Tethys Ocean. This is now represented by a rather discontinuous belt of ophiolitic rocks along the Bentong-Raub, Nan-Uttaradit, and Dien-Bien Phu sutures (Fig. 4.2). On the western side of the suture is an amalgamation of continental terranes called Sinoburmalaya (Gatinsky and Hutchison, 1986) or Sibumasu (Metcalfe, 1996). This includes the Shan States of Myanmar, western Thailand and Malaya, and part of Sumatra. On the eastern side is a similar terrane that comprises Indochina, eastern Malaya, and probably western Borneo (Fig. 4.2). Sundaland is therefore the eastern end of a major Late Triassic orogen, the "Cimmerides" (Sengor and Hsu, 1984) or "Indosinian" Orogen (Hutchison, 1989), that resulted from the closure of the Palaeo-Tethys Ocean.

There is also compelling evidence that suggests that all the Southeast Asian tectonostratigraphic terranes were an integral part of the supercontinent Gondwanaland during early Palaeozoic times (Hutchison, 1989; Metcalfe, 1996). Sibumasu and the Indochina-East Malaya blocks have distinct floral and faunal affinities that suggest they were mutually exotic. Metcalfe (1988, 1991) summarised the palaeontological evidence. East Malaya and Indochina contain late Palaeozoic-Mesozoic floras and faunas that bear no relation to Gondwanaland, but have affinities with South and North China (Cathaysia). According to some authors (e.g. Hutchison, 1989) these terranes must have rifted off Gondwanaland during the Ordovician and Silurian, and were probably attached to Eurasia by the Late Permian. Sibumasu is probably a composite of terranes that originated from NW Australian Gondwanaland during the late Palaeozoic, because of its distinctive Gondwanaland features of Carboniferous-Early Permian glaciomarine diamictites and Early Permian cool-water faunas. In order to have developed Gondwanaland floral signatures the Sibumasu terranes must have been attached, or were close to Gondwanaland, at least until Carboniferous-Permian times, to have been affected by glaciation (Hutchison, 1993). Sibumasu terranes became separated from Gondwanaland during the late Early Permian and probably collided with East Malaya-Indochina during the Late Triassic Indosinian Orogeny (Metcalfe, 1996) (Fig. 4.3).

The pre-Tertiary evolution of Southeast Asia involved the progressive suturing of these allochthonous continental terranes that had rifted off from Gondwanaland since the Late Palaeozoic (Gatinsky and Hutchison, 1986; Sengor and Hsu, 1984; Sengor, 1987; Metcalfe, 1988, 1996). Metcalfe (1996) recognised three main phases of rifting: Devonian, Carboniferous-Early Permian, and Late Triassic-Early Jurassic. These culminated in the closure of successive Tethys oceans and in the collision of terranes along major suture zones.

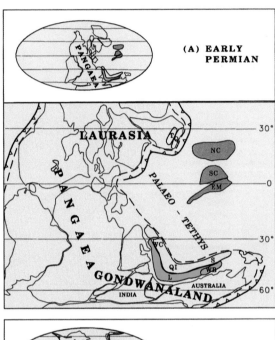

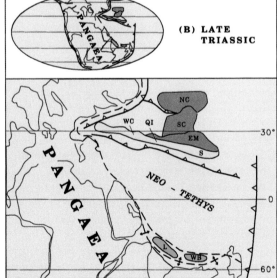

Fig. 4.3. Reconstructions of Pangaea and Tethys showing the distribution of major land-masses and oceans during Permo-Triassic times. (A) Early Permian: Sibumasu Block (yellow) still attached to northern Gondwanaland. East Malaya sutured onto South China in northern Palaeo-Tethys. (B) Late Triassic: Sibumasu has drifted northwards. Suturing of East Malaya and Sibumasu (of which western Malaya is part) and closure of Palaeo-Tethys, creating the Neo-Tethys behind it. After Metcalfe (1996). NC-North China, SC-South China, WC-Western Cimmerian Continent, QI-Qiangtang, L-Lhasa, S-Sibumasu, WB-West Borneo, EM-East Malaya.

The latter are the Song Ma Suture (Late Devonian), the Bentong-Raub Suture (Late Triassic), and the Woyla-Meratus-Boyan Sutures (Middle-Late Cretaceous) (Fig. 4.2). The Carboniferous-Permian rifting and drifting phase ended during the late Triassic with the closure of the Palaeo-Tethys Ocean. The consequent continental collision of Sibumasu and the Indochina-East Malaya Block resulted in the Indosinian Orogeny (Hutchison, 1989).

The Indosinian Orogeny caused the widespread emplacement of tin-rich granitoids throughout central Southeast Asia, from northern Thailand through Peninsular Malaysia to the Indonesian tin islands off Sumatra (Cobbing et al., 1992). This period must have been a phase of crustal thickening, uplift, and exhumation. Subsequently, during the Late Triassic to the Cretaceous the region underwent further post-orogenic uplift, crustal extension, strike-slip faulting, and extensional basin formation. The intermontane basins of Indochina and in the Central and Eastern belts of Peninsular Malaysia have been interpreted by some workers as being the result of late-orogenic extension (Metcalfe, 1989; Gabel et al., 1993). The Gagau Group (?Upper Jurassic-Lower Cretaceous) in Peninsular Malaysia consists of continental sediments and extrusive rocks in a wedge-shaped basin, bounded to the west by the east-dipping Lebir Fault (Burton, 1973), one of the major NW-trending strike-slip faults that may have been active during the Late Cretaceous-Tertiary (Tjia, 1972). The Late Cretaceous was also a time of widespread granite emplacement, for there are occurrences of Late Cretaceous granites on Peninsular Malaysia and in the surrounding South China Sea region (Hutchison, 1989, p. 314). There is also an important Late Cretaceous granite belt extending from Phuket in southern Thailand to Myanmar.

The continental terranes that constitute Sundaland probably extend into western Borneo as the West Borneo Basement (Haile, 1974) to form the relatively stable cratonic part of Borneo. The rest of Borneo consists of younger material that was accreted onto southern Sundaland during the Late Cretaceous and Tertiary. The western, eastern, and southern Sundaland margins, from Sumatra to northern Borneo, were the sites of subduction and continental accretion during the Late Cretaceous and early Tertiary. Ophiolites, melanges, and accretionary complexes of Late Cretaceous-early Tertiary age occur along the margins of Sundaland. Examples include the Woyla Group of Sumatra (Wajzer et al., 1991),

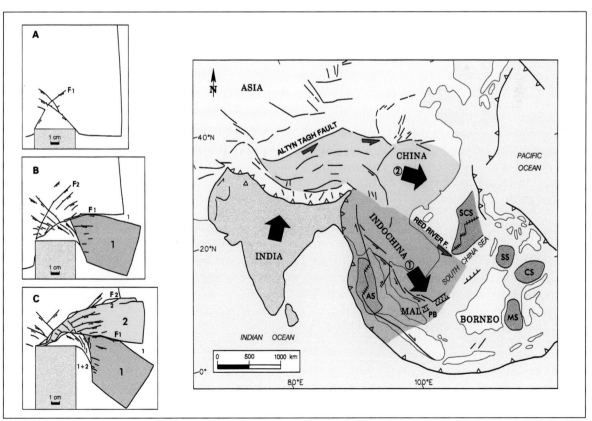

Fig. 4.4. Extrusion tectonics hypothesis as a model for the Tertiary tectonic evolution of Southeast Asia. Left: A to C represents the results of plane indentation experiment of Tapponnier et al. (1982), showing the effects of indentation by a rigid body into plasticine. Note the sequence of faulting (F1, F2) and extrusion/block rotation (1, 2). Right: Sketch of Southeast Asia tectonics, showing the broad resemblance with the experimental results. The first extrusion involved Indochina, resulting in left-lateral motion along the Red River Fault. The second extrusion involved the China Block and caused the Red River Fault to be reactivated in a dextral sense (after Tapponnier et al., 1982).

the Meratus Ophiolite of southern Borneo (Hamilton, 1979), and the Rajang Fold-Thrust Belt of northern Borneo (Hutchison, 1996b).

THE INDIA-ASIA COLLISION

While terrane accretion dominated the pre-Cenozoic evolution of Southeast Asia, processes of strike-slip tectonics, marginal basin formation, and arc collision were important during the Cenozoic. Two major events have shaped the region during the Cenozoic: (1) the India-Asia collision between 50 and 45 Ma (Dewey et al., 1989; Peltzer and Tapponnier, 1988) and (2) rifting in the South China Sea Basin between 32 and 17 Ma (Taylor and Hayes, 1980; Briais et al., 1993). Tectonic processes in the eastern part of Southeast Asia, involve arc-arc and arc-continent collision, back-arc extension, and sea-floor spreading. These, coupled with the arrival of the Australian passive margin, have also contributed to the complex history of Southeast Asia (Packham, 1996).

In a simple experiment of continental indentation using plasticine, Tapponnier et al. (1982) proposed that the Middle to Late Eocene India-Asia collision caused the extrusion of continental slivers along major strike-slip faults (Fig. 4.4). Tapponnier and his co-workers argued that continental blocks, such as Sundaland, were displaced to the southeast and east by almost a thousand kilometres away from the collision zone. The displacement between the extruded blocks was taken up along major strike-slip fault zones. Daines (1985) and Polachan and Sattayarak (1989) interpreted the Pattani Trough and Malay Basin as pull-apart basins that developed along NW-trending strike-slip faults as a result of the India-Asia collision. Daines (1985) named the zone of extensional basins from the Gulf of Thailand to Natuna the Malay-Natuna-Lupar Shear Zone.

Ongoing debate regarding the role of the India-Asia collision in controlling the Cenozoic tectonics in Southeast Asia centres upon whether the indentation of India is accommodated by underthrusting beneath Tibet, by lateral extrusion, or by crustal thickening. Strike-slip faults such as the Wang Chao, Three Pagodas, and Red River show left-lateral displacements of several hundred kilometres (Lacassin et al., 1993; Maluski et al., 1993; Scharer et al., 1990, 1993, 1994) and, thus, seem to support the extrusion model. On the other hand, Dewey et al. (1989) argued that the India-Asia collision was accommodated mainly by crustal thickening in Tibet, and that underthrusting and lateral extrusion of continental blocks were insignificant. Dewey and co-workers believe that the progressive indentation of India into Asia had resulted in an overall dextral megashear in Southeast Asia (Fig. 4.5) which could have reactivated pre-existing fault zones along which extensional and pull-apart basins formed. A dextral megashear model, as opposed to the extrusion model, takes account of the fact that Sundaland consists of smaller fault-bounded blocks that may have rotated and slid past each other during deformation. The internal deformation of Sundaland by distributed shear may have resulted in the complex palaeomagnetic rotations observed during the early Tertiary (Richter and Fuller, 1996).

A model of distributed deformation for Southeast Asia could well explain the lack of consistent and often contradicting palaeomagnetic results from different parts of the region. Thus, on a regional scale, Sundaland is a structurally heterogeneous crustal block containing pre-existing NW-trending faults that has undergone dextral shear as a result of the oblique collision of the Indian margin with Eurasia (Fig. 4.5). Consequently, these major NW-trending fault zones were reactivated as left-lateral shear zones during the Tertiary. Mazlan Madon (1997) used this idea in his model of distributed left-lateral shear for the development of the Malay Basin, offshore Peninsular Malaysia.

EVOLUTION OF THE SOUTH CHINA SEA

The Sarawak and Sabah basins in northern Borneo form a continuous sedimentary basin at the southern margin of the South China Sea marginal basin. Hence, the tectonic evolution of the South China Sea must have had a major impact on the structural evolution of the Sarawak and Sabah basins. The South China Sea Basin is almost triangular-shaped. The average water depth is about 4 km. Taylor and Hayes (1980) and Briais et al. (1993) recognised a symmetrical pattern of roughly east-west trending magnetic anomalies that indicate sea-floor spreading activity during the Middle Oligocene to Early Miocene (32-17 Ma). The average heat flow of between 85 and 115 mWm^{-2} also agrees with the theoretical heat flow of oceanic crust of the same age (Parsons and Sclater, 1977). There appears to have been three episodes of spreading, corresponding to magnetic anomalies 11-10, 10-7 and 6b to 5c (Briais et al., 1993).

Holloway (1981, 1982), Hutchison (1986), and Ru and Pigott (1986), among others, have attempted to reconstruct the plate tectonic evolution of the South China Sea. A rifting stage during the ?Late Cretaceous to Eocene times caused the formation of rift basins in the southern China margin, such as the Pearl River Mouth Basin. The existence of Palaeocene-Eocene rift basins in the South China margin (e.g. Wu, 1988; Yu, 1990) suggests that crustal extension was already widespread before the India-Asia collision and the initiation of movements on the Red River Fault (Briais et al., 1993). This extended continental margin later became the trailing edge of the South China Sea marginal basin. Sea-floor spreading in the South China Sea during the Oligocene caused several micro-continental blocks, such as the Reed Bank and Luconia Block, to drift southwards and collide with the northern Borneo margin (Holloway, 1981). In the south, a remnant ocean basin was being subducted beneath Borneo to form the Upper Cretaceous-Lower Miocene Rajang and Crocker Fold-Thrust Belts which include the West Crocker and Kudat formations in western Sabah (see Chapter 5). Subduction of that remnant ocean basin, the so-called "proto-South China Sea" (e.g. Hall, 1996), beneath Borneo, along the Lupar Line subduction zone may have provided the slab-pull force that drove rifting in the South China Sea Basin. The proto-South China Sea was an ocean basin that once separated the Luconia, Dangerous Ground, and Reed Bank carbonate platforms from Borneo but is now preserved as the Rajang-Crocker Fold-

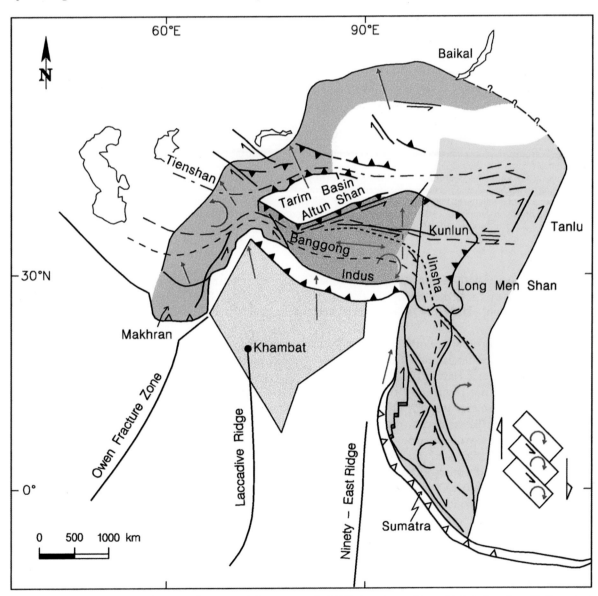

Fig. 4.5. Tectonic zones of the India-Asia collision zone (after Dewey et al., 1989). According to this model the indentation of India resulted in three main zones: a central region of crustal thickening in Tibet (red), sinistral megashear west of Tibet (green), and dextral megashear in Southeast Asia (yellow). SE Asian deformation may be viewed as rotating crustal blocks bounded by NW-trending faults in response to dextral megashear (shown schematically in right bottom corner).

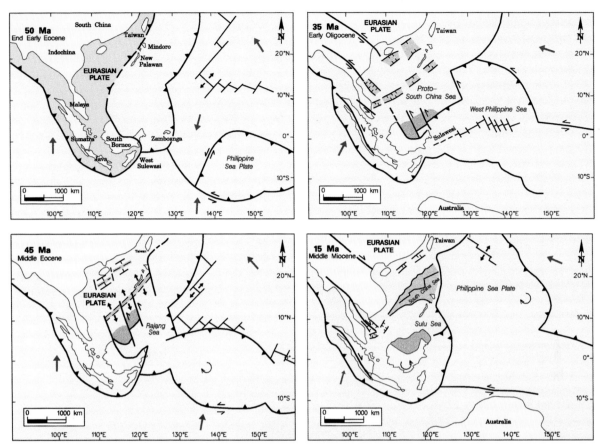

Fig. 4.6. Schematic reconstruction of the tectonic evolution of Southeast Asia since the Eocene (modified after Hall, 1996, 1997). 50 Ma- SE Asia peninsula, formed by Sundaland, surrounded by active margins. 45 Ma- Rifting of the South China margin to form extensional basins; rotation of the Philippine Sea Plate and arc magmatism associated with subduction in eastern Southeast Asia. 35 Ma- further extension in South China margin and central Sundaland, the latter along reactivated NW-trending strike-slip faults. Active sea-floor spreading in the South China Sea. Microcontinental blocks (blue areas) collide with northern Borneo to close the Rajang Sea. Opening of the Sulawesi Sea and W Philippine Basin completed. 15 Ma- South China Sea spreading completed with the suturing of microcontinental blocks to northern Borneo. Dextral reactivation of strike-slip fault zones and the arrival of Australia from the south resulted in basin inversion across Sundaland.

Thrust Belt. Because this ancient sea had no relation whatsoever with the present South China Sea, it is referred to as the "Rajang Sea" in the ensuing discussion. Haile (1974) suggested the term "Danau Sea".

Creation of new oceanic crust and sea-floor spreading, as determined from magnetic anomalies (Taylor and Hayes, 1980; 1983; Briais et al., 1993), did not begin until the Oligocene (32 Ma) with the southward-drift of the Luconia, Palawan and Reed Bank blocks from the South China margin towards Borneo. Peltzer and Tapponnier (1988) suggested that the collision of India and Asia had contributed partly to the extension in the South China Sea. Spreading occurred in a north-south and then northwest-southeast direction (Briais et al., 1993) until the late Early Miocene, about 17 Ma ago (Taylor and Hayes, 1983). The seafloor spreading in the South China Sea was concomitant with the oblique closure of the Rajang Sea to produce the Rajang-Crocker Fold-Thrust Belts of northern Borneo. The closure of the Rajang Sea was diachronous, with collision occurring first between the Luconia Block and the Sarawak margin during the Late Eocene and then between the Dangerous Grounds and NW Sabah during the Early Miocene (Hutchison, 1996b). The effect of the first collision during the Sarawak Orogeny is a major Late Eocene unconformity in onshore central Sarawak (Hutchison, 1996b). The second collision, termed the Sabah Orogeny (Hutchison, 1996b), between the Dangerous Grounds-Reed Bank area (the "NW Sabah Platform", see Chapter 5) and Sabah resulted in the Deep Regional Unconformity (DRU) in offshore NW Sabah (Levell, 1987) and in the Late Te5 (Lower-Middle Miocene) unconformity onshore in western Sabah (Liechti et al., 1960). The Sabah Orogeny resulted in the uplift of the deepwater Crocker Formation. The northwestwards thrusting and loading of pre-Middle Miocene rocks onto the NW Sabah Platform resulted in the formation of the Sabah foreland basin. The unfilled portion of this foreland depression is represented today as the Sabah Trough (Chapter 5).

The early Middle Miocene collision event

in offshore NW Sabah can be observed also in the outer shelf off Sarawak. It is represented by the "Green Unconformity" within the upper Cycle III (Chapter 19) which represents a major phase of tectonism and relative sea-level fall. The Sarawak and Sabah continental margins have continued to evolve as a passive margin since the early Middle Miocene collision event, with sedimentation characterised by northward progradation of clastic sediment derived from the uplifted fold-thrust belts.

EASTERN SE ASIA

Figure 4.6 shows schematically the tectonic evolution of eastern Southeast Asia, based on Hall (1996, 1997). In contrast to western SE Asia, eastern SE Asia is dominated by accreted crusts and marginal seas interspersed with microcontinental fragments (Fig. 4.2). The main feature of the reconstruction is the northward motion and clockwise rotation of the Philippine Sea Plate throughout the Cenozoic. The movements were apparently driven by the northwestward-motion of the Pacific Plate, the assembly of the Philippines archipelago, and the concomitant opening of a major system of marginal basins that were to evolve into the present-day West Philippine Basin, the Celebes Sea, and the Makassar Straits.

Between 50 and 40 Ma, the India-Australian Plate was undergoing subduction northwards beneath the Southeast Asian peninsula along an arc that extended from the Himalayas to Java (Fig. 4.6). North of the arc, in the Philippine Sea Plate, an oceanic spreading centre developed, bounded to the north and south by subduction zones. Meanwhile, the Rajang Sea was being subducted beneath the Borneo promontory and the Sulu arc. The indentation of India into Asia at 45 Ma, especially at the Assam-Yunnan syntaxis, is believed to have caused major effects on the tectonics of Southeast Asia. During this time, there were also major changes in plate motions of the Pacific Plate, concomitant with the opening of the West Philippine Basin, the Celebes Sea, and the Makassar Strait, and the cessation of spreading in the Wharton Basin of the Indian Ocean. The initiation of subduction of the Rajang Sea beneath the Luzon and the Sulu arc also triggered extension in the South China margin (Hall, 1997). The indentation also resulted in the southeastwards extrusion of Indochina and Malaya through the reactivation of the Red River and Wang Chao faults. Extensional basins developed along these shear zones, controlled by pre-existing structural fabrics (Khalid Ngah et al., 1996).

Subduction in Borneo may have contributed to continental extension of the South China margin and, eventually at 32 Ma, rifting and sea-floor spreading in the South China Sea marginal basin. Important developments in eastern Southeast Asia occurred around 25 Ma ago. Arc-continent collisions in the eastern Philippines and New Guinea, ophiolite obduction in Sulawesi, and the arrival of the Australian margin, caused rotations of blocks and accretion of continental fragments in the southeastern part of the region. Throughout this time the Philippine Sea Plate rotated clockwise, which may have also induced rotations of adjacent terranes such as Borneo. At around 25 Ma, the Australian passive margin collided with the island arcs in southeastern part of the region, resulting in major plate reorganisation and in the inversion of pre-existing extensional basins (e.g. Penyu Basin). Subduction in northern Borneo finally ceased around 17-15 Ma with the collision of the Dangerous Grounds with western Sabah. Since that time, the southern margin of the South China Sea has evolved essentially as a passive continental margin of northern Borneo. The Sulu Sea opened as a backarc basin south of the Cagayan Ridge during the Early Miocene (Schluter et al., 1996). Hutchison (1992a) interpreted the Sulu Sea opening as intra-arc rifting. The melanges of eastern Sabah are believed to be related to its opening.

TERTIARY SEDIMENTARY BASINS

The large number of Tertiary sedimentary basins in Southeast Asia indicates widespread extensional tectonism (Fig. 4.7). Many of these basins are rich in petroleum (Murphy, 1975; Soeparjadi et al., 1976; Beddoes, 1980). The basins fall broadly into two main categories: (1) those along the eastern margin of Sundaland, such as the forearc basins along the Myanmar-Sumatra-Java arc, (2) those in the interior of Sundaland, such as the intermontane basins of Thailand, and the extensional basins in the northern Sunda Shelf. The basins in Sumatra and the Java Sea developed by extension during Palaeocene-Eocene times, whereas the interior Sundaland basins, such as the Malay, West Natuna and Thailand basins, formed during the Late Eocene to Early Oligocene. The development

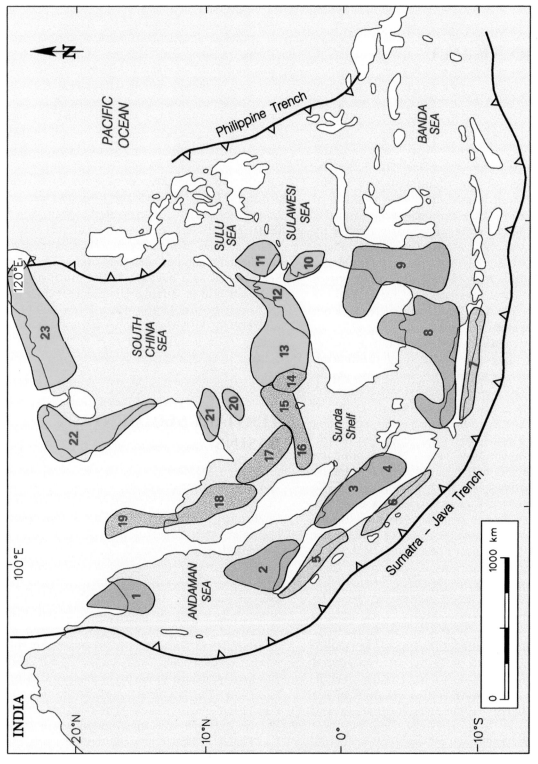

Fig. 4.7. Map showing the major Tertiary sedimentary basins of Southeast Asia. 1. Gulf of Martaban, 2. N Sumatra, 3. Central Sumatra, 4. S. Sumatra, 5. Sibolga, 6. Bengkulu, 7. S. Java, 8. E. Java/Sunda, 9. Makassar/Kutei, 10. SE Sabah/Tarakan, 11. NE Sabah (including Sandakan Sub-basin), 12. Sabah, 13. Sarawak, 14. Sokang, 15. W Natuna, 16. Penyu, 17. Malay, 18. Pattani, 19. Lower Central Plains, 20. Saigon, 21. Vung Tau, 22. Beibu Gulf/Yinghe Hei, 23. Pearl River Mouth. Colours represent different basin types: green - backarc basins, purple - interior extensional basins, pink - forearc basins, yellow - continental margin basins.

of these basins is interpreted, usually, in terms of the extrusion hypothesis (Tapponnier et al., 1982). Non-marine (lacustrine) sedimentation characterises the Oligocene (Gibling, 1988; Williams and Eubank, 1995) but by the Early Miocene, many basins such as the Malay and West Natuna basins became partly connected to the ancestral South China Sea (Beddoes, 1980; Daines, 1985).

The tectonic origin of the extensional basins in Sundaland has been the subject of many discussions. The plethora of models that exist suggest that the origin of these basins is still not well understood. Some of the proposed causes of basin formation are back-arc extension resulting from subduction and oblique convergence along the Sumatra-Java Trench (e.g. Kingston et al., 1983), formation by pull-apart along major left-lateral strike-slip faults by extrusion tectonics (Tapponnier et al., 1982), back-arc rifting combined with wrench faulting (Hamilton, 1979; Crostella, 1981), rifting associated with a mantle plume (Hutchison 1989; Khalid Ngah et al., 1996), extension in response to rotating stress field related to the collision of India and Eurasia (Harder et al., 1992; Huchon et al., 1994), dextral shear along pre-existing NW-striking faults (Polachan et al., 1989), and pull-apart basins related to oroclinal bending of the Andaman-Sumatra-Java Arc (Hutchison, 1992b).

Most Southeast Asian Tertiary basins underwent a phase of basin inversion during the Late Miocene. Basin inversion occurs when a pre-existing extensional basin (depositional low) is deformed by compression to form a structural high. In the basins of Sumatra and the Sunda Shelf, Neogene basin inversion is manifested by the "Sunda folds" (Eubank and Makki, 1981). In many places, the inversion folds are truncated by the regional Late Miocene-Pliocene unconformity (ASCOPE, 1981; Daines, 1985). Letouzey et al. (1990) and Ginger et al. (1993) describe the geometry of these inversion structures in the Sunda Shelf basins, which commonly form hydrocarbon traps. Inversion structures in the Malay and West Natuna basins were formed by dextral shear associated with basement wrench zones (Hamilton, 1979; Ginger et al., 1993; Tjia, 1994; Mazlan Madon, 1997). Basin inversion may have been caused also by intraplate stresses transmitted from nearby plate boundaries such as those generated by the northward subduction of Indian Ocean crust along the Sunda Trench (Letouzey et al., 1990) or the collision of the Australian craton with the arc terranes of eastern Indonesia (e.g. Hall, 1996).

CONCLUSION

The geological evolution of Southeast Asia involved the assembly of allochthonous terranes derived from the break-up of Gondwanaland from late Palaeozoic to Cretaceous times at the southern margin of eastern Eurasia. Several phases of rifting, drifting and orogeny have been identified from many regional geological studies. Present-day Southeast Asia essentially consists of a late Mesozoic continental core that formed during a major late Triassic collisional orogeny related to the closure of the Palaeo-Tethys Ocean. Surrounding the continental core (Sundaland) are many younger (Late Cretaceous to Cenozoic) tectonic features that include island arcs, marginal basins, continental margins, and accretionary complexes.

The evidence suggests that Sundaland has been undergoing extension since late Triassic times, probably as the result of uplift following the Indosinian Orogeny. This is manifested in the many late Mesozoic intermontane basins found throughout the region. It is against this background of overall extension that the Tertiary basins of Sundaland may have developed, but were partly influenced by the extrusion-related tectonism associated with the collision of India and Asia during the early Tertiary. The most important effect of the India-Asia collision was to reactivate some of the major strike-slip faults along which extensional and pull-apart basins developed.

Plate tectonics assumes that lithospheric deformation tends to occur within narrow zones along plate boundaries, and that very little deformation occurs within the plates. The many extensional basins of Sundaland, however, indicate that intracontinental deformation is pervasive, partly aided by the presence of pre-existing fracture systems. The tendency of older orogenic belts to undergo extensional collapse is due to forces that result from compensated uplift of thickened continental crust (Dewey, 1988). Hence, it is not surprising to find many Triassic continental basins developed on the Indosinian orogenic belt. The role of strike-slip tectonics in the development of these basins cannot be ruled out, as the structures of the Tertiary basins have shown. The highly oblique plate convergence between India-Australia and Southeast Asia along the Sumatra trench have resulted in partitioning of the deformation into strike-slip motion, as shown by the movement of arc-parallel faults such as the Semangko and Mentawai faults.

Hence, while plate tectonics is generally adequate in explaining the major features observed in Southeast Asia, it has not explained the complex intraplate continental deformation. It is probably inadequate to describe the Southeast Asian lithosphere as a single rigid plate that does not deform internally. The numerous faults and shear zones that are observed in the region (e.g. Wood, 1985) appear to have strongly influenced its deformational behaviour. Southeast Asia may be regarded as comprising several smaller entities that may have moved independently during deformation. Consequently, sedimentary basins and crustal uplifts are formed along fault zones depending on the relative motion between the deforming blocks.

REFERENCES

ASCOPE, 1981. Tertiary Sedimentary Basins of the Gulf of Thailand and South China Sea: Stratigraphy, Structure and Hydrocarbon Occurrences. ASEAN Council on Petroleum (ASCOPE) Secretariat, Bangkok.

Beddoes, L.R., 1980. Hydrocarbon plays in Tertiary basins of Southeast Asia. Proceedings of the Offshore South East Asia Conference, 26-29 February 1980 (preprint).

Briais, A., Patriat, P. and Tapponnier, P., 1993. Updated interpretation of magnetic anomalies and reconstructions of the South China Sea basin: implications for the Tertiary evolution of Southeast Asia. Journal of Geophysical Research, 98, 6299-6328.

Burton, C. K., 1973. Mesozoic. In: Gobbett, D.J. and Hutchison, C.S., eds., "Geology of the Malay Peninsula." Wiley Interscience, New York, 97-141.

Cobbing, E.J., Pitfield, P.E.J., Darbyshire, D.P.F. and Mallick, D.I.J., 1992. The granites of South-East Asian tin belt. Overseas Memoir, British Geological Survey, Keyworth.

Crostella, A., 1981. Malacca Strait wrench fault controlled Lalang and Mengkapan oil fields. Proceedings of the Offshore South East Asia Conference, 9 -12 February 1981, Exploration III - Geology Session, 1-12. (preprint).

Daines, S.R., 1985. Structural history of the W Natuna Basin and the tectonic evolution of the Sunda region. Proceedings of the Indonesian Petroleum Association 14th Annual Convention, Jakarta, 8-10 October 1985, 39-61.

Daly, M.C., Cooper, M.A., Wilson, I., Smith, D.G. and Hooper, B.G.D., 1991. Cenozoic plate tectonics and basin evolution in Indonesia. Marine and Petroleum Geology, 8, 2-21.

Daly, M.C., Hooper, B.G.D. and Smith, D.G., 1987. Tertiary plate tectonics and basin evolution in Indonesia. Proceedings of the Indonesian Petroleum Association 16th Annual Convention, Jakarta, 20-22 October 1987, 1, 399-428.

Dewey, J.F., 1988. Extensional collapse of orogens. Tectonics, 7, 1123-1139.

Dewey, J.F., Cande, S. and Pitman III, W.C., 1989. Tectonic evolution of the India-Eurasia Collision Zone. Eclogae geologicae Helvetiae, 82, 717-734.

Eubank, R.T. and Makki, A.C. 1981. Structural geology of the Central Sumatra Back-Arc Basin. Proceedings of the Indonesian Petroleum Association 10th Annual Convention, 26-27th May 1981, 153-196

Gabel, J., Trzeski, R., Zwingmann, H., Chonglakmani, C., Helmcke, D. and Meischner, D., 1993. Triassic extensional basins in northern Thailand. In: Late Orogenic Extension in Mountain Belts. Doc. BGRM Fr. No. 219, Abstracts of International Meeting, 4-6 March, Montpellier, 72.

Gatinsky, Y.G. and Hutchison, C.S., 1986. Cathaysia, Gondwanaland, and the Paleotethys in the evolution of continental Southeast Asia. Bulletin of the Geological Society of Malaysia, 20, 179-199.

Gibling, M.R., 1988. Cenozoic lacustrine basins of Southeast Asia, their tectonic setting, depositional environment and hydrocarbon potential In : Fleet, A.J., Kelts, K., and Talbot, M.R., eds., Lacustrine Petroleum Source Rocks. Geological Society of London Special Publication, 40, 341-351.

Haile, N.S., 1974. Borneo. In: Spencer, A.M., ed., Mesozoic-Cenozoic Orogenic Belts. Geological Society of London Special Publication, 4, 333-347.

Hall, R., 1996. Reconstructing Cenozoic SE Asia. In: Hall, R. and Blundell, D.J., eds., Tectonic Evolution of Southeast Asia. Geological Society of London Special Publication, 106, 153-184.

Hall, R., 1997. Cenozoic tectonics of SE Asia and Australasia. In: Howes, J.V.C. and Noble, R.A., eds., Proceedings of the IPA Petroleum Systems of SE Asia and Australasia Conference, Jakarta, Indonesia, May 1997, Indonesian Petroleum Association, Jakarta, 47-71.

Hamilton, W., 1979. Tectonics of the Indonesian region. United States Geological Survey Professional Paper, No. 1078.

Harder, S.H., McCabe, R.J. and Flower, M.F.J., 1992. A single mechanism for Cenozoic extension in and around Indonesia. Symposium Tectonic Framework and Energy Resources of the Western Margin of the Western Pacific Basin, Kuala Lumpur, 29 Nov-2Dec, 1992. Program and Abstracts, 34.

Hazebroek, H.P. and Tan, D.N.K., 1993. Tertiary evolution of the NW Sabah continental margin. In: Teh, G.H., ed., Proceedings of the Symposium on Tectonic Framework and Energy Resources of the Western Margin of the Pacific Basin. Bulletin of the Geological Society of Malaysia, 33, 195-210.

Holloway, N.H., 1981. The North Palawan Block, Phillippines: its relation to the Asian mainland and its role in the evolution in the South China Sea. Bulletin of the Geological Society of Malaysia, 14, 19-58.

Holloway, N.H., 1982. North Palawan Block, Philippines – its relation to Asian mainland and role in evolution of South China Sea. American Association of Petroleum Geologists Bulletin, 66, 1355-1383.

Huchon, P., Le Pichon, X. and Rangin, C., 1994. Indochina Peninsula and the collision of India and Eurasia. Geology, 22, 27-30.

Hutchison, C.S., 1986. Formation of marginal seas in Southeast Asia by rifting of the Chinese and Australian continental margins, and implications for the Borneo region. Bulletin of the Geological Society of Malaysia, 20, 201-220.

Hutchison, C.S., 1988. Stratigraphic-tectonic model for eastern Borneo. Bulletin of the Geological Society of Malaysia, 22, 135-151.

Hutchison, C.S., 1989. "Geological Evolution of South-East Asia". Oxford monographs on Geology and Geophysics, No. 13, Clarendon Press, Oxford.

Hutchison, C.S., 1992a. The Southeast Sulu Sea, a Neogene marginal basin with outcropping extensions in Sabah. Bulletin of the Geological Society of Malaysia, 32, 89-108.

Hutchison, C.S., 1992b. The Eocene unconformity on Southeast and East Sundaland. Bulletin of the Geological Society of Malaysia, 32, 69-88.

Hutchison, C.S., 1993. Gondwanaland and Cathaysian blocks, Palaeothethys sutures and Cenozoic tectonics in South-East Asia. Geologisches Rundschau, 82, 388-405.

Hutchison, C.S., 1996a. "South-East Asian Oil, Gas, Coal and Mineral Deposits". Oxford University Press.

Hutchison, C.S., 1996b. The "Rajang accretionary prism" and "Lupar Line" problem of Borneo. In: Hall, R. and Blundell, D.J., eds., Tectonic Evolution of Southeast Asia. Geological Society of London Special Publication, 106, 247-261.

Jolivet, L., Huchon, P. and Rangin, C., 1989. Tectonic settings of western Pacific marginal basins. Tectonophysics, 160, 23-27.

Khalid Ngah, Mazlan Madon and Tjia, H.D., 1996. Role of pre-Tertiary fractures in formation and development of the Malay and Penyu basins. In: Hall, R. and Blundell, D.J., eds., Tectonic Evolution of Southeast Asia. Geological Society of London Special Publication, 106, 281-289.

Kingston, D.R., Dishroon, C.P. and Williams, P.A., 1983. Global basin classification system. American Association of Petroleum Geologists Bulletin, 67, 2175-2193.

Lacassin, R., Leloup, P.H. and Tapponnier, P., 1993. Bounds on strain in large Tertiary shear zones of SE Asia from bounding restoration. Journal of Structural Geology, 15, 677-692.

Letouzey, J., Werner, P. and Marty, A., 1990. Fault reactivation and structural inversion; backarc and intraplate compressive deformations; examples of the eastern Sunda shelf (Indonesia). Tectonophysics, 183, 341-362.

Levell, B.K., 1987. The nature and significance of regional unconformities in the hydrocarbon-bearing Neogene sequence offshore West Sabah. Bulletin of the Geological Society of Malaysia, 21, 55- 90.

Liechti, P., Roe, R.W., and Haile, N.S., 1960. Geology of Sarawak, Brunei and western North Borneo. British Borneo Geological Survey Bulletin, 3, 1960.

Malod, J.A. and Kemal, B.M., 1996. The Sumatra margin: oblique subduction and lateral displacement of the accretionary prism. In: Hall, R. and Blundell, D.J., eds., Tectonic Evolution of Southeast Asia. Geological Society of London Special Publication, 106, 19-28.

Maluski, H., Lacassin, R., Leloup, P.H., Tapponnier, P., Briais, A., Bunopas, S., Hinthong, C. and Siribhaki, K., 1993. Mid-Oligocene left-lateral shear along the Wang Chao fault zone (NW Thailand). Terra Abstracts, EUG VII Strasbourg, 4-8 April 1993, p. 262.

Mazlan Madon, 1997. A kinematic model of extension and inversion in the Malay Basin. Petronas Research & Scientific Services Research Bulletin, 1, 1-6.

McCaffrey, R., 1996. Slip partitioning at convergent plate boundaries of SE Asia. In: Hall, R. and Blundell, D.J., eds., Tectonic Evolution of Southeast Asia. Geological Society of London Special Publication, 106, 3-18.

Metcalfe, I., 1988. Origin and assembly of Southeast Asian continental terranes. In: Audley-Charles, M.G., and Hallam, A., eds., Gondwana and Tethys. Geological Society of London Special Publication, 37, 101-118.

Metcalfe, I., 1989. Triassic sedimentation in the Central Basin of Peninsular Malaysia. In: Thanasuthipitak, T. and Ounchanum, P., eds., Proceedings of the International Symposium on Intermontane Basins: Geology and Resources. Chiang Mai University, Thailand, 173-186.

Metcalfe, I., 1991. Late Palaeozoic and Mesozoic palaeogeography of Southeast Asia. Palaeogeography, Palaeoclimatology, Palaeoecology, 87, 211-221.

Metcalfe, I., 1996. Pre-Cretaceous evolution of SE Asian terranes. In: Hall, R. and Blundell, D.J., eds., Tectonic Evolution of Southeast Asia. Geological Society of London Special Publication, 106, 97-122.

Murphy, R.W., 1975. Tertiary basins of Southeast Asia. Proceedings of the South East Asian Petroleum Exploration Society, 2, 1-36.

Packham, G., 1990. Plate motions and South East Asia: some tectonic consequences for basin development. Proceedings of the 8th Offshore SE Asia Conference, 4-7 December 1990, Singapore, 119-132 (OSEA-90175).

Packham, G., 1996. Cenozoic SE Asia: reconstructing its aggregation and reorganization. In: Hall, R. and Blundell, D.J., eds., Tectonic Evolution of Southeast Asia. Geological Society of London Special Publication, 106, 123-152.

Parsons, B. and Sclater, J.G., 1977. An analysis of the variation of ocean floor bathymetry and heat flow with age. Journal of Geophysical Research, 82, 803-827.

Peltzer, G. and Tapponnier, P., 1988. Formation and evolution of strike-slip faults, rifts, and basins during the India-Asia collision: an experimental approach. Journal of Geophysical Research, 93, 15085-15117.

Polachan, S. and Sattayarak, N., 1989. Strike-slip tectonics and the development of Tertiary basins in Thailand. Proceedings of the International Symposium on Intermontane Basins: Geology and Resources, Chiang Mai, Thailand, 30 January-2 February 1989, 243-253.

Rangin, C., Jolivet, L., and Pubellier, M., 1990. A simple model for the tectonic evolution of Southeast Asia and Indonesia region for the past 43 m.y. Bulletin de la Societe Geologique de France, 8 VI, 889-905.

Richter, B. and Fuller, M., 1996. Palaeomagnetism of the Sibumasu and Indochina blocks: implications for the extrusion tectonic model. In: Hall, R. and Blundell, D.J., eds., Tectonic Evolution of Southeast Asia. Geological Society of London Special Publication, 106, 203-224.

Ru, K. and Pigott, J.D., 1986. Episodic rifting and subsidence in the South China Sea. American Association of Petroleum Geologists Bulletin, 70, 1136-1155.

Scharer, U., Tapponnier, P., Lacassin, R., Leloup, P.H., Dulai, Z. and Shaocheng, J., 1990. Intraplate tectonics in Asia: a precise age for large-scale Miocene movement along the Ailao Shan - Red River shear zone, China. Earth and Planetary Science Letters, 97, 65-70.

Scharer, U., Tapponnier, P. and Xiachan, L., 1993. Precise dating of Tertiary strike-slip movement in the Red River shear belt. Terra Abstracts, EUG VII Strasbourg, 4-8 April 1993, p. 268.

Scharer, U., Zhang, L.-S. and Tapponnier, P., 1994. Duration of strike-slip movements in large shear zones: The Red River belt, China. Earth and Planetary Science Letters, 126, 379-397.

Schluter, H.U., Hinz, K. and Block, M., 1996. Tectono-stratigraphic terranes and detachment faulting of the South China Sea and Sulu Sea. Marine Geology, 130, 39-78.

Sengor, A.M.C., 1987. Tectonics of the Tethysides: orogenic collage development in a collisional setting. Annual Reviews of Earth and Planetary Science, 15, 213-244.

Sengor, A.M.C. and Hsu, K.J., 1984. The Cimmerides of eastern Asia: history of the eastern end of the Palaeo-Tethys. Memoires de la Societe Geologique de France, 147, 139-167.

Soeparjadi, R.A., Nayoan, G.A.S., Beddoes, L.R. and James, W.V., 1976. Exploration play concepts in Indonesia. Proceedings of the 9th World Petroleum Congress, 3, Exploration & Transportation, 51-64.

Tapponnier, P., Peltzer, G., Le Dain, A.Y., Armijo, R., and Cobbold, P., 1982. Propagating extrusion tectonics in Asia: new insights from simple experiments with plasticine. Geology, 10, 611-616.

Taylor, B. and Hayes, D.E., 1980. The tectonic evolution of the South China Basin. In: Hayes, D.E., ed., The Tectonic and Geologic Evolution of Southeast Asian Seas and Islands. American Geophysical Union, Geophysical Monograph, 23, 89-104.

Taylor, B. and Hayes, D.E., 1983. Origin and history of the South China Sea Basin. In: Hayes, D.E., ed., The Tectonic and Geologic Evolution of South-east Asian Seas and Islands, Part 2. American Geophysical Union, Geophysical Monograph, 27, 23-56.

Tjia, H.D., 1972. Strike-slip faults in West Malaysia. 24th International Geological Congress, Section 3, 255-262.

Tjia, H.D., 1994. Inversion tectonics in the Malay Basin: evidence and timing of events. Bulletin of the Geological Society of Malaysia, 36, 119-126.

Wajzer, M.R., Barber, A.J., Hidayat, S. and Suharsono, 1991. Accretion, collision and strike-slip faulting: the Woyla Group as a key to the tectonic evolution of North Sumatra. Journal of Southeast Asian Earth Sciences, 6, 447-461.

Williams, H.H. and Eubank, R.T., 1995. Hydrocarbon habitat in the rift graben of the Central Sumatra Basin, Indonesia. In: Lambiase, J.J., ed., Hydrocarbon Habitat in Rift Basins. Geological Society of London Special Publication, 80, 331-371.

Wood, B.G.M., 1985. The mechanics of progressive deformation in crustal plates - A working model for Southeast Asia. Bulletin of the Geological Society of Malaysia, 18, 55-99.

Wu, J.M., 1988. Cenozoic basins of the South China Sea. Episodes, 11, 91-96.

Yu, Hong-Sing, 1990. The Pearl River Mouth Basin: a rift basin and its geodynamic relationship with the southeast Eurasian margin. Tectonophysics, 183, 177-186.

Chapter 5

Basin Types, Tectono-Stratigraphic Provinces, and Structural Styles

Mazlan B. Hj. Madon

INTRODUCTION

This chapter reviews the tectonic framework and structural styles of the major structural provinces and sedimentary basins of Malaysia. The major petroleum-bearing and prospective sedimentary basins in Malaysia are Tertiary in age. They occur in three main offshore regions: the Straits of Melaka, the offshore area east of Peninsular Malaysia, and the continental margins of Sarawak and Sabah bordering the South China, Sulu, and Celebes (Sulawesi) seas (Fig. 5.1). Some basins in Sabah and Sarawak extend onshore. The Malay, Sarawak, and Sabah basins are filled with 10 km or more of Tertiary sediment. The Penyu Basin and the Sandakan Sub-basin are smaller, but nevertheless contain as much as 8 km of sediment. Tertiary sedimentary basins also occur onshore Peninsular Malaysia (Chapter 7).

The general geology of some of the larger sedimentary basins has been described by previous authors (Hamilton, 1979; Hutchison, 1989, 1996a) but the smaller ones are lesser known. Khalid Ngah et al. (1996) and Mazlan Madon et al. (1997) described the structure and stratigraphy of the Malay and Penyu basins. Scherer (1980) discussed the geology and exploration history of the Sarawak and Sabah offshore areas. The general geology of the central Sarawak shelf was described in more detail by Doust (1981). James (1984) and Sandal (1996) described the major structural features of the Sarawak Basin and the Baram-Champion delta complexes. The geology and stratigraphy of the Sabah Basin, offshore west Sabah, have been discussed by Levell (1987), Hazebroek and Tan (1993), and Hazebroek et al. (1994). Bell and Jessop (1974), Walker (1993), and Wong (1993) described the geology of the Northeast Sabah Basin (chapter 23) on the southwestern margin of the Sulu Sea, which includes the Sandakan Sub-basin. Field descriptions of the circular sub-basins in the Southeast Sabah Basin (chapter 24) were given by Tjia et al. (1990) and Allagu (1997).

BASIN NOMENCLATURE

It is necessary to review the nomenclature for describing sedimentary basins used in this book. A basin is basically a low area or depression on the earth's surface filled with sediment. In the stricter geological sense, a basin is an area that has subsided faster than its surroundings as a result of tectonic forces, long enough for a large thickness of sediment to accumulate in it. The term "basin" is also used commonly to represent a package of rock strata that was deposited during a geological time interval. A basin, therefore, should normally have stratigraphical (age) as well as geographical limits. Thus, a thick succession of deep marine turbidites that have been deformed into a mountain belt is sometimes also referred to as a basin. A basin is deposited over a basement, which is not necessarily composed of crystalline rocks but may comprise sediments (deformed or otherwise) that belong to an older succession deposited during a previous sedimentary cycle. In the Malaysian context, basement is generally defined as the pre-Tertiary rock substrate upon which Oligocene and younger sediments were deposited. It may be inappropriate, however, to use the word "basin" for an isolated outcrop or outlier of sedimentary rocks that may have been part of a larger, uplifted and deformed, sedimentary basin. The "circular basins" of Sabah, for example, may have been parts of a larger basin that covered eastern Sabah during the Miocene but are now preserved as isolated outcrops (Allagu, 1997). In this chapter the word "sub-basin" is used informally to denote an outlier of sedimentary rocks or a depocentre that appears to be part of a larger sedimentary basin.

The northern Borneo continental shelf, demarcated by water depth less than 200 m, is more than 300 km wide offshore Sarawak but only 100 km wide offshore Sabah. The Sarawak and Sabah continental margins are underlain by enormous thicknesses of Neogene sediment, in places exceeding 15 km. Part of this continental margin succession extends onshore as the unmetamorphosed and relatively mildly

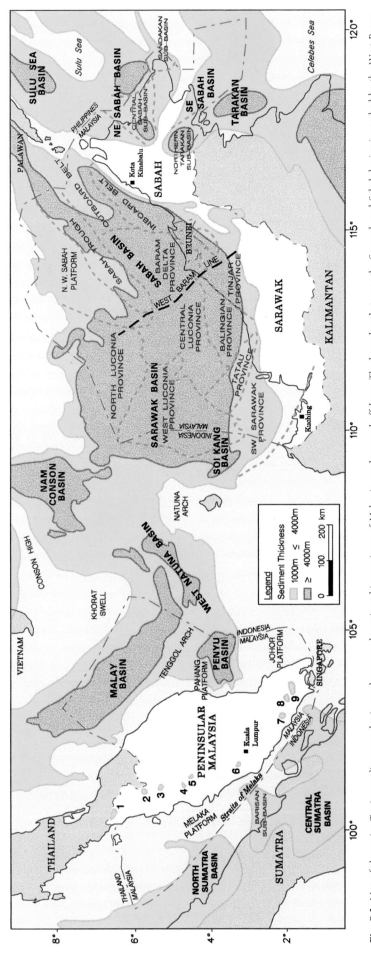

Fig. 5.1. Map of the major Tertiary sedimentary basins and structural-stratigraphic provinces of Malaysia, onshore and offshore. The boundary between the Sarawak and Sabah basins is marked by the West Baram Line. Numbers on Peninsular Malaysia represent location of Tertiary basins: 1. Bukit Arang, 2. Lawin, 3. Nenering, 4. Enggor, 5. Tanjong Rambutan, 6. Batu Arang, 7. Kg Durian Chondong, 8. Kluang-Niyor, 9. Layang-Layang.

deformed Upper Eocene and younger sedimentary rocks that unconformably overlie low-grade metamorphosed Upper Cretaceous-Lower Miocene turbiditic rocks. The term "Sarawak Basin" is commonly used in a geographical sense for the sedimentary basin underlying the continental shelf off Sarawak. Similarly, the Sabah Basin is the Neogene basin that lies offshore northwestern Sabah. In this chapter the Sarawak Basin is defined more strictly as the sedimentary succession overlying the Rajang Fold-Thrust Belt in Sarawak and is of Late Eocene to Recent age. The Sabah Basin is that succession overlying the West Crocker, Kudat, and equivalent formations, and is of Middle Miocene to Recent age. Together, the Sarawak and Sabah basins form the Tertiary continental margin basin of northern Borneo. The Baram Delta complex, which began prograding northwards onto this continental margin during

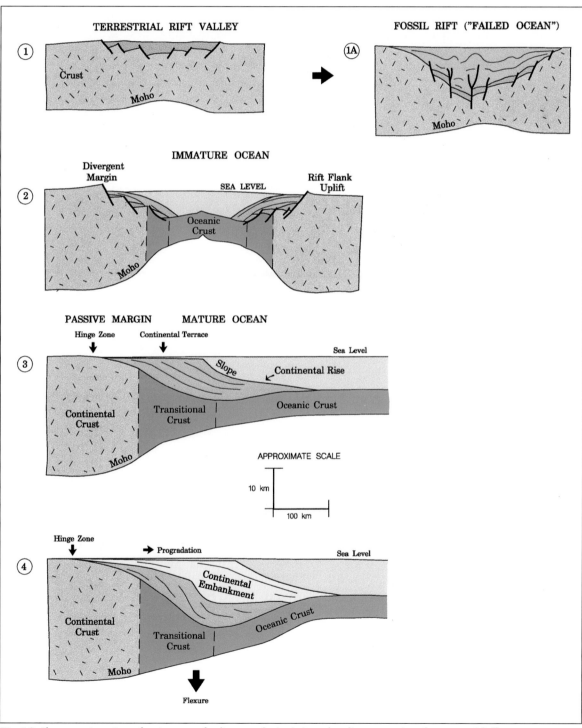

Fig. 5.2. Plate tectonic settings of major types of sedimentary basins in intraplate divergent settings, from terrestrial rift to mature ocean margins with well-developed continental terraces and embankments. Terrestrial rifts may remain land-locked as intracratonic rifts (1A). (Modified from Dickinson, 1974).

the Middle Miocene, straddles parts of Sarawak, Sabah, and intervening Brunei, and is usually treated as a separate basin although, strictly, it is an integral part of the continental margin of northern Borneo that stretches from Natuna to Palawan.

The term "province" has been used in offshore Sarawak mainly to represent a geographical area characterised by a certain tectonic and structural style that influenced its stratigraphic and sedimentation histories. The Balingian Province, for example, is characterised by basement-involved wrench-fault tectonics that affects Oligocene and younger sediment, while the Central Luconia Province is characterised by block-faulting or horst-graben structures upon which carbonate buildups developed during the Middle and Late Miocene. Hazebroek and Tan (1993) recognised "tectonostratigraphic provinces" in the NW Sabah margin. The qualifying term "tectonostratigraphic" is more commonly used for fault-bounded terranes of regional extent, characterised by a geological history that is distinct from that of neighbouring terranes (Howell et al., 1985). The provinces of offshore Sabah, however, are part of the same basin, despite having different structural styles and sedimentation histories. The term "structural province" is more appropriate.

BASIN TYPES

Efforts have been made in the past to classify Malaysian and Southeast Asian Tertiary basins (Murphy, 1975; Soeparjadi et al., 1976). Hutchison (1984, 1989, p. 63) discussed the difficulties in using these schemes to classify some Southeast Asian Tertiary basins. For example, Murphy's (1975) basin types, such as "shelfal" and "archipelagic", describe the present-day physiographic setting but do not necessarily relate to the tectonic origin of the basins. Because of their complex tectonic histories, only a general first-order basin classification can be achieved, perhaps using a plate tectonics approach (e.g. Bally and Snelson, 1980; Kingston et al., 1983), as a guide to understanding the origin of these basins. The extensional basins such as the Malay, Penyu, and Central Sumatra basins were rightly

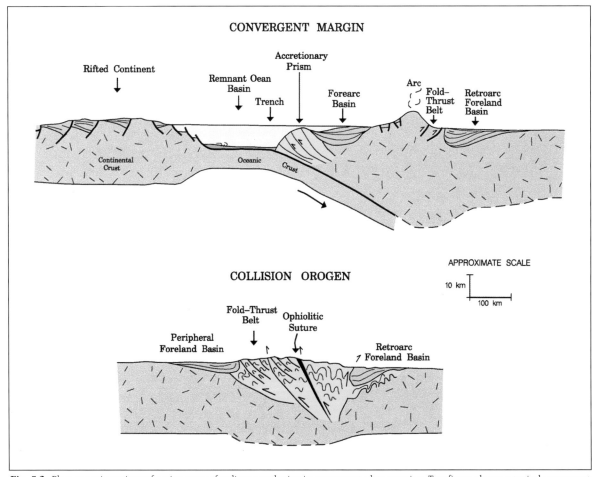

Fig. 5.3. Plate tectonic settings of major types of sedimentary basins in convergent plate margins. Top figure shows a typical convergent margin bordering a subduction zone. As a result of continental collision, this margin will eventually form a foreland fold and thrust belt and basin couplet. (Modified from Dickinson, 1974).

classified by Hutchison (1989) as intracontinental basins, although opinions differ whether the basins are wrench or shear basins (Kingston et al., 1983), back-arc rifts, or aulacogens (Hutchison, 1989). Such terms may serve as qualifiers to represent departures from the general or typical, based on local observations and interpretation.

The plate-tectonics classification of Dickinson (1974) is useful as a general classification scheme for the Tertiary basins of Malaysia. It is widely believed that ocean basins begin their life-cycle (the Wilson Cycle) as terrestrial rift valleys in divergent intraplate settings (Dickinson, 1974; Ingersoll, 1988). These rift basins may evolve, ultimately, into mature oceans whose margins are characterised by continental terraces and rises that are formed by thick accumulations of sediments (Fig. 5.2). In deltaic continental margins, which are usually underlain by "failed rifts" or aulacogens, sediment flux may be high enough to cause progradation of the shoreline over oceanic basement, and can result in continental embankments. Many rift valleys may, however, "fail" to become oceans and are fossilised as rift-sag couplets that are floored by continental basement. Rifts or extensional basins are typified by half-graben and tilted fault blocks, bounded by normal faults that form during extension. In strike-slip intraplate settings, rifting is commonly oblique to the regional stress field. The geometry of the resulting basin therefore depends on the orientation of pre-existing basement faults relative to the regional stress orientation. Hence, basins that form by a combination of extensional and strike-slip tectonics may have complex structural histories.

Convergent plate margins are characterised by subduction zones that may evolve into collisional suture zones (Fig. 5.3). In subduction zones, forearc basins form in arc-trench gaps, whereas retroarc foreland basins form behind arcs as a result of loading by the fold-thrust belt. Collision orogens result from continent-continent or continent-arc collision at convergent margins when nonsubductable continental crust is underthrust beneath the subduction complex. A continent-arc collision may result in the deformation and uplift of the arc-trench sedimentary complex into a fold-thrust belt, which loads onto the underplating continental block to form a peripheral foreland basin in front of the collision orogen (suture zone). Fragments of oceanic crusts may be incorporated into suture zones as ophiolites during arc-continent collisions.

The general classification described above can be applied to the major Tertiary basins of Malaysia (Fig. 5.1). The Malaysian basins have diverse origins and occur in different plate tectonic settings representing almost all stages of the Wilson cycle; from divergent (extensional), intraplate, to convergent margin settings (Figs. 5.2, 5.3). The North and Central Sumatra basins are examples of basins formed by crustal extension in a back-arc setting, but that were later subjected to strike-slip deformation associated with an oblique convergent margin. Compressional deformation has resulted in uplift of the Barisan volcanic arc and the formation of retroarc foreland basins in the Melaka Straits superimposed on the earlier-formed back-arc basins. These basins are asymmetrical; the sedimentary strata thicken towards the Barisan fold-thrust belt. In the Malaysian part of the Melaka Straits a relatively thin succession of Tertiary sediments occurs on the northeastern ramp (flexural) margin of the retroarc foreland.

The Malay and Penyu basins, located offshore east of Peninsular Malaysia, may be categorised as fossil intracontinental rifts that never made it to the oceanic stage. They are floored by continental crust formed during a previous Wilson Cycle that ended in the late Triassic Indosinian Orogeny (Chapter 4). The basins seem to have developed along a major intraplate shear zone that was reactivated as a result of the India-Asia collision during the early Tertiary (Tapponnier et al., 1982). Gravity anomalies indicate that the crust underneath the basins has been thinned by at least half its original thickness (Mazlan Madon, 1996; Mazlan Madon and Watts, 1998). Crustal thinning under the Malay Basin is probably responsible for the observed high geothermal gradients (>36°C km^{-1}) (Fig. 5.4).

The sedimentary basins onshore and offshore northwestern Borneo are difficult to classify because their tectonic origin is still poorly understood and controversial. Some of these basins may have evolved through different tectonic regimes. Hutchison (1984) interpreted the Neogene basins offshore Sarawak as basins that are "on or peripheral to continental fragments" based on the assumption that the central Sarawak shelf is underlain by an allochthonous continental fragment, the Luconia Block (Taylor and Hayes, 1980, 1983). Ismail Che Mat Zin and Swarbrick (1997) proposed that the Sarawak Basin was formed by strike-slip movement of major NW-SE faults during the Late Oligocene to Early Miocene. James (1984) interpreted the Sarawak Basin as a foreland basin that developed when the Luconia Block collided with Borneo in Late Eocene times. The latter model is preferred since the Sarawak Basin overlies the Rajang Group unconformably and

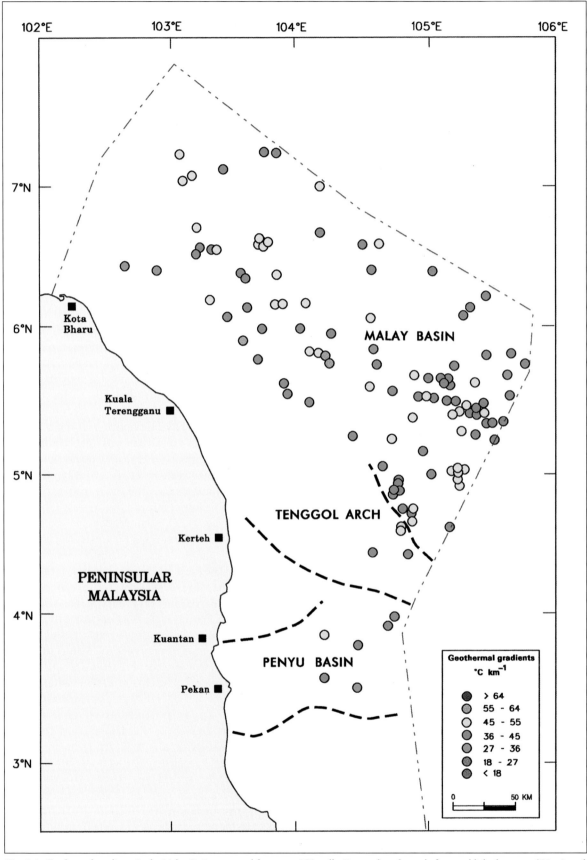

Fig. 5.4. Geothermal gradients in the Malay Basin measured from over 100 wells. Data gathered mostly from published sources (Wan Ismail W Yusoff, 1984, 1990; Mohd Firdaus Abdul Halim, 1994). The data show that the geothermal gradient in both basins is generally between 36 and 55°C km^{-1}, which is similar to the Sarawak Basin (Fig. 5.5).

therefore, must have been post-collisional. If the Sarawak margin is indeed underlain by the Luconia Block microcontinent, as commonly assumed (Hazebroek and Tan, 1993), the Sarawak Basin may have evolved initially as a peripheral foreland basin during the Sarawak Orogeny (Hutchison, 1996b) as a result of the closure of a remnant ocean basin, the Rajang Sea (Chapter 4). Deformation and uplift of the turbidite basin-fill produced the Rajang Fold-Thrust Belt which provides a source for sediment in the overlying Oligocene-Miocene succession in the Balingian and Tinjar provinces. Hence, the compressional and wrench deformation in these provinces could not have been caused directly by the collisional event (cf. James, 1984) but is post-collisional, after the Rajang Group has been deformed and uplifted. Wrench tectonics may have been important in modifying the structural development of the basin, but was probably not the sole cause of basin formation (cf. Ismail Che Mat Zin and Swarbrick, 1997). Cross-sections of the Sarawak continental margin are characterised by a terrace-slope-rise triad formed by thick prograding sedimentary packages (e.g. Epting, 1980; Doust, 1981; Ismail Che Mat Zin, 1996). The overall geometry resembles that of many continental terraces and rises in Atlantic-type passive margin settings (Bally et al., 1981). Hence, although the Sarawak Basin may have originated as a collision-related foreland basin, it subsequently developed as a passive continental margin.

The Sabah Basin is defined as the Middle Miocene to Recent sedimentary basin located east of the West Baram Line and, therefore, includes the entire Baram-Champion delta complex. The basin is perhaps best described as a peripheral foreland basin that formed after the collision of Reed Bank and Dangerous Grounds microcontinental fragments with northwestern Sabah. The collision led to a major phase of deformation at the end of the Early Miocene, called the Sabah Orogeny (Hutchison, 1996b). Progradational sedimentation and thrust-sheet propagation from the southeast appear to have loaded the extended continental basement underneath the Sabah margin to form a deep basin, including the Sabah Trough (Milsom et al., 1997). The northeastern limit of the Sabah Basin may be defined by the Strait of Balabac Fault (Beddoes, 1976), a transcurrent fault that runs NW-SE between the islands of Balabac and Banggi.

The Baram Delta Province, which appears to be confined by major crustal-scale basement faults, the West Baram Line and the Jerudong-Morris Faults, is an example of a continental embankment formed by excessive sediment loading on transitional-oceanic crust. It is interesting to note that the continental margin off NW Sabah is different from that of Sarawak; it is narrower (about 100 km wide) and remarkably straight. Geothermal gradients in the Sarawak margin are generally higher than in the Sabah margin (Fig. 5.5). Its outer limit is defined by the presence of a linear deepwater trough called the

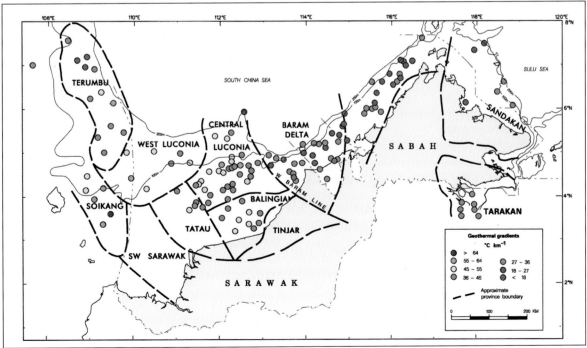

Fig. 5.5. Geothermal gradients at selected exploration wells on the continental margins of Sarawak and Sabah. The data indicate distinct thermal regimes in the basins which are separated by the West Baram Line. The NW Sabah margin is generally cooler (<36° C km^{-1}) than the Sarawak margin (> 36° C km^{-1}), which is similar to the Malay and Penyu basins. Updated from an unpublished map by OXY (Boodoo et al., 1989).

Sabah Trough. The truncation of this 2 km-deep bathymetric feature at the West Baram Line suggests that the West Baram Line is a major tectonic discontinuity that separates the Sabah Basin to the east and the Sarawak Basin to the west. Although the Baram Delta is usually regarded as a province of the Sarawak Basin, only about one third of the entire Baram Delta complex lies in Sarawak. The remainder is in Brunei Darussalam and Sabah. Hence, the Baram Delta Province lies within the same tectonic domain as the structural provinces of the Sabah Basin east of the West Baram Line, which are characterised by lower geothermal gradients, much higher post-Middle Miocene subsidence rates (Noor Azim Ibrahim, 1994), and a gravity signature indicative of basement formed by transitional crust (Milsom et al., 1997). It is more appropriate, therefore, to include the Baram Delta Province as a part of the Sabah Basin instead of the Sarawak Basin.

PENINSULAR MALAYSIA BASINS

Malay Basin

The Malay Basin is a northwest-trending elongate basin about 500 km long and 200 km wide (Fig. 5.6). It contains over 12 km or more of Oligocene to Recent sediments. The Oligocene sediments are generally terrestrial deposits with minor marine influence, whereas the Miocene-Recent sediments are coastal plain to shallow marine deposits. The sedimentary succession is subdivided into seismic stratigraphic units, referred to as "groups". These are labelled alphabetically (Fig. 5.7). Chapter 6 discusses the stratigraphy in more detail.

The structure and tectonic evolution of the Malay and Penyu basins have been studied by Tjia (1994), Liew (1994), Mazlan Madon (1995), Khalid Ngah et al. (1996), and Tjia and Liew (1996). The basin originated by extension during Late Eocene-Early Oligocene times, probably along a major left-lateral shear-zone and, subsequently, by thermal subsidence during the Miocene to Recent. Sedimentation during the Miocene was accompanied by structural inversion that caused major east-trending anticlines to grow in the axial part of the basin. A major Middle-Upper Miocene unconformity is associated with this inversion event. The anticlines are believed to be the result of right-lateral wrench deformation (Hamilton, 1979; Mazlan Madon, 1997a).

The Malay Basin is separated from the Penyu Basin by the Tenggol Arch, a shallow basement area generally at a depth of about 1.5 km below sea level (less than 1 second two-way time). Pre-Tertiary basement has been penetrated by many exploration wells on the basin margins, including the Tenggol Arch, and in the uplifted southeastern part of the basin (Fig. 5.6). Igneous, metamorphic, and sedimentary rocks have all been found. Fontaine et al. (1990) described Triassic limestones in the basement on the southwestern basin margin. Some wells have also penetrated volcanic rocks within the synrift section, indicating volcanic activity during basin extension.

Basement structures in the Malay Basin are dominated by pre-existing fault trends (Khalid Ngah et al., 1996). The northern part of the basin is characterised by N-trending faults and half-grabens (Liew, 1994), similar to those in the Pattani and other Gulf of Thailand basins. These late Mesozoic structures are similar to those documented on Peninsular Malaysia, such as the Lebir and Lupar faults. The western margin of the basin is slightly steeper than the eastern margin, and is marked by a series of *en echelon* normal faults, called the Western Hinge Fault Zone (Tjia, 1994). The axial part of the Malay Basin is characterised by very deep basement, which is poorly imaged by conventional seismic. Small pull-apart half-grabens also occur on the flexural margins of the basin (Liew, 1994).

The rift-sag basin geometry of the Malay Basin is typical of basins formed by lithospheric stretching (Fig. 5.7A). The initial basin geometry is that of half-grabens associated with the crustal extension. This synrift basin is overlain by a broad sag basin characterised by stratigraphic onlap onto the basin margins, which is produced by thermal subsidence and sediment loading (Watts et al., 1982). Gravity modelling has indicated that the Malay Basin is not a purely extensional basin; its development may have involved a significant strike-slip component of extension (Mazlan Madon, 1996; Mazlan Madon and Watts, 1998). During the Middle-Late Miocene the Malay Basin was subjected to basin inversion, which produced major anticlinal structures that trap large amounts of oil and gas in the basin (Tjia, 1994; Mazlan Madon, 1995, 1997a). The southeastern part of the basin was probably uplifted during this time also (Fig. 5.7B). Basin inversion is thought to have resulted from a dextral shear regime following a change in the regional stress field (Tjia and Liew, 1996; Mazlan Madon, 1997a).

Penyu Basin

The Penyu Basin (Oligocene-Recent) is characterised by normal faults and grabens that trend east and northeastwards and continue into the West Natuna Basin. The stratigraphy generally comprises a synrift half-graben fill overlain by relatively flat-lying postrift succession (Fig. 5.8). Chapter 6 and 9 give more details on the stratigraphy. Four main fault-bounded sub-basins have been identified from seismic data: the Kuantan, Pekan, Rumbia, and Merchong grabens (Chapter 9). These grabens/half-grabens are bounded by two main sets of faults trending ENE

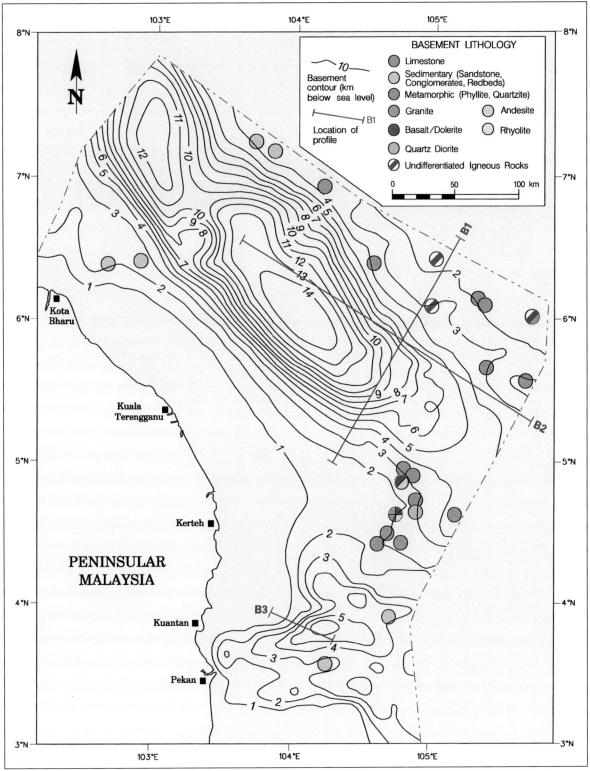

Fig. 5.6. Map showing the pre-Tertiary basement depth below sea level (contours in km) in the Malay and Penyu basins and basement lithologies encountered in exploration wells. B1, B2, and B3 are cross sections shown in Figs. 5.7 and 5.8. Contours based on an unpublished map by Esso (1985).

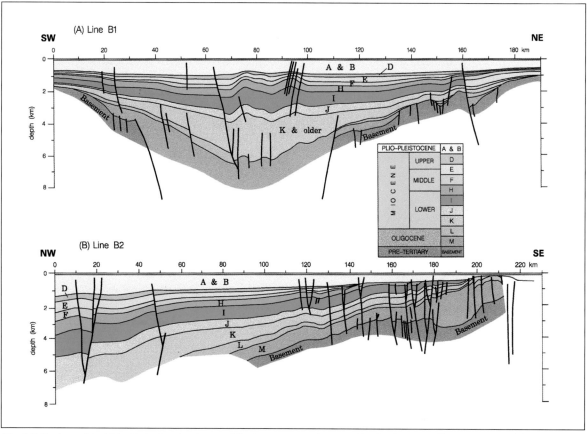

Fig. 5.7. Sections of the Malay Basin showing the overall basin geometry and structural styles. Location of lines in Fig. 5.6. (A) Cross section showing a steer's head geometry with deep, poorly resolved synrift basin at centre overlain by thickening and onlapping postrift strata. Note the inversion folds at centre. (B) Longitudinal section showing inverted basement in the southeastern corner and the erosional truncation of strata towards the southeast. Cross sections based on unpublished report by Esso (1985). The stratigraphy is explained in Chapter 6.

and NW. Most of the ENE-trending bounding faults are normal, whereas the NW-trending faults have had a significant strike-slip component of displacement. The synrift sedimentary fill of the half-grabens thickens towards the bounding faults, commonly with rollover anticline development (Fig. 5.8).

In the eastern part of the Penyu Basin, NW-trending faults define rhombic-shaped basins that indicate a pull-apart origin associated with strike-slip movement of these faults. The NW-trending bounding faults have the same trend as that of major NW-trending strike-slip faults onshore Peninsular Malaysia, and could have been older faults that were reactivated during basin development (Khalid Ngah et al., 1996).

Like the Malay and West Natuna basins, the Penyu Basin was inverted during the middle-late Miocene. Inversion was mild compared to that in the Malay and West Natuna basins, where some half-graben bounding faults were reactivated as reverse and thrust faults (Ginger et al., 1993). In the Penyu Basin, the half-graben bounding faults were reactivated only slightly to produce low-relief inversion anticlines.

Straits of Melaka

The Straits of Melaka, between Peninsular Malaysia and Sumatra, is a shallow seaway with water depths of 100 m or less, that straddles the Indonesia-Malaysia border. The Malaysian part of the straits is underlain by up to 1600 m of Tertiary sediments deposited on the southwestward-tilting Sundaland basement. These form the northeastern flexural margins of the highly petroliferous North and Central Sumatra basins (Fig. 5.9). The northern Straits of Melaka, offshore Kedah and Perak, is part of the North Sumatra Basin, while its southern part, between offshore Selangor and Johor, probably belongs structurally to the Central Sumatra Basin (see Fig. 5.1).

The geology of these basins is described in many published works by the Indonesian petroleum industry (Kamili and Naim, 1973; Mertosono and Nayoan, 1974; de Coster, 1974; Kingston, 1978; Eubank and Makki, 1981; Buck and McCulloh, 1994; Katz and Dawson, 1996). The North and Central Sumatra basins were formed by crustal extension in a back-arc setting associated with an oblique convergent margin.

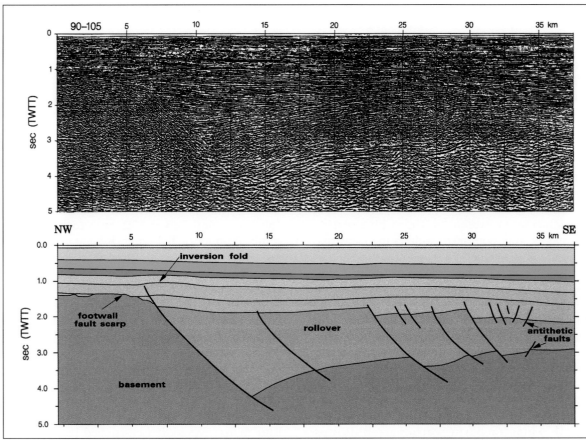

Fig. 5.8. Seismic section across a major half-graben normal fault in the Penyu Basin. Line drawing emphasises the major structural features such as the eroded footwall scarp, a slight fold in the postrift strata above the main bounding fault, and several antithetic faults some 30 km away from the bounding fault. Location in Fig. 5.6. From Mazlan Madon (1995).

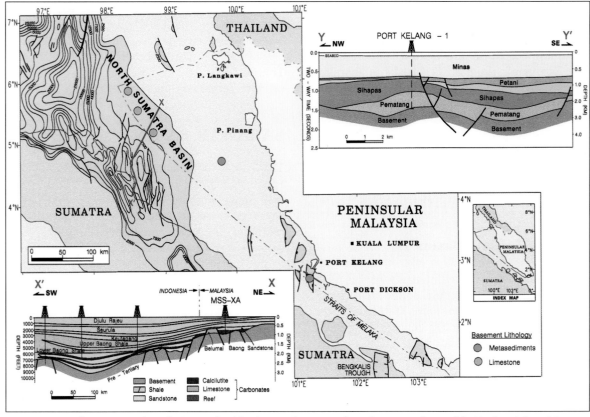

Fig. 5.9. Map and cross-section from the Straits of Melaka showing the stratigraphy and structure of parts of the North and Central Sumatra basins. Also shown are the basement lithologies encountered by exploration wells in the area. North Sumatra cross-section from Sun Oil (Hatch, 1987); Port Kelang structural cross-section from unpublished Sun Oil report (Port Kelang-1 well proposal, July 1991).

The basement structure in both basins is characterised by structural highs and grabens with a N-S orientation superimposed upon a pre-existing NW-SE basement grain. Wells drilled in offshore Kedah penetrated poorly fossiliferous and recrystallised limestone basement, probably equivalent to the Middle-Upper Permian Chuping Limestone of Langkawi and Perlis. Some basement limestones resemble the Ordovician Setul Limestone that crops out in NW Peninsular Malaysia.

Both basins underwent rifting during the Eocene-Oligocene, and a postrift phase during the Miocene to present, during which marine sediments were deposited. Nonmarine lacustrine basins that developed in the synrift extensional grabens became the sites of organic source-rock deposition, for example the Brown Shale in Central Sumatra Basin, contributing most of the oil accumulations in these basins (Katz and Dawson, 1996). North-south grabens have been mapped from seismic offshore Selangor and Johor, and are similar to those in the Central Sumatra Basin (Fig. 5.9). The postrift phase of basin development began with a marine transgression during the Early Miocene. In the North Sumatra Basin, carbonate buildups developed on structural highs and provide excellent hydrocarbon reservoirs, such as in the Arun gas field (Soeparjadi, 1982). A Late Miocene compressional phase related to uplift of the Barisan Mountains resulted in a retro-arc foreland depression east of the Barisan volcanic arc, deforming the North and Central Sumatra basin fills. The gentle slope of the Melaka Straits may have resulted from flexural loading of the Barisan Mountains on the Sundaland lithosphere. The oblique convergence between the Indian and Eurasian Plates along the Sumatra Trench have reactivated the pre-existing N-S and NW-SE basement faults to form many of the petroliferous inversion structures within the upper Miocene and Pliocene succession (Crostella, 1981).

Onshore Tertiary Basins

Most of the Tertiary sedimentary basins that underlie the offshore areas around Peninsular Malaysia do not extend onshore. The Penyu Basin appears to extend westwards beneath the coastal plain between Kuantan and Pekan. Small, isolated exposures of Tertiary sedimentary rocks do occur onshore in Peninsular Malaysia; the major ones are Bukit Arang, Enggor, Batu Arang, Kg. Durian Chondong, Kluang-Niyor, and Layang-Layang (Fig. 5.1). Stauffer (1973), Suntharalingam (1983), and Hutchison (1989) summarised the geology of some of these basins. A review of the tectonic evolution of the basins was published recently by Raj et al. (1998). Only a few detailed studies of these basins have been carried out, and some of the supposed basins have not been proven to be Tertiary in age. The Tertiary ages for the Nenering and Tanjong Rambutan deposits, for example, are still disputed (Uyop Said and Che Aziz Ali, 1997; Raj et al., 1998). It is also unclear whether these isolated outcrops of Tertiary sedimentary rocks represent relicts of much larger basins. They generally lie unconformably on folded and metamorphosed Palaeozoic and Lower Mesozoic rocks but are gently dipping or flat-lying. Basinal (synformal) geometry has been demonstrated in some basins (e.g. Batu Arang). Published work to date suggests that most of these basins formed as pull-aparts associated with NW-trending strike-slip faults (Raj, 1998; Raj et al., 1998). A gravity study by Vijayan (1990) on the Layang-Layang Basin indicates steep anomalies along the probably faulted northeastern basin edge.

The sedimentary rocks are mainly fluvial and lacustrine deposits, with significant lignite and coal seams in such places as Bukit Arang and Batu Arang. The most well-studied basin is Batu Arang, where the sediments have been dated by palynology as Oligocene to Miocene (Ahmad Munif Khoraini, 1993). Coal seams were mined at Enggor during the 1920s and at Batu Arang between 1915 and 1960 (Stauffer, 1973). The Nenering Tertiary deposits in Hulu Perak were described by Ibrahim Abdullah et al. (1991) and Teh and Sia (1992). Chapter 6 discusses the onshore Tertiary basins in greater detail.

SARAWAK BASIN

The Tertiary sedimentary basins of Sarawak and Sabah developed on the northern and eastern continental margins of Borneo (Fig. 5.10). The Sarawak Basin, of Late Eocene to Recent age, underlies the broad continental shelf and slope off northern Sarawak and parts of onshore Sarawak. The basin overlies unconformably the Rajang Group (Upper Cretaceous to Upper Eocene), which consists of highly deformed, low-grade metamorphosed deep marine shales, turbidites, some radiolarian chert, spilite, and dolerite. The Rajang Group includes the Lupar and Belaga formations that crop out in western and central Sarawak and continue into Sabah as the East Crocker, Sapulut and Trusmadi formations. Inliers of the Rajang Group occur in some parts of eastern Sarawak. Unlike the Rajang Group proper, these Upper Cretaceous to Eocene

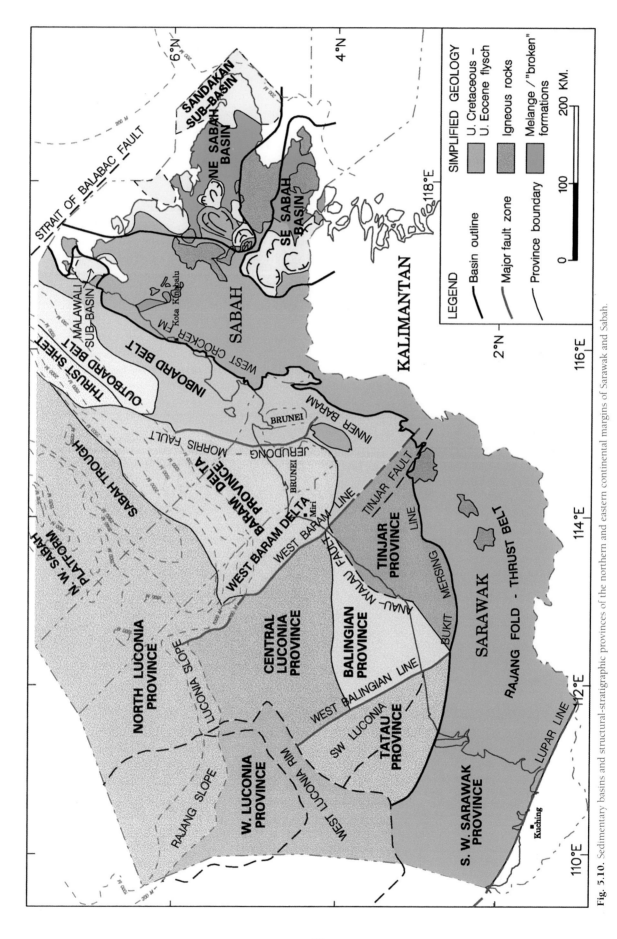

Fig. 5.10. Sedimentary basins and structural-stratigraphic provinces of the northern and eastern continental margins of Sarawak and Sabah.

Kelalan and Long Bawan formations consist of fluvial, deltaic, and shallow marine sediments (Hutchison, 1989). The Rajang Group and the younger West Crocker Formation in Sabah are thought to represent a Late Cretaceous-Early Tertiary accretionary complex that formed by subduction of the Rajang Sea along the NW Borneo subduction margin (James, 1984; Hutchison, 1996).

Although the Sarawak Basin resembles in general many continental terraces in cross section, a variety of structures occur in different parts of the basin. These are related to basement-involved extensional tectonics, strike-slip and wrench tectonics, and gravity-driven basement-detached tectonics. Several structural-stratigraphic provinces are recognised: the SW Sarawak, West Luconia, Tatau, Balingian, Tinjar, Central Luconia, and the West and North Luconia provinces (Fig. 5.10). The structural characteristics of these provinces are described below. The stratigraphic subdivision of the Sarawak Basin, discussed in more detail in Chapter 6, is based on the concept of the sedimentary cycle, in which progradational packages of sediment are bounded by regional transgressive surfaces during relative sea-level rise. A total of 8 sedimentary cycles (labelled I to VIII) are recognised in the Sarawak offshore (Ho, 1978).

In this chapter, the Baram Delta Province has been included in the Sabah Basin instead of Sarawak Basin primarily because a substantial portion of it lies in Brunei and Sabah (the latter as the East Baram Delta Province) eastwards from a major tectonic discontinuity, the West Baram Line. The Baram Delta is therefore discussed under the Sabah Basin later in the chapter.

SW Sarawak Province

The SW Sarawak Province is the offshore continuation of the Rajang Fold-Thrust Belt in central Sarawak westwards onto the continental shelf, in the Lupar River embayment (Fig. 5.1). Based on the few wells drilled to date, the pre-Oligocene basement consists of hard and dense lithic-feldspathic sandstone, siltstone and shale, correlated with the Rajang Group. The Rajang Group is overlain unconformably by a relatively thin succession (generally less than 700 m) of

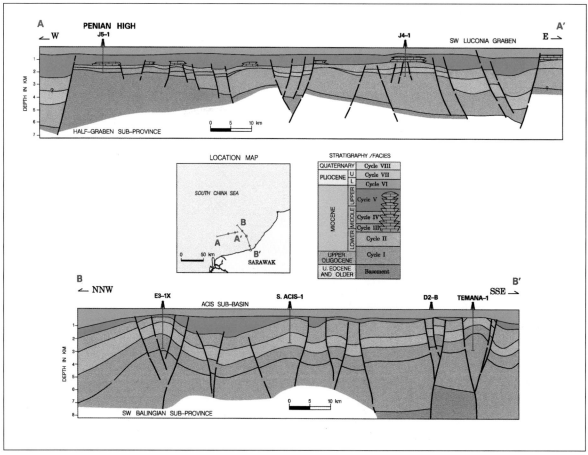

Fig. 5.11. Structural cross-sections from the Tatau and Balingian provinces showing the structural styles (after Shell, 1993) .(A) W-E section across the Penian High to the SW Luconia Graben, characterised by extensional faults. (B) NNW-SSE section across the Eastern Balingian sub-province, from the Acis Sub-basin to Temana-1, showing basement-involved compressional wrench-fault structures. The cycles are defined by Ho (1978); see also Chapter 6.

middle to late Miocene shallow marine sediments. The sediment thickness exceeds 2 km to the northwest of the province in the Soikang Sub-basin. The structural style of the SW Sarawak Province is dominated by NW-trending half-graben-bounding normal faults, similar to those in the neighbouring Tatau Province, which is described below.

Tatau Province

The Tatau Province is a triangular area located mainly offshore between the Balingian and SW Sarawak provinces (Fig. 5.10). It is subdivided into the Half-Graben and SW Luconia sub-provinces, separated by the Penian High. The structural style is characterised by major NW-trending fault-bounded grabens and horsts (Fig. 5.11). The overall trend of the faults is parallel to the that of the Rajang Fold-Thrust Belt. The grabens are smaller in the onshore area (e.g. Lemai, Mukah, and Igan-Oya grabens) and are generally deeper in the offshore area, some reaching about 7 km in the northwest of the province. Extension of the Tatau half-grabens took place during the Early Miocene. Compressional deformation at the end of the Early Miocene was followed by the development of carbonate buildups on the basement horsts (Fig. 5.11). Northward progradation of clastic sediments during the Middle Miocene has continued to the present.

Balingian Province

The Balingian Province lies mainly offshore central Sarawak between the West Balingian Line to the west, the Central Luconia Province to the north, and the Tinjar Province to the south (Fig. 5.10). Its boundary with the largely onshore Tinjar Province is marked by the NE-trending Anau-Nyalau Fault, which is a zone of northward-verging basement thrusts that continues southwestwards to form the northern boundary of the Tatau Horst in the southwestern Tinjar Province. The northern boundary of the Balingian Province is gradational with the Central Luconia Province and is defined generally as the southern limit of carbonate development during the Middle Miocene.

The Balingian Province consists of three sub-provinces that have different structural styles and histories (Swinburn, 1994): East, Southwest, and Northwest sub-provinces (Fig. 5.12). In addition, there are three major depocentres aligned in a north-south orientation at the centre

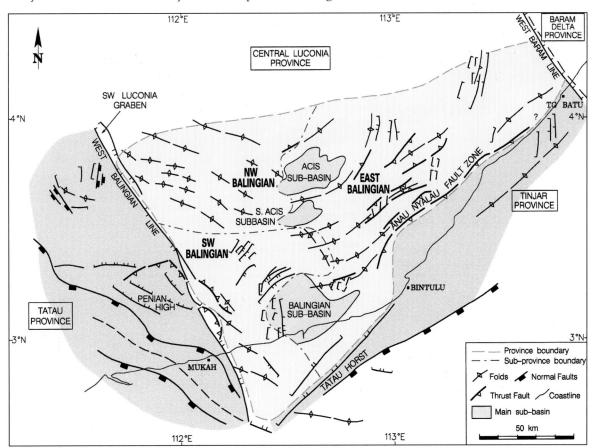

Fig. 5.12. Simplified map of the Balingian Province, showing the sub-provinces, depocentres, and structural features. Modified from Swinburn (1994).

of the Balingian Province. These depocentres or sub-basins, the Acis, South Acis, and Balingian sub-basins, represent areas of significantly higher subsidence rates and are thought to be important hydrocarbon kitchens (Chapter 14).

East Balingian is characterised by large high-amplitude NE-trending folds, commonly bounded by reverse faults and thrusts that are probably linked to Late Miocene to Pliocene strike-slip faults in the basement. SW Balingian is characterised by NE-trending basement normal faults flanking the Balingian Sub-basin and downthrowing to the southeast. *En echelon* NW-trending folds are a characteristic feature of NW Balingian. Near the West Balingian Line *en echelon* NW-trending folds were probably formed by dextral wrench movement of that fault zone during the Oligocene to Early Miocene. The folds were subjected to strong multiphase deformation during the Oligocene to Pliocene (Swinburn, 1994). Many of these folds are cut by N-S faults. Superimposition of N-S normal faults on east-west folds may result in complex structural traps, as documented in the D18 field (Almond et al., 1990a, b).

Tinjar Province

The Tinjar Province is situated mainly onshore central Sarawak (Fig. 5.10). It is bounded to the south by the contact between the Rajang Group and overlying post-Eocene succession which roughly coincides with the Tatau-Mersing Line, interpreted as an ophiolitic suture between the Luconia Block and the Rajang Fold-Thrust Belt (Hutchison, 1989, p. 267). The sedimentary succession consists of post-Eocene basal sediments of the Sarawak Basin deposited over the eroded surface of the Upper Cretaceous-Eocene Rajang Fold-Thrust Belt. These are generally coastal plain to shallow marine deposits.

The Tinjar Province was subjected to several phases of compressional deformation between Late Oligocene and Late Miocene times, which produced a structurally complex belt of NW- and NNE-trending anticlinal structures. The strata are generally tightly folded, with individual anticlines measuring up to 40 km long. Most of them are oriented NE-SW, parallel to the overall Rajang Fold-Thrust Belt. The

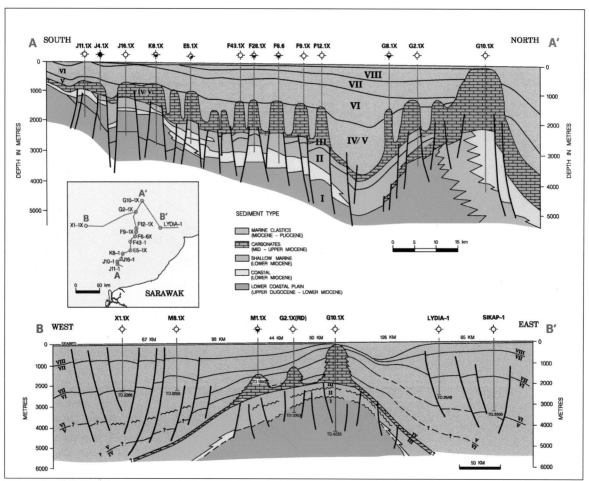

Fig. 5.13. Schematic cross-sections across the Central Luconia Province. (A) N-S cross section. (B) E-W cross section. Map shows location of profiles. Modified from OXY (1991).

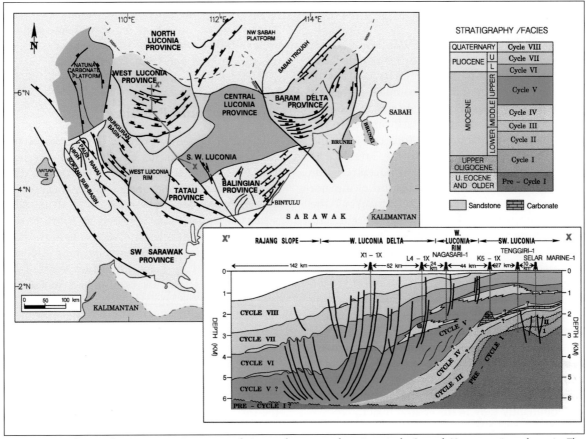

Fig. 5.14. Map showing the West Luconia Province in relation to other structural provinces in the Sarawak-Natuna continental margin. The West Luconia delta prograded northwards in a relatively low area between two major carbonate platforms, the Natuna and Central Luconia platforms. Cross-section shows the stratigraphy and structural style across the West Luconia Province. The cycles are defined by Ho (1978); see also Chapter 6. Modified from Triton (1994).

intensity of compressional deformation increases gradually inland in the Tinjar Province. Oligocene to Late Miocene deltaic progradation proceeded northwards from central Sarawak, after the most intense period of deformation that affected the underlying Upper Eocene Belaga Formation.

Central Luconia Province

The Central Luconia Province is an area of the Sarawak continental margin in which shallow marine carbonates developed since the Middle Miocene. Although many of the carbonate buildups have died out, some are still growing. Carbonate deposition appears to have been restricted to within a relatively narrow stratigraphic window between the Middle and Late Miocene. Epting (1980) described the general geology and sedimentology of the Central Luconia carbonates. To date, more than 200 carbonate buildups, made up of mainly corals and coralline-algal mounds, have been mapped. Many of these buildups have been drilled and found to contain substantial amounts of hydrocarbons (Fig. 5.13).

The pre-carbonate stratigraphy is similar to that observed in the Balingian Province to the south, with essentially a northward prograding siliciclastic depositional system characterised by coastal plain aggradation in response to continuing sea-level rise and sediment influx from the south. The area was subjected to extensional tectonism during the late Early Miocene which resulted in the formation of NNE-trending extensional grabens. Uplifted horsts created during that faulting episode acted as substrates for carbonate buildups to grow during a major marine transgression during the Middle to Late Miocene. Carbonate growth seemed to have started in the inboard areas and progressively developed seawards (Ismail Che Mat Zin and Swarbrick, 1997). A major sea-level drop occurred during the late Middle Miocene, ending carbonate deposition in the Central Luconia while northward deltaic progradation continued. The Central Luconia Province was subjected to two phases of faulting (Epting, 1980): during the Oligocene to Early Miocene and during the Early to Middle Miocene when the Balingian Province was undergoing a compressional phase.

West Luconia Province

The West Luconia Province is a large clastic depocentre, similar in size to the Baram Delta Province, developed between two areas of carbonate sedimentation. It is bounded to the east by the Central Luconia Province and to the west by the Natuna Carbonate Platform where the giant Terumbu carbonate is located (May and Eyles, 1985) (Fig. 5.14). Like the Baram Delta complex, this large siliciclastic depocentre is the result of the northward progradation of a major delta, the West Luconia Delta, across the southwestern Sarawak continental margin. The main phase of progradation occurred during the Middle Miocene to Pliocene, depositing more than 7 km of sediment beyond the shelf edge.

The West Luconia Delta is confined by major structural discontinuities that separate it from the carbonate platforms on both sides The structures are dominated by north-hading synsedimentary growth faults and associated rollover anticlines, similar to those in the Baram Delta. The fault traces are concave northwards, predominantly E-W in the central region, NE-SW in the east, and NW-SE in the west, following the overall shape of the delta. In some areas, the growth faults have been modified by basement wrench-fault movements. The outer rim of the delta complex, known as the West Luconia Rim, is less faulted and has a thick succession of Middle Miocene and younger sediments.

North Luconia Province

The North Luconia Province is the area of the Sarawak continental margin that occurs beyond the present shelf edge and includes the continental slope and abyssal plain (Fig. 5.10). Water depth generally exceeds 2000 m. The structural style is characterised by gravitational growth faults, particularly in the western half of the province, as a continuation of deltaic tectonics of the West Luconia Delta. Two subprovinces may be recognised based on differences in the structural styles: Rajang Slope in the west and Luconia Slope in the east. The Rajang Slope is the northward continuation of the West Luconia Delta, which has prograded onto transitional crust of the South China Sea marginal basin. This western North Luconia area is characterised by arcuate north-verging deltaic toe-thrusts formed by gravitational loading by the prograding delta front sediments. The thrusts appear to be listric and sole out along a major detachment fault (Fig. 5.15).

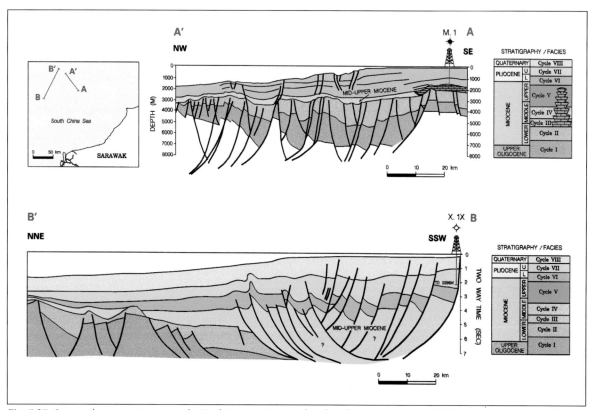

Fig. 5.15. Structural cross sections across the North Luconia Province, based on the interpretation by BHP Petroleum (Morris, 1992). Note the different structural styles between the eastern and western parts of the province. The eastern profile (AA') shows extensional structures in pre-Cycle III strata whereas the western profile (BB') shows gravity-driven deltaic growth faults in Cycle V and younger sequences overlying extensional structures in pre-Cycle III strata.

The Luconia Slope, located to the north of the Central Luconia Province, has a markedly different structural style. Here, the major faults are generally N- to NE-trending, west-dipping listric normal faults bounding en echelon half-grabens and tilted fault-blocks formed of pre-Middle Miocene sequences. A major erosional unconformity in the lower part of the Middle Miocene (referred to herein as the Middle Miocene Unconformity, MMU) truncates the crests of the tilted fault blocks containing steeply dipping mainly Oligocene-Lower Miocene strata (Fig. 5.15). To the south, in Central Luconia, uplifted fault blocks similar to those in the North Luconia have acted as the substrate for some Middle Miocene or younger carbonate buildups. A major NE-trending normal fault, the North Luconia Fault, seems to mark the northern limit of carbonate deposition in the Central Luconia Province.

The main phase of faulting in the Luconia Slope occurred during the Oligocene or earlier and was followed by minor faulting during the Early Miocene. Most faults do not cut the MMU but appear to sole out beneath the Oligocene synrift sequence at depths of about 8 to 10 km (Fig. 5.15). The MMU is probably related to tectonic movements during late Early Miocene that reactivated the pre-existing faults, causing uplift and erosion of the fault-blocks. Movement of the North Luconia Fault during this time appears to have controlled subsidence of the Luconia Slope.

SABAH BASIN

The Sabah Basin (Middle Miocene to Recent) is located on the northwestern continental margin of Sabah. Like the Sarawak Basin, the Sabah Basin also unconformably overlies deformed deepwater sediments that now form the Crocker Formation and Rajang Group rocks in western Sabah (Fig. 5.10). As pointed out by Hutchison (1996b), a distinction must be made between the Rajang Group as defined originally in Sarawak and the Rajang Group in Sabah. The latter comprises the East Crocker, Trusmadi, and Sapulut formations, whereas the former includes the Belaga and Lupar formations. In western Sabah, the Rajang Group as described by Hazebroek and Tan (1993) and previous authors, comprises an older Rajang Group, of Palaeocene-Oligocene age, which is equivalent to the Rajang Group in Sarawak. This older turbidite succession is overlain by a younger unmetamorphosed succession of turbidites belonging to the Oligocene-Lower Miocene West Crocker, Temburong, and Kudat formations. The inferred unconformity (Late Te_5 unconformity, Liechti et al., 1960) between these two successions of turbiditic rocks is probably related to the Late Eocene event that deformed and exhumed the Rajang Group into a fold-thrust belt which many authors believed to have been caused by the collision of Luconia Block with Borneo (e.g. James, 1984).

The structure and stratigraphic evolution of the NW Sabah continental margin have been discussed by Bol and van Hoorn (1980), Hinz et al. (1989), Tan and Lamy (1990), Hazebroek and Tan (1993), and Noor Azim Ibrahim (1994). Hinz et al. (1989) described the structure of the outer continental margin off NW Sabah based on seismic data. They recognised six structural zones of different structural complexity and style within the outer shelf and slope, including the Sabah Trough. Tan and Lamy (1990) subsequently published structural information from the inner shelf area. The Sabah Basin exhibits features indicative of compressional margins characterised by thrusts and wrench tectonics which reflect the strong influence tectonics has had over its structural evolution. Figure 5.16 shows some examples of the structural styles in the Sabah Basin, ranging from compressional, extensional (deltaic), to strike-slip and thrust tectonics.

The Sabah Basin is subdivided into provinces that are characterised by distinct structural styles and sedimentation histories (Tan and Lamy, 1990; Hazebroek and Tan, 1993). These are the Baram Delta, Inboard Belt, Outboard Belt, Sabah Trough, and the NW Sabah Platform (Fig. 5.10). Its sedimentation history involved basically the northwestward progradation of siliciclastic shelf sedimentation since the Middle Miocene after an earlier phase of deep marine sedimentation (West Crocker and Kudat formations). Sedimentation was punctuated by several tectonic events that resulted in several regional unconformities at the basin margin. These unconformities have been used as the basis for subdividing the stratigraphy of the Sabah Basin into "stages" IVA to IVG (Levell, 1987). Chapter 6 discusses the stratigraphy in more detail.

Based on a detailed analysis of the subsidence history using numerous well data, Noor Azim Ibrahim (1994) recognised two main phases of deposition in the Sabah Basin: a very rapid subsidence phase during early Middle Miocene to early Late Miocene, which resulted in deltaic aggradation, followed by a much slower

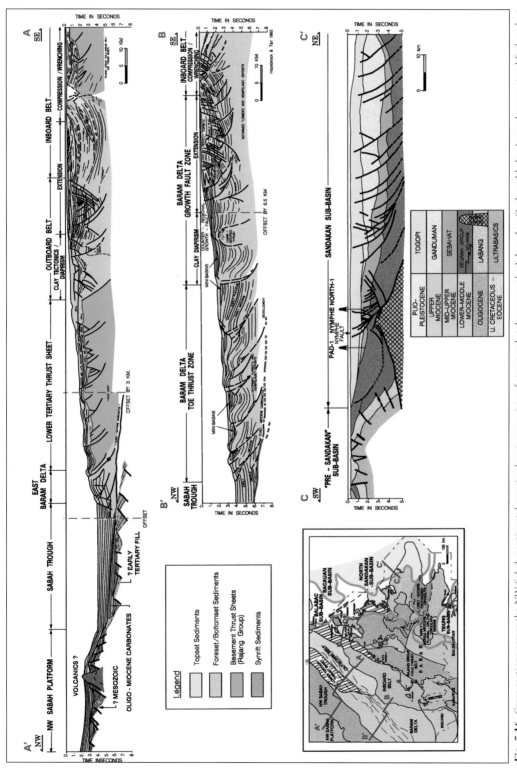

Fig. 5.16. Cross-sections across the NW Sabah continental margin showing a variety of structural styles across the Inboard to Outboard belts. Inset shows a simplified geological map of Sabah showing the major Tertiary sedimentary basins: NE Sabah Basin, which includes the Sandakan and Labuk Bay sub-basins, and the SE Sabah Basin, which includes the Maliau, Malibau, and Tidung sub-basins. Profile DD' is shown in Fig. 5.18. (A) Northern part of Sabah Basin across the Inboard and Outboard belts, the Thrust Sheet, the Sabah Trough, and the NW Sabah Platform. Modified after Hazebroek and Tan (1993). (B) Southern part of the Sabah Basin, from the Inboard Belt, through the Baram Delta growth fault zone, to the Baram Delta toe-thrust zone. Modified after Hazebroek and Tan (1993). (C) A schematic cross-section of the Sandakan Sub-basin, NE Sabah Basin.

subsidence phase accompanied by westward progradation of shelf-slope system as sediment accumulation rates exceeded the rate of increase in accommodation space. The transition between these two phases of subsidence was also observed to be younging in the northwestward direction, which suggests basin formation by foreland loading (Noor Azim Ibrahim, 1994). A foreland basin origin for the Sabah Basin is also indicated from gravity studies by Milsom et al. (1997).

Baram Delta Province

The Baram Delta Province takes its name from the present Baram Delta which is a wave-dominated micro/meso-tidal delta that is fed by the Baram River. The delta may have been tidally influenced in the past (Caline and Huong, 1992; Lambiase et al., 1997) and has been prograding northwards along a valley that was incised into the Lambir and Belait Hills since about 5.5 ka ago (Caline and Huong, 1992). Koopman (1996) recognised several successive phases of delta outbuilding northwestwards over an eroded Rajang Fold-Thrust belt: the Meligan Delta (Early Miocene), Champion Delta (Middle-Late Miocene), and the Baram Delta (Late Miocene to Quaternary). The Meligan and Champion deltas underlie the Inboard Belt in onshore Baram, Brunei, and westernmost Sabah, while the Baram Delta extends from the Baram coastal plain to the shelf edge.

The Baram Delta depocentre developed during the Early Miocene, probably as a fault-controlled depression formed at the intersection of two major crustal-scale faults, the West Baram Line to the west and the Jerudong-Morris Fault to the east (Fig. 5.17). The West Baram Line is regarded by many authors as a major strike-slip fault zone (e.g. Ismail Che Mat Zin and Swarbrick, 1997) or a transform fault (Milsom et

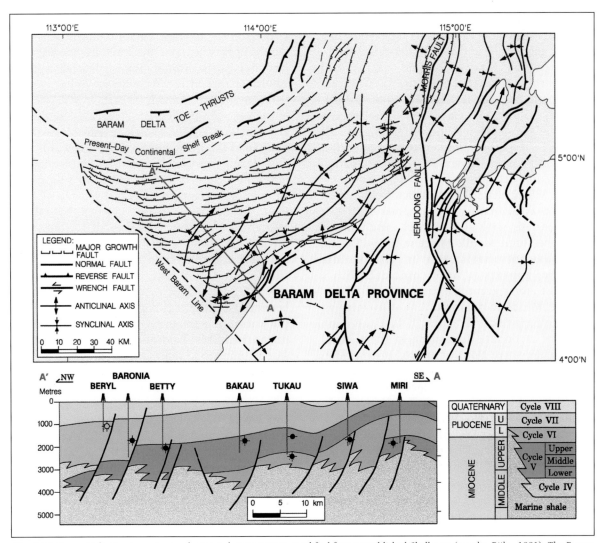

Fig. 5.17. Baram Delta Province structural map and cross-section, modified from unpublished Shell map (see also Rijks, 1981). The Baram Delta is confined between the West Baram Line and the Jerudong-Morris faults. Note the growth faults and superimposed anticlines in the central region. Cross section shows Middle Miocene (Cycle IV) and younger coastal-shelf progradational sediments overlying prodelta marine shale.

al., 1997). It is actually a series of *en echelon* east-dipping extensional faults that forms a major structural discontinuity between the Sarawak and Sabah basins. The Morris Fault is a major left-lateral fault zone that continues landwards into Brunei as the Jerudong Fault.

The Baram Delta Province straddles the northeastern part of Sarawak, Brunei, and western Sabah. The Sarawak part of the delta has been referred to as the West Baram Delta (Chow and Tan, 1997), whereas the Sabah part is known as the East Baram Delta (Tan and Lamy, 1990). This usage is informal and is based purely on political boundaries rather than on geological reasons. The geology of the central (Brunei) part of the province has been described by James (1984) and Koopman (1996).

West Baram Delta

The West Baram Delta was the first to be discovered, and is one of the most prolific, petroleum provinces in Malaysia. Malaysia's first oil field, the Miri Field, was discovered in 1910 in this province. The geological setting and hydrocarbon occurrences of the West Baram Delta have been described by Rijks (1981). Recently, Chow and Tan (1997) published a review of the geological understanding of this area and the prospects for future exploration efforts. Chapter 13 discusses the West Baram Delta in more detail.

Sedimentation in the Baram Delta started with a predominantly argillaceous sequence interbedded with limestones and minor sandstones until Middle Miocene times. Deltaic progradation, that began in the Middle Miocene, appears to have coincided with a regional tectonic event that gave rise to the lower Middle Miocene regional unconformity in offshore Sarawak and Sabah, before the main phase of carbonate deposition in the Central Luconia Province. Northwestwards progradation of the delta was punctuated by several transgressive phases that produced marine shale intervals, which are used to define the eight sedimentary cycles (Ho, 1978).

The structure of the Baram Delta Province is dominated by growth faults, the so-called "regional" growth faults that are downthrown basinwards to the north and northwest (Fig. 5.17). Antithetic (counter-regional) faults occur on the delta-slope where the sand-shale ratios change rapidly. Superimposed on the deltaic growth faults are the effects of basement-involved compressional wrench tectonics that occurred during the early Late Miocene, forming a series of NE-trending folds and some reverse faults. The compressional wrench movements may have resulted in clay diapirism, as mobile clay in the basal part of the sedimentary succession of the delta was reactivated (Koopman, 1996). Many oil accumulations in the province are trapped at the intersection between the E-trending growth faults and NE-trending compressional anticlines (Rijks, 1981). Reservoir sands occur mostly in Upper Miocene-Lower Pliocene middle Cycle V to Cycle VI topsets, stacked in rollover structures on the hangingwall of growth faults. An example is the Betty Field, described by Johnson et al. (1989).

East Baram Delta

The East Baram Delta is a continuation of the Baram Delta Province into Sabah (Tan and Lamy, 1990; Hazebroek and Tan, 1993) (Fig. 5.10). A large part of the East Baram Delta is in water depths of between 200 and 2500 m. Towards the boundary with Brunei, the structural style is similar to that in the West Baram Delta; growth faults and rollover anticlines with crestal collapse structures. Examples of structures with oil accumulation are the Samarang oilfield (Scherer, 1980) and the Kinabalu discovery. Eastwards, the East Baram Delta is characterised by SE-dipping imbricate thrust sheets, bounded by subparallel and partly overlapping faults that sole out along a major detachment surface formed by Oligo-Miocene carbonates (Hinz et al., 1989, Fig. 5.16A). The thrust sheets vary between 3 and 15 km wide and can be up to 190 km long. They form prominent ridges on the sea floor, separating intra-slope basins that are filled with hemipelagic sediments and minor turbidites. The thickness of sediments filling the intraslope basins increases landward up the continental slope, implying that thrust movement had occurred in pulses since probably the Middle Miocene (Hinz et al., 1989).

Tan and Lamy (1990) and Hazebroek and Tan (1993) believe that the thrust structures at the continental slope (Fig. 5.16B) were related to gravity sliding caused by the growth of the delta. However, the NE-SW orientation of these thrust faults, particularly in the northeastern part of the province, is parallel to the strike of the Rajang and Crocker fold-thrust belts and the Inboard Belt. This suggests that the imbricate thrusts may not be related to gravity-driven thrusting from the main Baram Delta, which is 200 km to the southwest, but were more likely caused by northward compressional force and thrust-slice stacking within the Inboard Belt and the Rajang/Crocker fold-thrust belt. Crustal thickening in the inboard region may have resulted in gravitational collapse and spreading, causing thrusting in a seaward direction.

Inboard Belt

The Inboard Belt occurs immediately seaward of the Rajang Fold-Thrust Belt. It is characterised by narrow NNE-trending anticlines, known as the Sabah Ridges (Bol and van Hoorn, 1980), which are separated by wide synclines such as the Labuan Syncline (van Vliet and Schwander, 1987). The structural grain of the Inboard Belt is NNE-SSW, parallel to that of the NW Borneo Trend in the Rajang Fold-Thrust Belt. This trend bends sharply in the northeast to NW-SE, which defines the Sulu Trend. The tight anticlinal structures that form the Sabah Ridges are believed to be the result of basement-wrench faulting during the Late Miocene to Early Pliocene (Bol and van Hoorn, 1980; Hazebroek and Tan, 1993; Koopman, 1996). Major left-lateral fault systems have been identified from seismic and surface outcrops in the Inboard Belt and adjacent onshore areas in Brunei and eastern Sarawak (Hazebroek and Tan, 1993; Koopman, 1996). One of these faults is the Morris Fault that crops out in Brunei to the southwest as the Jerudong Fault. Slump deposits and mud diapirs, related to the Morris Fault and Jerudong Line, indicate syndepositional movement on these old structural features (Levell and Kasumajaya, 1985; Morley et al., 1997).

Outboard Belt

The Outboard Belt occurs seaward of the Inboard Belt, in the outer shelf and upper slope region of the NW Sabah continental margin. The province represents an elongated Late Miocene to Pliocene depocentre, filled with prograding sediments (Hazebroek and Tan, 1993). It has a thin cover of deltaic and shelf (topset) sediments showing both extensional and compressional structural styles. Compressional features, such as high-relief anticlines with shale cores that resemble mud diapirs, are common in the Middle Miocene and deeper shale-prone sequences. Extensional features are represented by NE-trending normal faults downthrown basinwards. This zone is partly equivalent to Hinz et al.'s (1989) Zone V.

Thrust Sheet Zone

The Thrust Sheet Zone occurs at the extreme northeastern portion of the NW Sabah continental margin (Figs. 5.10, 5.16A). Hinz et al. (1989) proposed that this zone is formed by two major thrust sheets stacked over a rifted South China Sea continental basement of inferred Oligocene to Lower Miocene age that dips beneath the NW Sabah margin. The thrust sheets are overlain by Middle Miocene-Upper Pliocene clastics of the Sabah Basin. These chaotic seismic facies are bounded to the northwest and southwest by steep thrusts and to the southeast by normal faults. Hinz et al. (1989) and Hazebroek and Tan (1993) interpreted the thrust sheets as allochthonous masses of Crocker, Chert-Spilite, or equivalent sub-Middle Miocene formations that have been thrust over the NW Sabah margin when the Rajang Fold-Thrust Belt was uplifted. Similar structural relationships have also been documented in the Palawan area (Hinz and Schluter, 1985; Schluter et al., 1996).

Sabah Trough

The Sabah Trough (also known as the NW Borneo Trough, NW Sabah Trough, Palawan Trough, Borneo-Palawan Trough, or Nansha Trough) is a NE-trending deep bathymetric feature under 2800-2900 m of water. It separates the Dangerous Grounds submarine plateau to the northwest from the Sabah continental slope and shelf to the southeast. This 80 km-wide trough continues northeastwards as the Palawan Trough but ends abruptly southeastwards at the West Baram Line. The trough is characterised by a generally flat sea floor except for two volcanic seamounts that are capped by reef limestones (Fig. 5.16). The sediment fill consists of Miocene to Recent pelagic and hemipelagic sediments, thickening landwards from about 1 to 2.5 s two-way travel time (Hinz et al., 1989).

The Sabah Trough was interpreted by Hamilton (1979) as an inactive subduction trench formed during Eocene to Miocene times. Later workers (Hinz et al., 1989) have shown, however, that the trough is underlain by thinned continental crust of the NW Sabah Platform which has been underthrust beneath the Sabah Basin. The Sabah Basin could well represent a deep foreland basin formed by loading of the continental margin sediments and thrust stacking on extended continental basement.

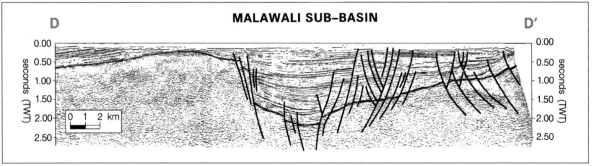

Fig. 5.18. East-west cross section of the Malawali Sub-basin, northeastern Sabah. For location see Fig. 5.16. The blue unconformity between poorly reflective rocks and overlying strata is late Middle Miocene in age.

NW Sabah Platform

The NW Sabah Platform is also known as the Southern South China Sea Platform, which includes the Dangerous Grounds, Reed Bank, and Northern Palawan. This structural province is characterised by half-graben and tilted fault blocks filled with Palaeocene to Lower Oligocene sediments, overlain unconformably by a thin succession of Upper Oligocene-Lower Miocene shallow-water platform carbonates forming the postrift blanket (Fig. 5.16B). The latter is overlain by hemipelagic/pelagic sediment that can be as thick as 0.5 to 1s two-way travel time. Half-graben normal faults are downthrown mainly to the northwest. The basement, of Triassic to Cretaceous age, has a continental origin, consisting of gabbros, diorites, and metamorphics, and Mesozoic plant-bearing redbeds (Kudrass et al., 1985; Xia and Zhou, 1993). The Oligo-Miocene carbonate has been seen on seismic to plunge landwards beneath the Sabah continental margin (Hinz et al., 1989). The NW Sabah Platform is part of the larger "North Palawan Block" that was interpreted by Holloway (1981) as a continental fragment that rifted off southern China during the Oligocene to collide with Sabah during the Early Miocene. Geophysical data indicate the presence of thinned continental crust underneath the region (Xia and Zhou, 1993). Hinz et al. (1989) presented seismic evidence to suggest that the continental crust of the NW Sabah Platform has been underthrust at the Sabah Trough beneath the Outboard Belt. Isostatic rebound of the subducted continental crust may have been the cause of uplift in Sabah.

Malawali and Other Sub-basins

Smaller depocentres or sub-basins occur near the northern tip of Sabah where the Crocker Formation changes its strike from NE-trending to ESE-trending (Fig. 5.10). The Malawali Sub-basin lies north of Marudu Bay between Banggi Island and Kudat Peninsula and may be grouped together with the Siagut Syncline, Kudat Platform, and other smaller depocentres such as Kindu-Mangayau and Tiga Papan, located to the west and northwest of Balambangan Island (see Chapter 22).

The Malawali Sub-basin is filled with Middle Miocene and younger clastics that reach a total thickness of up to 3600 m. The basin is relatively small (800 km^2 surface area), rather like a synclinal depocentre similar to the Labuan Syncline (Fig. 5.18). The basement surface of the Malawali Sub-basin is correlated with the Shallow Regional Unconformity (SRU) (Wong and Salahuddin Saleh Karimi, 1995). The Malawali Sub-basin is characterised by a conjugate system of NNE- and NW-trending normal (growth) faults in the main sub-basin, and NW and E-trending normal faults in the southern basement high area. A NW-SE seismic section across the basin (Fig. 5.18) shows a steep fault-bounded margin on the northern side of the basin whereas the southern flank is gently sloping, with numerous basement-involved faults cutting through the sediment. Sedimentation occurred by deltaic progradation from the southeast down the gently sloping basin flank. An intrabasinal unconformity occurs at the base of the Pliocene (approximately equivalent to Horizon III of Levell, 1987) and appears to have been caused by N-S compression during the Late Miocene-Pliocene.

BASINS IN EASTERN SABAH

NE Sabah Basin

Two major sedimentary basins have been recognised in eastern Sabah: the NE Sabah Basin and the SE Sabah Basin (Fig. 5.10). The NE Sabah Basin includes the circular outcrops of

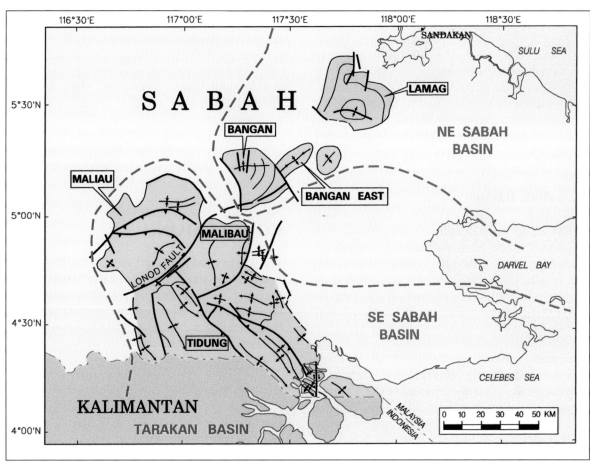

Fig. 5.19. Sketch map of the SE Sabah Basin (which includes the Maliau, Malibau, and Tidung sub-basins) and part of the NE Sabah Basin, which includes several circular sub-basins.

Lower-Middle Miocene shallow water sediments (Tanjong Formation) in central Sabah and extends offshore into the southwestern Sulu Sea margin as the Labuk Bay and Sandakan sub-basins. The structure and stratigraphy of the Sandakan Sub-basin has been described by Bell and Jessop (1974), Wong (1993), and Walker (1993). Graves and Swauger (1997) described the geology of the Sandakan Sub-basin on the Philippine side.

The basins of eastern Sabah are Early Miocene to Recent age. Like in western Sabah, the eastern Sabah basins unconformably overlie a basement composed of deepwater turbidites (Labang Formation). The turbidites rest on Lower Cretaceous ophiolite comprising the Chert-Spilite Formation. The basins also rest upon Middle Miocene marine olistostrome formations, such as the Garinono Formation, interpreted as synrift deposits formed during the opening of the Sulu Sea marginal basin (Hutchison, 1992; Clennell, 1996). The basins have been interpreted as aulacogens or rift basins, the so-called Sulu or Sandakan rift (Tjia et al., 1990; Hutchison, 1992), which were formed by a major rifting event related to the opening and sea-floor spreading in the SE Sulu Sea and the Makassar Straits. That rifting event was probably synchronous with the deformation of the West Crocker and Kudat formations in the Sabah Basin: the Sabah Orogeny of Hutchison (1996b). The rifting produced widespread olistostrome formations in eastern Sabah, such as the Garinono and Ayer formations, thought to be the initial rift infill upon which the shallow water Tanjong Formation was deposited (Hutchison, 1992). The olistostromes, which are also often referred to as mélanges in the literature, were mostly produced by submarine slope failures triggered by tectonic movements at the end of the Early Miocene (Clennell, 1991, 1996).

The structure of the offshore Sandakan Sub-basin is relatively well-known as a result of oil and gas exploration. Its northern part is characterised by NE-trending structural ridges, the Aguja and Pegasus ridges, which define a broad syncline with major NE-hading growth faults and associated rollover anticlines, such as the NE Benrinnes structure. The Pegasus Ridge continues on land as the Segama Ridge. The superimposition of NE-trending compressional folds on the growth faults has produced complex structures, such as the Nymphe structure (Fig. 5.16C). These structures appear to be positive flower structures related to wrench deformation. The southern part of the Sandakan Sub-basin is structurally less deformed, being dominated by NE-trending normal faults parallel to the Sulu Arc trend. Its sedimentation history is characterised by continuous southwestward-progradation of Middle Miocene and younger clastic sediments (Wong, 1993).

SE Sabah Basin

The SE Sabah Basin includes the circular outcrops of Lower-Middle Miocene Tanjong Formation, known separately as Maliau, Malibau, and Tidung sub-basins (Fig. 5.19), which are thought to represent the northern extension of the Tarakan Basin in Kalimantan (Leong, 1978; Hutchison, 1992). It comprises Lower-Middle Miocene shallow marine and deltaic mudstone-sandstone sequences (Tanjong and Kapilit formations) that rest, unconformably in part, upon deepwater mudstone and chaotic deposits of the Labang, Kuamut, and Kalabakan formations. The Tanjong and Kapilit formations have been dated by Allagu (1997) as Early-Middle Miocene and Middle to early Late Miocene, respectively. The structural style in the Maliau, Malibau, and Tidung sub-basins is dominated by NW-trending folds in Miocene and Pliocene sediments, with broad synclines separated by narrow faulted shale-cored anticlines. The Maliau sub-basin has been described by Tjia et al. (1990). This basin has an almost circular outline; its northern margin marked by arcuate north-verging thrusts. A major fault, the Lonod Fault, separates the Maliau and Malibau. The eastern margin of the Malibau Sub-basin is marked by normal faults. The Tidung sub-basin, the onshore extension of the northern Tarakan Basin, has been highly faulted into three thrust slices by two major SW-dipping thrusts. Subvertical reverse faults and flower structures occur in the offshore extension of the Tidung sub-basin southeastwards. Hutchison (1992) includes the Maliau and Malibau sub-basins in the Sulu Sea rift which he interpreted as an aulacogen. Detailed field mapping by Allagu (1997) suggests these sub-basins may have been originally one large basin that has been deformed during the Late Pliocene NE-SW compression, after being buried to great depths. The sub-basins have since undergone rapid uplift, for they now lie over a thousand metres above sea level.

The origin of the circular or saucer-shaped outliers of the Tanjong and Kapilit formations is enigmatic. Various hypotheses have been proposed, including mud-diapirism, differential subsidence, large-scale loading of sand-rich depocentres on muddy substrates formed by the olistostrome, and structurally-controlled depressions formed by the interference of folds and faults (Hutchison, 1989; Lee and Tham, 1989; Tongkul, 1993; Allagu, 1997). The circular outcrops of the NE and SE Sabah basins could be the remnants of a larger continuous sedimentary basin that has been intensely deformed and fragmented by late Tertiary tectonics (Clennell, 1996; Allagu, 1997).

CONCLUSION

The Tertiary sedimentary basins of Malaysia show a variety of structural styles that reflect their tectonic setting and origin. Basins around Peninsular Malaysia are extensional basins formed in continental crust. Hence, their structural styles are typified by basement normal faults, which often have significant strike-slip displacements. These basins are also characterised by high surface heat flow, perhaps because of the thinning of the lithosphere during their formation, which caused a thermal anomaly beneath the basins (Mazlan Madon, 1997b). Structural studies in the Malay and Penyu basins have shown that many extensional structures in the basement were influenced by pre-existing pre-Cenozoic faults that may have been reactivated by Cenozoic tectonic movements. Changes in the regional stress regime had caused these faults to be reactivated again during the course of basin filling and produced a variety of inversion structures.

Basins around Peninsular Malaysia are generally terrestrial extensional basins throughout most parts of their histories. In contrast, the sedimentary basins of Sarawak and Sabah developed a continental margin during the Late Eocene and Early Miocene when the Rajang and Crocker accretionary prisms were uplifted and eroded. Sedimentation in the Sarawak Basin was mainly in coastal to shallow marine environments during the Oligocene and Early Miocene. During the Middle Miocene, tectonic activity resulted in extensional faulting and the creation of depocentres and intervening basement highs. A relative rise in sea level during this event also resulted in marine conditions in offshore Sarawak and the deposition of shallow marine carbonates on partly submerged basement highs.

Landwards of the Central Luconia carbonate platform, especially in the Balingian and Tinjar provinces, the structures are more characteristic of compressional and wrench-related deformation. Such structures are also common in the Sabah Basin, where the effects of basement-involved wrench fault deformation is often superimposed upon earlier extensional and growth faults. The Sabah Basin appears to have developed in a similar tectonic setting as the Sarawak Basin. Hence, we would expect the structural styles to be similar. However, the main contrast between the two basins is that compressional thrusts and wrench structures, in places associated with mud diapirism, are more abundant in the Sabah Basin.

The basins in eastern Sabah probably developed as rifts during the Early Miocene but are now fragmented by younger tectonic activity. They comprise circular or saucer-shaped outcrops that appear to be remnants of a much larger sedimentary basin. The NE and SE Sabah basins seem to comprise at least two successive phases of sedimentation that may be separated locally by a hiatus. These may be interpreted as successor basins that resulted from switching of the position of the depocentre. In the NE Sabah Basin, a Lower-Middle Miocene Tanjong Formation appears to be succeeded to the northeast by a Middle Miocene-Pliocene Sandakan Formation, and in the Dent Peninsula by the Sebahat and Ganduman formations of the Dent Group. In the SE Sabah Basin, the Tanjong and Kapilit formations in the Maliau, Malibau and Tidung sub-basins are succeeded southwards by the Tarakan Basin proper in Kalimantan.

REFERENCES

Ahmad Munif Khoraini, 1993. Tertiary palynomorphs from Batu Arang, Malaysia. Warta Geologi, 19, 116.

Allagu, B., 1997. Sedimentology and structural development of the Malibau Basin, Sabah. In: Tectonics, Stratigraphy and Petroleum Systems of Borneo, Universiti Brunei Darussalam, 22-25 June 1997. (abstract).

Almond, J., Vincent, P. and Williams, L.R., 1990a. The application of detailed reservoir geological studies in the D18 Field, Balingian Province, offshore Sarawak. Bulletin of the Geological Society of Malaysia, 27, 137-159.

Almond, J., Mohd Reza Lasman, Vincent, P. and Williams, L., 1990b. The application of integrated 3D seismic and reservoir geological studies in a complex oilfield, D18 Field, Sarawak, Malaysia. Proceedings of the 8th Offshore South East Asia Conference, 4-7 December 1990, Singapore, 47-60. (OSEA 90185)

Bally, A.W. and Snelson, S., 1980. Realms of subsidence. In: Miall, A.D., ed., Facts and Principles of World Petroleum Occurrence. Canadian Society of Petroleum Geologists Memoir, 6, 9-94.

Bally, A.W., Watts, A.B., Grow, J.A, Manspeizer, W., Bernoulli, D., Schreiber, O. and Hunt, J.M., 1981. Geology of passive continental margins: History, Structure and Sedimentologic Record (with special emphasis on the Atlantic Margin). American Association of Petroleum Geologists, Education Course Note Series, 19, AAPG, Tulsa.

Beddoes, L.R., Jr., 1976. The Balabac Sub-Basin, southwestern Sulu Sea, Philippines. Proceedings of the Offshore South East Asia Conference, February 1976, Paper 15, 22pp.

Bell, R.M. and Jessop, R.G.C., 1974. Exploration and geology of the West Sulu Sea Basin. Australian Petroleum Exploration Association Journal, 1974, 21-28.

Bol, A.J. and van Hoorn, B., 1980. Structural styles in western Sabah offshore. Bulletin of the Geological Society of Malaysia, 12, 1-16.

Boodoo, W., Stevens, A.H., Toland, M.D. and Vlierboom, F.W., 1989. Occidental Malaysia Inc. North Borneo Regional Study, offshore Sarawak. Unpublished OXY report.

Buck, S.P. and McCulloh, T.H., 1994. Bampo-Peutu(!) Petroleum System, North Sumatra, Indonesia. In: Magoon, L.B. and Dow, W.G., eds., The Petroleum system - from source to trap. American Association of Petroleum Geologists Memoir, 60, 625-637.

Caline, B. and Huong, J., 1992. New insight into the evolution of the Baram Delta from satellite imagery. Bulletin of the Geological Society of Malaysia, 32, 1-13.

Chow, K.T. and Tan, D.N.K., 1997. The West Baram Delta, offshore Sarawak - new focus of exploration. Geological Society of Malaysia, Petroleum Geology Conference '97, 1-2 December 1997, Kuala Lumpur (abstract).

Clennell, B., 1991. The origin and tectonic significance of melanges in Eastern Sabah, Malaysia. Journal of Southeast Asian Earth Sciences, 63, 407-429.

Clennell, B., 1996. Far-field and gravity tectonics in Miocene basins of Sabah, Malaysia. In: Hall, R. and Blundell, D.J., eds., Tectonic Evolution of Southeast Asia. Geological Society of London Special Publication, 106, 307-320.

Crostella, A., 1981. Malacca Strait wrench fault controlled Lalang and Mengkapan Oil Fields. Proceedings of the Offshore SE Asia Conference, 9 - 12 February, Singapore, Exploration III session, 1-12.

de Coster, G.G., 1974. The geology of the Central and South Sumatra basins. Proceedings of the Indonesian Petroleum Association 3rd Annual Convention, 3-4 June 1974, 77-110.

Dickinson, W.R., 1974. Plate tectonics and sedimentation. Society of Economic Paleontologists and Mineralogists Special Publication, 22, 1-27.

Doust, H., 1981. Geology and exploration history of offshore central Sarawak. In: Halbouty, M.T., ed., Energy Resources of the Pacific Region. American Association of Petroleum Geologists, Studies in Geology Series, 12, 117-132.

Epting, M., 1980. Sedimentology of Miocene carbonate buildups, Central Luconia Province, offshore Sarawak. Bulletin of the Geological Society of Malaysia, 12, 17-30.

Esso., 1985. Petroleum Geology of the Malay Basin. Unpublished Report, Esso Production Malaysia Inc.

Eubank, R.T. and Makki, A.C., 1981. Structural geology of the Central Sumatra Back-Arc Basin. Proceedings of the Indonesian Petroleum Association 10th Annual Convention, 26-27 May 1981, 153-196.

Fontaine, H., Rodziah Daud and Updesh Singh, 1990. A Triassic 'reefal' limestone in the basement of the Malay Basin, South China Sea: regional implications. Bulletin of the Geological Society of Malaysia, 27, 1-25.

Ginger, D.C., Ardjakusuma, W.O., Hedley, R.J. and Pothecary, J., 1993. Inversion history of the West Natuna Basin: examples from the Cumi-Cumi PSC. Proceedings of the Indonesian Petroleum Association 22nd Annnual Convention, Jakarta, October 1993, 635-658.

Graves, J.E. and Swauger, D.A., 1997. Petroleum systems of the Sandakan Basin, Philippines. In: Howes, J.V.C and Noble, R.A, eds., Proceedings of the IPA Petroleum Systems of SE Asia and Australasia, Jakarta, 21-23 May 1997, Indonesian Petroleum Association, Jakarta, 799-813.

Hamilton, W., 1979. Tectonics of the Indonesian region. United States Geological Survey Professional Paper, No. 1078.

Hatch, G., 1987. Technical evaluation and project summary for PM-1 Block, Straits of Malacca, Malaysia. Sun Malaysia Petroleum Company, Unpublished Report. ER:SUN:1:87-01.

Hazebroek, H.P. and Tan, D.N.K., 1993. Tertiary tectonic evolution of the NW Sabah continental margin. In: Teh, G.H., ed., Proceedings of the Symposium on Tectonic Framework and Energy Resources of the Western Margin of Pacific Basin. Bulletin of the Geological Society of Malaysia, 33, 195-210.

Hazebroek, H.P., Tan, D.N.K. and Swinburn, P., 1994. Tertiary evolution of the offshore Sarawak and Sabah basins, NW Borneo. Abstracts of American Association of Petroleum Geologists International Conference & Exhibition, Kuala Lumpur, Malaysia, 21-24 August 1994, American Association of Petroleum Geologists Bulletin, 78, 1144-1145.

Hinz, K. and Schluter, H. U., 1985. Geology of the Dangerous Grounds, South China Sea, and the continental margin off southwest Palawan: results of Sonne Cruises SO 23 and SO 27. Energy, 10, 3/4, 297-315.

Hinz, K., Fritsch, J., Kempter, E.H.K., Mohammad, A.M., Meyer, J., Mohamed, D., Vosberg, H., Weber, J. and Benavidez, J. 1989. Thrust tectonics along the northwestern continental margin of Sabah/Borneo. Geologische Rundschau, 78, 705-730.

Ho, K.F., 1978. Stratigraphic framework for oil exploration in Sarawak. Bulletin of the Geological Society of Malaysia, 10, 1-14.

Holloway, N.H., 1981, The North Palawan Block, Philippines: its relation to the Asian mainland and its role in the evolution of the South China Sea. Bulletin of the Geological Society of Malaysia, 14, 19-58.

Howell, D.G., Jones, D.L. and Schermer, E.R., 1985. Tectonostratigraphic terranes of the Circum-Pacific Region. In: Howell, D.G., ed., Tectonostratigraphic terranes of the Circum-Pacific Region. Circum-Pacific Council for Energy and Mineral Resources Earth Science Series, No. 1, 3-30.

Hutchison, C.S., 1984. Is there a satisfactory classification for Southeast Asian Tertiary basins? Proceedings of the 5th Offshore South East Asia Conference, 21-24 February 1984, Singapore (preprint).

Hutchison, C.S., 1989. "Geological Evolution of South-East Asia." Oxford monographs on Geology and Geophysics no. 13, Clarendon Press, Oxford.

Hutchison, C.S., 1992. The Southeast Sulu Sea, a Neogene marginal basin with outcropping extensions in Sabah. Bulletin of the Geological Society of Malaysia, 32, 89-108.

Hutchison, C.S., 1996a. "South-East Asian Oil, Gas, Coal and Mineral Deposits." Oxford University Press.

Hutchison, C.S., 1996b. The "Rajang accretionary prism" and "Lupar Line" problem of Borneo. In: Hall, R. and Blundell, D.J., eds., Tectonic Evolution of Southeast Asia. Geological Society of London Special Publication, 106, 247-261.

Ibrahim Abdullah, Johari Mat Akhir, Abd Rashid Jaapur and Nor Azian Hamzah, 1991. The Tertiary basin in Felda Nenering, Pengkalan Hulu (Keroh), Perak. Warta Geologi, 17, 181-186.

Ingersoll, R.W., 1988. Tectonics of sedimentary basins. Geological Society of America Bulletin, 100, 1704-1719.

Ismail Che Mat Zin, 1996. Tertiary tectonics and sedimentation history of the Sarawak Basin, East Malaysia. Unpublished PhD Thesis, University of Durham, 277pp.

Ismail Che Mat Zin and Swarbrick, R.E., 1997. The tectonic evolution and associated sedimentation history of Sarawak Basin, eastern Malaysia: a guide for future hydrocarbon exploration. In: Fraser, A.J., Matthews, S.J. and Murphy, R.W., eds., Petroleum Geology of Southeast Asia. Geological Society of London Special Publication, 126, 237-245.

James, D.M.D., 1984. Regional Geological Setting. In James, D.M.D. (ed.). "The geology and hydrocarbon resources of Negara Brunei Darussalam." Muzium Brunei, 34-42

Johnson, H.D., Chapman, J.W. and Ranggon, J., 1989. Structural and stratigraphic configuration of the Late Miocene Stage IVC reservoirs in the St. Joseph Field, offshore Sabah, NW Borneo. Bulletin of the Geological Society of Malaysia, 25, 79-118.

Kamili, Z.A. and Naim, A.M., 1973. Stratigraphy of Lower and Middle Miocene sediments in North Sumatra Basin. Proceedings of the Indonesian Petroleum Association 2nd Annual Convention, June 4-5, 1973, 53-72.

Katz, B.J. and Dawson, W.C., 1996. Pematang-Sihapas Petroleum System of Central Sumatra. In: Howes, J.V.C and Noble, R.A, eds., Proceedings of the IPA Petroleum Systems of SE Asia and Australasia, Jakarta 21-23 May 1997, Indonesian Petroleum Association, Jakarta, 685-698.

Khalid Ngah, Mazlan Madon and Tjia, H.D., 1996. Role of pre-Tertiary fractures in formation and development of the Malay and Penyu basins. In: Hall, R. and Blundell, D.J., eds., Tectonic Evolution of Southeast Asia. Geological Society of London Special Publication, 106, 281-289.

Kingston, D.R., Dishroon, C.P. and Williams, P.A., 1983. Global basin classification system. American Association of Petroleum Geologists Bulletin, 67, 2175-2193.

Kingston, J., 1978. Oil and gas generation, migration and accumulation in the North Sumatra Basin. South East Asia Petroleum Exploration Society Proceedings, IV, 1977/78, 158-182.

Koopman, A., 1996. Structure. In: Sandal, D.T., ed., "The geology and hydrocarbon resources of Negara Brunei Darussalam". 2nd Edition, Syabas, Bandar Seri Begawan, 61-78.

Kudrass, H.R., Wiedecke, M., Cepek, P., Kreuzer, H. and Muller, P., 1985. Mesozoic and Cenozoic rocks dredged from the South China Sea (Reed Bank area) and Sulu Sea and their significance for plate tectonic reconstructions. Marine and Petroleum Geology, 3, 19-30.

Lambiase, J.J., Crevello, P.D., Morley, C.K. and Simmons, M.D., 1997. Miocene strata of the Baram Delta Province, Brunei Darussalam: rethinking controls on facies development. In: Tectonics, Stratigraphy and Petroleum Systems of Borneo, Universiti Brunei Darussalam, 22-25 June 1997. (abstract).

Lee, C.P. and Tham, K.C., 1989. Circular basins of Sabah. Abstracts of the Geological Society of Malaysia Petroleum Geology Seminar, Kuala Lumpur. Warta Geologi, 29, 54.

Leong, K.M., 1978. Sabah. In: ESCAP Atlas of Stratigraphy I. Burma, Malaysia, Thailand, Indonesia, Philippines. United Nations, Mineral Resources Development Series, 44, 26-31.

Levell, B.K., 1987. The nature and significance of regional unconformities in the hydrocarbon-bearing Neogene sequence offshore West Sabah. Bulletin of the Geological Society of Malaysia, 21, 55-90.

Levell, B.K. and Kasumajaya, A., 1985. Slumping at the late Miocene shelf-edge offshore West Sabah: a view of a turbidite basin margin. Bulletin of the Geological Society of Malaysia, 18, 1-29.

Liechti, P., Roe, R.W and Haile, N.S., 1960. Geology of Sarawak, Brunei and western North Borneo. British Borneo Geological Survey Bulletin, 3, 1960.

Liew, K.K., 1994. Structural development at the west-central margin of the Malay Basin. Bulletin of the Geological Society of Malaysia, 36, 67-80.

May, J.A. and Eyles, D.R., 1985. Well log and seismic character of Tertiary Terumbu carbonate, South China Sea, Indonesia. American Association of Petroleum Geologists Bulletin, 69, 1339-1358.

Mazlan B.H. Madon, 1995. Tectonic evolution of the Malay and Penyu basins, offshore Peninsular Malaysia. Unpublished D.Phil. Thesis, University of Oxford, 325pp.

Mazlan Madon, 1996. Gravity anomalies, isostasy, and the tectonic evolution of the Malay and Penyu basins. Geological Society of Malaysia Petroleum Geological Conference, 9-10 December 1996, Kuala Lumpur. (abstract).

Mazlan B. Hj. Madon, 1997a. The kinematics of extension and inversion in the Malay Basin, offshore Peninsular Malaysia. Bulletin of the Geological Society of Malaysia, 41, 127-138.

Mazlan B. Hj. Madon, 1997b. Analysis of tectonic subsidence and heat flow in the Malay Basin (offshore Peninsular Malaysia). Bulletin of the Geological Society of Malaysia, 41, 95-108.

Mazlan B. Madon and Watts, A.B, 1998. Gravity anomalies, subsidence history and the tectonic evolution of the Malay and Penyu basins. Basin Research, 10, 375-392.

Mazlan B. Hj. Madon, Azlina Anuar and Wong, R., 1997. Structural evolution, maturation history and hydrocarbon potential of the Penyu Basin, offshore Peninsular Malaysia. In: Howes, J.V.C and Noble, R.A, eds., Proceedings of the IPA Petroleum Systems of SE Asia and Australasia, 21-23 May 1997, Indonesian Petroleum Association, Jakarta, 403-424.

Mertosono, S. and Nayoan, G.A.S., 1974. The Tertiary basinal area of Central Sumatra. Proceedings of the Indonesian Petroleum Association 3rd Annual Convention, 3-4 June 1974, 63-76.

Milsom, J., Holt, R., Dzazali Bin Ayub and Ross Smail, 1997. Gravity anomalies and deep structural controls at the Sabah-Palawan margin, South China Sea. In: Fraser, A.J., Matthews, S.J. and Murphy, R.W., eds., Petroleum Geology of Southeast Asia. Geological Society of London Special Publication, 126, 417-427.

Mohd Firdaus Abdul Halim, 1994. Geothermics of the Malaysian sedimentary basins. Bulletin of the Geological Society of Malaysia, 36, 162-174.

Morley, C.K., Crevello, P. and Z. Hj. Ahmad, 1997. Outcrop example of shale diapir and growth fault development: the Jerudong Anticline, Brunei Darussalam. In: Tectonics, Stratigraphy and Petroleum Systems of Borneo, Universiti Brunei Darussalam, 22-25 June 1997. (abstract).

Morris, J.C., 1992. Malaysia, Offshore Sarawak: The hydrocarbon potential of deepwater blocks A & B. Unpublished report, BHP Petroleum.

Murphy, R.W., 1975. Tertiary basins of Southeast Asia. Proceedings of the South East Asian Petroleum Exploration Society, 2, 1-36.

Noor Azim Ibrahim, 1994. Major controls on the development of sedimentary sequences Sabah Basin, Northwest Borneo, Malaysia. Unpublished PhD Thesis, University of Cambridge, 254pp.

OXY, 1991. Malaysia SK6/SK8 Carbonate study. Unpublished Occidental Report, ER:OXY:3:91-02

Raj, J.K., 1998. Tectonic evolution of the Tertiary Basin at Batu Arang, Selangor Darul Ehsan, Peninsular Malaysia. Proceedings of the Seminar on Tertiary Basins of Peninsular Malaysia and Its Adjacent Offshore Areas. Geological Society of Malaysia, 21-22 February 1998, Kuala Lumpur.

Raj, J.K., Abdul Hadi Abd Rahman and Mustaffa Kamal Shuib, 1998. Tertiary basins of inland Peninsular Malaysia: review and tectonic evolution. Proceedings of the Seminar on Tertiary Basins of Peninsular Malaysia and Its Adjacent Offshore Areas. Geological Society of Malaysia, 21-22 February 1998, Kuala Lumpur.

Rijks, E.H.J., 1981. Baram Delta geology and hydrocarbon occurrence. Bulletin of the Geological Society of Malaysia, 14, 1-18.

Sandal, S.T. (ed.), 1996. "The geology and hydrocarbon resources of Negara Brunei Darussalam." 2nd edition, Syabas, Bandar Seri Begawan, 243pp.

Scherer, F.C., 1980. Exploration in East Malaysia over the past decade. In: Halbouty, M.T., ed., Giant oil and gas fields of the decade 1968-1978. American Association of Petroleum Geologists Memoir, 30, 423-440.

Schluter, H.U., Hinz, K. and Block, M., 1996. Tectono-stratigraphic terranes and detachment faulting of the South China Sea and Sulu Sea. Marine Geology, 130, 39-78.

Shell, 1993. SK5 Exploration Data Book 1994. Unpublished Sarawak Shell Berhad Report.

Soeparjadi, R.A, 1982. Geology of the Arun Gas Field. Proceedings of the Offshore SE Asia Conference, 9-12 February 1982, Singapore, 1-15.

Soeparjadi, R.A., Nayoan, G.A.S., Beddoes, L.R. and James, W.V., 1976. Exploration play concepts in Indonesia. Proceedings of the 9th World Petroleum Congress, 3, Exploration & Transportation, 51-64.

Stauffer, P.H., 1973. Cenozoic. In: Gobbett, D.J and Hutchison, C.S., eds., "Geology of the Malay Peninsula." Wiley Interscience, New York, 143-176.

Suntharalingam, T., 1983. Cenozoic stratigraphy of Peninsular Malaysia. Proceedings of the Workshop on Stratigraphic Correlation of Thailand and Malaysia. 8-10 September 1983, Haad Yai, Thailand, Geological Society of Malaysia, v. 1, 149-158.

Swinburn, P., 1994. Structural styles in the Balingian Province, Offshore Sarawak. Abstracts of American Association of Petroleum Geologists International Conference & Exhibition, Kuala Lumpur, Malaysia, August 21-24, 1994, American Association of Petroleum Geologists Bulletin, 78, 1164.

Tan, D.N.K. and Lamy, J.M., 1990. Tectonic evolution of the NW Sabah continental margin since the Late Eocene. Bulletin of the Geological Society of Malaysia, 27, 241-260.

Tapponnier, P., Peltzer, G., Le Dain, A.Y., Armijo, R., and Cobbold, P., 1982. Propagating extrusion tectonics in Asia: new insights from simple experiments with plasticine. Geology, 10, 611-616.

Taylor, B. and Hayes, D.E., 1980. The tectonic evolution of the South China Basin. In: Hayes, D.E., ed., The Tectonic and Geologic Evolution of South-east Asian Seas and Islands. American Geophysical Union, Geophysical Monograph, 23, 89-104.

Taylor, B. and Hayes, D.E., 1983. Origin and history of the South China Sea Basin. In: Hayes, D.E., ed., The Tectonic and Geologic Evolution of South-east Asian Seas and Islands, Part 2. American Geophysical Union, Geophysical Monograph, 27, 23-56.

Teh, G.H. and Sia, S.G., 1992. The Nenering Tertiary deposit, Keroh, north Perak - a preliminary study. Warta Geologi, 17, 49-58.

Tjia, H.D., 1994. Inversion tectonics in the Malay Basin: evidence and timing of events. Bulletin of the Geological Society of Malaysia, 36, 119-126.

Tjia, H.D. and Liew, K.K., 1996. Changes in tectonic stress field in northern Sunda Shelf basins. In: Hall, R. and Blundell, D.J., eds., Tectonic Evolution of Southeast Asia. Geological Society of London Special Publication, 106, 291-306.

Tjia, H.D., Ibrahim Komoo, Lim, P.S. and Tungah Surat, 1990. The Maliau Basin, Sabah: Geology and tectonic setting. Bulletin of the Geological Society of Malaysia, 27, 261-292.

Tongkul, F., 1993. Tectonic control on the development of the Neogene basins in Sabah, East Malaysia. In: Teh, G.H., ed., Proceedings of the Symposium on the Tectonic Framework and Energy Resources of the Western Margin of Pacific Basin. Bulletin of the Geological Society of Malaysia, 33, 95-103.

Uyop Said and Che Aziz Ali, 1997. Nenering continental deposits: its age based on palynological evidence. Geological Society of Malaysia Annual Conference, 1997, Kijal, Terengganu, 21-23 May, 1997, abstracts, p. 94.

van Vliet, A. and Schwander, M.M., 1987. Stratigraphic interpretation of a regional seismic section across the Labuan Syncline and its flank structures, Sabah, North Borneo. In: Bally, A.W., ed., Atlas of Seismic Stratigraphy. American Association of Petroleum Geologists, Studies in Geology, 27, 163-167.

Vijayan, V.R., 1990. Gravity survey of the Layang-Layang Tertiary Basin in Johor, Peninsular Malaysia - a preliminary report. Bulletin of the Geological Society of Malaysia, 26, 55-70.

Walker, T., 1993. Sandakan Basin prospects rise following modern reappraisal. Oil & Gas Journal, May 1993, 43-47.

Wan Ismail Wan Yusoff, 1984. Heat flow study in the Malay Basin. CCOP Technical Publication, 15, 77-87.

Wan Ismail Wan Yusoff, 1990. Heat flow in offshore Malaysian basins. CCOP Technical Publication, 21, 39-54.

Watts, A.B., Karner, G.D. and Steckler, M.S., 1982. Lithospheric flexure and the evolution of sedimentary basins. Philosophical Transactions of the Royal Society of London, A305, 249-281.

Wong, R.H.F., 1993. Sequence stratigraphy of the Middle Miocene-Pliocene southern offshore Sandakan Basin. In: Teh, G.H., ed., Proceedings of the Symposium on Tectonic Framework and Energy Resources of the Western Margin of Pacific Basin. Bulletin of the Geological Society of Malaysia, 33, 129-142.

Wong, R.H.F. and Salahuddin Saleh Karimi, 1995. Regional Interpretation of SBP93 Seismic Lines, Block SB5. PETRONAS Unpublished Report.

Xia, K-Y. and Zhou Di, 1993. The geophysical characteristics and evolution of northern and southern margins of the South China Sea. Bulletin of the Geological Society of Malaysia, 33, 223-240.

Chapter 6

Tertiary Stratigraphy and Correlation Schemes

Mazlan B. Hj. Madon,
Rashidah Bt. Abd. Karim
and Robert Wong Hin Fatt

INTRODUCTION

This chapter briefly describes the stratigraphic correlation schemes used in the Malaysian Tertiary sedimentary basins. Since most of these basins are located offshore (see Fig. 5.1, Chapter 5) and have no outcropping equivalents, their stratigraphies are based on seismic data and well logs. The traditional method of subdividing a sedimentary succession into lithostratigraphic units based on rock type and other lithological criteria has its limitations when applied to the subsurface. Correlation of borehole sections using wireline logs is often problematic because of the diachronous nature of lithostratigraphic surfaces. Although most conventional wireline logs can resolve beds that are a few tens of centimetres thick, the reliability of well-to-well correlation diminishes as well spacing increases. Seismic data, however, when used with high-resolution biostratigraphic zonation, provides a better tool for basin-wide stratigraphic correlation. Hence, biostratigraphy is an important tool that enables the lithostratigraphic and seismic stratigraphic units be placed in their chronostratigraphic framework. A reliable scheme for stratigraphic correlation is a major concern for the explorationist. It is often necessary to carry out a biostratigraphic analysis together with seismic interpretation of a given area.

The chapter reviews the stratigraphic nomenclature used in the major Tertiary sedimentary basins of Malaysia, and briefly discusses the biostratigraphic schemes used for stratigraphic correlation. Different stratigraphic nomenclatures are currently being used in the Malay, Penyu, Sarawak, and Sabah basins. Brief descriptions of the stratigraphy of these basins have been given by earlier workers (e.g. Armitage and Viotti, 1977; Ho, 1978; ASCOPE, 1981; Doust, 1981; Rijks, 1981; Md Nazri Ramli, 1987; Levell, 1987; Ismail Che Mat Zin and Swarbrick, 1997). In the Sarawak Basin and a large part of the Baram Delta Province, the succession is subdivided into sedimentary "cycles", while in the Sabah Basin each subdivision is called a "stage". In the Malay Basin, the basin fill is subdivided into "groups", while the seismic units in the Penyu Basin are given formation names. In eastern Sabah, where the offshore sedimentary basins are an extension of the onshore outcrops, the lithostratigraphic nomenclature (formations and groups) established by the Malaysian Geological Survey is used for the subsurface and offshore successions (e.g. Wong, 1993; Ismail Che Mat Zin, 1994). Figure 6.1 shows the different nomenclatures and their approximate correlation.

BIOSTRATIGRAPHY

Although much emphasis is placed on seismic stratigraphy, biostratigraphy plays an important role in providing the chronostratigraphic framework of exploration. Hence, it is appropriate to begin the chapter with a brief overview of the biostratigraphic methods for correlation of Malaysian Tertiary sequences. Three major groups of microfossils are used for biostratigraphic correlation of Malaysian Tertiary sequences: foraminifera, calcareous nannofossils, and palynomorphs. The utility of each microfossil group depends on the depositional environment of the sediments being investigated. Because the distribution of foraminifera and palynomorphs is controlled by depositional conditions and water salinity, it is useful for delineating subenvironments in terrestrial to marginal marine settings (Fig. 6.2). For age determination, the global standard planktonic foraminifera (Blow, 1969) and calcareous nannofossil zonation (e.g. Martini, 1971) schemes are used, generally with little or no modification. These are calibrated against the global timescale (Haq et al., 1988). In addition to these global biozonations,

Tertiary Stratigraphy and Correlation Schemes

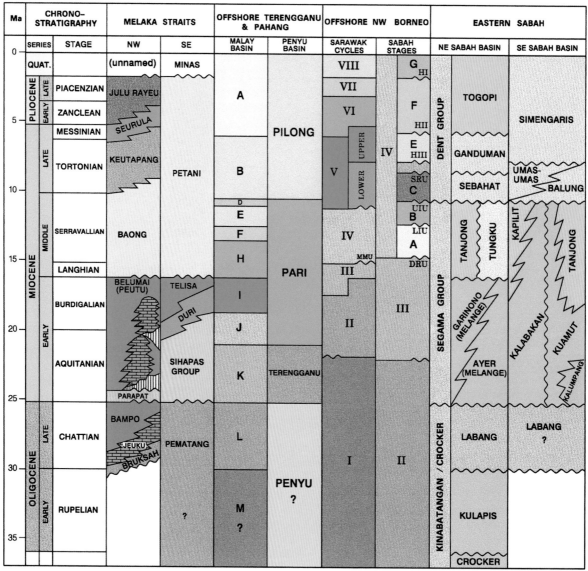

Fig. 6.1. Correlation of the stratigraphies in Malaysian Tertiary basins. Only the most frequently used schemes are shown.

there are also local biozonation schemes, especially for palynomorphs, and different zonation schemes have been established by the various laboratories that provide biostratigraphic services to oil companies (Fig. 6.3).

The paralic nature of Malaysian Tertiary sequences has made it difficult to find age diagnostic foraminifera and nannofossils. As a result, palynology has become the most widely used correlation tool. Morley (1991) gave a historical review of the development of palynological studies as applied to the Southeast Asian Tertiary. In the early days, the stratigraphic application of palynomorphs in the Malaysian basins involved the identification of the first appearance and last extinction datums in the botanical context. Palynomorph zones were established based on the last and first appearances of certain gymnosperms, angiosperms, and fern spores, and by identifying the sudden changes in abundance of any one or more stratigraphical indices. Since palynomorph distribution depends on local geographical factors (including depositional environments), there have been a number of local palynomorph zonation schemes established for different basins. PETRONAS Research (PRSS) has established a palynomorph zonation scheme for the Malay Basin (Azmi Mohd Yakzan et al., 1994). In the Sarawak and Sabah basins, offshore NW Borneo, a much improved palynomorph zonation introduced by Shell is being used by other operators. These palynological schemes are calibrated to the standard time scale indirectly via the planktonic and nannofossil zonation schemes.

Foraminifera and nannofossils are most useful in fully marine sequences. In the Malay Basin, marine sequences occur mainly in the Upper Pliocene and younger. In many parts

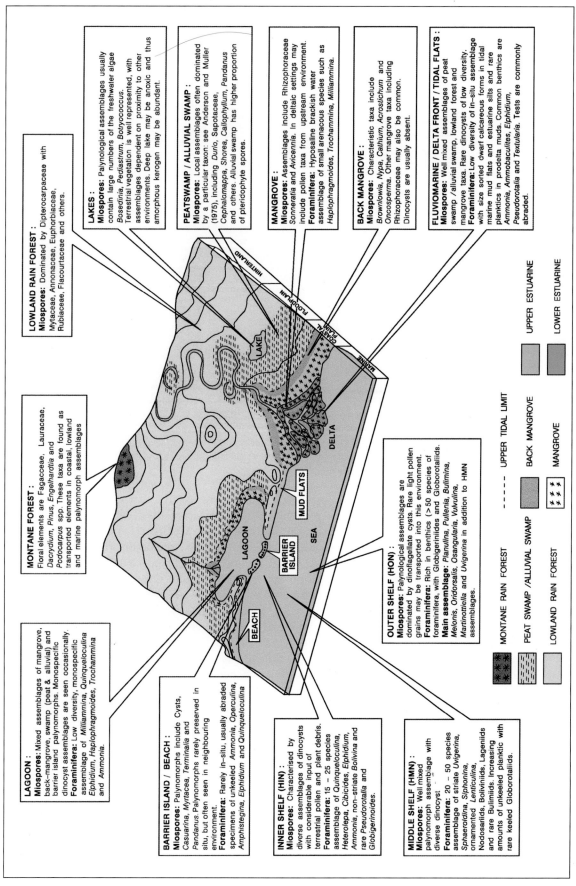

Fig. 6.2. Characterisation of depositional environments in Malaysian basins using foraminifera and palynomorphs.

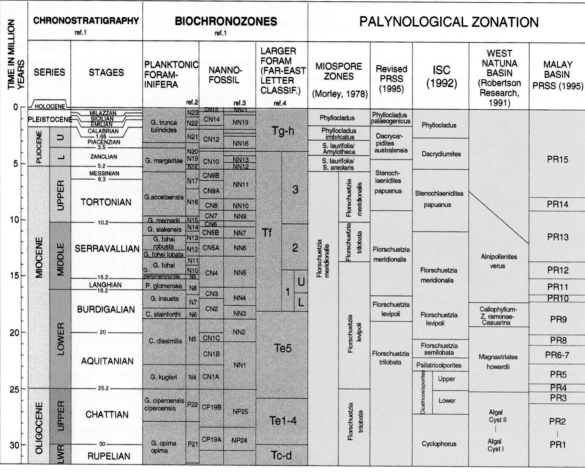

Fig. 6.3. Biozonation schemes used in the Malaysian Tertiary basins, which include the global planktonic foraminiferal and calcareous nannofossil zonations, calibrated to global geochronometric scale (Haq et al., 1988). Shown are the various local/regional palynomorph zonation schemes for basins offshore Peninsula Malaysia. Revised PRSS scheme after Morley (1978) in Azmi Mohd Yakzan et al. (1994). All schemes are calibrated to PRSS miospore zonation for Malay Basin (far right column) in Azmi Mohd Yakzan et al. (1994). References: 1. Haq et al. (1988), 2. Blow (1969), 3. Martini (1971), 4. Adams (1970), Chow (1996).

of the Sabah and Sarawak basins, calcareous nannofossils are more useful than foraminifera, especially in the subdivision of the Upper Miocene to Pleistocene succession. In the determination of epoch boundaries, the Pleistocene is usually demarcated close to and above the last occurrence (LO) of *Discoaster brouweri*, while the Lower-Middle Miocene boundary is usually demarcated at the LO of *Sphenolithus abies*. The first appearance datum (FAD) of *Ceratolithus acutus* has been used in the Mediterranean region to demarcate the Miocene-Pliocene boundary (Cita and Gartner, 1973), but is rarely observed in the Malaysian Mio-Pliocene. Evidence from the Mio-Pliocene of offshore Sarawak, however, suggests that the Miocene-Pliocene boundary occurs between the LO of *D. quinqueramus* and the first occurrence (FO) of *C. acutus*. The dearth of reliable material and the lack of opportunity to study the Pliocene marine sequences have hampered investigation into this problem.

A review of microfossil datums by Finch (1984) for offshore northeastern Sabah has shown that nannofossils can give accurate chronostratigraphic correlation when used with foraminifera. The Middle-Upper Miocene boundary, for instance, can be drawn below the LO of *C. coalitus* and at or near to the FO of the foraminifera, *Globorotalia siakensis*. Some index taxa used in Martini's (1971) zonation may not be directly applicable because of the differences in generic or specific concepts. For example, the Lower Pliocene index taxon, *Amaurolithus tricorniculatus*, is seldom found in the Malaysian Pliocene but, generically, it resembles the genus *Ceratolithus* rather than *Amaurolithus*. The preference so far has been not to rely on this taxon. A re-examination of the Sabah and Sarawak Pliocene has shown that the datums associated with this variant/homeomorphic nannofossil taxa are reliable when calibrated to foraminiferal datums. There is no significant variation in the total stratigraphic ranges of the major indices. This suggests that the zonations of both Martini (1971) and Okada and Bukry (1980) are generally applicable to the Malaysian Tertiary sequences.

MALAY BASIN

The stratigraphic subdivision of the Malay Basin was established by Esso Production Malaysia Inc. in the late 1960's. Using seismic stratigraphy, the sedimentary succession was subdivided into units called "groups" (Fig. 6.4), which are seismically defined packages of strata, bounded by unconformities or sequence boundaries. Here, the term "group" should not be confused with the formal lithostratigraphic unit "Group" (cf. Salvador, 1994, p. 35). The seismic groups are labelled A to M from top to bottom. Subsequently, the groups were given formation names by Armitage and Viotti (1977). The Pulai Formation, for example, is equivalent to Group K. This formational nomenclature is very rarely used now, as most operators prefer the alphabetical nomenclature.

Some operators developed their own nomenclature in their respective exploration acreages. During the late 1960's Conoco operated in the southern part of the Malay Basin and established a different stratigraphic nomenclature. Conoco's formations can be roughly correlated with Esso's seismic groups. The "Terengganu Shale", for instance, is a laterally extensive shaly formation in the southern Malay Basin, which is

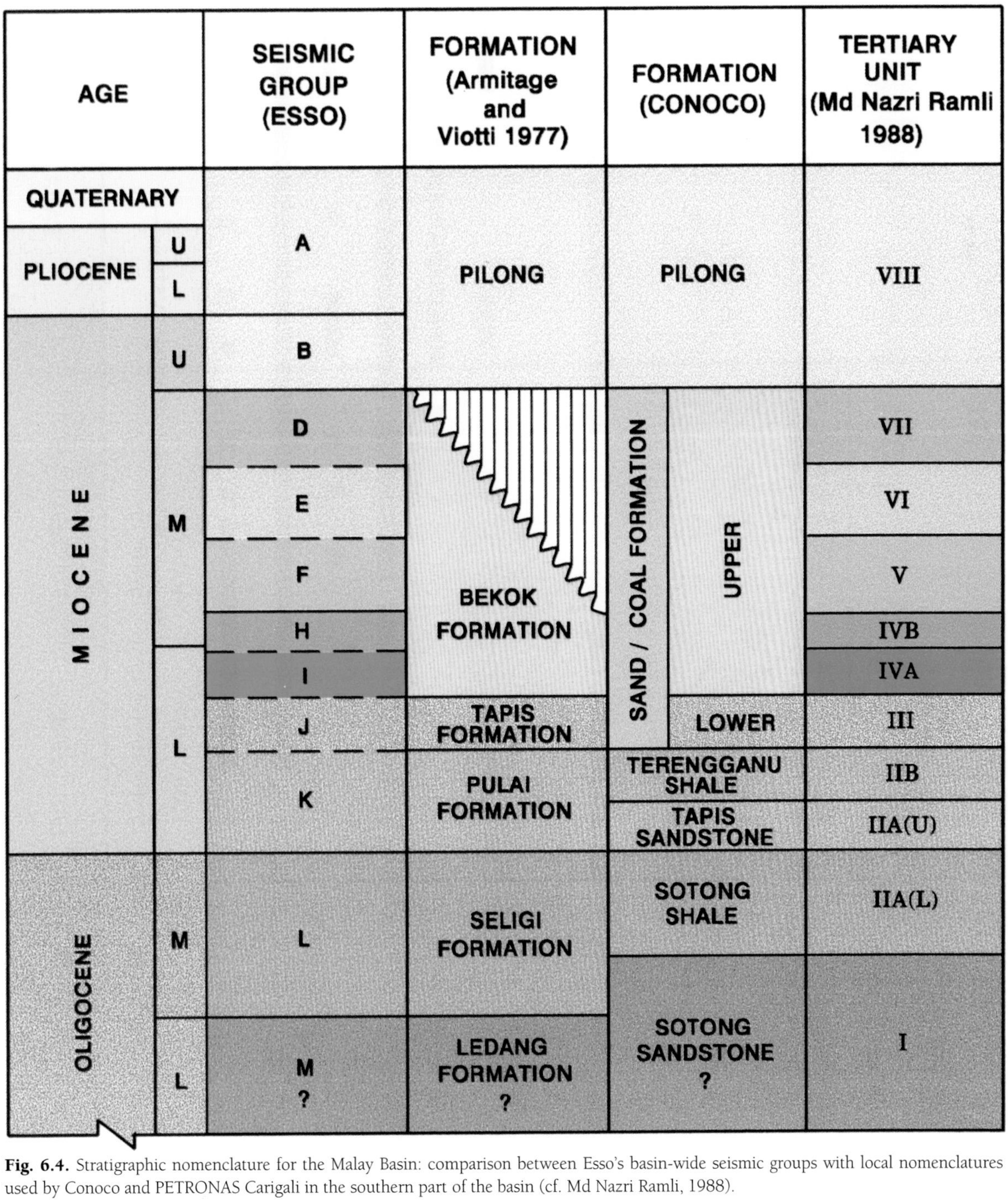

AGE	SEISMIC GROUP (ESSO)		FORMATION (Armitage and Viotti 1977)	FORMATION (CONOCO)		TERTIARY UNIT (Md Nazri Ramli 1988)
QUATERNARY						
PLIOCENE	U	A	PILONG	PILONG		VIII
	L					
MIOCENE	U	B				
	M	D	BEKOK FORMATION	SAND / COAL FORMATION	UPPER	VII
		E				VI
		F				V
		H				IVB
		I				IVA
	L	J	TAPIS FORMATION		LOWER	III
		K	PULAI FORMATION	TERENGGANU SHALE		IIB
				TAPIS SANDSTONE		IIA(U)
OLIGOCENE	M	L	SELIGI FORMATION	SOTONG SHALE		IIA(L)
	L	M ?	LEDANG FORMATION ?	SOTONG SANDSTONE ?		I

Fig. 6.4. Stratigraphic nomenclature for the Malay Basin: comparison between Esso's basin-wide seismic groups with local nomenclatures used by Conoco and PETRONAS Carigali in the southern part of the basin (cf. Md Nazri Ramli, 1988).

Tertiary Stratigraphy and Correlation Schemes

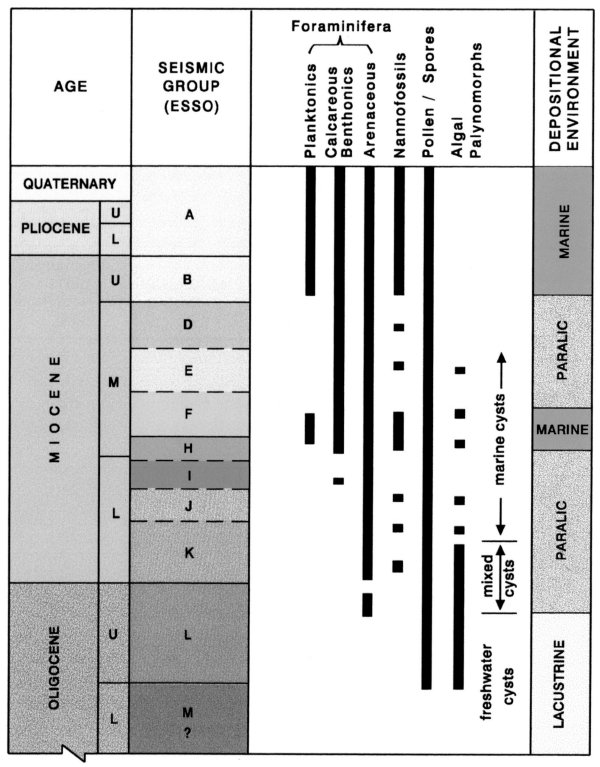

Fig. 6.5. Stratigraphic ranges of the main microfossil groups in the Tertiary succession of the Malay Basin (from Azmi Mohd Yakzan et al., 1994).

equivalent to the "K shale" at the top of Esso's Group K (Fig. 6.4). The "Tapis Sandstone", however, is not the same as the Tapis Formation defined by Armitage and Viotti (1977).

Conoco's nomenclature, however, has been superseded by a scheme introduced by PETRONAS Carigali when they assumed the operatorship of Conoco's exploration blocks in 1979 (Abu Samad Nordin and Md Nazri Ramli, 1985). In the Carigali scheme, described by Md Nazri Ramli (1988), the stratigraphic column is subdivided into 8 "Tertiary units", I to VIII, in ascending order (Fig. 6.4). The units are defined either by the top of regional shale markers (as in Units I and II) or by micropalaeontologically-defined marine transgressive pulses (as in Units

III and IV), or both. This scheme has been used only in PETRONAS Carigali's exploration blocks and in some neighbouring areas. Esso's nomenclature is used in most parts of the Malay Basin.

For biostratigraphic correlation, all the three major microfossil groups (foraminifera, nannofossils, and palynomorphs) are used, but their importance depends on the stratigraphic level (Fig. 6.5). In older Neogene successions, which are mainly nonmarine, well-to-well correlation relies heavily on palynomorphs, whereas in the marine-influenced Upper Miocene and younger strata, foraminifera and calcareous nannofossils are more useful. Since palaeo-water depths rarely reached beyond the middle neritic realm (~20 to 100 m depth), reliable index foraminifera and nannofossils are rarely recovered. The presence of planktonic foraminifera, however, is usually taken as indicative of a relative deepening of the environment. Rare but well-preserved planktonic specimens in a dominantly paralic assemblage also often indicate tidal influence during deposition. The presence of nannofossils is a useful indicator of a marine flooding event (Awalludin Harun et al., 1996). Even if a nannofossil assemblage is of low diversity and characterised by dissolution-resistant forms, it may contain index species that are useful for dating the flooding event.

Palynology is the main correlation tool for the Malay Basin sediments, which are dominated by coastal plain and marginal marine sequences. Despite the importance of palynomorphs, however, only a few palynological

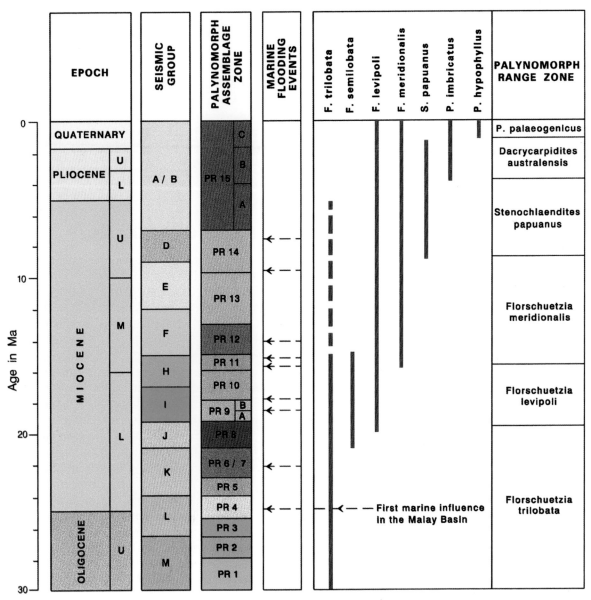

Fig. 6.6. Tertiary palynormorph assemblage zones (PR1 to PR15) established for the Malay Basin by PETRONAS Research, correlated with Esso's seismic groups. Also shown are the marine flooding surfaces and the extinction datums identified from key palynomorph index species (after Azmi Mohd Yakzan et al., 1994).

extinction events can be used for correlation, probably because of the relatively high sedimentation rate during the Neogene. Besides *Florschuetzia sp.*, other index taxa are poorly represented. Nevertheless, a thorough scanning for certain index taxa has helped to improve the recovery and applicability. An example is *Stenochlaenidites papuanus*, which has two morphological groups that can be used to distinguish between the Upper Miocene and Pliocene sections in the basin.

The quantitative analysis of palynomorph distributions has helped stratigraphers to overcome the problem of finding index species for correlation. A major development in stratigraphic palynology in Malaysian basins is the use of quantitative analysis of miospore frequencies, particularly in marginal marine sequences. Azmi Mohd Yakzan et al. (1994) evaluated 67 taxa from 16 wells in the Malay Basin and found 47 taxa that are stratigraphically diagnostic. Palynological datums and events were identified from the quantitative changes in the abundance of certain taxa such as *Pandaniidites sp., Myrtaceidites sp.*, and *Callophyllum*. Using these datums and events, 15 palynomorph zones were defined for correlation of the Upper Oligocene to Upper Miocene sequences (Fig. 6.6). The application of these biostratigraphic signatures in correlating depositional sequences in the basin was reported by Awalludin Harun et al. (1996).

PENYU BASIN

As in the Malay Basin, the stratigraphy of the Penyu Basin is based on seismic stratigraphy. In an early study, Khalid Ngah (1975) defined three major stratigraphic units based on seismic reflection characteristics and lithological data from two exploration wells. Overlying the pre-Tertiary basement is an Oligocene-Lower Miocene unit of continental deposits, which show relatively weak and transparent reflections. A middle unit, of Middle-Upper Miocene age, consists of coastal plain sandstone, mudstone, and lignite. This unit is characterised by very strong and continuous reflections. Capping the succession is a Pliocene to Quaternary marine unit of mainly soft mud, which shows weak and transparent seismic reflections.

The current nomenclature for the Penyu Basin was established by Texaco (1992) based on seismic data acquired during the early 1990's. Figure 6.7 shows the current nomenclature and a correlation with the neighbouring basins. The stratigraphy reflects a two-phase evolution that is typical of extensional basins: a synrift phase followed by a postrift phase (see Chapter 9). The Oligocene synrift sequence filling the extensional half-grabens is assigned to the Penyu Formation. A regional unconformity at the top of the Penyu Formation marks the end of the synrift phase during the Late Oligocene. This is overlain by a laterally extensive shale, the Lower Miocene Terengganu Formation, which is believed to be equivalent to the Barat Shale in the West Natuna Basin and to the Terengganu (K) Shale in the southern Malay Basin. The overlying Pari Formation (Lower to Upper Miocene) consists of nonmarine to marginal marine sediments deposited during the postrift phase. The uppermost unit, the Pilong Formation (Upper Miocene-Quaternary), consists of marine clays. Its name follows the nomenclature of Armitage and Viotti (1977) for the southern Malay Basin. The Pilong and Pari formations are separated by a regional Middle-Upper Miocene unconformity which is also present in the neighbouring basins. As in the Malay Basin, biostratigraphic zonation of the Penyu Basin sediments is based primarily on palynomorph assemblages. Figure 6.3 shows the palynomorph zonation scheme for the basin, established by International Stratigraphic Consultants (ISC).

STRAITS OF MELAKA

The Tertiary stratigraphic nomenclature for the Straits of Melaka follows that of the North Sumatra Basin (Fig. 6.8). In the northern part of the strait (exploration blocks PM320 and PM321, see Chapter 10), the stratigraphy is similar to that of the North Sumatra Basin, whereas in the southern part (Block PM322) the stratigraphy is correlated with that of the Central Sumatra Basin. In both basins, the stratigraphic subdivision is based on outcrop geology, which has been extended into the subsurface and offshore.

Stratigraphic information on the northern Melaka Strait (Block PM320/321) is derived from Mobil's MSS-XA and MG-XA wells and Sun Oil's Singa Besar-1, Dayang-1, and Langgun Timur-1 wells. Figure 6.8 shows a summary of the stratigraphy and its correlation with North Sumatra. The wells penetrated a pre-Tertiary basement composed of Permian to Cretaceous schists, phyllites, quartzites, dolomites, and limestones. The basement limestones in the Singa Besar-1 well are rich in Hemigordiopsidae foraminifera, such as *Shanita sp, Hemigordiopsis renzi, Globivalvulina*, and *Pachyphloia sp.*, which

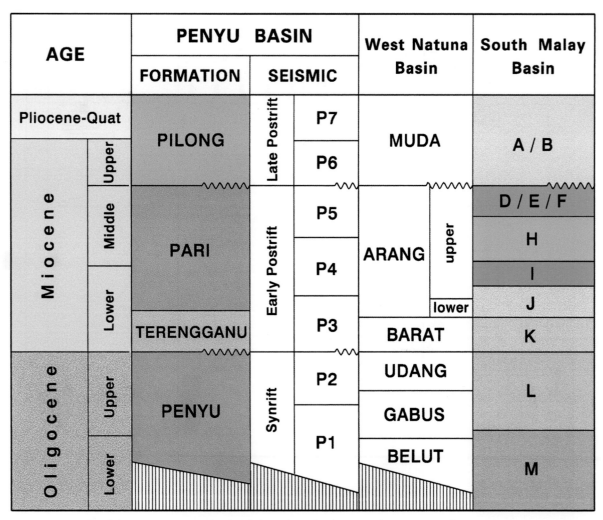

Fig. 6.7. Stratigraphy of the Penyu Basin, correlated with the Malay and West Natuna basins. Sources: formation names (Texaco, 1992), seismic units (Mazlan Madon et al., 1997), West Natuna stratigraphy (Phillips et al., 1997).

indicate a Middle Permian (Murghabian) age (Fontaine et al., 1992). These fusuline-poor limestones are similar to the Chuping Limestone (of Middle Permian-Triassic age) that crops out in the Langkawi Islands and in the state of Perlis. Other black limestones penetrated in the Langgun Timur-1 and Dayang-1 wells are fossil poor, and resemble the Setul Limestone (Silurian) of Langkawi.

Overlying the basement carbonates in the Dayang-1 and Langgun Timur-1 wells are Eocene dolostones and dolomitic limestones of the Tampur Formation. Since these rocks lack primary porosity, they are generally regarded as part of the economic basement in the North Sumatra Basin (Andreason et al., 1994), although sometimes they have become a secondary drilling objective on the Melaka Platform (Chapter 10). Both the pre-Tertiary rocks and the Tampur Formation are overlain unconformably by an Oligocene succession of coarse-grained sandstones and shales (Bruksah and Bampo formations), and some reefal limestones (Jeuku Limestone) (Murray, 1991; Buck and McCulloh, 1994). The synrift alluvial to lacustrine sediments of the Bruksah Formation onlap onto progressively shallower basement to the northeast. Synrift rocks are generally absent in Blocks PM320/321, except in the MG-XA well in the Central Graben area (Block PM321). Thus, the Neogene stratigraphy in the Malaysian part of the northern Melaka Straits (Blocks PM320/321) generally begins with the equivalent of the Lower Miocene Belumai Formation. The succession represents an Early Miocene period of deposition during a marine transgression across the Melaka Platform. In the North Sumatra Basin proper, the Belumai succession consists of calcareous sandstones and shales, and are locally interbedded or overlain by pinnacle and biohermal reef buildups of the Peutu Formation. In Block PM320, Peutu-type limestones are known as the "Melaka reefs", and are often difficult to differentiate from the pre-Tertiary carbonates because of dolomitization and recrystallization.

The Middle Miocene Baong Formation overlies and interdigitates with the Melaka

Tertiary Stratigraphy and Correlation Schemes

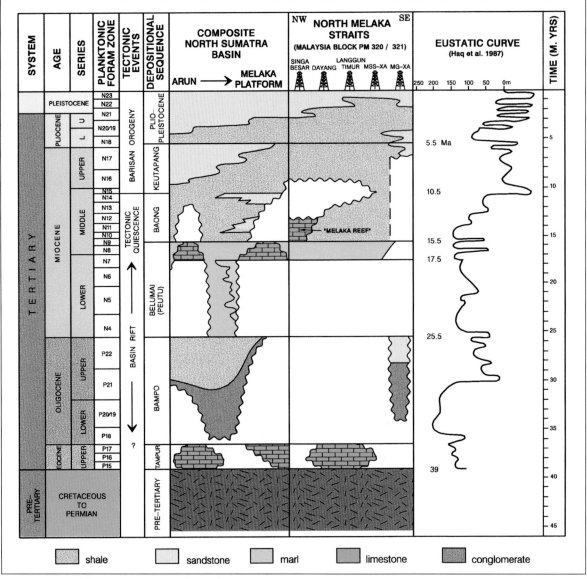

Fig. 6.8. Stratigraphic correlation across the North Sumatra Basin into the Malaysian blocks PM320/321, northern Straits of Melaka. Modified after Andreason et al. (1994).

limestones. Only a relatively thin Baong equivalent is present in Blocks PM320/321. It has been penetrated by the Dayang-1 and Langgun Timur-1 wells but is absent at Singa Besar-1 (Murray, 1991). The Upper Miocene post-Baong strata on the Melaka Platform consist of brown shales and fine-grained sandstones, equivalent to the upper part of the Keutapang Formation. These sediments were deposited in inner to middle shelf environments. A Plio-Pleistocene succession of silty claystones and sandstones unconformably overlies the Keutapang equivalent.

The only well drilled in the southern Melaka Straits (Block PM322) is Port Kelang-1 (Fig. 6.9). This well penetrated a pre-Tertiary basement of recrystallised, unfossiliferous limestone thought to be equivalent to the Setul Limestone or the Batu Caves limestone of Kuala Lumpur (Fontaine et al., 1992). An Upper Oligocene succession of sandstone, claystone, and lignite overlies the basement and is probably equivalent to the Pematang Group of Central Sumatra. The abundance of the freshwater alga, *Pediastrum spp.*, indicates deposition in a lacustrine environment. Overlying the Pematang interval are fine- to coarse-grained sandstones, claystones and, rarely, thin lignite, probably equivalent to the Sihapas Group. This sequence was deposited mainly in an upper estuarine to supratidal setting and is dominated by a back-mangrove flora.

In Central Sumatra, the Pematang-Sihapas boundary is an unconformity that represents the 25.5 Ma Oligo-Miocene sequence boundary (Katz and Dawson, 1997). There, the Sihapas is overlain unconformably by sandy-silty regressive marine deposits of the Petani Formation (Middle Miocene to Pliocene). The Petani Formation, however, is missing in the Port Kelang-1 well because of

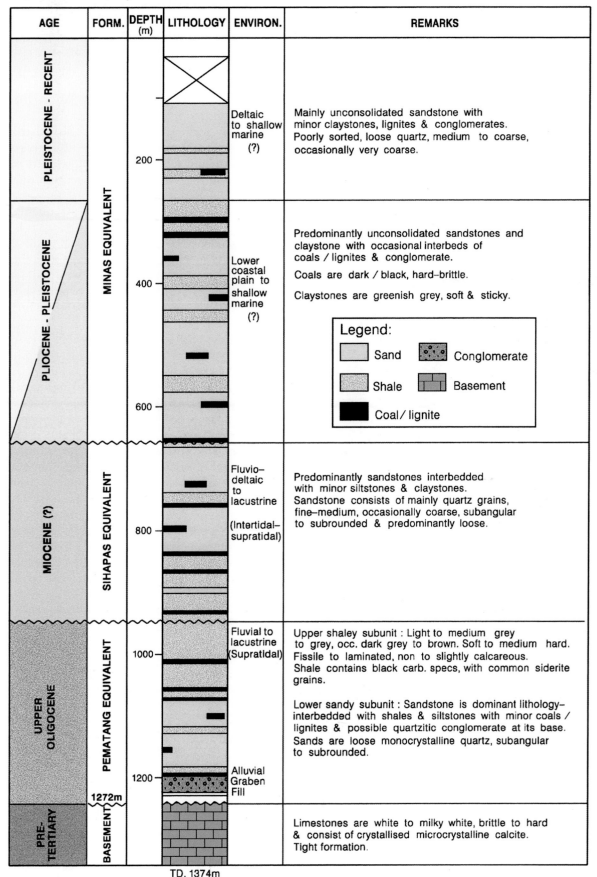

Fig. 6.9. Stratigraphy of the Port Kelang-1 well, southern Melaka Straits (PM321). After Murray (1991).

truncation at the base-Minas unconformity on the crest of the Port Kelang structure (see Chapter 10). The youngest formation in this well is the Minas equivalent, which consists of poorly consolidated medium-to coarse-grained quartzose sandstone with some interbeds of soft claystone and coal. The age of this sequence is poorly constrained, but is probably Pliocene or younger.

SARAWAK BASIN

In the Sarawak Basin, Shell geologists subdivided the Oligocene-Recent strata into sedimentary "cycles". Potter et al. (1984) and van Borren et al. (1996) summarised the historical development of the cycle concept in the Baram Delta Province (offshore NW Borneo). The concept was originally applied to the Baram Delta Province, where the sedimentary succession is characterised by alternating sandstone and mudstone. A cycle is defined as a regressive package of sediment bounded above and below by marine transgressive (or flooding) surfaces. The base of a cycle usually coincides with a regional transgressive marine shale on top of a fluvial/coastal sand belonging to the previous cycle. The cycles may be identifiable in well logs based on variation in sand content and wireline log patterns (Doust, 1981; Rijks, 1981). Dating and correlation of the cycles are based on biostratigraphic zonation using palynomorphs and foraminifera (Fig. 6.10). Where thick coastal plain deposits dominate the sedimentary succession, such as in the Balingian Province, correlation relies heavily on seismic interpretation and pollen zonation.

The cycle concept has been applied to the entire Sarawak shelf. A total of 8 regressive cycles have been recognised (Fig. 6.10). Doust (1981) summarised the characteristics of each cycle. Cycles I and II (Upper Eocene to Lower Miocene) are equivalent to much of the Setap Shale and Nyalau formations exposed onshore. These deposits are mainly fluvial and estuarine channel sands, overbank clays and coals. Minor carbonate deposition during this time is represented by the Subis and Melinau limestones.

Cycle III (Lower-Middle Miocene) consists of shale with thin limestone and sandstone beds, with patchy development of thicker limestones in the upper part. Cycle IV (Middle Miocene) represents open shallow marine conditions throughout most parts of the Sarawak shelf, with extensive development of carbonate buildups in the Central Luconia Province. During Cycle V times (Middle to Upper Miocene) the Baram and West Luconia deltas began to prograde, covering some parts of the carbonate shelf. Cycles VI and VIII (Upper Miocene to Pleistocene) consist of open marine to coastal clays and sands deposited during delta progradation, some completely covering the carbonate buildups in the Central Luconia Province.

As discussed by van Borren et al. (1996), the cycle concept also has its limitations, particularly when dealing with fully marine successions. Unlike in coastal-inner shelf settings, the depositional cycle is less well-developed in the open marine environment. Furthermore, since a cycle is bounded by marine flooding surfaces, the possible presence of a sequence boundary within it can easily be overlooked. In practice, sequence boundaries, particularly those associated with major erosional events at the basin margins, are more easily identified on seismic sections and well logs than cycle boundaries that are related to marine flooding surfaces.

The relationship between cycle stratigraphy and modern sequence stratigraphy has been illustrated by van Borren et al. (1996). A fundamental difference between the two concepts is in the selection of the boundary that defines a "cycle" or a "sequence". In sequence stratigraphic terminology (van Wagoner et al., 1990) cycle boundaries are represented by marine flooding surfaces, whereas sequence boundaries are recognised as major erosion surfaces and their correlative depositional surfaces which result from relative sea level falls. A cycle in offshore NW Borneo is therefore equivalent to the "genetic stratigraphic sequence" of Galloway (1989). The work by Agostinelli et al. (1990) is an example of the use of Galloway's alternative concepts in favour of the Exxon sequence stratigraphic model in the eastern Balingian Province, where Type 1 sequence boundaries are believed to be absent.

The local sequences identified from seismic and sequence stratigraphic analysis are dated using a biostratigraphic scheme. Shell has established a local palynomorph zonation scheme alongside planktonic foraminifera for offshore NW Borneo (e.g. Potter et al., 1984). Shell's palynological scheme has evolved and improved through the years; the latest version has zones with the prefix "P", shown in Fig. 6.10. Besides the use of index markers, palynomorph distribution is also used for the interpretation of eustatic events (Poumot, 1989) and climatic signals (Chow, 1996). Poumot's (1989) quantitative approach was applied to the Upper Miocene-Pliocene sequences, in which

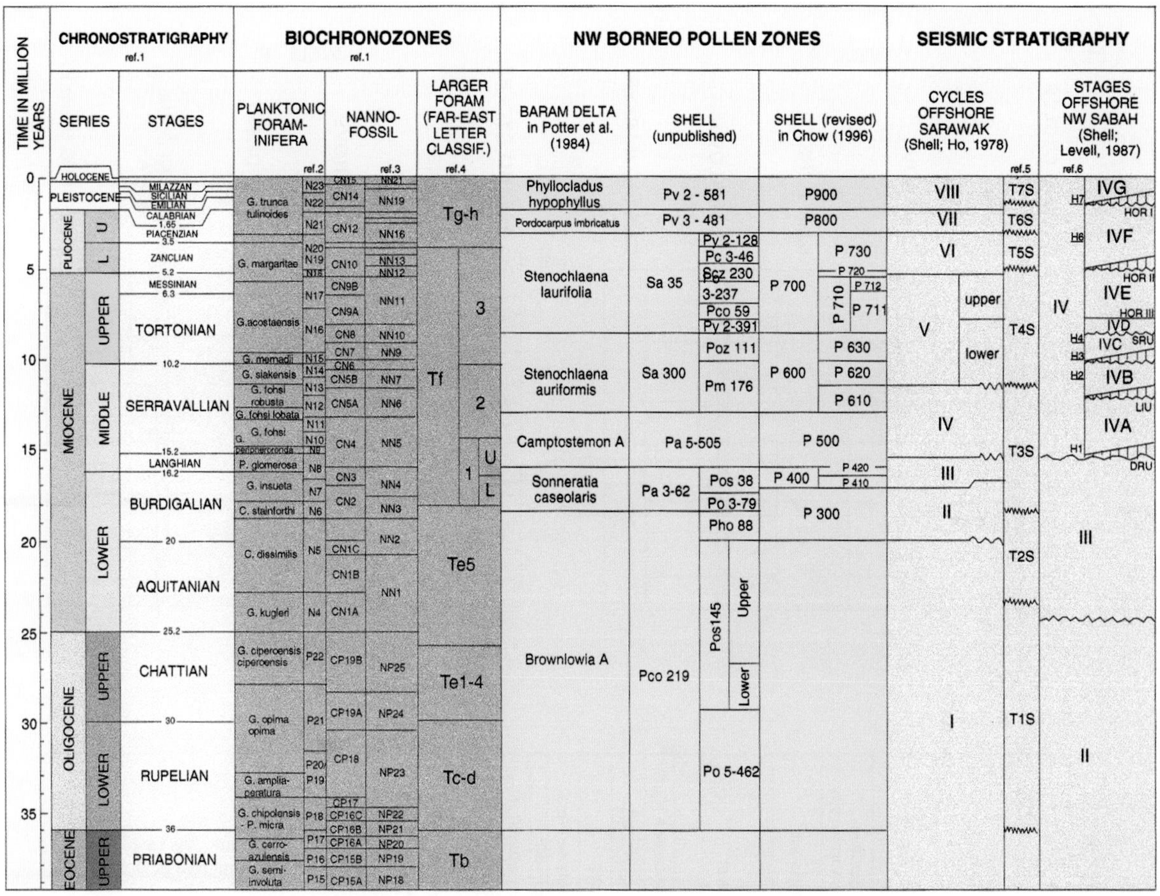

Fig. 6.10. NW Borneo pollen zonation schemes and seismic stratigraphic nomenclature for offshore Sarawak and Sabah, calibrated against the global chronostratigraphic scale and biochronozones, as in Fig. 6.3.

repetitive or cyclic changes in miospore assemblages, termed "palynocycles", were identified (Fig. 6.11). Palynocycles are attributed to palaeobotanical changes due to variations in sea level (eustasy) or climate. A total of 14 regional palynocycles, which were dated using foraminiferal and nannofossil datums, were used successfully for inter-well correlation.

Chow's (1996) approach takes advantage of the strong influence of climate on miospore assemblages. As a result, 16 climatic phases, representing ever-wet versus seasonal conditions, were identified on the basis of the miospore concentrations. The overall climatic pattern shown is remarkably similar to the global eustatic changes, implying a close link between climate and sea level change. Certain miospore assemblages were also observed to correspond with particular systems tracts as defined in logs and seismic data.

SABAH BASIN

The cycle concept is less applicable to the Sabah Basin (offshore NW Sabah) because of the strong influence of tectonics on the stratigraphy during the Neogene. Intense syn-depositional tectonic deformation has resulted in a complex stratigraphic architecture, characterised by regional unconformities and highly diachronous transgressive surfaces, particular at the landward basin margin (Levell, 1987; Noor Azim Ibrahim, 1994). The unconformities and their correlative conformities are used to define several unconformity-bounded stratigraphic units called "stages" (Fig. 6.10). The term "stage" as used in the Sabah Basin should not be confused with the formal chronostratigraphic unit "Stage" which refers to a package of rock strata deposited during a specified time interval (cf. Salvador, 1994, p. 78). A stage in the Sabah Basin is similar in some respects to a "sequence" in sequence stratigraphic terminology. Since Shell has been the major operator in the Sabah Basin, the planktonic foraminifera and pollen zones used in offshore Sarawak (described above) have also been used to date the Sabah stages (Fig. 6.10).

The strongest of the regional unconformities is the Deep Regional Unconformity (DRU), which separates a pre-Middle Miocene succession of deep marine rocks (Stages I to III) from the overlying

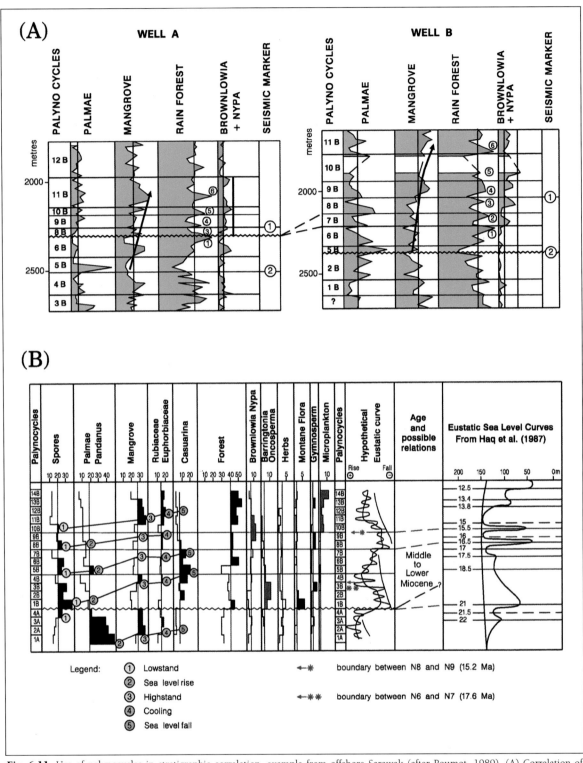

Fig. 6.11. Use of palynocycles in stratigraphic correlation, example from offshore Sarawak (after Poumot, 1989). (A) Correlation of palynocycles between two wells based on pollen diagrams. (B) Quantitative zonation with "megapalynocycles" and possible relation to eustatic sea level changes.

Middle Miocene to Quaternary prograding shelf-slope sediments. The latter is subdivided into 7 stages: Stage IVA to Stage IVG in younging order. These stages are bounded by major unconformities produced by tectonic movements and uplift of the southeastern margin of the basin. Each unconformity passes northwestwards into a conformable downlap surface.

EASTERN SABAH

The Tertiary basins of eastern Sabah has onshore and offshore parts. The continental margin off eastern Sabah is underlain by the offshore continuation of the NE and SE Sabah basins (Chapter 5). The NE Sabah Basin extends from Labuk Bay to the Dent Peninsula, and includes the circular outcrops of the Tanjong Formation in the Central Sabah Sub-basin. The offshore Sandakan Sub-basin on the southwestern Sulu Sea margin is also part of the NE Sabah Basin. The SE Sabah Basin underlies the onshore Kuamut, Segama and Tawau areas, and also includes the circular Tanjong Formation outcrops of the Maliau, Malibau, and Tidung sub-basins (Fig. 5.19, Chapter 5). These sub-basins are thought to represent the northern extension of the Tarakan Basin in Kalimantan (Leong, 1978; Hutchison, 1992).

The stratigraphy of eastern Sabah has been described in the Geological Survey memoirs based on field mapping and palaeontological analysis of outcrop samples (Haile and Wong, 1965; Leong, 1974). Traditionally, the outcrops were dated palaeontologically using the Far-East Letter Classification (Adams, 1970). Correlation with the standard European chronostratigraphic units are still rather broad. [Figure 6.12 shows a correlation based on Chow (1996)]. The letter scheme is based on larger foraminifera which are common in the shallow water shelf limestones in the region. All the published memoirs and reports of the geological surveys of Sarawak and Sabah before 1967 uses this scheme. The Tertiary is subdivided into "stages", each of which is characterised by an assemblage of larger foraminifera. Hence, it is not applicable to siliciclastic sequences that are devoid of larger foraminifera. This problem has been overcome partly by correlating the letter stages with planktonic foraminifera zones.

Figure 6.12 summarises the onshore lithostratigraphy of the NE and SE Sabah basins, derived from the latest Geological Map of Sabah. The lithostratigraphic units have been assigned ages based on palaeontological analysis, using the letter classification. Both basins unconformably overlie a basement comprising Eocene/Oligocene to Lower Miocene (Te$_{1-4}$) deep marine turbidites (Crocker, Kulapis, and Labang formations) that rest on Lower Cretaceous ophiolites (Chert-Spilite Formation). The Lower-Upper Miocene basin succession begins with deep marine, mainly siliciclastic, sequences that include the olistostromes and melanges (Ayer, Garinono), overlain by the shallow marine Tanjong and Kapilit formations. These are in turn overlain by an Upper Miocene-Quaternary sequence of progradational deltaic to shallow marine deposits of the Sandakan Formation and its equivalent on the Dent Peninsula, the Dent Group. Deposition of the Tanjong Formation appears to have been contemporaneous in part with the olistostromes (Clennell, 1996). Although the Sandakan Formation is generally younger than the Tanjong Formation, it is probably contiguous, in some places, with the Tanjong Formation. It is thought to represent a younger phase of deltaic progradation into the basin.

NE Sabah Basin

The stratigraphy of the Sandakan Sub-basin, offshore NE Sabah, is based on seismic interpretation and biostratigraphy (Wong, 1993; Walker, 1993). Previous operators (Aquitaine, WMC Petroleum) in this area have attempted to indirectly correlate the seismic units with the onshore formations. Uncertainties exist in the correlation, however, because of complex faulting, the presence of major unconformities, and the lack of age-diagnostic foraminifera and nannofossils. The current nannoplankton, foraminiferal, and palynological schemes give poor age control. The palynological zonation developed by ISC, although generally identifiable, is not accurate because of the long-ranging index markers used. The problem is also serious because of the poor microfossil content and preservation. Only some of the CN nannofossil zones can be recognised, but only in the Upper Miocene-Pliocene Dent Group section. Since marine microfossils are scarce, quantitative palynology, as applied in the other Malaysian basins, has the potential to be an important correlation tool in this basin.

The sequences in the Sandakan Sub-basin, offshore Dent Peninsula, were described by Wong (1993), Walker (1993), Ismail Che Mat Zin (1994), and Clennell (1996). Note that, while Walker (1993) and Wong (1993) assigned the Dent Group to the Upper Miocene-Pleistocene, Ismail Che Mat Zin (1994) interpreted the offshore Dent Group to the Pliocene-Pleistocene, slightly younger than previous authors, although the basis for this interpretation was not given. This discrepancy needs to be resolved.

It is also worth noting, that Line C-C' of Clennell (1996) and Line 1 of Ismail Che Mat Zin (1994) are the same line, but the stratigraphic interpretation is different. Clennell (1996) interpreted the deformed sediments of Unit 1 below the angular unconformity as a Middle-Upper Miocene prograding wedge, which he assigned to the Sandakan Formation and younger. Ismail Che Mat Zin (1994), however, called the same sequence the "Older Sebahat Formation". The strata below the unconformity

Tertiary Stratigraphy and Correlation Schemes

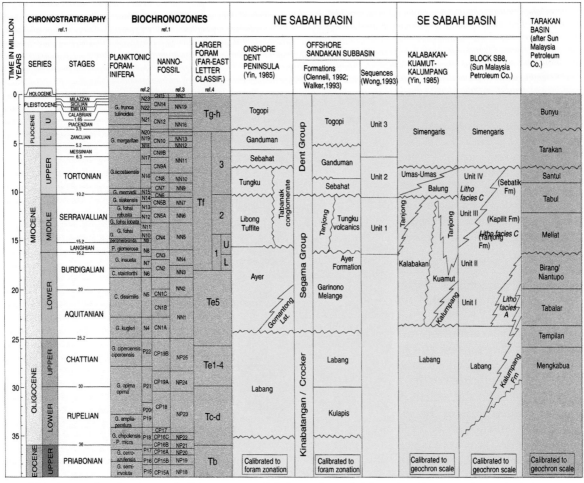

Fig. 6.12. Stratigraphy of onshore and offshore eastern Sabah (NE and SE Sabah basins), calibrated to the standard chrono- and bio-stratigraphic schemes as in Fig. 6.3. Note the revisions made upon the onshore stratigraphy (Lim, 1985) by workers in offshore Sandakan Sub-basin (e.g. Walker, 1993). Block SB8 in the SE Sabah Basin includes the Maliau, Malibau, and Tidung sub-basins (see Chapter 5 for details).

are highly deformed, and exhibit circular features like in the onshore sub-basins. They are most likely "pre-Dent", equivalent to the Tanjong Formation and the Segama Group. Furthermore, the onshore Sandakan Formation and the Dent Group are only gently dipping, in contrast with the tight folding of the sediments below the unconformity.

A reliable correlation between the onshore and offshore sequences is important because of its potential impact on exploration. In the Sandakan Sub-basin, the Dent Group, particularly the Sebahat Formation, forms the main reservoir target and has been found to contain hydrocarbons. Drilling of the Kuda Terbang-1 well by WMC in 1994, however, indicated that potential reservoirs may also occur in the pre-Dent section, probably the Tanjong Formation whose outcrops are widespread in central Sabah. At Kuda Terbang-1, the productive horizons occur below a major unconformity that spans the late Early Miocene to the Late Miocene. These rocks were assigned to the Segama Group by WMC (Bates, 1995), and may be correlated with the Tanjong Formation.

SE Sabah Basin

The stratigraphy of the SE Sabah Basin, covering much of the Kuamut and Tawau highlands and the Kalabakan Valley, was described by Collenette (1965). Figure 6.12 shows a summary of the stratigraphy. The basement here consists of the Cretaceous-Eocene Chert-Spilite Formation which is unconformably overlain by the Crocker, Kulapis, and Labang formations of Te5-f age (Lower-Middle Miocene), based on larger foraminifera found in the Malibau sub-basin (Collenette, 1965). These deep marine turbiditic rocks are overlain by Lower to Upper Miocene argillaceous sediments with marls, limestones and volcanic tuffs of the Kalabakan, Kuamut and Kalumpang formations. These are partly equivalent in age to the Segama Group in the NE Sabah Basin. The Tanjong Formation which crops out in the circular sub-basins also forms an important part of this succession.

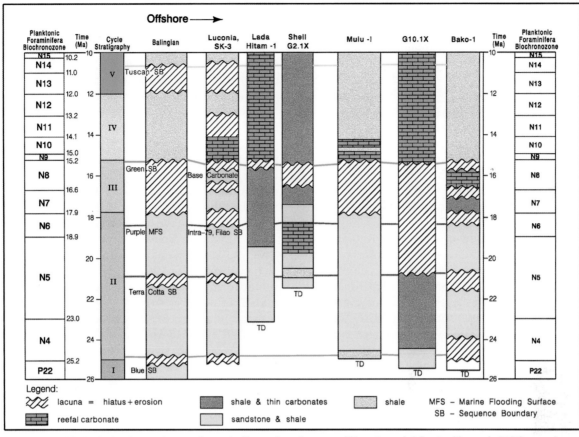

Fig. 6.13. Modified Wheeler diagram (time vs distance) of key wells and areas in offshore Sarawak (after Snedden et al., 1995).

The Dent equivalent in the SE Sabah Basin includes the Umas-Umas, Balung, and Simengaris formations which are Upper Miocene in age.

Several wells have been drilled on the coastal plain and in the offshore part of the SE Sabah Basin. Identification of the onshore formations in the subsurface is based on biostratigraphic evidence. As in offshore NW Sabah, the standard biozonation schemes for planktonic foraminifera and nannofossils are used here. Hence, the approximate equivalents of the onshore formations are deduced from the biostratigraphic ages of the subsurface formations. The Simengaris and Balung formations have been identified in the Tawau Offshore-1 and -2 wells. In the Tawau Barat-1, drilled by Sun Oil, a sequence of Upper Miocene to Pliocene rocks, probably equivalent to the Balung and Simengaris formations, were encountered. The ages were determined based on broad ranging palynomorphs (cf. Morley, 1978). Hence, it must be emphasised that these ages are very rough because of the poor recovery of microfossils from the sediments. Planktonic foraminifera and calcareous nannofossils are generally absent. A higher resolution palynomorph zonation is needed for better correlation between the onshore formations and offshore sequences.

SEQUENCE STRATIGRAPHIC STUDIES

Concepts in sequence stratigraphy (Mitchum et al., 1977; van Wagoner et al., 1990) are increasingly being applied to the stratigraphic analysis and correlation of Malaysian Tertiary sequences. Yap (1990) presented a sequence stratigraphic framework for the Malay Basin, in which the seismic group boundaries were correlated with the "global" coastal onlap chart of Haq et al. (1988). The major unconformity at the base of Group B was tied to the 5.5 Ma sequence boundary, which corresponds with the Miocene-Pliocene boundary. Esso has subsequently refined this interpretation (EPIC, 1994), and the unconformity has since been shifted to the 10.5 Ma sequence boundary, which is close to the Middle and Late Miocene boundary. Biostratigraphic work by Azmi Mohd Yakzan et al. (1994) suggests that the unconformity is about 7 Ma, which is closer to Esso's initial interpretation. This discrepancy has not been resolved.

Besides its application in regional correlation, sequence stratigraphy is also widely applied to oilfield development in the Malay Basin. The J sandstone reservoirs are probably the best studied (e.g. Ramlee and Bedingfield, 1996; Yap, 1996). Yap (1996) uses sequence stratigraphy to identify third-order sequences in the J sandstone reservoirs of the Tapis field. The gas discovery in the Bergading Deep-1 well has demonstrated the importance of sequence stratigraphy in assessing the deep reservoir potential of the northern Malay Basin (Wong and Salahuddin Saleh Karimi, 1996).

After the oil discovery at Rhu in the Penyu Basin, Texaco (1994) undertook a sequence stratigraphic study of the Penyu Formation to search for synrift hydrocarbon plays. In a seismic facies analysis of the same basin, Wong (1997) described three facies (lower, middle, and upper) in the synrift megasequence based on reflection characteristics. The lower seismic facies exhibits high-amplitude and continuous reflections that onlap onto the basement highs. This facies is interpreted as early-synrift fluvial sediments laid down when the sedimentation rate exceeded the subsidence rate. The middle facies is characterised by weak, discontinuous reflectors and is interpreted as lacustrine shales deposited when the subsidence exceeds sedimentation. The upper seismic facies, characterised by hummocky, semi-continuous to continuous reflection events, represent channel dominated fluvial-deltaic sediments filling the grabens when sedimentation exceeded subsidence during the late synrift phase.

Sequence stratigraphic concepts have been used in regional correlation in offshore Sarawak since the late 1980s and early 1990s, as exploration gradually moved into deeper waters (e.g. Abdul Manaf Mohammad and Wong, 1995). Some authors have attempted to revise the stratigraphy of offshore Sarawak in a sequence stratigraphic framework (e.g., in the Balingian Province, Wong, 1995a; Ismail Che Mat Zin and Jaafar Sipan, 1994). Ismail Che Mat Zin and Swarbrick (1997) proposed an alternative scheme to the cycle stratigraphy based on a basin-wide sequence stratigraphic analysis of the Sarawak Basin. They subdivided the basin strata into 7 unconformity-bounded sequences (T1S to T7S, Fig. 6.10). More recently, Shell and PETRONAS have carried out a sequence stratigraphic interpretation of the pre- and post-carbonate sequences in the Central Luconia Province in Sarawak (Taylor et al., 1997). In the study, the global "TB" sequences of Haq et al. (1988) were used for regional correlation across the whole Sarawak shelf. The results of the post-carbonate sequence stratigraphy were presented by Nasaruddin Ahmad et al. (1997). Snedden et al. (1995) studied the sequence stratigraphy of the Sarawak continental slope and correlated the Middle Miocene (Green) Unconformity with the top of Cycle III. Using a time-distance plot (Fig. 6.13), Snedden et al. (1995) showed how a sequence-bounding unconformity can be distinguished from a condensed section or maximum flooding surface. For a sequence boundary the lacuna widens landwards, but for a condensed section the lacuna widens seawards.

Detailed sequence stratigraphic studies have also been applied to the Sabah Basin. Noor Azim Ibrahim (1994) identified several unconformities, sequence boundaries, and marine flooding surfaces, and noted the seaward migration of successive unconformities (labelled H1- H7 in Fig. 6.10). The latter was a result of progressive uplift of the landward margin of the Sabah Basin in response to foreland thrust loading (Levell, 1987; Noor Azim Ibrahim, 1994). Other studies in the Sabah Basin include those on the Tembungo field and surrounding areas (Ismail Che Mat Zin, 1992), the Upper Miocene Stage IVC in the Labuan-Paisley Syncline (van Vliet and Schwander, 1987; Wong, 1996), and the Siagut Syncline (Wong, 1994). Marzuki Mohamad and Lobao (1997) used sequence stratigraphy to search for deepwater sands of the Upper Miocene-Lower Pliocene Lingan Fan complex in the southern Outboard Belt of the Sabah Basin.

On a field scale, sequence stratigraphy has also proved valuable for correlating between wells and predicting the distribution of reservoir facies. Wong (1995b) carried out a sequence stratigraphic interpretation of the SE Collins field in the Sabah Basin. The study concluded that the Lower Intermediate Unconformity (LIU), which was identified by previous workers as an onlap surface (e.g. Levell, 1987), is a downlap surface that has been uplifted during an earlier tectonic episode associated with the DRU. The LIU is now tied to the peak occurrence of *Globigerinoides subquadratus* in well records. The Middle Miocene Stage IVA was also found to consist of 2 third-order sequences separated by an Intra-Stage IVA sequence boundary. In the offshore NE Sabah Basin (Sandakan Sub-basin), Wong (1993) identified three sequences in the Middle Miocene-Quaternary strata. Unit 1 to Unit 3, from young to old (see Fig. 6.10), are bounded by Type 1 third-order sequence boundaries, and represent distinct phases of coastal-shelf progradation in response to differing rates of coastal progradation and aggradation.

CONCLUSION

The stratigraphy and correlation of sedimentary sequences in the Malaysian Tertiary basins are based on a combination of seismic/sequence stratigraphy and biostratigraphy. The different subsurface stratigraphic nomenclatures (groups, cycles, stages) in the Malay, Sarawak, and Sabah basins are the result of different operators exploring in these basins. In basins that have outcropping extensions or stratigraphic equivalents, such as in the Straits of Melaka and eastern Sabah, nomenclatures are based on onshore geology. Correlation between onshore formations and subsurface offshore stratigraphic units, however, are not straight-forward because one is based on lithostratigraphy and the other on seismic reflection characteristics. Biostratigraphy is, therefore, an important tool for correlation, not only for correlating between onshore and offshore stratigraphies, but for carrying out high-resolution sequence stratigraphic analysis within a given basin or oilfield.

The applicability of biostratigraphy in correlation depends on the recovery of microfossils from the rock samples, which is determined by depositional palaeoenvironments. The application of benthonic foraminifera in determining palaeoenvironments has been well established by Shell during the early days of exploration in the Baram Delta Province. The use of planktonic foraminifera and calcareous nannofossils in the Tertiary of offshore NW Borneo, particularly in marine sequences, have improved the interpretation and correlation of cycle and stage boundaries. It has also facilitated the recognition of marine flooding surfaces and condensed sections.

While there are already well-established biozonation schemes for correlation, the poor recovery of index planktonic foraminifera and nannofossils due to palaeoenvironmental factors poses a major problem. Biostratigraphers have resorted to establishing local biozonation schemes, especially in palynology. In some places, correlation is based on palynocycles, as shown by the work by Poumot (1989) in offshore Sarawak and the Malay Basin by PETRONAS Research (Azmi Mohd Yakzan et al., 1994). Because the palynological database is inadequate, some problems still remain, for example, in the dating of sequence boundaries. Hence, there are opportunities for improvement in palynostratigraphy, as more data become available through exploration activities. For biostratigraphy to be effective as a tool in sequence correlation, samples for analysis must be properly collected on the drilling rig. Laboratory techniques in sample processing and quantitative analysis must also be continuously improved and refined.

It is crucial for explorationists to accurately and reliably date sequences, and correlate them between wells, outcrops, and seismic data. In basins that have outcropping extensions, analogues for reservoir, source, and seal are derived from onshore investigations. Thus, a reliable correlation between onshore formations and offshore seismic stratigraphic units is needed. In the Sandakan Sub-basin (offshore NE Sabah Basin), for example, the major hydrocarbon occurrences are thought to be mainly in the Middle-Upper Miocene equivalents of the Ganduman and Sebahat formations (Dent Group) but recent drilling results have indicated that the Lower-Middle Miocene pre-Dent section (Tanjong Formation or Segama Group equivalents) is equally prospective. A biostratigraphic zonation scheme is desirable for accurate prediction of the stratigraphy ahead of the drill bit.

REFERENCES

Abdul Manaf Mohammad and Wong, R.H.F., 1995. Seismic sequence stratigraphy of the Tertiary sediments, offshore Sarawak deepwater area, Malaysia. Bulletin of the Geological Society of Malaysia, 37, 345-361.

Abu Samad Nordin and Md. Nazri Ramli, 1985. Stratigraphic scheme for Petronas Carigali's operating areas offshore Peninsular Malaysia. Abstracts of the Geological Society of Malaysia, Petroleum Geology Seminar, 6-7 December 1985, Kuala Lumpur, Warta Geologi, 11, 267.

Adams, C.G., 1970. A reconsideration of the East Indian Letter Classification of the Tertiary. Bulletin of the British Museum (Natural History), 19(3), 87-137.

Agostinelli, E., Mohamed Raisuddin Ahmad Tajuddin, Antonielli, E. and Mohamad Mohd Aris, 1990. Miocene-Pliocene palaeogeographic evolution of a tract of Sarawak offshore between Bintulu and Miri. Bulletin of the Geological Society of Malaysia, 27, 117-135.

Andreason, M.W., Smith, L.J. and Filewicz, M., 1994. Technical evaluation of Malaysia Block PM1. Unpublished Unocal report, November 30, 1994.

Armitage, J.H. and Viotti, C., 1977. Stratigraphic nomenclature - southern end Malay Basin. Proceedings of the Indonesian Petroleum Association 6th Annual Convention, May 1977, 69-94.

ASCOPE, 1981. Tertiary Sedimentary basins of the Gulf of Thailand and South China Sea: Stratigraphy, Structure and Hydrocarbon Occurrences. ASEAN Council on Petroleum (ASCOPE) Secretariat, Bangkok.

Awalludin Harun, Rashidah Abd Karim and Azmi Mohd Yakzan, 1996. Biostratigraphic signatures of depositional sequences in the Malay Basin. PRSS Technology Forum, 12-13 September 1996, Awana Genting, Kuala Lumpur.

Azmi Mohd Yakzan, Awalludin Harun, Bahari Md Nasib and Morley, R.J., 1994. Integrated biostratigraphic zonation for the Malay Basin. Abstracts of American Association of Petroleum Geologists International Conference & Exhibition, Kuala Lumpur, Malaysia, August 21-24, 1994, American Association of Petroleum Geologists Bulletin, 78, 1170-1171.

Bates, R., 1995. Geologic well evaluation report: Kuda Terbang-1. WMC Petroleum (Malaysia) Sdn Bhd, Unpublished report.

Blow, W.M., 1969. Late Middle Eocene to Recent planktonic foraminiferal biostratigraphy. Proceedings of the First International Conference on Planktonic Microfossils, 1967. Bronniman & Renz, Geneva, 199-422.

Buck, S.P., and McCulloh, T.H., 1994. Bampo-Peutu(!) Petroleum System, North Sumatra, Indonesia. In: Magoon, L.B. and Dow, W.G., eds., The Petroleum system - from source to trap. American Association of Petroleum Geologists, Memoir 60, 625-637.

Chow, Y.C., 1996. Palynological climatic biosignal as a high-resolution tool for sequence stratigraphical studies, offshore Sabah and Sarawak, Malaysia. PRSS Technology Forum, 12-13 September 1996, Awana Genting, Kuala Lumpur.

Cita, M.B. and Gartner., 1973. Studi sul Pliocene e sugli strati di passaggio dal Miocene al Pliocene IV. The stratotype Zanclean, foraminiferal and nannofossil biostratigraphy. Review Italian Palaeontology, 79, 503-58

Clennell, B., 1992. The melanges of Sabah, Malaysia. Unpublished PhD Thesis, University of London.

Clennell, B., 1996. Far-field and gravity tectonics in Miocene basins of Sabah, Malaysia. In: Hall, R. and Blundell, D.J., eds., Tectonic Evolution of Southeast Asia. Geological Society of London Special Publication, 106, 307-320.

Collenette, P., 1965. The geology and mineral resources of the Pensiangan and upper Kinabatangan areas, Sabah, Malaysia. Geological Survey, British Territories in Borneo, Memoir 9, Government Printing Office, Kuching.

Doust, H., 1981. Geology and exploration history of offshore central Sarawak. In: Halbouty, M.T., ed., Energy Resources of the Pacific Region. American Association of Petroleum Geologists, Studies in Geology, 12, 117-132.

EPIC (Esso-PETRONAS Integrated Collaborative Study), **1994.** Regional study of the Malay Basin-Final Portfolios. Unpublished report, Esso Production Malaysia Inc.

Finch, E.M. (compiler), 1984. A biostratigraphic review of the Carigali-BP Contract Area, offshore Sabah, Eastern Malaysia. Unpublished report, BP Research Centre, STR/43/84.

Fontaine, H., Asiah M Salih and Sanatul Salwa Hassan, 1992. Pre-Tertiary sediments found at the bottom of wells drilled in the Malacca Straits. Unpublished PRSS report, No: PRI/RP5/92-01.

Galloway, W.E., 1989. Genetic stratigraphic sequences in basin analysis I: Architecture and genesis of flooding-surface bounded depositional units. American Association of Petroleum Geologists Bulletin, 73, 125-142.

Haile, N.S. and Wong, N.P.Y., 1965. The geology and mineral resources of Dent Peninsula, Sabah. Geological Survey, Borneo Region, Malaysia, Memoir 16, Government Printing Office, Kuching.

Haq, B.U., Hardenbol, J. and Vail, P.R., 1987. Chronology of fuctuating sea levels since the Triussic. Science, 235, 1156-1167.

Haq, B.U., Hardenbol, J. and Vail, P.R., 1988. Mesozoic and Cenozoic chronostratigraphy and cycles of sea-level change. In: Wilgus, C.K., Hastings, B.S., Kendall, C.G.St.C., Posamentier, H.W., Ross, C.A. and van Wagoner, J.C., eds., Sea-level change: an integrated approach. Society of Economic Paleontologists and Mineralogists Special Publication, 42, 71-108.

Ho, K.F., 1978. Stratigraphic framework for oil exploration in Sarawak. Bulletin of the Geological Society of Malaysia, 10, 1-13.

Hutchison, C.S., 1992. The Southeast Sulu Sea, a Neogene marginal basin with outcropping extensions in Sabah. Bulletin of the Geological Society of Malaysia, 32, 89-108.

Ismail Che Mat Zin, 1992. Regional seismostratigraphy study of the Tembungo area, offshore West Sabah. Bulletin of the Geological Society of Malaysia, 32, 109-134.

Ismail Che Mat Zin, 1994. Dent Group and its equivalent in the offshore Kinabatangan area, East Sabah. Bulletin of the Geological Society of Malaysia, 36, 127-145.

Ismail Che Mat Zin and Jaafar Sipan, 1994. Application of sequence stratigraphic techniques on the non-marine sequences: An example from the Balingian Province, Sarawak. Bulletin of the Geological Society of Malaysia, 36, 105-117.

Ismail Che Mat Zin and Swarbrick, R.E., 1997. The tectonic evolution and associated sedimentation history of Sarawak Basin, eastern Malaysia: a guide for future hydrocarbon exploration. In: Fraser, A.J., Matthews, S.J. and Murphy, R.W., eds., Petroleum Geology of Southeast Asia. Geological Society of London Special Publication, 126, 237-245.

Katz, B.J. and Dawson, W.C., 1997. Pematang-Sihapas Petroleum System of Central Sumatra. In: J.V.C. Howes and R.A. Noble, eds., Proceedings of the IPA Petroleum Systems of SE Asia and Australasia Conference, Jakarta, Indonesia, May 1997, Indonesian Petroleum Association, Jakarta, 685-698.

Khalid Ngah, 1975. Stratigraphic and structural analyses of the Penyu Basin, Malaysia. Unpublished MS thesis, Oklahoma State University.

Leong, K.M., 1974. The geology and mineral resources of the Upper Segama Valley and Darvel Bay area. Geological Survey of Malaysia, Memoir 4 (Revised), Government Printing Office, Kuching.

Leong, K.M., 1978. Sabah. In: ESCAP Atlas of Stratigraphy I. Burma, Malaysia, Thailand, Indonesia, Philippines. United Nations, Mineral Resources Development Series, 44, 26-31.

Levell, B.K., 1987. The nature and significance of regional unconformity in the hydrocarbon-bearing Neogene sequence offshore West Sabah. Bulletin of the Geological Society of Malaysia, 21, 55-90.

Lim, P.S., 1985. Geological Map of Sabah 1:500,000 (3rd Edition), Geological Survey of Malaysia.

Martini, E., 1971. Standard Tertiary and Quaternary calcareous nannoplankton zonation. Proceedings of the 2nd Plankton Conference, Rome, 1969, v. 2, 739-785.

Marzuki Mohamad and Lobao, J.J., 1997. The Lingan Fan: Late Miocene/Early Pliocene turbidite fan complex, north-west Sabah. In: Howes, J.V.C and Noble, R.A, eds., Proceedings of the IPA Petroleum Systems of SE Asia and Australasia, 21-23 May 1997, Indonesian Petroleum Association, Jakarta, 787-798.

Mazlan B. Hj. Madon, Azlina Anuar and Wong, R., 1997. Structural evolution, maturation history and hydrocarbon potential of the Penyu Basin, offshore Peninsular Malaysia. In: Howes, J.V.C and Noble, R.A, eds., Proceedings of the IPA Petroleum Systems of SE Asia and Australasia, 21-23 May 1997, Indonesian Petroleum Association, Jakarta, 403-424.

Md Nazri Ramli, 1988. Stratigraphy and palaeofacies development of Carigali's operating areas in the Malay Basin, South China Sea. Bulletin of the Geological Society of Malaysia, 22, 153-188.

Mitchum, R.M., Vail, P.R. and Thomson III, S., 1977. Seismic stratigraphy and global changes of sea level, Part 2: the depositional sequence as a basic unit for stratigraphic analysis. In: Payton, C.E., ed., Seismic stratigraphy: applications to hydrocarbon exploration. American Association of Petroleum Geologists Memoir, 26, 53-62.

Morley, R.J., 1978. Palynology of Tertiary and Quaternary sediments in Southeast Asia. Proceedings of the Indonesian Petroleum Association 6th Annual Convention, 255-276

Morley, R.J., 1991. Tertiary stratigraphic palynology in Southeast Asia: current status and new directions. Bulletin of the Geological Society of Malaysia, 28, 1-36.

Murray, C., 1991. Review of the prospectivity of Block PM-1, Straits of Malacca, Malaysia. Sun Malaysia Petroleum Company, Unpublished Sun Oil report. ER:SUN:1:91-15.

Nasaruddin Ahmad, Newall, D., Ngau, A. and Powell, C., 1997. Seismic and sequence stratigraphy of the Upper Miocene/Pliocene siliclastics in the Central and SW Luconia provinces, offshore Sarawak. Abstract of the Geological Society of Malaysia Petroleum Geology Conference '97, Warta Geologi, 23, 375.

Noor Azim Ibrahim, 1994. Major controls on the development of sedimentary sequences Sabah Basin, Northwest Borneo, Malaysia.Unpublished PhD thesis, University of Cambridge, 254pp.

Okada, H. and Bukry, J., 1980. Supplementary modification and introduction of code numbers to low-latitude coccolith biostratigraphy zonation (Bukry, 1973; 1975). Marine Micropalaeontology, 5, 321-326.

Phillips, S., Little, L., Michael, E. and Odell, V., 1997. Sequence stratigraphy of Tertiary petroleum systems in the West Natuna Basin, Indonesia. In: Howes, J.V.C and Noble, R.A, eds., Proceedings of the IPA Petroleum Systems of SE Asia and Australasia, 21-23 May 1997, Indonesian Petroleum Association, Jakarta, 381-401.

Potter, T.L., Johns, D.R. and de Naris, T.B.G., 1984. Lithostratigraphy. In: James, D.M.D., ed., "The geology and hydrocarbon resources of Negara Brunei Darussalam." Muzium Brunei, 43-75.

Poumot, C., 1989. Palynological evidence for eustatic events in the tropical Neogene. Bulletin des Centres de Recherches Exploration-Production Elf-Aquitaine, 13, 437-453.

Ramlee, A.R. and Beddingfield, J.R., 1996. Mid J Group sequence stratigraphy and reservoir architecture in Block PM-9, Malay Basin. PRSS Technology Forum, 12-13 September 1996, Awana Genting, Kuala Lumpur.

Rijks, E.H.J., 1981. Baram Delta geology and hydrocarbon occurrence. Bulletin of the Geological Society of Malaysia, 14, 1-18.

Salvador, A. (ed.), 1994. International Stratigraphic Guide, International Subcommision of Stratigraphic Classification (ISSC), 2nd Edition: Geological Society of America, Boulder, Co.

Snedden, J.W., B.K. Rodgers and J.F. Sarg, 1995. Variation in stratigraphy from shallow to deepwater, Sarawak: sedimentary vs sequence stratigraphy. Unpublished MOBIL report, technical service job no. 6140-MBDALUA.

Taylor, G., Powell, C., Newall, M. and Ngau, A., 1997. Petronas and Sarawak Shell Berhad Joint Regional Study of the Pre-Carbonate Clastics, Central Luconia Province, Offshore Sarawak. Unpublished Internal Report, EXP.R.50793.

Texaco, 1992. The significance of the Rhu oil discovery, PM-14, offshore Peninsular Malaysia: a preliminary summary. TEXACO Exploration Penyu Inc., Unpublished report, ER:TEXACO:1:92-01.

Texaco, 1994. Sequence stratigraphic study of the Penyu Formation, Pahang Sub-basin. Unpublished TEXACO report.

van Borren, L., Koopman, A. and Schreurs, J., 1996. Stratigraphy. In: Sandal, D.T., ed., "The geology and hydrocarbon resources of Negara Brunei Darussalam." 2nd Ed., Syabas, Bandar Seri Begawan, 81-128.

van Vliet, A. and Schwander, M.M., 1987. Stratigraphic interpretation of a regional seismic section across the Labuan Syncline and its flank structures, Sabah, North Borneo. In: Bally, A.W., ed., Atlas of Seismic Stratigraphy. American Association of Petroleum Geologists, Studies in Geology, 27, 163-167.

van Wagoner, J.C., Mitchum, R.M., Campion, K.M. and Rahmanian, V.D., 1990. Siliciclastic sequence stratigraphy in well logs, cores, and outcrops. American Association of Petroleum Geologists, Methods in Exploration Series, No. 7, AAPG, Tulsa, 55pp.

Walker, T., 1993. Sandakan Basin prospects rise following modern reappraisal. Oil & Gas Journal, May 1993, 43-47.

Wong, R.H.F., 1993. Sequence stratigraphy of the Middle Miocene-Pliocene southern offshore Sandakan Basin. In: Teh, G.H., ed., Proceedings of the Symposium on Tectonic Framework and Energy Resources of the Western Margin of Pacific Basin. Bulletin of the Geological Society of Malaysia, 33, 129-142.

Wong, R., 1994. Seismic sequence stratigraphy of the Siagut Syncline, NW Sabah Basin. PETRONAS' in-house report.

Wong, R., 1995a. Seismic sequence stratigraphic interpretation of the East Balingian Basin, Sarawak. PETRONAS' in-house report.

Wong, R., 1995b. Seismic sequence stratigraphic interpretation enhances remaining hydrocarbon potential of the South East Collins field. Geological Society of Malaysia Petroleum Geology Conference, Kuala Lumpur, 1995.

Wong, R., 1996. Sequence stratigraphy of the Upper Miocene Stage IVC in the Labuan-Paisley Syncline, Northwest Sabah Basin. Geological Society of Malaysia, Petroleum Geology Conference, Kuala Lumpur, 1996. (submitted to Bulletin of the Geological Society of Malaysia).

Wong, R., 1997. Seismic facies analysis of the synrift sediments, Penyu Basin. Unpublished PETRONAS internal report.

Wong, R. and Salahuddin Saleh Karimi, 1996. Deep reservoir potential of the North Malay Basin. Geological Society of Malaysia, Petroleum Geology Conference, Kuala Lumpur, 1996. (abstract).

Yap, K.T., 1990. Sequence stratigraphy of the Group J in the Malay Basin and its impact on development opportunities. Geological Society of Malaysia, Petroleum Geology Conference, 27-28 November 1990, Kuala Lumpur. Warta Geologi, 16, 267.

Yap, K.T., 1996. Tapis Field: Lower Group J Reservoir. PRSS Technology Forum, 12-13 September 1996, Awana Genting, Kuala Lumpur.

Part 3

PENINSULAR MALAYSIA SEDIMENTARY BASINS AND PETROLEUM RESOURCES

Chapter 7

Geological Setting of Peninsular Malaysia

H.D. Tjia

Geological Setting Of Peninsular Malaysia

INTRODUCTION

Peninsular Malaysia consists of two geological terranes: the Western belt (or also West Domain) and the Core terrane (Fig. 26.1, Chapter 26). In more detail, Peninsular Malaysia can be divided into four domains (Fig. 7.1a). In the West Domain of Peninsular Malaysia crop out the oldest proven rocks consisting of Middle Cambrian (about 520 Ma) deltaic deposits similar to fossil-bearing beds in adjacent Tarutau island of Thailand. On the main Langkawi island, small outcrops of quartzitic sandstone below the Middle Cambrian sequence may be of Precambrian age (Fig. 26.2)

In terms of regional tectonics, Peninsular Malaysia, including the adjacent offshore areas, the Straits of Melaka and the South China Sea, are part of Sundaland (Fig. 7.1b). Peninsular Malaysia is elongated in north- northwest direction parallel to its main structural trend. This direction is superimposed upon structures with northeast and east strikes. In certain areas, such as the northwestern part of the country, the older strikes also possess surface expressions. The main NNW trend was developed in a Late Triassic - Early Jurassic deformational period. By the Early Tertiary, Sundaland had become tectonically stable. Younger major deformations have been confined to existing zones of weakness,

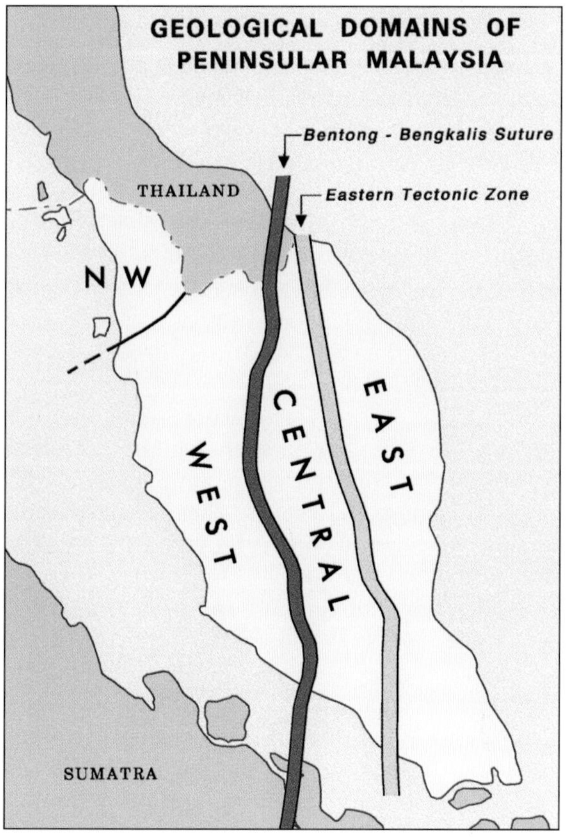

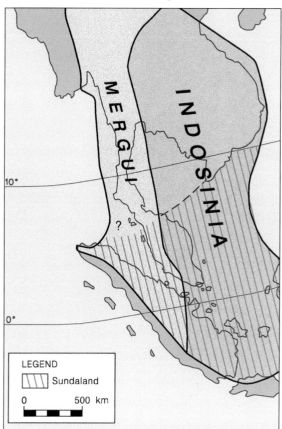

Fig. 7.1a The four geological domains of Peninsular Malaysia: Northwest (NW), West, Central and East. Two of the domain boundaries are distinct: the lower Triassic Bentong- Bengkalis suture and the Eastern Tectonic Zone as indicated by bouguer gravity anomalies.

Fig. 7.1b Sundaland combines the southern parts of two larger continental blocks: Indosinia and Mergui blocks. The latter is also known as Sibumasu. The northern boundary of Sundaland is ill-defined.

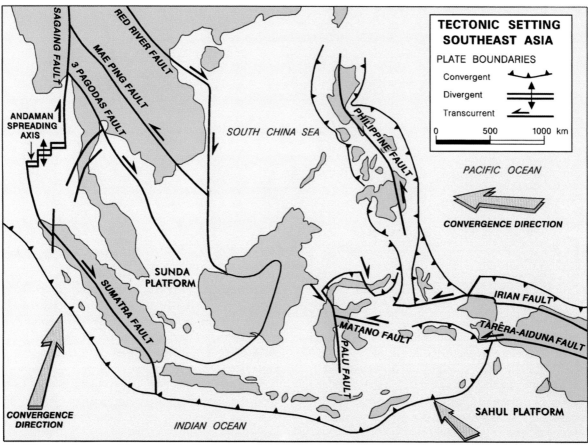

Fig. 7.2 Tectonic setting of Southeast Asia. The Southeast Asian continental slabs are being differentially extruded as a result of collision between the Indian subplate with the Eurasian plate (to the west of the figure).

and annual vertical crustal movements occurred at the sub-millimetre scale. However, these zones of weakness have most probably accommodated considerable lateral displacements, and offshore regional subsidence has formed the Tertiary basins that contain several kilometres of sediments. The largest, the Malay Basin accumulated at least 8 km of Tertiary sediments. Sundaland is the core of the Southeast Asian region and is experiencing convergence from all sides. The Indian Ocean - Australian plate is moving northward; the Pacific plate approaches from the east, while according to a widely accepted hypothesis the Southeast Asian portion (including Sundaland) of the Eurasian plate is being extruded toward the southeast (Fig. 7.2).

SEDIMENTARY BASINS AND STRATIGRAPHY

(1) Peninsular Malaysia Onshore

Throughout the Phanerozoic three or more major tectonic deformational periods, widespread granitic emplacements, and repeated large-scale denudation have blurred the outlines of the sedimentary basins of Peninsular Malaysia, except those of Jurassic and younger ages. The four-fold structural division (see below and Fig. 7.1a) and lithology can be used as a framework to differentiate the probable depocentres. A generalised stratigraphy is in Figure 7.3, which is based on Gobbett and Hutchison (1973), Foo (1983), Khoo (1983), Geological Survey of Malaysia (1985), Jabatan Penyiasatan Kajibumi Malaysia summary (1995), and other sources listed in the references.

Northwest Depocentre.-The oldest sedimentary rocks comprise the Middle Cambrian to Lower Ordovician and are only known from this depocentre. These are the deltaic-paralic Macincang Formation on the islands Langkawi and adjacent Tarutau (Thailand), and an equivalent unit in the Gunung Jerai area (Foo, 1983; Lee, 1983). The coarse-grained clastics were derived from a land mass to the west. During the remainder of the Lower Palaeozoic, depositional conditions generally deepened from shallow-marine Setul Limestone (Jones, 1978) to the bathyal fine-grained clastics and chert of the Mahang Formation (Burton, 1967). In the Langkawi islands, tectonic deformation separates the Lower and Upper Palaeozoic rocks (Koopmans, 1965). The Upper Palaeozoic is

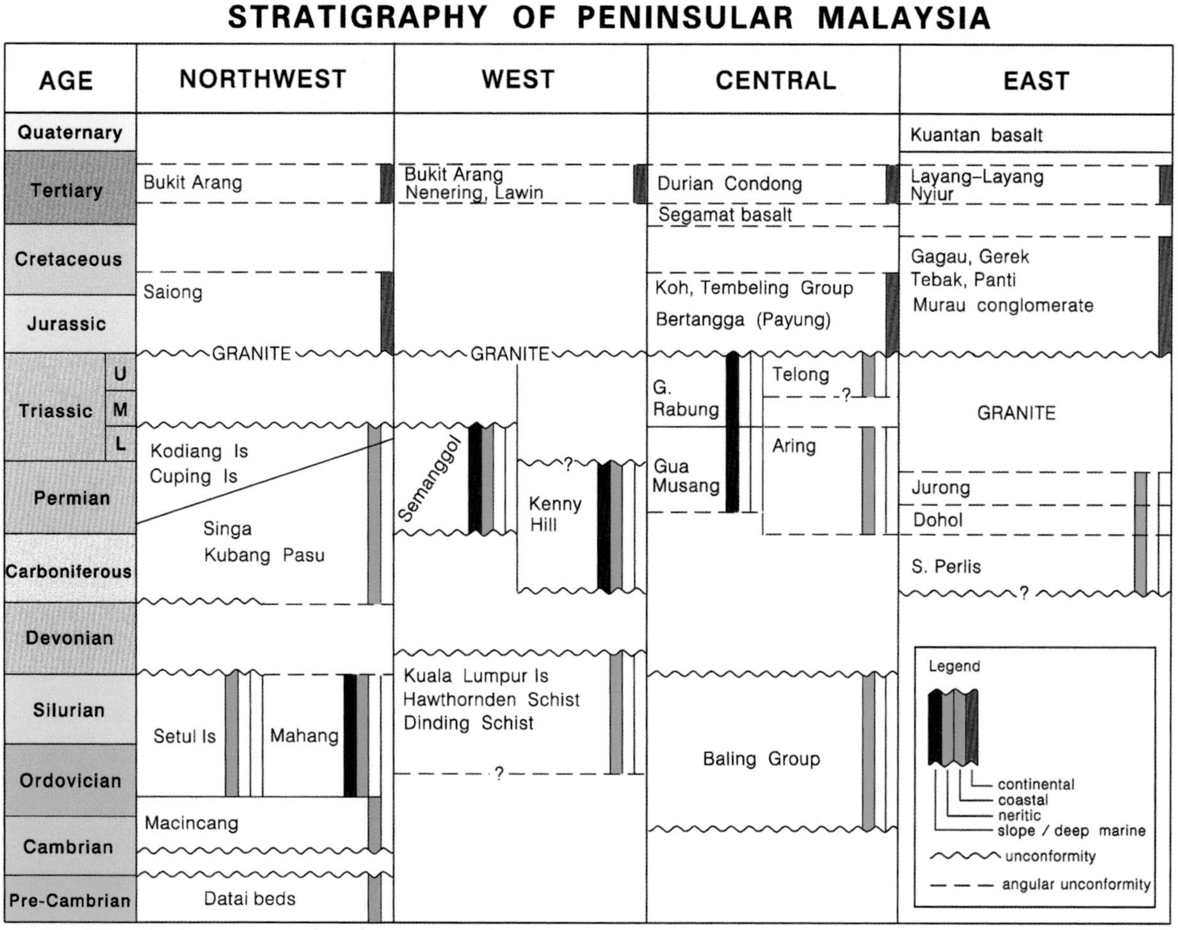

STRATIGRAPHY OF PENINSULAR MALAYSIA

AGE	NORTHWEST	WEST	CENTRAL	EAST
Quaternary				Kuantan basalt
Tertiary	Bukit Arang	Bukit Arang Nenering, Lawin	Durian Condong / Segamat basalt	Layang-Layang Nyiur
Cretaceous	Saiong		Koh, Tembeling Group	Gagau, Gerek Tebak, Panti Murau conglomerate
Jurassic	~GRANITE~	~GRANITE~	Bertangga (Payung)	
Triassic U/M/L	Kodiang ls / Cuping ls	Semanggol / Kenny Hill	G. Rabung / Gua Musang	Telong / Aring
Permian	Singa			GRANITE / Jurong / Dohol
Carboniferous	Kubang Pasu			S. Perlis
Devonian				
Silurian	Setul ls / Mahang	Kuala Lumpur ls / Hawthornden Schist / Dinding Schist	Baling Group	
Ordovician				
Cambrian	Macincang			
Pre-Cambrian	Datai beds			

Legend: continental, coastal, neritic, slope / deep marine, unconformity, angular unconformity

Fig. 7.3 Generalised stratigraphy of the four domains of Peninsular Malaysia. The depositional environments are colour-coded. Data are mainly from the Geological Survey of Malaysia (1985) and updated as noted in the text.

represented by shallow- marine limestone, clastics and pebbly mudstones. The presence of Upper Carboniferous to Lower Permian cool-water fauna and the pebbly mudstone (Singa Formation) suggest that these are periglacial deposits of Gondwanaland (Stauffer and Mantajit, 1981; Stauffer and Lee, 1986). Similar fauna-lithological associations are also known in Myanmar, Peninsular Thailand, and North Sumatra. Lower Permian pebbly beds have also been found near Kuala Lumpur in the West Domain (Tjia and Anizan Isahak, 1990). This extensive region of such deposits is regarded as a sliver of Gondwana, once probably located off NW Australia. From the Mesozoic onward the Northwest Depocentre has experienced geological development similar to that of the West Depocentre. Shallow-marine limestone at Kodiang straddles the era boundary. To the east, the Permian to Middle Triassic Semanggol Formation was deposited in slope and deep-marine surroundings. The regional deformational period at the Triassic-Jurassic transition is also apparent by a difference in structural style and absence of sedimentary rocks of that time. The Upper Jurassic- Lower Cretaceous Saiong Beds comprise a sequence of polymict conglomerate and conglomeratic redbeds at the Kedah-Thailand boundary (Khoo, 1983). Interbeds of red sandstone, shale and mudstone are present; no fossils have been found and the age is based on lithological correlation with the Tembeling and equivalent redbeds in the central and southern parts of the peninsula. Thickness is estimated at 1200 m; its top is eroded while the beds rest with apparent unconformity upon the Permian to Triassic Semanggol Formation.

West Depocentre. This depositional depression contains extensive Silurian to Permian limestone in the Kinta valley, while elsewhere lower Palaeozoic limestone and metasediments are separated by an angular unconformity from Palaeozoic clastics and from Lower-Middle Triassic shelf (Kodiang limestone), slope and deep-marine (Semanggol Formation) sedimentary beds. In the Kuala Lumpur area, also known as the Klang Valley, the lower Palaeozoic comprises Ordovician psammitic schist and Silurian-Lower Devonian pelitic schist and marble (the Kuala Lumpur Limestone). The Palaeozoic weakly metamorphosed clastics (Kenny Hill Formation) probably rest unconformably upon the older sequences that are characterised by a stronger degree of metamorphism and overprinting of structural trends. Large-scale granitoid emplacement in the Upper Triassic-Lower Jurassic created the Main Range or Banjaran Titiwangsa and associated granitic bodies. No younger Mesozoic rocks, except the isolated Cretaceous granite mass of Gunung Ledang (Mount Ophir), are known. About half a dozen small Tertiary basins developed in the West and Northwest Depocentres (Fig. 7.4a).

Central Depocentres. Only Upper Palaeozoic and younger sedimentary rocks are known in the Central Domain. Centres of Permian to Triassic volcanic activity have been identified in SW Kelantan and central Pahang, but elsewhere volcanism contributed only volcaniclastics to the marine sediments. Shallow-marine Permian-Lower Triassic carbonates crop out in the north-central part. Dark coloured limestone of probable Permian age was also encountered at shallow depth in the south. In the Lower-Middle Triassic a neritic sea covered the middle west of the Central Domain while elsewhere in the domain a slope to deep-marine condition prevailed. Turbiditic sequences with

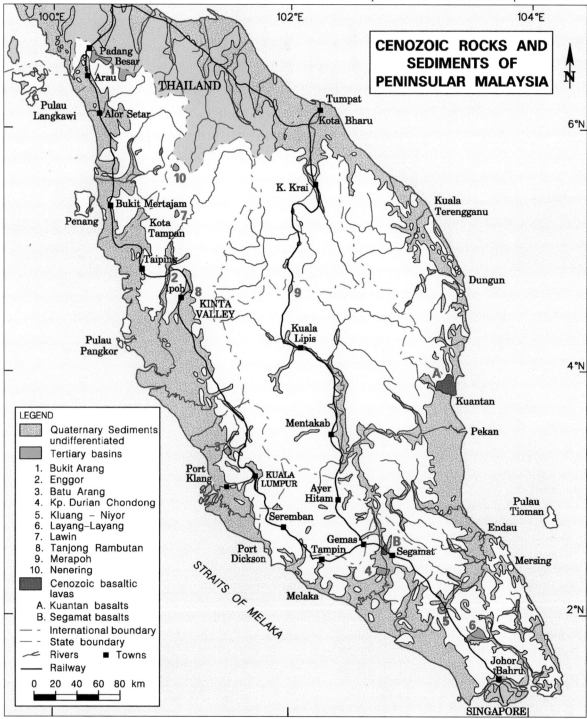

Fig. 7.4a Proven and suspected Tertiary basins onshore Peninsular Malaysia. After Geological Survey of Malaysia (1985) and Raj et al. (1998).

Geological Setting Of Peninsular Malaysia

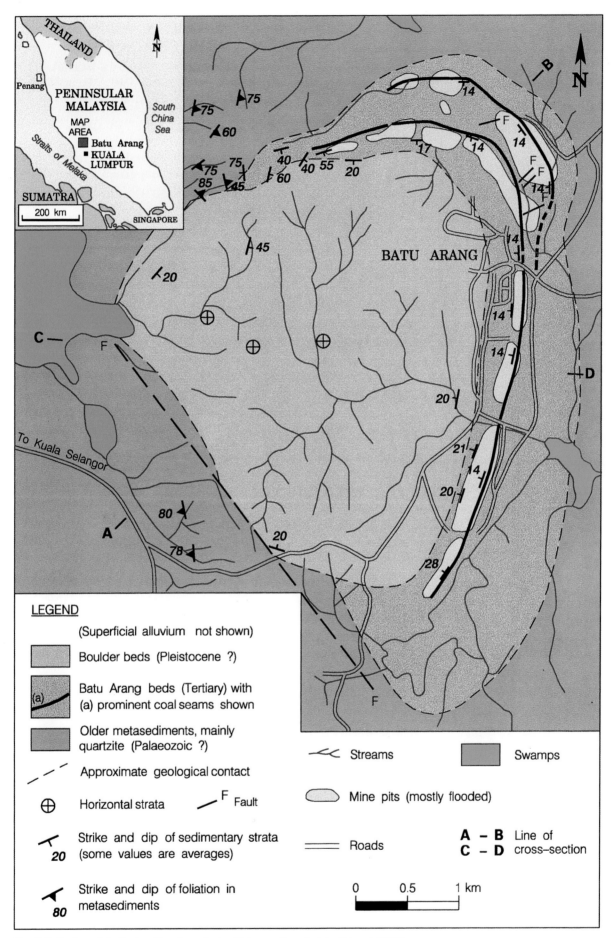

Fig. 7.4b Geology of the Tertiary Batu Arang basin of Selangor. Compiled by Raj et al. (1998).

Geological Setting Of Peninsular Malaysia

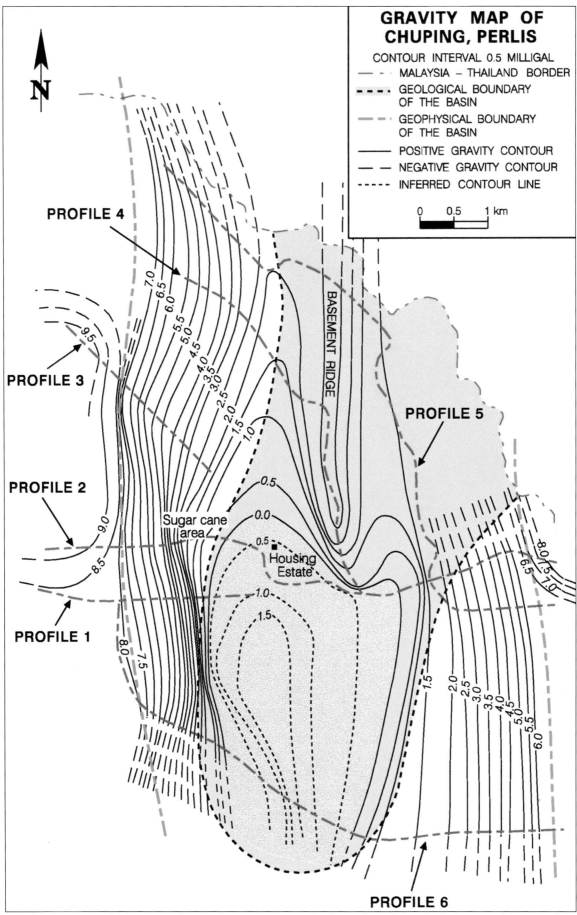

Fig. 7.4c Gravity map of the Tertiary Bukit Arang basin in Chuping, Perlis after Arafin et al. (1989). The known surface extent of the basin is smaller than indicated by the bouguer gravity anomalies.

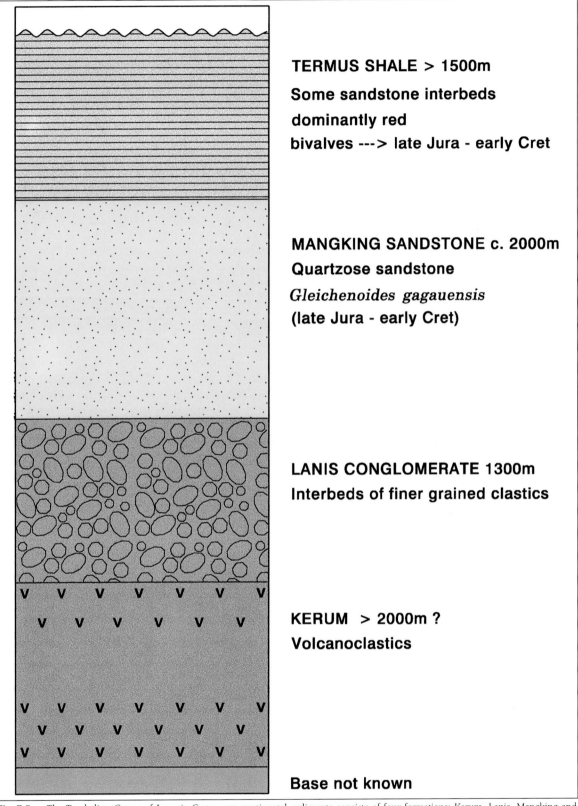

Fig. 7.5 The Tembeling Group of Jurassic-Cretaceous continental sediments consists of four formations: Kerum, Lanis, Mangking and Termus (Khoo 1983).

common slump intervals mark the Semantan and Gemas formations. The Upper Triassic-Lower Jurassic deformation brought to an end marine deposition in the Central Domain and in the entire peninsula. Regional southerly tilting may explain the fact that Permian-Lower Triassic sedimentary beds occupy the northern part, while the younger Triassic beds crop out in the south. The Central Domain is also depressed relative to the adjoining domains where plutonic rocks and older sedimentary units crop out over extensive areas.

The Upper Mesozoic Tembeling and equivalent continental deposits developed in a number of isolated basins, possibly grabens and half-grabens, that are elongated in NNW direction nearer to the east side of the Central Domain. Three of the larger outcrops are in the Central Domain, each being approximately 60 km long and 30 km wide. Results of fieldwork by the Geological Survey of Malaysia were compiled by Burton (1973), Rishworth (1974) and updated by Khoo (1983); Figure 26.7a. The Tembeling and Gagau groups are representative Mesozoic stratigraphic units. The 6800 m thick Tembeling Group of four formations (Khoo 1983) comprises in progressively younger sequences (1) Kerum Volcaniclastics, (2) Lanis Conglomerate, (3) Mangking Sandstone and (4) Termus Shale (Figs. 7.5 & 26.8a). The Kerum Formation consists of coarse to medium grained tuffaceous clastic beds with pillow lavas. The Lanis Formation is a well-cemented polymict boulder conglomerate occurring in thick to massive beds interbedded with coarse-grained subarkosic sandstone. The Mangking Formation consists of similar sandstone with intervals of reddish to purplish mudstone and shale. The latter is predominant in the Termus Shale. Stratification is good and normal-thick beds are the rule. Sahalan Abdul Aziz (in Liew et al., 1996) identified in a Tembeling-equivalent sequence certain facies types: trough-filling conglomerate, channel-fill sandstone, thick-bedded siltstone and mudstone, thick to massive bedded sandstone, ripple-laminated clastics, and alternating layers of reddish coloured sandstone, siltstone and mudstone. Foraminifers in calcareous rocks collected from the Tekai river in the Tembeling area suggest some marine influence, but it has not been ascertained if the samples originated from *in situ* outcrops or from floats. In the same general area freshwater gastropod limestone was identified. A semi-arid palaeoclimate is indicated by plants, pollen and the occurrence of fresh feldspar. Stratification, occasional ripple marks, cross beds, and scour-and-fill structures suggest deposition from running water.

The larger Upper Mesozoic basins began as pull-aparts controlled by N-S wrench faults (Fig. 26.9a). The infilling sediments were laid down in lakes and in floodplains with braided-

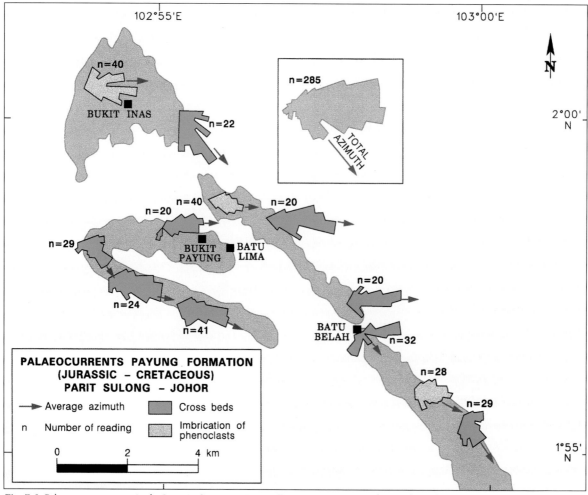

Fig. 7.6 Palaeocurrent pattern in the Jurassic-Cretaceous Payung Formation near Parit Sulong, Johor. The palaeocurrents indicate a general western provenance. Azimuths shown are for current readings after the containing beds were restored to original position. Location of the Payung Formation is shown on Figure 7.7. After Tan Yong Phang (1988).

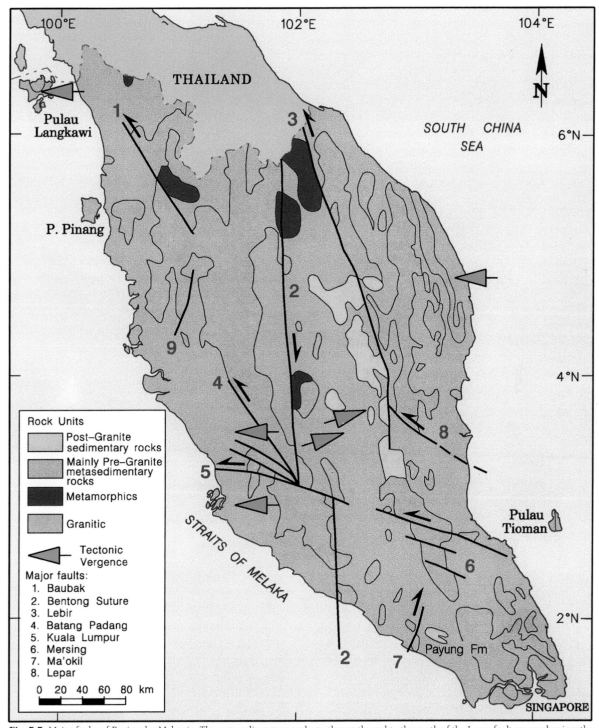

Fig. 7.7 Major faults of Peninsular Malaysia. The metasedimentary rocks to the north and to the south of the Lepar fault are predominantly Carboniferous and Permian, respectively. The green-tipped arrows show tectonic vergence. Geology of the base map is simplified after the Geological Survey of Malaysia (1985).

meandering channel systems. Palaeocurrent patterns imply that at least the larger contiguous Upper Mesozoic outcrops (Koh, Tembeling, Bertangga) are separate depocentres. While topographic highs probably existed around each of the depocentres, a dominant mountain range in the west supplied material to all the basins in the domain. Figure 7.6 shows the palaeocurrent system in the Jurassic-Cretaceous continental Payung Formation, Johor State (location shown on Fig. 7.7). At the beginning of the Cenozoic, basaltic lava flows developed in the Segamat area.

Burton (1973) pointed to the altitudinal difference between the base of Cretaceous sediments in the north and their equivalents in Johor in the south. Regional slopes of about 1: 200 to 1 :450 gradients are suggested.

East Depocentres. Two main elongated depocentres occupy the East Domain, being separated by the WNW-ESE striking Lepar fault zone (Fig. 7.7). In addition are several Upper Mesozoic, continental deposits that are named as Gagau Group and equivalent units. Type sections of the lower Cretaceous Gagau Group were established by Rishworth (1974) in the Gagau uplands. The two formal formations of the group, the Badong Conglomerate and the Lotong Sandstone totalling 1500 m thickness are separated by an "eruptive" interval of pyroclastics and lava flows (Fig. 7.8). The group forms cuestas and mesas reflecting the gentle dips of its beds. The Badong Conglomerate is massively bedded, polymictic-lithic and is associated with unweathered feldspar grains and biotite flakes set in a generally argillaceous groundmass. The younger Lotong Sandstone displays better stratification and its conglomerate is marked by the prevailing presence of quartz and smaller proportion of lithic clasts. Thick to massive siltstone intervals are typical. Rishworth (1974) grouped as Gagau-equivalents a dozen or so smaller outcrops of similar lithology at high hill and mountain summits in the Central Domain and in the southern part of the East Domain. From study of the type area, Rishworth concluded that volcanic activity and faulting were significant in the Late Jurassic to Early Cretaceous period.

The two larger depocentres are occupied by Permian, clastic marine sequences to the south of the Lepar fault zone, and by metasedimentary rocks of similar origin and of dominantly Carboniferous age to the north of the fault zone (Geological Survey of Malaysia, 1985). A Permian limestone complex occurs to the west of Kuantan and in small isolated patches in SE Johor. All have been regionally metamorphosed and intruded by mainly Upper Triassic granitic plutons. Volcanic rocks accompany the Permian sequences. There are also younger Middle-Upper Cretaceous dioritic dykes and Quaternary basaltic flows in the Kuantan area.

Tertiary Basins Onshore. Raj et al. (1998) compiled information on ten established and suspected Tertiary basins onshore Peninsular Malaysia (Figs. 7.4a, 7.4b & 7.4c) and concluded that only seven have been proven (Table 7.1). The interpreted Cenozoic deposits at Tanjung Rambutan and at Merapoh are likely of Quaternary age considering their 'young' appearance and the lignite character of the carbonaceous bed (Tanjung Rambutan). The Nenering beds, most recent description by Qalam Azad Rosle and Teh (1998), are most probably of Tertiary age, although Uyop Said and Che Aziz Ali (1997) reported the presence of Aptian-Albian palynomorphs. These, however, may have been reworked. The Tertiary age is based on the semi-consolidated state (unlike the consolidated and indurated Tembeling beds) and generally moderate dips (rarely exceeding 20 degrees) of right-side up beds. A North-South basin axis and 2 normal faults with about 20 m throw affect the Nenering beds. Renewed faulting appears to have

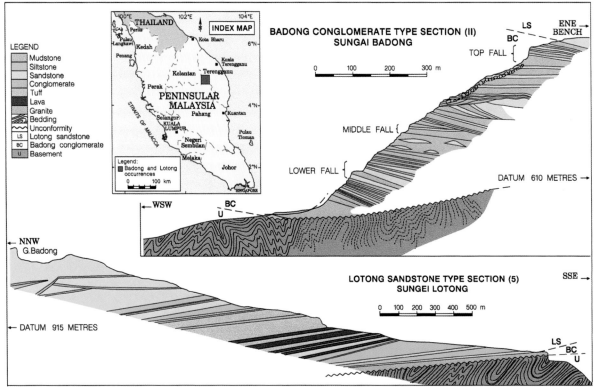

Fig. 7.8 Type sections of the Gagau Group: Badong Conglomerate and Lotong Sandstone. After Rishworth (1974).

LOCALITY	STRATIGRAPHY	LITHOLOGY	COAL
Batu Arang Selangor	Boulder Bed unconformable upon Coal Measures	**Boulder Beds** Poorly sorted gravels and boulders in sandy to gravelly matrix; Weakly consolidated to semi-consolidated; At least 300 m thick; Dipping 20° to 45° towards basin centre; Alluvial fans; Age unknown **Coal Measures**; Shale, sandstone & Clay with some lignite seams; Weakly consolidated to consolidated; Thickness 265 m or more; Strike E-W to N-S with dips of 10° to 40° towards S and W Lake deposit Eocene to Oligocene	Two thick coal seams: upper 15 m lower 8 m Sub-bituminous to lignite
Kampung Durian Condong Northwest Johor	Alluvium unconformable over Lower Sequence	**Alluvium** Mainly sand with some clay; Unconsolidated; At least 5 m thick; Horizontal bedding; Fluvial deposit ?Pleistocene (=Older Alluvium) **Lower Sequence** Shales, Volcanic ash & clay with some lignite seams; Weakly consolidated to semi-consolidated; Maximum thickness known 195 m; Dipping 20° to 25° towards West Lake deposit Age is Late Tertiary	A few lignite seams of limited lateral extent;l one seam is 6 m thick
Kluang-Nyior Central Johor	Alluvium unconformable over Lower Sequence	**Alluvium** Mainly sand with some clay; Unconsolidated; At least 5 m thick; Horizontal bedding Fluvial deposition ?Pleistocene (=Older Alluvium) **Lower Sequence** Shale & clay with some lignite seams; Weakly consolidated to semi-consolidated; Maximum 67 m thick; Dipping 6° to 20° towards N and S Late Tertiary	A few lignite seams of limited lateral extent, one is up to 10 m thick
Bukit Arang N. Perlis and Kedah	Boulder Bed unconformable over Lower Sequence	**Boulder Bed** Poorly sorted gravels and boulders in sandy to clayey matrix; Loose to weakly consolidated; About 90 m thick; Gently dipping towards N; Fluvial deposits Age unknown **Lower Sequence** Layers of sand and clay, with or without thin lignite seams; Weakly consolidated to semi-consolidated; Minimum thickness 130m; Strike NE or NW, Dips 10° - 35° towards North Lake deposits Late Neogene or Quaternary	Thin lignite seams of limited lateral extent
Enggor Central Perak	Thin layer of Surface Wash over Coal Bearing Strata	**Surface Wash** Quartzite pebbles in sandy matrix; Unconsolidated; About 1.2m thick; Horizontal bedding; Fluvial deposit Age unknown **Coal Bearing Strata** Shale, sandstone and clay with some lignite seams; Weakly consolidated to semi-consolidated; Maximum thickness 63m; Reported dipping 10° towards NW; Same age as Coal Measures of Batu Arang	One 1-m thick lignite seam and another of poor quality 1.5 m thick
Layang-Layang South Johor	Pengeli Sand Member conformable over Badak Shale Member	**Pengeli Sand Member** Sand, clayey sand & clay; Unconsolidated; At least 3.5 m thick; Horizontal bedding Fluvial to deltaic deposits ?Pleistocene (=Older Alluvium) **Badak Shale Member** Mudstone & clayshale; Weakly consolidated to semi-consolidated; Exposed thickness 95 m; Dips<15° Lake Deposit Miocene, probably older	No lignite at the surface, but likely present at depth
Lawin North Perak	Boulder Beds	**Boulder Beds** Poorly graded sand, grit and pebble beds; Weakly consolidated to semi-consolidated; At least 300 m thick; Dipping 20° to 45° towards basin centre; Cut by normal faults, throws up to 14 m; Alluvial fan deposits; Age similar as Boulder Beds at Batu Arang	Lignite seams absent
Nenering NE Perak	Alluvial fan interfingering with conglomerate	**Upper Unit** (probably 50 m thick) of braided channel sands and finer-grained clastics grading laterally into **Middle Unit** (40-70 m thick) of thick-bedded boulder/gravel beds, pebbly sandstone unconformable over Palaeozoic rocks **Lower Unit** of massive bedded conglomerate with unknown contact with the other units ; Age most likely Tertiary; reported Cretaceous pollen may be reworked	No coal or lignite

Table modified after Raj et.al. 1998

Table 7.1 Geology of Tertiary basins Onshore Peninsular Malaysia.

been facilitated by old faults in the Silurian basement underlying the basin.

A negative bouguer anomaly of more than 10 mgal marks a 20 km x 13 km area below the coastal plain at Teluk Datuk, Selangor (Fig. 7.9). The anomaly has N-S elongation and may represent a buried Tertiary basin similar in strike to that of the Kelang graben just offshore.

(2) Straits of Melaka

The generalised Cenozoic stratigraphy of the Straits of Melaka in Figure 7.10 uses formation names established in the adjacent North Sumatra and Central Sumatra basins. Fifteen grabens or half-grabens are known in the basement in the Malaysian waters of the strait. In the northern grabens (in the region to the north of the Perak river mouth) the Cenozoic began with continental Upper Oligocene beds that rapidly changed into marine deposits. Lower Miocene carbonate buildups formed on structural highs. Shallow marine to coastal conditions were maintained during the rest of the Cenozoic, except for minor subaerial exposures during the various Quaternary glacial episodes when the strait and the entire Sundaland formed a vast Southeast Asian continent. In the North Sumatra Basin, hydrocarbon production comes from intra-Baong and older post-Baong sands, while gas is produced from the Arun reef limestone (Courteney et al., 1989). The depressions in the strait to the south of the Perak river mouth began as lakes in the Late Oligocene, later also received fluvio-deltaic deposits. From the Pliocene onward the environment changed into lower coastal plain to shallow-marine conditions. The Quaternary glacial episodes also caused subaerial exposure of the southern Straits of Melaka. In the Central Sumatran Bengkalis Trough, major hydrocarbon producing reservoirs are contained in the

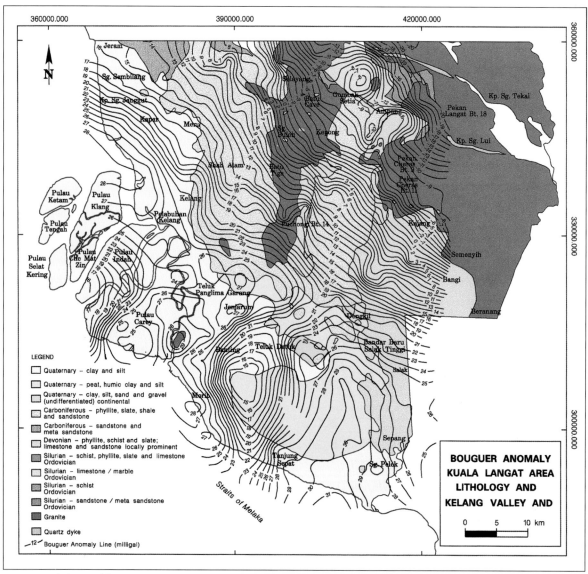

Fig. 7.9 Lithology and bouguer gravity anomaly map of the Klang (Kelang) valley and Kuala Langat area. Note the gravity low between Teluk Datuk and Tanjung Sepat. After an unpublished map by Jamaludin Osman, Geological Survey of Malaysia. Inserted with permission by its Director-General.

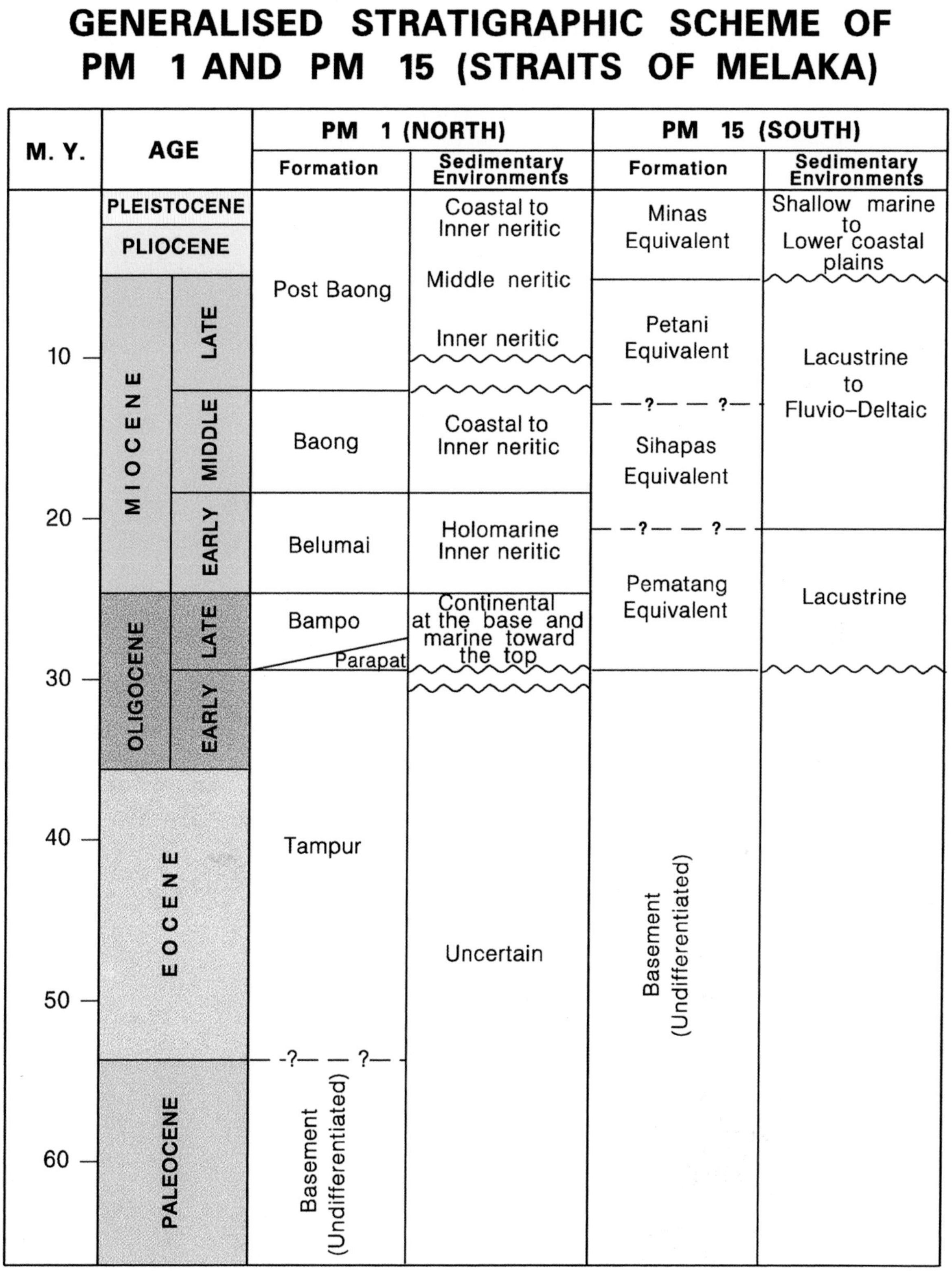

GENERALISED STRATIGRAPHIC SCHEME OF PM 1 AND PM 15 (STRAITS OF MELAKA)

M.Y.	AGE		PM 1 (NORTH)		PM 15 (SOUTH)	
			Formation	Sedimentary Environments	Formation	Sedimentary Environments
	PLEISTOCENE		Post Baong	Coastal to Inner neritic	Minas Equivalent	Shallow marine to Lower coastal plains
	PLIOCENE			Middle neritic		
10	MIOCENE	LATE	Post Baong	Inner neritic	Petani Equivalent	Lacustrine to Fluvio–Deltaic
		MIDDLE	Baong	Coastal to Inner neritic	Sihapas Equivalent	
20		EARLY	Belumai	Holomarine Inner neritic	Pematang Equivalent	Lacustrine
30	OLIGOCENE	LATE	Bampo / Parapat	Continental at the base and marine toward the top		
		EARLY				
40	EOCENE		Tampur	Uncertain	Basement (Undifferentiated)	
50						
60	PALEOCENE		Basement (Undifferentiated)			

Fig. 7.10a Stratigraphic schemes used in the northern and southern parts of the Straits of Melaka. After Liew (1994).

Sihapas and pre-Sihapas units (Courteney et al., 1991).

(3) Malay Basin

This is the largest among the hydrocarbon-bearing Tertiary basins of the Sunda Shelf, being about 500 km long and 200 km wide (Fig. 7.11). Its southern and central portions strike NW, but this changes into northerly in the northern part where the Malay Basin adjoins the north-striking Pattani Basin in the Gulf of Thailand.

Several stratigraphic schemes have been used for the Malay Basin (Fig. 7.12). The EPMI (Esso Production Malaysia Incorporated) seismostratigraphic scheme is based on information from the northeastern and

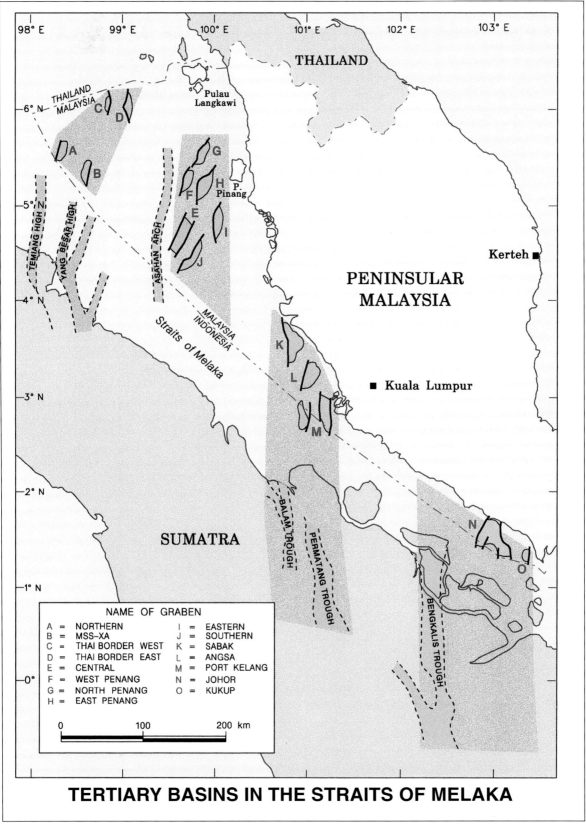

Fig. 7.10b Four N-S belts contain the known Tertiary basins in the Straits of Melaka according to Liew (1994).

southeastern parts of the basin and is now most commonly used. In its 1994 update incorporating micropalaeontological markers, absolute ages were assigned to the stratigraphic 'groups' by correlation with so called global eustatic sea-level changes. It has become increasingly clear that this type of correlation is untenable (see e.g. Miall, 1997), although critics also conceded that the other sequence-stratigraphic principles have great practical merit. PETRONAS Carigali is the other important operator in the basin. Its nomenclature is based on the recognition of regional shale intervals and/or marine transgressive pulses and is preferred for the northern and southwestern parts (Md Nazri Ramli, 1988).

The initial Oligocene-Lower Miocene synrift deposits were mainly non-marine. The Lower Miocene - Lower Middle Miocene sequences were laid down under stable geological conditions. These were terminated by strong regression, followed by a major marine transgression in the Middle Miocene. Regression again prevailed in the Upper Middle Miocene that was soon succeeded by transgression towards the

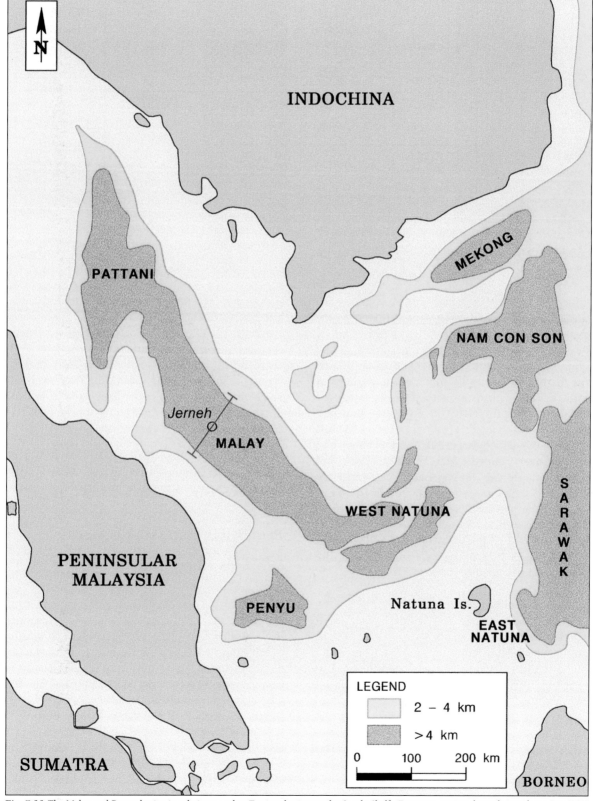

Fig. 7.11 The Malay and Penyu basins in relation to other Tertiary basins on the Sunda Shelf. For cross-section through Jerneh see Fig. 7.17.

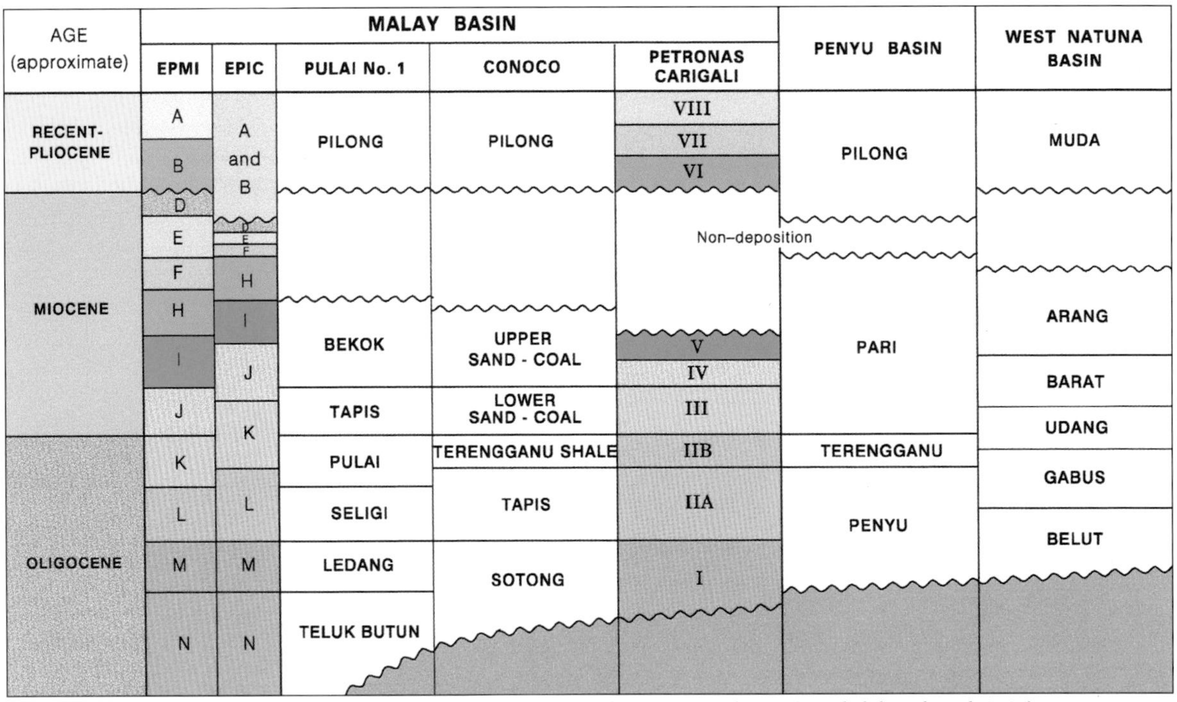

Geological Setting Of Peninsular Malaysia

AGE (approximate)	MALAY BASIN					PENYU BASIN	WEST NATUNA BASIN
	EPMI	EPIC	PULAI No. 1	CONOCO	PETRONAS CARIGALI		
RECENT-PLIOCENE	A	A and B	PILONG	PILONG	VIII	PILONG	MUDA
	B				VII		
					VI		
MIOCENE	D				Non–deposition		
	E	E F					
	F	H					ARANG
	H	I	BEKOK	UPPER SAND - COAL	V	PARI	
	I	J			IV		BARAT
	J	K	TAPIS	LOWER SAND - COAL	III		UDANG
	K		PULAI	TERENGGANU SHALE	IIB	TERENGGANU	GABUS
OLIGOCENE	L	L	SELIGI	TAPIS	IIA	PENYU	BELUT
	M	M	LEDANG	SOTONG	I		
	N	N	TELUK BUTUN				

Fig. 7.12 Correlation of the Tertiary stratigraphy in the Malay, Penyu and West Natuna basins (see Khalid Ngah et al. 1996).

end of the Middle Miocene. A major unconformity within the Upper Miocene separates younger, essentially undisturbed, sequences from the underlying deformed units. The stratigraphy of the younger beds is better known from exploration for placer tin (Aleva et al., 1973). Repeated subaerial exposures during the major glacials produced stacked channels in the Upper Miocene to Pleistocene intervals. Figure 7.12 compares the stratigraphy of the Malay Basin with the adjacent Tertiary basins.

It is widely accepted that the Oligocene and younger Cenozoic beds reach at least 8 km thickness in the Malay Basin. A specially processed regional seismic line across the basin suggests the presence of yet older stratified units beneath the Oligocene sediments. On this cross section, stratified intervals are observed to reach more than 14 km thickness. Figure 17.7 is the interpreted geoseismic section.

(4) Penyu Basin

The sediments in the Penyu Basin have been divided into four formations (Fig. 7.13). These are in younging order: (1) Penyu Formation of very fine-medium grained sandstone and reddish brown claystone interbeds; separated by an unconformity from the succeeding (2) Terengganu Formation of dominantly claystone and fine to coarse-grained sandstone. This is conformably overlain by the (3) Pari Formation, composed of fine to coarse-grained thin to massive bedded very poorly consolidated sands, interbedded with minor claystone and thin lignite laminae. This is separated by another unconformity from the (4) Pilong Formation composed of soft claystone with stringers of coarse-grained, poorly consolidated sandstone. The environment changed from predominantly continental (Penyu-1 well: braided stream system) via predominantly marginal marine (Terengganu and Pari) to predominantly marine (Pilong). The basement at the Rhu high consists of Cretaceous fine-grained clastics and volcanics (Texaco, 1992a). The maximum thickness of the Cenozoic sediments is about 6 km.

REGIONAL STRUCTURES

Peninsular Malaysia Onshore

Peninsular Malaysia consists of four structural domains: the Northwest, West, Central and East domains (Fig. 7.1a and Fig. 26.5). The boundaries are respectively: an arbitrary line from Kerian to the Betong salient at the international border with Thailand, the Bentong-Raub suture zone, and the Lebir lineament (or Eastern Tectonic Zone).

The NW Domain is characterised by NE and northerly striking structures, by distinct contact aureoles around the granitoid bodies of the Langkawi islands, and by presumed continuous deposition throughout the Palaeozoic (Foo, 1983; Hutchison, 1977). The arbitrary Kerian-Betong domain boundary encloses the only area of NE-striking structures.

Geological Setting Of Peninsular Malaysia

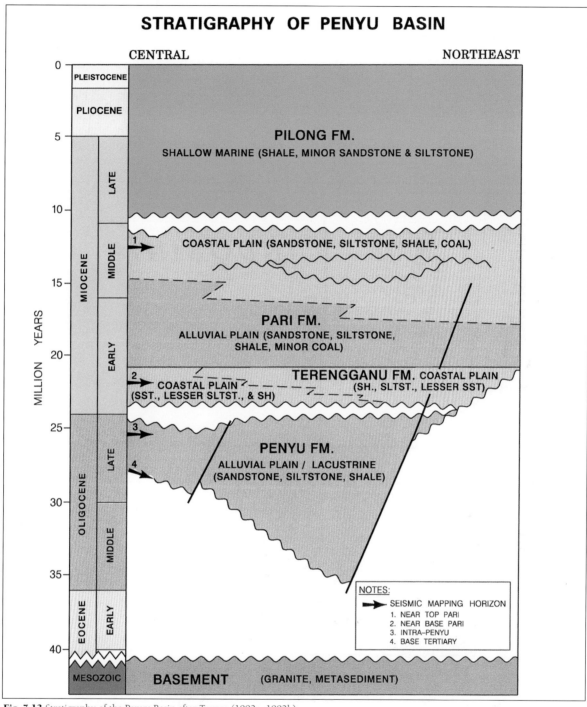

Fig. 7.13 Stratigraphy of the Penyu Basin after Texaco (1992a, 1992b).

The West Domain possesses extensive granitoids; their Rb-Sr and K-Ar dates are commonly discordant and contact aureoles are generally absent. The evidence taken together with the coarse-grained porphyritic texture has been interpreted to indicate meso-zonal emplacement (Hutchison, 1977). Tin-ore occurs in pegmatite and in skarn; malayaite also may be present; its cassiterite may be strongly pleochroic, and there are no tin-iron skarns (Hosking, 1977). Structures trend NNW and North (Geological Survey of Malaysia, 1985), while the dominant tectonic transport was westward.

The Bentong-Raub line was first recognised as an ophiolite belt by Hutchison (1975). Olistostromes were later identified and studied in some detail at several localities (Tjia, 1987, Tjia and Syed Sheikh Almashoor, in press). This Bentong suture is up to 18 km wide, and runs generally North-South, and most probably crosses the Straits of Melaka to continue as the petroliferous Bengkalis Trough in Central Sumatra. Others believe that from Negeri Sembilan state onward the Bentong-Raub line curves into a SE direction. They have projected the line to run as far as the sea strait between the tin islands Bangka and Billiton (Fig. 7.14). The olistoliths in Peninsular Malaysia contain Upper Ordovician up to Lower Triassic fossils. The presence of these fossils indicates that the West and Northwest domains docked with the rest of the peninsula by that time.

In the Central Domain there are a few small, characteristically fine-grained strongly sodic granitoids. There are also Late Cretaceous granitoids with high-emplacement characteristics (Hutchison, 1977). Economic minerals are typified by occurrences of gold and base metals (Chung and Yin, 1980). The Central Domain also contains the larger known outcrops of amphibolite facies metamorphics: the Taku Schist, the Setong Migmatite, and other bodies in western Pahang (Hutchison, 1973, Jaafar Ahmad, 1979). The Middle to Upper Triassic rocks were transported eastward (Jaafar Ahmad, 1976; Tjia, 1986) in contrast to the general

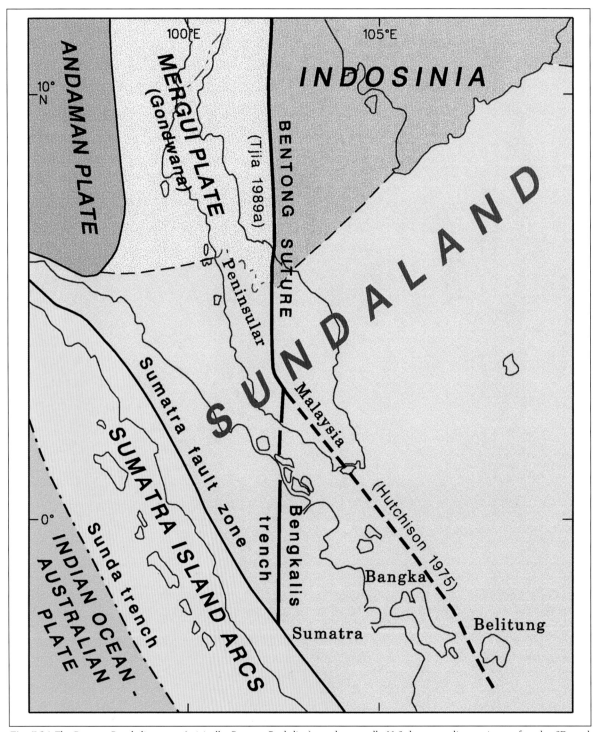

Fig. 7.14 The Bentong-Bengkalis suture (originally: Bentong-Raub line) trends generally N-S, but an earlier version preferred a SE-ward continuation. Indosinia-Sundaland-Mergui are continental lithosphere; the Andaman and Indian Ocean-Australian plates are oceanic; island-arc lithosphere underlies western Sumatra and the islands to the west.

westward transport in the other domains. The Central Domain also contains the larger outcrops of Jurassic-Cretaceous terrestrial sedimentary rocks. The structural grain in the Central Domain varies between North and NNW.

The Eastern Tectonic Zone (ETZ, Fig. 7.15) runs roughly North-South separating the Central Domain from the East Domain. Associated with the ETZ are three major fault zones: the NNW Lebir, the N-S Cini and the Lepar, the latter offsetting the ETZ left-laterally by about 35 km. The 125-km northerly elongated Upper Triassic granitoid of Ulu Kemapan - Gunung Besar - Bekok in Johor is also within the ETZ. Its whole rock Rb-Sr date is 222 ± 5 Ma or Late Triassic (Bignell and Snelling, 1977). The recently recognised ETZ is marked, in addition to the Lebir and Cini faults, by a distinct zone of negative to low-positive (-60 mgal to +15 mgal) bouguer gravity anomalies (H.D.Tjia, in Liew et al.,1996).The major Jurassic-Cretaceous depocentres of the Central Domain are closely associated with the ETZ. Some small serpentinite bodies and chaotic deposits (olistostrome?; Lepar fault zone) are associated with the ETZ, which may possibly represent another suture within the Laurasian block.

In the East Domain there are extensive outcrops of dominantly Upper Triassic granitic rocks that are mostly equigranular with some porphyritic varieties. Well-developed contact aureoles surround several of the granitoids. Their Rb-Sr and K-Ar dates are generally concordant, from which fact epizonal emplacement has been deduced (Hutchison 1977). Mineralisation produced tin-iron skarns; pleochroic cassiterite is rare, while malayaite has not been reported from among the tin minerals (Hosking 1977). The regional structural trend is NW to NNW.

The regional fault zones in the peninsula are shown on Figure 7.7. The (latest?) wrench components on the regional fault zones correspond to a regional compressive stress directed approximately 70° - 250°. Regional fault zones striking within the WNW - NNW sector are sinistral, while those striking North to NE are dextral. Actual displacements have only reliably been determined for the Kuala Lumpur (> 20 km sinistral) and the Baubak (or Bok Bak; 55 km sinistral) fault zones. A number of the small Tertiary basins shows close association with these fault zones, such as the Batu Arang (Kuala Lumpur f.z.) and Lawin (Baubak f.z.). Radiometric dates of mylonite indicate ages as young as Early-Middle Eocene, which suggest the probable time of the youngest fault activity (Zaiton Harun 1992).

Straits of Melaka

Liew (1994), based on Sun Oil early 1990s reports of the Straits of Melaka, stated the following. The Tertiary basement of the strait slopes gently towards Southwest. At discrete locations, the basement hosts 15 north-trending grabens (Fig. 7.10b). The grabens range from 825 m to almost 4000 m in depth. They have been grouped into (1) Bengkalis Trough related, (2) Pematang-Balam Trough, (3) Asahan Arch-Kepulauan Arua Nose, and (4) Tamiang-Yang Besar High related depressions. The groupings are in N-S zones, probably representing regional fracture zones, that are separated from each other by zones of regional highs. Liew (1994) considers the grabens and graben groupings to have begun developing in the ?Late Oligocene by regional dextral shearing of the NW-SE trending Straits of Melaka belt. On major, but non-regional faults evidence of wrenching is compelling. Seismic lines show flower structures on some of the basement-involved faults. On time structure maps N-S *en echelon* normal faults define some of the graben boundaries (Figs. 7.10b, 26.11a, 26.11b, 26.12a, 26.12b, 21.13 & 21.14). Although altered by subsequent tectonic deformation, many of the graben outlines at basement level resemble rhombic or sigmoidal pull-aparts, such as the Central, North Penang, East Penang, Port Kelang (Klang), and Johor grabens.

Malay Basin

The Malay Basin is a large epicontinental Tertiary basin located in the northern Sunda Shelf. It has recently been proposed that the Malay Basin is an aulacogen that, together with the Penyu and West Natuna (in Indonesia) rifts, originated on top of an Upper Cretaceous dome, termed the Malay Dome (Fig. 7.16; Tjia, 1994). The geological age is indicated by widespread occurrence of Late Cretaceous granitic plutons in the northern Sunda Shelf region. These basins form a triple junction that even today still possesses a moderately high geothermic gradient. The junction also coincides with thicker crust which may represent the root of differentiated mantle-plume material. The Malay Dome is estimated to be 1000 km across and even today one notes the height of the pre-Tertiary basement of Peninsular Malaysia, of the islands of Natuna, Anambas, Tambelan, Con Son, and of the adjacent shallow sea floor. Below 10 km very steep to vertical contacts between basin-fill and pre-Tertiary basement may represent the rift surfaces of the Malay Basin (Fig. 7.17). Subsidence continued from the Oligocene towards the end of Early Miocene in a transtensional stress field. In the same period a NW-trending regional fault zone along the basin's axis, named the Axial Malay fault zone, moved in a left-lateral sense. This wrench motion was transtensional and created E-W half-grabens that became discrete depocentres accumulating about 6 km of pre-Oligocene deposits and at least 8 km of Oligocene and younger sediments. Towards the northwest, the Axial Malay fault zone joins the Three Pagodas fault zone. These fault zones, together with several other NW trending regional

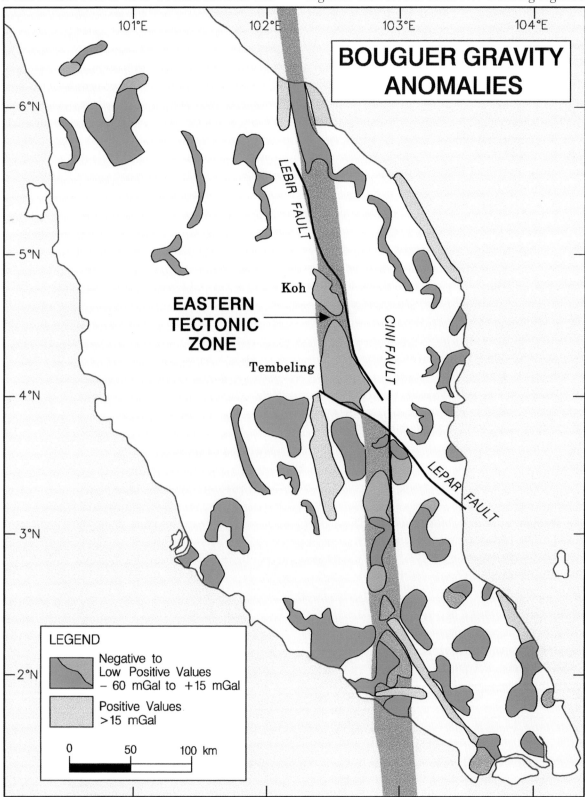

Fig. 7.15 Bouguer gravity anomaly patterns simplified after ARK's bouguer gravity map at a scale of 1 : 500,000 (ARK Geophysics, 1995). Geological interpretation by H.D.Tjia (in Liew et al., 1996).

fault zones across Indosinia, have facilitated extrusion of the Southeast Asian continent in response to collision of the Indian sub-plate with the Eurasian plate (Tapponnier et al., 1982). The Middle-Late Miocene stress regime in the Malay Basin was transpressional, possibly as result of reorientation of plate movements in the Indonesian region *sensu lato*. Reversals of wrench motion were common. The Axial Malay fault zone moved in a dextral sense, and this caused structural inversion that began at the south and extended progressively northwestward. Stronger compression in the southeast is expressed in reverse faulting, that is also reported from the adjacent West Natuna Basin. The reverse faults have partly (by up to a kilometre) restored downthrows (as much as 2.5 km separation) on former growth faults (Fig. 7.18). Half-graben infills were deformed into anticlines whose E-W strikes were determined by the original orientation of the half-grabens. After reaching a peak at about 10 Ma, regional compression or transpression has declined but may persist today. This would account for tightening of the

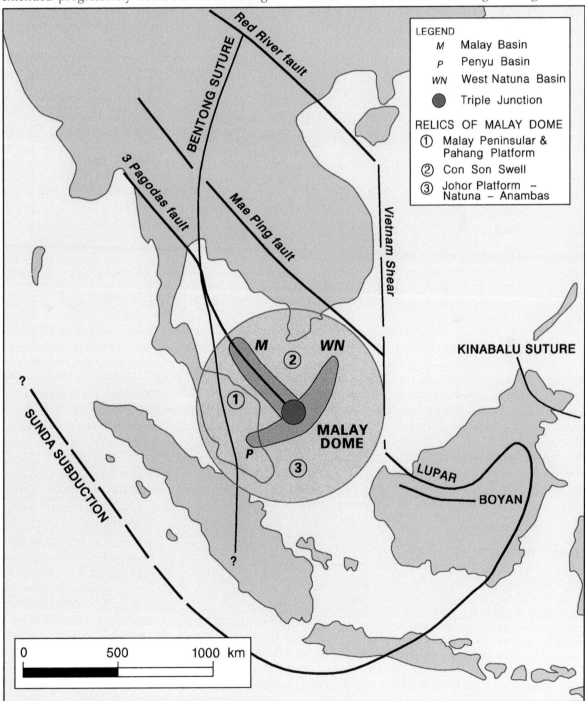

Fig. 7.16 The widespread occurrence of Upper Cretaceous granitoids in the northern Sunda Shelf probably represents the Malay Dome, believed to mark the site of a mantle plume. The Malay, Penyu and West Natuna basins are thus considered as failed rift arms of the Malay Dome.

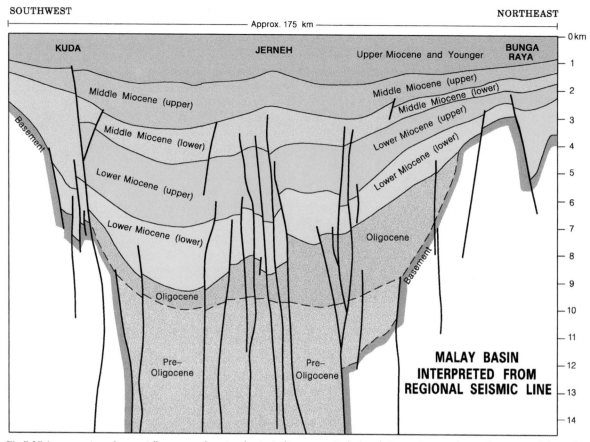

Fig.7.17 Interpretation of a specially processed, regional seismic line across Kuda-Jerneh-Bunga Raya in the northern part of the Malay Basin reveals structures at depths exceeding 14 km. The vertical boundaries of the basin at depth resulted from rifting, the gentler pre-Tertiary basement surface at shallower depth indicates thermal subsidence. The vertical exaggeration is about 13 times. This NE-SW section is the line across Jerneh shown on Figure 7.11.

anticlines whose crests have fractured along the extensional direction perpendicular to strike. The probable rift origin of the Malay Basin implies that its initially restricted environment of deposition could favour hydrocarbon generation prior to that believed to have been responsible for the oil and gas being extracted. These pre-Oligocene hydrocarbons could have migrated into fractured basement highs. More recent discussions on the tectonics of the Malay Basin are in Khalid Ngah et al. (1996), and Tjia and Liew (1996).

Penyu Basin

The Penyu Basin is another failed rift arm of the Malay Dome. The basin plan is 160 km x 200 km, somewhat elongated in E-W direction. The geology of the basin has been described by Khalid Ngah (1975), Texaco (1992a, 1992b), and Mazlan Madon (1995), Mazlan Madon et al. (1997). The surrounding basement highs have thin sedimentary cover, except for the Tenggol Arch in the northeast which possesses about a kilometre thin sedimentary cover. The pre-Tertiary basement was faulted into ten major half-grabens. A major 100 km NW-SE Rumbia fault, cuts the Penyu Basin into two parts. The western part is marked by E-W half-grabens, the eastern part has WNW-ESE orientated depressions (Fig. 7.19). Heave on the Rumbia fault exceeds 2 km and left-lateral separation of once-contiguous structures suggest 20 km displacement. An *en echelon*, sigmoidal fault pattern adjacent to and parallel with the Rumbia supports sinistral wrenching. Another similar E-W orientated fault pattern, as seen near the lower border of Figure 7.20, also indicates sinistral wrenching. The E-W striking relatively smaller faults and half-grabens are interpreted as secondary structures developed by sinistral wrenching along the Rumbia fault and other, as yet unknown, major NW-SE faults in the basement of the Penyu Basin area. Continued left-slip motion on the Rumbia fault is postulated to have rotated the E-W faults and half-grabens in the eastern part of the basin into their present WNW-ESE orientation, involving 25 degrees clockwise rotation. Anticlinal features above the deepest parts of the half-grabens show up on seismic, but the effect of compressional tectonics

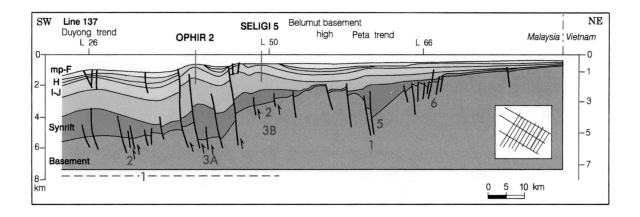

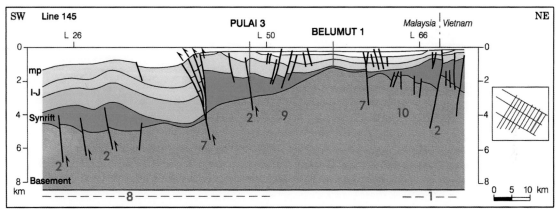

Fig. 7.18 Interpreted regional seismic sections across the Malay Basin showing structural inversion, reverse faulting, and possible flower structures. Compare with Figure 7.12 for the geological ages of alphabetically named intervals. Insets show the grid of regional seismic lines covering the Malaysian part of the Malay Basin; the regional lines in Figure 7.18 are in red.

1. Inversion
2. Reverse fault (in I-F units)
3A. Inversion; J-top down 2 km restored 1 km by inversion
3B. Inversion; J-top up 0.8 km restored 0.3 km
4. Half graben
5. Inversion; down 0.75 km restored 0.3 km
6. Step faults in basement
7. Flower structure
8. Inversion; basement up 0.4 km
9. Normal fault, basement detached
10. Broken basement

is less obvious than in the Malay Basin (Fig. 7.20). The known Cenozoic development of the Penyu Basin began with an Oligocene (trans?)tensional regime that formed the half-grabens floored by pre-Tertiary basement. Lake and river deposits of the Penyu Formation filled the depressions. Major growth faulting is apparent. A regional unconformity at the top of the Penyu Formation indicates tectonic uplift at the end of the Oligocene. Renewed subsidence in Early Miocene time began with the deposition of the coastal plain Terengganu and the Middle Miocene Pari formations. Some of the existing faults were reactivated and normal faulting continued into the Middle Miocene. At the end of the Middle Miocene another unconformity indicates a probable transpressional event. This is overlain by the shallow-marine Pilong Formation (Upper Miocene - Pliocene).

TECTONIC EVOLUTION

Until the beginning of the Mesozoic, the West (including the Straits of Melaka) and Northwest domains developed as a tectonic plate separate from the Central and East domains. The latter two domains are part of the Indosinian block. Metcalfe (1998) suggests that both the Indosinian block and that of the NW and W domains (Mergui block; also referred to as Sibumasu block) were derived from northeast Gondwana, but arrived at different times at their SE Asian locations. The blocks combined along the Bentong-Bengkalis (originally named Bentong-Raub line) suture during the lower Triassic. The present dominant NW-SE structural grain of Peninsular Malaysia became established by the Upper Triassic-Lower Jurassic deformation period, accompanied by large-scale granitic emplacement. In the East Domain granitic

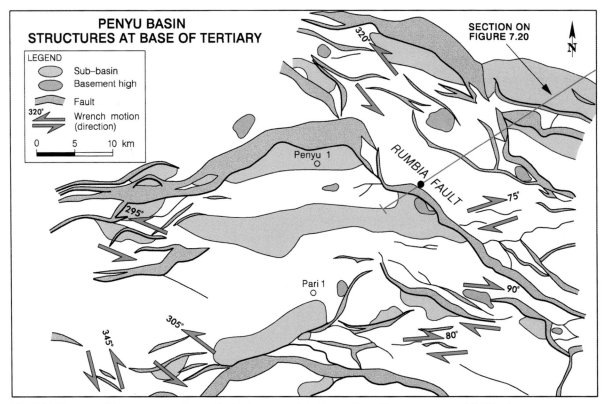

Fig. 7.19 Plan-view of the structures in the pre-Tertiary basement of the Penyu Basin. The sub-basins trend E-W in the western part, but are orientated WNW-ESE to the east of the Rumbia fault. Note the en echelon sigmoidal faults indicating sense of fault motion.

intrusions occurred earlier; in the Permian in the north and in the Middle-Upper Triassic in the south. Early sedimentation was prevalently marine, but from the Late Mesozoic onward, the depositional environment had become terrestrial. Moderately sized pull-apart depressions, associated with N-S wrench faults and early-phase volcanism, were filled by several kilometres of Upper Jurassic-Cretaceous 'redbeds'. By the Mesozoic-Cenozoic boundary the crust was largely stabilised.

Northwest and West Domains

A major low-angle lag fault zone separates suspected Precambrian coarse-grained quartz-rich clastics from the overlying Middle Cambrian-Lower Ordovician, deltaic Macincang Formation. It is not known if this faulting was pre-Macincang or if it is a manifestation of the Middle Devonian deformational episode. The Macincang strata are succeeded, probably conformably, by shallow-marine Setul Limestone and clastic intervals. In the Middle Devonian these were overfolded and thrusted westward onto the moderately deformed Macincang sediments. The latter exhibit large open folds and are therefore considered to be located upon a relatively stable tectonic platform. Farther south in the West Domain, in Selangor and Negeri Sembilan states, this Middle Palaeozoic deformation caused SE-verging isoclinal folds and thrust faults in the ?Silurian-Lower Devonian strata. Superimposed and varied structural styles attest to the strata having been subjected to multiple deformations. In the same area the Upper Palaeozoic-Lower Triassic sequences were also deformed into asymmetrical folds without superposed structures showing westward tectonic transport. Marine sedimentation ceased with the widespread emplacement of the Upper Triassic-Lower Jurassic Titiwangsa granitoids. Subsequently Jurassic-Cretaceous fluvial redbeds and Lower-Middle Tertiary lacustrine-paludal sediments with coal intervals were deposited across the West Domain.

Central Domain

The shallow-marine Baling Group are the oldest strata in the Central Domain. Some of its formations contain significant volcanic constituents, presumably products of arc volcanism. The Baling Group strata were deformed into complex refolded folds and faults of varied directions of tectonic transport. The first deformational episode took place in the Middle Palaeozoic to ?Early Carboniferous. Carbo-Permian-Triassic sedimentation occurred in deeper marine environments in the west and in

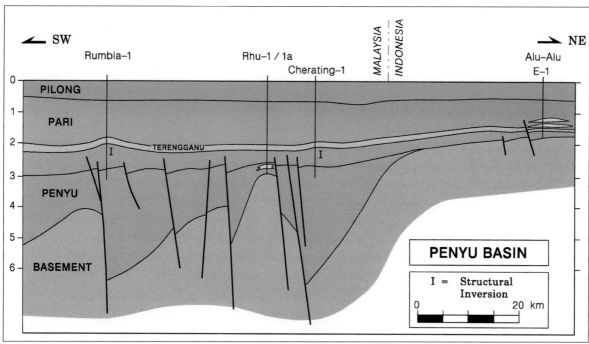

Fig. 7.20 A structural section across the Penyu Basin. Approximate section line is on Figure 7.19. Compare the lower degree of inversion with that in the Malay Basin (Figure 7.18).

neritic conditions in the east. The second major deformation was associated with the Titiwangsa granitoid emplacement (Late Triassic-Early Jurassic) that caused the entire Peninsular Malaysia to acquire continental conditions. The younger Palaeozoic-Triassic strata of the Central Domain were deformed into asymmetrical folds and were cut by reverse faults verging mainly eastward. Locally complex structures were produced by the tectonic deformation of syndepositional slumps, such as in the Semantan Formation. The post-granitoid continental strata of the Central Domain were deformed mainly into large basinal sags. The larger of these basins are pull-aparts associated with northerly and NW-striking strike-slip faults. Radiometrically dated mylonites suggest that wrench faulting in Peninsular Malaysia persisted into the Eocene. The stronger deformation close to these fault zones manifests as compressive folds and reverse faults, but the Jurassic-Cretaceous continental beds are generally regional homoclines and synclinal basins. There are Early Cenozoic alkali basalt flows in the Segamat area of this domain.

East Domain

The oldest known rocks north of the Lepar fault zone are Carboniferous, in which some intervals show relative abundance of coalified plant remains that indicate at least some coastal if not continental setting. These rocks were intruded by Permian granitoids. To the south of the Lepar fault zone, the oldest strata are Permian, as suggested by the fossiliferous limestone at Sumalayang mountain. All the Upper Palaeozoic strata have been metamorphosed and subjected to at least three deformational episodes, two being coaxial with general N to NNW strikes, and the younger episode resulting in east-west open warps. In the southern part of the domain, acid volcanism occurred at the Permian-Triassic boundary. Granitoid intrusions took place in the Middle and Upper Triassic. The Jurassic-Cretaceous is represented by continental redbeds that show evidence of semi-arid conditions in the Peninsula. Except for the Gagau and Gerek stratigraphic units, the other redbed occurrences are isolated patches. The beds are almost undeformed, and regionally tilted to the south. Compressive folds and reverse faults characterise these beds in the proximity of the major strike-slip faults: Lebir, Lepar and others. The less than 2 million-year old alkali basalt near Kuantan may be a product of impact with an extraterrestrial body rather than an expression of diastrophism. Shocked, lamellar quartz grains were identified in the granitic rocks below the lava.

Malay Basin and Penyu Basin

A mantle plume developed beneath the current junction of the Malay-Penyu-West Natuna basins and manifested as the Malay Dome. The three named basins originated as rifts on this dome. Hard collision of the Indian sub-plate with the Eurasian plate that began in the Middle Eocene, differentially extruded elongated crustal slabs of the Southeast Asian lithosphere towards the southeast along three major NW-SE striking wrench fault zones. One of these, the Axial Malay fault zone, runs in the pre-Tertiary basement of the Malay Basin. In pre-Late Oligocene time, the Axial Malay fault zone moved sinistrally, thereby creating E-W half-grabens within the greater basin. This transtensional stress regime was of regional character. Smaller predominantly N-S pull-aparts developed in the pre-Tertiary basement in the Melaka Straits. The entire strait as a NW-SE trending zone is thought to have experienced dextral wrenching. Onshore, several of the small Tertiary basins are closely associated with WNW to NW strike-slip faults. A genetic relationship is suggested, but more structural data are needed. Then in the early to middle part of the Miocene, reorientation of the regional stress field into compressional caused many of the wrench faults in the Malay Basin to reverse their motion sense. The Axial Malay fault zone became dextral and resulted in structural inversion. A number of large growth faults also accommodated compression by reverse motion involving several hundred metres. Apart from the Malay Basin, response to this younger compression was relatively mild in the other Tertiary basins described in this chapter. The change into a compressive stress regime in the Middle-Upper Miocene reflects the reorientation of movement patterns of the neighbouring plates. By that time the Philippine Sea and the South China Sea plates ceased spreading, thus enabling the westward thrust of the Pacific plate to be felt in the region of Peninsular Malaysia. Furthermore, the Indian Ocean-Australian plate had approached closer to the region. This probably obstructed or hampered the ongoing extrusion of SE Asian crustal slabs.

REFERENCES

Aleva, G.J.J., Bon, E.H., Nossin, J.J., and Sluiter, W.J., 1973. A contribution to the geology of part of the Indonesian tin belt: The sea areas between Singkep and Bangka islands and around the Karimata islands. Bulletin of the Geological Society of Malaysia, 6, 257-272.

Arafin, M.S., Lee, C.Y. and Kaur, G., 1989. Gravity delineation of the Tertiary beds of Chuping in Perlis, Malaysia. Sixth Regional Congress on Geology, Mineral and Energy Resources of Southeast Asia, GEOSEA 6, Jakarta, July 1987, Proceedings, 493-504.

ARK Geophysics, 1995. Bouguer gravity [map], scale 1 : 500 000. PETRONAS, Kuala Lumpur (unpublished).

Basir Jasin, 1996. Discovery of Early Permian radiolaria from the Semanggol Formation, northwest Peninsular Malaysia. Warta Geologi 22 (4), 283-287.

Bignell, J.D. and Snelling, N.J., 1977. Geochronology of Malayan Granites. Institute of Geological Sciences, Overseas Geology and Mineral Resources No. 47, 70 p.

Burton, C.K,. 1967. Dacryonconarid tentaculites in the Mid-Palaeozoic euxinic facies of the Malayan geosyncline. Journal of Palaeontology 41, 449-454.

Burton, C.K,. 1973. Chapter 5. Mesozoic. In Gobbett, D.J. and Hutchison, C.S., eds., "Geology of the Malay Peninsula". John Wiley, New York, 97-141.

Chung, S.K. and Yin, E.H., 1980. Regional geology: Peninsular Malaysia. Geological Survey of Malaysia, Annual Report 1980, 82-99.

Courteney, S., Cockcroft, P., Lorentz, R.and Miller, R., editors, 1989. Oil & Gas Fields Atlas Volume 1: North Sumatra and Natuna. Indonesian Petroleum Association, Jakarta.

Courteney, S., Cockcroft, P., Lorentz, R. and Miller, R. and others, editors, 1991. Oil & Gas Fields Atlas Volume II: Central Sumatra. Indonesian Petroleum Association, Jakarta.

Foo, K.Y. ,1983. The Palaeozoic sedimentary rocks of Peninsular Malaysia: stratigraphy and correlation. Proceedings Workshop on Stratigraphic Correlation of Thailand and Malaysia. Geological Society of Thailand and Geological Society of Malaysia I, 1-19.

Geological Survey of Malaysia, 1985. Geological Map of Peninsular Malaysia 8th edition, 1 : 500 000. Geological Survey of Malaysia, Ipoh: 2 sheets.

Gobbett, D.J. and Hutchison, C.S. (Editors), 1973. "Geology of the Malay Peninsula" West Malaysia and Singapore. John Wiley, New York, 438 pp.

Hosking, K.F.G., 1977. Known relationships between the 'hardrock' tin deposits and the granites of Southeast Asia. Geological Society of Malaysia Bulletin 9: 141-157.

Hutchison, C.S., 1973. Chapter 9 Metamorphism. In Gobbett, D.J. and Hutchison, C.S., eds., "Geology of the Malay Peninsula". John Wiley, New York: 253-303

Hutchison, C.S., 1975. Ophiolite in Southeast Asia. Bulletin of the Geological Society of America, 86, 797-806.

Hutchison, C.S., 1977. Granite emplacement and tectonic subdivision of Peninsular Malaysia. Bulletin of the Geological Society of Malaysia, 9, 187-207.

Ibrahim Amnan and Fontaine, H., 1996. New palaeontological data of the limestones in northwest Kelantan and north Pahang, Peninsular Malaysia. Warta Geologi 22 (2), 88-89.

Jaafar Ahmad, 1976. The Geology and Mineral Resources of the Karak and Temerloh Areas, Pahang. Geological Survey of Malaysia, District Memoir 15, 138 p.

Jaafar Ahmad, 1979. The Petrology of the Benom Igneous Complex. Geological Survey of Malaysia, Special Paper 2.

Jabatan Penyiasatan Kajibumi Malaysia, 1995. Laporan tahunan - Annual report 1995. Kementerian Perusahaan Utama, Kuala Lumpur.

Jones, C.R., 1978. The Geology and Mineral Resources of Perlis, North Kedah and the Langkawi Islands. Geological Survey of Malaysia, District Memoir 17, 257 pp.

Khalid Ngah, 1975. Stratigraphic and Structural Analyses of the Penyu Basin, Malaysia. MSc thesis, Oklahoma State University.

Khalid Ngah, Mazlan Madon and Tjia, H.D., 1996. Role of pre-Tertiary fractures in formation and development of the Malay and Penyu basins. In Hall, R., and Blundell, D., eds., Tectonic Evolution of Southeast Asia, Geological Society special publication No. 106, 281-289.

Khoo, H.P. 1983. Mesozoic stratigraphy in Peninsula Malaysia. Proceedings Workshop on Stratigraphic Correlation of Thailand and Malaysia. Geological Society of Thailand and Geological Society of Malaysia I, 370-383

Koopmans, B.N., 1965. Structural evidence for a Palaeozoic orogeny in northwest Malaya. Geological Magazine 102, 501-520.

Lee, C.P., 1983. Stratigraphy of the Tarutao and Machinchang formations. Proceedings Workshop on Stratigraphic Correlation of Thailand and Malaysia. Geological Society of Thailand & Geological Society of Malaysia I, 20-38.

Liew, K.K., 1994. Structural patterns within the Tertiary basement of Straits of Malacca. PETRONAS Research & Scientific Services Sdn Bhd, Report No. PRSS-RP5-94-02, 20 May 1994 (unpublished).

Liew, K.K., Mohd Fauzi Abdul Kadir, Mohd Jamaal Hoesni, Sahalan Abdul Aziz and Tjia, H.D., 1996. Hydrocarbon Potential of Mesozoic Strata Onshore Peninsular Malaysia, Results of Field & Laboratory Studies 1995. PETRONAS Research & Scientific Services Sdn Bhd, Report PRSS-RP5-96-10, 31 March 1996 (unpublished).

Mazlan Madon 1995. Tectonic evolution of the Malay and Penyu basins, offshore Peninsular Malaysia. D.Phil. Thesis, University of Oxford, 325 pp (unpublished).

Mazlan Madon, Azlina Anuar and Wong, R. 1997. Structural evolution, maturation history, and hydrocarbon potential of the Penyu basin, offshore Peninsular Malaysia. Proceedings of the Petroleum Systems of SE Asia and Australasia, Indonesian Petroleum Association, Jakarta, May 1997.

Md.Nazri Ramli, 1988. Stratigraphy and palaeofacies development of Carigali's operating areas in the Malay Basin, South China Sea. Bulletin of the Geological Society of Malaysia, 22, 153-187.

Metcalfe, I., 1998. The Palaeo-Tethys in East Asia (abstract). Ninth Regional Conference on Geology, Mineral and Energy Resources of Southeast Asia - GEOSEA '98, Kuala Lumpur, 27-28.

Miall, A.D., 1997. "The Geology of Stratigraphic Sequences". Springer Verlag, Berlin: 433 pp.

Qalam Azad Rosle and Teh, G.H., 1998. The stratigraphy, structure and significance of the Nenering Tertiary beds, Pengkalan Hulu (Keroh), Hulu Perak. In Geological Society of Malaysia, Seminar publication on Tertiary basins of Peninsular Malaysia and its adjacent offshore areas, Kuala Lumpur 21-22 February 1998, 2.1-2.26.

Raj, J.K., Abdul Hadi Abd. Rahman and Mustaffa Kamal Shuib, 1998. Tertiary basins of inland Peninsular Malaysia: Review and tectonic evolution. In Geological Society of Malaysia, Seminar publication on Tertiary basins of Peninsular Malaysia and its adjacent offshore areas, Kuala Lumpur 21-22 February 1998, 1.1-1.21.

Rishworth, D.E.H. 1974. The Upper Mesozoic Terrigenous Gagau Group of Peninsular Malaysia. Geological Survey of Malaysia, Special Paper 1: 78p.

Stauffer, P.H. and Mantajit, N., 1981. Late Palaeozoic tilloid of Malaya, Thailand and Burma. In Hambrey, M.J., and Harland, W.B., eds., Earth's pre-Pleistocene Glacial Record. Cambridge University Press, Cambridge: 331-337.

Stauffer, P.H. and Lee, C.P., 1986. Late Palaeozoic glacial marine facies in Southeast Asia. Bulletin of the Geological Society of Malaysia, 20, 363-397.

Tan Yong Phang 1988. Stratigrafi - Geologi Kawasan Parit Sulong, Johor. MSc thesis, Universiti Kebangsaan Malaysia, Bangi: 168 pp. (unpublished).

Tapponnier, P., Peltzer, G. LeDain, A., Armijo, R. and Cobbold, P., 1982. Propagating extrusion tectonics in Asia: new insights from simple experiments with plasticine. Geology 10, 611-616.

Texaco, 1992a. The significance of the Rhu oil discovery, PM-14, offshore Peninsular Malaysia: a preliminary survey, Volume 1, text. Rhu Assessment Team, Internal report, May 15, 1992. Texaco Exploration Penyu Inc.

Texaco, 1992b. Malaysia PM-14, Exploration well drilling recommendation Rumbia No. 1, PETRONAS Carigali Sdn Bhd, Texaco Exploration Penyu Inc., Clyde Expro Plc, Santos Ltd., Internal report.

Tjia, H.D., 1986. Geological transport directions in Peninsular Malaysia. Bulletin of the Geological Society of Malaysia, 20, 149-177.

Tjia, H.D., 1987. Olistostrome in the Bentong area, Pahang. Warta Geologi 13 (3): 105-111.

Tjia, H.D., 1989a. Tectonic history of the Bentong-Bengkalis suture. Geologi Indonesia 12 (1): 89-111.

Tjia, H.D., 1989b. Major faults of Peninsular Malaysia on remotely-sensed images. Sains Malaysiana 18, 101-114.

Tjia, H.D., 1994. Origin and tectonic development of Malay-Penyu-West Natuna basins. In PETRONAS Research & Scientific Services "Research for Business Excellence Seminar", Kuala Lumpur, 20-21 June 1994 (limited distribution).

Tjia, H.D. and Anizan Isahak, 1990. Permian glacigenic deposits at Salak Tinggi, Selangor. Sains Malaysiana 19 (1), 45-64.

Tjia, H.D. and Liew, K.K., 1996. Changes in tectonic stress field in northern Sunda Shelf basins. In Hall, R. and Blundell, D.J., eds., Tectonic Evolution of Southeast Asia, Geological Society special publication No.106, 291-306.

Tjia, H.D. and Syed Sheikh Almashoor, in press. The Bentong suture in southwest Kelantan, Peninsular Malaysia. Bulletin of the Geological Society of Malaysia (submitted 1996).

Uyop Said and Che Aziz Ali, 1997. Nenering continental deposits: its age based on palynological evidence (abstract). Warta Geologi 23 (3): 170-171.

Zaiton Harun, 1992. Anatomi Sesar-sesar Utama Semenanjung Malaysia. PhD thesis, Universiti Kebangsaan Malaysia, Bangi, Malaysia (unpublished).

Chapter 8

Malay Basin

*Mazlan B. Hj. Madon, Peter Abolins,
Mohammad Jamaal B. Hoesni
and Mansor B. Ahmad*

INTRODUCTION

The Malay Basin is situated in the southern part of the Gulf of Thailand, between Vietnam and Peninsular Malaysia (Fig. 8.1A). The basin covers an area of about 80,000 km², and is filled with up to 14 km of sediment (Abd Rahim Md Arshad et al., 1995). The basin continues northwestwards to merge with Thailand's Pattani Trough and southeastwards with the Indonesia's West Natuna Basin. Its northeastern flank lies partly in

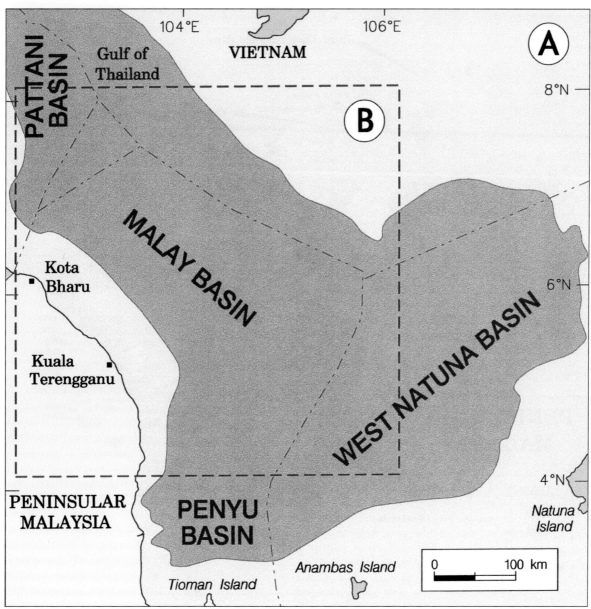

Fig. 8.1. (A) Location of the Malay Basin in relation to neighbouring Penyu and West Natuna basins. Boxed area shown in Fig. 8.1(B).

Vietnamese territory.

Petroleum exploration in the Malay Basin began in 1968 and is now in a relatively mature stage. Many oil and gas accumulations have been discovered and 13 of these are currently producing (Fig. 8.1B). Early exploration activities (pre-1980s) in the basin were reviewed by Armitage (1980), Ahmad Said (1982), and Chua and Wong (1997). Data collected by the petroleum industry over the years have resulted in numerous studies that have helped improve our understanding of the petroleum systems in the basin. These studies have dealt with a variety of topics, from tectonics, structure, regional geology, stratigraphy, and petroleum geology and exploration, to field-scale studies of reservoirs and their sedimentological characteristics.

Although the basin has been explored for over 3 decades, most of the studies carried out by oil companies are unpublished. Before the mid-1980s most publications dealt with reservoir engineering aspects of the major oil fields. These include studies on the Tapis (Goh et al., 1982, 1983), Pulai (Flores and Kelm, 1982), Semangkok (Khandwala et al., 1984), Tinggi (Brami et al., 1984), Jerneh (Thambydurai et al., 1988), and the Guntong and Tabu fields (Hui and Pillai, 1988; Amran Nong Chik et al., 1996). Only brief accounts of the geology of the basin can be found in Hamilton (1979), Armitage (1980), and ASCOPE (1981). Recently, more detailed tectonic and structural studies have been published. These include, among others, Khalid Ngah et al. (1996), who described the pre-Tertiary basement faults and their control on the development of the basin, and Tjia (1994a, 1994b) who discussed the

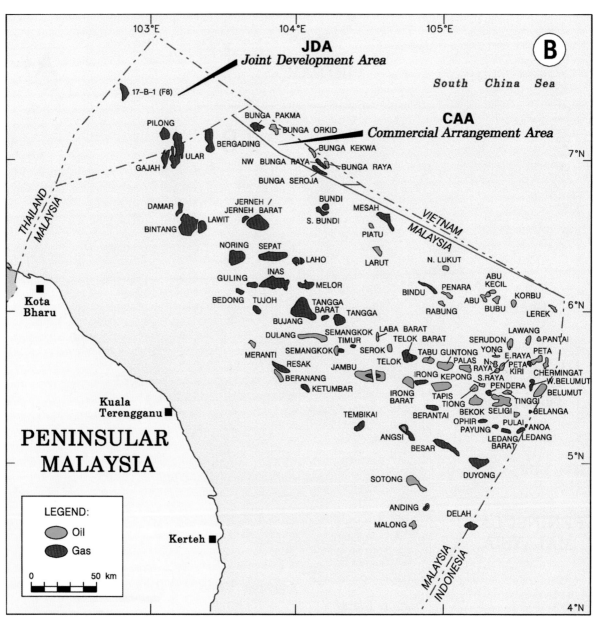

Fig. 8.1. (B) Oil and gas fields/discoveries in the Malay Basin. JDA - Malaysia-Thai Joint Development Area, CAA - PM3 Commercial Agreement Area (Malaysia-Vietnam).

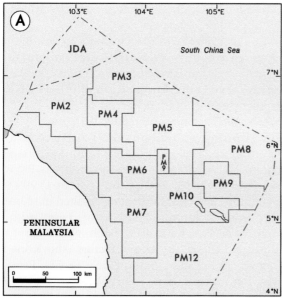
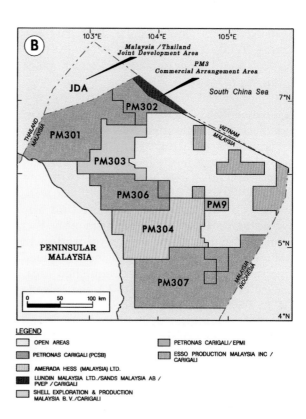

Fig. 8.2. Malay Basin production-sharing contract areas. (A) 1985 PSC round. (B) Current (1999).

major fault zones, basin inversion, and tectonic origin of the basin. Mazlan Madon and Watts (1998) discussed the implications of gravity anomalies on the tectonic evolution of the basin.

A number of papers on petroleum geology have also been published, including those on the hydrocarbon trap styles by Ng (1987) and Mohd Tahir Ismail and Rudolph (1992). Creaney et al. (1994) and McCaffrey et al. (1998) described the geochemical characteristics of the oil families in the basin. Waples and Mahadir Ramly (1995) discussed the geochemical characteristics of gases in the basin. Comprehensive studies on the petroleum geology of the basin have been carried out by oil companies, and these are presented in major reports by Esso (1985), PETRONAS Carigali (Leslie et al., 1994), and the joint Esso–PETRONAS Integrated Study (EPIC, 1994).

EXPLORATION HISTORY

Before 1976 oil companies operated in Malaysia under various exploration licenses. In 1968 the first concessions in the offshore area east of Peninsular Malaysia were awarded to Esso Production Malaysia Inc. (EPMI) and Conoco. Esso operated in the area north of the 5° N latitude while Conoco was given the area to the south, which includes the Penyu Basin. After PETRONAS was incorporated on 17 August 1974, exploration areas were awarded to oil companies under production-sharing contracts (PSCs) with PETRONAS. Figure 8.2 shows the block classification during the 1985 PSC round and also the current (1999) blocks.

In July 1969 Esso drilled its first well, Tapis-1, in the southern part of the basin. Tapis is a large E–W-trending anticline cut by N–S normal faults. The Tapis-1 well found gas in Group J and K sandstones. Tapis-2, drilled in 1974, discovered oil in the J sandstones. The discovery of Tapis was followed by the drilling of two wells on the Jerneh structure in the central part of the basin. In 1970 Esso drilled the Jerneh, Pilong, Sepat, Bintang, and Belumut structures. Continuous exploration effort resulted in the first significant oil discoveries at Seligi and Bekok in 1971. The first oil production came from the Pulai and Tapis fields in 1978. The Tapis Field came on stream in March 1978. The largest oil field in the basin is Seligi, with over 550 MMSTB estimated ultimate recovery (EUR). Further drilling resulted in the discovery of the Angsi, Besar, Palas, Guntong, Irong, Irong Barat, Semangkok, Tinggi, and Dulang fields. Most of these fields are currently on production.

The introduction of PSCs at the end of 1976 marked the beginning of a highly successful exploration campaign by Esso in the Malay Basin (Ahmad Said, 1982). During the first three years, there were 10 oil discoveries (Palas, Tiong, Guntong, Liang, Irong, Ophir, Irong Barat, Tabu, Serok, and Kepong) and 9 gas discoveries

(Ledang, Lawit, Tujoh, Damar, Inas, Bedong, Noring, Telok/Telok Barat, and Berantai). Four more oil and 3 more gas discoveries were made during the next two years.

In 1970 Conoco spudded its first exploration well, Duyong-1, and discovered gas. In 1973 more oil and gas were discovered in the Sotong and Anding structures in the southwestern part of the basin. The Anding and Sotong structures were appraised in 1974, and are being developed. The Feri-1 wildcat well, drilled during that year, did not find any hydrocarbons. In 1978 Conoco relinquished its acreage and withdrew from exploration in Malaysia. PETRONAS Carigali then assumed operatorship of Conoco's relinquished area. Carigali drilled the Duyong 5H-2.1 and Duyong 6H-32.1 wells in 1980 and continued to drill appraisal and wildcat wells between 1982 and 1984. Duyong was put on stream in 1984. Further oil and gas discoveries were made at Dulang, Malong, Beranang, Resak, and Meranti. Malong is the only discovery on the Tenggol Arch, and was appraised successfully by a second well. The Anding Barat, Keladang, Kempas, Jelutong, and Delah structures were also drilled by Carigali but without success.

In 1985 PETRONAS introduced new PSC terms and subdivided offshore Peninsular Malaysia area into several blocks (Fig. 8.2). The Japex-Taiyo Operating Company (JTOC) acquired the exploration interest in Block PM10 in July 1987, but relinquished it in July 1993 after finding sub-commercial oil and gas accumulations at Jambu and Tembikai. Esso acquired the exploration interests in blocks PM5 and PM8 in 1988, and is the current (1999) operator in these blocks. Drilling resulted in more discoveries including Melor, Larut, East Larut, South Raya, East Raya, Laba Barat, North Larut, Bundi, North Lukut, Abu, Lawang, Abu Kecil, Chermingat, Serudon, Bubu, Lerek, Penara, Peta, Inas, Piatu, Yong, and Pasir. Western Mining Company (WMC) operated in Block PM7 from May 1989 to May 1994, and relinquished the block after making a minor discovery in the Ketumbar structure.

Home Oil of Canada signed a PSC for Block PM2 in August 1988, but Lasmo Oil later bought Home Oil's assets, and made two

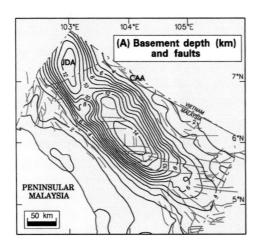

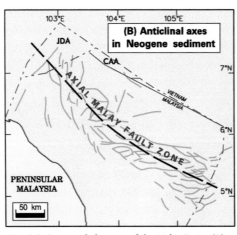

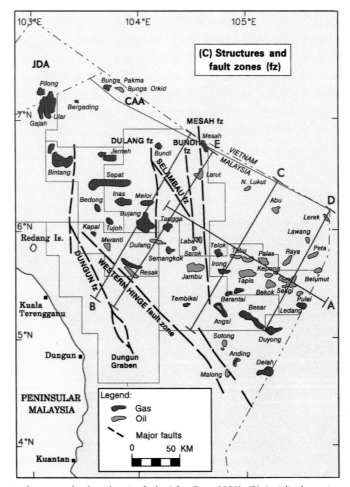

Fig. 8.3. Structural elements of the Malay Basin. (A) pre-Tertiary basement depth and major faults (after Esso, 1985). (B) Anticlinal axes in Neogene sediment fill (after Esso, 1985). (C) Major fault zones identified by Khalid Ngah et al. (1996). Blue lines are the locations of cross-sections shown in Fig. 8.4 (lines A to D) and Fig. 8.34 (line E).

discoveries (Ular and Gajah) near the JDA area. Despite the encouraging results, Lasmo withdrew from the block shortly after drilling Ular-1 in 1991. In 1991 Esso took over the operatorship of Block PM2 from Lasmo. In 1989 a consortium led by International Petroleum Corporation (IPC) Malaysia Ltd (now known as Lundin Malaysia Ltd) won the rights to operate in the adjacent Block PM3. IPC farmed out an interest to Hamilton Oil shortly after acquiring the block. Hamilton Oil discovered the Bunga Raya, Bunga Orkid and Bunga Pakma oil and gas fields. Shortly after these discoveries, Hamilton/BHP pulled out and IPC took over operatorship. The discovered fields fall under a designated Commercial Arrangement Area between Malaysia and Vietnam. Bunga Kekwa, discovered in 1994 by IPC, is currently on production. Recently, there have been many gas discoveries north of the Gajah and Ular structures in the Malaysia-Thai Joint Development Area (JDA, Fig. 8.2), particularly in Block A-18 which is being operated by Carigali-Triton Operating Company (CTOC).

By the end of 1997, some 330 exploration wells were drilled and 410,000 line-km of 2D and 3D seismic were acquired in the Malay Basin. The exploration effort to date has resulted in the discovery of some 50 oil and 30 gas accumulations (Fig. 8.1B). In 1996 PETRONAS re-blocked the relinquished areas and signed another round of PSCs with multinational oil companies to explore for hydrocarbons in the basin. Figure 8.2 shows the current (1999) operators.

TECTONIC FRAMEWORK

Structure

The Malay Basin is located at the centre of Sundaland, the cratonic core of Southeast Asia. It is one of the deepest continental extensional basins in the region, and is believed to have formed during early Tertiary times. The Tenggol Arch separates the Malay Basin from the Penyu Basin, while the Narathiwat High separates the Malay Basin from Thailand's Pattani Basin. The Malay Basin is an elongate NW-SE trending basin, about 500 km long and 250 km wide (Fig. 8.3A), underlain by a pre-Tertiary basement of metamorphic, igneous, and sedimentary rocks. These are thought to be the offshore continuation of the geology of eastern Peninsular Malaysia. Some of the basement lithologies encountered are described in Fig. 5.6 of Chapter 5. The basin is bounded by relatively shallow (<1.5 km) basement areas: the Terengganu Platform and Tenggol Arch to the southwest, the Narathiwat High to the northwest, and the Con Son Swell to the northeast. The basement represents the late Mesozoic continental landmass that existed before the basins were formed.

The Malay Basin is asymmetrical along its length and in cross section. Its southwestern flank is slightly steeper than its northeastern flank (Fig. 8.3A). Basement faults in the southeastern and central parts of the basin mostly trend E-W, oblique to the overall basin trend. These basement faults appear to have influenced the geometry of compressional anticlines in the Neogene sedimentary fill as a result of dextral motion along the Axial Malay Fault Zone (AMFZ, Fig. 8.3B). The southwestern margin is marked by the Western Hinge Fault (WHF, Fig. 8.3C), which is a zone of *en echelon* normal faults and associated fault-bounded, pull-apart basins. The latter have been interpreted as resulting from right-lateral wrench motion along the WHF (Tjia, 1994b). To the south of the WHF the Tenggol Fault marks the northeastern edge of the Tenggol Arch. This major fault zone was active during the Late Oligocene–Early Miocene, and is believed to have acted as the shelf edge during an Early Miocene transgressive episode (Mazlan Madon, 1992). The Dungun Fault is a splay of the WHF that cuts across the Terengganu Platform on the southwestern flank of the Malay Basin (Fig. 8.3C). A number of small pull-apart grabens occur along the Dungun Fault, the largest being the Dungun Graben which is about 35 km long and less than 10 km across. Liew (1994) interpreted the Dungun Graben as a pull-apart graben formed by right-lateral wrench movement of the Dungun Fault.

The Malay Basin is a complex rift composed of numerous extensional grabens. Most of these grabens have not been penetrated because of their great depths, but were interpreted from magnetic, gravity, and seismic data. Some smaller grabens have been identified from seismic mapping on the west-central margin, including the Dungun Graben, described above, and the Tok Bidan Graben (Liew, 1994). The pre-Tertiary basement shallows to the southeast as a result of late Middle Miocene tectonic deformation and uplift (Fig. 8.4A). The deformation also resulted in numerous compressional anticlines (cf. Fig. 8.3C) that are usually bounded by reactivated normal faults on their southern sides. The intensity of deformation increases southeastwards (compare Lines B, C and D, in Fig. 8.4).

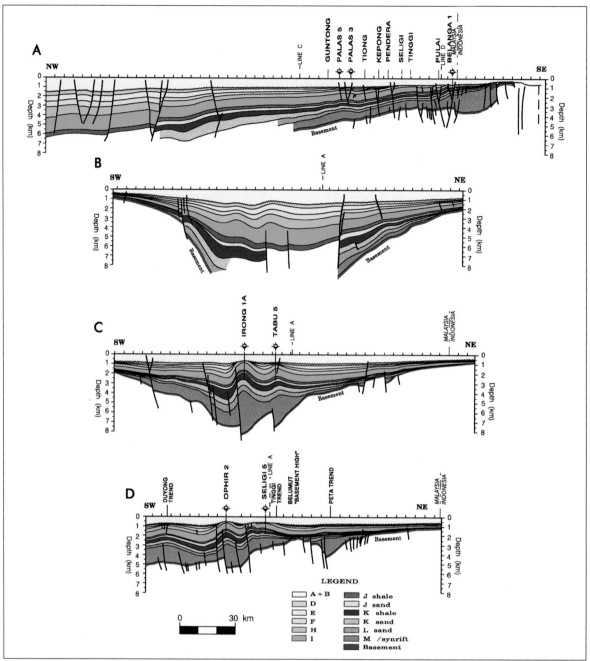

Fig. 8.4. Cross sections of the Malay Basin. (A) NW-SE section. (B, C, D) NE-SW sections. Modified from Esso (1985). For locations, see Fig. 8.3. All sections have the same scale.

Tectonic Origin

Various tectonic models have been proposed for the origin of the Malay Basin. Some authors have interpreted the basin as a back-arc basin (e.g. Kingston et al., 1983; Mohd Tahir Ismail et al., 1994), while others have proposed that the basin originated as a pull-apart along a major strike-slip fault zone that was reactivated by the mid-late Eocene collision between India and Asia (Tapponnier et al., 1982). Back-arc extension is probably not the main cause of extension, since the basin is located more than 1000 km away from the Sumatra Trench. White and Wing (1978) postulated that the Malay Basin may have developed by regional thinning of the continental crust. Crustal thinning is manifested by high surface heat flows (85-100 mWm^{-2}), as interpreted from well data (Wan Ismail Wan Yusoff, 1984, 1990), and the low-amplitude, negative free-air gravity anomalies which indicate thinning by a factor of at least 2 (Mazlan Madon and Watts, 1998) (Fig. 8.5). Other tectonic models include crustal extension above a hot spot (Hutchison, 1989, p. 180-183; Tjia, 1994a; Khalid Ngah et al., 1996) and distributed shear deformation of a pre-existing basement fault zone (Mazlan Madon, 1997a).

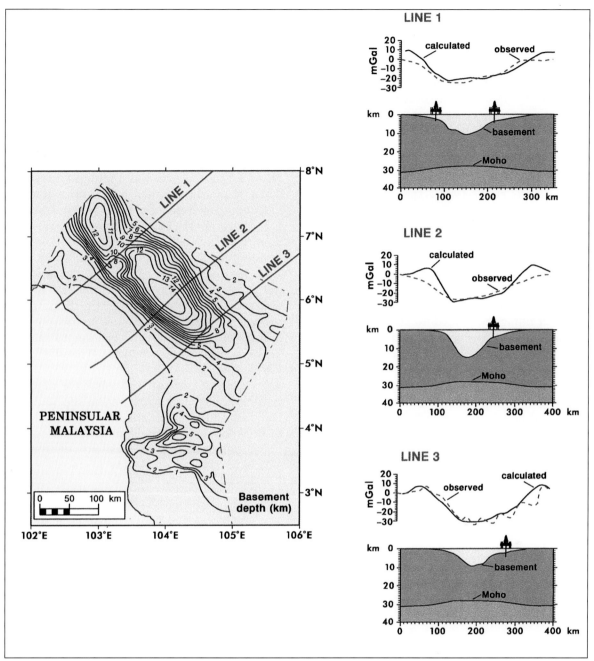

Fig. 8.5. Gravity modelling across the Malay Basin. The results indicate that crustal thinning beneath the basin is responsible for the negative free-air anomalies of -20 to -30 mGal (after Mazlan Madon and Watts, 1998).

The cause of the crustal thinning is still uncertain. A widely used model is Tapponnier et al.'s (1982) extrusion hypothesis, whereby the India-Asia collision has caused the re-activation of major strike-slip fault zones and the formation of extensional basins (see Fig. 4.4, Chapter 4). This model is able to explain the key observations: (1) the almost synchronous development of extensional basins in Sundaland at the time of the collision, and (2) crustal extension occurring within a relatively narrow deformation zone emanating from the India-Asia collision zone. Consequently, the extrusion model has been used as a working hypothesis for the Malay, West Natuna, and Thai basins (Daines, 1985; Polachan and Sattayarak, 1989). The Malay Basin appears to be at the southeastern end of a major strike-slip fault zone, the Three Pagodas Fault, which continues into Thailand. Although there is no seismic evidence of a through-going strike-slip fault, structural evidence seems to suggest that the basin has developed by transtension of a NW-trending shear zone, which Tjia (1994a) referred to as the Axial Malay Fault Zone (Fig. 8.3B). The Malay Basin has also been compared with the structurally similar Yinggehai Basin, offshore southern China, which was formed at the distal end of the Red River Fault Zone (Zhang and Zhang, 1991).

Cross sections of the Malay Basin (e.g.

Fig. 8.4B) show the "steer's head" geometry of a typical rift-sag couplet (cf. Dewey, 1982). This characteristic geometry is widely attributed to a mechanism of lithospheric stretching (McKenzie, 1978); initial rifting of the brittle upper crust results in fault-controlled subsidence, whereas the later sag phase represents thermal subsidence as the thermal anomaly associated with lithosphere stretching decays. The progressive thinning and onlap of the sedimentary units onto the basin margins are also characteristic of rift basins (Watts et al., 1982). The gentle inward-tilting of the basin flanks, however, probably indicates a combination of thermal and flexural subsidence of the basin flanks. Mazlan Madon (1997b) analysed the subsidence and thermal histories of the basin, and showed that they are consistent with a lithospheric stretching model. The observed heat flow may be explained using a model of rifting that started 40 Ma ago for about 10 Ma (Fig. 8.6). The subsidence history derived from backstripped well data shows a typical pattern of extensional basins — rapid initial subsidence followed by slower thermally-induced subsidence due to cooling of the lithosphere.

Although the Malay Basin superficially resembles many rift basins, it does not appear to have been formed by orthogonal NE-SW extension, for most of the basement faults along the basin axis strike roughly east-west, approximately 35° oblique to its NW-SE trend (Fig. 8.3A). These faults are actually the border faults of major E-trending extensional half-grabens. Khalid Ngah et al. (1996) interpreted the rectilinear basement fault pattern as inherited from pre-existing basement inhomogeneities. The kinematic development of the basin was discussed by Mazlan Madon (1997a), and is shown schematically in Fig. 8.7. The basin is believed to have formed as a result of a regional dextral mega-shear caused by the indentation of India into Asia during the early Tertiary (see Fig. 4.5, Chapter 4). The regional dextral shear resulted in sinistral shear along pre-existing NW-trending fault zones, including the AMF. This local shearing caused the E-trending faults within the fault zone to rotate, which led to the formation of the E-trending half-grabens. These half-grabens were then subjected to transpression during the Late-Middle Miocene due to changes in the regional stress field (cf. Tjia and Liew, 1996).

Other tectonic models have invoked crustal extension without strike-slip faulting. Hutchison (1989, p. 180-183) postulated that thermal doming of central Sundaland during Late Cretaceous times may have caused crustal uplift and extension in the Malay Basin. This is a

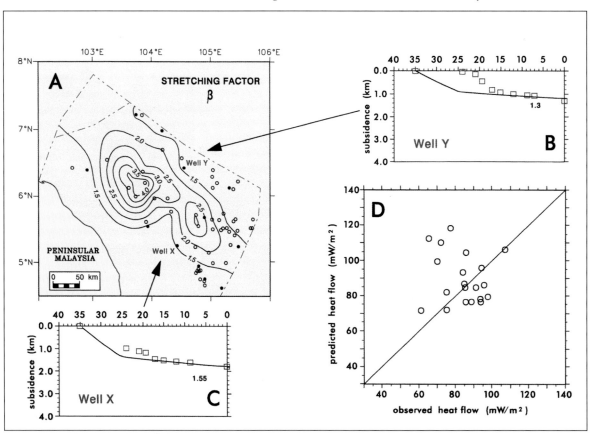

Fig. 8.6. Map of stretching factor (β) for the Malay Basin (A) based on backstripped subsidence curves, examples of which are shown for wells X and Y (B, C). Note that the subsidence histories are typical of extensional (rift) basins. Comparison between predicted and observed surface heat flows are shown in (D). (After Mazlan Madon and Watts, 1998).

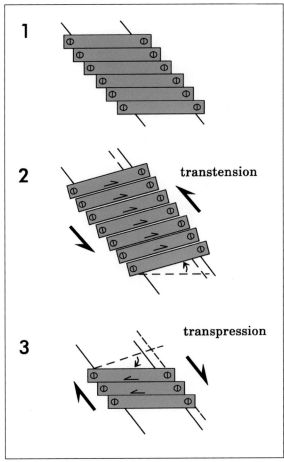

Fig. 8.7. Shear model for the kinematic evolution of the Malay Basin (Mazlan Madon, 1997a), inspired by the block rotation model of McKenzie and Jackson (1986).

possibility because, although many basins began to form during the early Tertiary, the region may have undergone post-orogenic extensional collapse since Jurassic-Cretaceous times as a consequence of the Late Triassic Indosinian Orogeny (Mazlan Madon, 1995). Jurassic-Cretaceous intermontaine extensional/strike-slip basins in Peninsular Malaysia and Thailand provide evidence for this phase of post-orogenic extension. Alternatively, Tjia (1994a) and Khalid Ngah et al. (1996) proposed a model in which active extension due to thermal up-doming is superimposed upon strike-slip faulting. In their hypothesis the authors proposed that the Malay, Penyu, and West Natuna basins represent the three arms of a failed rift that formed above a Late Cretaceous thermal dome (see also Chapter 7 for details).

Basin Inversion

The Malay Basin underwent inversion during the Middle–Late Miocene. The structural evidence for basin inversion was described by Tjia (1994b). He identified compressional anticlines, inverted and uplifted half-grabens, and reverse throws on half-graben normal faults. The intensity of inversion is generally greater at the centre than on the flanks of the basin. Inversion also increases in intensity towards the southeast. Hence, in the West Natuna Basin inversion structures are more commonly associated with thrust and reverse faults (Ginger et al., 1993).

The lower part of Fig. 8.7 shows a mechanism by which the half-grabens may have been inverted. If the originally sinistral shear is reversed, so that the crustal blocks bounded by the east-trending basement faults rotate clockwise, sinistral oblique slip will occur along their bounding faults. This would induce a N-S shortening across the half-grabens, resulting in reverse dip-slip re-activation of the faults. In the Malay Basin, dextral shearing has resulted in regional basement uplift in its southeastern part, and more intense inversion in the West Natuna Basin. The Natuna basement ridge may have acted as a buttress that resisted the dextral motion of the AMFZ, resulting in compression and inversion in the West Natuna Basin (Mazlan Madon, 1997a).

Basin inversion resulted in the development of large wrench-induced compressional anticlines, mostly within the axial region of the basin. Many of the anticlines are oriented roughly E-W, and are *en echelon*, parallel to the basement normal faults that bound major synrift half-grabens (Figs. 8.3A, B). The location and geometry of the inversion anticlines appear to have been strongly controlled by the basement faults, which were produced by sinistral slip of the AMFZ during basin inception (Fig. 8.7). The inversion anticlines then formed over the half-grabens by a NW-trending dextral shear during the basin inversion phase (Tjia, 1994b; Mazlan Madon, 1997a). In the southeastern part of the basin, the structures are typified by tightly folded postrift strata that once filled the half-grabens. The bounding faults to the half-grabens have been reactivated as reverse and/or strike-slip faults. They have complex upward-diverging geometries, typical of positive flower structures associated with wrench faulting. The flower structures may have formed by sinistral oblique-slip re-activation of their E-trending half-graben border faults, as a result of the reversal in the shear direction.

The timing of growth of the inversion anticlines, which is critical for hydrocarbon entrapment, were determined during the EPIC (1994) study from the changes in stratigraphic thickness across the inversion structures. Because the inversion structures are growth features formed during deposition, the thicknesses of the

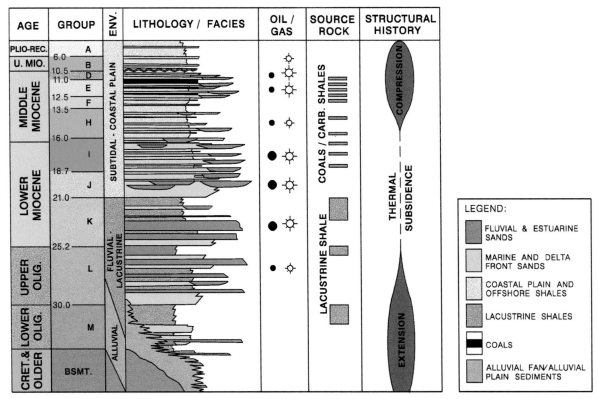

Fig. 8.8. Generalised stratigraphy, hydrocarbon occurrences, and structural history of the Malay Basin (EPIC, 1994).

syn-inversion stratigraphic units decrease towards the crestal region as the depositional surface is being deformed and eroded. The EPIC study showed that, although the timing of structural growth is generally synchronous across the whole basin, the peak of fold growth is earlier in the south than in the north. The onset of compression coincides with the top of Group H (Fig. 8.8), whereas the time of peak deformation varies from near the top of Group E in the south to the present day at Jerneh. However, although basin inversion caused wholesale uplift and formation of flower structures, on a basin scale the subsidence accelerated.

STRATIGRAPHY AND PALAEOENVIRONMENTS

The Malay Basin strata are subdivided informally into seismostratigraphic units (Fig. 8.8). Chapter 6 discusses the stratigraphy in some detail. Each unit, known as a Group, is bounded by basin-wide seismic reflectors. Some of these reflectors represent major sequence boundaries and are erosional unconformities on the basin flanks. The groups are designated alphabetically, in order of increasing age, from A down to M. This nomenclature, which was introduced by Esso in the late 1960s, is generally preferred over those established by other operators.

The stratigraphic development of the Malay Basin is directly related to its structural evolution, which occurred in 3 phases: 1. a pre-Miocene (Oligocene or possibly earlier) extensional or synrift phase, 2. an Early to Middle Miocene thermal/tectonic subsidence phase, which was accompanied by basin inversion, and 3. a Late Miocene-Quaternary subsidence phase, which represents a tectonically quiescent period. Note that, although the age of the oldest sediment (Group M) is Oligocene, the presence of older sediments cannot be ruled out.

The pre-Miocene phase represents the extensional phase of basin development, during which subsidence was controlled by faulting. Initially, sedimentation in isolated half-graben depocentres (Fig. 8.9) deposited thick synrift successions of alternating sand-dominated and shale-dominated, fluviolacustrine sequences. Groups M to K, which fill the extensional sub-basins, comprise the deposits of braided streams, coastal plains, lacustrine deltas, and lakes. These deposits show increasing lacustrine influence towards the basin centre (Mohd Tahir Ismail et al., 1994).

Extensional faulting ceased during the Late Oligocene. Continued thermal subsidence, however, resulted in the deposition of Groups L to D. The basin was probably at or near sea level by Early Miocene times, as indicated by the abundance of coal-bearing strata in the

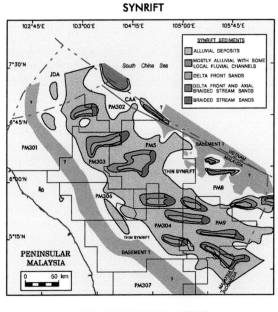
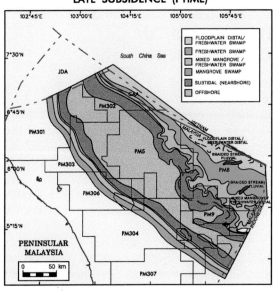
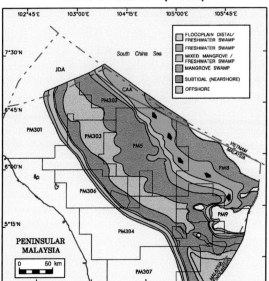
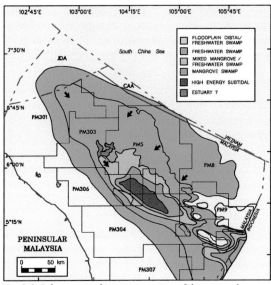

Fig. 8.9. Palaeogeographic reconstruction of depositional environments for the Malay Basin. Based on EPIC (1994).

succession. The first signs of marine inundation were recognised from micropalaeontology within the Lower Miocene strata (Azmi Mohd Yakzan et al., 1994; Mohd Tahir Ismail et al., 1994). A cyclical succession of offshore marine, tidal-estuarine, coastal plain, and fluvial sediments was deposited in the Lower to Middle Miocene. Groups I and J consist of progradational to aggradational fluvial to tidally-dominated estuarine sands. Groups H and F are dominantly marine to deltaic sediments with fluvial/estuarine channels, deposited during an overall sea-level rise. Groups E and D were deposited by the progradational stacking of dominantly fluvial/estuarine channels, and culminated with a localised erosional unconformity.

The Early-Middle Miocene period of thermal/tectonic subsidence was accompanied by compressional deformation, which resulted in local inversion of half-grabens by re-activation of their bounding faults, and a major uplift in the southeastern part of the basin. Seismic data show a prominent unconformity (the Middle-Upper Miocene regional unconformity) that truncates folded and uplifted strata as old as Group H in the southern part of the basin. The unconformity is overlain by undeformed marine sediment of Groups A and B. Deformation was contemporaneous with sedimentation, such that erosion and non-deposition on the crests of the structures occurred simultaneously with deposition on the flanks. It is estimated that up to 1200 m of sediment may have been eroded off the crests of some structures (Murphy, 1989). Inversion is more severe in the southeastern part of the basin, while sedimentation of Groups D, E, and F in the central and northern parts of the basin was relatively continuous. Sediments in the north may have been derived partly from erosion of pre-existing sediment in the south.

The Late Miocene to Quaternary was a phase of gentle subsidence without significant tectonic activity. Fully open-marine conditions were established only during this phase. The resulting strata, Groups A and B, above the Middle-Upper Miocene unconformity, consist of predominantly marine clays and silts deposited during an overall marine transgression in nearshore to shallow marine environments.

Figure 8.9 is a schematic reconstruction of the palaeogeographic development of the Malay Basin, based on the EPIC (1994) regional study. Only the palaeoenvironments for selected (key) seismic groups are shown, i.e. Groups L and M (representing synrift deposition), K, J, H, E, and A/B. The reconstruction shows a progression from mainly nonmarine (alluvial to coastal plain) environments during the Oligocene (Groups L and M), which is the synrift phase of basin development, to increasingly marine depositional environments (coastal fluviomarine to inner neritic) during the Miocene and later (Groups K to A/B). The palaeogeographic maps (Fig. 8.9) show that the basin was connected to the ancestral South China Sea via the West Natuna Basin (Nik Ramli, 1986; Mazlan Madon, 1994). The Malay Basin was effectively a narrow seaway or gulf that received sediment from its northeastern and southwestern flanks. Nik Ramli (1988a) described fan-delta complexes that might have developed along the coasts of the seaway during Late Oligocene times to deposit the thick braided-stream deposits in Group K. Similarly, subtidal sandbar deposits in Group J appear to be aligned along the basin axis, which indicates the influence of tidal currents in the deposition of these sandstones (Yap, 1996).

HYDROCARBON OCCURRENCES

The major oil and gas occurrences, including producing fields, in the Malay Basin are shown in Fig. 8.1B. A total of 13 oil fields are presently (1999) in production. The southern part of the basin contains most of the oil reserves, including several giant fields such as Seligi and Tapis (both have EURs of almost 600 MMSTB). The producing non-associated gas fields are Jerneh (EUR of ~3900 BSCF) and Duyong (930 BSCF). The hydrocarbons occur in reservoirs from Group L to Group D. Groups E, I, J, and K are the most prolific (Fig. 8.8). There is a broad geographical subdivision of the basin into a northern gas-prone province and a southern oil-prone province, although there are exceptions to this generalisation. For example, a major gas trend occurs in the southwestern margin of the basin (Duyong–Besar–Angsi), while an oil trend occurs on the northeastern flank (Bunga Orkid–Bunga Kekwa–Bunga Raya) (Fig. 8.1B).

On a regional scale, the geographic and stratigraphic distribution of oil and gas appears to be controlled primarily by basin morphology. The basin is considerably deeper in the north, and has a much thicker, hence more deeply buried, sedimentary succession. Having been buried to greater depths, the sediments would have reached gas-window maturity or greater. Hence, there is a great tendency for oil to be flushed out by high-maturity gas. In contrast, the Middle-Late Miocene tectonic uplift in the south has resulted in shallower burial and lower thermal maturities. The main source intervals are mostly in the oil

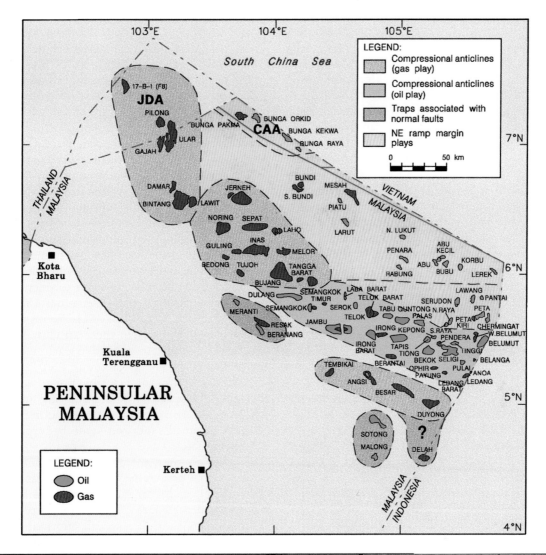

Fig. 8.10. Top: Hydrocarbon occurrences and distribution of trap styles/play types in the Malay Basin. Bottom: Close-up of the southern part of the basin, showing the major oil and gas trends.

window and, therefore, there is less possibility of oil being flushed by gas from below the oil window.

On a semi-regional scale, the main factors controlling oil and gas distribution include source-rock quality and maturity, and the relative timing of generation and structuration. The effective source rocks are coaly in the north (mainly in Group I, see below) and lacustrine in the south (Groups K, L, and M). Mohd Tahir Ismail (1995) suggested that the relative timing between structuration and hydrocarbon migration varies from south to north. Structures were formed earlier in the south, and thus were able to trap oil, whereas late structuration in the north resulted in more gas being trapped.

HYDROCARBON PLAYS AND TRAP STYLES

The hydrocarbon occurrences in the Malay Basin may be categorised according to the structural style of the traps. Previous authors have described the main structural trap styles in the basin (Ng, 1987; Mohd Tahir Ismail and Rudolph, 1992), which are mostly compressional anticlines and fault-dip closures. In the following section, trap styles and hydrocarbon plays are categorised based on structural features, geographic and stratigraphic distribution, and source-reservoir relationships.

Compressional Anticlines

By far the most prolific trap style is the compressional anticline (Fig. 8.10). This comprises E-W trending anticlines formed by inverted half-grabens. The compressional anticlines are located mainly along the central/axial part of the basin. They are formed during the basin inversion phase in the Middle Miocene (at the beginning of Group F times, Fig. 8.8). Structural maps by Esso (1985) indicate that most of the anticlines are the result of wrench movement associated with transpressional deformation of the underlying fault-bounded half-grabens. This is clearly observed in the southern part of the basin, where the anticlines are bounded to the south by major basement normal faults (Fig. 8.4). In the central part of the basin, basement faults are more difficult to identify because of poor seismic resolution.

Many major discoveries in the Malay Basin are of this play type. Examples include Tabu, Irong, Jambu, Liang, Guntong, Tapis, Palas, Bekok, Tiong, Tinggi, Seligi, and Pulai. Many of these are now producing. Based on the hydrocarbon distribution map, several major oil and gas trends can be recognised (Fig. 8.10). A trend usually comprises a series of anticlinal structures associated with a major E-trending deep-seated basement fault. Such faults are observed on the southern side of the Liang–Belumut and Dulang–Palas oil trends.

The hydrocarbon distribution map also indicates that compressional anticlines in the south are more oil-prone, while those in the northern part of the basin are gas prone. In the south, most of the anticlines are either domal or asymmetrical, and are often compartmentalised by normal faults. The main reservoirs are shallow marine and fluvial sandstones of Groups H, I, J and K. Hydrocarbons in these structures were sourced from the interbedded carbonaceous shales and coals, mainly in Group I, or from the lacustrine shales of Groups K and L. The structures are sealed by intra-group claystone and shale beds. In some faulted anticlines, the juxtaposition of sand and shale beds provide lateral seal across faults.

In contrast, the compressional anticlines in the central part of the basin involve reservoirs in Groups D and E sands. Most traps are formed by 4-way dip closures in domal structures or asymmetrical anticlines, and normal fault-bounded structures (Fig. 8.11). The reservoirs are generally formed by shallow marine sandstones of Groups D and E. Group F is usually overpressured and, thus, its reservoir potential has not been fully tested. Interbedded claystone and shale units within Groups D and E, as well as in the upper Group D, provide the top seals for the reservoirs. Hydrocarbons are thought to be charged from deeply buried source rocks, mainly in Group I, migrating vertically through fault conduits. Examples of this type of play include Bintang, Lawit, Jerneh, Noring, Sepat, Inas, Bujang, Dulang, Tangga, Melor, Ular, and Bergading.

In addition to the main oil and gas trends in the central/axial Malay Basin, there is a major gas trend in the southwestern part of the basin, close to the Tenggol Fault. This is the Angsi–Duyong trend, which includes the major gas discoveries Angsi and Besar (Fig. 8.10). These large compressional anticlines are structurally similar to those in the main oil province to the north, and are underlain by synrift half-grabens controlled by normal faults. Only Duyong is currently producing.

Traps Associated with Normal Faults

In the southwestern margin of the Malay Basin, traps are formed by closures associated with normal faults, such as fault/dip closed structures and fault-bend folds. Examples are Meranti, Resak, Beranang discoveries along the Western Hinge Fault Zone (Fig. 8.11). The Tembikai and Sotong discoveries occur in similar structures associated with the Tenggol Fault system. The main sandstone reservoirs are from Groups H, I, J, and K. The top seals are provided by the interbedded claystone and shale within Groups H, I, and J, and also by Group K shale. Lateral fault seals are formed either by sand-shale juxtaposition or by shale smear along fault planes. Hydrocarbons were probably derived mainly from the interbedded carbonaceous shale and coal, and also from the lacustrine Group K shale.

Tenggol Arch Plays

The Tenggol Arch is a relatively shallow and flat, NW-trending pre-Tertiary basement feature that separates the Malay and Penyu basins. Its northeastern boundary is marked by the Tenggol Fault, a major normal fault zone with a maximum throw of about 2500 m (Ng, 1987). This basement feature has remained a positive area until the late Early Miocene, and was a source of sediment for the half-grabens in the Malay Basin to the north. The arch is relatively featureless except for isolated basement mounds, probably relics of the pre-Tertiary erosion surface. Some of these basement topographic anomalies result in structural closures when Tertiary sediments are draped over them upon compaction. These basement drapes form a unique trap style (Fig. 8.12).

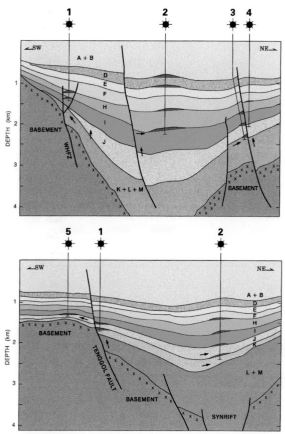

Fig. 8.11. (A) Schematic cross sections of the Malay Basin showing the different trapping styles. 1. Traps associated with normal faults. 2. Compressional anticlines, including deep reservoir play in the lower figure. 3 & 4. NE ramp margin play (normal fault associated traps). 5. Tenggol Arch play (basement-drape).

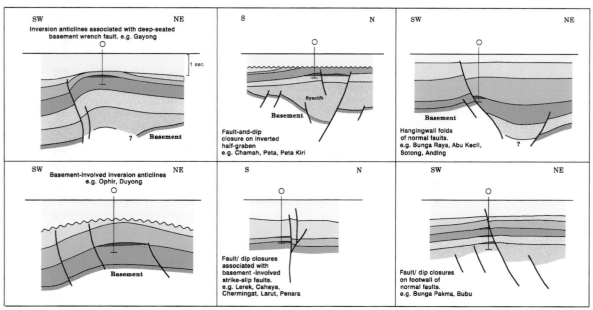

Fig. 8.11. (B) Some examples of trap styles based on actual examples.

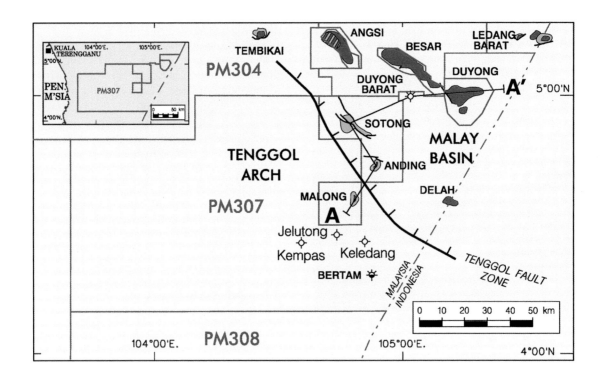

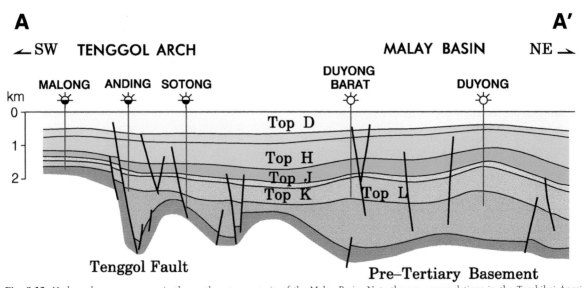

Fig. 8.12. Hydrocarbon occurrences in the southwestern margin of the Malay Basin. Note the gas accumulations in the Tembikai-Angsi-Besar-Duyong trend and the more oil-prone Tenggol Arch play (Anding, Sotong, Malong, and Bertam). Shown is a schematic cross-section from Malong to Duyong. Line of cross-section is 115 km

Oil has been discovered in one of these structures, at Malong, in 1983 (Fig. 8.12), where it occurs in the Group J shallow marine sandstones. Noor Azim Ibrahim and Mazlan Madon (1990) described the sedimentology of the Malong reservoirs. Hydrocarbons at Malong are probably charged from the Group K or older lacustrine shales in the basin deep across the Tenggol Fault, and have migrated up-dip onto the Tenggol Arch. Interbedded shales provide the top seal for the drape structures.

Because the sedimentary succession on the Tenggol Arch is generally less than 1500 m thick, any potential source rocks are thought to be immature. The Malong oil, therefore, may have come from the deeper half-grabens, situated to the northeast of the Tenggol Fault. The oils at Anding, Sotong, and Feri were probably sourced from these half-grabens along the Tenggol Fault. The migration distance to the Malong structure is, at most, about 10 km. Nevertheless, the Malong discovery has spurred interest in other basement-drape features on the arch. The assumption then was that oil from the basinal grabens had migrated over long distances (30–60 km) from the Malay Basin to fill up these structures. Some of these structures were tested: Kempas 1984, Jelutong 1987, and Keledang 1991, and found to be dry. This suggests that the long-distance migration model may need to be revised.

NE Ramp Margin Plays

During the 70s and 80s exploration efforts were directed towards finding the more obvious structural traps, i.e. the compressional anticlines. Between 1968 and 1985 about 70 new field wildcats were drilled, resulting in 15 oil and 18 gas discoveries. With much improved exploration technology and a better geological understanding of the basin, the emphasis has gradually shifted towards the higher risk stratigraphic and combination traps, particularly on the northeastern flank (ramp margin) of the basin. After the introduction of the 1985 PSC, 19 oil and 17 gas discoveries were reported from the northeastern flank of the basin. Chua and Wong (1997) summarised the different trap types that have been explored in this part of the basin. Figure 8.11 shows some examples of the major trap styles on the NE ramp margin. These include subtle stratigraphic/fault traps in Groups I, J, and K, such as Larut and Bunga Raya.

Faulted anticlinal traps in Groups I, J, and K reservoirs occur in the northeastern and eastern parts of the basin. Closure occurs on either side of a normal fault, but is usually associated with fault-bend folds developed in the hangingwall of the fault (Fig. 8.11). Lateral seal is provided by sand-shale juxtaposition, while top seal is provided by the interbedded shale. Hydrocarbons are sourced either from *in situ* interbedded source beds or from beds down-dip (Groups K and L shales) via long-range migration. Examples of this trap style are the Bunga Orkid, Bunga Kekwa, and Bunga Raya oilfields in the PM3 CAA area (Fig. 8.10), and the Lerek, Pantai, Lumut, East Belumut, Larut, and Abu oilfields in Blocks PM5 and PM8, to the south.

In addition to the more conventional trap styles, there are also examples of traps formed by basement drapes, much like those observed on the Tenggol Arch. The South Raya Field is such a trap, where oil and gas have been found in Groups I, J, and K reservoirs. Stratigraphic channel plays have also become drilling targets in recent years. The Bindu discovery, for example, was identified from seismic amplitude anomalies. Some 330 BSCF of gas was found in mid-Group I sandstone. Bunga Seroja is an example of a Group H channel play. There is also potential for stratigraphic pinch-out and onlap traps on the northeastern ramp margin.

Deep Reservoir Play

This play type involves reservoirs within or below the overpressured zone, thought to be present beneath existing discoveries/fields. The Bergading structure is an example of a normal fault bounded N-trending anticline, located in the northern part of the Malay Basin (near the JDA, Fig. 8.10). The Bergading-1 well was drilled by Esso in 1979 and found gas in Groups B, D, and E. In 1995, PETRONAS tested the potential of deeper reservoirs (Groups H and I). The Bergading Deep-1 well was drilled to about 3100 m, and successfully penetrated the overpressured zone in Group F. Substantial amounts of gas and condensate were found in Groups H and I reservoirs. The success of Bergading Deep-1 has opened a new play for future exploration in the basin.

RESERVOIR ROCKS

Depositional Environments and Reservoir Architecture

Hydrocarbons in the Malay Basin are found in sandstone reservoirs of Groups D down to K (Fig. 8.8). The depositional environment of the sandstones vary with the stratigraphy. In the older groups (K, L, M) reservoirs are formed, mainly fluvial channels in a nonmarine-lacustrine setting. In the J and younger groups, the sandstones are predominantly shoreface and subtidal shelf sands (particularly in the J group), and fluvial-deltaic to estuarine channel complexes (I group and younger). Some detailed sedimentological studies of reservoir sandstones in the basin have been published. Among them are the studies of the Upper Oligocene–Lower Miocene reservoir sandstones in Groups J and K in the southeastern part of the basin (Nik Ramli, 1986, 1987, 1988a, 1988b; Noor Azim Ibrahim and Mazlan Madon, 1990; Khalid Ngah, 1990a), and the E sandstones in the Jerneh gas field (Mazlan Madon, 1994). The sandstones of Group K are mainly deltaic and braided stream deposits whereas those in Group J are mainly shallow marine sediments. Upper Miocene reservoir sandstones of Groups D and E have been interpreted as tidal, deltaic to

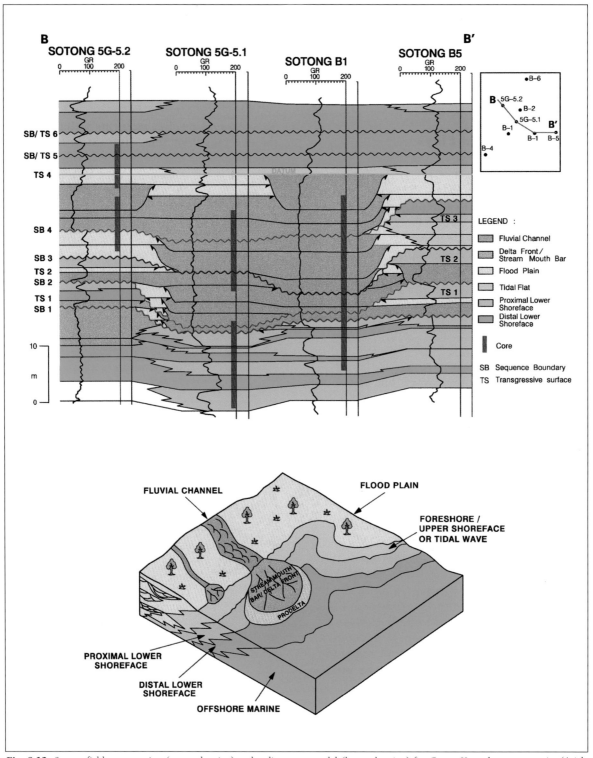

Fig. 8.13. Sotong field cross section (upper drawing) and sedimentary model (lower drawing) for Group K sandstone reservoirs (Asiah Mohd Salih and Mohd Fauzi Abdul Kadir, 1995). See Fig. 8.12 for location.

lower coastal plain deposits (Khandwala et al., 1984; Thambydurai et al., 1988; Mazlan Madon, 1994).

The K reservoirs consist of braided stream sands deposited in alluvial fans and fan delta complexes (Nik Ramli, 1988a). Oligocene sediments were derived from the flanks of the basin via rivers flowing across the steeply faulted basin margins. Detailed reservoir architectural studies have been carried out by Asiah Mohd Salih and Mohd Fauzi Abdul Kadir (1995) on the K2 sandstone in the Sotong discovery (Fig. 8.13). The reservoirs consist mainly of stacked fluvial channel complexes in a lowstand systems tract. The channel complexes are 5-15 m thick, with individual channels measuring about 5-6 m thick. Non-reservoir facies are interpreted as flood plain, tidal flat, and lower shoreface deposits. The transition from lacustrine (K shale) to marine (J sandstone) is a significant event in the depositional history in the basin. This first marine incursion was recorded in Group J (18.7–21 Ma) from palynological analysis (Azmi Yakzan et al., 1994). This Lower Miocene interval consists of coastal plain-estuarine to tidally-dominated shallow marine sandstones deposited in a lowstand systems tract (Ramlee and Beddingfield, 1996).

Group J contains about 40% of the total reserves in the basin. The majority of the reservoirs occur in the middle Group J (Ramlee and Bedingfield, 1996) estuarine and high-energy subtidal deposits. The lower and upper J sequences have lower reservoir quality, having been deposited in low to moderate energy subtidal environments. Group J estuarine channel and subtidal sands, with their long axis parallel to the tidal current direction, form the best quality reservoirs. Up-dip of the estuaries, where the depositional energy is lower, poor quality reservoir sands were developed.

In the PM9 block the J19/21 interval contains reservoirs ranging from estuarine, proximal tidal bar sandstones to distal, subtidal bar sandstones. The sediment supply was predominantly from the eastern and northeastern margins of the basin. The lower Group J facies were described by Yap (1996) as subtidal sand bars aligned WNW-ESE, parallel to the palaeo-shoreline. Deposition occurred during a sea level fall that narrowed the basin and enhanced the tidal effect. Goh et al. (1982, 1983) described the lower J sandstone reservoirs in the Pulai field. These consist of coarsening-upward sequences of bioturbated sandstones overlain by cleaner, coarser-grained, and better sorted glauconitic sandstones. The sediments represent shoreface and bar sandstone facies that grade laterally into bay, lagoon and tidal flat mudstone facies. Previous workers, such as Nik Ramli (1986) and Noor Azim Ibrahim and Mazlan Madon (1990), interpreted the J sandstones as shoreface and shelf sediments deposited in a wave and storm-influenced environment, with minor tidal influence. Figure 8.14 shows some core logs from the Malong area.

Group I reservoirs host many large accumulations in the southern Malay Basin, e.g. at Tabu, Guntong, Palas, Kepong, and Seligi (Fig. 8.10). The results of an EPR-PRSS joint study on reservoir architecture and 3D geological modelling of the Guntong Field were compiled by Campion (1997). In this field, the hydrocarbon reservoirs are mainly in Groups I and J. The J reservoir sandstones are mainly subtidal deposits

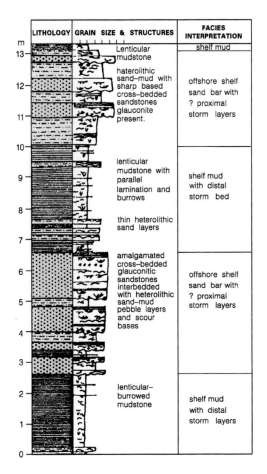

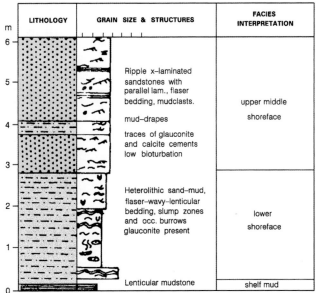

Fig. 8.14. Sedimentary logs from the Malong discovery, Tenggol Arch (see Fig 8.12 for location), showing (left) two shallow marine parasequences representing offshore sandbars and (right) a shoreface parasequence. After Noor Azim Ibrahim and Mazlan Madon (1990).

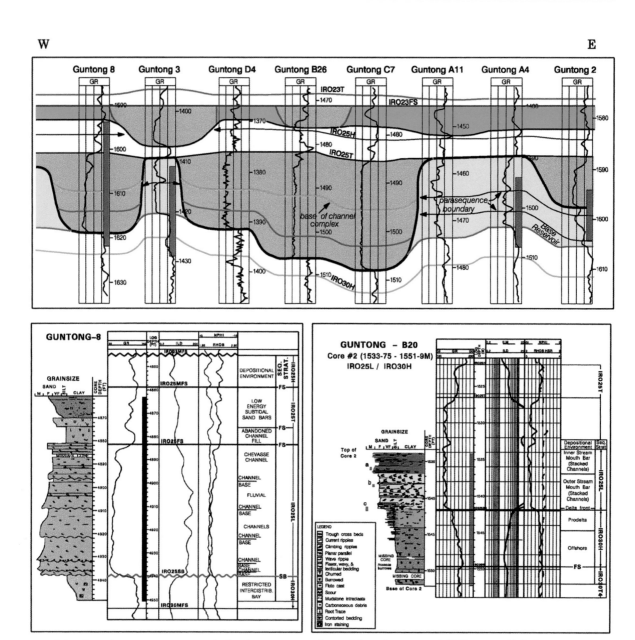

Fig. 8.15. Group I reservoir cross section (top) and core sequences (bottom) from the Guntong field (from unpublished EPR-PRSS report: Campion, 1997).

forming east-west trending, moderate-high energy, subtidal bars. The I fluvial and tidal reservoirs occur at depths of between 1300 and 2500 m. A sequence stratigraphic cross section of the I reservoirs in the field is shown in Fig. 8.15. The two main reservoir intervals are a lower fluvial-deltaic facies above a major sequence boundary, and an upper reservoir of sandy tidal estuarine deposits. The fluvial deltaic deposits consist of braided fluvial channels, with sharp-based, blocky log signature, which rest sharply on a coarsening upward, offshore-delta front to stream mouth bar succession. The contact between the two main facies was interpreted as a major sequence boundary which represents a relative sea-level fall.

The Group E sandstones form an important group of reservoirs in the central part of the Malay Basin. Mazlan Madon (1994) studied the reservoirs in the Jerneh gas field, which occur at depths of 1250 to 2000 m. The sandstones were interpreted as deltaic, and include distributary mouthbar, shoreface, and channel sediments formed during a Middle-Late Miocene regression. Different reservoir sandbodies characterise the delta front and delta plain facies association. The former is typified by distributary mouthbar and shoreface sandstones whereas the latter is associated with channel point bar deposits. The overlying transgressive deposits are typified by laterally extensive shallow marine sheet sandstones. The lateral continuity and architecture of the Group E reservoirs in the Jerneh field therefore vary as a result of the different depositional environments.

Reservoir Quality

There have been several reservoir quality studies in the Malay Basin, viz: J sandstones (Nik Ramli, 1988b; Noor Azim Ibrahim and Mazlan Madon, 1990), K sandstones (Nik Ramli, 1987; Khalid Ngah, 1990b), and E sandstones (Mazlan Madon, 1994). Regional reservoir quality and diagenetic studies include Chu (1992) and Hill et al. (1992), and the unpublished work by EPIC (1994). Most studies indicate that two main factors control reservoir quality: depositional environment and burial diagenesis. Compaction during burial is the most important diagenetic factor controlling reservoir quality. Ignoring burial diagenetic effects, however, depositional environment has the strongest control on reservoir quality. This is well demonstrated by the K sandstones in the Sotong Field (Fig. 8.16A). The best reservoir quality is shown by Group J estuarine sandstones, and mid-lower Group K braided stream deposits. Shoreface sandstones in the lower Group J in the Tapis field have permeabilities of 200–300 mD (Goh et al., 1982), whereas Group K sandstones in the Pulai field have 18–31% porosity and 300–3000 mD permeability (Flores and Kelm, 1982). Low energy subtidal and tidal flat reservoirs have the poorest quality.

Nik Ramli (1988b) also concluded that the reservoir properties of the J sandstones are primarily determined by depositional factors; high-energy facies storm bar sandstones tend to be of better quality than low-energy lower shoreface to inner shelf sandstones. In addition, it was noted that meteoric water flushing of the sandstones associated with the Middle-Upper Miocene regional unconformity in the southern Malay Basin has generated secondary porosity.

Textural parameters, such as sorting and grain size, also exert some control on reservoir properties. Cleaner and coarser grained sandstones tend to have the best reservoir properties. The reservoirs of groups M to K are mostly braided stream deposits, which are very coarse-grained and conglomeratic, whereas the younger groups tend to be very fine to fine grained. This generally fining trend with decreasing stratigraphic age could be the result of decreasing energy of deposition, the coastal plain and tidal settings being dominant in the younger groups. Chemical diagenesis is thought to have played a minor role in altering reservoir properties, and its effect is usually local (Mazlan Madon, 1994; EPIC, 1994). Authigenic minerals constitute 0–25% of the bulk volume, with an average of about 5%. The major minerals are silica (SiO_2) and carbonates, the former being dominant.

Apart from the direct impact of burial diagenesis on porosity, there is the indirect effect of high geothermal gradients on the rapid destruction of porosity with depth. Figure 8.16B shows the porosity data from Block PM 307 (formerly Block PM12) in the southwestern part of the basin. The data indicate that porosity declines very rapidly between 1500 m and 3000 m. Reasonable porosity values (>10%) can still be maintained down to depths of about 3000 m. A study of Group E sandstones by Mazlan Madon (1994) indicated that compactional effects during shallow burial (less than 1 km) have resulted in up to 50% reduction of the original porosity (now 10 to 25%). The sandstones have undergone a rapid decline in porosity, at a rate of about 3.8% of the original porosity per 300 m of burial. This is comparable to that in the neighbouring Pattani Basin (cf. 3.5% per 300 m, Trevena and Clark, 1986). This has been attributed to the high geothermal gradients in both basins.

Although clay minerals potentially have a negative impact on reservoir quality, very little work on clay minerals has been published. A regional study of clay minerals in Malaysian basins by Hill et al. (1992) provides some idea on the role of clays in determining reservoir behaviour. Authigenic kaolinite occurs in varying sizes, but the large plates generally pre-date the smaller ones. Large kaolinite is common in fluvial sandstones, whereas small kaolinite is more common in marine sandstones. Migration of small kaolinite may clog pore throats and result in formation damage during production. Detrital illite is most common in bioturbated sandstones. Authigenic illite distribution is dependent on depth of burial/temperature. Pore-bridging filamentous illite is common between 2000 and 2600 m. Smectite or smectite rich mixed-layer clays are relatively uncommon in the Malay Basin.

SOURCE ROCKS

Overview of Malay Basin Source Rocks

The abundant oil and gas reserves in the Malay Basin give testimony to the presence of effective source rocks. Geochemical studies of the oils and condensates (EPIC, 1994; McCaffrey et al., 1998) have indicated two main depositional settings for the source rocks — lacustrine and fluviodeltaic, with varying degrees of mixing between these two end-members. These source rock types reflect the

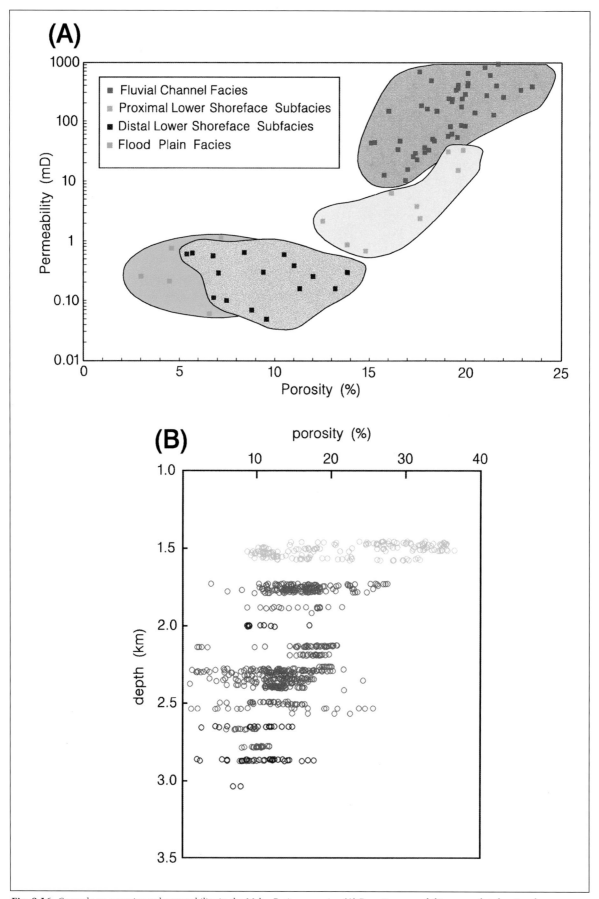

Fig. 8.16. Controls on porosity and permeability in the Malay Basin reservoirs. (A) Porosity-permeability cross-plot showing the poroperm characteristics of different sedimentary facies in the K2 sandstones at Sotong (after Asiah Mohd Salih and Mohd Fauzi Abdul Kadir, 1995). (B) Porosity-depth plot of data from wells in PM307, southwestern Malay Basin, showing rapid porosity reduction with depth. Data compiled from core-plug analyses.

sedimentary evolution of the basin, which changed gradually from lacustrine, through paralic and fluviodeltaic, to open marine (Fig. 8.8). The lacustrine source rocks consist of shales rich in algal components. They occur in the Oligocene/Early Miocene K, L, M and pre-M/synrift groups. Fluviodeltaic source rocks are found mainly in the Lower-Middle Miocene I and E groups. They are mainly coastal plain shales and coaly/carbonaceous shales. Figure 8.17 shows examples of the oil-prone material present in the source rocks.

Lacustrine source rocks have only been penetrated at shallow depths on the flanks of the basin. The fluviodeltaic source rocks are encountered in the basin centre. This situation is also reflected in the distribution of oil families in the basin — lacustrine oils occur mainly on the flanks while fluviodeltaic oils predominate in the basin centre (Fig. 8.18). Since the lacustrine sequences and the associated good quality source rocks occur mainly in the synrift (pre-M), M, L, and K groups, most of the oils in these groups are generally lacustrine in origin. The major source rock types are further discussed in the next section.

A large volume of source-rock screening data has been gathered by present and past operators, including Esso, PETRONAS Carigali,

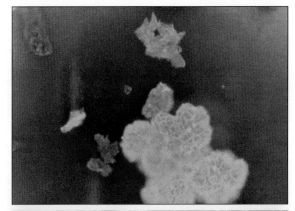
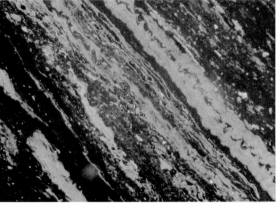

Fig. 8.17. Photomicrographs of source rocks in the Malay Basin, taken under ultra-violet light. Top: Lacustrine shale containing Type I organic matter, *Pediastrum* (upper) and *Botryococcus* (lower). Bottom: Fluviodeltaic coaly shale, containing fluorescent organic matter, mainly suberinite, cutinite, and sporinite.

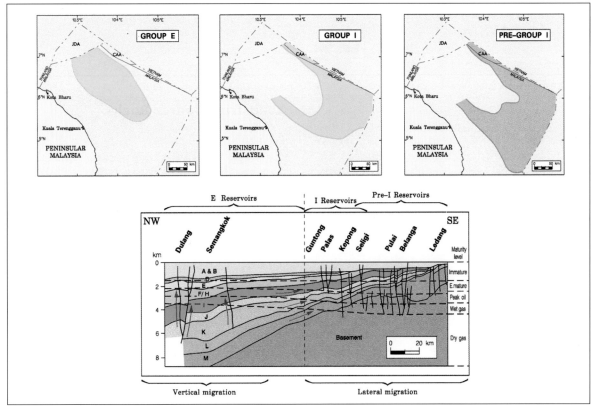

Fig. 8.18. Distribution of oil families in the Malay Basin. Top: Maps showing the approximate geographical limits of the three oil families. Bottom: Schematic NW-SE section showing the distribution of reservoirs for the three main oil families. Section based on Line A in Fig. 8.3. Also shown are the iso-maturity lines representing the maturity levels (right). Red arrows show the dominant mode of hydrocarbon migration: vertical in the northwest and lateral in the southeast.

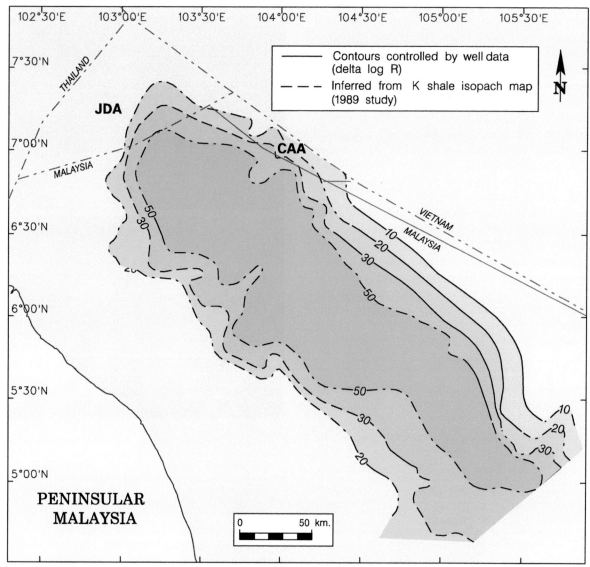

Fig. 8.19. Isopachs of lacustrine source rocks in the K Group identified from log response. Contours in metres.

JTOC, and BHP/Hamilton. The data consist of traditional source rock analyses such as total organic carbon (TOC) determination and Rock-Eval pyrolysis. Although these techniques are proven quantifiers of source-rock quality, they can be carried out only on discrete horizons, or composites of those horizons. They do not represent a continuous down-hole record of source-rock quality. Several operators, most notably Esso, have used electric logs to assess the source-rock potential continuously down-hole. One particular technique, the dLogR, has been published by Esso (Creaney et al., 1994), and is particularly powerful when calibrated locally with measured TOC data. The dLogR technique has been used to determine net source-rock thicknesses per group across the Malay Basin. This involves overlaying appropriately scaled porosity (usually the sonic) and resistivity (usually the deep) curves. A wide separation between the two curves indicates good quality, mature source rocks.

Lacustrine Source Rocks

Lacustrine shales are now recognised as perhaps the single most oil-prone facies, especially in Southeast Asian Tertiary basins (Cole and Crittenden, 1997). Since lacustrine source rocks in the axial regions of the Malay Basin are at great depths, their presence can only be inferred from seismic data. Figure 8.19 shows the isopachs of lacustrine source rocks (K shale) identified from log response. Although lacustrine source rocks may exist at depth, their nature and quality can only be inferred. The oldest lacustrine source rocks were deposited in isolated half-grabens developed during the early transtensional (synrift) phase of basin evolution. Although these lakes were perhaps small initially, they may have evolved into larger lakes as the basin subsided more rapidly. The lateral and temporal extents of these large lake systems are not known, but periods of low and high stands during K, L, and M times, identified in the EPIC (1994) study, must have been associated with large lakes. Open lacustrine environments are conducive for algae to flourish. Furthermore, if the lakes are stratified, sub-oxic bottom conditions may result in good preservation of algal-rich, Type I organic matter that contributes to high-quality, oil-prone source rocks.

Figure 8.20 is a schematic summary of all the Malay Basin source rocks and compares the lacustrine source rocks of Groups K, L, and M with the younger fluviodeltaic source rocks. The lacustrine sequences are generally shaley. Coals are rare, and are therefore not considered here, as their overall contribution to the lacustrine hydrocarbon charge would be low. Note that lacustrine shales are not ranked as highly as the younger coals and shales of Groups E and F. This is largely because of their lower organic richness (original TOC) rather than poor kerogen quality (original hydrogen index). Figure 8.21 shows the data from the K shale, the richest and most oil-prone lacustrine shale in the basin. Interestingly, few of the samples plot in the Type I kerogen field, as expected of lacustrine shales. Since the samples are from the basin flanks, there may have been dilution by lesser quality terrigenous material or partial oxidation due to a shallower water depositional environment. It is possible that source-rock quality may improve towards the basin centre.

Since much of Group M is unpenetrated, an accurate assessment of its source potential cannot be made. The relatively small data-set from Group M shales, however, suggests that they are mediocre in terms of oil source potential because of low organic richness and quality (Fig. 8.22). The lacustrine source rocks in Groups J, K, L and M are generally within the oil window in much of the southern and southeastern parts of the basin and on the basin margins (Fig. 8.23). The K shales, in particular, become rapidly overmature northwards and towards the basin centre (Fig. 8.18). It is expected that some of the oils generated from these northern source rocks may have been cracked and contributed to late gas flushing on the flanks.

A majority of the lacustrine source-rock extracts studied are from the K and L shales. These shales have similar geochemical characteristics. They are characterised by low Pr/Ph ratios (<2.5), indicative of low oxicity depositional conditions, and moderately waxy to waxy fingerprints. Such features are typical of lacustrine shales. Higher-plant biomarkers (oleananes and bicadinanes), in marked contrast to the coal-driven petroleum systems (e.g. in Group I), are generally low in quantity (Fig. 8.29). This feature alone does not imply a low higher-plant input but, when coupled with a predominance of C_{27} steranes over C_{29} steranes, a low higher-plant input is likely. An interesting feature of some extracts is the above-average amounts of the biomarker gammacerane. Occasionally, elevated amounts of gammacerane, seen particularly in the L shales, are suggestive of saline conditions, although never becoming

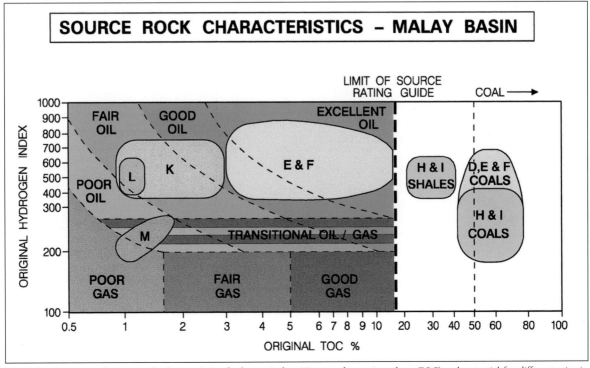

Fig. 8.20. Summary of source rock characteristics (hydrogen index, HI vs total organic carbon, TOC) and potential for different seismic groups in the Malay Basin. After EPIC (1994).

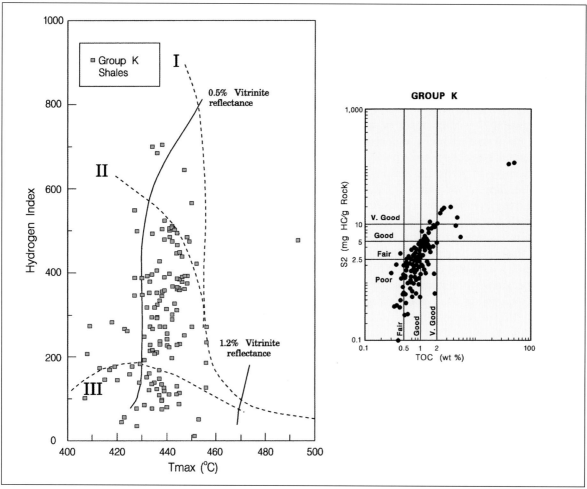

Fig. 8.21. Source rock screening data for Group K shales. Left: Hydrogen index (HI) vs Tmax (°C). Right: S_2 vs total organic carbon (TOC).

hypersaline. They also indicate that the depositional conditions varied laterally and temporally. The presence of C_4-methylated steranes, in varying relative abundances, indicates an algal contribution. The variation in their relative abundance further confirms the changing depositional conditions, although consistently lacustrine. Tricyclic terpanes tend to be low or absent, a feature that has been observed in non-saline lakes where the main algal contribution is from *Botryococcus* and *Pediastrum* (Wan Hasiah and Abolins, 1998).

In conclusion, the source-rock potential of the K and L lacustrine sequences is good, with considerable oil generative capacity. Group M, on the other hand, is less promising, and is only marginally oil prone in some places.

Fluviodeltaic Source Rocks

Fluviodeltaic environments became increasingly important in the Malay Basin from Group I times (Lower Miocene). Fluviodeltaic source rocks are generally best developed in Groups I and E. In most parts of the basin, however, Group E is

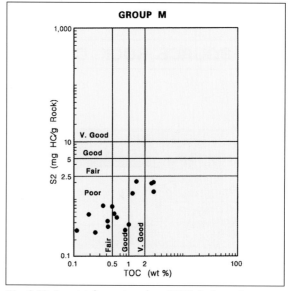

Fig. 8.22. S_2 vs total organic carbon (TOC) for lacustrine source rock in Group M.

immature for hydrocarbon generation (see Fig. 8.23) and, therefore, is rarely an effective source rock. Group E source rocks are likely to have been effective in the northernmost part of the basin (in the JDA and Thai waters). Because of their general immaturity, Group E source rocks are not discussed further. Emphasis is given to the fluviodeltaic Group I source rocks which, unlike those of Group E, have been sufficiently buried in many parts of the basin to have generated hydrocarbons. The Group I oil window extends as a ring around the entire basin (Fig. 8.23). In the basin centre, much of Group I is within the gas window. These very mature Group I kitchens are likely to have been the source for some overmature gases in the northern and central parts of the basin.

The good regional seal that occurs above Group I (i.e. the Group H shales and siltstones) and below (Group J shale), and the good correlations that can be made between Group I reservoired oils and Group I source rocks (Fig. 8.28), have led operators (notably Esso) to consider Group I as a closed petroleum system. Much of what is known about the source rocks that drive this prolific system has been learnt from well penetrations that, with a few exceptions, are geographically restricted to the east and southern flanks of the basin.

During Group I times, widespread coastal plain environments in the Malay Basin resulted in the deposition of thick, coal-bearing coastal plain sequences (Fig. 8.9). The organic matter in these sequences is dominated by terrigenous-derived material, which contrasts markedly with the algae-derived organic matter in the lacustrine shales of Groups K, L, and M. Group I source rocks consist of coals, coaly shales, and shales that occur as a continuous band running sub-parallel with the palaeocoastline. Up to 40 m of net source-rock occur in Group I, approximately half of which is coal. The greatest amount of coal occurs in the middle of Group I (EPIC, 1994). Fluctuations in the depositional conditions during Group I times are manifested in the variation of source-rock quality and biomarker distribution.

Coals are increasingly being accepted as effective source rocks by petroleum geochemists. Two decades ago it was generally believed that coals could generate only gas. This was based primarily on geochemical studies of coals in western Europe. Recent studies on coals from Southeast Asia and Australasia, using modern geochemical techniques such as pyrolysis, biomarkers, and petrography, have shown that coals may form excellent sources for oil and gas. Although many geochemists now believe that coals can generate significant quantities of oil, current debate now centres on whether the oils can be expelled from the coals. More detailed discussion on coals as source rocks for oil can be found in Scott and Fleet (1994).

Figure 8.24 summarises the available source-rock screening data from Group I in the form of an S_2 vs TOC plot and a HI vs Tmax plot. The organic richness (TOC) ranges widely, from very lean shales (TOC < 0.50 wt%) to clean coals (TOC approaching 80 %wt). Poor and mediocre source rocks (TOC < 1.00 wt%) show a range of kerogen composition, although generally dominated by Type III, gas-prone material. The richer source rocks, coals in particular, consistently have a more oil-prone kerogen assemblage, composed of Type II, as well as Type III, kerogen.

The HI-Tmax plot generally shows that the coals tend to have higher HI and lower Tmax values than the shales. Considering that the coals and shales are interbedded, the distinction between coals and shales based on Tmax (a maturity parameter) requires explanation. It is known that SE Asian coals can generate abundant hydrocarbon-like material at low maturity (Cook and Struckmeyer, 1986; Khorasani and Murchison, 1988; Wan Hasiah, 1997). The nature of this early-generated material is not well-defined, although it will contribute to the S_2 of the coal during pyrolysis. Although contributing to the S_2 (hence HI) of the coal, this early generated material is not considered to contribute volumetrically to the total expelled hydrocarbons. For this reason, the HI of coals tends to be misleadingly higher than the HI for shales. An alternative method should therefore be used when assessing the generative potential of coals. Furthermore, since Tmax is, by definition, the temperature of maximum hydrocarbon generation, the early generated material from coals will have the effect of reducing the Tmax value. More advanced techniques, such as pyrolysis–gas chromatography (PyGC), are more reliable indicators of source rock quality of organic rich shales and coals. PyGC traces of selected Group I coals and shales are shown in Fig. 8.25, which demonstrates the range of source-rock quality and the inadequacies of Rock-Eval-based source-rock assessment of coals. Isaksen et al. (1998) similarly pointed out the advantages of PyGC.

The biomarker distributions of Group I source rocks mostly indicate a significant land plant input, as suggested by geochemical characteristics such as high Tm/Ts, dominant C_{29} steranes, and abundant oleananes and bicadinanes. This is in line with a predominantly lower coastal plain setting in which copious amounts of higher plant material were deposited across the entire area. Occasionally, some Group I biomarker distributions show features that are indicative of algal contribution, suggestive of a lacustrine setting. Such features include a C_{27}

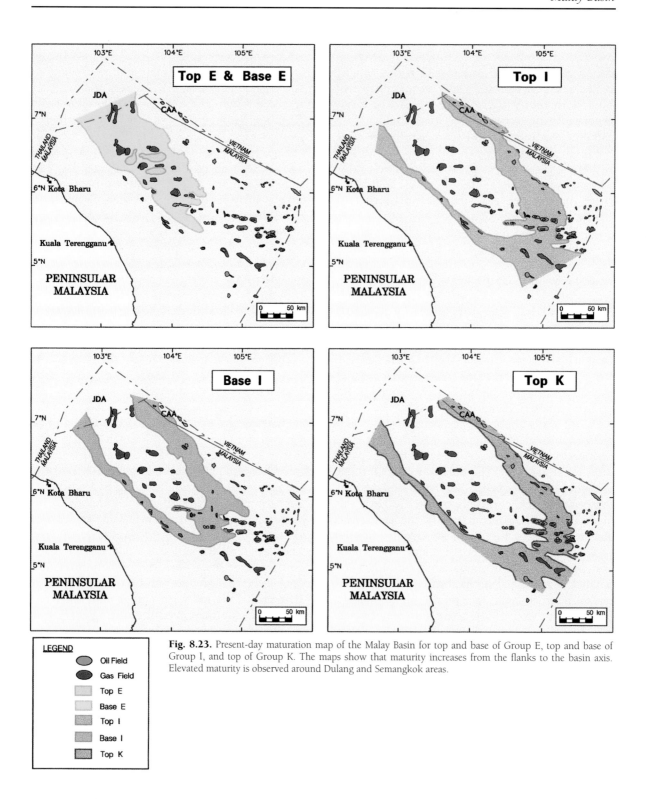

Fig. 8.23. Present-day maturation map of the Malay Basin for top and base of Group E, top and base of Group I, and top of Group K. The maps show that maturity increases from the flanks to the basin axis. Elevated maturity is observed around Dulang and Semangkok areas.

sterane preference, low amounts of oleanane and bicadinanes, and low Tm/Ts ratio, and can occur in both coals and shales. This tends to suggest a lacustrine influence. An algal input will generally enhance the oil-generative capacity of the source rock, as shown by their elevated HI values. The lacustrine influence in some Group I source rocks is attributed to the existence of marsh lakes on the coastal plain. Some of these lakes may have been sufficiently large and long-lasting to have accumulated the source sediment that generated some of the Group I oils. This suggestion is primarily based upon geochemical analysis of a small proportion of Group I oils that indicate lacustrine influence (see later).

Overall, therefore, the Group I source rock facies is characterised by consistently high levels of higher plant input, with varying amounts of algal contribution. This variation in source facies is manifested in the geochemical characteristics of the Group I oils (discussed in the next section).

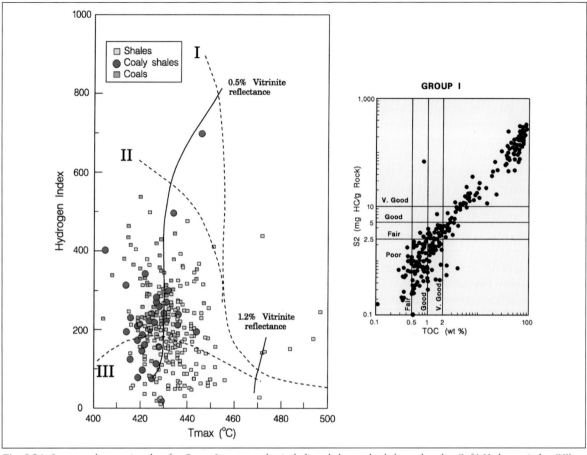

Fig. 8.24. Source rock screening data for Group I source rocks, including shales, coaly shales and coals. (Left) Hydrogen index (HI) vs Tmax (°C). (Right) S_2 vs total organic carbon (TOC).

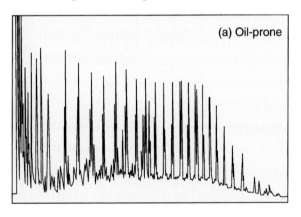

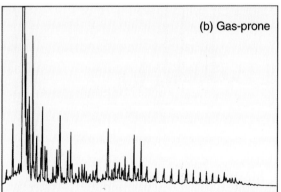

Fig. 8.25. Pyrolysis-gas Chromatography (PyGC) traces of selected Group I coals. (a) Oil-prone. (b) Gas-prone. Both coals possess HI values of 330, but this value is clearly misleading for the lower coal.

OIL TYPES

Oils in the Malay Basin may be classified into three groups, based on a combination of geochemical characteristics and age of the host reservoir. These are the Group E, Group I, and Pre-Group I oils.

Group E Oils

Oils in Group E are geographically restricted to the basin centre, with the Dulang and Semangkok fields accounting for the majority of E oils. Large oil accumulations in Group E, such as at Dulang, tend to coincide with areas prone to deep-seated faults. As Group E is immature over much of the basin (only the lowest part of E reaches the oil window in some areas), much of the hydrocarbons in the group are thought to have vertically migrated up these faults from stratigraphically deeper, more mature, intervals. Although the source sequences that are actually charging the E reservoirs have not been determined, the geochemical characteristics of the E oils (described below) suggest that the source is not pre-I, as they do not have the lacustrine signature of Pre-I oils. Since Groups F and H are generally considered as non-source, it is likely that Group I source rocks are charging the E reservoirs. This is supported by the similarity between the Group E oils in Dulang and the Group I oils in Guntong (Fig. 8.26).

A majority of Group E oils are moderately waxy to waxy, although in the far north of the basin (within Malaysian waters), such as at Ular, condensates are more common. This is a consequence of the great thickness of sediment in the north and, consequently, the high maturity of the source sequences. Most Group E oils and condensates typically have high Pr/Ph and Pr/nC_{17} ratios, suggesting oxidising depositional conditions for the source rock. Like the stratigraphically deeper I oils, the E oils show some variations in biomarker distribution, although a majority tend to have a terrigenous, rather than a lacustrine, signature. A typical E (fluviodeltaic) oil biomarker distribution is shown in Fig. 8.26. This is characterised by moderate quantities of higher plant biomarkers such as bicadinanes and oleanane, Tm/Ts >1.00, low levels of tricyclic terpanes, moderate quantities of diasteranes, and a sterane distribution that is generally skewed towards C_{29}. Such features are typical of oils receiving significant higher plant input. The low Tm/Ts ratio and moderate quantities of diasteranes probably indicate that coal was not a major contributor to the E oils. As none of the Group E oils have the very diagnostic fingerprint of the Pre-I oils (cf. Fig. 8.27), cross-stratal migration from below Group I does not seem to be an effective migration pathway. Lateral migration is probably the dominant mechanism (Fig. 8.18).

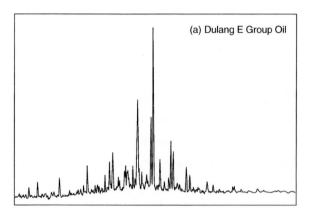

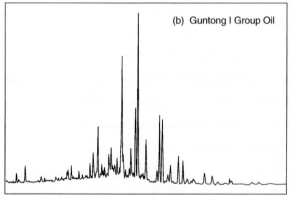

Fig. 8.26. Geochemical similarity between (a) Group E oil at Dulang and (b) Group I oil at Guntong. Both traces show m/z 191 distributions.

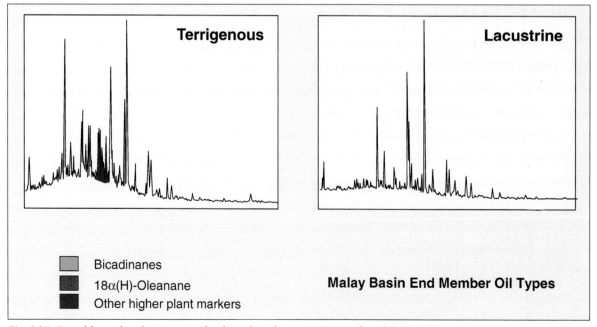

Fig. 8.27. Typical biomarker characteristics of end-member oils, i.e. terrigenous (fluviodeltaic) vs lacustrine, in the Malay Basin. Traces show m/z 191 distributions.

Group I Oils

Group I oils are common in the southern half of the Malay Basin. They are found over much of the width of the basin (except on the undrilled western flank), such as in Tembikai-1 and Meranti-1 in the west, the compressional structures such as Guntong and Tabu in the central region, and at Abu and Serudon in the east. In the deeper northern part of the basin, oils in Group I are restricted to the eastern flank, such as in the Bunga Kekwa field.

Group I oils show variable compositions at both bulk and molecular level. They range from being high-API condensates to moderate-API waxy oils. Such a variation may occur in the same well, in sands only a few 10's of metres apart. The Piatu well is an example where compositional variations at the molecular level is clearly indicated in the biomarker distributions. Considering the degree of variation observed in the Group I source rocks (discussed earlier), these variations in oil composition are expected, provided that all the source rocks have been effective. The oils range from very terrigenous, as demonstrated by large quantities of higher plant markers such as oleananes and bicadinanes (e.g. at Piatu), to the more lacustrine-influenced, such as Penara, which have low higher-plant input and relatively larger amounts of C_{27} steranes.

The close spatial and temporal relationships between different source rock facies during Group I times have resulted in the mixing of the oils they generated (Fig. 8.28). Thus, although there are many similarities between the source rock and oil biomarker distributions, correlating any oil to a particular source rock may not be possible. Over much of the basin, with the exception of some flank areas, there is sufficient evidence to support Esso's conclusion (EPIC, 1994) that all Group I oils were sourced from Group I source rocks. However, on the basin flanks, particularly the northeastern flank, Group I oils were probably sourced from stratigraphically deeper levels, such as the Pre-I lacustrine sequences.

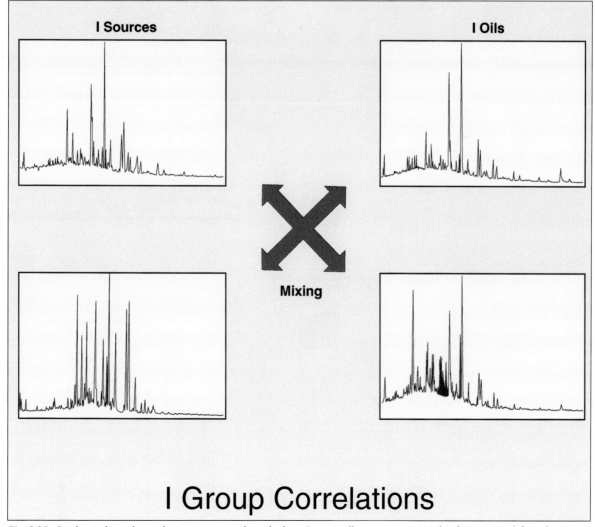

Fig. 8.28. Geochemical correlations between source rocks and oils in Group I. All traces are m/z 191 distributions. Top left is a lacustrine-influenced I source, bottom left is a fluviodeltaic-influenced I source. The resulting oils on the right can therefore be of various types.

Pre-Group I Oils

Pre-I oils occur in Groups J, K, and L. There are no known occurrences in Group M. With only few exceptions, the Pre-I oils are restricted to the southern end of the basin, primarily on the eastern side. This distribution is a direct consequence of the maturity of the pre-I source rocks. The marginal areas, being relatively shallow, are the only parts of the basin where pre-I source rocks are not over-mature. Pre-I oils are also known to occur on the western flank, such as at Resak and Beranang, but the occurrence is sparse. Whether this scarcity is real, or due to the low number of wells drilled, remains to be seen. Amerada Hess recently acquired Block PM304, which has a number of flank leads that need to be worked up and which may provide some answers. The northeastern flank of the basin, such as at Bunga Kekwa, also contains some Pre-I oils.

An interesting feature of Pre-I oils is that, despite a wide range of bulk properties, their geochemical characteristics vary only little. The oils vary from very waxy crudes, such as Bunga Kekwa, to light condensates, such as at Tembikai. This variation is considered to be due to a combination of differing maturity of the oils and of post-accumulation processes such as evaporative fractionation. Source facies is not considered to have been a factor, as most of the oils appear to have been derived from similar source rocks, as suggested by the similarities in biomarker distributions of the majority of Pre-I oils, as discussed below.

A common feature of all the Pre-I oils is a moderately low Pr/Ph ratio (generally between 2.0 and 3.0) and a low Pr/nC_{17}, (generally between 0.3 and 0.5). Such features are typical of mildly oxidising or non-oxidising depositional conditions. Biomarker distributions are characterised by low Tm/Ts and high quantities of C_{29}Ts and diahopane. Higher plant markers, such as bicadinanes and oleananes, typical of many southeast Asian oils (such as those in Group I) are rare or absent. Regular sterane distribution tends to be evenly balanced between C_{27} and C_{29} steranes, while C_4-methyl steranes are common although generally not in high abundance. Tricyclic terpanes are generally absent or in low abundance, as is gammacerane. Such features are characteristic of many southeast Asian oils sourced from freshwater lacustrine sequences.

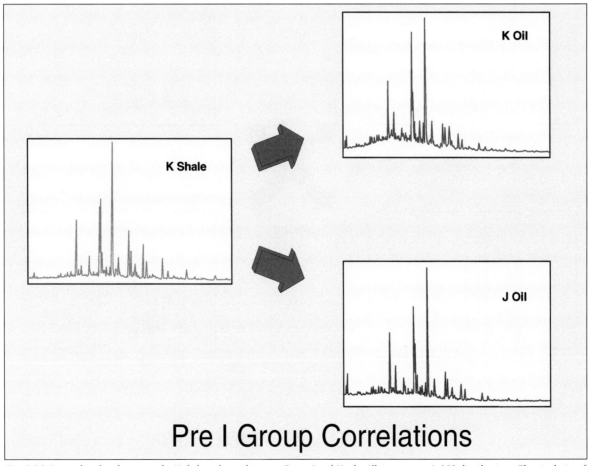

Fig. 8.29. Biomarker distribution in the K shale and correlation to Group J and K oils. All traces are m/z 191 distributions. The similarity of J and K oils is striking, and a mutual correlation to the K shale is likely.

The geochemical characteristics outlined above are seen in the Group J and Group K oils, suggesting that they were derived from the same source. Such biomarker distributions correlate very well with those described earlier for the K shale. This similarity is demonstrated in the respective biomarker distributions shown in Fig. 8.29. It is therefore believed that the K shale lacustrine sequence has been an effective source over a large area of the basin, and has been responsible for charging the J and K reservoir sands.

NATURAL GASES

The Malay Basin is not purely an oil province but also contains several large gas accumulations. In fact, natural gas is estimated to represent more than half of the hydrocarbon reserves of the basin. The total (associated and non-associated) gas reserve in place is estimated to exceed 60 TSCF. A large proportion of this is in the Jerneh, Lawit, Duyong, and Seligi fields. The largest gas accumulations occur mainly in the reservoirs of Groups D, E, I and J. Smaller accumulations have been found in the reservoirs of groups B, H, F, K, and L. In the northwestern part of the basin, gas is found in stratigraphically younger reservoirs. Gas accumulations occur in stratigraphically deeper/older reservoirs towards the south. The possible reasons for this distribution were discussed earlier. This observed gas distribution, however, may not represent a complete picture because of the lack of well penetrations of the older units in the north/central parts of the basin.

Geochemical data suggest that the natural hydrocarbon gases (methane, ethane, propane, etc) in the Malay Basin were derived from either a biogenic or a thermogenic source (Curry, 1992; EPIC, 1994; Waples and Mahadir Ramly, 1995; McCaffrey et al., 1998). Because of the high mobility of gases in the subsurface, mixing of gases tends to be common. Mixing is often recognised from the geochemical composition of the gas.

Biogenic gas, consisting mainly of methane, forms at shallow depths and low temperatures (less than 75°C). It is generated by fermentation and reduction reactions caused by bacterial degradation of organic matter. Biogenic gas has a distinctive stable carbon isotope composition, being isotopically very light ($\delta^{13}C <$ -55‰). In the Malay Basin, biogenic gas occurrences are mainly confined to the eastern flank, within Group H and older units. These occurrences, however, are probably not purely biogenic, as some degree of mixing with thermogenic gas is likely. Generally, biogenic gas does not contribute significantly to the total gas reserve of the Malay Basin.

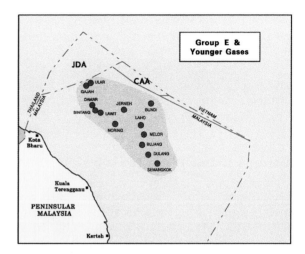

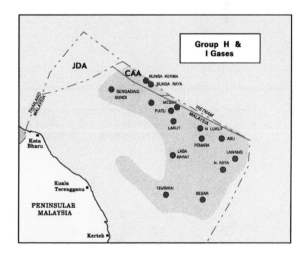

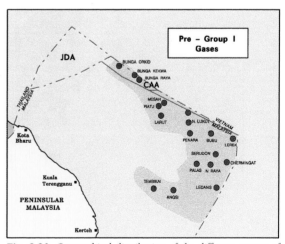

Fig. 8.30. Geographical distribution of the different groups of gases in the Malay Basin. (Top) Groups E & post-E gases. (Middle) Group H & I gases. (Bottom) Pre-Group I gases. Outlined areas represent the extent of the occurrences for the different groups, while the red circles are data points for the $\delta^{13}C$ values plotted in Fig. 8.31.

Thermogenic gas forms at higher temperatures than biogenic gas. It may be generated directly by kerogen decomposition or derived from the cracking of oils. Stable carbon isotope compositions ($\delta^{13}C$) of methane, ethane, and propane indicate that the thermogenic gases in the Malay Basin were generated from source rocks with maturities ranging from 0.7% to more than 2.0% vitrinite reflectance.

The Malay Basin gases may be classified into three stratigraphic groupings based on the observed compositional and isotopic trends. These groupings are described below.

Groups E and Younger Gases

Gases in the Group E and younger units are mainly confined to the axis of the basin (Fig. 8.30). They are clearly distinguishable from gases of the other two groupings by their heavier methane and ethane isotopic values (Fig. 8.31). The isotopic values indicate high-maturity gases, probably derived via vertical migration from deeply buried, mature sources. Based on wetness and the isotopic difference of methane and ethane, all gases in this grouping can be classified as dry thermogenic gas (Fig. 8.31A). These gases are commonly associated with high levels of inorganic CO_2, which probably originated from carbonates or metasediments in the basement at depth. The much heavier methane and ethane isotopic values also indicate a coaly percursor. It is therefore likely that the Group E and younger gases were derived from the thermal breakdown of kerogen/coals at very high thermal maturities.

Groups H and I Gases

Gases in groups H and I are generally found towards the basin flanks (Fig. 8.30). They are isotopically and compositionally similar to the stratigraphically deeper gases in the pre-I groups. This is especially true for those in Group I. Gases from Group H show less spread in their ethane stable carbon isotope values, and are geochemically confined between the Group E and younger gases and the pre-Group I gases. This suggests a lesser degree of mixing between the Group H and older gases, probably due to the presence of a good seal.

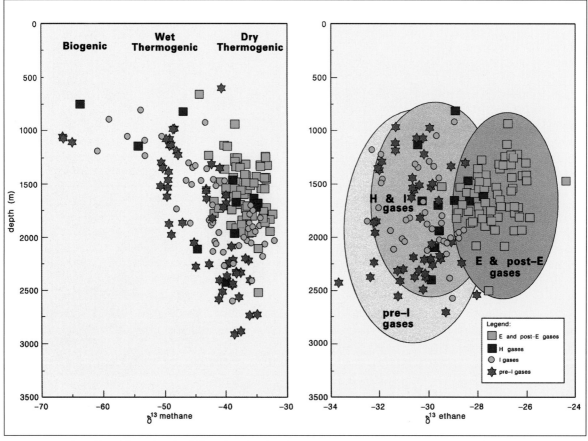

Fig. 8.31. Depth profiles of $\delta^{13}C$ values for (left) methane (CH_4) and (right) ethane (C_2H_6) depth in the Malay Basin gases, showing distinct stratigraphic and compositional groupings. The methane isotope plot (left) shows the separation between biogenic, wet thermogenic, and dry thermogenic gases. The ethane isotope plot (right) shows that gases in Group E & younger reservoirs (clustered to the right of the plot) are distinct from those in the older reservoirs (Groups H & I and pre-Group I gases).

The Group I gases show a larger spread in isotopic composition, which indicates mixing of gases from different source facies and maturity. These consist of mixtures of biogenic, wet thermogenic, and dry thermogenic gases in varying proportion (Fig. 8.31). The wet components are those that have been generated from within Group I, whilst the dry component probably originated from the underlying, mature, lacustrine shales of Group J and older.

maturity trend with depth. The CO_2 content also increases with depth.

Non-Hydrocarbon Gases

In the Malay Basin, carbon dioxide (CO_2) occurs in significant quantities (Fig. 8.32). Although CO_2 can reach significant concentrations in Malay Basin gases, very high CO_2 levels tend to be the

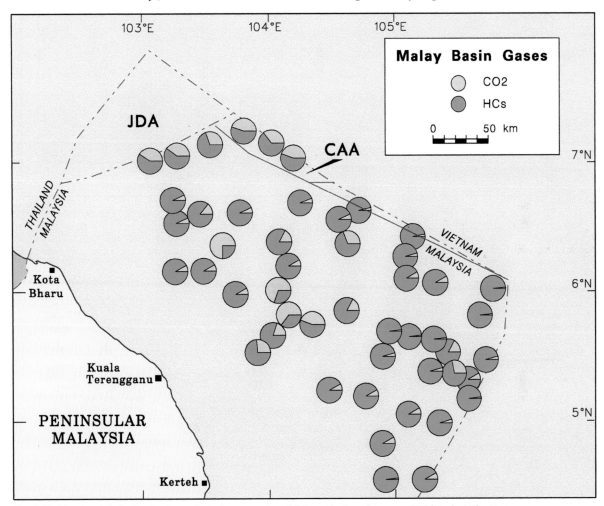

Fig. 8.32. Map showing the distribution and relative proportion of CO_2 and hydrocarbon gases (HCs) in the Malay Basin.

Pre-Group I Gases

Figure 8.30 (bottom) shows the distribution of pre-Group I gases, which consist of mixtures of biogenic and wet and/or dry thermogenic gases (Fig. 8.31). Examples of biogenic, wet thermogenic, and dry thermogenic gases are Lerek, Penara, and Bunga Kekwa, respectively (their locations are shown in Fig. 8.30). The wet thermogenic gases are believed to have been generated from oil-prone lacustrine shales and have migrated laterally, albeit over relatively short distances. The dry component was derived from the more mature source rocks down dip. The pre-I gases show a consistent

exception rather than the rule. CO_2 concentrations range between 5 and 40 mol%. High-CO_2 occurrences are also geographically restricted to certain locations, particularly in the northern part of the basin (Fig. 8.32) and, occasionally, in the central region, such as in the Dulang Field. CO_2 in the Malay Basin originates from two possible sources: organic, which involves primarily the breakdown of kerogen in coals, and inorganic, which is generally the thermal breakdown of carbonates, presumably in the basement. Most occurrences of CO_2 in the basin show a mixture of both organic and inorganic origin. In some instances, however, where no mixing has occurred, the

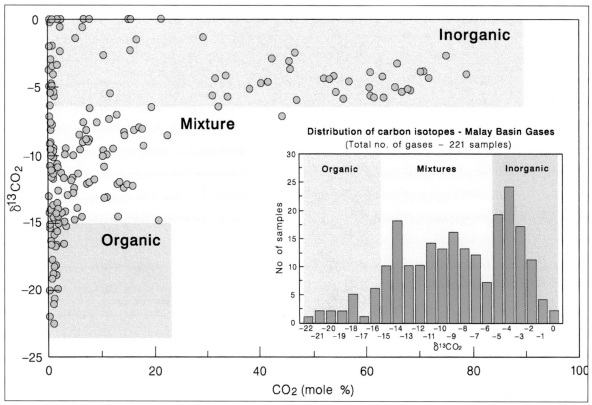

Fig. 8.33. Crossplot of $\delta^{13}C$ against mole % for CO_2 in the Malay Basin gases, to distinguish between organic and inorganic CO_2. Inset shows the frequency distribution of $\delta^{13}C$ values for CO_2.

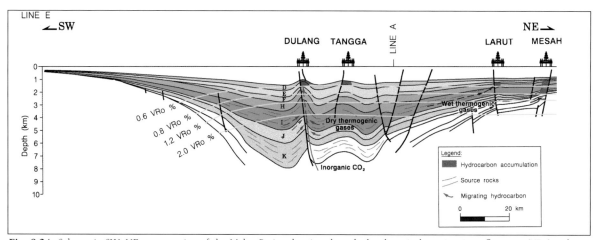

Fig. 8.34. Schematic SW–NE cross section of the Malay Basin, showing the calculated equivalent vitrinite reflectance (VRo) isolines, postulating the wet and dry thermogenic gas generation zones and migration pathways. Based on Line E in Fig. 8.3.

two sources can be clearly distinguished isotopically.

Figure 8.33 shows plots of both CO_2 content and CO_2 isotopic composition for gases in the basin. High concentrations of CO_2 always have compositions indicative of inorganic derivation from the thermal metamorphism of carbonates in the pre-Tertiary basement. In these gases, isotopic values range from 0 to −5‰. Such gases are found in wells located in the basin centre and are associated with large petroleum gas accumulations (Fig. 8.34). These gases are thought to have migrated upwards along fault systems, and were subsequently mixed with shallower thermally-generated gas. Organic-derived CO_2 has isotopic values ranging from −15 to −25‰, but rarely constitutes more than 5 mol% of the total gas composition.

Although the geographical distribution of CO_2 in the Malay Basin is probably well understood, the stratigraphic distribution of CO_2 still poses many questions. As yet, no predictive technique is available to estimate CO_2 concentrations on a reservoir basis, to all but the broadest levels. Clearly, further work is required on this aspect.

OVERPRESSURE

The deeper strata in the Malay Basin are usually overpressured. It is estimated that more than 80% of wells drilled during the earlier phase of exploration were terminated because of overpressure (Mohd Shariff Kader, 1994). The onset of abnormal pressures occurs in progressively older sedimentary units from NW to SE, and from the centre to the flanks (Singh and Ford, 1982; Mohd Shariff Kader, 1994). In the basin centre, the onset of overpressure occurs in stratigraphically younger (Groups E and F) and shallower (about 1200–2000 m) formations (Fig. 8.35). Towards the basin flanks, the onset of overpressure occurs in stratigraphically older and vertically deeper horizons because of the increasing sand percentage away from the centre (Fig. 8.36). The top of overpressure usually coincides with the top of the oil window, which suggests that there is a link between hydrocarbon generation and the onset of overpressure.

Overpressure in the basin is thought to be the result of one or more factors. Singh and Ford (1982) attributed the widespread occurrence of overpressures to mainly a combination of rapid burial of certain stratigraphic units (especially Groups D to H) and uplift of initially normally-pressured strata (Groups J to M). Disequilibrium compaction due to high sedimentation and burial rates is thought to be the most common cause of overpressure in the basin (Wan Ismail Wan Yusoff, 1993). In the Malay Basin, a very thick (13-14 km), largely shaly, sedimentary succession accumulated over a relatively short time span (probably less than 35 Ma). The sedimentation rates for some seismic units, e.g. Groups H and F, may be as high as 1000 mMa^{-1} in the basin centre (Leslie et al., 1994). Such high rates resulted in an effective pressure seal that retains the excess pressure. The thick undercompacted shale sequences of Groups E, F, and K provide such a seal.

In the EPIC (1994) report, it was argued, based on Bower's sonic velocity versus effective stress analysis, that compaction disequilibrium alone could not explain the high overpressure in the basin. Therefore, other contributory factors must have played a role. In the southeastern part of the basin, tectonic compression and uplift during the Middle Miocene may have been the cause of overpressure in Group J and K reservoirs (Singh and Ford, 1982). Other factors that contribute to overpressure in many basins worldwide, such as hydrocarbon generation, clay de-watering, and aquathermal pressuring (Spencer, 1987; Buhrig, 1989; Osborne and Swarbrick, 1997), could also have contributed to the overpressure in the Malay Basin. Furthermore, the high geothermal gradients in the central and northern parts of the basin could have easily caused the cracking of oil to gas. This would result in the increase in fluid volume and, hence, pressure (Hunt et al., 1994).

Overpressure exerts a strong control on hydrocarbon generation and migration pathways. In overpressured zones, restricted fluid movement prevents oil expulsion. The retained oil may subsequently be cracked to gas as the basin subsides. The high-maturity, wet gases in the Malay Basin may have formed this way. In the north/central parts of the basin, pore pressures are close to, or exceed, the formation fracture pressure (70–90% of lithostatic pressure). As a result, the hydrocarbons, including high-maturity gases, migrated vertically and were trapped in

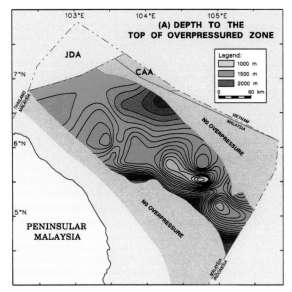
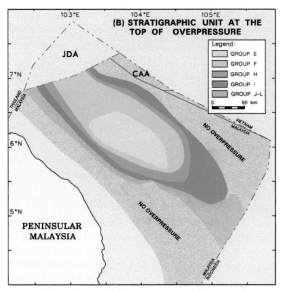

Fig. 8.35. Overpressure in the Malay Basin (from Leslie et al., 1994). (A) Depth to the top of the overpressured (transition) zone. (B) Stratigraphic unit at the top of the overpressured zone.

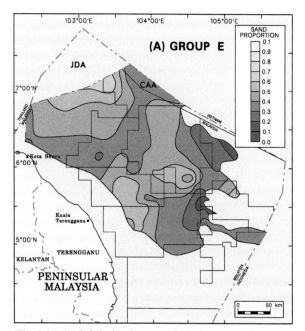
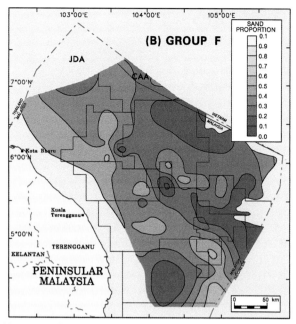

Fig. 8.36. Sand-shale distribution in the Malay Basin. (A) Group E and (B) Group F (from Leslie et al., 1994).

shallower reservoirs. Below the top of overpressure, hydrocarbon saturations decrease as temperature and pressure increase. On the basin flanks, lateral migration of hydrocarbons from the basin deep is driven by high formation-pressure gradients. There is a greater chance of finding hydrocarbons below the onset of overpressure because of better seal integrity. As future exploration effort is towards deep reservoirs, more discoveries are expected within overpressured zones.

MIGRATION AND ENTRAPMENT

The earlier sections discussed how the source rocks of the Malay Basin can be categorised as lacustrine and fluviodeltaic, and that these facies occur (broadly) in the Pre-I and I groups, respectively. Group E also has excellent source rock potential but is generally immature over much of the basin, the only exception being in occasional deep synclines in the basin centre and in the far north of the basin.

The Malay Basin is a relatively young Tertiary basin, probably less than 40 Ma old, which explains the significantly high present-day surface heat flows, especially in the northern and central parts of the basin. Geothermal gradients vary from about 32 °Ckm^{-1} on the flank to 53 °Ckm^{-1} in the basin centre (Fig. 8.37). High heat flows of around 105 mWm^{-2} are recorded in the axial region, decreasing towards the basin flanks. Wan Ismail Wan Yusoff (1993) suggested that some high heat flow areas are associated with subsurface fluid movements. Other areas, such as around Belumut, are still experiencing low heat flow conditions.

A large database exists for vitrinite reflectance, spore colour, and Tmax measurements in the Malay Basin. On the whole, the data are reliable for maturity interpretation. In some wells, however, vitrinite reflectance values are lower than those predicted from maturity modelling (Waples et al., 1994). This is inconsistent with the high present-day geothermal gradients and subsurface temperatures. Suppressed vitrinite reflectance is also commonly observed in other SE Asian Tertiary basins (Okui et al., 1997; Watts, 1997). To explain this phenomenon, Waples et al. (1994) suggested that a heat pulse during Pleistocene times may have been responsible for the present high heat flow, and that the heat flow may have been lower in the past. However, most of the low reflectance values were obtained from kerogen in shales instead of from coals, which would give more acceptable values (EPIC, 1994). Waples et al. (1994) applied the Flourescence Alteration of Multiple Macerals (FAMM) technique to obtain equivalent vitrinite reflectance measurements that can be used for maturity modelling. Using FAMM data, the heat flow history may be explained by a model in which the crustal heat flow was constant or decaying exponentially, without invoking a Pleistocene heat pulse. Using the same data set, Mazlan Madon (1997b) has also shown that the subsidence and thermal history of the basin are consistent with a model in which rifting occurred at 35 Ma for a period of 10 Ma.

A hydrocarbon generation/accumulation

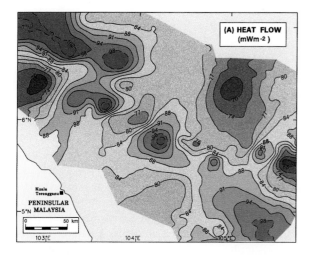

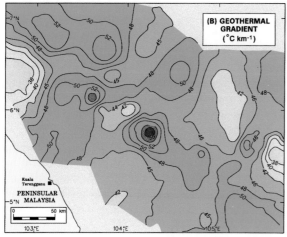

Fig. 8.37. Maps of (A) heat flow and (B) geothermal gradient in the Malay Basin (after Wan Ismail Wan Yusoff, 1993 and Mazlan Madon, 1997b). Heat flow varies from 85 to 105 mWm^{-2}. Geothermal gradients range between 32 and 53 °Ckm^{-1}, and have a similar regional trend as the heat flow.

model for the Malay Basin generally involves the K shale source rocks charging the J and K sands, and the I source rocks charging the I sands. In these two hydrocarbon systems, the main migration mechanism is up-dip lateral migration. Cross-stratal migration appears to have been insignificant in most instances. The volume of hydrocarbon charge into some structures is often greater than the capacity of the first encountered trap, so that the fill-and-spill process (tertiary migration) takes place. This involves reservoirs being filled with hydrocarbons to the spill point, before being spilled out to migrate into the next up-dip structure. According to EPIC (1994), the fill-and-spill process was probably responsible for the chain of fields from Tabu to Palas (see Fig. 8.10), whose oils have similar geochemical characteristics. To map the lateral migration pathways, structural reconstructions are required for the relevant formations at the time of expulsion/migration. As migration distances increase, distinct shadow zones develop. These shadow zones are areas around which laterally migrating fluids are diverted because of the dominant structural style. Such shadow zones, however, could still be filled if vertical drainage can be effective.

The main exception to the lateral migration and filling of structures are the Group E reservoirs. Because the Group E source rocks are generally immature, the oils in the Group E reservoirs must have come from a deeper, more mature, source. Cross-stratal migration has to be the main migration mechanism for the Group E oils. The geochemical similarity between some Group E oils (such as at Dulang) and some Group I oils (such as at Guntong) supports this interpretation. The presence of deep faults associated with the Group E structures also implies that vertical migration is an important migration process. The deep faults may have acted as conduits for migrating fluids. Furthermore, the high concentration of inorganically-derived CO_2 (probably from the basement) in many Group E reservoirs is also a strong evidence for vertical migration. This is probably the most important migration mechanism in the heavily-faulted northern/central parts of the Malay Basin (Fig. 8.18).

Lateral migration or vertical migration are probably never mutually exclusive. There are probably many instances where both elements are present. An example is Lerek, where cross-stratal migration has also occurred (EPIC, 1994). Other examples also exist around the flanks of the basin. The broad vertical distribution of geochemically distinct oil types (stratigraphically deep lacustrine versus stratigraphically shallow fluviodeltaic) suggests that cross-stratal migration is limited to areas affected by heavy faulting, where leaked hydrocarbons have charged the stratigraphically shallower Group E reservoirs.

CONCLUSION

The Malay Basin is in a mature phase of exploration, with close to 1000 exploration and development wells drilled, and over 150 oil and gas accumulations discovered since the 1960s. There are two main hydrocarbon systems on the basis of source rock geochemistry: lacustrine and fluviodeltaic. The basin underwent an extensional (constructive) phase during the Oligocene during which lacustrine source rocks were deposited. The Miocene was a time of fluviodeltaic source and reservoir deposition. A compressional phase in Late Miocene times resulted in structural inversion and the formation of many of the major hydrocarbon traps in the basin. Rapid sedimentation, and high heating rates resulted in widespread overpressure development and large accumulations of thermally generated gas, particularly in the central/northern parts of the basin. Hydrocarbon migration is dominantly lateral up-dip along stratigraphic carrier beds, while in some areas, particularly in the Group E reservoirs, vertical migration via deep-seated fault conduits seems to be predominant.

Hydrocarbon distribution in the basin is dependent on two main factors: 1. source quality and maturity, and 2. the relative timing of generation/migration and structuration. The most prolific structural trap style is the inversion anticline. The J sandstones are by far the most productive reservoirs in the basin. Reservoir porosities are generally depth-dependent. Besides burial depth, depositional environment (facies) is the main factor controlling reservoir properties. Diagenesis plays a relatively minor role in determining reservoir quality. Hence, reasonable porosity values (>10%) can still be found at depths down to about 3000 m.

There are at least 3 major hydrocarbon systems in the basin, linking source rock to the reservoir rocks. These correspond to the three main oil types: Group E oils, Group I oils, and the pre-Group I oils. Natural hydrocarbon gases also fall into these general categories on the basis of their geochemical characteristics. The hydrocarbon systems are fed by two main source rock systems: lacustrine shales of Oligocene-Lower Miocene age in Groups K, L, M, and pre-M (synrift), and fluviodeltaic shales (and often coals), mainly in Group I. The lacustrine source rocks charge the Groups J, K, and L oil reservoirs, and are generally sealed off by the regional Group J shale from the younger reservoirs. The fluviodeltaic source rocks charge the Group I and younger oil reservoirs. In addition, the lacustrine source rocks also charge the gas reservoirs throughout the stratigraphic column from L up to B. Generally, source rocks in Group E and younger rocks are thermally immature. Deep-seated faults play an important role in providing migration routes for many gas accumulations in the basin.

Although the basin is in a relatively mature stage of exploration, oil companies are continuously looking for, and testing, new hydrocarbon play concepts. An example of a play concept that has emerged out of these efforts is the deep reservoir in the northern part of the basin. The Bergading Deep-1 well was drilled by PETRONAS in late 1995 to test this idea. The well was drilled into the overpressured zone in Group F before encountering a sweet spot in Group H. The success of Bergading Deep-1 has opened a new play in this mature basin.

REFERENCES

Abd Rahim Md Arshad, Dashuki Mohd and Tjia, H.D., 1995. A deep seismic section across the Malay Basin: processing of data and tectonic interpretation. Abstracts of the Geological Society of Malaysia Petroleum Geology Conference 1995. Warta Geologi, 21, 412.

Ahmad Said, 1982. Overview of exploration for petroleum in Malaysia under the Production Sharing Contracts. Offshore South East Asia 82, 9-12 February 1992, Singapore, 1-14.

Amran Nong Chik, Samsuddin Selamat, Mohd Rohani Elias, White, J.P. and Wakataka, M.T., 1996. Guntong Field: development and management of a multiple-reservoir offshore waterflood. Journal Petroleum Technology, December 1996, 1139-1143.

Armitage, J.H., 1980. A decade of exploration and development by EPMI off the east coast of Peninsular Malaysia. South East Asia Petroleum Exploration Society (SEAPEX) Conference.

ASCOPE, 1981. Tertiary Sedimentary basins of the Gulf of Thailand and South China Sea: Stratigraphy, Structure and Hydrocarbon Occurrences. ASEAN Council on Petroleum (ASCOPE) Secretariat, Bangkok.

Asiah Mohd Salih and Mohd Fauzi Abdul Kadir, 1995. Sedimentological study of K2 reservoir in Sotong Field. Unpublished PRSS Report No: PRSS TCS10-97-21.

Azmi M. Yakzan, Awalludin Harun, Bahari Md Nasib and Morley, R.J., 1994. Integrated biostratigraphic zonation for the Malay Basin. Abstracts of AAPG International Conference & Exhibition, Kuala Lumpur, Malaysia, August 21-24, 1994. American Association of Petroleum Geologists Bulletin, 78, 1170-1171.

Brami, J.B., Khairuddin, K. and Muhammad, Y.M., 1984. History and geology of the Tinggi Field, offshore Peninsular Malaysia. Petro-Asia Conference 1984, October-December.

Buhrig, C. 1989. Geopressured Jurassic reservoirs in the Viking Graben: modelling and geological significance. Marine and Petroleum Geology, 6, 31-48.

Campion, K.M. (editor), 1997. Malay Basin Field Studies: Guntong, Seligi, and Tapis Fields. EPR-PRSS Collaborative Research Report, EPR.14EX.97.

Chu, Y.S., 1992. Petrographic and diagenetic studies of the reservoir sandstone of the Malay Basin. Bulletin of the Geological Society of Malaysia, 32, 261-283.

Chua, B.Y. and Wong, R., 1997. Some possible new exploration ideas in the northern and western Malay Basin of Peninsular Malaysia. Proceedings of the ASCOPE '97 Conference, Kuala Lumpur, 24-27 November 1997.

Cole, J.M. and Crittenden, S., 1997. Early Tertiary basin formation and the development of lacustrine and quasi-lacustrine/marine source rocks of the Sunda Shelf of SE Asia. In: Fraser, A.J., Matthews, S.J. and Murphy, R.W., eds., Petroleum Geology of Southeast Asia. Geological Society of London Special Publication, 126, 147-183.

Cook, A.C. and Struckmeyer, H., 1986. The role of coal as a source rock for oil. In: Glennie, R.C., ed., Second South-Eastern Australian Oil Exploration Symposium. Petroleum Exploration Society of Australia Symposium, 14-15 November 1985, 419-432.

Creaney, S., Abdul Hanif Hussein, Curry, D.J., Bohacs, K.M. and Redzuan Hassan, 1994. Source facies and oil families of the Malay Basin. Abstracts of AAPG International Conference & Exhibition, Kuala Lumpur, Malaysia, August 21-24, 1994. American Association of Petroleum Geologists Bulletin, 78, 1139.

Curry, D.J., 1992. Geochemistry and source facies relationships of gases from the Malay Basin. Unpublished Report, Esso Production Malaysia Inc.

Daines, S.R., 1985. Structural history of the W Natuna Basin and the tectonic evolution of the Sunda region. Proceedings of the Indonesian Petroleum Association 14th Annual Convention, 8-10 October 1985, 39-61.

Dewey, J.F., 1982. Plate tectonics and the evolution of the British Isles. Journal of the Geological Society of London, 139, 317-412.

EPIC, 1994. Regional study of the Malay Basin-Final Portfolios. Esso-PETRONAS Integrated Collaborative Study, Unpublished report, Esso Production Malaysia Inc.

Esso, 1985. Petroleum Geology of Malaysia. Unpublished report, Esso Production Malaysia Inc.

Flores, D.J. and Kelm, C.H., 1982. Development and reservoir engineering studies of the Pulai Field, offshore Peninsular Malaysia. Offshore South East Asia 82, 9-12 February 1992, Singapore.

Ginger, D.C., Ardjakusuma, W.O., Hedley, R.J. and Pothecary, J., 1993. Inversion history of the West Natuna Basin: examples from the Cumi-Cumi PSC. Proceedings of the Indonesian Petroleum Association 22nd Annual Convention, Jakarta, October 1993, 635-658.

Goh, S.T., Heacock, D.W. and Loveless, D.E., 1983. Exploration, development, and reservoir engineering studies of the Tapis Field, offshore Peninsular Malaysia. Journal of Petroleum Technology, June 1983, 1051-1060.

Goh, S.T., Loveless, D.E. and Heacock, D.W., 1982. Exploration, development and reservoir engineering studies for the Tapis Field, offshore Peninsular Malaysia. Offshore South East Asia 82, 9-12 February 1982, Singapore.

Hamilton, W., 1979. Tectonics of the Indonesian region. United States Geological Survey Professional Paper, No. 1078.

Hill, J.A., Soo, D.K.Y. and Verriah, T., 1992. Clay mineralogy in subsurface sandstones of Malaysia and the effects on petrophysical properties. Bulletin of the Geological Society of Malaysia, 32, 15-43.

Hui, S.K. and Pillai, H., 1988. Waterflood development of the Guntong and Tabu Fields. Proceedings of the 7th Offshore South East Asia Conference, 2-5 February 1988, 746-757.

Hunt, J.M., Whelan, J.K., Eglinton, L.B. and Cathles, L.M., 1994. Gas generation – a major cause of deep Gulf Coast overpressures. Oil & Gas Journal, 59-63.

Hutchison, C.S., 1989. "Geological Evolution of South-East Asia". Oxford monographs on Geology and Geophysics no. 13, Clarendon Press, Oxford.

Isaksen, G.H., Curry, D.J., Yeakel, J.D. And Jenssen, A.I., 1998. Controls on the oil and gas potential of humic coals. Organic Geochemistry, 29, 23-44.

Khalid Ngah, 1990a. Deposition and diagenesis of Oligocene-Lower Miocene sandstones in the southern Malay Basin. Unpublished Ph.D. Thesis, University of London.

Khalid Ngah, 1990b. Porosities in Pulai-II Sandstone – Implication for hydrocarbon exploration in older reservoir. Abstracts of the Geological Society of Malaysia Petroleum Geology Seminar, Kuala Lumpur, 26-27 November 1990. Warta Geologi, 16, 267.

Khalid Ngah, Mazlan Madon and Tjia, H.D., 1996. Role of pre-Tertiary fractures in formation and development of the Malay and Penyu basins. In: Hall, R. and Blundell, D.J., eds., "Tectonic Evolution of Southeast Asia". Geological Society of London Special Publication, 106, 281-289.

Khandwala, S.M., Abd Malek Abd Rani and M Nasir Abd Rahman, 1984. Field development planning for the Semangkok field, offshore Peninsular Malaysia. 5th Offshore South East Asia Conference, 21-24 February 1984, Singapore, 8.35-8.49.

Khorasani, G.K. And Murchison, D.G., 1988. Order of generation of petroleum hydrocarbons from liptinitic macerals with increasing thermal maturity. Fuel, 67, 1160-1162.

Kingston, D.R., Dishroon, C.P. and Williams, P.A., 1983. Global basin classification system. American Association of Petroleum Geologists Bulletin, 67, 2175-2193.

Leslie, W.C., Mohd Khair Abd Kadir, Muzammal Abdul Ghani, Mohd Shariff Kader, Abdullah Adli Zakaria, Che Shaari Hj Abdullah and Rodziah Daud, 1994. Malay and Penyu Basins Regional Study 1992-1994. Unpublished report, PETRONAS Carigali, 4 volumes.

Liew, K.K., 1994. Structural development at the west-central margin of the Malay Basin. Bulletin of the Geological Society of Malaysia, 36, 67-80.

Mazlan B. Hj. Madon, 1992. Depositional setting and origin of berthierine oolitic ironstones in the lower Miocene Terengganu shale, Tenggol Arch, offshore Peninsular Malaysia. Journal of Sedimentary Petrology, 65, 899-916.

Mazlan B. Hj. Madon, 1994. Depositional and diagenetic histories of reservoir sandstones in the Jerneh Field, central Malay Basin. Bulletin of the Geological Society of Malaysia, 36, 31-53.

Mazlan B. H. Madon, 1995. Tectonic evolution of the Malay and Penyu Basins, offshore Peninsular Malaysia. Unpublished D.Phil. thesis, University of Oxford, 325pp.

Mazlan B. Hj. Madon, 1997a. The kinematics of extension and inversion in the Malay Basin, offshore Peninsular Malaysia. Bulletin of the Geological Society of Malaysia, 41, 127-138.

Mazlan B. Hj. Madon, 1997b. Analysis of tectonic subsidence and heat flow in the Malay Basin (offshore Peninsular Malaysia). Bulletin of the Geological Society of Malaysia, 41, 95-108.

Mazlan B. Madon and Watts, A.B, 1998. Gravity anomalies, subsidence history, and the tectonic evolution of the Malay and Penyu basins. Basin Research, 10, 375-392.

McCaffrey, M.A., Abolins, P., Mohammad Jamaal Hoesni and Huizinga, B.J., 1998. Geochemical characterisation of Malay Basin oils: some insight into the effective petroleum systems. Ninth Regional Congress on Geology, Mineral and Energy Resources of Southeast Asia - GEOSEA '98, 17-19 August 1998, Kuala Lumpur, Programme and Abstracts, 149.

McKenzie, D., 1978. Some remarks on the development of sedimentary basins. Earth and Planetary Science Letters, 40, 25-32.

McKenzie, D.P. and Jackson, J.A., 1986. A block model of distributed deformation by faulting. Journal of the Geological Society of London, 143, 349-353.

Mohd Shariff B. Kader, 1994. Abnormal pressure occurrence in the Malay and Penyu basins, offshore Peninsular Malaysia - a regional understanding. Bulletin of the Geological Society of Malaysia, 36, 81-91.

Mohd Tahir Ismail, 1995. Relationship of structural timing and hydrocarbon migration in the Malay Basin. Abstracts of the Geological Society of Malaysia Petroleum Geology Conference, 1995. Warta Geologi, 21, 396-397.

Mohd Tahir Ismail and Rudolph, K.W., 1992. Structural trap styles of the Malay Basin. Abstracts of the Symposium on the Tectonic Framework and Energy Resources of the Western Margin of the Pacific Basin. Kuala Lumpur, 29 November–2 December 1992. Warta Geologi, 18, 267.

Mohd Tahir Ismail, Shahrul Amar Abdullah and Rudolph, K.W., 1994. Structural and sedimentary evolution of the Malay Basin. Abstracts of AAPG International Conference & Exhibition, Kuala Lumpur, Malaysia, August 21-24, 1994. American Association of Petroleum Geologists Bulletin, 78, 1148.

Murphy, R.W., 1989. Inversion tectonics – a discussion. Geological Society of London Special Publication, 44, 336-337.

Ng, T.S., 1987. Trap styles of the Tenggol Arch and the southern part of the Malay Basin. Bulletin of the Geological Society of Malaysia, 21, 177-193.

Nik Ramli, 1986. Depositional model of a Miocene barred wave-and storm-dominated shoreface and shelf, southeastern Malay Basin, offshore West Malaysia. American Association of Petroleum Geologists Bulletin, 70, 34-47.

Nik Ramli, 1987. Petrology, diagenesis and quality of K sandstone (Pulai Formation) reservoirs in the southeastern part of the Malay Basin. Abstracts of the Geological Society of Malaysia Petroleum Geology Seminar, Kuala Lumpur, 7-8 December 1987. Warta Geologi, 13, 284.

Nik Ramli, 1988a. Development of a humid tropical fan-delta system: the middle Tertiary 'K' sandstone in the southeastern Malay Basin, offshore West Malaysia. In: Nemec, W. and Steel, R.J., eds., Fan Deltas: "Sedimentology and Tectonic Settings". Blackie, Glasgow, 341-353.

Nik Ramli, 1988b. Characteristics of J sandstone (Tapis Formation) reservoirs in the southeastern part of the Malay Basin, offshore West Malaysia. 7th Offshore South East Asia Conference, Singapore, 2-5 February 1988 (preprint).

Noor Azim Ibrahim and Mazlan Madon, 1990. Depositional environments, diagenesis, and porosity of reservoir sandstones in the Malong Field, offshore West Malaysia. Bulletin of the Geological Society of Malaysia, 27, 27-55.

Okui, A., Imayoshi, A. and Tsuji, K., 1997. Petroleum system in the Khmer Trough, Cambodia. In: Howes, J.V.C and Noble, R.A, eds., Proceedings of the IPA Petroleum Systems of SE Asia and Australasia, 21-23 May 1997, Indonesian Petroleum Association, Jakarta, 365-380.

Osborne, M.J. and Swarbrick, R.E., 1997. Mechanisms for generating overpressure in sedimentary basins: a reevaluation. American Association of Petroleum Geologists Bulletin, 81, 1023-1041.

Polachan, S. and Sattayarak, N., 1989. Strike-slip tectonics and development of Tertiary basins in Thailand. Proceedings of the International Symposium on Intermontaine Basins: Geology and Resources, Chiang Mai, Thailand (30 Jan - 2 Feb, 1989), 243-253.

Ramlee, A.R. and Bedingfield, J.R., 1996. Mid J Group sequence stratigraphy and reservoir architecture in Block PM-9, Malay Basin PRSS Technology Forum, 12-13 September 1996, Awana Genting, Kuala Lumpur (unpublished proceedings).

Scott, A.C. and Fleet, A.J. (eds.), 1994. Coal and coal-bearing strata as oil-prone source rocks? Geological Society of London Special Publication, 77.

Singh, I. and Ford, C.H., 1982. The occurrence, causes and detection of abnormal pressure in the Malay Basin. Offshore South East Asia 82, 9-12 February 1992, Singapore.

Spencer, C.W., 1987. Hydrocarbon generation as a mechanism for overpressuring in Rocky Mountain Region. American Association of Petroleum Geologists Bulletin, 71, 368-388.

Tapponnier, P., Peltzer, G., Le Dain, A.Y., Armijo, R. and Cobbold, P. 1982. Propagating extrusion tectonics in Asia: new insights from simple experiments with plasticine. Geology, 10, 611-616.

Thambydurai, R., Mustaffa, A.F., Mueller, K.H. and Dixon, M.R., 1988. Jerneh gas field development planning. Proceedings of the 7th Offshore South East Asia Conference, Singapore, 2-5 February 1988, 859-870 (preprint).

Tjia, H.D., 1994a. Origin and tectonic development of Malay-Penyu-West Natuna basins. Paper presented at PRSS Technology Day, 21 June 1994, Kuala Lumpur. (unpublished manuscript).

Tjia, H.D., 1994b. Inversion tectonics in the Malay Basin: evidence and timing of events. Bulletin of the Geological Society of Malaysia, 36, 119-126.

Tjia, H.D. and K.K., Liew, 1996. Changes in tectonic stress field in northern Sunda Shelf basins. In: Hall, R. and Blundell, D.J., eds., Tectonic Evolution of Southeast Asia. Geological Society of London Special Publication, 106, 291-306.

Trevena, A.S. and Clark, R.A., 1986. Diagenesis of sandstone reservoirs in the Pattani Basin, Gulf of Thailand. American Association of Petroleum Geologists Bulletin, 70, 299-308.

Wan Hasiah, A., 1997. Evidence of early generation of liquid hydrocarbons from suberinite as visible under the microscope. Organic Geochemistry, 27, 591-596.

Wan Hasiah, A. and Abolins, P., 1998. Organic petrological and organic geochemical characterisation of the Tertiary coal-bearing sequence of Batu Arang, Selangor, Malaysia. Journal of Asian Earth Sciences, 16, 351-367.

Wan Ismail Wan Yusoff, 1984. Heat flow study in the Malay Basin. CCOP Technical Publication, 15, 77-87.

Wan Ismail Wan Yusoff, 1990. Heat flow in offshore Malaysian basins. CCOP Technical Publication, 21, 39-54.

Wan Ismail Wan Yusoff, 1993. Geothermics of the Malay Basin, offshore Malaysia. Unpublished M.Sc. thesis, University of Durham, 213 pp.

Waples, D.W., Leslie, W. and Mahadir Ramly, 1994. A thermal model for the evolution of the Malay Basin. Abstracts of AAPG International Conference & Exhibition, Kuala Lumpur, Malaysia, August 21-24, 1994. American Association of Petroleum Geologists Bulletin, 78, 1169.

Waples, D.W. and Mahadir Ramly, 1995. Geochemistry of gases in the Malay Basin. Abstracts of the Geological Society of Malaysia, Petroleum Geology Conference 1995. Warta Geologi, 21, 413.

Watts, A.B., Karner, G.D. and Steckler, M.S., 1982. Lithospheric flexure and the evolution of sedimentary basins. Philosophical Transactions of the Royal Society of London, A305, 249-281.

Watts, K.J., 1997. The northern Nam Con Son basin petroleum system, based on exploration data from Block 04-2 Vietnam. In: Howes, J.V.C and Noble, R.A, eds., Proceedings of the IPA Petroleum Systems of SE Asia and Australasia, 21-23 May 1997, Indonesian Petroleum Association, Jakarta, 481-498.

White, J.M. Jr. and Wing, R.S., 1978. Structural development of the South China Sea with particular reference to Indonesia. Proceedings of the Indonesian Petroleum Association 7th Annual Convention, Jakarta, June 1978, 159-178.

Yap, K.T., 1996. Tapis Field: Lower Group J Reservoir. PRSS Technology Forum, 12-13 September 1996, Awana Genting, Kuala Lumpur.

Zhang, Q.M. and Zhang, Q.X., 1991. A distinctive hydrocarbon basin - Yinggehai Basin, South China Sea. Journal of Southeast Asian Earth Sciences, 6, 69-74.

Chapter 9

Penyu Basin

Mazlan B. Hj. Madon and Azlina Anuar

INTRODUCTION

The Penyu Basin is a small extensional basin located offshore east of Peninsular Malaysia and is separated from the much larger Malay Basin by the Tenggol Arch (Fig. 9.1). The basin appears to be structurally contiguous with the West Natuna Basin of Indonesia but is relatively under-explored. Very little work has been done or published on the basin. The structure and stratigraphy of the basin was first studied by Khalid Ngah (1975). Brief accounts of the basin can be found in Hamilton (1979), ASCOPE (1981), Hutchison (1989) and Khalid Ngah et al. (1996). A more detailed structural analysis was done by Mazlan Madon (1995). Mazlan Madon et al. (1997) discuss the structural evolution and hydrocarbon potential of the basin. This chapter is a review of the geology of the basin based on previous works and incorporates the results of recent exploration efforts.

EXPLORATION HISTORY

Exploration in the Penyu Basin is still at an early stage. After almost three decades of exploration, only 8 wells have been drilled in the basin (Fig. 9.2). The first phase of exploration in the Penyu Basin took place in the mid-60s and early 70s, during which Continental Oil Company (Conoco) acquired about 2000 line-km of seismic data. Two wildcat wells were drilled by Conoco. Penyu-1 was the first well drilled in 1970 to test a large inversion fold structure formed within a relatively shallow Upper Miocene postrift sequence. Although there is good reservoir development in the targeted section, no hydrocarbons were found. In 1973, Pari-1 was drilled to test a four-way dip closure formed by a basement-drape feature. No hydrocarbons were found because of poor reservoir development in the objective section.

In 1978 PETRONAS Carigali acquired from Conoco the exploration rights for the concession area. Between 1981 and 1983 Carigali acquired about 3000 line-km of seismic but did not drill any wells. In 1990 a new operator, Texaco Exploration Penyu Inc., acquired an additional 4000 line-km of seismic data. During late 1991 Texaco drilled three wells, Cherating-1, Merchong-1, and Rumbia-1, to test postrift inversion plays, similar to that drilled at Penyu-1. The structures are basically Miocene folds, which are transpressional features formed during the Early Miocene to Middle Miocene but affecting both the Oligocene and Miocene sections. The primary reservoir intervals were sandstones above and below the Terengganu Formation with intraformational shales as seals. The shallower Miocene sandstones of the Pari Formation and the deeper Penyu Formation sandstones were secondary objectives. All the three wells were unsuccessful.

After failing to find hydrocarbons in the postrift inversion plays, Texaco acquired a further 3500 line-km of seismic to assess the prospectivity of synrift plays. Rhu-1/1A was drilled in late 1991 with the main objective being late synrift sediments that form a drape structure over an intrabasinal basement ridge. The potential reservoirs in the faulted anticlinal closure were Oligocene intra-Penyu Formation sandstones, thought to be sealed by intraformational shales. Deep-graben lacustrine source rocks were believed to be the source for the hydrocarbons. The Rhu play concept was a success; oil was discovered in the upper Penyu Formation at depths between 2600 and 2670 m.

In 1993 Texaco drilled a Rhu-like structure in the southeastern corner of the basin. The Soi-1 well was to target the "ramp play", analogous to the discoveries at Rhu and on the northeastern margin of the Malay Basin. The well was drilled to a depth of 2630 m but did not find any hydrocarbons. During that same year, an appraisal well, Rhu-2, was drilled on the Rhu structure to test the

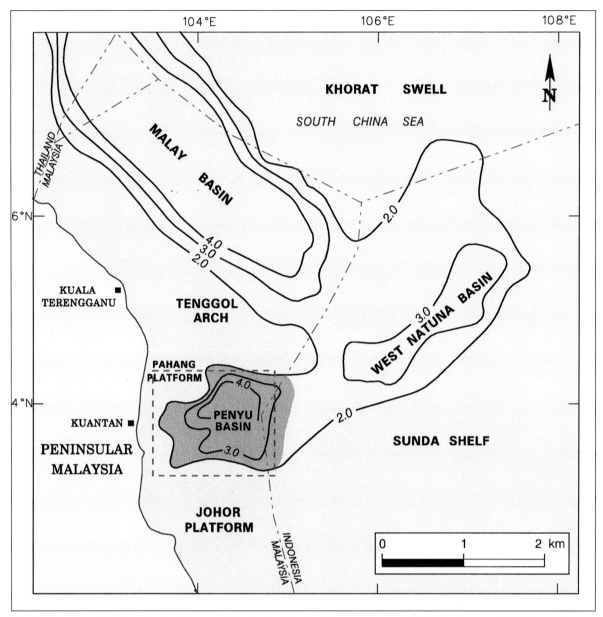

Fig. 9.1. Location of the Penyu Basin offshore east of Peninsular Malaysia. Contours represent pre-Tertiary basement depth in seconds two-way travel time. Modified from ASCOPE (1981).

stratigraphic continuity and structural geometry of the hydrocarbon column discovered at Rhu-1/1A. Unfortunately, the well failed to produce any oil from the targeted zones. Texaco left the area in 1995 after the Rhu discovery was found to be subcommercial.

STRATIGRAPHY

As in the neighbouring Malay and West Natuna basins, the sediments in the Penyu Basin are virtually entirely siliciclastic, consisting of interbedded shale, siltstone, sandstone and, in the middle part, coal. The sedimentary succession is categorised generally into synrift and postrift sequences, ranging in age from Oligocene to Recent. The distinction between synrift and postrift sedimentation phases is characteristic of basins formed by lithospheric extension (e.g. McKenzie, 1978). The synrift sequence represents Oligocene sedimentation during the extensional phase of basin development while the postrift sequence represents Miocene to Recent sedimentation when extensional fault activity has ceased. Whereas the synrift sequence occurs as half-graben fills whose thicknesses are determined by the amount of extension along the bounding faults, the postrift sequence has a more uniform thickness across the entire basin as a result of gentle sagging due to nonfault-related subsidence.

Figures 9.3 shows the stratigraphic

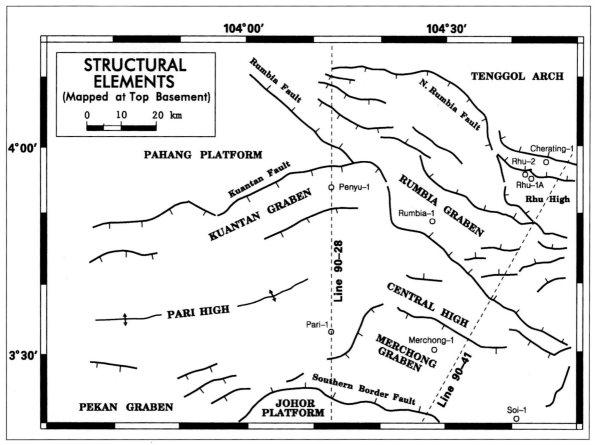

Fig. 9.2. Structural elements (faults and grabens) in the Penyu Basin identified from seismic data (after Mazlan Madon, 1995). Bold lines with barbs represent major faults; barbs indicate the downthrown side. Shown are the location of wells and the two seismic profiles in Fig. 9.5 (Lines 90-28 and 90-41).

subdivision of the Penyu Basin. An informal four-fold seismo-lithostratigraphic subdivision was established by Texaco in the early 1990's. Four formations were identified, in ascending order: Penyu, Terengganu, Pari, and Pilong. Further subdivision into seismic units, P1 to P7, was proposed by Mazlan Madon (1995) based on the recognition of seismic markers, H1 to H8, some of which are unconformities that represent major tectonic events. The most important markers are the top of basement (H8), intra-Penyu (H7), base-Pari (H6), and top-Pari (H3). The seismic units are dated mainly by palynology, which is also used in determining the environments of deposition (Fig. 9.4).

The Penyu Formation basically represents the synrift deposition during the Oligocene. It consists of interbedded sandstone and shale deposited in alluvial and lacustrine environments. By Late Oligocene times the basin was near sea level and was affected by marine incursions. Palynomorph assemblages in the formation indicate that there was increasing marine influence as the basin evolved (Figs. 9.3, 9.4). The influence of faulting on subsidence during the synrift phase is reflected in the increase in the thickness of seismic unit P1 from Pari-1 to Penyu-1 well (Fig. 9.4).

The top of the synrift succession is marked by the Base-Pari Unconformity, which is the result of a Late Oligocene basin inversion. This inversion event caused minor folding and truncation of synrift strata in some parts of the basin. Subsequently, during the early to Middle Miocene, postrift sequences belonging to the Terengganu and Pari formations were deposited in alluvial and coastal plain environments. Coals are particularly abundant in the Pari Formation at Penyu-1 well (Fig. 9.4).

The top of the Pari Formation is marked in places by an erosional unconformity, the Top-Pari Unconformity, which truncates part of the uplifted and folded Pari Formation that overlies inverted half-grabens. The Late Miocene-Pliocene regional unconformity in the Malay and West Natuna basins (Armitage and Viotti, 1977; Daines, 1985, Tjia, 1994) is probably the result of the same basin inversion event.

The Pilong Formation consists of sandstone and shale deposited in nearshore to shallow marine environments following a marine transgression during the Late Miocene to Early Pliocene times. The undeformed nature of the post-inversion sequence suggests that the Pliocene was a phase of relative tectonic quiescence.

STRUCTURE

The Penyu Basin is bounded to the north by the Pahang Platform and Tenggol Arch, and to the south by the Johor Platform (Figs. 9.1, 9.2). These areas of shallow pre-Tertiary basement, from which sediments in the basin were probably derived, are generally less than 1 km deep. The top of pre-Tertiary basement is recognised in most places by high-amplitude reflections, overlain by poorly reflective synrift half-graben fill, in marked contrast with the postrift sediments (Fig. 9.5).

The Penyu Basin is not a simple basin but comprises several grabens or sub-basins. Mazlan Madon (1995) identified four major fault bounded sub-basins: the Kuantan, Pekan, Rumbia, and Merchong grabens (Figs. 9.2, 9.5). The Kuantan and Pekan grabens, in the eastern part of the basin, are bounded by the Pahang and Johor platforms, which are the offshore extensions of the late Mesozoic basement of Peninsular Malaysia. The two grabens form a conjugate pair that are separated by the Pari High, which is essentially the interference zone between their hangingwalls. The Kuantan Graben is the deepest graben in the Penyu Basin, with more than 7 km of sediment. The Rumbia and Merchong grabens are characterised by linear and steep bounding faults. They are separated by the Central High.

Pre-existing basement faults seem to have had a major influence on basin development. Two fault trends dominate the basement structure: 1. ENE-trending, mostly half-graben bounding, normal faults and 2. NW-trending combined dip- and strike-slip fault. The two largest faults are the ENE-trending Kuantan Fault, which is the border fault to the Kuantan Graben, and the NW-trending Rumbia Fault, which is a left-lateral strike-slip fault that extends northwestwards near the Terengganu coastline. The orientation of the NW-trending bounding faults suggests that the faults have a similar origin to the major NW-trending strike-slip faults on Peninsular Malaysia.

Contrasting structural styles occur in different parts of the Penyu Basin (Figs. 9.2, 9.5). The Kuantan Graben is an asymmetrical half graben bounded by a single border fault to the north, whereas the Merchong Graben is

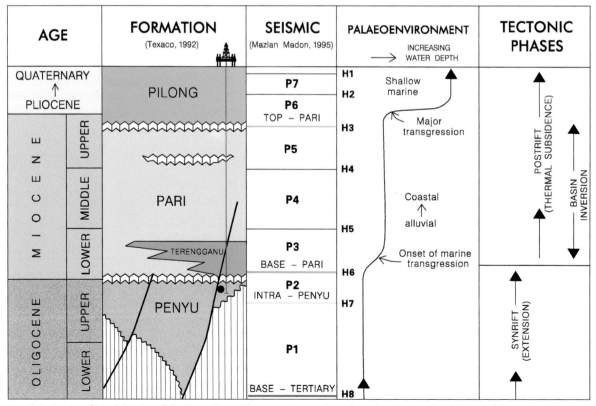

Fig. 9.3. Stratigraphic subdivision of the Penyu Basin. The succession consists of synrift (Oligocene) and postrift (Miocene and younger) sequences. Formation names follow Texaco's (1992) nomenclature. Seismic units, P1 to P7, are based on Mazlan Madon (1995), identified by markers H1 to H8.

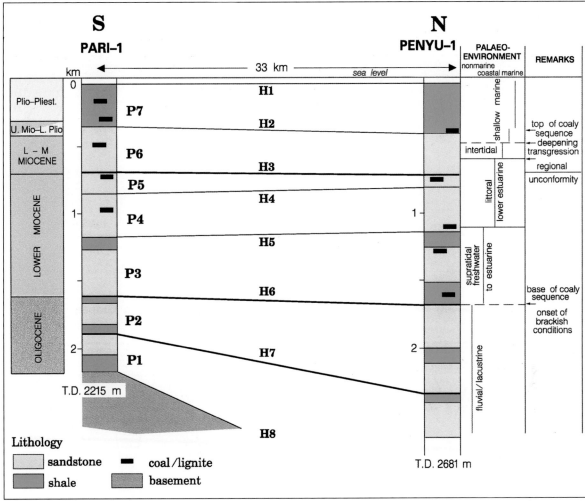

Fig. 9.4. Correlation of stratigraphy in Penyu-1 and Pari-1 wells. P1 to P7 are stratigraphic units defined by seismic picks H1 to H8. Palaeoenvironmental interpretation in Penyu-1 is based on unpublished palynological and foraminiferal analyses.

symmetrical and bounded by steep, strike-slip border faults to the northeast and southwest. The orientation of the half-graben border faults indicates predominantly north-south extension direction. Partitioning of strain by these faults may have caused strike-slip motion that resulted in a complex internal structure in the Rumbia and Merchong grabens.

The basin fill shows a variety of internal structural styles as a result of deformation during its evolution. Synrift wedges filling the half-grabens are characterised by rollover-like structures associated with synsedimentary movement of their listric border faults. An example is shown in Fig. 9.5A where the hangingwall sediment in the Kuantan Graben has developed a rollover structure associated with movement on the Kuantan Fault. Synthetic faults in the hangingwalls are generally more common than antithetic faults. These faults, however, seem to affect only the synrift strata. The strike-slip faults such as the Rumbia and Merchong border faults are associated with wrench-related features such as positive and negative flower structures overlying reactivated basement faults. The steep-sided asymmetrical geometry of the Merchong Graben is typical of pull-apart basins.

BASIN EVOLUTION

The Penyu Basin is one the three major extensional basins of interior Sundaland. These basins are believed to have formed as pull-aparts along a major NW-trending shear zone that was reactivated by the India-Asia collision during the Late Eocene (Tapponnier et al., 1982; Daines, 1985; Polachan et at.,1991). The exact timing of basin formation is unknown, but analogy with the West Natuna Basin (Daines, 1985; Wongsosantiko and Wirojudo, 1984; Ginger et al., 1993) suggests that the basin was formed during the Late Eocene to Early Oligocene (between 40 and 35 Ma).

Figure 9.6 is a kinematic model for the Penyu Basin based on the structural features observed in seismic (Mazlan Madon, 1995). The

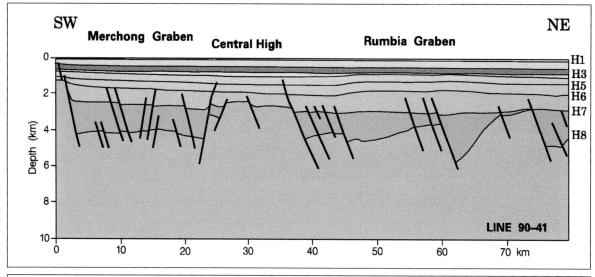

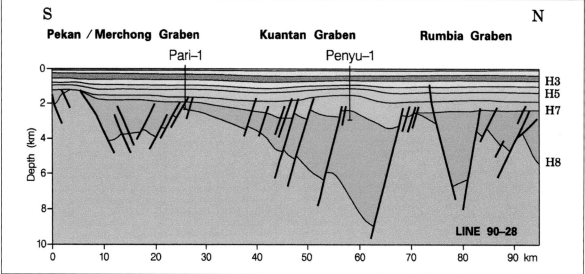

Fig. 9.5. Seismic profiles from the Penyu Basin showing the structural styles. See Fig. 9.2 for location. (A) Line 90-28, oriented N-S through the Kuantan and Merchong grabens. (B) Line 90-41, across the western part of the basin, showing the structural styles in the Merchong (SW) and Rumbia (NE) grabens. Note the steep bounding faults and complex intragraben faulting in the Merchong Graben.

basin appears to have been formed as a result of local transtension caused by regional stresses acting on pre-existing basement faults (Fig. 9.6A). The orientation and geometry of the half-grabens were controlled by the orientation of the faults relative to the regional extension direction. Besides crustal extension, deformation has also resulted in internal rotation of the fault-bounded crustal blocks. One of the major basement faults is the Rumbia Fault, which appears to be a reactivated basement fault. During the Early to Middle Miocene, the Penyu Basin underwent an inversion phase probably as a result of change in the regional stress field (e.g. Huchon et al., 1994; Tjia and Liew, 1996). North-south shortening across the basin resulted in the formation of E-trending, doubly-plunging anticlines, the so-called "Sunda folds", above the major half-grabens (Fig. 9.7). The largest inversion anticline occurs above the deepest part of the Kuantan Graben and has been tested by the Penyu-1 well (Fig. 9.5A). The geometry of the inversion structures suggests that they were strongly controlled by basement faults that bound the underlying half-grabens. Orthogonal extension and contraction in the Kuantan Graben have resulted in a simple inverted half-graben, whereas oblique extension/ contraction in the Rumbia and Merchong grabens produced wrench-fault structures.

The geometry and stratigraphic architecture of the Penyu Basin is typical of basins formed by lithospheric stretching. The average heat flow in the basin is about 85 mWm^{-2} (Mazlan Madon, 1995), which is similar to that of the West Natuna Basin (cf. Thamrin, 1985). Such a high heat flow implies the presence of a thermal anomaly beneath the basin that may have been caused by lithospheric stretching. The high heat flow may contribute to maturation of potential source rocks in the half-grabens. During the stretching (synrift) phase of basin development the half-grabens or subbasins evolved as separate depocentres controlled by basement faults. Rapid subsidence of the footwall near the half-graben bounding faults resulted in the accumulation of thick synrift sequences. During the postrift phase, however, the basin subsided regionally as the extended and heated crust cooled to attain the equilibrium temperature. This two-stage subsidence history is characteristic of basins formed by lithospheric stretching (McKenzie, 1978). Mazlan Madon (1995) estimated the amount of extension, or stretching factor (β), of 1.5 to 2.5 from subsidence analysis.

PLAY TYPES

Postrift Sunda Fold Play

The inversion (or Sunda) fold is the most obvious structure in the Penyu Basin and was the exploration target in the past. Besides the Penyu structure, the Cherating structure (Fig. 9.8) is an example of this type of play. Trapping is achieved by four-way dip closure of the elongate or domal anticlinal structures in the postrift section above half-graben depocentres. Potential reservoirs are within the Pari Formation, deposited in coastal plain environments. The source rocks are likely to be synrift lacustrine shales in the half-graben deeps. Migration of hydrocarbons may occur through vertical conduits such as faults, particularly the half-graben normal faults that propagate upwards into the postrift strata. Most of the major Sunda fold plays have been tested by drilling and turned out to be dry. Some of the reasons for the failure of this play type is that few faults connect the source facies with the postrift reservoirs, and that the Top-Pari Unconformity which truncates the Sunda folds may have caused hydrocarbons to be lost through the surface.

Late Synrift Basement-Drape Play

The basement drape play is a proven play in the late synrift sequence, which either onlaps or drapes over pre-rift basement high. The oil discovery at Rhu is an example of this type of play (Fig. 9.8). The main reservoirs consist of thick fluvial sandstones in the Penyu Formation. Interbedded claystone/shale intervals in the Terengganu Formation act as seals.

The Rhu oil appears to have migrated from lacustrine source facies in the deep half graben up the gently dipping ramp margin, into a structurally high anticlinal trap that was formed by late synrift sediment drape over a basement ridge. Anticlinal closure near the Base-Pari seismic horizon seems to persist downwards through the Terengganu and Penyu formations. The Rhu Ridge became the focus

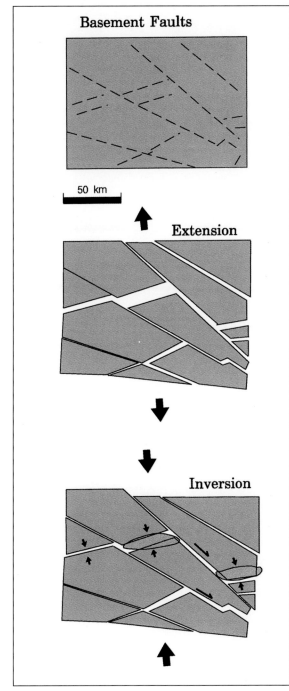

Fig. 9.6. Kinematic model for the development of the Penyu Basin by N-S extension. (A) Basin development was influenced by the presence of older E- and NW-trending basement faults (dashed lines). (B) Oligocene north-south extension caused opening of half-grabens along lines of weakness. (C) Middle Miocene compression caused inversion of half-grabens and folding of the graben fills to form Sunda-type folds.

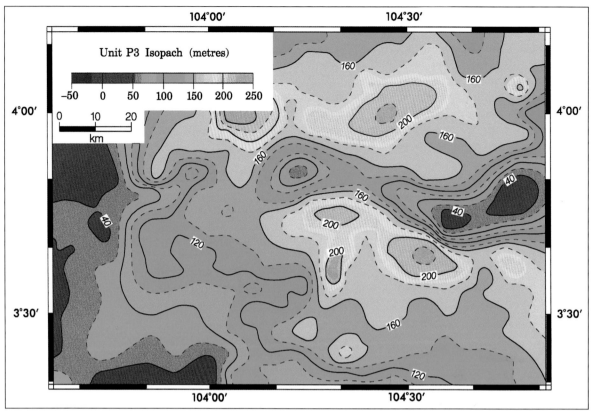

Fig. 9.7. Structural map of the Penyu Basin showing isopach of seismic unit P3, showing the effect of erosion/truncation at the Mid-Miocene Unconformity on thickness. Areas where Unit P3 is thin (less than 80 m) through the centre of the map (blue regions) represent inversion anticlines, which overlie major half-grabens. The domal structure in the centre overlies the Kuantan Graben and has been tested by the Penyu-1 well.

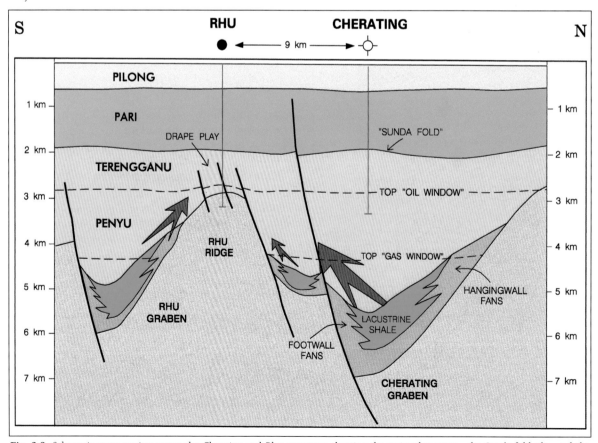

Fig. 9.8. Schematic cross section across the Cherating and Rhu structures showing the main play types - the Sunda-fold play and the basement-drape play. Oil that is trapped in the Rhu structure was generated from lacustrine source shales in the deep half-grabens (green shaded). Hangingwall and footwall fan sediments (orange shaded) in the synrift section provide potential new plays in the basin.

of migrating hydrocarbons because the sediment drape and the associated fault traps overlying the ridge were in existence before the onset of hydrocarbon migration. Hydrocarbons are sourced from mature oil-prone lacustrine sediments in the depocentre to the south of the Rhu Ridge. Hence, bed-parallel migration rather than vertical migration via fault conduits is the key element of this play.

Potential New Plays

Potential new plays which remain to be tested include fluvial deposits of the lower synrift graben-fill that onlap onto basement highs (Fig. 9.8). Both hangingwall and footwall fan deposits, sealed by impervious basement and intraformational shales, have the potential to trap large amounts of hydrocarbons that are

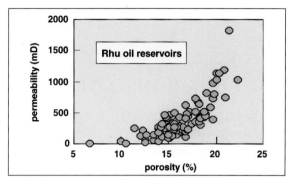

Fig. 9.10. Plot of sandstone porosity and permeability measurements from core samples of the Rhu oil reservoirs.

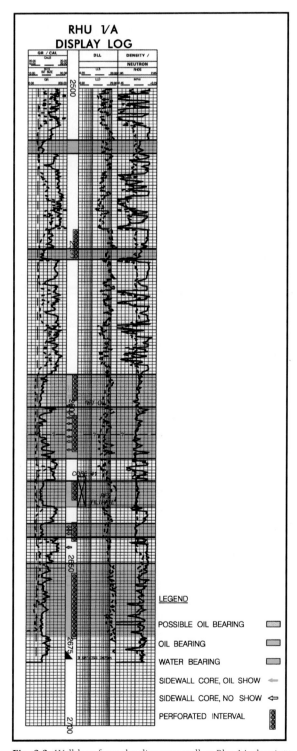

Fig. 9.9. Well-logs from the discovery well at Rhu-1A showing the main reservoir sandstones and perforated intervals. Depth in metres.

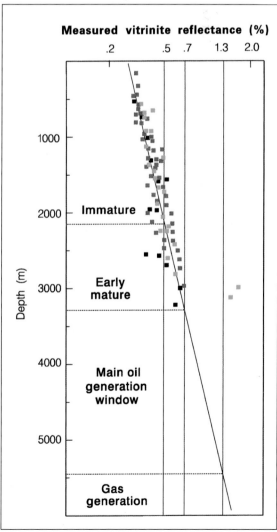

Fig. 9.11. Vitrinite reflectance measurements from Penyu Basin wells plotted against depth, showing a linear trend. The main oil-generation window is expected to occur between 3500 m and 5500 m.

generated by shales in the graben centre. Alluvial fans along the border faults may consist of coarse-grained sandstones and conglomerates which can have excellent reservoir quality. Because these plays are deeper (> 3 km) than the Rhu-type reservoirs, enough porosity must be preserved at depth. At Rhu, secondary porosity from fractures in the sandstones has upgraded the prospectivity of the play. Careful mapping of the seismic facies will help identify drilling targets.

THE RHU OIL DISCOVERY

The oil discovery at Rhu is proof of the existence of an oil system in the Penyu Basin. The discovery well Rhu-1/1A produced oil from the upper Penyu Formation with a cumulative production exceeding 6000 BOPD from reservoir sandstones at depth of between 2544 and 2815 m (Fig. 9.9). The reservoirs have porosities ranging between 10% and 22% and some have permeabilities of over 1000 mD (Fig. 9.10). Some of the reservoir zones have secondary fracture porosity. The well section is normal-pressured and encountered only minimal gas shows. The Rhu oil is moderately waxy (35° API gravity), has a high pour point (44°C), and a low sulphur content (<0.1 wt%). Table 9.1 compares the bulk properties of the Rhu oil with those of selected Malay and West Natuna basin oils.

The potential source rocks for the Rhu oil have not been penetrated and occur, probably, deep in the half-grabens. Vitrinite reflectance data from seven wells show a linear trend with the top of oil window (early mature 0.5-0.7% VRo) at about 2100 m depth (Fig. 9.11). If this trend is extrapolated downwards, the main oil generation window (0.7-1.3% VRo) should be at about 3200 m while the top of the gas generation window (<1.3% VRo) is at 5500 m (Fig. 9.8). This implies that the Oligocene synrift succession (Penyu Formation) is generally mature. The Rhu oil must, therefore, have come from source rocks within the Penyu Formation that have been buried deeper than 2km. Biomarker characteristics of the Rhu oil suggest that the source rock from which the oil was derived contains both algal and terrigenous organic matter and, therefore, was deposited in a relatively reducing lacustrine environment with a high terrigenous input.

Source-rock analyses show that the penetrated Penyu Formation shales contain generally less than 1% Total Organic Carbon (TOC) (Fig. 9.12), with the predominant kerogen type being Type III land-plant-derived organic matter (Fig. 9.13). Coals within the formation have greater TOC (10-26 wt %) and moderate to excellent generating potential. The younger postrift succession (Terengganu, Pari, and Pilong formations) could not have generated the Rhu oil because they are immature.

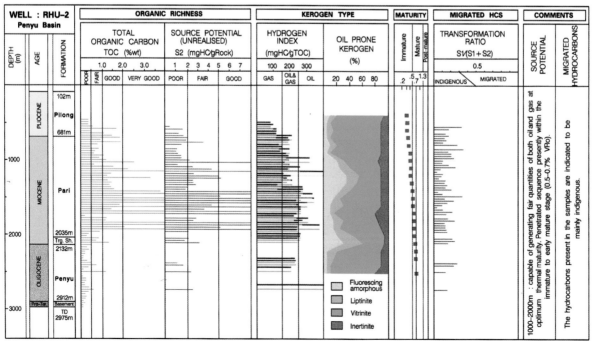

Fig. 9.12. Geochemical summary log of Rhu-2 well, Penyu Basin. The Miocene Pari Formation shows fair hydrocarbon generating potential, however the penetrated section is only within the immature to early mature stage.

Basin	Field	Gravity (°API)	Pour point (°C)	Wax (wt%)	Sulphur (wt%)
Penyu	Rhu-1A	35	44	27.8	0.1
Malay	Bertam	34	36	9.57	0.07
West Natuna	Kepiting	39	37.8	17.4	0.1
	Udang	39	37.8	17.4	0.1
	Ikan Pari	47	18.8	15.4	0.1
	KH, Kakap	46	23.3	19	0

Table 9.1. Bulk properties of Rhu oil compared with selected Malay Basin and West Natuna oils (source: IPA, 1989)

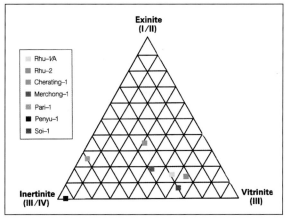

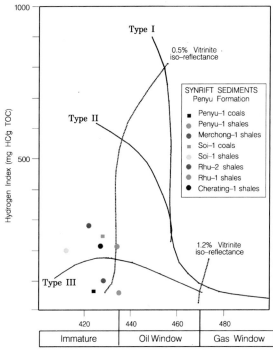

Fig. 9.13. Source-rock characteristics of the synrift sediments in the Penyu Basin. **Left:** Ternary plot showing the average kerogen assemblages. **Right:** Type of kerogen as indicated by the Hydrogen Index vs pyrolysis Tmax crossplot.

CONCLUSION

The Penyu Basin evolved from a nonmarine extensional basin during Oligocene times, brackish in Early-Middle Miocene, to coastal-shallow marine during late Miocene-Recent. Its structure is dominated by ENE-trending normal faults and NW-trending combined dip-and strike-slip faults. Fault geometry and orientation suggest basin formation by regional N-S extension. Pre-existing basement faults have exerted a strong control on basin development.

The main exploration play types in the Penyu Basin are: 1.Sunda fold play, 2. Late synrift basement-drape play, 3. Synrift stratigraphic play. Most of the Sunda folds have been tested by drilling but without success. The basement drape play has been proven by the oil discovery at Rhu. The more speculative synrift play holds promise, but detailed understanding of reservoir distribution is essential for identifying drilling targets.

Much of the penetrated section in the Penyu Basin is immature; the top of the oil window is about 2100 m depth. The source of the Rhu oil must therefore occur in the deeper parts of the basin. Biomarker characteristics indicate that the Rhu oil was probably derived from a source rock that was deposited in a reducing, saline, lacustrine environment, with a high terrestrial organic matter influx.

REFERENCES

Armitage, J.H. and Viotti, C., 1977, Stratigraphic nomenclature - southern end Malay Basin. Proceedings of the Indonesian Petroleum Association 6th Annual Convention, May 1977, 69 94.

ASCOPE, 1981, Tertiary Sedimentary Basins of the Gulf of Thailand and South China Sea: Stratigraphy, Structure and Hydrocarbon Occurrences. ASEAN Council on Petroleum Secretariat, Bangkok.

Daines, S.R., 1985. Structural history of the W Natuna Basin and the tectonic evolution of the Sunda region. Proceedings of the Indonesian Petroleum Association 14th Annual Convention, 8-10 1985, 39-61.

Ginger, D.C., Ardjakusuma, W.O., Hedley, R.J., and Pothecary, J., 1993. Inversion history of the West Natuna Basin: examples from the Cumi-Cumi PSC. Proceedings of the Indonesian Petroleum Association 22nd Annnual Convention, Jakarta, October 1993, 635-658.

Hamilton, W., 1979, Tectonics of the Indonesian region. U.S. Geological Survey Professional Paper No. 1078.

Huchon, P., Le Pichon, X., and Rangin, C., 1994. Indochina Peninsula and the collision of India and Eurasia. Geology, 22, 27-30.

Hutchison, C.S., 1989. "Geological Evolution of South East Asia". Oxford monographs on Geology and Geophysics No 13, Clarendon Press, Oxford.

IPA, 1989. Indonesia - Oil & Gas Field Atlas, Vol. 1: North Sumatra and Natuna. Indonesian Petroleum Association.

Khalid Ngah, 1975. Stratigraphic and structural analyses of the Penyu Basin, Malaysia. Unpublished MS thesis, Oklahoma State University.

Khalid Ngah, Mazlan Madon and H.D. Tjia, 1996. Role of pre-Tertiary fractures in formation and development of the Malay and Penyu basins. In: Hall, R. and Blundell, D., eds., "Tectonic Evolution of Southeast Asia". Geological Society of London Special Publication No. 106, 281-289.

Mazlan B. H. Madon, 1995. Tectonic evolution of the Malay and Penyu Basins, offshore Peninsular Malaysia. D.Phil. Thesis, University of Oxford, 325pp.

Mazlan B. Hj. Madon, Azlina Anuar, and Wong, R., 1997. Structural evolution, maturation history, and hydrocarbon potential of the Penyu Basin, offshore Peninsular Malaysia. In: Howes, J.V.C. and Noble, R.A., eds., Proceedings of the IPA Petroleum Systems of SE Asia and Australia, 21-23 May 1997, Indonesian Petroleum Association, Jakarta, 403-424.

McKenzie, D.P., 1978. Some remarks on the development of sedimentary basins. Earth Planetary Science Letters, 40, 25-32.

Polachan, P., Pradidtan, S., Tongtaow, C., Janmaha, C., Intarawirti, K., and Sangsuwan, C., 1991. Development of Cenozoic basins in Thailand. Marine and Petroleum Geology, 8, 84-97.

Tapponnier, P., Peltzer, G., Le Dain, A.Y., Armijo, R., and Cobbold, P., 1982. Propagating extrusion tectonics in Asia: new insights from simple experiments with plasticine. Geology, 10, 611-616.

Texaco, 1992. The significance of the Rhu oil discovery, PM-14; offshore Peninsular Malaysia: a preliminary summary. Unpublished Texaco Report, ER : TEXACO : 1-92-01

Thamrin, M., 1985. An investigation of the relationship between the geology of Indonesian sedimentary basins and heat flow density. Tectonophysics, 121, 45-62.

Tjia, H.D., 1994. Inversion tectonics in the Malay Basin: evidence and timing of events. Bulletin Geological Society of Malaysia, 36, 119-126.

Tjia, H.D. and Liew, K.K., 1996. Changes in tectonic stress field in northern Sunda Shelf basins. In: Hall, R. & Blundell, D.J., eds., Tectonic Evolution of Southeast Asia. Geological Society of London Special Publication No., 106, 291-306.

Wongsosantiko, A. and Wirojudo, G.K., 1984. Tertiary tectonic evolution and related hydrocarbon potential in the Natuna area. Proceedings of the Indonesian Petroleum Association 13th Annual Convention, Jakarta, August 1984, 202-211.

Chapter 10

Basins in the Straits of Melaka

*Mazlan B. Hj. Madon
and Mansor B. Ahmad*

INTRODUCTION

The Straits of Melaka, offshore Peninsular Malaysia, is one of the least explored sedimentary basinal areas in Malaysia. Figure 10.1 shows the current blocks and operator for PM320 and 322. Geologically, it represents the northeastern ramp margin of the North and Central Sumatra basins. The Tertiary sedimentary section here is more than 4000 m thick in some graben depocentres. This chapter briefly discusses the geology and hydrocarbon potential of this area. Very little has been published on the Malaysian part of the Melaka Straits. Most of the existing literature comes from the Indonesian side where petroleum exploration has been very successful. The stratigraphy of the northern part of the straits was first described by Nik Ramli Nik Hassan (1978). Liew (1994a, 1994b, 1997) described the occurrences of many fault-bounded grabens in the straits and the structural patterns in the basement. Much of the geological information discussed in this chapter was also derived from unpublished reports by oil companies, mainly Mobil and Sun.

EXPLORATION HISTORY

Exploration activity in the Malaysian Straits of Melaka began in 1971, when the Malaysian government granted a concession in the northern part of the straits (offshore Kedah and Perak) to Mobil Malaysia Exploration Company. In 1972 Mobil drilled the first wildcat well, MSS-XA (TD 1294 m subsea), to test the calcarenite and sandstone reservoirs below the Middle Miocene Baong Shale or its equivalent. The well encountered about 1100 m of Tertiary sediments above pre-Tertiary brecciated dolomite. In 1974 a second well, MG-XA (TD 1651 m), was drilled in the Central Graben by Mobil and Teijin. The well encountered a thick succession of coarse-grained sandstone, conglomerate, and red claystone, filling a deep half-graben. These sediments were interpreted as nonmarine alluvial deposits, overlain by 670 m of Pliocene marine sands and clays. Both the Mobil wells were drilled on poorly understood stratigraphic traps and failed to encounter any hydrocarbon shows. About 6,157 line km of seismic data were acquired by Mobil during this exploration phase before the concession was relinquished in 1976.

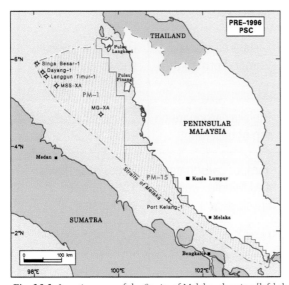
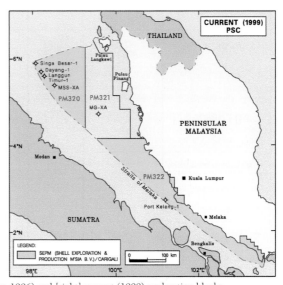

Fig. 10.1. Location map of the Straits of Melaka, showing [left] the old (pre-1996) and [right] current (1999) exploration blocks.

In 1985 PETRONAS subdivided the Straits of Melaka area into Blocks PM1 and PM15 (Fig. 10.1), and acquired about 1,958 line km of seismic data. In December 1987 Sun Malaysia Petroleum Company, in partnership with Champlin, Gulf Canada, Kerr McGee, and PETRONAS Carigali, signed a production-sharing contract (PSC) for Block PM1. A PSC for Block PM15 was awarded to the same consortium in June 1988. Sun acquired an additional 8,573 line km of seismic data in the two blocks. A geochemical "sniffer" survey, totalling 950 km of line traverses, was carried out in the northwestern corner of Block PM1 (Mansor Ahmad and Geneau, 1989; Geneau, 1990). Sun also carried out a gravity survey in Block PM15.

Three wells, Singa Besar-1, Langgun Timur-1 and Dayang-1, were drilled by Sun in Block PM1. The wells are located in the northwestern corner of the block which represents the northeastern ramp margin of the North Sumatra Basin (Fig. 10.1). The drilling objectives were the Lower Miocene Melaka carbonate reefs. The first well, Dayang-1 (TD 1142 m), was drilled in 1989. It encountered a pre-Tertiary basement high instead of the prognosed Lower Miocene reef. The overlying Lower Miocene section consisted of only calcareous and glauconitic sandstones with minor limestone beds. Minor gas shows were recorded in the sandstones. The second well, Singa Besar-1 (TD 844 m), tested another Lower Miocene reef play, but the Lower Miocene section is absent. Instead, the well found some gas in fractured pre-Tertiary limestone and dolomite at 769 m. The well produced at a rate of 3.7MMSCF of gas per day, with 39% CO_2. The Singa Besar discovery was declared non-commercial because of a high water flow during testing, believed to have been from deep-seated basement faults. In 1990 Sun drilled its last well in Block PM1, Langun Timur-1 (TD 2057 m), which was to test a reefal buildup along the same trend as the previous wells. The well found a pre-Tertiary basement of fractured dolomite and some patchy oil stains in the Lower Miocene Belumai section. Block PM1 was relinquished in 1991.

Sun drilled only one well in Block PM15 before relinquishing it in 1992. The Port Kelang-1 well, was drilled in 1991 to test Oligocene synrift sediments of the Pematang Group equivalent but found only sub-commercial traces of oil. The well was drilled to a TD of 1974 m subsea and penetrated a pre-Tertiary limestone basement at a shallower depth than expected. In 1996 PETRONAS re-classified the Melaka Straits area into three new blocks: PM320, PM321, and PM322 (Fig. 10.1). Table 10.1 summarises the drilling results in these areas.

WELL	OPERATOR	YEAR	TD (m)	OBJECTIVES	RESULTS
MSS-XA	Mobil	1972	1294	calcarenites and sandstones in below Baong Shale	Dry
MG-XA	Mobil	1974	1651	Synrift graben fill reservoirs	Dry
Dayang-1	Sun	1989	1142	Melaka carbonate/ Bampo Formation	Dry. Trace gas/oil shows in Baong sandstones
Singa Besar-1	Sun	1989	844	Melaka carbonate on top of basement	3.3 m NGS, flowing at 3.7 MMSCF/day
Langgun Timur-1	Sun	1989/ 1990	2028	Tampur carbonates and Middle Graben Fill clastics	Minor gas shows TD in Tampur carbonates
Port Kelang-1	Sun	1991	1374	Upper Pematang sandstones	Dry

Table 10.1. Summary of well results in the Straits of Melaka. Total depth (TD) in metres below sea level.

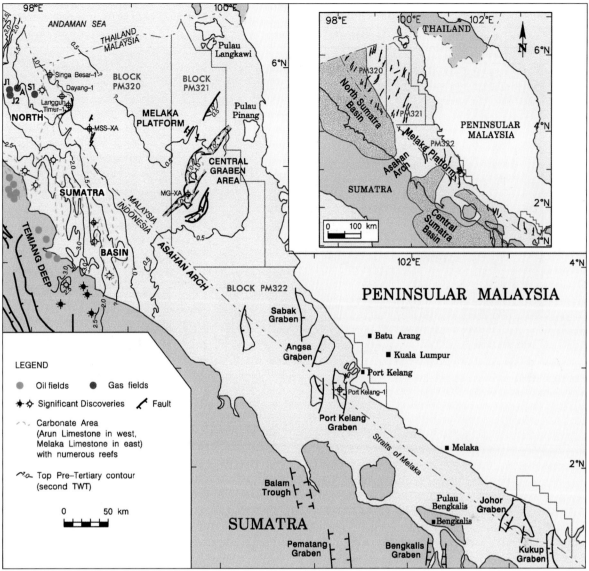

Fig. 10.2. Structural elements of the Straits of Melaka. The Malaysian part of the straits is regarded as the continuation of the North and Central Sumatra basins. Most major basement faults in the southern part are aligned N-S, similar to the grabens of the Central Sumatra Basin.

REGIONAL GEOLOGY AND STRUCTURE

The geological setting of the Straits of Melaka has been described in Chapter 5. The northwestern part of the straits (offshore Perlis, Kedah, and Perak) is underlain by up to 1600 m of Tertiary sediments. The pre-Tertiary basement geology of Peninsular Malaysia continues uninterrupted into Sumatra with no major structural offset (Hutchison, 1993). The basement structure is dominated by a combination of NW-trending horsts and grabens and N-S grabens (Moulds, 1989). The overlying sediments were deposited on this southwestward-tilting Sundaland basement (the Melaka Platform) which forms the northeastern flexural margins of the highly petroliferous North Sumatra Basin (Fig. 10.2). The southern part, stretching from offshore Selangor to offshore Johor, is characterised by fault-bounded grabens that are structurally similar to those in the Central Sumatra Basin. The geology of the North and Central Sumatra basins has been described in many publications by the Indonesian petroleum industry (e.g., de Coster, 1974; Kamili and Naim, 1973; Mertosono and Nayoan, 1974; Kingston, 1978; Eubank and Makki, 1981; Buck and McCulloh, 1994; Katz and Dawson, 1997). Both basins are believed to have formed by crustal extension in a back-arc setting associated with an oblique convergent margin along the Sumatra trench.

The major structural element in the northern straits (Blocks PM320 and PM321) is the gently west- to southwest-dipping Melaka Platform, which forms part of the northeastern flank of the North Sumatra Basin. Liew (1994a) identified at least 4 major N-S grabens in Block PM320, which he believed to be the northward continuation of major N-S

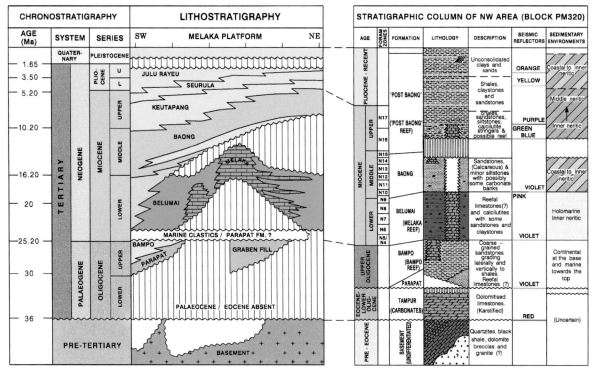

Fig. 10.3. Stratigraphic correlation between the North Sumatra Basin [left] (after Anderson et al., 1993) and [right] the NW area of Melaka Straits (after Murray, 1991).

structural features in Sumatra, namely the Tamiang and the Yang Besar highs. In most parts of the area, however, shallow pre-Tertiary rocks are overlain by an onlapping succession of upper Tertiary sandstone and claystone that thins progressively towards the northeast. Very little structural deformation has affected the succession. Miocene carbonates, locally formed by reefal buildups, occur in the northwestern part of the block. In the central-southern part of Block PM321, there is a series of NNE-trending grabens filled with a mainly continental succession of upper Tertiary sediments. Apart from the Central Graben, shown in Fig.10.2, Liew (1994a) recognised at least 5 other grabens in this area, including several major ones just west of Pulau Pinang.

Block PM322 (formerly PM15) is a narrow strip between southern Peninsular Malaysia and the Malaysia/Indonesia border. This area has a shallow pre-Tertiary basement and is part of the eastern margin of the Central Sumatra Basin. The Asahan Arch separates the North from the Central Sumatra basins. Several N-S trending fault-bounded grabens occur in the area; the major ones are the Port Kelang and Johor grabens (Fig. 10.2).

STRATIGRAPHY

In the northern straits, west of the 100°E meridian, there are 2 areas where relatively thick Tertiary sediments occur (Fig. 10.2). One is in the western edge of Block PM320, where the sediment thickness increases gradually westwards, towards the North Sumatra Basin. Another is in the Central Graben area in Block PM320. The southern straits (Block PM322) have a number of grabens that may have the potential for petroleum exploration. The stratigraphy of these three areas is discussed below.

NW Area (PM320)

The western edge of Block PM320 forms part of the northeastern margin of the North Sumatra Basin, where oil and gas have been found in Miocene reefs, such as in the Arun gas field (Soeparjadi, 1982). The geology of the area is described in unpublished reports by Sun Malaysia Petroleum Company (Hatch, 1987; Murray, 1991). Figure 10.3 shows the stratigraphy of the area. In this part of the North Sumatra Basin, the oldest sediments are Oligocene synrift siliciclastic sediments overlying pre-Tertiary basement of diverse lithologies, which include metasedimentary and igneous rocks. Overlying the basement in places are pre-rift carbonates of ?Cretaceous-Eocene age (Tampur Formation). Major extension and graben formation occurred during the Oligocene, with deposition initially in a terrestrial environment. This resulted in the deposition of sandstones and conglomerates at the graben margins that grade laterally into, and

Play elements	North Sumatra Basin (Buck and McCulloh, 1994)	Central Sumatra Basin (Katz and Dawson, 1997)
Source	Multiple sources, all marine shale with Type III land plant-derived kerogen. Generally lean (<1wt% TOC) but can be very thick in graben deeps (>1.8 km). Principal source: Oligocene Bampo Fm shales/claystones. Secondary: Peutu Fm basinal shales and marls. Baong Fm shales may be important source in E and SE parts.	Pematang Group "Brown Shale", middle synrift lacustrine facies. Oil prone Type I/II kerogen, 20-47° API oil, 0.2 wt% S, 4-46°C pour point. separate lacustrine depocentres, and hence source-rock kitchens
Reservoir	Primary: Miocene Peutu Fm reefal carbonates. e.g. Arun carbonates, average ϕ 16%. Secondary porosity due to meteoric water flushing.	Primary: Basal Sihapas Group "Menggala sandstone", representing early postrift transgressive shallow marine/intertidal, ϕ 25%, k 400-450 mD Secondary: Upper Pematang redbeds, synrift fluvial deposits, ϕ 15-20%, k <100 mD
Seal	Overpressured shales 1. Baong Fm top seal 2. Lateral clastic/fine carbonate equivalent of Peutu reefs 3. Overpressured Bampo Fm shale as bottom seal.	Telisa Fm shale as regional seal, mid-outer shelf sediment
Trap	Mainly stratigraphic: carbonate reefs encased in shale	Mainly structural: fault/structural closure + top seal, Miocene/later strike-slip related structures, drag folds, inversions, thrusts. Drapes over palaeohighs (e.g. Minas) Some stratigraphic traps.
Timing of generation/migration	Generation since Middle Miocene to present, peaked 12-4 Ma. Focussed migration through faults, basal clastics, and unconformity surfaces.	Generation during the Miocene to present.

Table 10.2 Comparison of the main hydrocarbon systems in the North and Central Sumatra basins (sources: Buck and McCulloh, 1994; Katz and Dawson, 1997).

are overlain by, lacustrine to lagoonal organic-rich shales of the Bampo Formation. These shales are the major source rocks in the North Sumatra Basin (Table 10.2).

Because their distribution is controlled by the basement topography, the synrift rocks are known by various names in different parts of the basin. The Bruksah Formation contains coarse clastics, derived locally from basement highs, and grades upwards and interfingers with the shallow-water bioclastic Jeuku Limestone, and with shales and claystones of the Bampo Formation. The latter contains deepwater foraminifera and, rarely, nannofossils of Early to Late Oligocene age (Buck and McCulloh, 1994).

More marine conditions were established during latest Oligocene and Early Miocene for carbonate reefs to develop, initially as fringing reefs (Bampo reefs) and later as buildups on structural highs (Melaka reefs). The latter grew to over 240 m high and have become the primary drilling objective. Reef growth was terminated during the Middle Miocene by the influx of coarse siliciclastics (Baong Formation). In Blocks PM320/321 the Baong Formation includes a laterally extensive unit of calcareous sandstone that acts as a conduit for hydrocarbon migration from the basin centre to the west. The top of the Baong Formation is marked by a regional Upper Miocene unconformity which, in places, cuts down into the underlying carbonates, increasing the porosity and permeability locally by epidiagenetic solution.

The overlying post-Baong Formation sediments were deposited in a shallow marine environment during a regional transgressive episode. Post-Baong pinnacle reefs developed on topographic highs but were eventually drowned by marine inundation. The non-carbonate sediments consist of interbedded sandstone, shale, and claystone that range in age from Late Miocene to Quaternary.

Central Graben Area (PM321)

The Central Graben is located in the central part of the PM321 block, east of the intersection between the Melaka Platform and the Asahan Arch (Fig. 10.2). It lies to the north of, and is on trend with, the Central Sumatra Basin. Hence, its stratigraphy is believed to be similar to that of the Central Sumatra Basin and of the grabens in PM322, but markedly different from that of the NW area (Block PM320). Only a limited amount of stratigraphic data is available from Mobil's MG-XA well.

The geology of the area was reviewed by Greenaway and Goh (1989). Figure 10.4 shows the stratigraphy interpreted from seismic data, compared with that established in the Central Sumatra Basin. The pre-Tertiary basement consists of metamorphic quartzites, shale and limestones of indeterminate age. The early graben fill comprises continental to lacustrine sediments of Upper Eocene-Oligocene age, equivalent to the Pematang Group. Some reservoir-quality sandstones are developed, while the interbedded lacustrine brown shales are very rich source rocks that have provided the source for the Central Sumatra oilfields. The upper part of the graben-fill sequence comprises fluvio-deltaic sediments (upper Pematang and lower Sihapas equivalents) which consist mainly of sandstones and minor shale interbeds. Following a widespread Early to Middle Miocene inversion event (represented by the base-Minas unconformity), a lower coastal plain to shallow marine sequence (Telisa and Minas formations equivalent) was deposited during Late Miocene to Recent times.

Structurally, the Central Graben area consists of a number of smaller grabens, formed by Oligocene extension. Later structural modification associated with right-lateral wrenching occurred during the Middle-Late Miocene. The MG-XA well penetrated only the

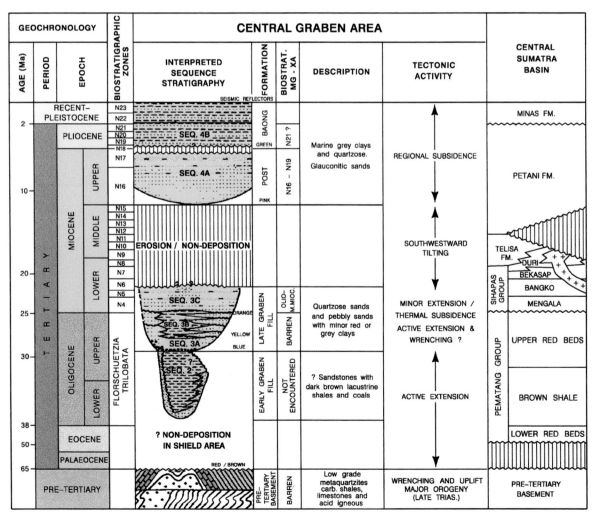

Fig. 10.4. Sequence stratigraphy of the Central Graben area (after Greenaway and Goh, 1989) correlated to the Central Sumatra stratigraphy based on Koning and Darmono (1984) and Williams et al. (1985).

younger (late) graben fill sequences of predominantly sandy facies. The well reached a total depth of 1651 m in pre-Tertiary basement. Seismic data indicate, however, that the sediment thickness may be as much as 3900 m thick. Organic shales in the Pematang-equivalent section are likely to be thermally mature and to have generated hydrocarbons.

Southern Area (PM322)

The southern Straits of Melaka is much less explored. This area has a relatively shallow basement, overlain by a thin veneer of sediment. Thicker sedimentary sections occur in several major grabens (Fig. 10.2). These grabens are bounded by NE- and N-trending faults, which are parallel in trend to the grabens in central Sumatra and the Central Graben area (Block PM321). Some grabens appear to extend on land into Peninsular Malaysia. An example is the Batu Arang basin, which is filled with a coal-rich fluvio-lacustrine sequence (Batu Arang Beds) overlain unconformably by the conglomeratic Boulder Beds (Stauffer, 1973; Raj, 1998). Raj (1998) correlated the coal measures of the Batu Arang beds (Upper Eocene-Lower Oligocene) with the Parapat/Graben Fill sequence of the North Sumatra Basin and with the Pematang Group of the Central Sumatra Basin. The Boulder Beds (Middle Miocene-Pliocene) are thought to be equivalent to the Baong and Minas formations of the North and Central Sumatra basins, respectively (Raj, 1998).

The geological evolution of the southern Melaka Straits is probably similar to that of the Central Sumatra grabens (cf. Williams and Eubank, 1995; Katz and Dawson, 1997). Rifting during the ?Eocene to Oligocene resulted in nonmarine clastic sedimentation (Pematang Group) in N- to NW-trending half-graben depocentres. Deposition was accompanied by folding, faulting, and uplift. During the Early to Middle Miocene, the basin underwent regional thermal subsidence, accompanied by predominantly marine sedimentation (Sihapas Group). Another phase of uplift, folding and dextral wrench faulting during the late Early to Middle Miocene resulted in the formation of traps for many of the major hydrocarbon accumulations in Central Sumatra (Williams and Eubank, 1995). These structures were later subjected to a Plio-Pleistocene deformation which produced a strong NW-SE structural overprint over Central Sumatra.

The grabens of the southern Melaka Straits are thought to be related to major graben

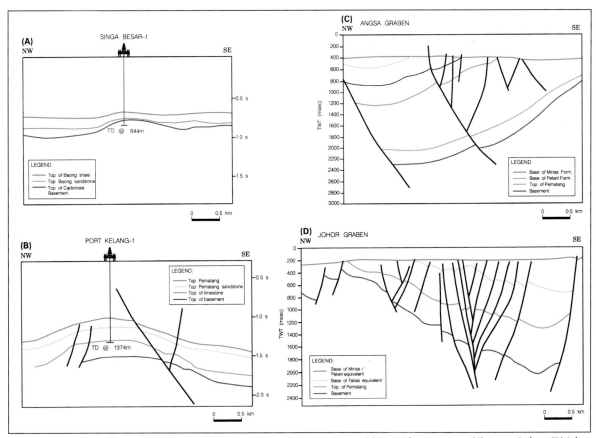

Fig. 10.5. Structural styles in the southern Melaka Straits. (A) Singa Besar structure. (B) Port Kelang structure. (C) Angsa Graben. (D) Johor Graben. See Fig. 10.2 for location of profiles. Total depth (TD) of wells in metres below sea level.

systems identified in Central Sumatra to the south (Liew, 1994a). The Sabak, Angsa, and Port Kelang grabens are probably part of the Pematang-Balam system, which includes the Pematang Graben, shown in Fig. 10.2. The Johor and Kukup grabens are probably related to the Bengkalis graben system. Among these offshore grabens, the Angsa Graben has the thickest sedimentary fill, of up to 2.4 s two-way time. Figure 10.5 shows cross-sections across the Angsa and Johor grabens, which have the structural style typical of extensional half-graben basins. Compressional strike-slip faulting seems to have occurred in some places.

Liew (1994a) described the Port Kelang Graben, which is a NNW-trending graben located to the west of Port Kelang on the west coast of Peninsular Malaysia. The graben is about 15 km wide and 4 km deep. It is bounded by N-trending normal faults. Some of these basement faults have been reactivated by a ?Middle-Late Miocene wrench deformation to form anticlinal, sometime flower, structures. One of these have been drilled into by the Port Kelang-1 well (location in Fig. 10.2). Except for this deformation event, the area seems to have undergone a relatively uninterrupted regional subsidence as a result of the gentle southwestward tilting of the Melaka Platform. Figure 6.5 (Chapter 6) summarizes the stratigraphy of the Port Kelang-1 well, drilled by Sun in 1991. The succession consists of interbedded sandstone, shale, and coal deposited in environments ranging from alluvial, through lacustrine, to coastal plain and fluviomarine to shallow marine.

HYDROCARBON PLAYS AND PROSPECTIVITY

Two main types of hydrocarbon play have been tested (Fig. 10.6): the carbonate play (pre-Tertiary, Eocene, Lower Miocene) and the synrift siliciclastics play (Oligocene). These are described below.

Carbonate Play (North Melaka Straits)

The pre-Oligocene carbonate reservoir has been the main drilling objective in the western part of Block PM320. Sun drilled 3 wells to test this play type: the Dayang-1, Singa Besar-1, and Langgun Timur-1 wells. Fractured pre-Miocene shelfal carbonates were penetrated in all the three wells. The first well was dry, but minor gas shows were found in the next two wells (Table 10.1). The Singa Besar-1 well found gas in dolomitised limestone, thought to be the equivalent of the Eocene Tampur limestone. Only a thin unit of the Miocene Melaka reef was found in this well. The Tampur carbonates are platform limestones that have been subjected to subaerial exposure. Leaching and erosion could have resulted in the generation of secondary porosity, thereby enhancing its reservoir properties. A Miocene marine transgression resulted in the deposition of the Baong Formation siliciclastics in the area, which provide a secondary reservoir objective. The Baong Shale provides the top seal, whereas the interbedded sandstones act as conduits for hydrocarbon migration from the west (Fig. 10.6). Long-distance migration of hydrocarbons is necessary for this play to work. Faulting during Plio-Pleistocene times may have created migration routes that bypass the North Sumatra area and charged the Singa Besar discovery. No source rock has been identified in the area. Any source rock would be immature because of the shallow depth of burial.

Besides the Eocene Tampur carbonates, there is potential for finding hydrocarbons in the Lower Miocene Melaka reefs equivalent to the Peutu carbonates of North Sumatra. Figure 10.7 shows the location of Lower Miocene reefal buildups in the Belumai Formation that have been mapped from seismic data by Hatch (1987). The Dayang-1 and Singa Besar-1 wells were also targetted at these buildups.

Synrift Clastics Play (Central/Southern Melaka Straits)

Grabens in the Melaka Straits (see Figs. 10.2, 10.5) were formed as a result of crustal extension during early Tertiary times. The main reservoir objectives in these grabens are the synrift siliciclastic sediments of the Eocene-Oligocene Pematang Group. Pematang fluvial sandstones are proven hydrocarbon reservoirs in the Central Sumatra Basin. Interbedded lacustrine shales act as the top seal and source rock for the hydrocarbons. The synrift play in the Central Graben (PM321) was tested by Mobil's MG-XA well.

The only well drilled in the southern Straits of Melaka (PM321) is located in the Port Kelang Graben, some 20 km southeast of Port Kelang (Fig. 10.2). The Port Kelang-1 well tested an inverted fault-bounded structure in the Pematang Group. The drilling objectives were the Upper Oligocene to Middle Miocene fluvial and lacustrine sandstones in the Pematang and

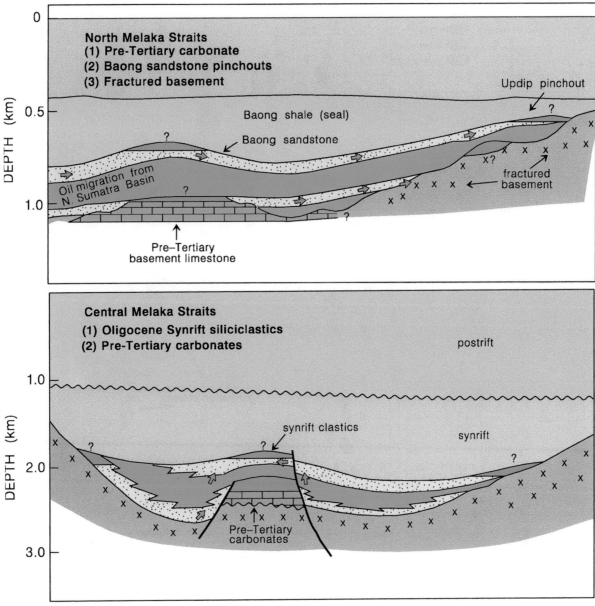

Fig. 10.6. Schematic illustrations of the main types of hydrocarbon play in the Straits of Melaka. Top: Possible play types in the northern Melaka Straits include pre-Tertiary carbonate and fractured basement, and the updip pinchouts of the Baong sandstone. Pre-Tertiary carbonates are abundant on land, e.g., Chuping Limestone. The traps are expected to be charged by long-distance migration from the west. Bottom: Play concepts in the central Melaka Straits include the Oligocene synrift siliciclastics and the pre-Tertiary basement carbonates in horst blocks. Unlike in the north, pre-Tertiary carbonates are scarce on land, e.g., the Silurian Kuala Lumpur Limestone. A local source rock kitchen formed by synrift lacustrine shales within the grabens is expected.

Sihapas equivalents, which are directly overlain by the Minas equivalent. The Petani equivalent is missing, because of truncation by the base-Minas unconformity (see Fig. 5.9, Chapter 5). Port Kelang-1 found traces of oil show, which indicate that there is a working hydrocarbon system. Geochemical analysis of the Pematang shales indicate the presence of organic-rich oil-prone source rocks (Fig. 10.8). Geothermal gradient (43.5 °C km^{-1}) and vitrinite reflectance data indicate that shales in the Pematang section may be thermally mature in the graben deep (Fig. 10.9). Burial and temperature history modelling (Wong, 1991) indicates that the oil window is at a depth of between 2200 m and 3200 m (equivalent VRo 0.64%-1.27%). Sealing may be a problem, however, because of the truncation of the regional seal (Petani Formation) by the base-Minas unconformity.

Other Plays

During the early stages of exploration in the early 1970s, the emphasis had been to search for stratigraphic traps in the NW Area (PM320). The Baong sandstone, for example, is a known reservoir in some fields of the North Sumatra Basin, and may form possible traps in its up-dip pinchouts against the pre-Tertiary basement to

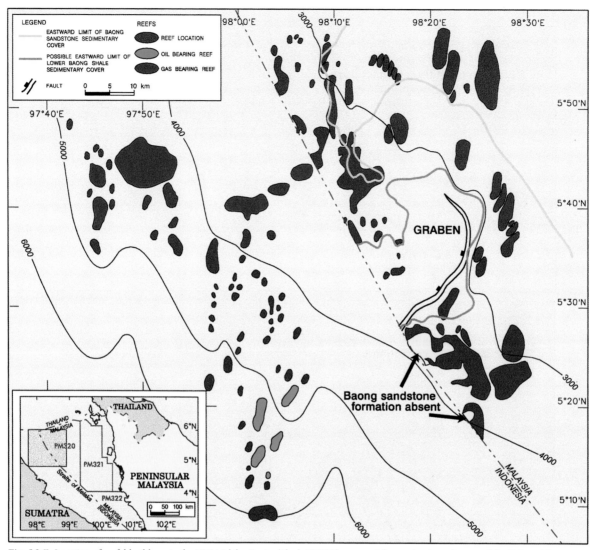

Fig. 10.7. Location of reefal buildups in the NW Melaka Straits (Block PM320), mapped from seismic at the top of the Belumai Formation equivalent. Also shown are the eastern limits of the Baong shale and sandstone units. Note the oil- and gas-bearing reefal buildups on the Sumatran side of the straits. Modified from Hatch (1987).

the northeast (Fig. 10.6). The MSS-XA well was drilled to test these stratigraphic traps. This well was risky, however, because of poor source potential of the sedimentary section at the location. The accumulation of hydrocarbons in these traps relies on the long-distance migration from deeper parts of the basin to the west. The uncertain presence of an effective top seal was also a major risk factor. Since then, exploration focus has shifted towards the synrift graben fill in the Central Graben area and in the southern part of the straits. This resulted in the drilling of the MG-XA well in the Central Graben, and more recently the Port Kelang-1 in the southern part of the straits.

Mansor Ahmad et al. (1998) described possible hydrocarbon plays that remain to be tested. These include some stratigraphic as well as structural traps. Stratigraphic traps are formed by transgressive sands of the Baong Formation and the onlap synrift sediment in grabens. There are also the basement-drape play, graben structural play, and fractured basement carbonate play (Fig. 10.6). In the NW area, compression-related structural traps may also form potential plays in the Upper Miocene-Pliocene post-Baong section, which are the equivalents of the Keutapang and the Julu Rayeu formations.

CONCLUSION

The Malaysian part of the Straits of Melaka is still poorly explored. Structurally, the PM320/321 area is dominated by a series of N-S-trending grabens, formed on a westward tilting Sundaland basement. The major depocentre lies to the west in Indonesia while the grabens become shallower to the east. Prospective traps are formed by carbonate reefs, low-relief structural closures in the Baong sandstones, and stratigraphic pinch-outs of the Baong and post-Baong sandstones. The first two wells by Mobil in the NW area (now

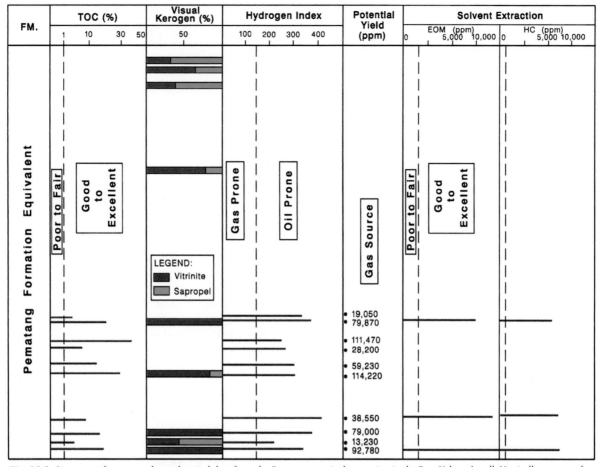

Fig. 10.8. Summary of source rock geochemical data from the Pematang-equivalent section in the Port Kelang-1 well. Vertically not to scale.

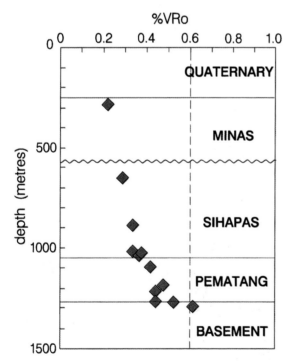

Fig. 10.9. Vitrinite reflectance (VRo) versus depth in the Port Kelang-1 well. Note that the Minas Formation rests unconformably upon the Sihapas Group. The Petani-equivalent section is missing.

Block PM320) in 1972 and 1974 (MSS-XA and MG-XA, respectively) were designed to test potential stratigraphic traps but neither encountered hydrocarbons due to the lack of a top seal, good source, or reservoirs.

Significant amounts of mature source rocks are present in the North Sumatra Basin, being more gas-prone in the northwest and oil-prone in the southeast. The westernmost grabens in PM320 and PM321 may contain mature sediments, at least locally. Hydrocarbon charge is likely to occur via the laterally extensive Baong sandstones from the basin depocentre or vertically from the local graben depocentres. Numerous direct hydrocarbon indicators (DHIs) have been identified from seismic data. The presence of gas in Singa Besar-1 indicates that a viable petroleum system is at work.

The major risks in this area include the presence of valid traps and the maturity of source rocks in the grabens where sediment is only about 2 seconds two-way time in apparent thickness. Furthermore, many structures are small and complexly faulted. The erosion of the regional top at the base-Pliocene unconformity may inhibit the trapping of hydrocarbons.

REFERENCES

Anderson, B.L., Bon, L. and Wahono, H.E., 1993. Re-assessment of the Miocene stratigraphy, palaeogeography and petroleum geochemistry of the Langsa Block in the offshore North Sumatra Basin. Proceedings of the Indonesian Petroleum Association 22nd Annual Convention, Jakarta, 169-190.

Buck, S.P. and McCulloh, T.H., 1994. Bampo-Peutu Petroleum System, North Sumatra, Indonesia. In: Magoon, L.B. and Dow, W.G., eds., The Petroleum system - from source to trap. American Association of Petroleum Geologists Memoir, 60, 625-637.

de Coster, G.G., 1974. The geology of the Central and South Sumatra basins. Proceedings of the Indonesian Petroleum Association 3rd Annual Convention, 3-4 June 1974, 77-110.

Eubank, R.T. and Makki, A.C., 1981. Structural geology of the Central Sumatra Back-Arc Basin. Proceedings of the Indonesian Petroleum Association 10th Annual Convention, 26-27 May 1981, 153-196.

Geneau, M.E., 1990. A discussion of "Sniffer" geochemical surveying offshore Malaysia. Bulletin of the Geological Society of Malaysia, 25, 57-73.

Greenaway, P. and Goh, K., 1989. Geoscience report Central Graben study area, Block PM1, Malaysia. Sun Malaysia Petroleum Company, Unpublished Report. ER:SUN:1:89-30.

Hatch, G., 1987. Technical evaluation and project summary for PM-1 Block, Straits of Malacca, Malaysia. Sun Malaysia Petroleum Company, Unpublished Report. ER:SUN:1:87-01.

Hutchison, C.S., 1993. Gondwanaland and Cathaysian blocks, Palaeothethys sutures and Cenozoic tectonics in South-East Asia. Geologisches Rundschau, 82, 388-405.

Kamili, Z.A. and Naim, A.M., 1973. Stratigraphy of Lower and Middle Miocene sediments in North Sumatra Basin. Proceedings of the Indonesian Petroleum Association 2nd Annual Convention, June 4-5, 1973, 53-72.

Katz, B.J. and Dawson, W.C., 1997. Pematang-Sihapas Petroleum System of Central Sumatra. In: Howes, J.V.C. and Noble, R.A., eds., Proceedings of the IPA Petroleum Systems of SE Asia and Australasia Conference, Jakarta, Indonesia, May 1997, Indonesian Petroleum Association, Jakarta, 685-698.

Kingston, J., 1978. Oil and gas generation, migration and accumulation in the North Sumatra Basin. South-East Asia Petroleum Exploration Society Proceedings, IV, 1977/78, 158-182.

Koning, T. and Darmono, F.X., 1984. The geology of the Beruk Northeast Field, Central Sumatra: oil production from pre-Tertiary basement rocks. Proceedings of the Indonesian Petroleum Association 13th Annual Convention, 29-30 May 1984, 385-406.

Liew, K.K., 1994a. Timing of Cenozoic basin formation in northern Sundaland, Southeast Asia. American Association Petroleum Geologists International Conference and Exhibition, Kuala Lumpur, Malaysia, 21-24 August 1994. Bulletin of the Geological Society of Malaysia, 37, 231-251

Liew, K.K., 1994b. Structural patterns within the Tertiary basement of the Straits of Malacca. (abstract). Warta Geologi, 20(3), 233-234.

Liew, K.K., 1997. Structural patterns within the Tertiary basement of the Straits of Melaka. PETRONAS Research Quarterly, 3 (11), 9-10.

Mansor Ahmad and Geneau, M.E., 1989. A discussion of Sniffer surveying, offshore Malaysia. [abstract]. Petroleum Geology Seminar, Kuala Lumpur, 4-5 December 1989. Warta Geologi, 15(6), 292.

Mansor Ahmad, Hamdan, A., Wong, R.H.F. and Muzammal, A.G., 1998. Hydrocarbon prospectivity of the Straits of Melaka. Ninth Regional Congress on Geology, Mineral and Energy Resources of Southeast Asia - GEOSEA '98, 17-19 August 1998, Kuala Lumpur, Programme and Abstracts, 146.

Mertosono, S. and Nayoan, G.A.S., 1974. The Tertiary basinal area of Central Sumatra. Proceedings of the Indonesian Petroleum Association 3rd Annual Convention, 3-4 June 1974, 63-76.

Moulds, P., 1989. Development of the Bengkalis Depression, central Sumatra, and its subsequent deformation - a model for other Sumatran grabens? Proceedings of the Indonesian Petroleum Association 18th Annual Convention, 1, 217-245.

Murray, C., 1991. Review of the prospectivity of Block PM-1, Straits of Malacca, Malaysia. Sun Malaysia Petroleum Company, Unpublished Report. ER:SUN:1:91-15.

Nik Ramli Nik Hassan, 1978. The stratigraphy of the northern part of the Straits of Melaka [abstract]. Geological Society of Malaysia, Petroleum Geology Seminar, Petaling Jaya, Selangor, 11 December 1978. Warta Geologi, 5(4):53.

Soeparjadi, R.A., 1982. Geology of the Arun Gas Field. Proceedings of the Offshore South East Asia 82, 9-12 February 1992, Singapore.

Stauffer, P.H., 1973. Cenozoic. In: Gobbett, D.J. and Hutchison, C.S., eds., "Geology of the Malay Peninsula". Wiley-Interscience, New York, p. 143-176.

Raj, J.K., 1998. Tectonic evolution of the Tertiary Basin at Batu Arang, Selangor Darul Ehsan, Peninsular Malaysia. Proceedings of the Seminar on Tertiary Basins of Peninsular Malaysia and its adjacent offshore areas. Geological Society of Malaysia, 21-22 February 1998, Kuala Lumpur.

Williams, H.H., Kelley, P.A., Janks, J.S. and Christensen, R.M., 1985. The Palaeogene rift basin source rocks of Central Sumatra. Proceedings of the Indonesian Petroleum Association 14th Annual Convention, 8-10 October 1985, 14/2, 57-90.

Williams, H.H. and Eubank, R.T., 1995. Hydrocarbon habitat in the rift graben of the Central Sumatra Basin, Indonesia. In: Lambiase, J.J., ed., "Hydrocarbon Habitat in Rift Basins". Geological Society of London Special Publication, 80, 331-371.

Wong, D., 1991. Review of the prospectivity of PSC Block PM15 (Post Port Kelang-1), Straits of Melaka, Malaysia. Sun Malaysia Petroleum Company, Unpublished report.

Chapter 11

Petroleum Resources, Peninsular Malaysia

Robert Wong Hin Fatt

INTRODUCTION

This chapter describes the petroleum resources of Peninsular Malaysia with reference to the intensity of exploration activity and the most significant plays in the 3 offshore Tertiary basins.

The first exploration well was drilled in 1969 in the Malay Basin. By the end of 1997, 158 wildcat and 184 appraisal wells had been drilled (Fig. 11.1). This figure shows the peak in the number of wells drilled following the awards of the 1976 and 1985 PSCs. A new peak will soon follow the awards of the 1995 R/C PSC. Each round of exploration is triggered by more progressive fiscal terms.

These exploration efforts resulted in the discovery of 12.5 BSTB OIIP with EUR of 4.3 BSTB in addition to 57.1 TSCF GIIP with EUR of 39.4 TSCF excluding small oil and gas fields of EUR less than 8 MMSTB and 50 BSCF respectively. These reserves are contained in 53 oil fields and 28 gas fields and they are located almost entirely in the Malay Basin. These include 14 'giant' oil fields and 6 'giant' gas fields, each with recoverable reserves of more than 100

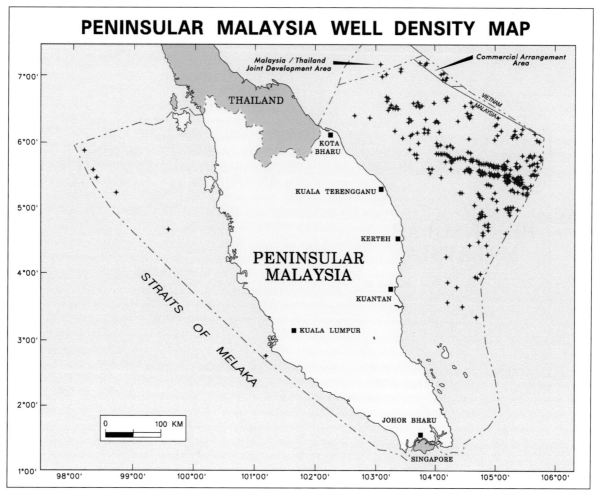

Fig. 11.1 Map showing well location/density in Peninsular Malaysia. Most of the wells are located in the oil-prolific Malay Basin.

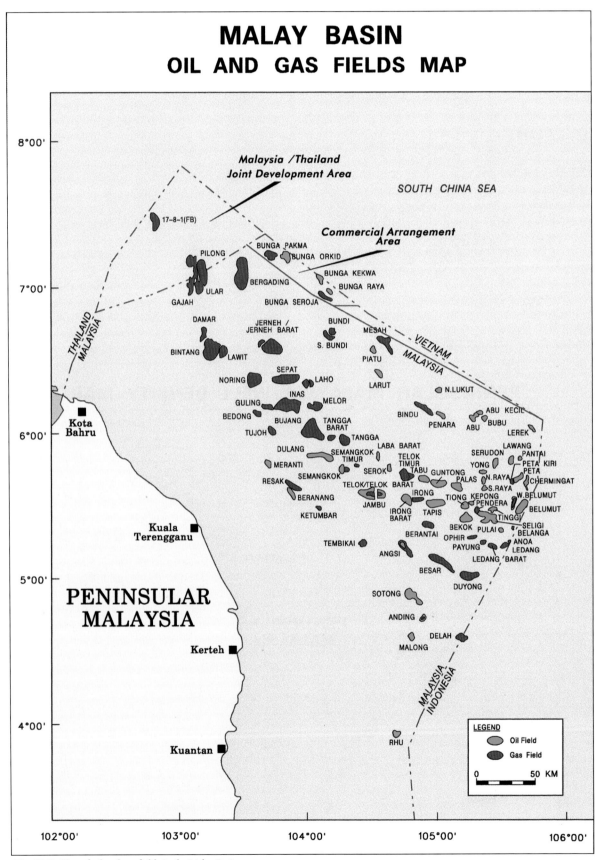

Fig. 11.2 Map of oil and gas fields in the Malay Basin.

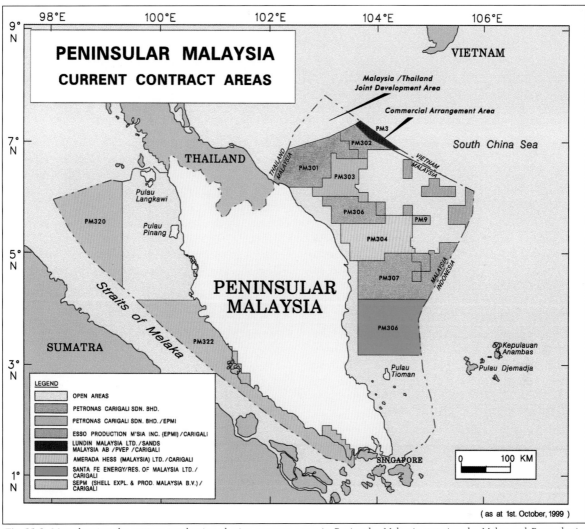

Fig. 11.3 Map showing the current production sharing contract areas in Peninsular Malaysia covering the Malay and Penyu basins and Melaka Straits.

MMSTB oil and 1 TSCF gas. The petroleum from the 15 producing oil fields and the 4 producing gas fields are piped to the onshore facilities at Kerteh, Terengganu (Fig. 11.2). Overall annual field growth of around 5% has been achieved due to discovery of new oil pools and better reservoir management. Up to 1.10.1999 the following oil fields came on stream : Raya, Yong, Malong-Sotong-Anding (MASA).

As at 1.10.1999, there are 13 active PSCs in Peninsular Malaysia area (Fig. 11.3). PETRONAS Carigali Sdn Bhd (Carigali) operates Blocks PM301, PM302 and PM306/307 under the R/C PSC. Carigali is also the operator of the PM9 PSC under the global arrangement, the Ular Gas Field in Block PM2 and 3 oil fields in Block PM6/12 under the 1985 PSC terms. Esso operates Blocks PM5 and PM8 under the 1985 PSC. Esso is also the operator of 7 producing oil fields in the 1995 Global PSC and 6 gas fields under the recently signed Gas PSC. Under the 1976 PSC, Esso still holds 3 gas fields.

Amerada Hess operates Block PM304 under the R/C PSC. Another independent, Santa Fe Energy, operates Block PM308 also under the R/C PSC. Finally, the PM3 Commercial Arrangement Area (PM3 CAA) is held by a third independent, Lundin Oil, under the 1985 PSC. Shell Exploration and Production Malaysia operates the Melaka Straits and Malay Basin blocks.

OIL AND GAS RESOURCES BY BASIN

There are three main sedimentary basins within the Peninsular Malaysia region. The oil and gas-prolific Malay Basin is the largest which contains more than 12,000m of sediments and occupies an area of about 78,000 sq. km. This is followed by the Penyu Basin of 14,000 sq. km. with more than 5,000m of sediments. Both these basins are located offshore east of Peninsular Malaysia. The Melaka Straits encompasses the northeastward extension of both the North and Central Sumatra basins. The sedimentary cover is generally thin apart from thicknesses of up to 3,000m. in the few grabens located within Malaysian territory.

2D seismic acquisition began in 1968 offshore Terengganu. By end 1997, more than 194,000 line-km of 2D seismic had been acquired in the three main sedimentary basins of Peninsular Malaysia (Fig. 11.4). 3D seismic acquisition began in 1984. By end 1997 more than 247,000 line-km of 3D seismic data had been acquired within the Malay Basin (Fig. 11.5).

By 1.1.1998, 410,000 km of seismic data (including 20 3D seismic surveys) had been completed, and 328 exploration wells drilled in the Malay Basin; 14,000 km of 2D seismic shot and 8 exploration wells drilled in the Penyu Basin; and 17,000 km of 2D seismic concluded and 6 exploration wells drilled in the Melaka Straits. It should be noted that the resource figures relate to activities as at 1.1.1998. During 1998, 6 exploration wells were drilled in the Malay Basin and 3901km 2D and 56018km 3D seismic were shot.

Malay Basin

The Malay Basin can be divided into six regions namely; Northeast, Southeast, North, West, Central and South based on their respective play types and geographical locations (Fig. 11.6).

In terms of exploration activity intensity, the Southeast Region (Fig. 11.7) of the Malay Basin ranks highest, with 46 wildcat and 91 appraisal wells, followed by Northeast Region, and the Central Region. Intense exploration in the Northeast Region only began in the late 80s following the introduction of the 1985 PSC fiscal terms. Numerous discoveries have been made in the Northeast Region. The North Region (Fig. 11.8) with 12 wildcats is essentially a gas province. As gas resources are now becoming more important, a new surge in exploration activity in the North Region is expected. The West Region (Fig. 11.8), although covering a large part of the Malay Basin, is only tested by 12 wildcats and 4 appraisals. This region is probably under-explored and future large discoveries can be expected.

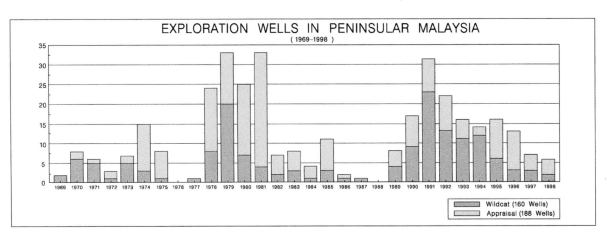

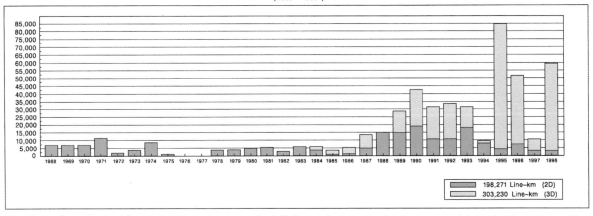

Fig. 11.4 Charts showing the number of exploration wells drilled and the amount of seismic acquired in Peninsular Malaysia. The increases in wells drilled follow new rounds of PSCs. The increase in the line-km of seismic in the late eighties and nineties is largely due to the voluminous appraisal and development 3D seismic.

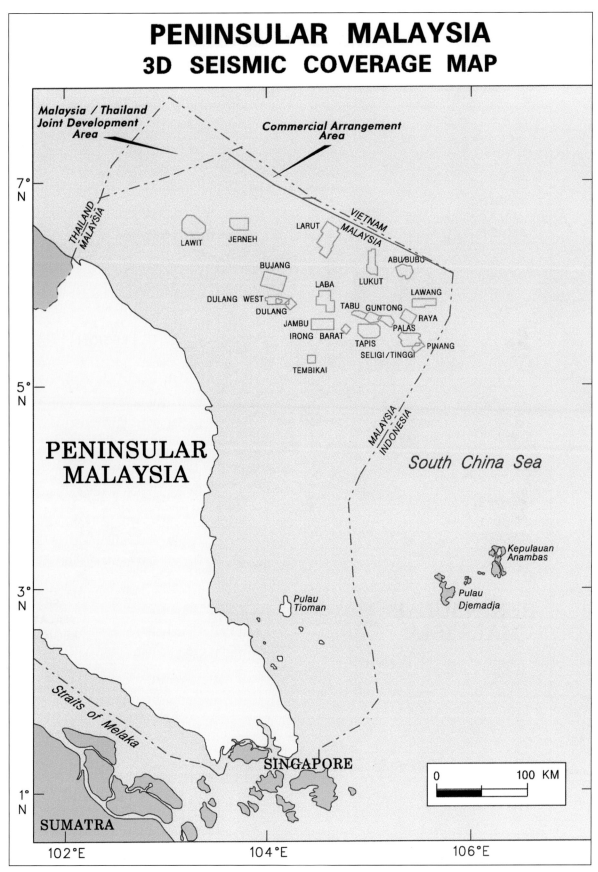

Fig. 11.5 Map showing the 3D seismic coverage in Peninsular Malaysia. They are confined to the 20 oil and gas fields to be appraised and developed.

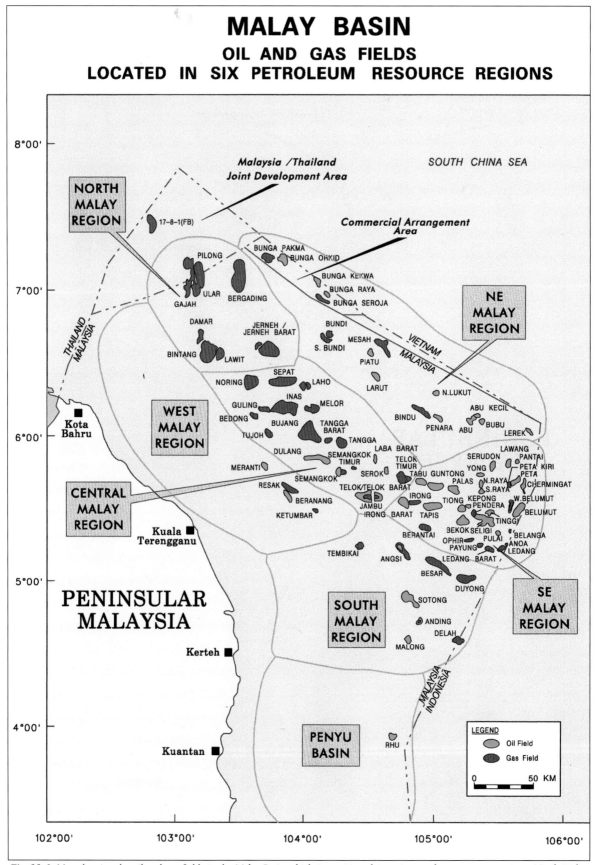

Fig. 11.6 Map showing the oil and gas fields in the Malay Basin which is partitioned into six petroleum resource regions- North Malay, West Malay, Central Malay, South Malay, SE Malay and NE Malay.

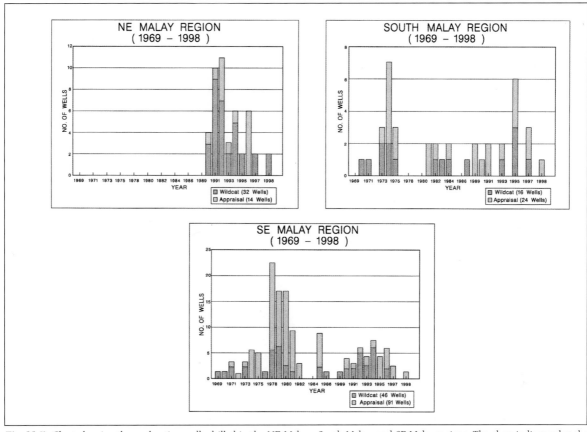

Fig. 11.7 Chart showing the exploration wells drilled in the NE Malay, South Malay and SE Malay regions. The chart indicates that the NE Region only came into prominence during the 1985 PSC, whereas the SE Region was the highlight of the 1976 PSC.

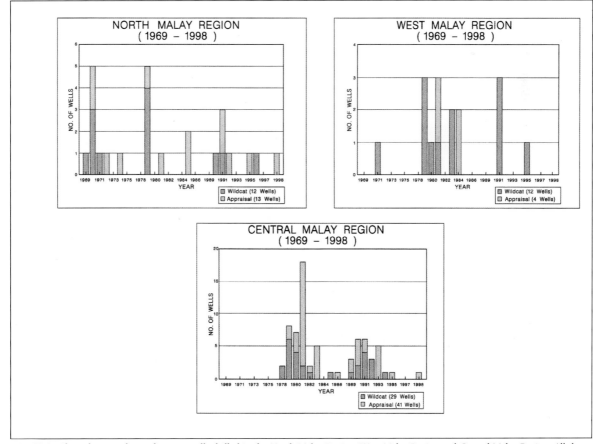

Fig. 11.8 Chart showing the exploration wells drilled in the North Malay Region, West Malay Region and Central Malay Region. All three regions were explored from the concession period through the 1976 PSC period to the 1985 PSC. In comparison with other regions, the West Region and the North Region appear to be under-explored.

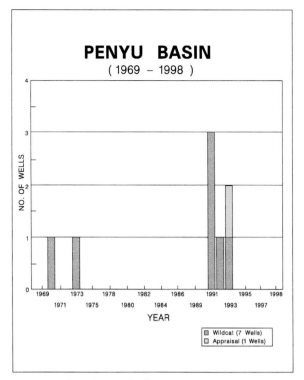
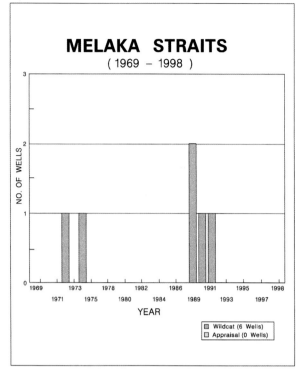

Fig. 11.9 Chart showing the few exploration wells drilled in Penyu Basin and Melaka Straits. Melaka Straits has no discoveries so far. Penyu Basin has one oil discovery. The next generation of large discoveries could possibly come from the Penyu Basin.

Southeast Region

For discovered oil resources to date, the Southeast Region ranks the highest with 69 % of EUR of oil in the entire Malay Basin (Fig. 11.10). This region contains several producing oil fields which are of giant size (EUR > 100 MMSTB) - Seligi (589 MMSTB), Guntong (573 MMSTB), Tapis (484 MMSTB), Bekok (296 MMSTB), Tiong/Kepong (232 MMSTB), Irong Barat (152 MMSTB), Tabu (121 MMSTB), Tinggi (125 MMSTB), Pulai (112 MMSTB) and Palas (115 MMSTB).

The most important play type in the Southeast Region is the J sands, particularly the J20 fluvial-braided sands trapped in large compressional anticlines. Other significant reservoirs within the same play type are the K and L sands. These are followed by Group I sands deposited in coastal plain to tidal settings and also trapped in large compressional anticlines.

The Southeast Region contains 62 % of the entire Peninsular Malaysia gas reserves (Fig. 11.11). 86 % of all associated gas, and 15 % of all non-associated gas in the Malay Basin occurs in the Southeast Region (Figs 11.12, 11.13).

Northeast And Central Region

In spite of more wildcats being drilled in the Northeast Region than in the Central Region, the latter area has higher volumes of discovered oil and gas resources (Fig. 11.10, Fig. 11.11). However, for both regions, the volume of discovered oil and gas resources are far below those of the Southeast Region. The Central Region contains several oil and gas fields; two giant oil fields are in production (Semangkok and Dulang). The larger gas fields (Non-Associated Gas) are Noring, Inas, Bujang, Tangga Barat and Telok. The most significant reservoirs in the Central Region are the Groups E and H tidal to fluvial sands. The Dulang Field reservoirs are Group E sands.

Numerous oil and gas fields have been discovered in the Northeast Region of the Malay Basin. The significant oil fields are Larut, Bunga Kekwa, Bunga Raya, Bunga Orkid and Bunga Pakma. Bunga Kekwa came on stream in 1997. This field are situated in the Malaysia-Vietnam commercial arrangement area. The most significant play type in the Northeast Region is the faulted structures of Group I and J fluvial to tidal sands. Other important plays occur in stratigraphic and structural traps of H channel sands and onlaps of K and L fluvial sands onto basement highs.

South Region

The South Region of the Malay Basin (Fig. 11.6) saw a new surge of exploration activities recently with the latest oil discovery in Angsi. The results

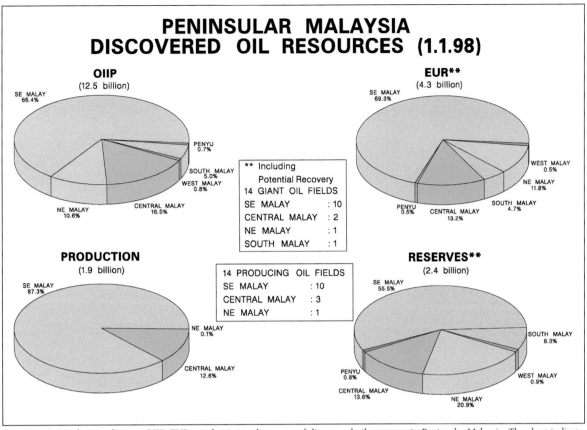

Fig. 11.10 Pie charts indicating OIIP, EUR, production and reserves of discovered oil resources in Peninsular Malaysia. The chart indicates the most oil-prolific region is the SE Region of the Malay Basin.

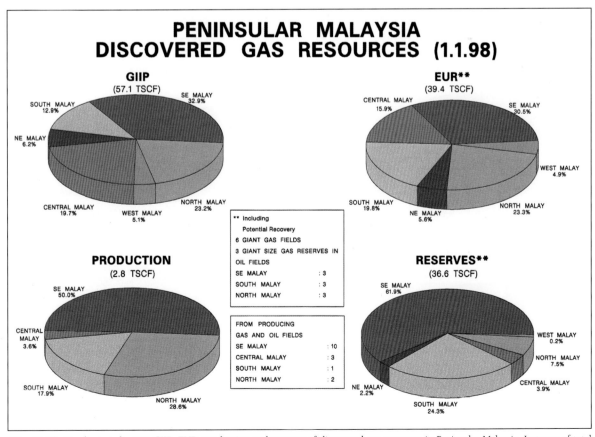

Fig. 11.11 Pie charts indicating GIIP, EUR, production and reserves of discovered gas resources in Peninsular Malaysia. In terms of total gas resource, the SE Region of the Malay Basin stands out.

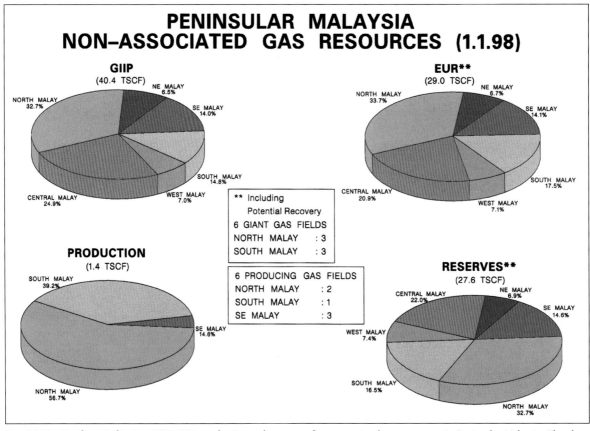

Fig. 11.12 Pie charts indicating GIIP, EUR, production and reserves of non-associated gas resources in Peninsular Malaysia. The chart depicts that the North Region of the Malay Basin yields the most non-associated gas resource

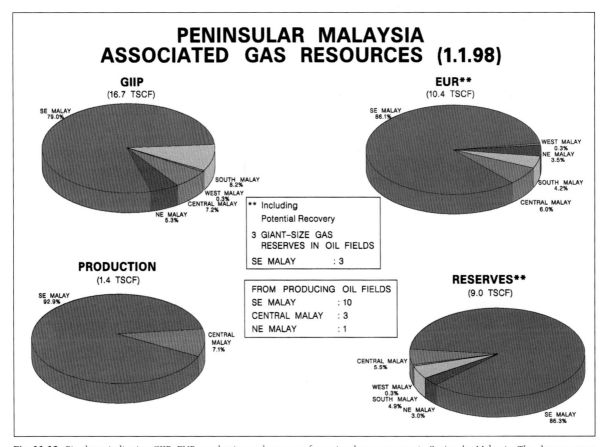

Fig. 11.13 Pie charts indicating GIIP, EUR, production and reserves of associated gas resources in Peninsular Malaysia. The chart suggests that the SE Region of the Malay Basin contains the most associated gas resource.

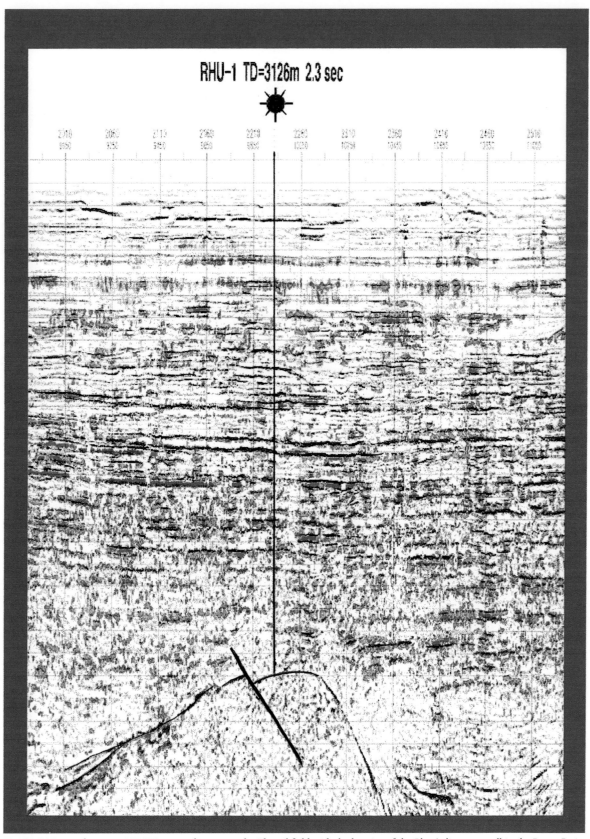

Fig. 11.14 Figure showing a NE-SW seismic line across the Rhu oil field with the location of the Rhu-1 discovery well in the Penyu Basin. The productive zones are situated just above the basement horst block at depths of 2500-2700m.

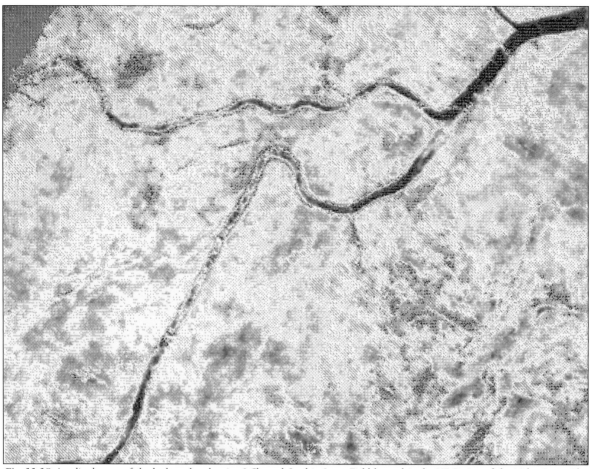

Fig. 11.15 Amplitude map of the hydrocarbon bearing I Channel Sand in Larut Field located in the NE Region of the Malay Basin. The meandering nature of the bright coloured channel is clearly seen.

indicate that a large oil discovery is still possible if hydrocarbon exploration continues with new geological ideas. The Angsi Field was first considered a gas field when the deeper Group K sands were tested gas bearing. The shallower and more elusive Group I channel sands only encountered oil shows initially. Only with the second Angsi well was the oil proven in the Group I channel sands. These were subsequently delineated by 3D seismic and tested by more appraisal wells. The South Region incorporates the Tenggol Arch area which includes the Malong oil discovery. There are several oil and gas discoveries in this region, with the Duyong gas field in production, and an integrated Malong-Sotong-Anding (MASA) field now in production. Resak gas field is under development.

West Region

The West Region of the Malay Basin contains a number of oil and gas discoveries, notably the Resak gas field. Only 12 wildcats have been drilled in this large area, thus making this region relatively unexplored. The main play type is Groups I, J, K and L sands in extensional fault play traps. The H and I channel sands, although tested by a few wells, and shown to contain sub-commercial quantities of oil, need further exploration efforts, such as 3D seismic, in order to delineate these sands more accurately. Another Angsi look-alike hidden in the haphazard coverage of seismic and well data may be awaiting discovery by the astute explorationist.

North Region

The North Region of the Malay Basin is essentially a gas province containing 34% of non-associated gas found in the basin (Figs. 11.11, 11.12) with several giant gas discoveries, notably the producing Jerneh (4.9 TSCF) and Lawit (2.1 TSCF) gas fields. Total GIIP and EUR in this region are 14 TSCF and 10 TSCF respectively. Although 12 wildcats have been drilled, a new surge in gas exploration activity in the North Region is anticipated following the discovery well Bergading Deep (Fig. 11.20). The main exploration play in this region is the faulted anticlines of Groups D and E fluvial to coastal

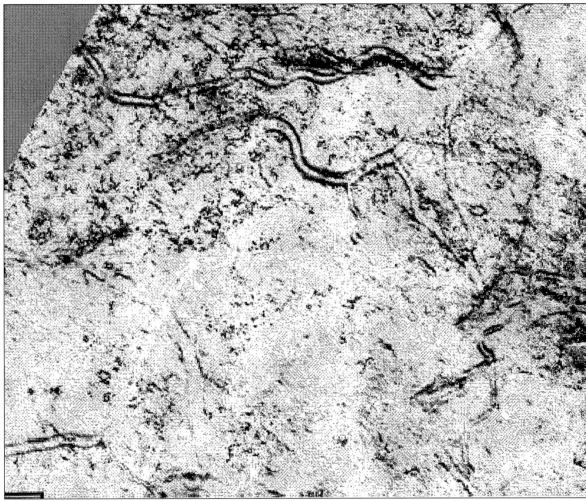

Fig. 11.16 Coherency cube slice of the same I Channel Sand in Figure 11.15. The black and white coloured display indicates additional smaller channels attached to the main channel in the south. The channel sands are shown in white and the edge of the channels are indicated by the dark curved lines.

sands. Other plays include fault traps of Groups H and I meandering channel sands below the overpressured Group F zone. These plays underlie the shallower Groups D and E gas reservoirs and the overpressured Group F shales.

Penyu Basin

The Penyu Basin has only 8 wells (Fig. 11.9) drilled in it during the last 20 years of exploration. The sole discovery is a marginal oil field (Rhu) reservoired in the Oligocene synrift sand (Fig. 11.14). Many deep half-grabens in this basin are receptacles of oil-prone source rocks. Although oil shows have been encountered in the Sunda folds, i.e. post-rift sediments that have undergone mild compression, this play is perceived to be inadequately tested. Based on the number of wells drilled (Fig. 11.9), the basin is also grossly under-explored like the West Region of the Malay Basin. The next big discovery could possibly come from this basin.

Melaka Straits Basins

A total of 6 wells were drilled in the Melaka Straits (Fig. 11.9). Four were drilled in the shallow platform area of the North Sumatra Basin. One well encountered gas within the Pre Tertiary carbonates. Effective long range migration is critical for accumulation of hydrocarbon in this play. Two wells were drilled in the graben play within the Central Sumatra Basin. One well encountered oil shows. The validity of the graben play depends on thick source facies and the adequate burial depth, as most of the grabens are small and shallow. The renewed exploration in this region is encouraging.

Onshore Basins

No exploration drilling has ever been carried out in the onshore basins. The petroleum potential of the onshore Pre-Tertiary basins is given in Chapter 26. A description of several small coal-bearing Tertiary basins is given in Chapter 7.

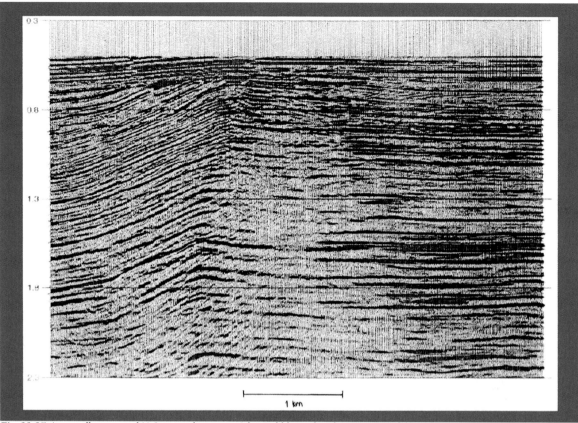

Fig. 11.17 A normally processed N-S seismic line across Seligi Field located in the SE Region of the Malay Basin. Notice the poor quality of the data and the sagging of the seismic events at the crestal area probably due to the gas accumulations.

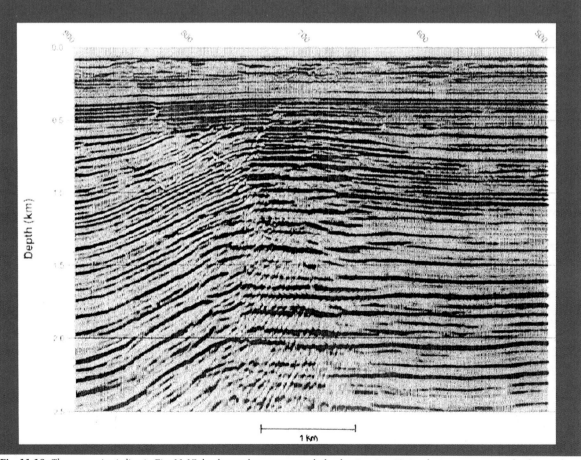

Fig. 11.18 The same seismic line in Fig. 11.17 that has undergone pre-stack depth migration. Notice the improvement in the data quality. The gas sag has been almost completely removed.

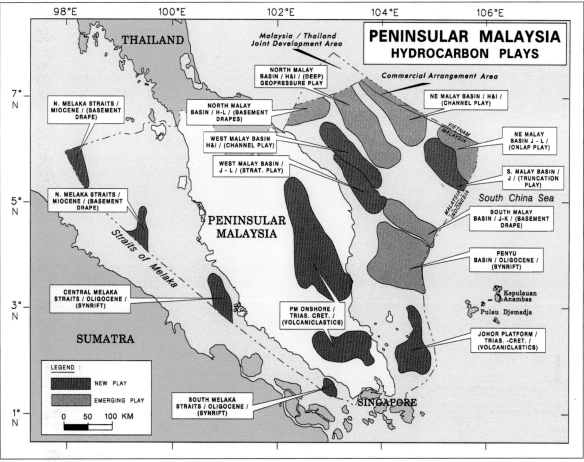

Fig. 11.19 Map showing the location of the 9 new and 6 emerging hydrocarbon plays in Peninsular Malaysia. 5 of the emerging plays are situated in the Malay Basin and 1 in Penyu Basin. 3 of the new plays are in the Malay Basin, 1 in Johor Platform, 1 onshore, and 4 in the Melaka Straits.

FUTURE OUTLOOK

Prospects for new discoveries

The recent entry of new oil companies, e.g. Amerada Hess and Santa Fe Energy, augurs well for future discoveries in previously explored areas, albeit unsuccessful.

Prospects for reserves growth in known fields/discoveries

Efforts towards enhancing reserves growths have been spearheaded by PETRONAS under the Low Resistivity/Low Contrast Project. Under this project, several fields have been identified in Peninsular Malaysia where the water saturations have been over-estimated. Such fields include Seligi, Tapis, Jerneh and Guntong. These by-passed reservoirs are now included in the reserves figure thereby making the reserves in each field larger. On-going study continues in the minor reservoirs of the Irong Barat Field. Enhanced Oil Recovery studies are also on-going. Reliance on more sophisticated interpretation tools is needed to realise the potential of the known fields and the new plays. Figure 11.15 exhibits a 3D seismic amplitude map of I Channel Sand in the Larut Field located in the Northeast Region of the Malay Basin, with the corresponding coherency cube slice shown in Figure 11.16. Figure 11.17 demonstrates a normally processed seismic line across the Seligi Field located in the Southeast Region of the Malay Basin. This line has undergone pre-stack depth migration processing (Fig. 11.18) to minimise the gas sag effect on the seismic data.

New and emerging plays

Studies by PETRONAS and Production Sharing Contractors on the future exploration potential leading to estimates of undiscovered petroleum resources, have unveiled two categories of hydrocarbon plays - new and emerging hydrocarbon plays. A new play is defined as a geological concept based on experience of the explorationists, or on a geological analogue derived from comparative study of the same or an adjacent basin. A lot more work has to be carried out on the source rocks, reservoirs and seals to reduce the exploration risk. An emerging play is a geological idea that is considered to be matured for drilling, or that has been proven to be valid by limited drilling. Each play is described in the following format: Location/Reservoir-Type/Play Type.

Fifteen hydrocarbon plays comprising 6 emerging and 9 new plays have been outlined in Peninsular Malaysia (Fig. 11.19).

The 6 emerging plays consist of 5 plays in the Malay Basin and one in the Penyu Basin. The 5 emerging plays in the Malay Basin are:

1) North Malay Basin/H-L/Basement Drape
2) North Malay Basin/H&I/Deep Geopressure
3) NE Malay Basin/H&I/Channel Play
4) NE Malay Basin/J-L/Onlap Play
5) South Malay Basin/J-K/Basement Drape

The emerging play located in the Penyu Basin is the Penyu Basin/Oligocene/Synrift.

An example of a new emerging play is illustrated in Figure 11.20. This is a ENE-WSW seismic line across the Bergading Gas Field within the North Malay Basin/H&I/Deep Geopressure play. This field was discovered by the Bergading Deep-1 well drilled in late 1995. The well went through the overpressure zone within Group F before it encountered the

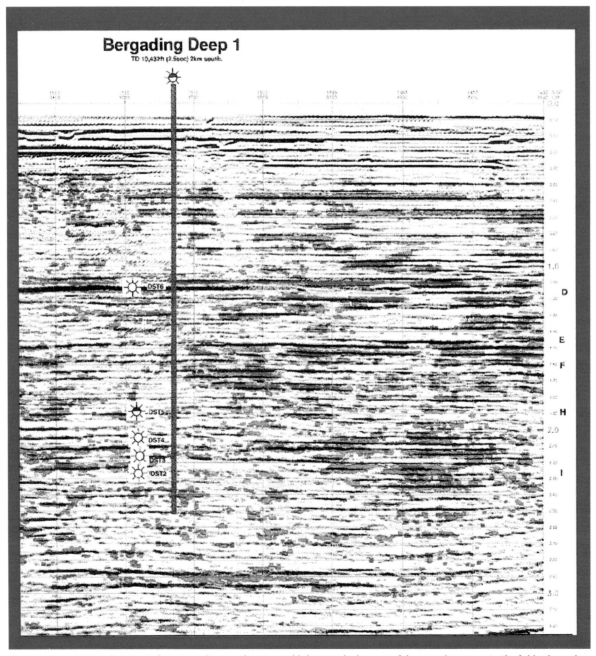

Fig. 11.20 An ENE-WSW seismic line across the Bergading Gas Field showing the location of the Bergading Deep-1. The field is located in the North Region of the Malay Basin. This line is extracted from the Bergading 3D seismic cube.

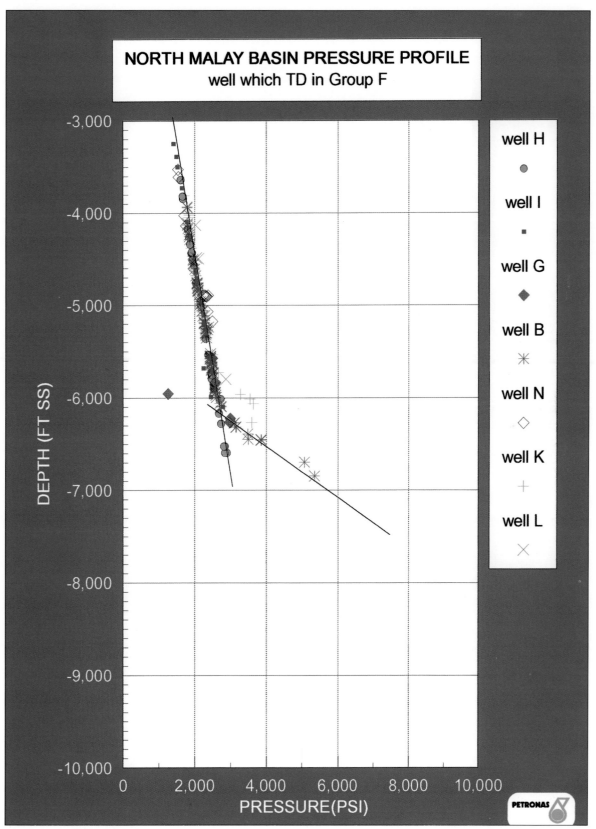

Fig. 11.21 North Malay Region pressure profile of wells which bottomed in Group F. The data portrays a rapidly increasing trend in Group F.

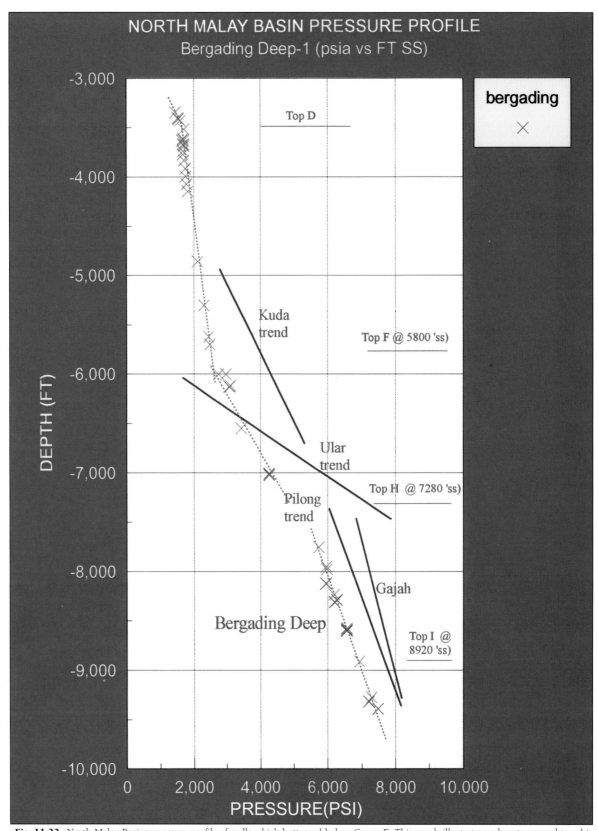

Fig. 11.22 North Malay Region pressure profile of wells which bottomed below Group F. This graph illustrates a close to normal trend in the Groups H and I below the overpressured Group F shales.

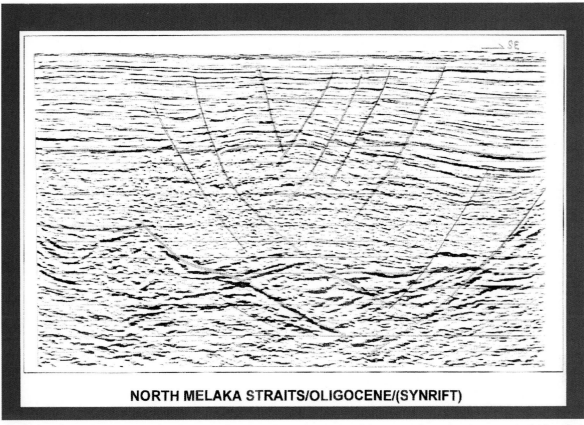

Fig. 11.23 A seismic line to illustrate the new hydrocarbon play in the North Melaka Straits. This is the Oligocene Synrift Play which contains a faulted anticline within the synrift sediments.

sweet spot in Group H. Formation pressure from wireline test was plotted against depth in 7 wells drilled in the North Malay Basin which bottomed in Group F (Fig. 11.21). A similar plot from wells drilled beyond the overpressured Group F shales is shown in Fig. 11.22. Notice the slight return to a more controllable formation pressure trend once the wells have penetrated the more sand-prone Group H. This is especially true in the Bergading Deep.

The 9 new plays identified are:

1) NW Malay Basin/H&I/Channel Play
2) West Malay Basin/J-L/Stratigraphic Play
3) South Malay Basin/J/Truncation
4) Johor Platform/Oligocene/Synrift
5) PM Onshore/Triassic-Cretaceous/Volcaniclastics
6) South Melaka Straits/Oligocene/Synrift
7) Central Melaka Straits/Oligocene/Synrift
8) North Melaka Straits/Oligocene/Synrift
9) North Melaka Straits/Miocene/Basement Drape

An example on the new play is displayed in Figure 11.23. This seismic line crosses the central graben of the Melaka Straits, and shows an inversion structure in the synrift sediments. This is the North Melaka/Oligocene/Synrift play.

CONCLUSIONS

Oil exploration in the Peninsular Malaysia region has progressed rapidly since it started in the late nineteen sixties. By end 1997 a total of 342 exploration wells were drilled and 441,000 km. of seismic data completed including 20 3D seismic surveys. These intense exploration efforts have led to the discovery of OIIP of 12.5 BSTB and GIIP of 57.1 TSCF almost entirely in the Malay Basin. This indicates that the Malay Basin is certainly a prolific oil and gas basin.

The exploration scene has undergone three rounds of PSCs starting with the 1976 PSC, followed by the 1985 PSC and the 1995 R/C PSC. 13 PSCs are currently in operation in the Peninsular Malaysia Region as at 1.1.1999.

The Malay Basin has been divided into six regions based on the reservoirs/play types and geographical locations. The Southeast Region contains the most hydrocarbons with OIIP of 8.3 BSTB. This region contains 10 giant oil fields. The main play type is the Group J fluvial-braided sands trapped in compressional anticlines.

The Penyu Basin has only 8 exploration wells drilled and the West Region of the Malay Basin has 16 exploration wells drilled but both yield only 0.5% each of the EUR of oil in the Peninsular Malaysia region. Based on the number of wells drilled and the hydrocarbons discovered, it is surmised that these two regions are grossly under-explored considering their substantial sizes. The possibility of future large discoveries in these two regions cannot be ruled out.

Proven recoverable reserves for fields have been increased by using more advanced techniques such as Low Resistivity/Low Contrast project for improved reservoir characterisation and Enhanced Oil Recovery mechanisms. The use of ocean bottom cable for the acquisition of 3D seismic data below the oilfield platforms, and specialised seismic processing, such as Pre-stack Depth Migration to remove gas sag and improve fault delineation, have located untested reservoirs and unswept zones.

In-house and petroleum sharing contractors' studies have also revealed future hydrocarbon plays in the Peninsular Malaysia region. Out of the 15 plays identified, 6 are in the emerging category and 9 are new plays.

In conclusion, new discoveries can be found in the Peninsular Malaysia region, even in the mature Malay Basin, through new exploration ideas and advanced seismic acquisition, processing and interpretation techniques. In addition, field reserves growth can be maintained and /or increased through improved seismic, reservoir and engineering studies.

Part 4

SARAWAK SEDIMENTARY BASINS AND PETROLEUM RESOURCES

Chapter 12

Geological Setting of Sarawak

Mazlan B. Hj. Madon

INTRODUCTION

Petroleum in Sarawak is produced from the Neogene rocks that occur mostly in the offshore areas of the Sarawak Basin. There has been more exploration and production in the eastern part of the basin, particularly in the West Baram Delta, where it all started, and in the Balingian and Central Luconia provinces. These provinces are therefore better understood than the rest. Although the Tatau, West Luconia, and North Luconia provinces have been explored to some extent, and with varying degrees of success, more work needs to be carried out. The SW Sarawak Province is the least explored of the eight provinces and thus little data is available from it. The geology and hydrocarbon habitats of each structural province in the Sarawak Basin (Fig. 12.1) are described in subsequent chapters. All the provinces are described, except SW Sarawak which is mentioned briefly in Chapter 18, together with the West Luconia Province. The aim of this chapter is to discuss the geological setting and evolution of Sarawak and its continental margin as a framework for discussion in the ensuing chapters on the hydrocarbon habitats in the geological provinces of the Sarawak Basin.

The Sarawak continental margin forms part of the Sunda Shelf which structurally connects Borneo with Peninsular Malaysia and the rest of continental Southeast Asia. The Sarawak Basin is of Late Eocene to Recent age. Part of it continues on land to be exposed

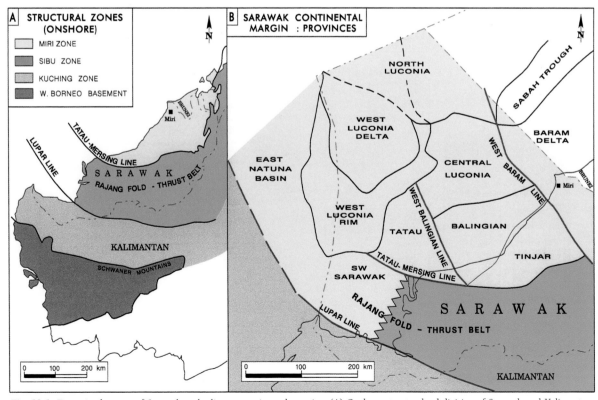

Fig. 12.1. Tectonic elements of Sarawak and adjacent continental margins. (A) Onshore structural subdivision of Sarawak and Kalimantan (Indonesian Borneo). After Liechti et al. (1960). (B) Structural provinces of Sarawak Basin. Onshore geology simplified from Geological Survey of Malaysia (1986).

between Sibu and Miri and further northeast into the Inboard Belt of the Sabah Basin. The pre-Oligocene rocks that form the basement of the Sarawak Basin are exposed to the south in the crescentic Rajang Fold-Thrust Belt, of Late Cretaceous to Late Eocene age (Fig. 12.1). These low-grade metamorphosed deep-marine rocks, together with older rocks in the western continental core of Borneo, comprise the economic basement with regard to petroleum exploration.

The Sarawak Basin continues westwards into Indonesia as the East Natuna Basin, and is separated from the Sabah Basin to the east by a major tectonic discontinuity called the West Baram Line. Some authors (e.g. James, 1984, p. 38; Agostinelli et al., 1990) have interpreted the West Baram Line as a major transform fault. The evidence suggests that this fault has had a significant influence on the sedimentary evolution of the NW Borneo margin, separating an area of carbonate sedimentation (Central Luconia Province) from an area that was dominated by deltaic siliciclastic sedimentation (Baram Delta Province). Although the Baram Delta Province has been traditionally regarded as part of the Sarawak Basin, the tectonic significance of the West Baram Line suggests that the Baram Delta Province belongs tectonically to the Sabah Basin. Scherer (1980) and Doust (1981) have given general summaries of the geology and exploration history of the Sarawak Basin. A detailed description of the geology of Sarawak and the neighbouring areas was given by Liechti et al. (1960) while more recent compilations are given in James (1984), Hutchison (1989), and Sandal (1996).

THE STRUCTURE AND STRATIGRAPHY OF SARAWAK

Sarawak may be subdivided into three tectonostratigraphic zones that represent decreasing stratigraphic and structural complexity towards the east (Fig. 12.1A). Deposition and deformation young generally from west to east and from south to north (Fig. 12.2). In westernmost Sarawak, west of Batang Lupar, the Kuching Zone is believed to be the peripheral part of the continental basement of Borneo, the West Borneo Basement, which extends southwards into Kalimantan. Palaeozoic and early Mesozoic rocks crop out mostly in this part of the state. The oldest rocks are believed to be the phyllites and schists of the Kerait Schist and Tuang Formation, which are probably pre-Late Carboniferous in age. In many places these rocks are covered by thick successions of Tertiary continental strata.

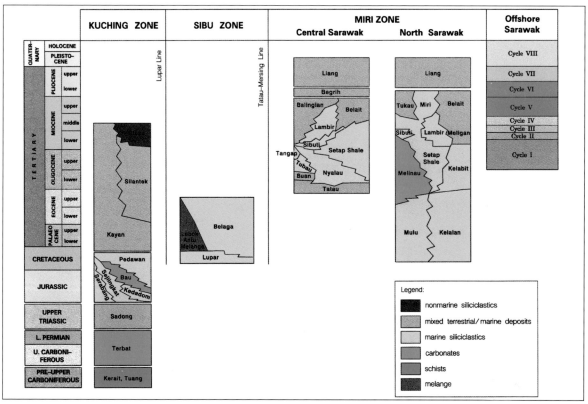

Fig. 12.2. Generalised stratigraphic columns for the onshore structural provinces of Sarawak and correlation with the offshore sedimentary cycles. Onshore stratigraphy modified from Geological Survey of Malaysia (1995).

In central and northern Sarawak, east of Batang Lupar, the rocks are almost exclusively Late Mesozoic and Cenozoic in age. The Sibu Zone is essentially formed by Upper Cretaceous to Upper Eocene deepwater sediments of the Rajang Group which have been deformed into a fold and thrust belt, referred to as the Rajang Fold-Thrust Belt. Further to the east and north is the Miri Zone, which is underlain by Upper Eocene to Recent strata, representing the youngest of the three tectonostratigraphic zones. These zones are described further below.

West Borneo Basement

The "West Borneo Basement" is the pre-Tertiary continental core of Borneo and is believed to be the eastwards extension of Sundaland (Hutchison, 1989). This basement forms the Schwaner Mountains in the southwestern part of the island in Indonesian East Kalimantan, and appears to extend eastwards to underlie the Barito and Kutei basins as well as northwards beneath the Tertiary Melawi and Ketungau basins within the Kuching Zone. The geology of the West Borneo Basement has been summarised by Hutchison (1989, p. 181; 1996, p. 252-253). The oldest rocks are Middle–Upper Carboniferous (post-Namurian) to Permian mica schists, hornfels, and metaquartzites. K-Ar dating on biotite granites gave ages of 201–320 Ma (Williams et al., 1988). These rocks are all intruded by Middle Jurassic to Late Cretaceous plutonic suites comprising quartz diorite, tonalite, granite, some gabbro, diorite, and norite, with K-Ar ages (75–115 Ma) ranging from Campanian to Barremian (Late to Early Cretaceous).

Kuching Zone

The Kuching Zone is probably the northward extension of the West Borneo Basement into Sarawak. Hutchison (1989) included the Kuching Zone and the West Borneo Basement in the "West Borneo Block" which he interpreted as the Sundaland continental terrane. The geology of this zone has been studied by Tan (1982). This zone comprises sub-Upper Cretaceous formations, including Upper Carboniferous to Triassic marine limestones in the central region south of Kuching. In close association with the limestones are basic to intermediate volcanic rocks of Late Triassic age. The Carbo-Permian and Upper Triassic limestones are overlain by Jurassic-Cretaceous mainly siliciclastic sequences and minor carbonates. Cretaceous granitoids intrude the Cretaceous and older rocks. All these rocks are overlain unconformably by mixed terrestrial-marine siliciclastic rocks of Late Cretaceous to Middle Miocene age that crop out extensively in the western and eastern parts of the region.

The eastern boundary between the Kuching Zone and the Sibu Zone is marked by the Lupar Line, a NW-trending zone of melange (Lubok Antu Melange). This is overlain by flysch-like deep-sea sediments (Lupar Formation) that are interbedded with pillow basalt and intruded by gabbro. The Lubok Antu Melange consists of blocks of sandstone, radiolarian chert, basic igneous rocks, and limestone in sheared mud matrix, and has been interpreted as a tectonic melange formed during the Eocene (Tan, 1982; see also Haile et al., 1994). The Lupar Formation is apparently overlain by the basal member of the Belaga Formation (Rajang Group) which forms the Sibu Zone. Foraminiferal evidence (Tan, 1982) suggests that the Lupar Formation, and hence the gabbros and pillow lavas, are probably Santonian-Maastrichtian (Upper Cretaceous) in age.

Sibu Zone

The Sibu Zone is underlain predominantly by the Rajang Group, which consists of intensely folded, thrust, and low-grade metamorphosed Late Cretaceous to Eocene turbidites, with radiolarian chert, spilite, and dolerite (Haile, 1969). The Rajang Group includes the Lupar and Belaga formations of West Sarawak, and part of the Danau Formation and Embaluh Group in adjacent Kalimantan (Hutchison, 1996). The Belaga Formation has been dated using foraminifers as being Upper Cretaceous to Upper Eocene (Wolfenden, 1960). It forms a large outcrop belt almost 200 km wide and continues eastwards in a crescentic trend into Sabah as the Sapulut, East Crocker, and Trusmadi formations (Fig. 5.10, Chapter 5). The succession is fairly uniform except for two large patches of Plio-Pleistocene rhyolitic/dacitic rocks that crop out in the Hose Mountains and the Usun Apau Plateau, and minor andesitic/basaltic flows of similar ages in the Neuwhenhuis Mountains. Flysch-like turbiditic rocks of the Selangkai Formation in Kalimantan are thought to be equivalent to the Rajang Group (Hutchison, 1996). Outliers of the overlying Oligo-Miocene Nyalau Formation also occur within the Sibu Zone, such as in the Usun Apau, Hose Mountains, and Merit-Pila areas. The latter is where the Merit-Pila coal field is located.

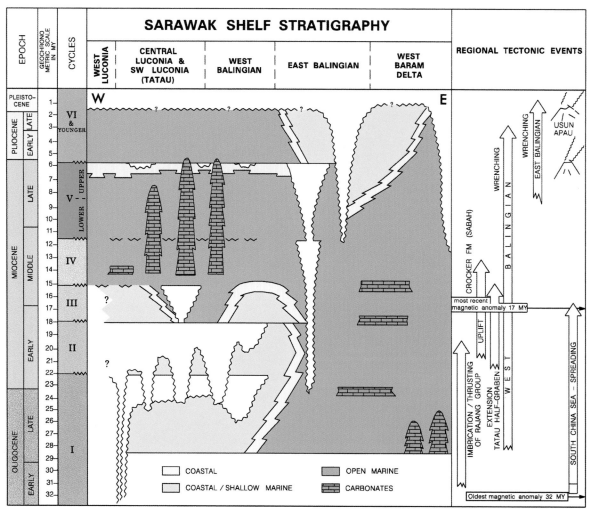

Fig. 12.3. Schematic stratigraphic summary for the Sarawak continental shelf, showing the major unconformities, sediment types, palaeoenvironments, and regional tectonic events (modified after Hazebroek et al., 1994).

Miri Zone

The boundary between the Sibu Zone and the Miri Zone is represented by the Tatau-Mersing Line. This is a structurally complex zone consisting of Palaeocene to Eocene ophiolitic rocks, including spilite, basalt, tuff, and radiolarian chert (Hutchison, 1989). The Tatau-Mersing Line also represents a major unconformity between the Belaga, Mulu and Kelalan formations (Rajang Group) and the overlying Upper Eocene-Recent sediments of the Miri Zone (Fig. 12.2). The unconformity is Late Eocene in Sarawak and Middle–Late Miocene in Sabah, suggesting that the deformation associated with the collision of the Luconia Block with the northern margin of Borneo was diachronous northward. An important hiatus has also been documented within the Melinau Limestone (Liechti et al., 1960). This Late Eocene unconformity represents a major phase of tectonism that deformed the deep marine rocks of the Rajang Fold-Thrust Belt. Overlying these deformed rocks are Oligocene-Miocene, mainly shallow water, deltaic-marine sediments that crop out in central Sarawak (Tatau, Nyalau, Setap, and younger formations). Further to the northeast towards Miri, similar shallow water Miri, Lambir and Tukau formations occur. The Oligo-Miocene sediments are predominantly siliciclastics except for isolated outcrops of carbonates such as the Melinau Limestone. Inliers of the Rajang Group occur in the Tinjar Province to the north where they are known as the Kelalan and Mulu formations. Rocks in the Miri Zone represent the lower part of the Sarawak Basin succession which extends offshore into the continental margin. Figure 12.3 shows the Oligocene-Recent stratigraphy of the Sarawak shelf.

PRE-TERTIARY TECTONIC HISTORY

Before the advent of plate tectonics theory, the tectonic evolution of Sarawak was interpreted in the context of the geosynclinal theory (e.g. Haile, 1969). Haile (1973) interpreted the deepwater turbiditic rocks in the Sibu Zone as "eugeosynclinal" (deepwater basinal) deposits whereas the Upper Eocene–Recent neritic and coastal sediments of the Miri Zone were interpreted as "miogeosynclinal" (shelfal) deposits. The geosynclinal concepts have now been superseded by plate tectonics theory. Nevertheless, Haile's tectonic subdivision is still useful because it emphasises the major tectonic elements of an evolving continental margin.

The pre-Tertiary tectonic development of Sarawak can be traced back in time in the context of the break-up of Gondwanaland and the evolution of the Tethys Ocean (Chapter 4). The West Borneo Basement, of which western Sarawak is part, may have been contiguous with the East Malaya Block, which is part of the Sundaland craton (Hutchison, 1989). The Late Carboniferous to Early Permian was a time of shallow marine sedimentation (Terbat Formation) when the West Borneo Basement was probably still attached to Gondwanaland. Sedimentation was interrupted by a major tectonic event during the Middle Permian to Early Triassic caused by the suturing of East Malaya to Cathaysia. This was followed by continental to shallow marine sedimentation (Sadong Formation). Widespread andesitic to basaltic lavas and pyroclastics are indicated by the Upper Triassic Serian Volcanics (probably correlateable with the Sedili Volcanics in Johor, Peninsular Malaysia), while elsewhere marine sedimentation continued into the Jurassic and Cretaceous, and is associated with basic volcanism (Bau Limestone and Kedadom and Pedawan formations).

During the Early-Middle Cretaceous, deposition was again locally interrupted by folding. Deposition of the Pedawan Formation ended with folding and igneous (granitic) activity during the Late Cretaceous. East of Batang Lupar deep marine turbidite sedimentation occurred in a remnant ocean basin until the Late Eocene (Belaga Formation), probably associated with a subduction zone that existed along the Lupar valley, called the Lupar Line. Similar turbiditic rocks occur in the inner Baram-Tinjar area, northeastern Sarawak, as the Mulu and Kelalan formations, indicating the large extent of this turbidite sedimentation across Sarawak. South of the subduction zone there may have been the accretionary complex, forearc basin, and associated volcanic arc (Hutchison, 1996). By Late Cretaceous times, western Sarawak was already uplifted, resulting in essentially continental sedimentation (Kayan Sandstone) that persisted into the Miocene with the deposition of the Silantek Formation and Plateau Sandstone, probably mainly intermontane basins in a back-arc setting (Hutchison, 1996).

The Late Mesozoic to Early Cenozoic evolution of Sarawak is a manifestation of subduction and collision processes at the northern margin of western Borneo associated with the closure of the Rajang Sea. The Kuching and Sibu Zones may be interpreted in this context as Mesozoic–early Cenozoic accreted crustal material that has been added to the continental core of Borneo (the West Borneo Basement) as the Rajang Sea was being subducted southwestwards beneath West Borneo. The Miri Zone represents the onshore extension of the Sarawak Basin that developed during the Late Eocene on the uplifted active continental margin formed by the Kuching and Sibu Zones (e.g. James, 1984). Since the Middle Cretaceous, the West Borneo Basement has acted as a unit together with Peninsular Malaysia, south of the Khlong Marui Fault (Fuller et al., 1991) and may have been part of the Sundaland landmass. Palaeomagnetic data from the island, however, have resulted in two contrasting interpretations: one that favours a counter-clockwise rotation of Borneo (reviewed by Fuller et al., 1991) and another that argues for no rotation (Lee and Lawver, 1994). Schmidtke et al. (1990) provided data from Sarawak that suggest counter-clockwise rotation of Borneo during the Cretaceous and Cenozoic, supporting previous interpretations (Haile and Briden, 1982). Further work is needed to resolve the palaeomagnetic problem.

SARAWAK BASIN

Tectonic Setting

The Sarawak continental shelf represents the easternmost part of the Sunda Shelf. Drilling and geophysical data suggest that the sedimentary succession on the shelf is in excess of 12 km thick and consists mainly of siliciclastic sediments. The sediments are thought to have been derived from the erosion of the uplifted Rajang Fold-Thrust Belt, to the south. Several phases of tectonism and uplift of the hinterland during the Neogene has also contributed to the sediment supply into the basin. In Chapter 5 the Sarawak Basin was defined as the Upper Eocene–Recent sedimentary succession unconformably overlying the Rajang Group. The lower part of the Sarawak Basin crops out as the onshore Miri Zone (Fig. 12.1) while the remainder lies mostly in the continental shelf and slope of Sarawak. The Sarawak Basin is subdivided into several tectono-stratigraphic provinces based on structural styles and sedimentation histories. Chapter 5 describes these provinces in some detail.

Geological Setting of Sarawak

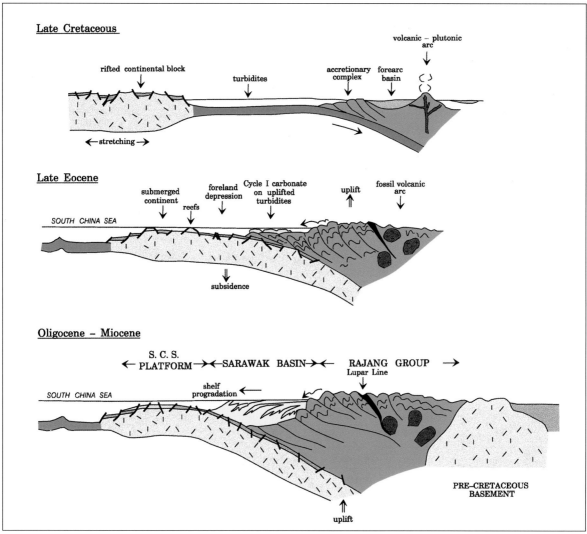

Fig. 12.4. Cartoon depicting the evolution of the Sarawak Basin, from subduction-accretion phase, through foreland basin phase, to passive margin phase. The bottom drawing is a schematic regional cross section of continent-ocean transition, from the Schwaner Mountains to the South China Sea, showing the salient features of Sarawak geology and tectonics.

The Sarawak Basin is most widely believed to have originated as a foreland basin that formed after the collision of the Luconia Block with the West Borneo Basement, and the closure of the Rajang Sea during the Late Eocene (James, 1984; Hazebroek and Tan, 1993; Hazebroek et al., 1994) (Fig. 12.4). Deformation and uplift of the Rajang Group accretionary prism to form the Rajang Fold-Thrust Belt, which forms the Sibu Zone, provide the sediment supply to the Sarawak Basin. The Sarawak Basin was a foreland basin during the Oligocene to Early Miocene, subjected to active extensional and strike-slip tectonics, but later underwent a phase of coastal-shelf progradation and passive continental margin outbuilding during the Middle Miocene to Recent.

Some authors have interpreted the Sarawak Basin as a strike-slip basin based on structural styles and subsidence history (Ismail Che Mat Zin, 1996, 1997; Ismail Che Mat Zin and Swarbrick, 1997). Although strike-slip tectonics may have been a major basin-modifying process, the evidence suggests that the basin originated as a foreland depression in a collisional setting. In particular, the presence of a large fold-and-thrust belt across Borneo, on the northern margin of the West Borneo Basement (WBB), implies a major compressional orogen, which formed by collision between the WBB and another continental block north of the orogen. The deepwater turbiditic sediments (flysch) of the Rajang Fold-Thrust Belt suggest that there was subduction of oceanic lithosphere prior to the collision. The time span of that subduction process, i.e., the age of that ocean, must have been between Late Cretaceous and Late Eocene, the age range of the Rajang Group (Liechti et al., 1960). The plutonic rocks in the Schwaner Mountains, along with the volcaniclastics in the Selangkai Formation of Kalimantan, have been recognised as a volcano-plutonic arc associated with the subduction (Hutchison, 1996), which also suggests that there had been a north-facing subduction zone along the northern margin of the WBB, where the Rajang Group now crops out. The continental block north of this subduction zone must therefore lie underneath the present Sarawak continental shelf, which many authors believe is the Luconia Block (Taylor and Hayes, 1980).

The Luconia Block has been interpreted as analogous to the NW Sabah Platform of the Sabah Basin which, together with the North Palawan Block, Reed Bank, and Dangerous Grounds, are composed of continental crust that had rifted off the South China margin (Holloway, 1981). Exploration wells on the North Palawan Block encountered Lower Cretaceous marginal marine sandstone, conglomerate and coal, unconformably overlain by Palaeocene shelf limestones (Taylor and Hayes, 1980). This suggests that it is underlain by continental crust. The continental composition of Reed Bank has been proven by drill-hole and dredge samples (Hamilton, 1979; Taylor and Hayes, 1980; Kudrass et al., 1985). The Luconia Block is therefore a continental terrane, thought to underlie the central Sarawak shelf. The collision of this terrane with the West Borneo Basement caused the uplift and deformation of the subduction accretionary prism to form the Rajang Fold-Thrust Belt.

Subsidence History

The question remains as to why the Luconia Block, which is believed to be a continental fragment, has subsided so that more than 12 km of sediments overlie it. Thinning of the continental crust prior to rifting at the southern China margin is one explanation for the subsidence of the Luconia Block. Modelling of free-air gravity data (Xia and Zhou, 1993) suggests that the Dangerous Grounds-Reed Bank area is underlain by transitional crust 20–25 km thick. A similar situation is likely for Luconia. Although the Luconia Block may have been thinned before collision, inherited from the rifting tectonics at the southern China margin, from which it is postulated to have originated (Taylor and Hayes, 1980), the subsidence is still too excessive to be explained simply by sediment loading.

One explanation for the extra subsidence is the flexural loading of the South China Sea lithosphere by the Rajang Group orogen during and after the suturing of the two blocks since Late Eocene times. A useful test of this hypothesis would be to study the subsidence history of the Sarawak Basin. Sedimentary basins subside as a result of one, or a combination of three, main tectonic mechanisms: crustal thinning, lithospheric cooling, and tectonic loading (Dewey, 1982). To determine which of these tectonic driving forces operates, the subsidence due to sediment loading has to be first removed. The backstripping technique (Sclater and Christie, 1980) is used to remove the effects of sediment loading on the total subsidence of sedimentary basins, so that only the tectonically induced subsidence is considered in the interpretation. The backstripped tectonic subsidence is due to either thermal (cooling), crustal/lithospheric thinning, or tectonic (supracrustal) loading. The technique, however, does not differentiate whether a basin forms by rifting or strike-slip or compressional tectonics. Visual inspection of tectonic subsidence curves may serve only as a very rough guide for interpreting basin origin, but tectonic factors such as multiple rifting events, thermal (igneous) events, tectonic uplifts/erosion, may complicate the history and easily cause misinterpretation. Without these complicating factors, the dominant mechanism of subsidence may be inferred from the subsidence history. Generally rift basins form by thinning of the crust and lithosphere (McKenzie, 1978). The subsidence history of rift basins is thus typified by rapid initial subsidence followed by much slower subsidence due to cooling of the lithosphere. The Malay Basin shows the characteristics of rift basins (Mazlan Madon and Watts, 1998). Foreland basins, however, show slow subsidence initially and a more rapid subsidence later; the subsidence rate is controlled by the flexural rigidity of the lithosphere (e.g. Beaumont, 1981). Strike-slip basins have a much more rapid initial subsidence (Pitman and Andrews, 1985) as a result of uncompensated or detached crustal thinning during their formation.

Figure 12.5 shows a series of tectonic subsidence curves from selected wells that penetrate the pre-Cycle I basement (Rajang Group) in the Sarawak Basin. The subsidence curves were obtained by removing the effects of sediment loading and, therefore, represent the tectonic component of subsidence. [The reader is referred to Sclater and Christie (1980) for a detailed description of the backstripping method.] The figure shows that many wells in the Sarawak Basin have subsidence patterns that are similar to that of a foreland basin, i.e. slow initial subsidence and a much faster subsidence later. The subsidence history, therefore, also supports a model of foreland basin for the Sarawak Basin, as deduced from the evidence above.

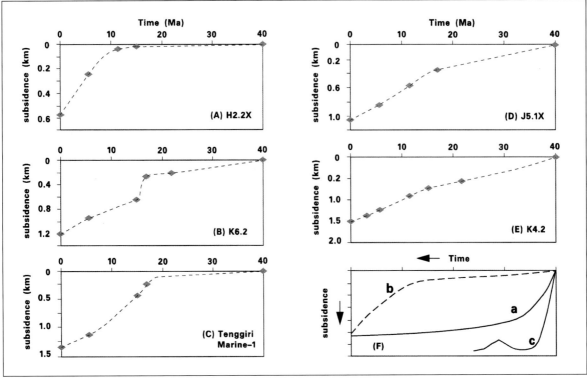

Fig. 12.5. Tectonic subsidence curves for selected offshore wells in the Sarawak Basin (A to E), showing the general increase in subsidence rates with time, indicative of a foreland basin. Sea water (density 1030 kg m^{-3}) is assumed to fill the basin. Details of methodology in Sclater and Christie (1980). (F) Schematic theoretical subsidence curves for different basin types: a.- rift basin, b.- foreland basin, c.- strike-slip/pull-apart basin.

Summary of Tectonic Evolution

In summary, the geological evidence suggests that the Luconia Block underlies a large part of the central Sarawak shelf, and is one of a series of continental fragments that had rifted off and drifted from southern China during the Late Cretaceous to Eocene (e.g. Ru and Pigott, 1986). The highly-deformed Rajang Group turbidites and deepwater sediments in the Kuching and Sibu zones represent the remnants of the Rajang Sea that once separated the Luconia and West Borneo continental blocks. The Rajang Fold-Thrust Belt is interpreted as a collisional foreland thrust belt that formed by the suturing between Luconia and West Borneo. The tectonic evolution of the Sarawak continental margin may be summarised as having involved five main stages:

- Late Cretaceous to Eocene southward subduction of the Rajang Sea oceanic crust beneath the northern margin of the West Borneo Basement.
- Continental collision and suturing of the Luconia Block with the West Borneo Basement (James, 1984; Hazebroek et al., 1994).
- Creation of a peripheral foreland basin north of the Rajang Group collisional orogen, and deposition of Cycle I and II sediments offshore.
- Extensional deformation of the Sarawak margin during the Early Miocene (end of Cycle III times).
- Deposition of Cycles IV and younger sequences, punctuated with periods of compressional wrench deformation.

Figure 12.6 shows a reconstruction of events leading to the suturing of the Luconia Block with West Borneo. The plutonic rocks in the Schwaner Mountains probably represent a volcano-plutonic arc associated with a northwards-facing accretionary complex that form the older (Cretaceous) part of the Rajang Group exposed in the Kuching Zone to the north (Hutchison, 1996). In the original reconstruction by James (1984), the Luconia Block was shown to be drifting towards West Borneo during the Oligocene, although it is generally accepted that the two terranes collided, and the Rajang Group was uplifted, by Late Eocene times (Hutchison, 1996). Figure 12.6 predicts that the Rajang Group (Sibu Zone) must be underlain by trapped Rajang Sea ocean crust. Uplift and erosion of the Rajang Fold-Thrust Belt resulted in the deposition of primarily coastal to shallow marine sediment during the Late Eocene to the present. Sediments of Cycle I and younger were probably derived from uplifted Rajang Group.

From a convergent margin phase during the Late Cretaceous to Late Eocene, the Sarawak margin evolved into a foreland basin phase during the Oligocene to Early Miocene, and to a passive margin phase from the Middle Miocene to the present (Hamilton, 1979; James, 1984; Hutchison, 1989; Chapter 5). During the foreland basin phase (Oligocene–Early Miocene) the Sarawak margin subsided fairly rapidly as the Rajang Group was being thrust northwestwards over the southern leading edge of the Luconia Block and providing sediment to fill up the foreland basin. The "passive margin" phase of continental margin evolution involves gradual subsidence of the Sarawak margin as the shelf prograded outwards to the north, accompanied by phases of deformation involving extensional, wrench faulting, and thrusting in different parts of the margin.

Palaeogeographic Evolution of Offshore Sarawak Since the Oligocene

After the Rajang Sea closed during the Late Eocene, the Sarawak margin began to form initially as a foreland basin, and then into a passive margin which gradually filled up by coastal progradation northwards from the uplifted Rajang Fold-Thrust Belt. Reconstruction of the palaeogeographic evolution of the Sarawak continental margin has been made possible because of data gathered through petroleum exploration. Several versions of palaeogeographic maps of Sarawak Basin for the Neogene have been published earlier, mainly by Shell geologists (Scherer, 1980; Doust, 1981; Ismail Che Mat Zin and Swarbrick, 1997). Understanding the palaeogeographic evolution of the Sarawak Basin is crucial for predicting the source-reservoir facies distribution. These reconstructions show the depositional environments of the Sarawak margin since the Late Oligocene (Fig. 12.7). Generally, the coastline was thought to have changed from being NW-oriented during the Late Oligocene–Early Miocene to NE-oriented during the Middle Miocene. A regional review undertaken recently by Shell and PETRONAS (Taylor et al., 1997) using newly acquired data suggests that this interpretation may be too simple.

Because of the active extensional tectonics during the Late Oligocene and Early Miocene, the coastline may have been controlled by tectonically-produced topographic relief, in the form of coastal embayments and basement highs. During the Early Miocene (Cycle I times) the coastline was still roughly east-west oriented, with deltas prograding northwards. Two areas of relatively deeper water (outer neritic to bathyal depths) occurred in the northwestern part of the area, and in the southeastern part where the Subis Limestone was deposited. The location of embayments and intrabasin highs was controlled by major basement faults. These faults had remained active, although not all at the same time, until early Middle Miocene times. As a result, the structural style during the deposition of Cycles I, II and III was dominated by extensional basement faults which greatly influenced the basin relief, morphology, and sedimentation pattern.

During Cycle I times, there was a roughly E–W-trending coastal plain with fluvio-deltaic tongues depositing siliciclastic sediment in the basin. Coastal plain environments dominated the southern half of the Sarawak margin, with the coastline trending almost E–W (Fig. 12.7). The central shelf area, which is now the Central Luconia Province, was already a high area during this time, underlain by shallow shelf sea with major shoals or banks. There were two deep embayments with bathyal water depths in the Sarawak margin during the Late Oligocene; one in the northwest, west of the Central Luconia high, and another in the southeast, near the Miri and Suai areas. These embayments appear to have been controlled by major NE–trending basement faults. Sediment is supplied to the basin through deltaic progradation from the south across the shelf. Meanwhile, the West Baram Line appears to have played a major role in controlling sedimentation, forming the palaeoshelf edge in the eastern part of the Sarawak margin. East of the West Baram Line, but still within Sarawak, is the West Baram Delta. This area is part of the larger Baram Delta complex that extends into Brunei and Sabah. The tectonic setting of the Baram Delta is described in Chapter 5.

During Cycles I and II times (Late Oligocene to Early Miocene), the western Sarawak margin was a relatively high area subjected to extensional tectonism. The tectonic movements resulted in the uplift of horst blocks (such as the Penian High) and subsidence of intervening grabens, which culminated in the Base Cycle II and Base Cycle III semi-regional unconformities in the SW Luconia and the Western Balingian sub-provinces (Fig. 12.3). To the north, different horst blocks were active at different times, as a result of continuing movements on their bounding faults. The topography created by the extension may have been an important determining factor in the distribution of reservoir-quality sandstones versus non-reservoir mudstones.

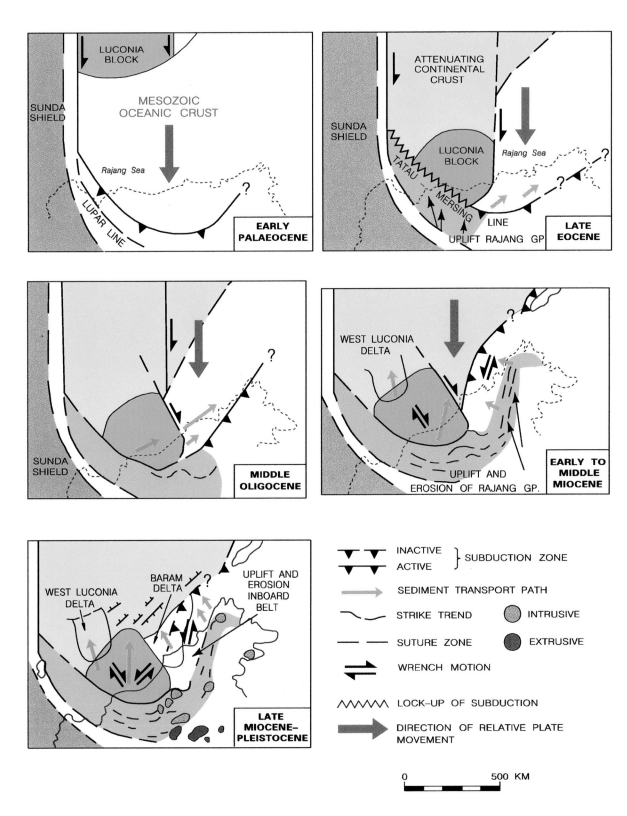

Fig. 12.6. Geological reconstruction of the events leading to the collision of the Luconia Block with the West Borneo Basement, to form the Sarawak foreland basin. Based on Hazebroek and Tan (1993) and Hazebroek et al. (1994).

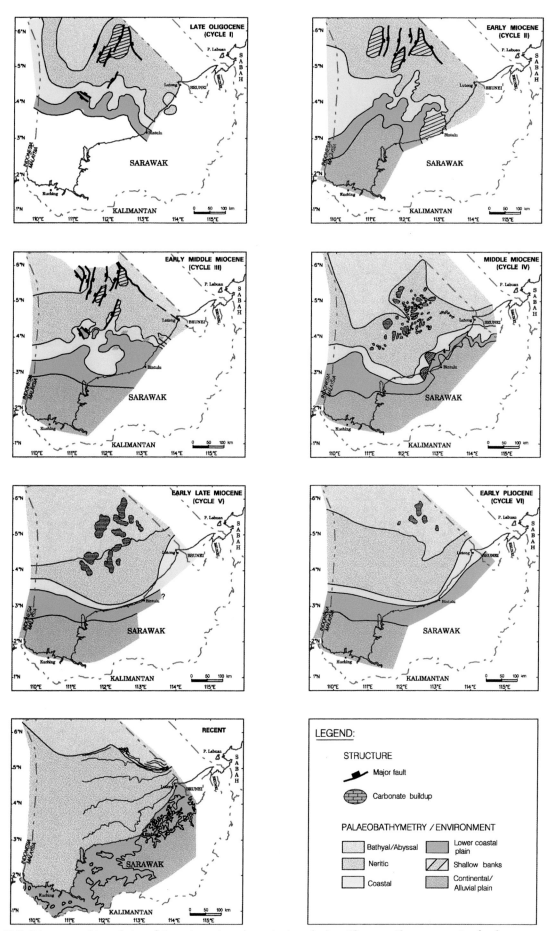

Fig. 12.7. Palaeogeographic evolution of Sarawak continental margin since the Late Oligocene. The reconstruction for the westernmost portion of the Sarawak shelf is speculative because Cycles I to III sediments are too deep. Modified from Taylor et al. (1997).

Widespread development of carbonate buildups occurred in the central Sarawak shelf during the Early to Middle Miocene, resulting in the large carbonate platform of the Central Luconia Province. Carbonate growth in Central Luconia was contemporaneous with deltaic deposition in the West Luconia and Baram Delta provinces to the west and east, respectively. The earliest carbonate growth occurs during the Early Miocene on the G2–G10 High, the northernmost horst structure on the margin (Fig. 12.7). The growth of the carbonate buildups, whose geometries range from platforms to pinnacles, seem to have coincided with a second-order eustatic sea level rise. The results of the joint study by Shell and PETRONAS (Taylor et al., 1997) indicate that the end of carbonate sedimentation in the Central Luconia Province was the result of a combination of eustatic sea level drop towards the end of the Middle Miocene, coupled with the start of deltaic progradation from the south and southeast. This idea is in contrast with Epting (1980) who postulated that carbonate growth was terminated by sea level rise (drowning). The sea level drop resulted in two contrasting carbonate environments: low-relief buildups in the south, which are subjected to subaerial exposure and burial by shallow marine siliciclastics, and high-relief buildups in the central and northern part of the Central Luconia Province, which experienced prolonged subaerial exposure and karstification before being swamped by siliciclastic sediments from the Baram Delta.

The end-Middle Miocene sea-level drop ended carbonate deposition in Central Luconia and marked the beginning of a regressive phase that was to continue until Early Pliocene times with the deposition of Cycles V to VII sequences. The Sarawak continental shelf continued to prograde and continuously fill the basin with coastal-fluviomarine sediments. The Baram Delta began to fill up the bathymetric depression east of the West Baram Line while the West Luconia Delta prograded northwards from SW Sarawak into the North Luconia Province. Cycle V sediments are seen on seismic to onlap onto the carbonate buildups, implying the passive infilling of the inter-buildup depressions by siliciclastic sediments, controlled in places by movements on some of the major normal faults bounding the horsts beneath the buildups. Depositional environment during Cycles V to VIII times ranges from deltaic/coastal through inner neritic from south to north, with strong fluvial input through major river systems from the uplifted Rajang and Tinjar foldbelts (Fig. 12.7).

The Sarawak shelf was under open marine conditions during Cycle V times. At the end of Cycle V times, however, a major uplift occurred throughout offshore and onshore Sarawak and resulted in the Base Cycle VI Unconformity. Throughout almost the entire history of the Sarawak continental margin, the West Baram Line remained a major structural feature that controlled sedimentation in the eastern part of the margin.

HYDROCARBON OCCURRENCES

Sarawak accounts for about 23% of Malaysia's total oil reserves and about 51% of its natural gas reserves. Oil and gas are being produced from 13 oil fields and 5 gas fields, all of which are located offshore. Indeed, most occurrences of hydrocarbons in Sarawak are offshore (Fig. 12.8). Hydrocarbons have been found in almost all the offshore structural provinces of the Sarawak Basin, particularly in the Central Luconia, Balingian, and West Baram Delta provinces. Lesser quantities have been encountered in the Tatau and West Luconia provinces. Hydrocarbons have not yet been found in the relatively poorly explored North Luconia Province. In the onshore areas, hydrocarbons occur predominantly in the Miri area, onshore West Baram Delta. Minor occurrences have been reported in the Suai area of the Tinjar Province and in the Igan-Oya Graben of the Tatau Province (Chapters 16 and 17).

The main drilling objectives in the Sarawak Basin are the Cycles I and II sandstones in anticlinal traps (Balingian), Cycles IV and V reefal carbonates in Central Luconia, and Cycles V and VI sandstones in the West Baram Delta. The carbonate accumulations are generally gas prone, whereas the others are oil prone. Some of the carbonate buildups have significant oil rims below the gas columns. Most of them are stratigraphic traps, with some faulting on the flanks. The Balingian oil and gas fields are complex faulted anticlines. The Baram Delta fields occur in faulted rollover anticlines associated with growth faults.

The Miri Field was the only onshore oil field in Sarawak. It was shut down in October 1972 after a cumulative production of 80 MMSTB. A more detailed description of the geology and history of the Miri Field can be found in Chapter 13. Nine out of the 13 offshore oil fields are in the West Baram Delta, while the

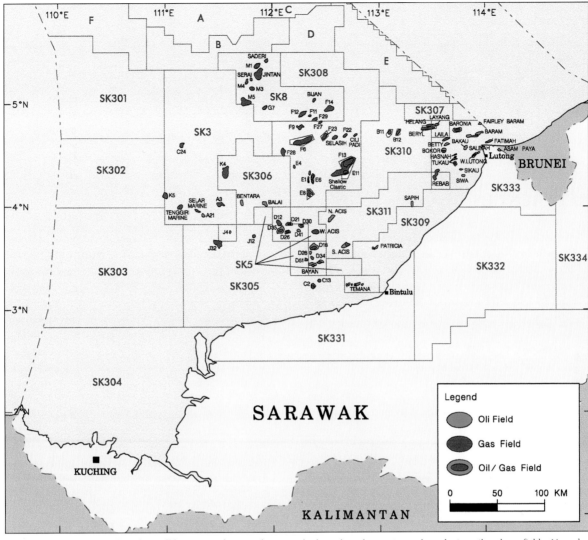

Fig. 12.8. Map of Sarawak and its offshore area, showing the major hydrocarbon discoveries and producing oil and gas fields. Note the 1998 PSC block nomenclature.

remaining four (Bayan, D18, D35, Temana) are in the Balingian Province. Over 1200 MMSTB of oil has been produced from these fields (the Miri Field included), and more than 1000 MMSTB came from oil fields in the West Baram Delta. The largest oil field in Sarawak is the Baram Field, which has an estimated ultimate recoverable (EUR) of about 280 MMSTB. The oil fields of Sarawak have also produced more than 1.6 TSCF of associated natural gas. Some of the major oil fields currently under development in Sarawak are J4 and D28.

Besides oil, Sarawak contributes a significant percentage (78%) towards Malaysia's natural gas production (associated and non-associated), with a combined cumulative production of over 5.5 TSCF non-associated gas from 5 gas fields: E11, F6, F23, M1, and M3. The first 3 fields account for about 95% of the total cumulative production of natural gas (associated and non-associated). The first gas field developed by Sarawak Shell Berhad was E11, which came on-stream in 1982. F6 is the largest gas field in Malaysia, with a EUR of close to 5.0 TSCF. Currently under development are the M4, Jintan, and Helang gas fields.

REFERENCES

Agostinelli, E., Mohamed Raisuddin Ahmad Tajuddin, Antonielli, E. and Mohamad Mohd Aris, 1990. Miocene-Pliocene palaeogeographic evolution of a tract of Sarawak offshore between Bintulu and Miri. Bulletin of the Geological Society of Malaysia, 27, 117-135.

Beaumont, C., 1981. Foreland basins. Geophysical Journal of the Royal Astronomical Society, 65, 291-329.

Dewey, J.F., 1982. Plate tectonics and the evolution of the British Isles. Journal of the Geological Society of London, 139, 317-412.

Doust, H., 1981. Geology and exploration history of offshore central Sarawak. In: Halbouty, M.T., ed., Energy Resources of the Pacific Region. American Association of Petroleum Geologists, Studies in Geology, 12, 117-132.

Epting, M., 1980. Sedimentology of Miocene carbonate buildups, Central Luconia Province, offshore Sarawak. Bulletin of the Geological Society of Malaysia, 12, 17-30.

Fuller, M., Haston, R., Lin, J.-L., Richter, B., Schmidtke, E. and Almasco, J., 1991. Tertiary palaeomagnetism of regions around the South China Sea. Journal of Southeast Asian Earth Sciences, 6, 161-184.

Geological Survey of Malaysia, 1986. Geological Map of Sarawak and Sabah, Malaysia. 4th Edition. Geological Survey Department, Ministry of Primary Industries, Kuala Lumpur.

Geological Survey of Malaysia, 1995. Annual Report 1995. Geological Survey Department, Ministry of Primary Industries, Kuala Lumpur.

Haile, N.S., 1969. Geosynclinal theory and the organizational pattern of the Northwest Borneo Geosyncline. Quarterly Journal of the Geological Society of London, 124, 171-194.

Haile, N.S., 1973. The recognition of former subduction zones in Southeast Asia. In: Tarling, D.H. and Runcorn, S.K., eds., Implications of Continental Drift to the Earth Sciences, 2. Academic Press, London, 885-892.

Haile, N.S. and Briden, J.C., 1982. Past and future palaeomagnetic research and the tectonic history of East and Southeast Asia. In: Palaeomagnetic Research in Southeast and East Asia. UN/ESCAP, CCOP Technical Publication, Bangkok, 13, 25-46.

Haile, N.S., Lam, S.K. and Banda, R.M., 1994. Relationship between gabbro and pillow lavas in the Lupar Formation, West Sarawak: Implications for interpretation of the Lubok Antu Melange and the Lupar Line. Bulletin of the Geological Society of Malaysia, 36, 1-9.

Hamilton, W., 1979. Tectonics of the Indonesian region. United States Geological Survey Professional Paper, No. 1078.

Hazebroek, H.P. and Tan, D.N.K., 1993. Tertiary tectonic evolution of the NW Sabah continental margin. In: Teh, G.H., ed., Proceedings of the Symposium on Tectonic Framework and Energy Resources of the Western Margin of Pacific Basin. Bulletin of the Geological Society of Malaysia, 33, 195-210.

Hazebroek, H.P., Tan, D.N.K. and Swinburn, P, 1994. Tertiary evolution of the offshore Sarawak and Sabah Basins, NW Borneo. Abstracts of the American Association of Petroleum Geologists International Conference & Exhibition, Kuala Lumpur, Malaysia, 21-24 August 1994, American Association of Petroleum Geologists Bulletin, 78, 1144-1145.

Holloway, N.H., 1981. The North Palawan Block, Philippines: its relation to the Asian mainland and its role in the evolution in the South China Sea. Bulletin of the Geological Society of Malaysia, 14, 19-58.

Hutchison, C.S., 1989. "Geological Evolution of South-East Asia". Oxford monographs on Geology and Geophysics, no. 13, Clarendon Press, Oxford.

Hutchison, C.S., 1996. The "Rajang accretionary prism" and "Lupar Line" problem of Borneo. In: Hall, R. and Blundell, D.J., eds., Tectonic Evolution of Southeast Asia. Geological Society of London Special Publication, 106, 247-261.

Ismail Che Mat Zin, 1996. Tertiary tectonics and sedimentation history of the Sarawak Basin, East Malaysia. Unpublished PhD Thesis, University of Durham, 277pp.

Ismail Che Mat Zin, 1997. Subsidence history of Sarawak Basin. Proceedings of the ASCOPE '97 Conference, 24-27 November 1997, v. 1, 107-127.

Ismail Che Mat Zin and Swarbrick, R.E., 1997. The tectonic evolution and associated sedimentation history of Sarawak Basin, eastern Malaysia: a guide for future hydrocarbon exploration. In: Fraser, A.J., Matthews, S.J. and Murphy, R.W. (eds.), Petroleum Geology of Southeast Asia. Geological Society of London Special Publication, 126, 237-245.

James, D.M.D., 1984. Regional geological setting. In: James, D.M.D., ed., "The geology and hydrocarbon resources of Negara Brunei Darussalam". Muzium Brunei, 34-42

Kudrass, H.R., Wiedecke, M., Cepek, P., Kreuzer, H., Muller, P., 1985. Mesozoic and Cenozoic rocks dredged from the South China Sea (Reed Bank area) and Sulu Sea and their significance for plate tectonic reconstructions. Marine and Petroleum Geology, 3, 19- 30.

Lee, T. and Lawver, L., 1994. Cenozoic plate reconstruction of the South China Sea region. Tectonophysics, 235, 149-180.

Liechti, P., Roe, R.W. and Haile, N.S., 1960. Geology of Sarawak, Brunei and western North Borneo. British Borneo Geological Survey Bulletin, 3, 1960.

Mazlan B. Madon and Watts, A.B., 1998. Gravity anomalies, subsidence history and the tectonic evolution of the Malay and Penyu basins. Basin Research, 10, 375-392.

McKenzie, D.P., 1978. Some remarks on the development of sedimentary basins. Earth and Planetary Science Letters, 40, 25-32

Pitman III, W.C. and Andrews, J.A., 1985. Subsidence and history of small pull-apart basins. In: Biddle, K.T. and Christie-Blick, N., eds., Strike-slip Deformation, Basin Formation and Sedimentation. Society of Economic Paleontologists and Mineralogists Special Publication, 37, 45-49.

Ru, Ke and Pigott, J.D., 1986. Episodic rifting and subsidence in the South China Sea. American Association of Petroleum Geologists Bulletin, 70, 1136-1155.

Sandal, S.T. (ed.), 1996. "The geology and hydrocarbon resources of Negara Brunei Darussalam". 2nd Edition, Syabas, Bandar Seri Begawan, 243pp.

Scherer, F.C., 1980. Exploration in East Malaysia over the past decade. In: Halbouty, M.T., ed., Giant Oil and Gas Fields of the Decade 1968-1978. American Association of Petroleum Geologists Memoir, 30, 423-440.

Schmidtke, E.A., Fuller, M.D. and Haston, R.B., 1990. Palaeomagnetic results from Sarawak, Malaysia Borneo, and the late Mesozoic and Cenozoic tectonics of Sundaland. Tectonics, 9, 123-140.

Sclater, J.G. and Christie, P.A.F., 1980. Continental stretching: an explanation of the post-mid-Cretaceous subsidence of the central North Sea basin. Journal of Geophysical Research, 85, 3711-3739.

Tan, D.N.K., 1982. The Lubok Antu Melange, Lupar Valley, West Sarawak: a Lower Tertiary subduction complex. Bulletin of the Geological Society of Malaysia, 15, 31-46.

Taylor, B. and Hayes, D.E., 1980. The tectonic evolution of the South China Basin. In: Hayes, D.E., ed., The Tectonic and Geologic Evolution of Southeast Asian Seas and Islands. American Geophysical Union, Geophysical Monograph, 23, 89-104.

Taylor, G., Powell, C., Newall, M. and Ngau, A., 1997. PETRONAS and Sarawak Shell Berhad Joint Regional Study of the Pre-Carbonate Clastics, Central Luconia Province, Offshore Sarawak. Unpublished Internal Report. EXP.R.50793.

Williams, P.R., Johnston, C.R., Almond, R.A. and Simamora, W.H., 1988. Late Cretaceous to Early Tertiary structural elements of West Kalimantan. Tectonophysics, 148, 279-297.

Wolfenden, E.B., 1960. The geology and mineral resources of the Lower Rajang Valley and adjoining areas, Sarawak. Geological Survey Department for the British Territories in Borneo, Kuching, Memoir, 11.

Xia K. Y. and Zhou D., 1993. The geophysical characteristics and evolution of northern and southern margins of the South China Sea. In: Teh, G.H., ed., Proceedings of the Symposium on Tectonic Framework and Energy Resources of the Western Margin of Pacific Basin. Bulletin of the Geological Society of Malaysia, 33, 223-240.

Chapter 13

West Baram Delta

*Denis N.K. Tan,
Abdul Hadi B. Abd. Rahman,
Azlina Anuar, Boniface Bait
and Chow Kok Tho*

INTRODUCTION

The West Baram Delta is the western part of the Upper Tertiary Baram Delta province which is roughly triangular in shape, with its apex occurring onshore and centred in Brunei and the northeastern coastal area of Sarawak.

It is the birthplace of the Malaysian oil and gas industry, with the first oil field (the Miri Field) discovered in 1910. Since that time, a total of some 50 exploration wells have been drilled in the West Baram Delta, resulting in the discovery of some 1.4 billion barrels of oil and 5.7 trillion cubic feet of gas in 18 oil and gas fields (Fig. 13.1). Nine of these oil fields are currently producing some 100,000 BOPD.

This chapter gives a review of the petroleum geology and hydrocarbon habitat of the West Baram Delta. In view of its historical significance, a detailed description of the Miri Field is provided. The onshore geology of Miri is analogous to the hydrocarbon-bearing reservoirs in the offshore fields, and is discussed in some detail.

EXPLORATION HISTORY

Onshore West Baram Delta

Exploration for oil in the Baram Delta commenced in 1909. The first exploration well, Miri-1, was drilled in 1910. This well resulted in the discovery of the Miri Field that ultimately produced about 80 million barrels of oil before it was abandoned in 1972.

A total of 27 exploration and appraisal wells were drilled by Shell until the concession was relinquished in 1981. Except for Miri-1, none of the other wells encountered significant hydrocarbons. Most of these exploration wells were located based on surface geology and/or single fold seismic.

In 1987, Malaysian Baram Oil Development Company (MBODC, as operator) and PETRONAS Carigali (Carigali) were awarded a PSC to explore Block SK14. MBODC acquired 738 km of 2D seismic and drilled 5 wells, resulting in the Asam Paya oil discovery. The acreage was relinquished in 1993 without development of the Asam Paya field.

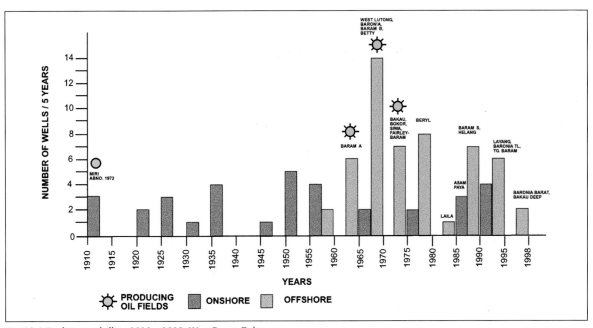

Fig.13.1 Exploration drilling 1910 – 1998: West Baram Delta.

In 1990, Idemitsu (operator), AGIP and Carigali were awarded Block SK17 which is located south of Block SK14. A total of 1018 km of 2D seismic was acquired and two exploration wells drilled, without success. The block was relinquished in 1995.

Offshore West Baram Delta

The first offshore exploration well, Siwa-1, was drilled in 1957 from a fixed platform constructed in shallow water. Unfortunately, the well did not encounter any hydrocarbons. The first offshore field was discovered in 1963 by the Baram-1 well drilled with the jack-up rig Orient Explorer. This success led to an increase in activities and opened the way to further discoveries in the offshore Baram Delta during the late 1960s and 1970s. The earlier exploration play was the Cycle V and VI topset sands in large anticlinal features, and drilling was terminated upon encountering overpressures. Recent exploration effort resulted in more gas discoveries in footwall closures along the nose of anticlinal features.

From 1955 to 1988 a total of 11 oil and gas discoveries were made by Shell in the offshore West Baram Delta. Siwa North-1 was the last well drilled before final relinquishment in 1988.

In 1988 a new PSC was awarded to Carigali (Operator) and Shell to produce the 9 oil fields, namely, Bakau, Baram, Baronia, Betty, Bokor, Fairley-Baram, Siwa, Tukau and West Lutong fields under Baram Delta Operations (BDO). The Laila and Beryl gas fields fall under the MLNG-Dua PSC which is held by Shell (operator) and Carigali.

The Block SK15 PSC, excluding the BDO fields, was operated by Carigali from 1988 to 1996. During this period, Carigali acquired more than 1100 km of 2D seismic, some 18,000 line-km 3D seismic and drilled 3 exploration wells, Baronia Timurlaut-1, Tanjung Baram-1 and Bokor South-1. The first 2 wells were oil and gas discoveries, and Bokor South-1 found minor gas.

Nippon Oil (operator) and Carigali were awarded Block SK10 in 1987. Nippon Oil drilled 11 exploration and appraisal wells resulting in the discovery of the Helang gas field in 1990 and the Layang oil and gas accumulation in 1991. In 1997, Nippon Oil and Carigali relinquished Block SK10, retaining only the Helang and Layang fields.

Shell (operator) and Carigali were awarded the deepwater Block E (Fig. 13.2) on 5 April 1996 under the deepwater PSC terms. Shell

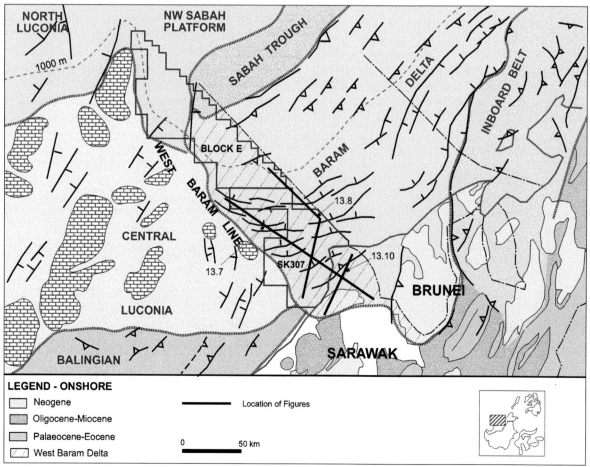

Fig. 13.2 West Baram Delta Regional Geological Setting.

acquired a total of 5200 km of long-cable 2D seismic data in 1996 and 1997. Evaluation of the acreage is in progress.

On 27 June 1997, Carigali (operator) and Shell were awarded Block SK307 (Fig. 13.2) under the Revenue-over-Cost (R/C) PSC terms. Carigali acquired 3300 km of long-cable 2D seismic in 1997 and 1998. In 1998, Carigali drilled two exploration wells, Baronia Barat-1 and Bakau Deep-1, and made a minor oil discovery in Baronia Barat, and a gas and condensate discovery in Bakau Deep.

MIRI OIL FIELD

The Miri-1 Discovery

Crude oil, collected from surface seepages in the Miri area, was originally used by the local inhabitants as an excellent remedy for aching bones and muscles, and was probably used for lamp and trade. By the 1880s, probably due to increasing demand, a more noticeable and commercially-oriented type of production system was set up. Claude Champion de Crespigni, the Resident of the Baram district of Marudi, reported in 1882 to the Brooke Government that at least eighteen simple oil wells were in production. The era of European oil technology then slowly followed. In 1902, Dr. Charles Hose, the new Resident who succeeded de Crespigni, prepared a map of the Baram district indicating the locations of surface oil seepages. This early map attracted the interest of the larger oil companies, and an exploration licence was issued about five years later.

In 1909, the first Sarawak Oil Mining Lease was granted to the Anglo Saxon Petroleum Company, an offshoot of the Royal Dutch/Shell Group. Dr. Josef Theodor Erb evaluated the prospectivity of the various locations on Hose's map and identified a number of structures that combined probable closures with surface oil shows. These structures included the Bulak-Malang-Setap anticline, Buri anticline, the Miri Hill, and some monoclines with gas seepages between Miri and Sibuti. The Miri Hill was ranked as the best prospect and a well location was selected on the top of the hill, with the hope of finding dip-closed reservoirs at deeper levels in the buried part of the hill. Erb mapped the Miri Hill as the top of an asymmetric anticline, with a gentle northwest flank and a steep, partly overturned southeast flank. Dip closure was expected to be present in the deeper levels of the anticline.

According to Miri legend, the erection of the derrick on top of the hill gave rise to barely concealed merriment amongst the local people. In their experience, the place to find oil was the marshland area between Pujut and Miri where they had seen a steady yield from hand-dug wells for generations. Nevertheless, the Miri-1 well was spudded on 10 August 1910 and proved Erb right; upon completion on 22 December 1910, the well produced 88 BOPD of light oil from a sand at a depth of 138 m (452 ft). The results of the well were initially thought to confirm Erb's model but later drilling results proved this structural model wrong (see below).

Stratigraphy

The first stratigraphic subdivision of the Miri Field dates from before 1916 and consisted of:

- An Upper Level, rather a group of Upper Levels to which the wells of the shallow pool C belong,
- The Shallow Level, embracing the majority of the wells, and
- The Deep Level or *thousand foot sand*, struck only in a few wells (including Miri-1).

Later drilling results, however, proved the stratigraphy and Erb's structural model untenable. Large reservoirs were discovered in the lowlands at the foot of the hill. Although the asymmetric anticline exists, closures appeared to be due to the fault system cutting the structure. Correlation of the very similar sands across faults, based on their lithologic characteristics and oil gravity, left too many uncertainties. Before a workable structural model could be established, a breakthrough in palaeontology was required. A very detailed study by L.C. Artis (*in* Schumacher, 1941) established a biostratigraphic zonation based on benthonic foraminifera, resulting in the familiar lithostratigraphic scheme (Fig. 13.3) which is still in use today. The study solved many correlation problems and showed, for instance, that the southern part of the field had consistently been mis-correlated to the northern part: each sand in the south had been correlated to the next higher sand in the north.

In 1979, the Miri-611 well was drilled to a depth of 1127.8 m (3700 ft) in the northern part of the Miri Field to evaluate the feasibility for enhanced oil recovery and appraise the hydrocarbon potential of the deeper reservoirs. The typical log profile of the various sands in the Miri Field, based on this well is given in Figure 13.4.

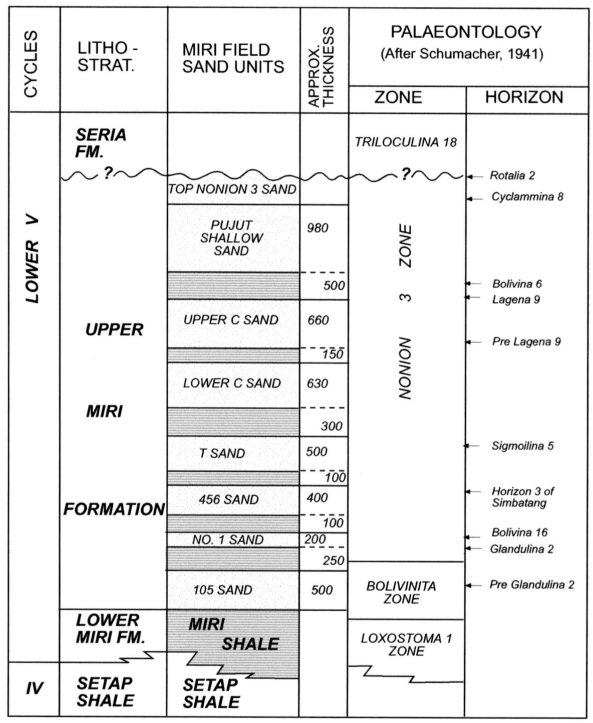

Fig. 13.3 Stratigraphic Framework of the Miri Field.

The Miri Structure

Combining the results of the new correlation and those of various detailed fault studies, Schumacher (1941) developed the current structural model of the Miri Field (Fig. 13.5). The Miri Hill structure is an asymmetric, slightly overturned, northeast-trending anticline, bounded to the southeast by a set of northwest-hading steep normal faults, of which the Shell Hill Fault, with a maximum throw of some 750 m, is the most important. The structure is cut by a set of flat, southeast-hading antithetic faults and a set of northwest-hading listric faults, with a combined throw of some 300 m, and bounded to the southeast by a set of merging northwest-hading reverse faults, the Canada Hill Thrust (Fig. 13.5). The Miri Hill structure underwent two phases of deformation; namely a Late Miocene extension, which resulted in the Shell Hill Fault and the associated listric and antithetic normal faults, followed by a Pliocene compression which resulted in development of the anticlinal feature and the Canada Hill Thrust.

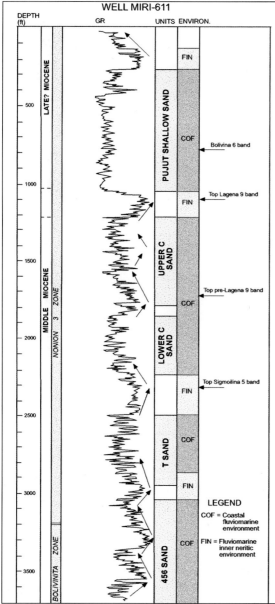

Fig. 13.4 Typical log profile of Miri Field.

Hydrocarbon Distribution

The hydrocarbons in the Miri Field appear to be trapped by at least 18 faults. It appears that the steep antithetic faults form more efficient seals than the listric faults. Many of these pools however, were very small, especially in the central part of the field where listric and antithetic faults intersect to form numerous small fault compartments. Some 80 separate accumulations have been recognised but only 48 contained sufficient reserves to justify one or more wells. Each accumulation is given a double name, e.g. Upper C, E9, which combines the sand unit (Upper C sand) and the fault trap (E9 fault). Although 48 accumulations were produced at some stage, 84% of the total production came from 10 accumulations (Fig. 13.6).

Two general types of oil are known from the Miri Field: a waxy light crude of about 40° API and a relatively heavy oil of about 20° API with low wax content. Within these two gross oil types, considerable distinction could still be made. For example, the Pujut Shallow, E11 was characterised by a typical 18° API oil, whereas the Upper C, E11 produced consistently 21° API oil. Scanty information available on the composition of some oil types suggests that the heavy oils were depleted in a few but specific components, indicating that the oil had been altered by near surface processes, e.g. bacterial action, oxidation or, perhaps, differential leaking.

Field History

By February 1916, 50 wells had been drilled and the Miri Field had reached a cumulative production of 1.25 million barrels, with a daily production rate of some 2700 BOPD. However, water and sand problems affected field development for a few years between 1916 and 1919. These problems were resolved and, by 1929, some 500 wells had been drilled, and peak production of over 15,000 BOPD was reached. In the 1930s, activities and production declined. Production dropped to less than 11,000 BOPD in 1931, 7000 BOPD in 1933, and 2400 BOPD in 1941, shortly before the Japanese invasion.

During the war, the Japanese produced a total of 700,000 barrels. After the war, it took 2 years to bring the field back on stream. However, the post-war history of the Miri Field was characterised by two conflicting forces, namely the uphill struggle to keep the dying field alive and the necessity to release increasing areas of the field for town expansion and development. Despite efforts to boost production, including field rehabilitation and water injection projects, the field continued to decline. By the end of 1971, the field was producing only some 675 BOPD (and more than 10 times as much water) from 98 wells. The battle was finally lost and the Miri Field was totally abandoned on 20 October 1972.

REGIONAL GEOLOGY

The Baram Delta Province is roughly triangular in shape, with its apex occurring onshore and centred in Brunei and the northeastern coastal area of Sarawak (Fig. 13.2). The province expands offshore to cover the whole width of

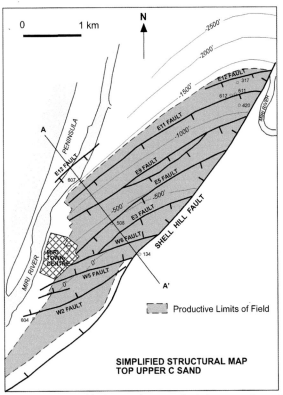
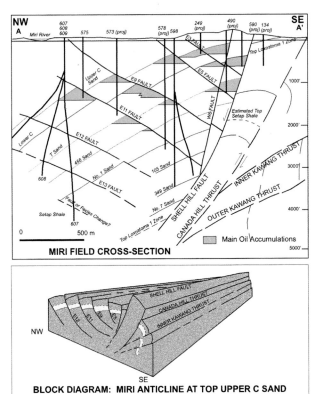

Fig. 13.5 Miri Field: Structural map, geological cross-section and block diagram.

Brunei waters, and encroaches southwest into Sarawak (where it is known as the West Baram Delta) and northeast into offshore NW Sabah (where it is named the East Baram Delta). The western margin of the Baram Delta is marked by the West Baram Line, a system of large, northeast-hading, down-to-the-basin faults that separates the delta from the older Balingian and Central Luconia provinces to the west. The eastern margin of the Baram Delta is defined by the Morris Fault - Jerudong Line which separates the delta from the older, intensely tectonised Inboard Belt of offshore NW Sabah.

The West Baram Delta is characterised by the deposition of a northwestward prograding delta since Middle Miocene times. Periods of delta outbuilding were separated by rapid transgressions, represented by marine shale intervals that form the base of eight sedimentary cycles. The regressive sequences of each depositional cycle grade northwestwards from coastal-fluviomarine sands to neritic, marine shales. The depositional cycles become successively younger due to the overall outbuilding nature of the shelf systems (Figs. 13.7 and 13.8). The geological evolution, stratigraphy, and trap configuration of the Baram Delta are well documented by geologists from Shell and other oil companies who have worked in the area.

Since the Middle Miocene, the Baram Delta has been subsiding relative to the more stable Central Luconia and Balingian provinces to the west. Within the Baram Delta, major increases in sedimentary thickness occur across growth faults, which generally trend NE-SW in the main depocentre but swing towards the NW-SE direction, on trend with the West Baram Line to the west. The West Baram Delta comprises up to 9 - 10 km of Miocene to Recent siliciclastic sediments derived from the south-southeast, along the trend of the present-day Baram river, and from the west and southwest across the West Baram Line.

The distal part of the Baram Delta is situated on the continental slope and extends into the Sabah Trough. The stratigraphy in this deepwater area (> 200 m) consists of hemipelagic sediments with thin interbedded turbiditic sands of possible Late Miocene - Pliocene (Cycles VI and VII) age. Correlation with wells on the shelf is not possible because of the presence of large growth faults. The nearest wells are the Z1 and Merpati wells in Brunei (Sandal, 1996). It is possible that Oligocene - Early Miocene Cycles I - III sediments, present in the Central Luconia and Balingian provinces, may extend to the distal Baram Delta and underlie the younger deltaic sequence. The dominant structures are large elongate anticlines formed by toe thrusts at the distal end of the Baram Delta. The amplitude of these anticlines decreases away from the shelf edge.

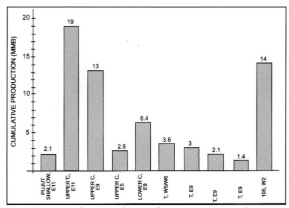

Fig. 13.6 Cumulative Production from top 10 accumulations in Miri Field.

STRATIGRAPHY AND SEDIMENTOLOGY

Onshore

The Tertiary rocks of onshore northwest Sarawak, adjacent to the West Baram Delta, consist of a thick succession of sand-shale sequences with subordinate carbonates. These rock units can be subdivided, primarily based on their degree of deformation, into two parts, namely (1) an older, deformed sequence of clastics and subordinate carbonates, ranging in age from Late Cretaceous to Early Miocene, and (2) a younger, more gently deformed series of progradational deltaic sediments, of Middle Miocene to Quaternary age. The older sequences (Palaeogene to early Neogene) occur in the more interior parts of Sarawak, whereas the upper Neogene deltaic series crop out extensively in the coastal areas and extend into the offshore. The onshore lithostratigraphy is described in detail by Liechti et al. (1960). A schematic relationship between the onshore formations and the offshore stratigraphy is shown in Figure 13.9. The younger Neogene succession, comprising the Setap Shale, Lambir, Tukau and Miri formations, is summarised below.

Following an Early Miocene tectonic event, uplift and erosion were accompanied by the deposition of a thick pile of clastic sediments which prograded seaward throughout Neogene times. Relatively coarse sediments, predominantly sandstones, were deposited in coastal plain, deltaic, and coastal environments. A thick argillaceous succession which underlies the sandstone sequence was deposited in outer shelf, slope, and base-of-slope environments. Sedimentation was strongly influenced by tectonic activity and very thick sequences (up to 6000 m) accumulated in sub-basins such as the Belait Syncline (James, 1984).

The Neogene formations crop out in onshore northwest Sarawak to form an integral part of the clastic wedge, with the arenaceous Lambir, Miri, and Tukau formations passing laterally basinward into the Setap Shale Formation. Their stratigraphic relationships are depicted in Figure 13.10.

Setap Shale Formation: The Setap Shale Formation (Early Miocene - Quaternary) consists mainly of shales with occasionally less indurated, relatively monotonous clay intervals and thin, mainly turbiditic sandstone beds. South of the Lambir Hills, the Setap Shale comprises mobile, undercompacted clays and forms mud volcanoes (Fig. 13.11). The Setap Shale overlies the Melinau Formation, and its contact with the Rajang Group, Kelalan and Mulu formations is postulated to be an unconformity or hiatus (Liechti et al., 1960). The characteristic facies persists into the offshore area, where it gets progressively younger in the northwesterly direction.

Lambir Formation: The Lambir Formation (Middle - Late Miocene) consists predominantly of sandstones and shales with minor limestones and marls. The Lambir, in general, has a slightly diachronous transitional contact with the underlying Setap Shale. In the Lambir Hills area, however, an erosional contact with the Setap Shale is observed. The Lambir transgresses rapidly into the more argillaceous Miri and Tukau formations. The Lambir Formation is up to 2100 m thick and is a time-stratigraphic equivalent of the Belait Formation in neighbouring Brunei.

Tukau Formation: The Tukau Formation (Middle - Late Miocene) consists of alternations of softer sandstones and shales rich in lignitic materials. The formation reaches up to 2700 m thick and grades laterally into the uppermost Lambir Formation as well as the Lower Miri Formation.

Miri Formation: The Middle Miocene Miri Formation crops out around Miri, providing an invaluable surface analogue for the subsurface sediments of the offshore West Baram Delta. Its outcrops are restricted to the narrow coastal region around Miri. Liechti et al. (1960) described the formation as a predominantly marine arenaceous succession, based on observations of outcrops of the Miri Anticline and examination of subsurface material from the Miri Field. The basal contact with the underlying Setap Shale Formation is a gradual transition from an arenaceous succession downward into a predominantly argillaceous succession. Based on

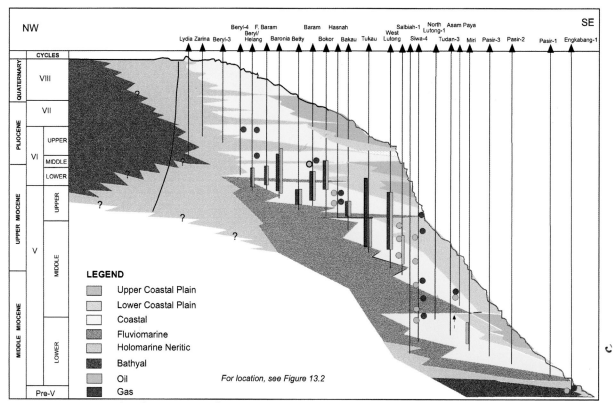

Fig. 13.7 West Baram Delta Schematic Stratigraphy.

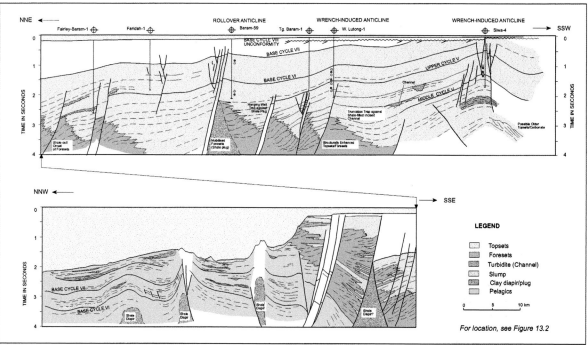

Fig. 13.8 Regional Geological Cross-section of West Baram Delta.

lithological differences and small benthonic foraminifera assemblages, Liechti et al. (1960) and Wilford (1961) subdivided the formation into a Lower and Upper unit, as follows:

The maximum total thickness of the Miri Formation is estimated to exceed 1830 m (6000 ft). The Lower Miri unit consists of interbedded shales and sandstones, and passes downwards into the underlying Setap Shale Formation. The Upper Miri unit is more arenaceous, consisting of rapidly recurrent and irregular sandstone-shale alternations, with the sandstone beds passing gradually into clayey sandstone and sandy or silty shale. Based on marine microfauna and lithological characteristics, Liechti et al. (1960) concluded that these sediments were deposited in a littoral to inner neritic shallow marine environment.

The Miri Formation crops out around Miri town as shown in Figure 13.12. Shell assigned a Middle Miocene age to the Miri Formation successions (Hamid Mohamad et al., 1986).

On the basis of lithology, bed geometry, sedimentary structures and bioturbation, 10 different facies are defined for the Miri Formation (Abdul Hadi Abd. Rahman, 1995):

1. *Medium-scale trough cross-bedded facies,* with a maximum composite thickness of about 1 m, consists of fine- to medium-grained quartzarenite displaying medium- to large-scale trough cross-bedding. Generally, the trough wavelength ranges between 0.1 m to 3 m. There are 3 extreme varieties of this facies. The first variety is large-scale cross-bedded units with mud-draped bottom sets and foresets (Fig. 13.13a). The second type displays granule- to pebble-sized mudclast-filled troughs, and the third variety exhibits clean, trough cross-bedding without mud drapes (Fig. 13.13b). This facies may also contain isolated clay pebbles and peat lump and detritus, and, in places, displays fill structures. The fill sediments show undulatory to cross-laminated structures of fine sand and shaly material. Bioturbation is common, and in places abundant. Trace fossils recorded from this facies include *Ophiomorpha nodosa, Ophiomorpha irregulaire, Palaeophycus, Teichichnus* and *Planolites.*

These bedforms have been interpreted as sand waves commonly associated with estuaries and shallow sub-tidal areas. Mud drapes are distinct features of trough cross-stratified structures, indicating pauses in bedload movement. The deposition and preservation of mud drape on sand waves are favoured by large sand-wave asymmetry, a high bottom concentration of suspended mud, large time-velocity asymmetry and low strength of tidal currents, and a high eccentricity of the tidal-current eclipse (Allen, 1980, 1982).

The overall stratification suggests that the trough cross-bedded sandstones of the Miri Formation were deposited in shallow sub-tidal areas of a tide-dominated estuary during periods of major, sub-tidal bedload transport.

The fill structures could represent sediment filling pre-existing depressions (Terwindt, 1971) or scour-and-fill structures that are generally produced on channel bottoms, under certain flow conditions over an unconsolidated sediment surface (Reineck and Singh, 1986). The trace fossil assemblage is typical of shallow marine environment.

2. *Small-scale trough cross-bedded facies* is a subordinate facies consisting of mixed mud-clasts and mud-draped trough cross-bedded facies. Figure 13.13c shows a small-scale, 15 cm thick, pronouncedly mud-draped trough cross-bedded layer sandwiched between lenticular-bedded layers. This thin, small-scale, mud-draped cross-bedded facies most probably represents shallow ebb-flow runoff on muddy tidal flats, indicating periods of emergence.

3. *Herringbone cross-bedded facies* does not occur as a distinct facies but within planar to tabular cross-bedded units. Figure 13.14a shows a special type of herringbone cross-bedding where two opposite-dipping cross-bedded units is separated by a thin mud

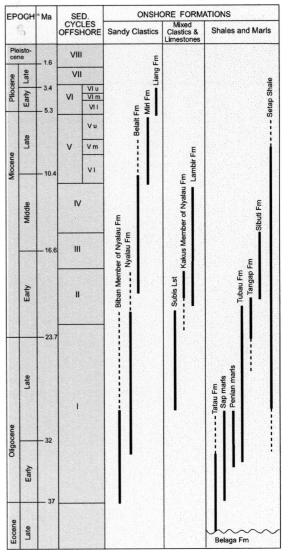

Fig. 13.9 Schematic Correlation between onshore formations and offshore stratigraphy.

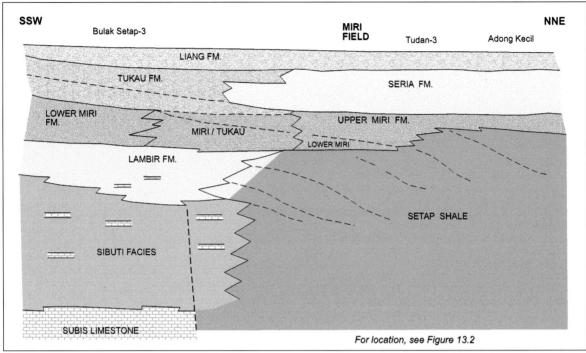

Fig. 13.10 Schematic Stratigraphic relation of Neogene formations, Miri area.

Fig. 13.11 Mud volcanoes, Setap Shale Formation, Peninjau area, Kilometre 44, Miri-Bintulu Road.

layer. Flat-top ripples are also preserved in some herringbone cross-beds. Bipolar dip direction in adjacent sets of cross-bedding in herringbone cross-bedding has been widely used as an indication of tidal deposition (Singh, 1969; Collinson and Thompson, 1988; Dalrymple, 1992). It results from reversal in current directions that can most readily occur in a tidally influenced environment. Terwindt (1988) considered flat-top ripples as one of the specific characteristics of inter-tidal deposits. The close association of these structures with flat-top ripples and flaser-bedded horizons confirms deposition in very shallow sub-tidal to inter-tidal environments.

4. *Flaser-bedded facies* is a cross-bedded sands with numerous intercalated mud flasers. Several varieties of flaser-bedded deposits

are observed in two outcrops of the Miri Formation. Figure 13.14b shows a bifurcated wavy flaser-bedded succession. The preserved discontinuous mud drapes outline the crestal shape of the ripples. The flaser-bedded facies passes gradually into wavy-bedded and lenticular-bedded horizons. Round-crested asymmetrical ripples result from current flowing in one direction only (Collinson and Thompson, 1988). The flaser-bedding indicates that sand and mud are continuously supplied into the environment, implying alternating periods of current activity and periods of quiescence. Flaser-bedding results from the presence of remnant mud layers in the ripple troughs, which has been deposited over the ripple during slack water, but eroded from the crest during the next tidal current flow (Reineck and Wunderlich, 1968). This bedform indicates a condition in which the deposition and preservation of sand are more favourable than for the mud (Reineck and Singh, 1986). The environments are shallow sub-tidal zones and inter-tidal zones.

5. *Wavy-bedded facies* comprises alternation of continuous, wavy layers of mud draping on top of rounded asymmetrical sand ripples. The sand layers are about 5 to 10 cm thick while the shale layers range from a lamina thick to about 2 cm (Fig. 13.14c). The formation of wavy bedding requires an environment where the deposition and preservation of both sand and mud are possible. Dalrymple (1992) suggested that wavy bedding could be produced by deposition on mixed inter-tidal flats.

Miri Formation	Foraminiferal zone
Estimated thickness ranges from 472 to 1310 m (1550 to 4300 ft) Upper Miri	Zone of *Nonion 3*
	Main zone of *Bolivinita 1*
Estimated thickness ranges from 168 to 1015 m (550 to 3330 ft) Lower Miri	Zone of *Loxostoma 1*

Table 13.1 Subdivisions of the Miri Formation

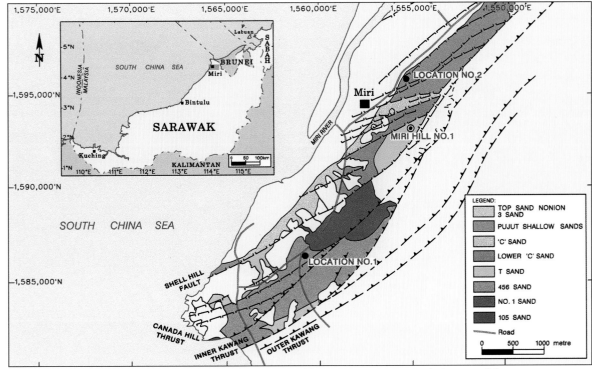

Fig. 13.12 Geological map of Miri Hill showing Shell's subdivision of the various reservoir units. The locations of the two major outcrops cited in the text are indicated (modified from Hamid Mohammad et al., 1986).

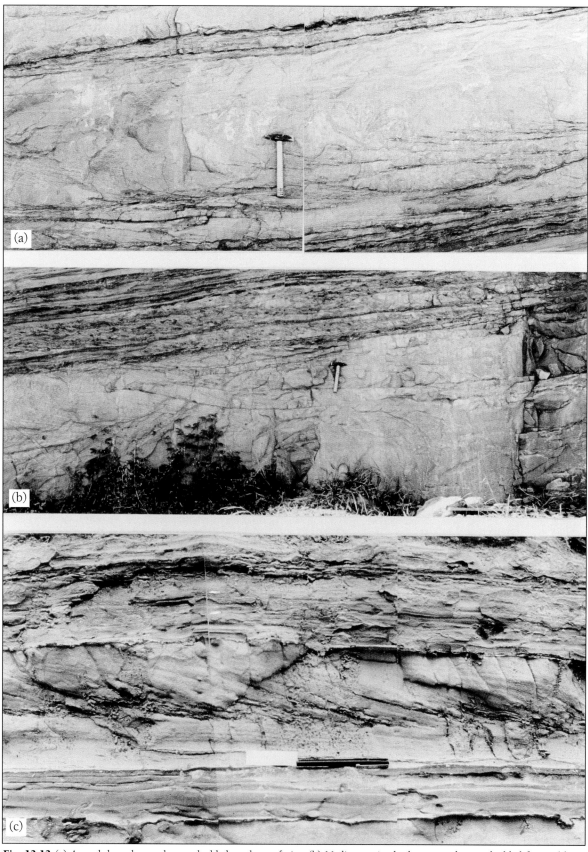

Fig. 13.13 (a) A mud-draped, trough cross-bedded sandstone facies; (b) Medium-grained, clean, trough cross-bedded facies; (c) An example of a small-scale, strongly-draped, trough cross-bedded facies sandwiched between lenticular-bedded layers. This facies is interpreted to be the product of a surface run-off. Airport Road outcrop (Location 1 of Fig. 13.12) of the Middle Miocene Miri Formation.

6. *Sand-clay alternation facies* comprises regularly interbedded fine-grained, thin-bedded sand and mud. The sand layers exhibit horizontal and ripple cross-laminations (Fig. 13.14d). The thickness of sand may range from about 1 to 23 cm, while the clay layer ranges from 1 to 5 cm. The boundaries between the sand and clay layers are commonly very sharp. Individual layers may persist over several hundreds of metres. The facies thickness ranges between 15 to 90 cm. Load casting effects is common in the clay intercalations. This facies grades upward into lenticular-bedded facies. Trace fossils are rare; species recorded are *Ophiomorpha nodosa* and *Palaeophycus*. This facies is interpreted to have formed under low-current action.

7. *Lenticular-bedded facies* consists of subordinate sand lenses of irregular form embedded in mud (Fig. 13.13c). The percentage of sand by thickness is usually not more than 25%, giving the appearance of sand lenticles floating in mud. The boundary between sand and clay layers is sharp. Lenticular bedding requires conditions of low current or wave action depositing minimum sand, and dominant slack water conditions for mud deposition. Terwindt (1971) suggested that the sand layers represent isolated small-scale ripples that have travelled over a clay bed and were covered by clay subsequently. The interpreted environments are sub-tidal and inter-tidal zones.

8. *Mudcrack surfaces and associated mudstones* were recognised at one outcrop (Fig. 13.15). The presence of these surfaces is significant because they indicate that the associated intertidal mudstones experienced periods of exposure in between deposition. The mudcrack surfaces also establish that the fining-upward facies succession of the Miri Formation implies successive shallowing, a typical feature of tidal flat sequences.

9. *Hummocky cross-stratified sandstone facies* consists of fine-to very fine-grained sheet sandstone occurring either as solitary beds between predominantly interbedded sand and shale of tidal origin, or in close association with massive coarse sandstone. The sandstone displays low angle, undulating cross-stratification (Fig. 13.16). The sandstone bed ranges from about 15 to 90 cm thick. Minimum lateral extent of this facies measured in the field is about 750 to 1000 m. Mudstone clasts are found at the base of some thin beds. Bioturbation is quite common, but not abundant and, commonly, restricted to the upper part of this facies. Trace fossils include *Chondrites*, *Planolites* and *Ophiomorpha nodosa* and indicate a shallow marine environment. Hummocky cross-stratified (HCS) structure was first defined by Harms (1975), and is commonly used as an indicator of shallow-marine sedimentation. The long, low-angle undulating hummocky cross-stratification can be produced by storm waves, below fair-weather wave base (Harms, 1975; Leckie, 1988).

The abrupt bases of these beds of the Miri Formation indicate sudden emplacement of sand into the environment. Mudstone intraclasts indicated that the already existing clay beds and laminae were actively reworked. The stratigraphic position of these beds immediately overlying shallow sub-tidal deposits indicates a sudden increase in water depth of the environment.

10. *Massive coarse sandstone facies* consists of poorly-sorted, medium- to coarse-grained sandstone with no internal structures. The sandstone occurs either as individual beds separated by interbedded thin sand, silt and clay of facies 6 above, or as basal layers virtually amalgamated with overlying HCS sandstones (Fig. 13.17a). The most characteristic feature of this facies is the dewatering and associated collapse structure, suggesting deposition in a fluidised condition. The loose, grain-supported framework suggests that the sandstone has undergone very little compaction (Fig. 13.17b). Hamblin and Walker (1979) suggested that a storm of hurricane proportion could rework and transport sand in very shallow water, creating a density current. The deposited turbidite may be subsequently reworked and laid down as HCS beds.

This massive, coarse-grained sandstone facies is interpreted as a transgressive storm deposit, probably formed immediately after an abrupt relative sea-level rise in a shallow, tidally influenced environment. The bedding character of the facies would suitably fit the storm-induced density flow deposition.

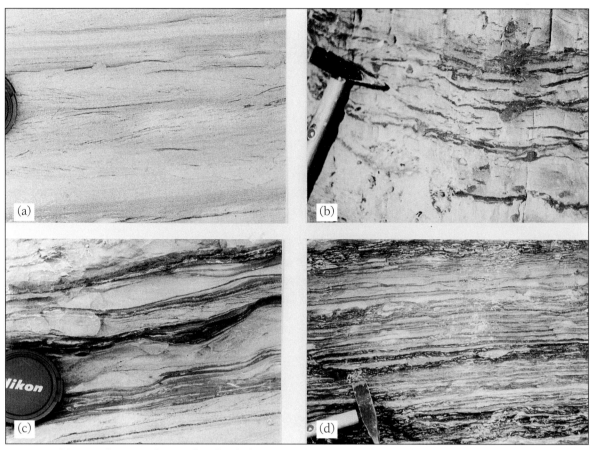

Fig. 13.14 (a) Herringbone cross-laminated sand with thin, intermediate shaly streak; (b) A flaser-bedded sandstone facies; (c) Wavy-bedded facies showing continuous mud layer draping asymmetrical sand ripples; (d) Thinly-bedded, sand-shale alternation unit. Airport Road outcrop (Location 1 of Fig. 13.12) of the Middle Miocene Miri Formation.

Outcrops and facies succession

The lithological-stratigraphic logs for the exposed sections of the Miri Formation are shown in Figures 13.18, 13.19 and 13.20. On the basis of the stratigraphic succession of the different facies, 2 different facies associations, namely tidal facies and wave- and storm-dominated facies, are recognised.

1. *Tidal facies association* is the most dominant facies association in the Miri Formation and is well displayed at the left flank of the Airport Road outcrop (Location 1; Figs. 13.19 and 13.20). The complete logged section shows repeated upward progression of subtidal-intertidal-tidal flat cycles. Each cycle suggests a gradual shallowing of the environment of deposition. This kind of succession is common in progradational macro-tidal depositional systems, where the accumulation of sand is faster than the rising high tide elevation.

    The ideal, complete single fining-upward succession of this tidal facies association is shown in Figure 13.21. The multiple stacking of this tidal-influenced sandstone and mudstone suggests progressive basinal subsidence and/or relative sea-level rise occurring with sedimentation. Figure 13.22 shows a block diagram of a siliciclastic coastal tide-dominated environment, showing the different sub-environments producing the different facies. This succession represents periods and sites of active deposition, with large volumes of sediment transported landwards to accumulate in shallow sub-tidal and inter tidal settings at the head of an estuarine embayment.

2. *Wave-and storm-dominated facies* association in which HCS and associated coarse-grained massive sandstones are products of storm- and wave-dominated inner shelf sedimentation. The presence of these sandstones, interbedded within the predominantly tide-dominated sandstone and mudstone succession, suggests abrupt periods of sea-level rise within a background of gradual transgressive, tidal estuarine system. Figure 13.23 shows the field (vertical and lateral) organisation and geometry of massive, coarse-grained

sandstone and HCS beds and their relationship with tidal deposited units. Its vertical organisation suggests that it is a transgressive facies succession. The succession begins with the deposition of massive coarse sandstone and HCS sandstone in storm-dominated, lower offshore areas at depths of about 20 m, probably tens of km from the coast. This could happen at the beginning of a marine transgression into a tide-dominated environment.

The upward-fining massive, coarse-grained sandstone, HCS sandstone and associated mudstone facies association represent wave- and storm-dominated sedimentation during the transgression of a coastal, tide-dominated estuary.

Facies model

Based on outcrop study, a sedimentological model that can best explain the palaeogeography and depositional processes of the Miri Formation would be a tide-dominated estuary, comparable to the model proposed by Dalrymple et al. (1992). Tate's (1976) interpretation of estuarine sedimentation in the Middle Miocene of Brunei lends further support to this hypothesis, and suggests that estuarine sedimentation extended to areas in Brunei, although it may not occur as a continuous system.

Estuarine conditions are initially created at the beginning of a transgression and migrate landwards as the transgression proceeds (Dalrymple et al., 1992). The interaction between the rate of sediment supply and accommodation determines the morphological variation within an estuary (Jervey, 1988; Dalrymple et al., 1992). In tide-dominated estuaries, sediment supplied by both river and marine sources are readily distributed by tidal currents, resulting in rapid infilling of the deeper and wider parts and development of the classic funnel-shaped geometry and facies distribution (Dalrymple et al., 1992). The vertical variation and spatial distribution of the facies within the estuary can be related to relative sea-level changes.

Summary

1. Outcrops of the Miri Formation around Miri provide some realistic insights into the geometry of the two main different types of sands. Cross-bedded tidal sands

Fig. 13.15 Mudcracks surface coated with rusty-coloured iron oxide. Northwest Miri Hill outcrop (Location 2 of Fig. 13.12) of the Middle Miocene Miri Formation.

Fig. 13.16 Hummocky cross-stratified sandstone facies with basal mudstone intraclasts. Airport Road outcrop (Location 1 of Fig. 13.12) of the Middle Miocene Miri Formation.

have minimum composite thickness (amalgamated beds) of 1 m, and extend laterally several kilometres with little variation. Hummocky cross-stratified sandstones have average thicknesses of around 45 cm, with good lateral continuity in thickness and texture. Minimum lateral extent measured in the field is 1 km.

2. The stratification of sand and shale of the Middle Miocene Miri Formation suggests the existence of a macro-tidal estuary during that period. Tide-dominated estuarine sedimentation is represented by sub-tidal trough cross-bedded sandstone, inter-tidal flaser and wavy-bedded facies and tidal mud-flat lenticular beds.

3. The massive coarse-grained sandstone and hummocky cross-stratification facies of the Miri Formation constitute a rare example of 'transgressive storm-beds' which forms on top of shallow subtidal-intertidal-tidal flat deposits.

4. The Middle Miocene Miri Formation probably constitutes a classic example of the sensitivity of tidal environments to abrupt sea-level rise, as is illustrated by the coarse-grained sandstone and HCS sandstone association sandwiched within the shallow, tide-dominated deposits.

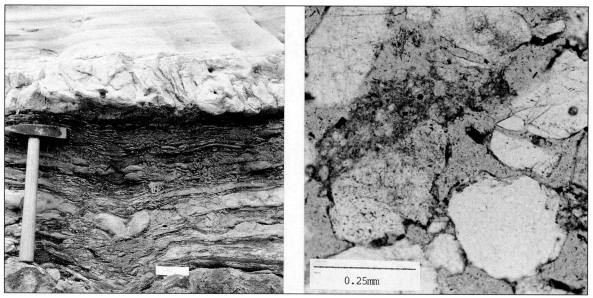

Fig. 13.17 (a) A succession of massive coarse-grained sandstone at the base followed by sand-shale alternation, overlained by another coarse-grained sandstone, and capped by a clean, fine-grained, hummocky cross-stratified sandstone; (b) A photomicrograph of the coarse-grained sandstone, showing poorly-sorted texture in a loose, grain-supported framework. Airport Road outcrop (Location 1 of Fig. 13.12) of the Middle Miocene Miri Formation.

Estuarine, tidal sedimentation during the Middle Miocene was punctuated by large and abrupt relative sea-level rise, resulting in the deposition of massive coarse-grained sandstone and hummocky cross-stratified sandstone.

Offshore

The offshore stratigraphy of the West Baram Delta is characterised by the occurrence of coastal to coastal - fluviomarine sands which have been deposited in a northwestwards prograding delta since the Middle Miocene (from Cycle IV onwards). Periods of delta outbuilding are separated by rapid transgressions, which are represented by marine shale intervals at the base of the sedimentary cycles (Ho, 1978). The regressive sands of each cycle grade northwestwards into neritic, mainly shaly sediments. Three major eustatically-controlled cycles of regression and transgression have been described (Rijks, 1981). In the West Baram Delta, Cycle V (Middle to Upper Miocene) to Cycle VII (Upper Pliocene) are well developed, prograding over thick diachronous pro-delta shales which have been mapped as the Setap Shale Formation onshore. Each cycle develops in a coastal plain environment to the south, dominated by deposition of sands, silts and clays, and grades northwards into holomarine neritic to bathyal environments with deposition of mainly clays, silts, minor sands and, in places, turbidites.

The following are summaries of the stratigraphy and lithofacies of the reservoirs in the Betty, Bokor and Baram fields which have been studied by various workers (Johnson et al., 1989; Sering and Ramli, 1994; Abdul Hadi Abd. Rahman, 1995).

Betty Field: The reservoirs comprise a stacked sequence of shallow-marine sands and shales of Late Miocene age (Upper Cycle V), occurring at depths of between 2194 and 2941 m (7200 and 9650 ft). The detailed sedimentological and reservoir characteristics of these sands were studied by Johnson et al. (1989), based on some 350 m (1150 ft) of continuous cores from well BE-5. Four main facies types, namely sandstone, sandstone-dominated heterolithic, mudstone-dominated heterolithic, and mudstone facies, were recognised. These main facies types were further subdivided into 10 sub-facies, based on variations in texture, sedimentary structure, bioturbation, mud content, porosity and permeability.

The *sandstone facies* is divided into 3 sub-facies: poorly-stratified sandstone, bioturbated sandstone, and low-angle/parallel laminated to hummocky cross-stratified sandstone. The *poorly-stratified sandstone* is mainly fine to medium grained, well sorted, friable, and either structureless or fairly stratified, with good reservoir quality (porosity 24%, permeability 1200 mD) and deposited in a high-energy, wave-dominated, shallow marine (nearshore) environment. The *bioturbated sandstone sub-facies* consists of fine-grained, well-sorted sandstone with abundant vertical and horizontal burrows, slightly reduced reservoir quality (porosity 22%,

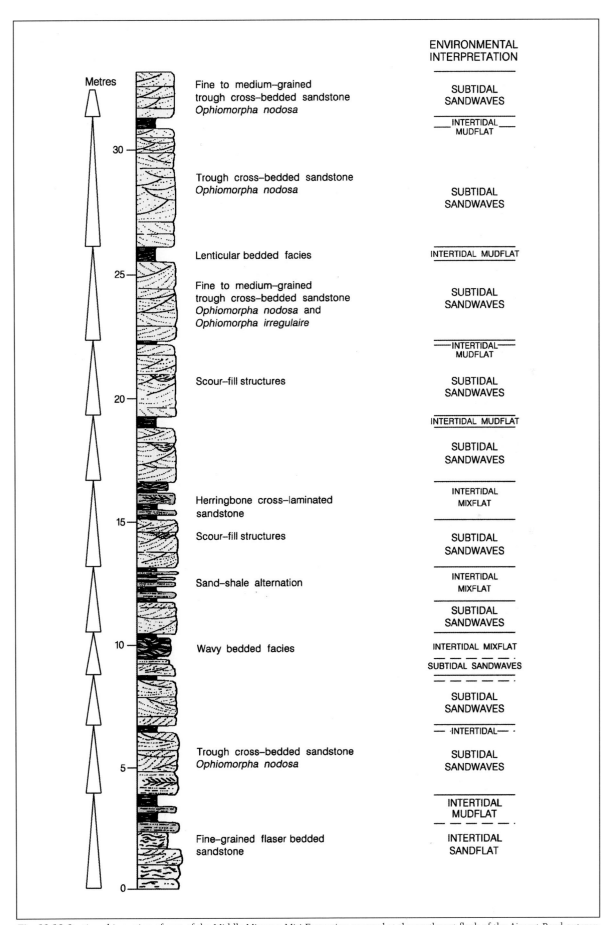

Fig. 13.18 Stratigraphic section of part of the Middle Miocene Miri Formation exposed at the southwest flank of the Airport Road outcrop (Location 1 of Fig. 13.12), showing repeated fining-upward tidal successions (modified from Abdul Hadi Abd. Rahman, 1995).

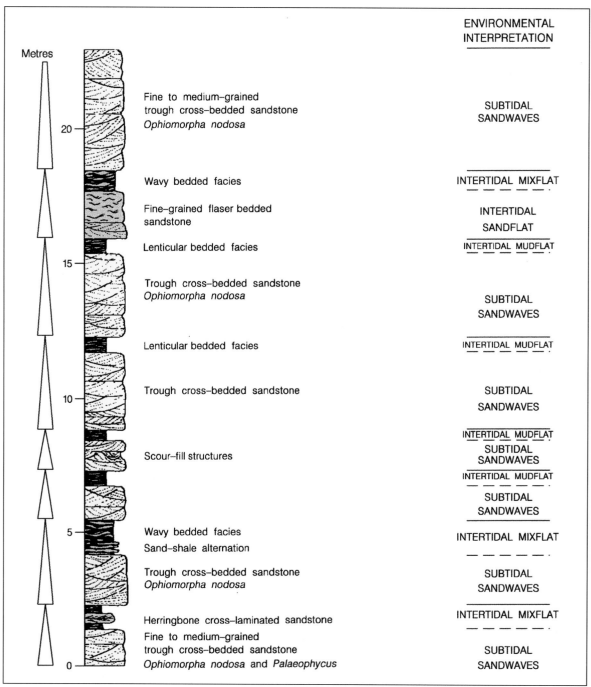

Fig. 13.19 Another stacked, fining-upward tidal successions of the Middle Miocene Miri Formation from the central part of the Airport Road outcrop, Miri (modified from Abdul Hadi Abd. Rahman, 1995)

permeability 475 mD) and deposited in moderate-energy, nearshore environment in which the rate of bioturbation exceeded the rate of deposition. The low-angle/parallel laminated to hummocky cross-stratified sandstone is fine grained, moderately sorted and characterised by well-developed lamination which ranges from parallel to low angle (<10°). The fine grain size and lamination result in relatively moderate reservoir quality (porosity 19%, permeability 90 mD). The sandstone is interpreted to be high-energy storm deposits that were deposited rapidly, probably below fair-weather wave-base, in a nearshore to inner neritic environment.

The *sandstone-dominated heterolithic facies* comprises sandstones with significant proportions of either interstitial or interbedded clays and can be divided into 2 sub-facies. The *bioturbated heterolithic sandstone sub-facies* consists of fine-grained, slightly argillaceous sandstone which has been completely homogenised by extensive bioturbation and relatively poor reservoir quality (porosity 17%, permeability 52 mD) due to the high proportion of dispersed clay. The sandstone was deposited

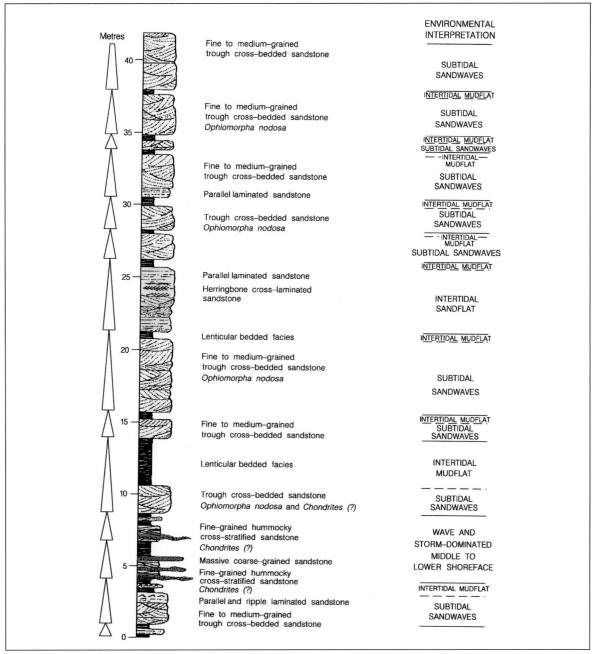

Fig. 13.20 Stratigraphic section of the Middle Miocene Miri Formation at the hill-cut at northwest Miri Hill (Location 2 of Fig. 13.12), showing wave and storm-dominated lower shoreface facies towards the bottom part (modified from Abdul Hadi Abd. Rahman, 1995)

in a low-energy inner neritic environment. The second sub-facies is *interbedded sandstone and shale* with individual sandstone beds (12 – 90 cm thick) displaying erosive base, clay clasts, low-angle to ripple lamination, and bioturbated or sharp tops, and occurring in single or amalgamated units and may be overlain by cm-thick mudstone layers. Reservoir quality is highly variable but generally moderate with porosity of 17% and permeability of 140 mD. This sub-facies is interpreted as an alternation of storm-generated sandstone beds interbedded with fair-weather mudstone.

The *mudstone-dominated heterolithic facies*, comprising various mudstone lithologies (ranging from laminated to bioturbated) with up to 50% sandstone intercalations, typically form the intra-reservoir shale layers and is generally non-reservoir, although minor porosity/ permeability may occur in some of the sandier layers. This facies was deposited in a fluviomarine coastal to inner neritic environment.

The *mudstone facies* forms the intraformational seal in the Betty field and was deposited in a fluviomarine coastal to inner neritic environment.

From their study, Johnson et al. (1989) concluded that the Betty reservoirs contain a broad, hierarchical range of vertical facies

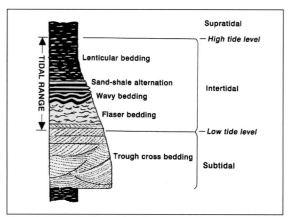

Fig. 13.21 An ideal, complete fining-upward succession of tidal estuarine sedimentation. This type of sedimention is commonly develop by the progradation of a tidal flat (modified from Dalrymple et al., 1992).

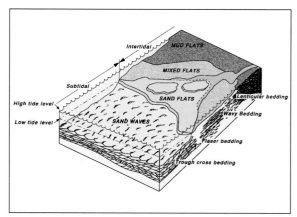

Fig. 13.22 A block diagram showing the different sub-environments in a siliciclastic tide-dominated coast (based on Dalrymple et al., 1992).

successions characterised by repeated progradational/regradational units of which three main types are apparent:
- Amplified sequence in which the progradational unit contains well-developed, high-energy sandstone;
- Stacked sequence characterised by a composite progradational unit with intercalated shale layers; and
- Single sequence comprising a single symmetrical unit with relatively low-energy facies.

Bokor Field: Sering and Ramli (1994) identified 6 lithofacies, namely poorly stratified sandstone, cross-laminated sandstone, bioturbated sandstone, sand-dominated heterolithic, mud dominated heterolithic, and mudstone, from 98 m (323 ft) of core from the Late Miocene (Upper Cycle V) reservoirs of the Bokor Field between 590 and 744 m (1940 and 2440 ft). These facies were considered to be deposited in a fluvial-dominated delta which prograded into a muddy marine shelf. Based on their reservoir quality, these facies were grouped into 3 units: good (porosity 35 – 40%, permeability 1000 – 2000 mD), moderate (porosity 30 – 35%, permeability 100 – 300 mD), and poor (porosity < 25%, permeability < 10 mD).

Baram Field: Abdul Hadi Abd. Rahman (1995) studied some 125 m (410 ft) of cores from the Upper Cycle V (Late Miocene) reservoir section in two wells, BA-8 and BA-16, and recognised 12 lithofacies based on lithology, sedimentary structures, degree of bioturbation, and fossil and trace fossil assemblage. These lithofacies are:

1. *Hummocky cross-stratified (HCS) sandstone* comprises fine- to very fine-grained, thinly-bedded (< 30 cm) sandstones, characterised by fine, parallel to low angle (< 10°) laminations, and interbedded with thicker beds of bioturbated and/or laminated mudstone (Fig. 13.24a). Most beds have sharp bases and the upper parts show slight bioturbation. Trace fossils present include *Ophiomorpha nodosa, Chondrites, Planolites* and *Skolithos linearis*(?). These HCS beds probably developed under the combined influence of storm waves and a unidirectional storm-generated geostropic current in a storm dominated lower shoreface or inner shelf areas.

2. *Swaley cross-stratified (SCS) sandstone* consists of very fine- to fine-grained sandstone, occurring commonly as 3 to 10 m thick amalgamated beds without interbedded mudstone (Figs. 13.24b and 13.24c). Mudstone intraclasts are common at the base of individual layers, but may also be scattered throughout the facies. Low angle (< 15°) parallel to slightly divergent stratification dominated this facies. Bioturbation is minimal. Trace fossils include *Ophiomorpha nodosa, Ophiomorpha irregulaire, Planolites, Skolithos, Teichichnus* and *Terebellina*.

This facies was deposited in a middle to upper shoreface environment, in 5 to 10 m water depths, dominated by storms and fair weather waves. The small-scale wave ripples represent waning of storm phases, or fair-weather conditions, with long quiet periods for biogenic activity, as indicated by the bioturbations.

3. *Massive coarse sandstone* is a well-sorted, angular quartzarenite showing distinct, massive texture in contrast to other sandy facies, and is interbedded with laminated

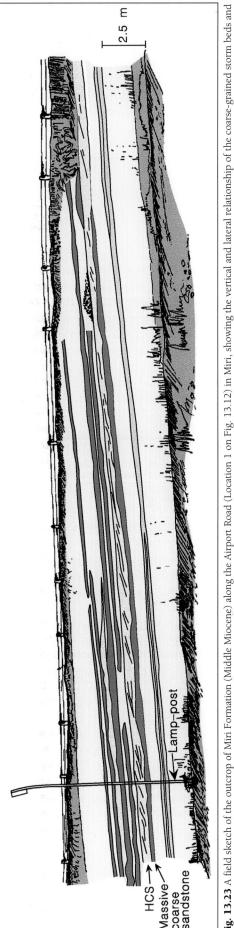

Fig. 13.23 A field sketch of the outcrop of Miri Formation (Middle Miocene) along the Airport Road (Location 1 on Fig. 13.12) in Miri, showing the vertical and lateral relationship of the coarse-grained storm beds and hummocky cross-stratified sandstone (modified from Abdul Hadi Abd. Rahman, 1995).

mudstone and bioturbated silty mudstone. These are coarse-grained storm beds. The thickness and stratigraphic position suggest that they are probably the tail end sediments, deposited in deeper waters during a very stormy event.

4. *Lenticular-bedded ribbon sand facies* comprises fine- to very fine-grained sandstone ribbons and lenses (1 to 5 cm thick), interbedded with subordinate mudstone (Fig. 13.24d), and occurs in close association with HCS and SCS sandstones. The bases of the sand layers are in most cases sharp. Internally, the sandstones are either parallel laminated or show a mixture of asymmetrical and symmetrical ripple cross-lamination. Bioturbation is rare. This facies was possibly deposited during less turbulent periods in a wave- and storm-dominated shelf, when the amount of sand reworking and transport was greatly reduced.

5. *Interbedded laminated sandstone and bioturbated mudstone facies* comprises thin laminated sandstone (2 - 20 cm thick) interbedded with mottled, bioturbated mudstone of similar thicknesses (Fig. 13.25a). The sandstones are usually fine- to very fine-grained, with horizontal to low angle parallel laminations. Bioturbation is apparent but not common; traces recognised include *Ophiomorpha nodosa, Ophiomorpha irregulaire, Skolithos* and *Palaeophycus*. The mudstones are dark grey and mostly intensely bioturbated. *Teichichnus, Chondrites* and *Palaeophycus heberti* are some of the identifiable traces. The sandstones show sharp, non-erosive or bioturbated contact with the overlying mud; the mudstones are commonly erosively overlain by sandstones. Mudstone clasts horizon is common, usually occurring immediately on top of a mudstone layer, or interbedded within a sandstone bed. Tube-like escape structures (*Skolithos?*), large bore-type *Ophiomorpha nodosa* escape structure, and associated collapse structure are also preserved at many intervals.

The highly bioturbated nature of the mudstone layers implies deposition largely from suspension in a low-energy environment characterised by intense biogenic activity. Deposition most probably took place below fair-weather wave-base. This, however, is in contrast to the lithology and character of the sandstone, which suggest deposition in a wave- and storm-dominated environment. This facies was probably deposited in a marginal environment close to storm-wave base.

6. *Laminated mudstone facies* is characterised by distinct, very fine sand to silty laminations of millimetre to centimetre thickness. The boundaries between the sandy streaks and the background mud are mostly sharp. Internally, these thin sands may show parallel lamination or may contain low-amplitude wave ripples (Fig. 13.25b). The background mudstone is dark grey or greyish black. In some intervals, distinctive subtle colour banding that reflects mm- to cm-scale graded siltstone-mudstone bed is preserved. This facies is interpreted to be the result of intermittent, storm-related deposition beneath a dysaerobic water column.

7. *Silty, bioturbated mudstone facies* consists of dark grey to black mudstone with irregular light-coloured, silty patches which reflect extensive bioturbation (Fig. 13.25c). Average thickness is about 1.5 m. Dominant trace fossils include *Chondrites, Palaeophycus heberti, Teichichnus rectus* and *Terebellina*. Intense biogenic reworking of some intervals makes further identification impossible.

    The muddy background lithology and high degree of bioturbation of this facies imply deposition from suspension in a low energy environment with intense animal activity, probably a holomarine inner neritic environment with water depths of around 50 m. The lack of well-preserved primary sedimentary structures suggests deposition below fair weather-wave base.

8. *Muddy bioturbated siltstone facies* consists entirely of a pervasively bioturbated, light to medium grey mudstone, with silt and very fine sand "stirred" in by the organisms (Fig. 13.25d). Trace fossils recognised include *Ophiomorpha nodosa, Chondrites, Teichichnus* and *Palaeophycus*(?). This facies was probably deposited fairly close to shore but still below fair weather-wave base

9. *Shelly, laminated fining-upward sandstone* displays distinctive internal stratification. The basal part is planar laminated, medium- to coarse-grained sand with interbedded thin horizons of calcareous shell fragments, grading upward into a thin layer of light grey, low-angle parallel laminated, fine-grained sand, and capped with ripple cross-laminated sand (Fig. 13.26a). This facies overlies erosively a thin layer of laminated mudstone and is overlain sharply by a thicker laminated mudstone. This facies is interpreted as a gutter cast, and is thought to form during storms related to a relative sea-level fall, when oscillatory wave and wind-forced currents scour the shelf and cut deep gutter casts. It commonly occurs in close association with HCS and SCS facies.

10. *Mud-draped, ripple-laminated sandstone* comprises small- to medium-scale (8 - 13 cm long), ripple-laminated mounds enveloped by thin, black mud drapes (Fig. 13.26b). This facies is interpreted as a sub-tidal sand deposited in an environment with alternating periods of current activity and quiescence, allowing the deposition and preservation of sand and mud.

11. *Heterogeneous sandstone facies* displays intermixing of clay clasts, irregular clay laminations and bioturbation (Fig. 13.27a). The basal contact, commonly either with laminated or bioturbated mudstone, is erosive; at the top, this facies either gradually diffuses into a flaser-bedded horizon or shows bioturbated contacts with mudstone. Trace fossils identified include *Ophiomorpha nodosa, Ophiomorpha irregulaire, Palaeophycus* and *Thalassinoides* (?).

    The lithology, presence of irregular, bioturbated mud layers, and close association of this facies with flaser-bedded facies suggest deposition in a tide-dominated environment. Under fair-weather conditions, tidal and longshore currents operate to depths of about 10 to 20 m, and are able to move sand around within these shallow areas.

12. *Flaser-bedded sandstone facies* comprising fine- to medium-grained sandstones with simple to bifurcate flaser bedding (Fig. 13.27b), similar to the flaser-bedded facies of the Miri Formation. The single horizon of flasered bed occurs as the top part of a heterogeneous sandstone. The sedimentary structures and stratigraphic position of this facies suggest deposition in sub-tidal environments.

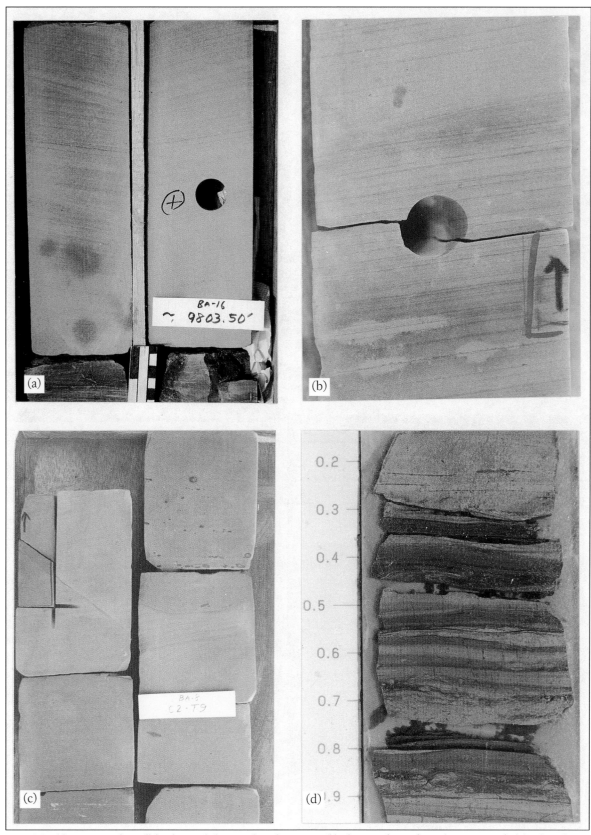

Fig. 13.24 (a) Fine-grained, parallel to low angle laminated mudstone unit; (b) Close-up of part of a swaley cross-stratified (SCS) sandstone unit, showing fine, parallel to low-angle laminations; (c) Another SCS unit, showing small mud clasts lining parallel to low angle laminations; (d) Lenticular-bedded ribbon sand facies. Core widths are ~ 10 cm {Cores are from the Upper Cycle V (Late Miocene) reservoir sections from wells BA-8 and BA-16 of Baram Field}.

Fig. 13.25 (a) Part of an interbedded, laminated sandstone and bioturbated mudstone facies; (b) Laminated mudstone, with a thin horizon of low-amplitude ripple laminated sandstone; (c) Dark mudstone with silty streaks and silt-filled burrows; (d) Muddy, bioturbated, sideritised siltstone. Core widths are - 10 cm {Cores are from the Upper Cycle V (Late Miocene) reservoir sections from wells BA-8 and BA-16 of Baram Field}.

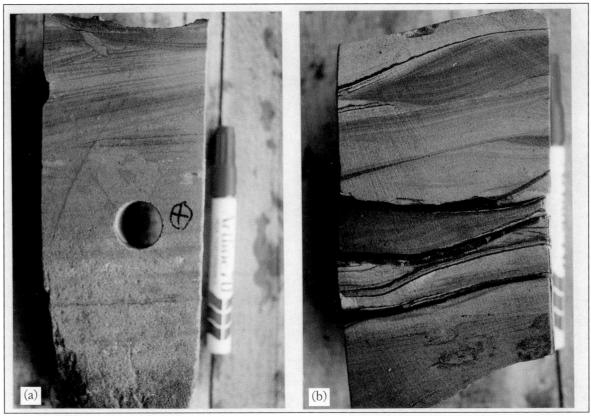

Fig. 13.26 (a) Shelly, laminated and fining-upward sand unit interpreted as a gutter cast; (b) Mud-draped, ripple-laminated sandstone. Core widths are ~ 10 cm {Cores are from the Upper Cycle V (Late Miocene) reservoir sections from well BA-8 of Baram Field}.

Facies succession and depositional environment

Based on the descriptions of the different facies above, 3 facies successions were mapped in the Baram Field (Figs. 13.28a, b and 13.29a, b):

1. *Sharp-based, upward-coarsening SCS succession* is characterised by thick (ca. 10 m) amalgamated SCS sandstone abruptly overlying thinly interbedded HCS sandstones and mudstones. The thick SCS unit comprises several amalgamated units of fine-to very fine-grained HCS sandstone, punctuated in places by massive coarse-grained sandstone and interbedded mud clasts conglomeratic layers. The sharp-based SCS sandstone corresponds to a very sharp decrease in the gamma-ray log readings. The thick SCS sandstone is interpreted to be deposited during a rapid fall of relative sea level, resulting in shoreface progradation into mud-dominated inner shelf, thus giving the sharp-based characteristics.

2. *Gradational-based, upward-coarsening HCS-SCS succession* begins with laminated or bioturbated mudstones and interbedded, cm-thick, fine to very fine sandstone beds. This lower succession may contain thin layers of facies 1, 3, 5, 9 and cm-thick planar laminated or ripple cross-laminated sandstones. The sandstone beds thicken upwards, the proportion of HCS beds increases and may show amalgamated HCS beds. Further upward, mudstone interbeds decrease in thicknesses and SCS sandstones predominate. The gradational vertical facies organisation suggests a progressive, gradual shallowing of the environment of deposition in a prograding, wave-dominated shoreface.

3. *Fining upward tidal succession* characterised by an upward-fining, upward-shallowing succession of tide-dominated facies, e.g. thick, fine-to medium grained heterogeneous sandstone capped at the top by a flaser-bedded horizon. This succession is interpreted as the result of a slight shallowing of a sub-tidal, tide dominated environment. Its thickness and stratigraphic position suggest periods of localised and restricted tide-dominated sedimentation within a 'temporally' more usual wave-and storm-dominated environment.

STRUCTURE

Onshore

The Neogene delta system of northwest Sarawak prograded over the active convergent margin of Northwest Borneo, resulting in the amalgamation of basement-rooted structures with deltaic growth faulting (Sandal, 1996). The structural basement of the Baram Delta Province is inferred to be the Rajang Group/West Crocker Formation, exposed in the uplifted and deeply truncated hinterland of Sarawak and Sabah. Although the basement has not been drilled into in the Baram Delta area, depths of up to 12 km had been estimated (Taylor and Hayes, 1980). The delta substratum thus comprises deep-marine to prodelta clays of the Setap Shale Formation. This unit forms a plastic substratum of varying thickness to the overlying delta systems and it maximises de-coupling of the deformation styles generated in the underlying basement with the overlying deltaic overburden. Thus the dominant structural styles in the West Baram Delta are primarily gravitational and depositional loading through the outbuilding deltaic systems. Additional complexities generally result from their interrelation with the basement-controlled structural element.

The main structural styles observed in the onshore part of the West Baram Delta are growth faults (e.g. Shell Hill, Kuala Baram, North Lutong and North Rasau Faults), normal antithetic and synthetic faults, and north-dipping reverse faults (e.g. Canada Hill, Kawang and Rasau Thrust Faults). The displacement and strike-slip movement of these reverse faults is not clear but is estimated to be some 600 m in the Miri field area. All these faults trend ENE - WSW and hade to the northeast (Fig. 13.30).

Compressional tectonics during the Late Miocene - Pliocene resulted in numerous folds, including the Riam - Buri Anticline, the Liku Badas Syncline, and the Miri - Rasau Anticline. The Liku - Badas Syncline is a relatively simple and unfaulted structure. Some of the growth faults (e.g. Shell Hill Fault) had been reactivated as reverse faults during these compressional periods. Various companies have explored for oil trapped against these structures. The only successes, so far, are the Miri and Asam Paya fields in Sarawak, and the Seria and Rasau fields in Brunei.

The West Baram Line which marks the western margin of the West Baram Delta with the Central Luconia Province could not be traced onshore but may be represented by the right-lateral Tinjar Fault.

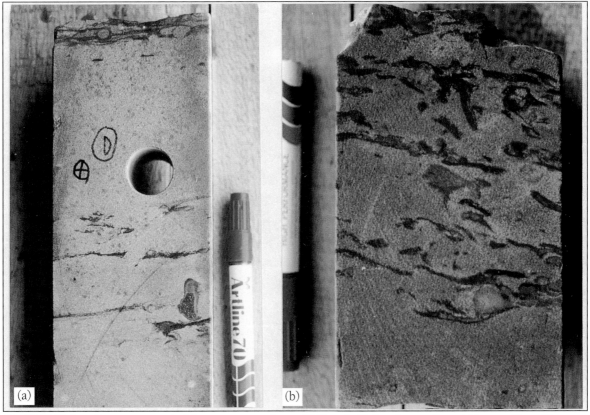

Fig. 13.27 (a) A medium-to coarse-grained heterogeneous sandstone; (b) Flaser-bedded sandstone. Core widths are ~ 10 cm {Cores are from the Upper Cycle V (Late Miocene) reservoir sections from wells BA-8 and BA-16 of Baram Field}.

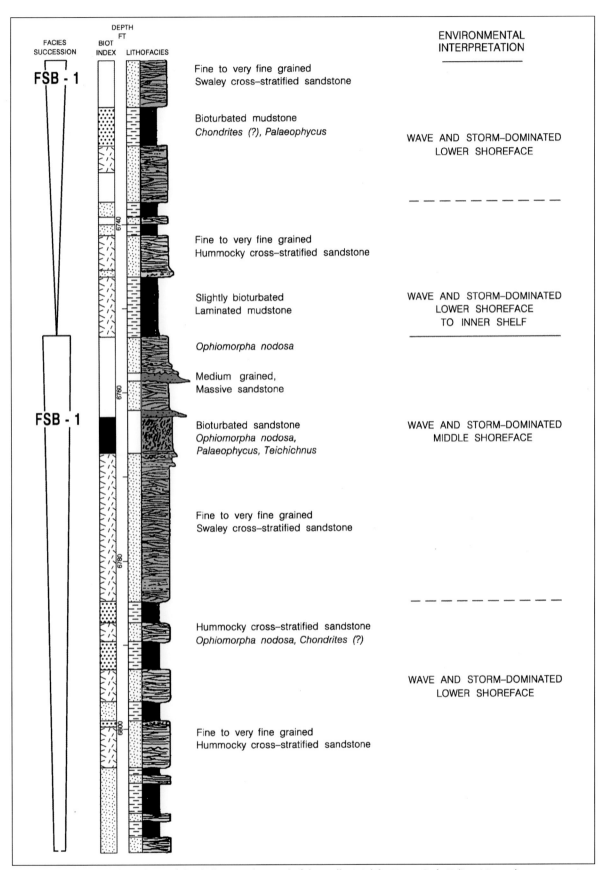

Fig. 13.28 (a) Stratigraphic column of the shallow cored interval of the Well BA-8 (The Upper Cycle V (Late Miocene) reservoir sections from Baram Field).

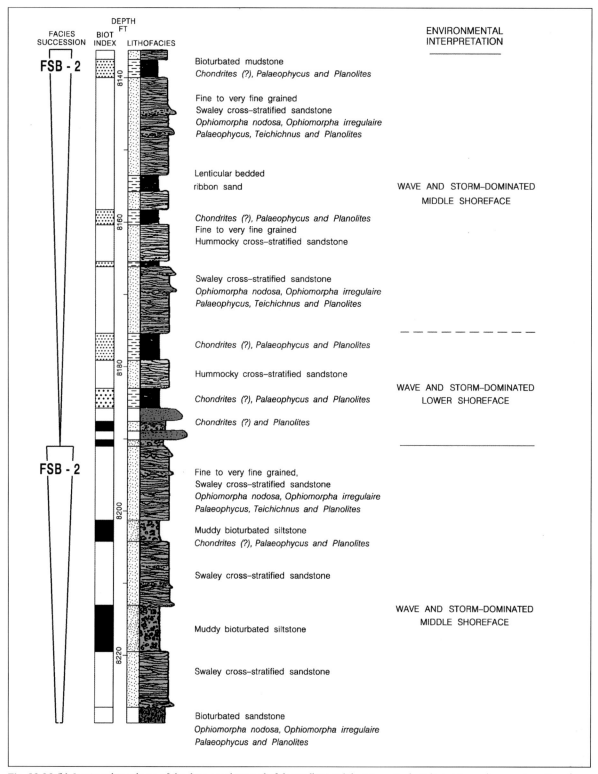

Fig. 13.28 (b) Stratigraphic column of the deep cored interval of the Well BA-8 {The Upper Cycle V (Late Miocene) reservoir sections from Baram Field}.

Offshore

The offshore West Baram Delta is characterised by delta tectonics, exemplified by families of growth faults forming macro- and mega-structures (tectonostratigraphic units), eventually grading into a delta 'toe', characterised by clay deformation and asymmetric folds in the deeper offshore.

The presence of a very thick sedimentary pile (> 6000 m) affected by growth faulting implies the existence of an underlying subsiding, weakened crust. This roughly triangular portion of attenuated crust is bounded to the SW by an irregular NW-SE lineament, the West Baram Line. This 'line' is expressed on seismic

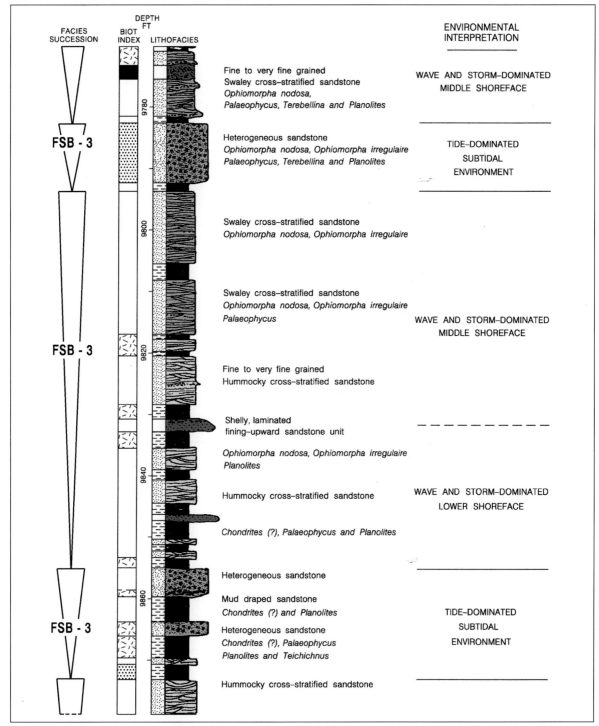

Fig. 13.29 (a) Stratigraphic column of the shallow cored interval of the Well BA-16 (The Upper Cycle V (Late Miocene) reservoir sections from Baram Field).

by a hinge bounding the NE extension of the Central Luconia 'platform' (carbonate palaeoshelf edge). This hinge is most likely step-faulted, *en echelon*, and probably has a transform character.

In addition to the growth faulting, the West Baram Delta was affected by later, wrench-induced folding along NE-SW trending axes. Compressional deformation culminated during Cycle VII. Quaternary uplift resulted in the emergence of the land area and the erosion of the younger cycles from the crestal areas in the south, e.g. Cycles VI and VII from the Miri and Siwa anticlines, and most of Cycle VII from Bakau and Bokor. This strong basinward tilt, as shown by emergence, erosion and excessive dips of non-tectonised strata has the effect of destroying, partially to almost totally in places, the counter-regional dips created by rollover into a number of major growth faults.

The West Baram Delta can be subdivided, in a proximal to distal direction, as follows:

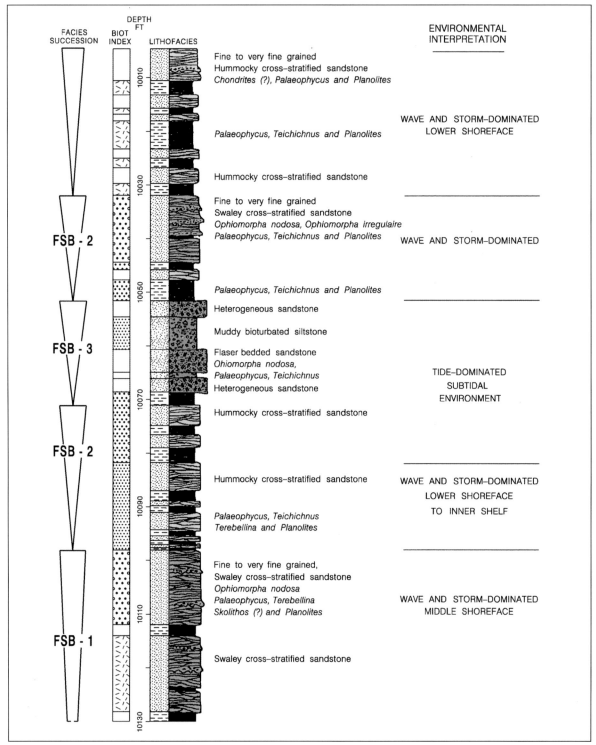

Fig. 13.29 (b) Stratigraphic column of the deep cored interval of the Well BA-16 {The Upper Cycle V (Late Miocene) reservoir sections from Baram Field}.

1. An updip region, bounded to the NW by the Siwa-Miri-Tudan and the Ensalai-Rasau trends. This is characterised by the fact that former growth faults, if any, have become unrecognisable due to later tectonics: the tectonic style (broad synclines and narrower, reverse faulted anticlines) is largely basement induced. The updip region is almost entirely onshore;

2. A mid-dip (central) region, starting in the south at the Siwa-Miri-Rasau trend and bounded to the NW by the so-called Outer Shelf growth fault system. This realm is characterised by structures that were initially growth fault-induced. Seven mega-structural trends (Fig. 13.31) can be recognised:
   - Siwa-Miri-Rasau trend;
   - Tukau-West Lutong trend;

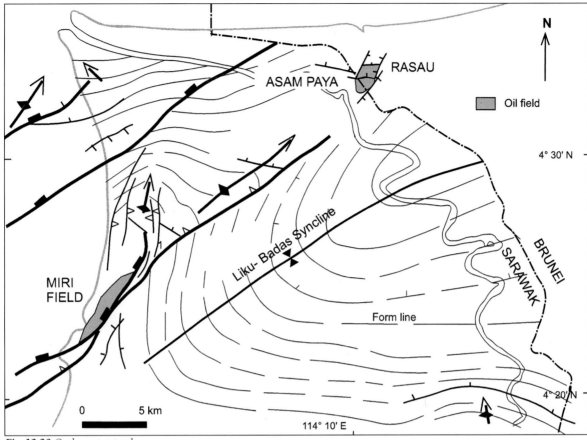

Fig. 13.30 Onshore structural map.

- Bokor-Bakau trend;
- Laila-Baram trend;
- Baronia-Fairley-Baram trend;
- Lydia-Beryl trend, and
- Outboard trend.

A number of these trends have been deformed by basement-controlled compression and wrench faulting. This is particularly evident along the Siwa to Rasau trend, where compression and wrenching have structurally obliterated the initial growth faults. Further downdip, and extending to the Baronia mega-structure boundary fault, wrench-induced broad anticlines running obliquely to the growth fault trends have significantly deformed the original structures, destroying some pre-existing traps in the process, but concomitantly resulting in the focussing of hydrocarbon migration (e.g. Tukau, Bokor, Bakau, West Lutong, Baram-B and -A). The influence of these superimposed deformations almost ceases at the Baronia-Fairley-Baram trend where it is expressed only by a slight along-strike tilt of the pre-existing rollover. From the Baronia-Fairley-Baram trend and northwards, large, well-defined counter-regional faults occur, thus defining the mega-structures unambiguously (further updip counter-regional faults do occur, but are relatively minor).

3. A downdip region, bounded by the Outer Shelf growth fault system. It is characterised by very large boundary faults that reach or almost reach the sea bottom. It grades downdip, via structures affected by clay deformation, into a series of sub-parallel, asymmetric, overthrust anticlines. This delta 'toe' extends into very deep waters until it reaches and forms the edge of the Sabah Trough abyssal plain.

The tectonic style of the West Baram Delta described above, shows the interaction of two main types of deformation. Gravity-induced growth faults are intersected by wrench-induced, compressional, NE-SW trending folds of Late Miocene or early Pliocene age. This resulted in a broad style of both fault- and dip-dependent closures which formed traps for the major hydrocarbon accumulations discovered to date. The intensity of wrench-induced deformation decreases basinward and growth faulting is the dominant tectonic style in the outer part of the shelf.

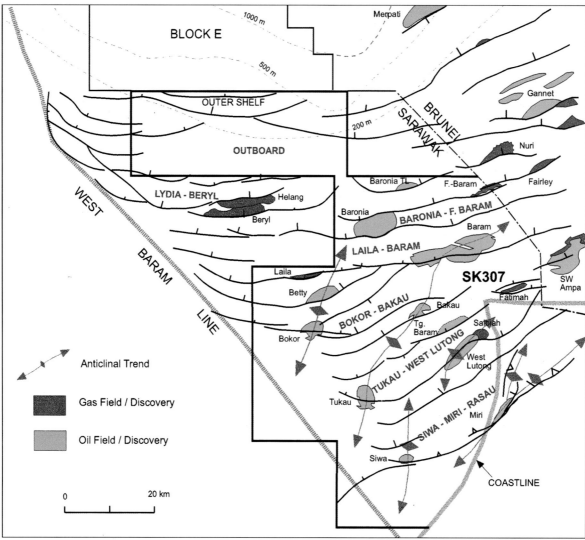

Fig. 13.31 West Baram Delta mega-structural trends.

DELTA EVOLUTION

The SE margin of the Baram Delta is marked by the Morris Fault, in offshore Sabah, which extends southwards into the Jerudong Line, in Brunei, and the SW margin is marked by the West Baram Line. In the late Miocene, the Morris Fault - Jerudong Line was a tectonically active shelf edge, downthrown to the west into the deepwater Baram depocentre. Tan and Lamy (1990) considered this deep depocentre to be a pull-apart basin associated with deep-seated, major N-S strike-slip faults. This depocentre was extensively filled by the Miocene Setap Shale, and the Baram Delta was built out to form a fan-shaped delta. The outermost NW margin of this delta is defined by the Sabah Trough. The Baram Delta can be subdivided into two delta systems; namely Middle - Late Miocene East Baram Delta (Champion delta in Brunei), and Late Miocene - Quaternary West Baram Delta (Baram delta in Brunei). These delta systems were probably supplied by separate river systems from the east and southeast across the Morris Fault – Jerudong Line, and south and southwest across the West Baram Line. The beginning of the West Baram Delta started onshore Sarawak and is well exposed in the southern foothills of the Lambir Hill (Hutchison, 1996) where the pro-delta Setap Shale is abruptly overlain by the sandy Middle Upper Miocene Lambir Formation. Hutchison (1996) speculated that the West Baram Delta sediments could have been derived from erosion of the Upper Oligocene – Lower Miocene Nyalau Formation.

Although the Baram Delta exhibits all the tectonic characteristics of other deltaic areas, e.g. Niger and Mississippi, which are situated on passive continental margins, the Baram is relatively small in areal extent and was formed on an active continental margin. Its shape and size suggest that it may have developed initially as a pull-apart basin whose length and width were pre-determined by its bounding faults.

HYDROCARBON OCCURRENCES

Onshore

The West Baram Delta is dominated by gravity tectonics. The resulting growth faults and shale diapirs are responsible for forming most of the structural traps. Onshore and in the shallow offshore area, structural inversion has tightened the anticlines developed in response to compression and shale diapirism (e.g. Miri, Asam Paya and Siwa fields). Several attempts by oil companies to explore for hydrocarbons in the onshore areas have had mixed results. The main targets are the Cycles III, IV and V sandstones which are stratigraphic equivalents of the Lambir and Miri formations. The main reasons for the disappointing exploration results are that the growth faults and their associated normal faults have been re-activated by subsequent compressional tectonics and the faults are either not sealing or the traps have been breached. This is indicated by the numerous surface hydrocarbon seeps observed near the coastal areas and by residual hydrocarbons reported from the exploration wells.

Offshore

The conventional play in the West Baram Delta is Middle Cycle V to Cycle VI stacked topset sands in rollover and fault-dependent structures on the downthrown blocks of growth faults. The interference of post-depositional tectonics with the growth fault systems results in a broad style of both fault- and dip-dependent closures. Trapping may be enhanced by stratigraphic factors, such as lateral shale-outs and shale-filled channels, e.g. the Nuri accumulation in Brunei.

The shallow marine sands are generally continuous and decrease in quality with depth and towards the distal parts of the delta. Up to 74

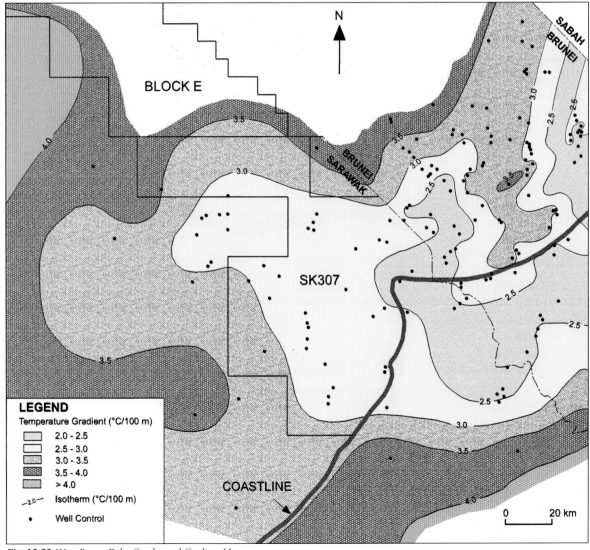

Fig. 13.32 West Baram Delta Geothermal Gradient Map.

stacked hydrocarbon-bearing sands, sealed by intraformational shales, have been observed. The source rock intervals have not been positively identified, although it is generally believed that the hydrocarbons were derived from dispersed land-plant organic matter. In general, there is an abundance of charge that appears to become more gas prone with depth and in the more distal parts of the delta. In Brunei, most oil is found at depths down to 2000 m while gas is generally more dominant at depths below 2000 - 2500 m. This phenomenon is partly attributed to differential retention of oil and gas below partly sealing shales, with sealing potential and capacity increasing with depth and basinwards (Sandal, 1996). Geothermal gradients range from 1.8 to 3.2°C/100 m, with an average of 2.8°C/100 m (Fig. 13.32), which is lower than in the contiguous Central Luconia and Balingian provinces to the west. Due to the low geothermal gradients, economic porosities (> 15%) are generally present to a depth of about 3050 to 3350 m (Rijks, 1981).

In the distal parts of the West Baram Delta proximal turbidites offer a hitherto unexplored target. Flank plays with turbiditic sands trapped against toe-thrusts, shale diapirs, or large counter-regional faults may be present. The risk of no or poorly-developed reservoir is high. Turbiditic sands are likely to be thin, probably up to 10 - 20 m thick, because of the poor sand development on the outer Baram Delta

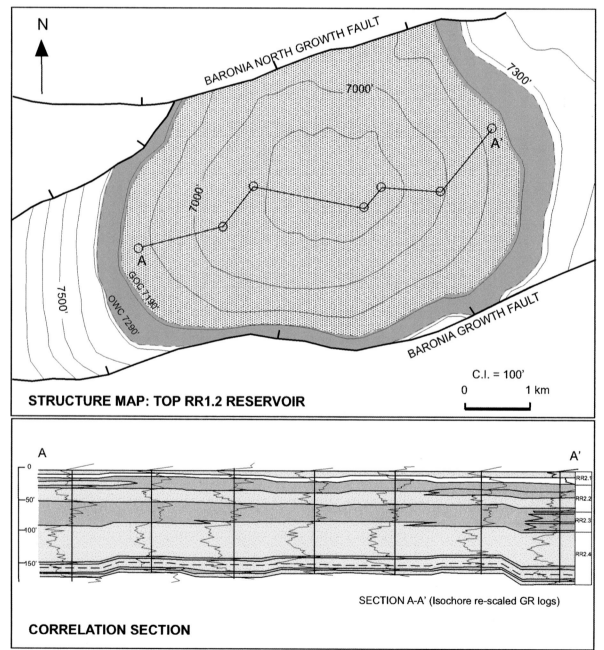

Fig. 13.33 Baronia Field: Structural map and cross-section.

shelf, e.g. Z1, Merpati, and Perdana in Brunei. The presence of gas chimneys and amplitude anomalies on seismic suggests the presence of gas charge, although the presence of oil cannot be ruled out. Condensate-rich gas has been found in the Z1/Merpati accumulation in neighbouring Brunei. The most likely explanation for the charge is that it originated from mature source rocks rich in land plant-derived organic matter that has been transported, e.g. by turbidity currents, to a position downdip from the structures.

RESERVOIR, SEAL AND TRAP STYLES

During the overall regressive Cycles V to VIII times, a thick sequence of regressive coastal plain, coastal, and fluviomarine sediments, with generally good reservoir characteristics, were deposited. These reservoir sequences are best developed in Cycles V and VI, and can reach a thickness of up to 3050 m (10,000 ft) and 1829 m (6000 ft), respectively. The hydrocarbon accumulations are predominantly found in fault- and dip-closed structures. Although a large proportion of the hydrocarbons is trapped by later faults, the major growth faults provide the principal trapping mechanism. Their syn-sedimentary nature and very large throws juxtaposed thick paralic sand/shale sequences, on the downthrown blocks, against marine shale sequences, on the upthrown blocks.

To illustrate the above, three West Baram Delta fields are described briefly below:

The **Baronia Field** (Scherer, 1980) is a simple unfaulted domal structure, without internal faulting, located between two major east-west trending growth faults (Fig. 13.33). The main prospective sequence consists of sandstones interbedded with siltstones and shales of Late Miocene age (Upper Cycle V – Lower Cycle VI) at depths of 1615 to 2410 m (5300 to 7900 ft). The sandstones range in thickness from 3 to 75 m (10 to 245 ft). The shales range in thickness from 1.5 to 90 m (5 to 295 ft) and form the intra-formational seals. Porosities range from 13 to 25% in the poorly developed sandstones, and 26 to 30% in the better developed sandstones. Most of the sandstones are fine grained, with permeabilities ranging from 100 to 350 mD. The hydrocarbons in Baronia are found in stacked reservoirs. The discovery well, Baronia-1, logged a total of 81 m (265 ft) NOS and 162 m (531 ft) NGS in 13 different reservoirs. The main reserves are distributed over 10 sandstone reservoirs with at least 8 separate oil-water contacts. The Baronia oil is a light (38.5° – 41° API), waxy (3 – 4 wt%), land plant-derived oil with low sulphur content (0.07 wt%).

The **Betty Field** (Fig. 13.34) is a gently-dipping, domal, rollover anticline located at the intersection of the NE – SW trending Baronia - Betty – Bokor anticlinal trend and the E – W trending Betty growth fault (Johnson et al., 1989). The Betty reservoirs consist of Upper Miocene (Upper Cycle V) sandstones occurring at depths of 2194 to 2941 m (7200 to 9650 ft). Details of the stratigraphy and sedimentology of the Betty reservoirs have been described above. The hydrocarbons are trapped within at least 21 stacked sand bodies separated by sealing intra-formational shales. The main hydrocarbon accumulation occurs on the downthrown block of the Betty growth fault. Only minor hydrocarbons are found on the upthrown block, confirming the effectiveness of the fault seal.

The **Baram Field** is an elongated, ENE – WSW trending, complexly faulted, anticlinal structure formed at the intersection of the Baram growth fault and the later Tukau – Bakau- Baram compressional uplift. The Baram Field consists of three separate accumulations; namely, Baram 'A' which was discovered in 1963, Baram 'B' which occurs west of and separated from Baram 'A' by a saddle, and Baram South which is a relatively unfaulted structure on the footwall of the Baram growth fault. The Baram 'A' and 'B' accumulations on the hanging wall of the Baram growth fault are heavily dissected by generally E – W trending listric and antithetic faults, resulting in a complex structural configuration. The hydrocarbon-bearing reservoirs (Upper Cycle V and Cycle VI) occur at depths of 760 to 2925 m (2500 to 9600 ft). The Upper Cycle V sediments were deposited in fluviomarine, coastal, and inner neritic environments whereas the Cycle VI sediments were deposited in coastal to inner neritic environments. Individual sands rarely exceed 15 m in thickness and are moderately sorted, very fine grained to silty. Porosities range from 16 to 30% and permeabilities are in the order of 20 to 300 mD. The oil is light (38° to 48° API), except in the shallow reservoirs (19° API) and in the very deep reservoirs (33° API).

SOURCE ROCKS

Because of the high sedimentation rate, most of the higher land-plant debris fed into the West

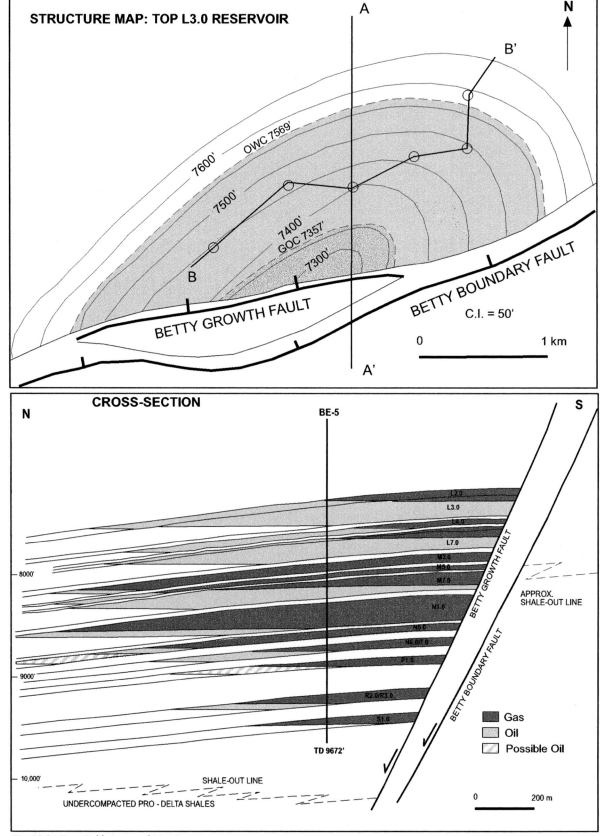

Fig. 13.34 Betty Field: Structural Map & cross-section.

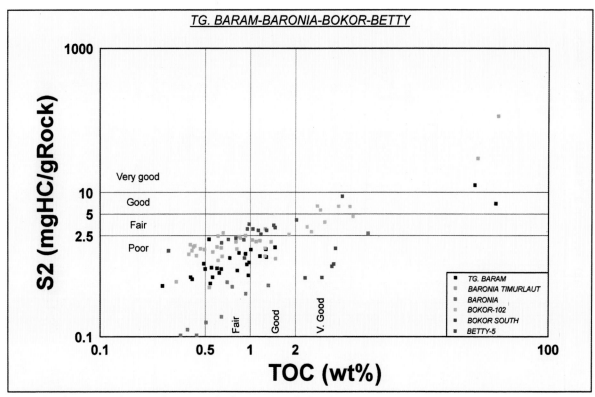

Fig. 13.35 The source rock potential summary plot for a number of selected West Baram Delta wells highlights the marginal hydrocarbon potential of the sediments.

Baram Delta would have been preserved by virtue of their rapid removal from the oxidation zones. The large amounts of clastic input, however, may have the effect of diluting the organic concentration of the sediments. Source rock evaluation has indeed confirmed the presence of organically lean intervals within Cycles V and VI (Rijks, 1981; Seah et al., 1987; Sandal, 1996). The total organic matter concentrations rarely exceed 2.0 wt%, while the occurrences of coal and/or coaly intervals with TOC values as high as 45wt% are exceptions (Fig. 13.35). Sedimentation rates as high as 950 m/Ma that prevailed between 10 Ma and the present day (Noor Azmi Ibrahim, 1994) seem to preclude the development of coal-rich layers in the delta. In wells that penetrated overpressured sequences, e.g. Bokor-2 well (Mohammad Yamin Ali et al., 1995), TOC concentration and hydrocarbon source potentials were higher in the overpressured zone than in the overlying hydrostatic section (Figs. 13.36a and 13.36b). This may be a result of better organic matter preservation within parts of the overpressured interval due to their rapid removal from the oxidation zone through rapid sedimentation and burial.

Today, it is generally accepted that the hydrocarbons in the West Baram Delta were generated mainly from transported terrigenous organic matter. Visual organic matter typing of the source rock intervals confirms that, in most

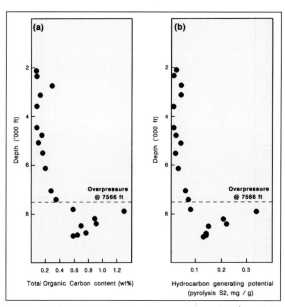

Fig. 13.36 (a) Total Organic Carbon content variations and (b) pyrolysis S_2 variations with depth in the well Bokor-2 (RDR), West Baram Delta. Below the overpressure boundary, there seems to be slight enhancement in organic matter concentration and hydrocarbon generation potential (S_2 yield).

cases, vitrinite is the most predominant maceral type, followed by inertinite. Between 5 to 20% of the total kerogen assemblage is made up of fluorescing liptinitic and amorphous organic matter. To further characterise these source rock intervals, a selected number of samples from Middle - Upper Cycle V and Lower Cycle VI have been solvent-extracted and analysed for their

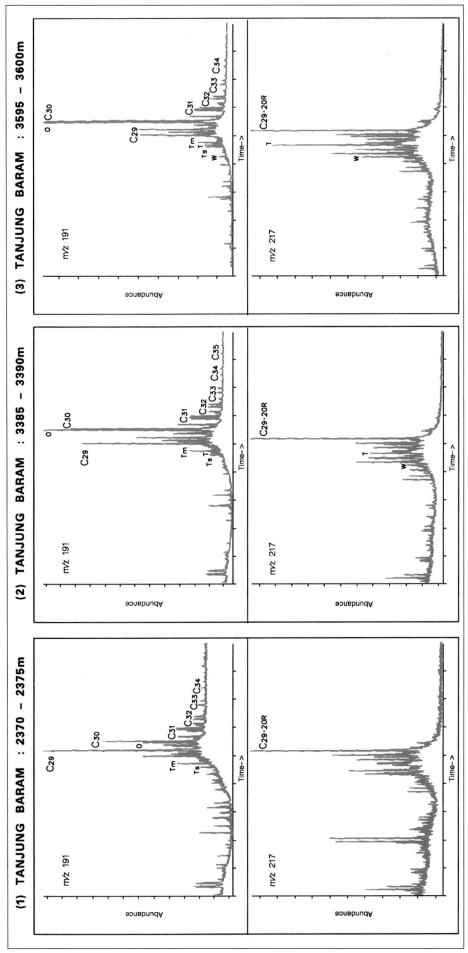

Fig. 13.37 Gas chromatograph-mass spectrometry traces of Tanjung Baram-1 extract samples. The appearance of the bicadinanes (peaks W, T and R) with increasing depth suggests that, apart from being source-related, the presence (or absence) of bicadinanes may also be affected by increasing maturity levels.

biomarker distributions (Zulkifli Salleh and Nazri Mokhtar, 1995a, b). The majority of extract samples have not reached the hydrocarbon generation window and, thus, display immature biomarker traces that are often very difficult to correlate with the oils. It was still feasible, however, to conclude that the main organic matter contributors in the source rock extracts were higher land plants, as suggested by the presence of oleanane and C_{24} tetracyclics, the lack of tricyclic biomarkers (Fig. 13.37), and the persistent dominance of C_{29} over C_{27} and C_{28} regular steranes (Fig. 13.38). Bicadinanes were absent in the least mature samples but, as the oil window is approached, they began to appear in the biomarker traces. This suggests that the presence of bicadinanes may be influenced by maturation levels, apart from being obviously source-related. Tm/Ts ratios of less than 2 and the presence of diasteranes attest to a non-coal source rock in the West Baram Delta (Abolins, 1998).

Maturity

Based on vitrinite reflectance data (Fig. 13.39) and a geothermal gradient of 2.7°C/100 m (1.5°F/100 ft), the required depth of burial within a normally pressured sequence for hydrocarbon generation to commence is around 4000 to 4600 m (13,000 to 15,000 ft). As mentioned earlier, this phenomenon is partly

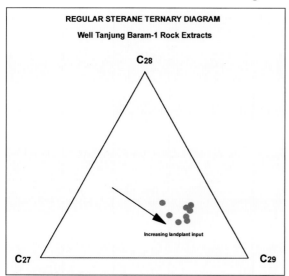

Fig. 13.38 The ternary diagram illustrates the relative distribution of C_{27}, C_{28} and C_{29} regular steranes in the Tanjung Baram-1 extracts, exhibiting the predominance of C_{29} over the C_{27} and C_{28} regular steranes. This predominance is in line with hydrocarbons originating from land-derived organic matter.

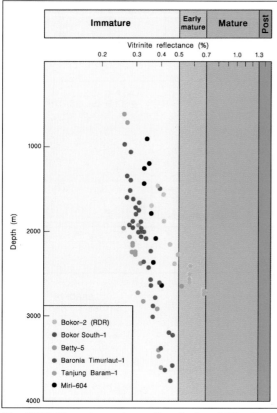

Fig. 13.39 Thermal maturity profiles of selected West Baram Delta wells in general suggest that the top of the oil window (early mature) will be reached between 4000-5000 m; however, sediments in the Bokor-2 (RDR) well hit the top of the oil window at around 2300 m.

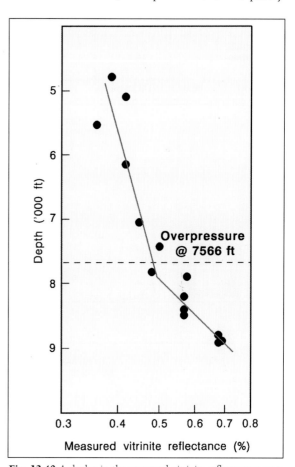

Fig. 13.40 A dogleg in the measured vitrinite reflectance versus depth profile of the Bokor-2 (RDR) well seemed to coincide with the top of the overpressure zone. This indicates the possible role of overpressure in enhancing organic matter maturation.

Well/Field	Gravity (°API)	Wax	Sulphur (wt%)	Pr/Ph ratio	$\delta^{13}C$ (oil)	$\delta^{13}C$ (sat)	$\delta^{13}C$ (aro)
Asam Paya-1	40.3	Semi-waxy	0.1	6	–	–	–
Baram-8	34.7	Low	0.2	4.9	-28.1	–	–
Baronia-29							
1736.7 m	34.5	Waxy	–	–	–	–	–
2985.8 m	31.7	Waxy	–	–	–	–	–
Baronia Barat-1							
2730 m	34.5	Low	–	–	–	–	–
2777 m	31.7	Waxy	–	–	–	–	–
Betty-5							
2301.2 m	34.4	Non-waxy	0.1	5.9	-28.1	–	–
2698 m	48.5	Non-waxy	0.2	6.3	-28.0	–	–
Engkabang-1	–	Semi-waxy	-0.1	6.2	–	-27.7	-26.5
Engkabang-2	31.8	Low		3.1	-26.9	–	–
Laila-1	–	Non-waxy	–	5.3	–	-28.4	-26.9
Lembuk-1							
DST#3	–	–	–	4.4	–	–	–
DST#4	–	–	–	5.8	–	–	–
Miri	40.3	Low	0	6.8	-28.1	–	–
Siwa-5	–	Waxy	–	4.2	–	-28	-27.6
Tanjung Baram-1	35	–	0.2	4.6	–	–	–
Tukau-11	–	Non-waxy	–	–	–	-27.7	-27.5
W. Lutong-7	31.3	Low	0.1	5.8	-27.7	–	–
W. Lutong-11	35.7	Low	0.1	6.4	-27.8	–	–

Table 13.2: West Baram Delta Oil Properties.

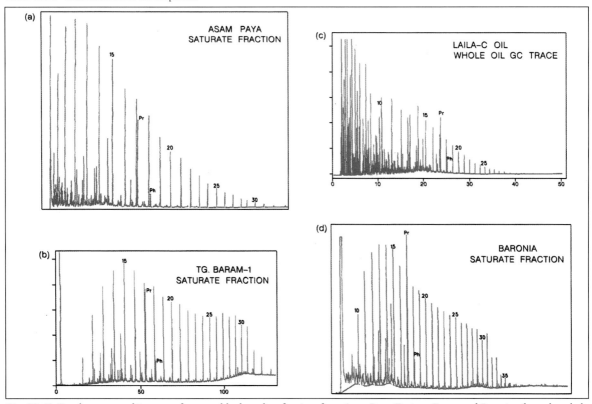

Fig. 13.41 Gas chromatographic traces of saturated hydrocarbon fractions from Asam Paya, Tanjung Baram and Baronia oils, and a whole oil trace from Laila. All four exhibit characteristics consistent with a higher land plant origin.

attributed to the differential retention of oil and gas below partly sealing shales, with sealing potential and capacity increasing with depths and basinwards. Schaar (1976) documented the prospectivity of overpressured sections in the West Baram Delta, and concluded that hydrocarbons are found only in reservoirs experiencing overpressures due to inflation rather than undercompaction.

The vitrinite reflectance profile of the Bokor-2 well (Mohammad Yamin Ali et al., 1995) exhibits a slight dogleg between the hydrostatic- and over-pressured intervals (Fig. 13.40). As a consequence of the increased temperature within the overpressured zone, the top of the oil window occurs at a shallower depth. This discordance in maturity level between over-and hydrostatic- pressured zones may be taken as an indication of slightly enhanced maturation within the former. The subject of organic matter maturation within overpressured zones has been addressed by many workers (Swarbrick, 1997) but it is still unclear whether overpressure enhances or retards the maturation process.

In the offshore areas, hydrocarbon kitchens containing mature Cycles V and VI sediments are expected to be present in the synclinal areas associated with the mega structural trends. In addition to the synclinal area north of the Siwa-Miri-Rasau trend, the onshore Miri field could have been sourced from the Liku - Badas syncline.

HYDROCARBON CHARACTERISTICS

The West Baram Delta oils exhibit light to medium gravity ranging from 31° to 49° API (Table 13.2). Distinctly low sulphur contents (< 0.2 wt%) indicate the possibility that these oils were sourced from non-marine source rocks.

The gas chromatographic traces of whole oils from Laila and Lembuk (Abolins, 1998) and saturate fraction samples from Asam Paya, Tanjung Baram, Laila and Baronia (van der Veen and Posthuma, 1980; Vlierboom, 1991) are remarkably consistent in their general characteristics: high Pr/Ph ratios, an abundance of higher molecular weight n-alkanes going up to at least nC_{33}, and a detectable preference of odd carbon numbers (Fig. 13.41). These attributes are

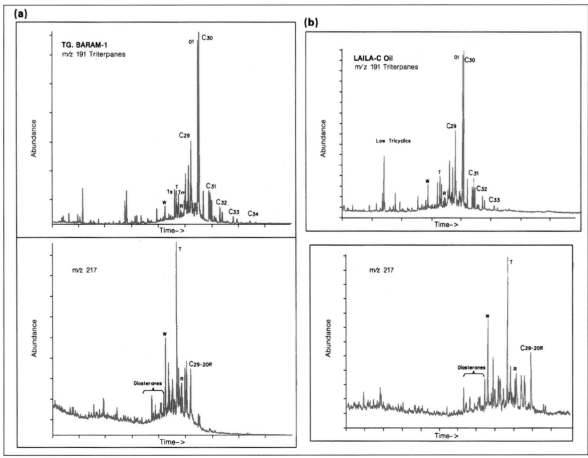

Fig. 13.42 The triterpane m/z 191 and sterane m/z 217 traces from (a) Tanjung Baram and (b) Laila oils further confirms their terrigenous nature.

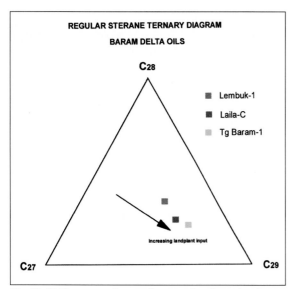

Fig. 13.43 The ternary diagram illustrates the relative distribution of C_{27}, C_{28} and C_{29} regular steranes in the West Baram Delta oils. It exhibits the predominance of C_{29} over the C_{27} and C_{28} regular steranes, similar to the extract hydrocarbons as shown in Figure 13.38, again indicating major contributions from land-derived material.

common for oils generated from higher land-plant materials.

Gas chromatography-mass spectrometry confirms the terrigenous nature of the source rocks by displaying ubiquitous occurrence of bicadinanes and oleanane in the tritepane m/z 191 traces, low tricyclic terpanes, moderately abundant diasteranes (Fig. 13.42), predominance of C_{29} regular steranes (Fig. 13.43), and high hopane to sterane ratios. The Tm/Ts ratios of less than 2 in these oils suggest the absence of any major coal influence.

Oil-oil and Oil-source Correlations

In general, the West Baram Delta oils correlate very well with each other, indicating the very similar nature of the source rocks that generated them. Oil-source rock correlations prove to be somewhat problematic, mainly because of the immature state of most of the extracts.

In the Tanjung Baram-1 well, the oil-like source extract correlates quite well with the encountered hydrocarbons (Fig. 13.44). Parameters such as the Extractable/Total Organic Matter (EOM/TOC) ratio and the Production

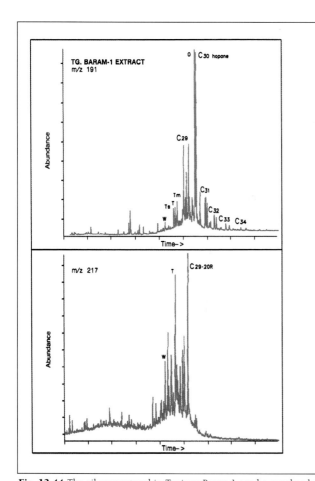
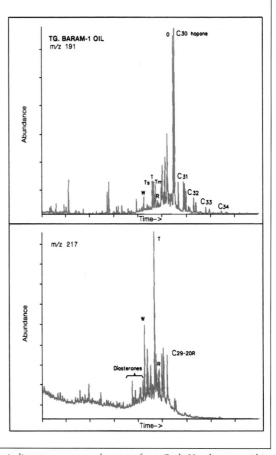

Fig. 13.44 The oil encountered in Tanjung Baram-1 can be correlated to an indigenous source rock extract from Cycle V sediments within the same well, thus suggesting that Cycle V sediments may be the source rock responsible for generating some of the West Baram Delta oil accumulations.

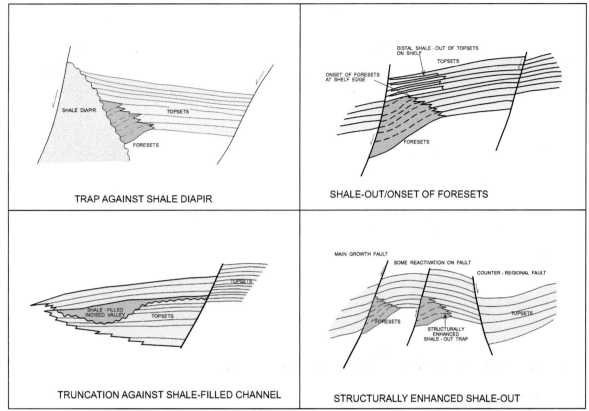

Fig. 13.45 Schematic combined structural and stratigraphic traps.

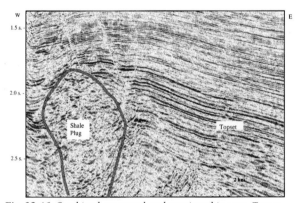

Fig. 13.46 Combined structural and stratigraphic trap: Topset trapped against shale plug.

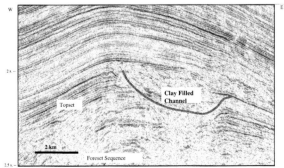

Fig. 13.47 Combined structural and stratigraphic trap: Topset truncated and trapped against shale-filled channel.

Index (S_1/S_1+S_2) were used to ensure that the extract was not an oil stain. Furthermore, the slight differences between the source extract and oil traces, such as that seen in the norlupane distribution, the Tm/Ts ratio, the relative oleanane abundance to the C_{30} hopane, and the higher C_{29}-20R regular sterane abundance in the source extract, all attest to its authenticity.

An operator in the onshore part of the West Baram Delta claimed that, for the first time, the oils encountered in the Penipah-1 well have been positively correlated with extracts from the Cycle III claystones of the same well. This correlation is, at best, only tentative because there seemed to be some migrated hydrocarbons (i.e. staining) mixed with the indigenous Cycle III claystone source rock extracts.

It is thus concluded that the West Baram Delta oils were generated from land-derived organic materials deposited and preserved in a deltaic setting. These source rocks began accumulating during the Late Miocene with the deposition of the Middle and Upper Cycle V sediments (e.g. Tanjung Baram-1) and continued into the Pliocene with the deposition of Cycle VI sediments (e.g. Beryl-1). The influence of coals as hydrocarbon source rocks is not perceived as important in the West Baram Delta. Holomarine sequences encountered by the West Baram Delta wells are not the source of hydrocarbons.

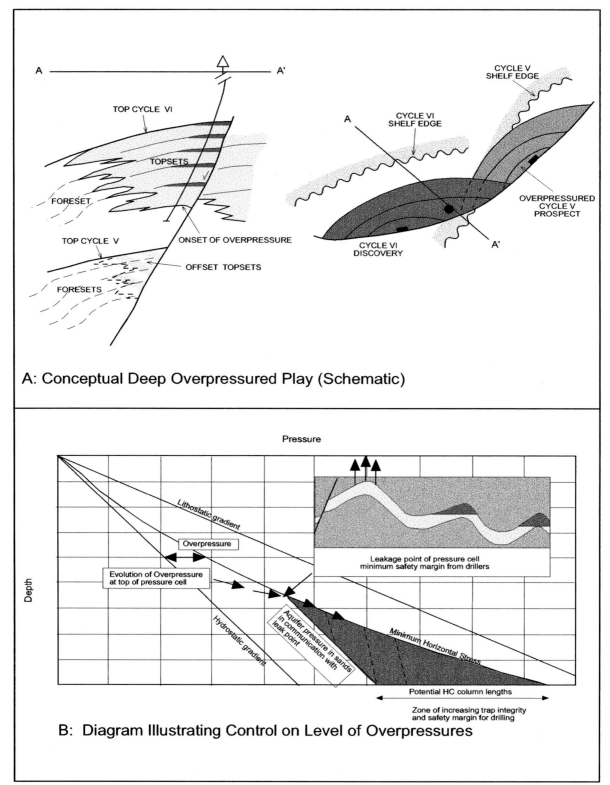

Fig 13.48 Overpressured play concept.

POTENTIAL NEW PLAYS

Onshore

Potential new play types in the onshore West Baram Delta include Cycles III and IV topset sands and lowstand fan deposits in either combined stratigraphic and structural, or solely stratigraphic traps, and Cycle V topsets in sub-thrust trapping configuration on the downthrown blocks of the major thrust faults. These conceptual play types will require good quality 3D seismic to image and develop into viable prospects.

Offshore

Potential new plays in the offshore West Baram Delta are those that have essentially not been tested in Sarawak but have been drilled in Brunei. These potential new play types (Chow and Tan, 1997) are as follows:

1. *Combined Structural and Stratigraphic Traps* (Fig. 13.45)
   A number of play types have been identified where a stratigraphic element forms a key part of the trap. The stratigraphic element may be due to 4 principal causes:
   - Stratigraphic traps against shale diapir (Fig. 13.46),
   - Shale-out traps: distal shale-out of topsets on shelf, or at shelf/slope break,
   - Truncation against shale-filled incised channels (Fig. 13.47), and
   - Structurally enhanced shale-out traps.

   Combined structural and stratigraphic traps have not been tested in Sarawak, although they have been drilled in Brunei, e.g. the Bugan field (Sandal, 1996) and the SW Ampa 21 Area (James, 1984).

2. *Overpressured Play* (Fig. 13.48)
   There is evidence for overpressures in the West Baram Delta (Schaar, 1976; Mantaring et al., 1994), e.g. in the Baram, Betty, and Baronia fields, and in the wells Hasnah-1, Zarina-1, Laila-1, Bakau-5, Beryl-2, Sikau-1 and -2. In the south in Siwa and Miri-604 overpressures were encountered at the top of the Setap Shale. There is, however, limited penetration of the overpressured sequences, except in Siwa-1 which penetrated some 1060 m of overpressured Setap Shale.

   Commonly, in a deltaic setting where depositional systems and structuration change with time, structures with shallow objectives are likely to be offset relative to structures with deeper objectives. It is likely, therefore, that some of the older wells may have tested their deeper objectives off-structure because these wells were located on poor quality and/or limited seismic data.

   In the West Baram Delta there is evidence for the presence of semi-regional seals retaining overpressures. Such seals would also retain hydrocarbons. Overpressure estimation is important for the risking of trap integrity. If fluid pressure exceeds the minimum horizontal stress, then the trap may fail. The difference between the minimum horizontal stress and formation pressure is the minimum effective stress (MES), which provides a measure of trap integrity. Traps with low MES cannot retain substantial hydrocarbon columns and are also hazardous to drilling because of the narrow margin to contain kicks. If a pressure cell can be identified, the shallowest point within the cell may act as a pressure valve which would enhance the integrity of deeper structures. Besides being a function of the local stress pattern, trap integrity is also dependant on the degree and timing of faulting as well as the relative position of the trap with respect to the leak point in the effective pressure cell (Gaarenstroom et al., 1993).

   Although inconclusive at this time, there is some evidence of porosity enhancement below the top overpressure in the West Baram Delta. There may be potential, therefore, for the occurrence of reservoir-quality sands at depths.

   Inflation overpressures may be caused by shale diapirism/intrusion into a topset sequence (Kho et al., 1994). If overpressures are due to shale intrusion, deeper hydrostatically-pressured reservoir sequences may be left untested.

3. *Turbidite Play*
   The potential for turbidite play in the West Baram Delta has not been evaluated. This play may be present in the more distal and deep water parts of the delta, e.g. Block E.

SUMMARY

The West Baram Delta is the western part of the greater Baram Delta Province that extends eastwards into Brunei and offshore Sabah. It is characterised by the deposition of a northwestward prograding delta since Middle Miocene times. Periods of delta outbuilding were separated by rapid transgressions representing various sedimentary cycles. The regressive sequences of each cycle grade northward from coastal – fluviomarine sands to neritic shales, minor sands, and, in places, turbidites. Sediment supply was derived from the south-southeast and southwest. The tectonic style shows the interplay between two main deformational styles, namely, NE – SW to NW – SE trending, syndepositional growth faults, and Late Miocene or Early Pliocene NE – SW trending, wrench-induced compressional folds.

The main hydrocarbon play is Lower Cycle V to Cycle VI topset sands stacked in rollover and fault-dependent structures. The interference of post-depositional tectonics with the growth faults resulted in a broad style of both fault- and dip-dependent closures. The first oil field, Miri Field, was discovered onshore in 1910. Since that time, a total of some 50 exploration wells have been drilled, resulting in the discovery of some 1.4 billion barrels of oil and 5.7 TSCF of gas in 18 oil and gas fields. Nine of these oil fields are producing, with current production of some 100,000 BOPD oil.

The hydrocarbons are reservoired in stacked coastal – fluviomarine sands that are generally continuous. Reservoir quality generally deteriorates with depth and towards the distal part of the delta. The oils are generally light to medium (31° to 49° API), with low sulphur content (< 2 wt%). Geochemical studies indicate that the oils were generated from land-derived organic matters deposited and preserved in a deltaic setting.

Potential new plays onshore include Cycles III and IV topset sands and lowstand fan deposits in combined stratigraphic and structural, or stratigraphic traps, and Cycle V topsets in sub-thrust trapping configuration. Offshore, potential new plays include Middle Cycle V and Cycle VI topsets in combined structural and stratigraphic traps, deeper topsets within the overpressured zones, and turbidite play in the more distal and deep water parts of the delta.

REFERENCES

Abdul Hadi Abd. Rahman, 1995. Reservoir characterisation of the Upper Cycle V (Late Miocene) of Baram Field, Baram Delta, offshore Sarawak, East Malaysia. Unpublished PhD Thesis, University of Reading, England, 333 pp.

Abolins, P., 1998. A Hydrocarbon Characterisation Study of Oils and Condensates from the West Baram Delta Area. PETRONAS Report No. PRSS-TCS08-98-25.

Allen, J.R.L., 1980. Sand waves: A model of origin and internal structure. Sedimentary Geology, 26, 281 - 328.

Allen, J.R.L., 1982. Mud drapes in sand-wave deposits: A physical model with application to the Folkestone Beds (Early Cretaceous, Southeast England). Philosophical Transactions of the Royal Society of London, A306, 291-345.

Chow, K.T. and Tan, D.N.K., 1997. The West Baram Delta, offshore Sarawak: New Focus of Exploration. Abstract GSM Petroleum Geology Conference, December 1997.

Collinson, J.D., and Thompson, D.B., 1988. "Sedimentary structures". Second Edition, George Allen & Unwin, London, 207 pp.

Dalrymple, R.W., 1992. Tidal depositional systems. In: Walker, R.G. and James, N.P. (eds.), Facies models - response to sea level change. Geological Association of Canada Publication, 195-218.

Dalrymple, R.W., Zaitlin, A.B., and Boyd, R., 1992. Estuarine facies model: Conceptual basis and stratigraphic implications. Journal of Sedimentary Petrology, 62, 1130-1146.

Duke, W.L., 1990. Geostropic circulation or shallow marine turbidity currents? The dilemma paleoflow patterns in storm-influenced prograding shoreline systems. Journal of Sedimentary Petrology, 60, 870-883.

Gaarenstroom, L., Tromp, R.A.J., de Jong, M.C., and Brandenburg, A., 1993. Overpressures in the central North Sea: Implications for trap integrity and drilling safety. In: Parker, J.R. (ed.), Petroleum geology of northwest Europe. Proceedings of 4th Conference London, Published by Geological Society of London, 1305 - 1313.

Hamblin, A.P., and Walker, R.G., 1979. Storm-dominated shallow marine deposits: The Fernie-Kootenay (Jurassic) transition, southern Rocky Mountains. Canadian Journal of Earth Sciences, 16, 1673-1690.

Hamid Mohamad, A., Budding, M.C., and Johnson, H.D., 1986. Miri Hill field trip. Reservoir Geology/Sedimentology Team, Sarawak Shell Berhad, 18 pp.

Harms, J.C., 1975. Stratification and sequence in prograding shoreline deposits. In: Harms, J.C., Southard, J.B., Spearing, D.R. and Walker, R.G. (eds.), Depositional Environments as Interpreted from Primary Sedimentary Structures and Stratification Sequences. Society of Economic Paleontologists and Mineralogists, Short Course No. 2 Lecture Notes, 81-102.

Ho, K.F., 1978. Stratigraphic Framework for Oil Exploration in Sarawak. Bulletin of the Geological Society of Malaysia, 10, 1-13.

Hutchison, C.S. (1996). "South-East Asian Oil, Gas, Coal and Mineral Deposits". Oxford Monographs on Geology and Geophysics No. 36, Clarendon Press, 265 pp.

James, D.M.D.(Ed.), 1984. "The geology and hydrocarbon resources of Negara Brunei Darussalam". Muzium Brunei Special Publication, 169 pp.

Jervey, M.T., 1988. Quantitative geological modeling of siliciclastic rock sequences and their seismic expression. In: C.W., Wilgus et al., (Eds.), Sea level changes: An integrated approach. Society of Economic Paleontologists and Mineralogists, Special Publication, 47-69.

Johnson, H.D., Kuud, T. and Dundang, A., 1989. Sedimentology and Reservoir Geology of the Betty Field, Baram Delta Province, offshore Sarawak, NW Borneo. Bulletin of the Geological Society of Malaysia, 25, 119-161.

Kho, S.C., Hemmings, and ten Have, T., 1994. Overpressures in the Baram Field, offshore Sarawak, East Malaysia. 1994 AAPG International Conference, Kuala Lumpur Abstracts.

Leckie, D., 1988. Wave-formed, coarse-grained ripples and their relationship to hummocky cross-stratification. Journal of Sedimentary Petrology, 58, 607-622.

Liechti, P., Roe, F.N., Haile, N.S., and Kirk, H.J.C., 1960. The geology of Sarawak, Brunei and the Western part of North Borneo. British Borneo Geological Survey, Bulletin 3, 360 pp.

Mantaring, A., Matsuda, F., and Okamoto, M., 1994. Analysis of Overpressure Zones at the Southern Margin of the Baram Delta Province and their Implications to Hydrocarbon Expulsion, Migration and Entrapment. 1994 AAPG International Conference, Kuala Lumpur Abstracts.

Mohammad Yamin Ali, Wan Ismail Wan Yusoff, Azlina Anuar, Idrus Mohd Shuhud, Mohd Fauzi Abdul Kadir and Mohd Nezam Mansor, 1995. Hydrocarbon Prospectivity Within and Below the Overpressured Zones in Baram Delta Province. PETRONAS Report No. PRSS-RR-95-09, Volumes 1-3.

Noor Azim Ibrahim, 1994. Major Controls on the Development of Sedimentary Sequence, Sabah Basin, Northwest Borneo. Unpublished PhD Thesis, University of Cambridge, 255 pp.

Reineck, H.E., and Wunderlich, F., 1968. Classification and origin of flaser and lenticular bedding. Sedimentology, 11, 99-104.

Reineck, H.E., and Singh, I.B., 1986. "Depositional sedimentary environments", 2nd Edition. Springer-Verlag, Berlin, 551 pp.

Rijks, E.J.H., 1981. Baram Delta Geology and Hydrocarbon Occurrence. Bulletin of the Geological Society of Malaysia, 14, 1-18.

Sandal, S.T.(Ed.), 1996. "The geology and hydrocarbon resources of Negara Brunei Darussalam"; 2nd Edition 1996. Brunei Shell Petroleum & Brunei Museum Publication, 243 pp.

Schaar, G., 1976. The Occurrence of Hydrocarbon in Overpressured Reservoir of the Baram Delta (Offshore Sarawak, Malaysia). Proceedings of Indonesian Petroleum Association, June 1976, 163-169.

Scherer, F.C., 1980. Exploration in East Malaysia over the last decade. In: Halbouty, M.T. (ed.) Giant Oil and Gas Fields of the decade 1968 - 1978: AAPG Mem. 30, 423 - 440.

Schumacher, P.v., 1941. The Miri Field. Shell Oilfields Internal Report R. 306.

Seah, E., Zulkifli Salleh and Awang Sapawi Awang Jamil, 1987. Rock Eval Analyses and Total Organic Carbon Analyses of Core and Surface Samples from Sarawak. PETRONAS Report No. MGK/08/87.

Sering, L. and Ramli, A., 1994. Reservoir Geological Modeling of the A3/A6.0 Viscous Crude Reservoirs of Bokor Field, Offshore Sarwak, East Malaysia. 1994 AAPG International Conference, Kuala Lumpur Abstracts.

Singh, I.B., 1969. Primary sedimentary structures in Precambrian quartzites of Telemark, southern Norway, and their environmental significance. Norsk Geol. Tidsskr., 49, 1-31.

Swarbrick, R.E., 1997. Characteristics of Overpressured Basins and Influence of Overpressure on the Petroleum System. In: Howes, J.V.C. & Nobles, R.A. (eds.), Proceedings IPA Petroleum Systems of SE Asia and Australasia Conference, Jakarta, Indonesia, May 1997, Indonesia Petroleum Association, Jakarta, 859-866.

Tan, D.N.K. and Lamy, J.M., 1990. Tectonic Evolution of the NW Sabah Continental Margin since the Late Eocene. Bulletin of the Geological Society of Malaysia, 27, 241-260.

Tate, R.B., 1976. Palaeo-environmental studies in Brunei. SEAPEX Proceedings, Volume III, 102-124.

Taylor, B., and Hayes, D.E., 1980. The Tectonic Evolution of the South China Basin. In: Hayes, D.E. (ed.) The Tectonic and Geologic Evolution of Southeast Asian Seas and Islands. Geophysical Monograph 23, 89-104.

Terwindt, J.H.J., 1971. Litho-facies of inshore estuarine and tidal-inlet deposits. Geologie en Mijnbouw, 50, 515-526.

Terwindt, J.H.J., 1988. Palaeo-tidal reconstructions of inshore tidal depositional environments. In: de Boer, P.L., van Gelder, A. and Nio, S.D. (eds.) "Tide-influenced sedimentary environments and facies". D. Reidel Publishing Company, Dordrecht, 233-263.

van der Veen, F.M. and Posthuma, J., 1980. Geochemical Analysis of Five Crude Oils from Wells Betty-5, Baronia-29 and J4-1, Sarawak. Shell External Report No. RKER.80.076, PETRONAS Exploration Dept. No. ER:SSB:3:80-01.

Vlierboom, F.W., 1991. Geochemical Analyses of 29 Crude Oil Samples, offshore Sarawak and Sabah. PETRONAS Exploration Dept. No. ER:OXY:4:91-04.

Wilford, G.E., 1961. The geology and mineral resource of Brunei and adjacent parts of Sarawak with descriptions of Seria and Miri oilfields. British Borneo Geological Survey, Brunei Government, Memoir 10, 319 pp.

Zulkifli Salleh and Nazri Mokhtar, 1995a. Organic Geochemical Study of Tanjung Baram-1 well, offshore Sarawak. PETRONAS Report No. PRSS SR6-95-38

Zulkifli Salleh and Nazri Mokhtar, 1995b. Organic Geochemical Study of the Baronia Timurlaut-1 well, offshore Sarawak. PETRONAS Report No. PRSS SR6-95-40

Chapter 14

Balingian Province

Mazlan B. Hj. Madon and Peter Abolins

INTRODUCTION

The Balingian Province is a proven hydrocarbon province in Sarawak and has been actively explored since the early 1920s. Current oil production comes from four fields, Temana, Bayan, D18, and D35 (Fig. 14.1), which are reservoired in Cycles I, II, and III sands, Cycle II sands being the most prolific. Several other discoveries, such as D12, D26, D28, and D34, are slated for future development.

The Balingian Province has had a complex structural history, being located in a collisional/wrench zone between the Central Luconia Province to the north, and the onshore Tinjar Province and the Rajang Fold-Thrust Belt to the south (Chapter 5). The presence of major hydrocarbon accumulations in the province is proof of a viable petroleum system and has therefore attracted several exploration companies during the 1990s. This chapter reviews some of the key features of the geology and hydrocarbon habitat of the Balingian Province.

The general geology and petroleum systems of the Balingian Province have been described in the literature. Scherer (1980) described the exploration activities in the offshore Sarawak region during the late 1960s and 1970s. Doust (1981) and Ahmad Said (1982) gave a

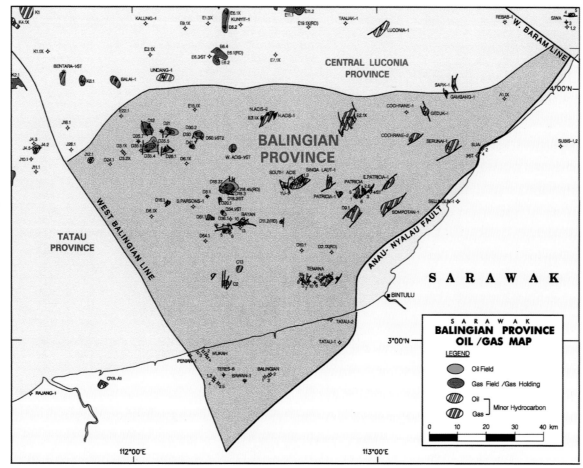

Fig. 14.1. Location of the Balingian Province, its major oil and gas fields and exploration wells.

general account of the exploration history of offshore central Sarawak, which includes the Balingian Province, until about 1981. There has not been any later account of the exploration history of the area since then. The structural styles in Balingian were described by Swinburn (1994). Ismail Che Mat Zin and Swarbrick (1997) discussed the tectonic and sedimentary development of the offshore Sarawak Basin. Agostinelli et al. (1990) described the palaeogeographic development of eastern Balingian based on seismic and well studies. Previous studies concerning the petroleum systems of Balingian include Awang Sapawi Awang Jamil et al. (1991) and Swinburn et al. (1994). Almond et al. (1990a, b) described the geology and reservoir characteristics of the D18 field in NW Balingian.

EXPLORATION HISTORY

Exploration in the onshore part of the Balingian Province by Shell started in the 1920s. After World War I, geological and gravity surveys were extended westwards from the Baram-Tinjar region to the Mukah River. Three wells drilled during the late 1920s — Tatau-1, Balingian-1 and Balingian-2 — found some indications of

WELLS	YEAR DRILLED	TOTAL DEPTH (m)	OBJECTIVES	HYDROCARBON INDICATIONS	REMARKS
Balingian-1	1929	166	Cycle V/VI	Slight indication in core 52-162m	
Balingian-2	1929	249	Cycle V/VI	Gas pockets common below 77m ss	
Balingian-3 Basin.	1956	1982	Pre-Cycle III Cycle VI	Fluorescence in cuttings at several horizons	Stratigraphic test of pre-Liang Fm. in the East Balingian
Mukah -1	1938	2049	Cycle I/III	Traces of oil at 1432m and 1440m. Oil stains and smell in numerous horizons between 224 - 2028m.	Blowout at 1914m. Gas produced at 488m used to drill Mukah 2,3,4. Gave out ca. 5 bbls/mmscf condensate during drilling Mukah/Teres 5.
Mukah -2	1938	1805	Cycle I/III	Oil shows and free oil at 407-410m. Gas shows at 406-421m. Oil and gas smell and indications in several horizons at 352-1396m.	Gas blowout at 177m.
Mukah -3	1938	918	Cycle I/III	Oil indication between 480 and 640m.	
Mukah -4	1938	850	Cycle I/III	Oil shows at about 633m.	
Penian -1	1939	617	Cycle I/II	Nil	No hydrocarbon indications.
Tatau -1	1928	159	Cycle I/II	Minor gas shows at 16.7m.	Gas blowout at 19 m.
Tatau -2	1956	1526	Cycle I/II	Faint fluorescence at 1402m.	Stratigraphic test.
Teres -1	1934	139	Cycle I/II	Minor oil and gas shows	
Teres -2	1934	206	Cycle I/II	Minor oil and gas shows at 178m.	
Teres -3	1937	627	Cycle I/II	Nil	No hydrocarbon indications.
Teres -4	1936	555	Cycle I/II	Slight traces of gas	
Teres- 5	1938	2129	Cycle I/III	Free oil in sandstones between : 1224.4 - 1231.4m 1076 - 1265m 1373.4 - 1314m 1385.3 - 1398.4m Gas shows between 1161-1176.5m	
Teres-6	1991	2474.00	Cycle I/II	Traces of oil from a DST at 1222-1232m/1267-1275m. Very good flourescence in swc at 963-1420m and 1704-1754m.	Residual oil
Bawan-1	1991	1820.00	Cycle I/II	Fluorescence in swc at 1598-1720m.	Residual oil

Table 14.1. Summary of drilling results in onshore Balingian Province.

hydrocarbons (Fig. 14.1, Table 14.1). Between 1930 and 1940, after the discoveries of oil sands in shallow core holes, exploration efforts were directed to the Teres area where 11 wells were drilled (4 Mukah wells, 6 Teres wells, and Penian-1). Oil shows, as well as free oil, were found in the Mukah and Teres wells but none of the discoveries were developed. With the exception of Balingian-3 and Tatau-2 wells, drilled during the 1950s, there has been little exploration activity in the onshore area since the World War II.

During the 1960s exploration efforts in the Balingian Province shifted offshore. Although more seismic data were acquired in the Balingian Sub-basin and surrounding areas during this decade, no wells were drilled onshore. Offshore drilling began in 1961 with Shell's Cochrane-1, which found marine shales of poor source quality in Cycle II. The Temana Field was discovered in 1962 and appraised in 1971. Temana-4 showed that the oil-bearing upper Cycle I sandstones in the eastern block are relatively tight. Production rates in Temana's eastern block were less than 300 BOPD. In the western block, however, the shallower Cycle II sands produced 2584 BOPD at Temana-5. Further drilling results showed that the production behaviour was due to complex reservoir architecture, faulting, and sedimentological factors.

In 1971 Shell drilled 19 offshore wells but no significant finds were reported. By this time major discoveries were being made in the Baram Delta Province. The Balingian wells included South Acis-1, Patricia-1, E15.1X, D2.1X, E3.1X, and West Patricia-1. Well A1.11X in the eastern part of the province found tight Cycle I limestone (Subis Limestone equivalent). A tighter seismic grid was shot over the Patricia structure in 1973, followed by the drilling of Patricia-1 and -2. Although the wells encountered oil in a few thin sandstones, reservoir development is inferior compared to Temana. More seismic data were acquired between 1980 and 1985. In 1984, the D9 structure, a few kilometres south of Patricia, was drilled. The well found gas and undifferentiated hydrocarbons. Four more wells were drilled on Patricia during 1985. Patricia-4 encountered substantial gas sands but a side-track failed to locate an oil rim.

In 1989 a consortium led by Overseas Petroleum and Investment Corporation (OPIC) signed a production-sharing contract (PSC) in the onshore part of the Balingian and Tatau provinces. Extensive exploration activities were carried out during the initial phase, including gravity, magnetics, and Synthetic Aperture Radar (SAR) surveys (Chiu and Mohd Khair Abd Kadir, 1990). In 1991 OPIC drilled two wells in onshore southern Balingian, the Teres-6 and Bawan-1 wells, and found traces of oil. Meanwhile, Agip was exploring in the offshore area in the easternmost part of Balingian, where Shell had drilled a total of 7 wells up to 1978. Subsequently, between 1990 and 1992, another 6 wells drilled by Agip in that area led to a minor oil discovery at Sapih-1 and a gas discovery at Rebab-1. The drilling results seem to suggest that there is more gas than oil in East Balingian. More oil has been found in the West Balingian, as shown by Shell's discoveries, such as Bayan, D18, D30, and D35.

STRUCTURE AND STRATIGRAPHY

The Balingian Province has an area of about 27000 km^2, of which about 30% is on land. It consists of siliciclastic sediments of Cycles I to VIII (Oligocene to Recent) overlying the Rajang Group, a tightly folded Upper Cretaceous to Upper Eocene flysch succession. The structure of the Balingian Province has been described briefly in Chapter 5. The province is bounded to the north by the Central Luconia Province, to the west by the West Balingian Line, and to the south by the NW-trending Anau-Nyalau Fault Zone which connects with the Tatau Horst. A small part of the province extends into onshore central Sarawak between Mukah and Balingian. Its eastern boundary is marked by the West Baram Line.

The Balingian Province may be subdivided into the East and West Balingian sub-provinces (Chapter 5, Fig. 5.10), each having a distinctive structural style. Swinburn (1994) subdivided West Balingian further into two sub-provinces: NW and SW Balingian, which have had slightly different structural histories. In addition, there are two major sub-basins or depocentres in the Balingian Province: the Balingian Sub-basin and the Acis Sub-basin (Fig. 14.2). A third smaller sub-basin, the South Acis Sub-basin occurs near the Acis Sub-basin. These three N-S aligned sub-basins separate the East and West Balingian sub-provinces. The sub-basins represent areas of almost continuous post-Eocene deposition; the Balingian Sub-basin, for example, contains more than 6 km of sediment infill. They are believed to be important hydrocarbon kitchens for the oil and gas fields located on the surrounding intrabasinal structural highs and on their up-dip margins.

Generally, tectonic movements in the Balingian Province occurred earlier in the west than in the east (Swinburn, 1994). The western part of Balingian (NW and SW sub-provinces) was subjected to a phase of folding, faulting, and erosion during Cycles I and II times (Late Oligocene-Early Miocene) but became relatively stable later, with only thinly developed Cycles III-V and minor tectonic activity during Cycles IV and V times (Fig. 14.3). A thin Cycle VI succession unconformably overlies much of the offshore area. In contrast, sedimentation in East Balingian was not significantly affected by tectonism throughout Cycles I to IV times. Folding during Cycle III culminated in large reverse-faulted anticlines that trend NE-SW, extending eastwards into the Baram Delta Province. Erosion was particularly severe in the onshore southeastern area where the entire post-Cycle I succession has been removed (Fig. 14.3).

The SW Balingian Sub-province is characterised by Oligocene to Early Miocene (end of Cycle I to end of Cycle II) wrench deformation of varying intensity. As a result, the Cycle I and Cycle II structural axes trend NW-SE. The basement in this sub-province is relatively shallow; large basement highs under the C5 and C8 structures are expressed as high-amplitude reflections on seismic. The faults trend mainly NE-SW and extend into relatively undisturbed overburden. A prominent angular unconformity occurs at the base of Cycle II over most of the sub-province, while a more subtle unconformity occurs at the base Cycle III (Fig. 14.3). Both unconformities pass into correlative conformities towards the centre of the Balingian Sub-basin. The flanks of the Balingian Sub-basin are characterised by down-to-the-basin normal faults that locally show growth in the Cycle II succession.

Multiple phases of deformation in the NW Balingian Sub-province from the Oligocene to Pliocene have also resulted in major unconformities at the bases of Cycles II, III, V, and VI (Fig. 14.3). The top of the basement is only weakly expressed on seismic as the basin becomes deeper to the north. The structural axes that resulted from these deformation episodes are characterised by *en echelon* NW-SE trending folds with complex fault patterns. A pervasive set of small N-S oriented normal faults, confined largely to the Cycles III and IV strata, occurs throughout

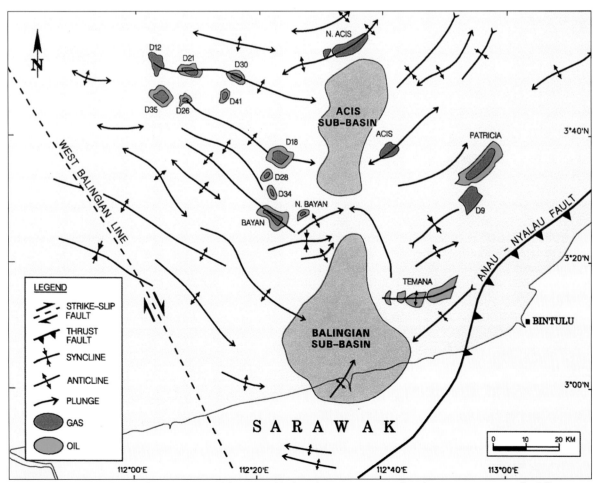

Fig. 14.2. Sketch map showing the major structural features and hydrocarbon occurrences in the Balingian Province. Note the differences in the structural trends between the West and East Balingian. Based on Shell (1994)

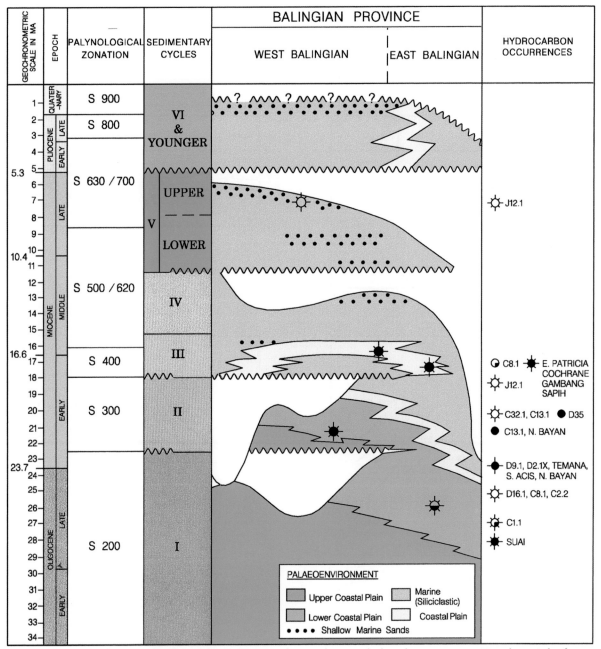

Fig. 14.3. Generalised stratigraphy of the Balingian Province showing the major hydrocarbon occurrences. Note the periods of non-deposition/erosion represented by hiatuses. Based on Shell (1994).

much of NW Balingian. These faults are probably related to the Cycles III/IV extensional phase that occurred further to the west in the Tatau Province (Chapter 17).

The East Balingian Sub-province was subjected to strong wrench deformation during the Late Miocene to Pliocene, between the end of Cycle V to Cycle VI, which produced NE-SW trending structures. Individual structures are bounded, typically, by reverse faults that converge at depth (Fig. 14.4). These are probably related to oblique or strike-slip faults in the basement. In the northern part of East Balingian, there are large-amplitude folds (e.g. E3/North Acis and South Acis) and, less commonly, reverse faults. The overall style of deformation in East Balingian is consistent with NE-SW oriented sinistral wrench deformation that decreases in intensity northwestwards away from the Patricia trend. Large basement highs occur under the D1 prospect and between D2 and Temana.

East Balingian is characterised by a thick post-Cycle I (post Oligocene) succession. Deposition of Cycle I took place in deep water environments until the area was uplifted towards the end of Cycle I times to form shallow water Cycles I/II carbonates, such as the Subis and Melinau limestones, and those found in the A1 structure. A major phase of uplift and erosion of all the pre-Cycle V strata occurred southeast of Temana, adjacent to the Tatau Horst, and resulted in a major angular unconformity. Elsewhere, especially on structural crests, an angular unconformity occurs at the base of Cycle VI, while a less angular unconformity occurs at the base of Cycle VII. Subsidence in East Balingian was punctuated with periods of transpressional strike-slip deformation that resulted in major unconformities at the mid-Cycle V, base-Cycle VI, and base-Cycle VIII levels.

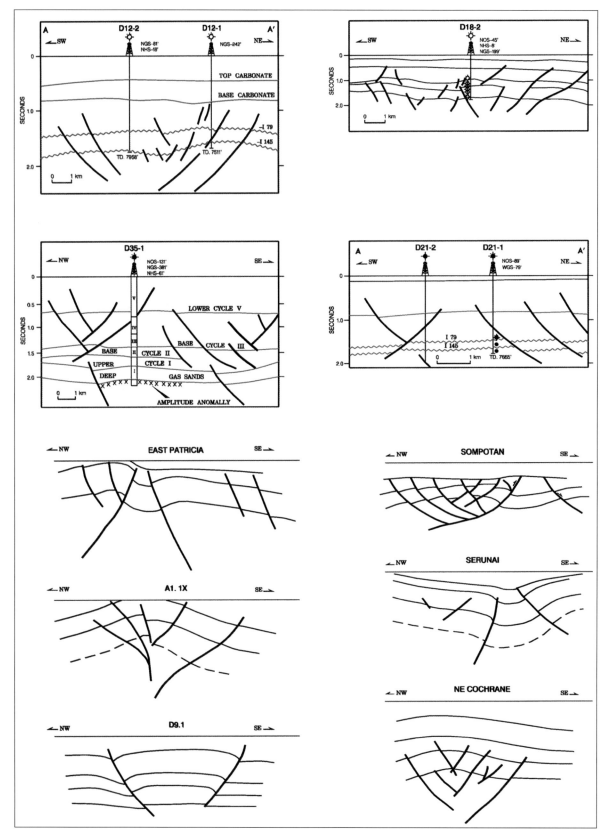

Fig. 14.4. Some of the common structural trap styles in the Balingian Province. Based on OXY (1989) and Shell (1994).

HYDROCARBON OCCURRENCES

Hydrocarbons in the Balingian Province occur in Upper Oligocene to Lower Miocene siliciclastic reservoirs of Cycles I, II and III. Most occurrences are in the offshore areas. Major oil and gas accumulations occur in the western part of the province, whereas minor hydrocarbon accumulations occur in the eastern part (Fig. 14.1). The largest of these accumulations, including the currently producing fields, occur in a NW-trending fairway running from the Temana Field, near the town of Bintulu, to the D35 Field. To date, only four fields are in production: Bayan, D18, D35, and Temana.

The Bayan Field was discovered by Shell in 1981 with the drilling of the Bayan-1 well. The well was drilled on a NE-trending faulted anticline, with reservoir sandstones in the Lower Miocene Cycles I and II at depths of about 1200-1500 m. The trap is a closed anticline bounded by an E- and a NE-trending fault on the northern and southern sides, respectively. The Bayan oil has an API gravity of about 37°. The field has an estimated ultimate recovery (EUR) of about 110 MMSTB. Cumulative production to date (1998) is about 80 MMSTB.

The D18 Field has been described in the literature by Almond et al. (1990a, b). It was discovered in 1981 by the crestal exploration well D18.1, and came on stream in 1986 at an initial rate of about 5300 BOPD. The D18 structure is a NE-trending anticline cut by numerous N-S and NNE-WSW trending normal faults. The reservoirs are sandstones of Cycle II, deposited in lower coastal and river-dominated delta plain environments, and having average porosities of about 25%. The field has a complex reservoir architecture because of numerous cross-cutting faults. The field has an EUR of about 70 MMSTB. The cumulative production to date (1998) is about 36 MMSTB.

The D35 Field was discovered in 1984 with the drilling of the D35.1 well. The structure is a domal anticline, cut by N-S normal faults. Cycles I and II coastal plain sandstones form the reservoirs for the oil, which has an API gravity of about 34°. The field has an EUR of about 80 MMSTB.

The Temana Field is one of the earliest discoveries in the offshore Balingian Province. The discovery well, Temana-1, was drilled in 1962 and found oil in Cycles I and II shallow marine and transitional marine sandstones in a structurally complex anticlinal trap. The anticline is compartmentalised by N-trending normal faults. Temana oil has an API gravity of about 35°. The field started production in December 1979 with an initial flow rate of 12000 BOPD. It has an EUR of about 140 MMSTB and, to date (1998), has produced about 90 MMSTB.

SOURCE ROCKS

Depositional Environments

The Balingian Province provides a very good demonstration of how palaeogeography controls source rock deposition (hence, quality) and, subsequently, the hydrocarbon distribution. It also provides strong supportive evidence for coal as an important source rock for oil in the long-standing "coal as an oil source" controversy. Using the basic geological principle of "the present is the key to the past", we may view the present-day Sarawak coastline from Kuching to Miri as an analogue of the palaeocoastline during the Early Miocene. This may provide us with some important insights into the source and reservoir depositional environments of the Balingian Province and its petroleum systems.

Today's coastline is characterised by large deltaic systems such as the Baram Delta, and major estuarine features such as the mouths of the Lupar and Rajang rivers. Major Early Miocene river systems, comparable to the present-day Baram and Rajang rivers, existed along the palaeocoastline and were responsible for the deposition of large amounts of sediment in the offshore area to the east. Reservoir facies, such as fluvial channels and offshore mouthbars, were deposited in these river systems, the recognition of which is the key to exploration success in the Lower Miocene succession of Balingian.

The present-day coastline also provides clues to the nature of potential source facies that could have been deposited during the Early Miocene. Thick mangrove belts, similar to those that presently fringe the coastline along many kilometres of northern Sarawak, may trap sediment and organic matter in their complex root systems. Landward of the mangrove forests, large areas of peat swamps occur, which have accumulated tens of metres of peat, the precursor of coal. Marsh lakes are numerous, although often of a rather ephemeral nature. Occasionally such lakes will be long standing, allowing accumulation of organic material. Such settings would act as ideal habitats for algal bloom development.

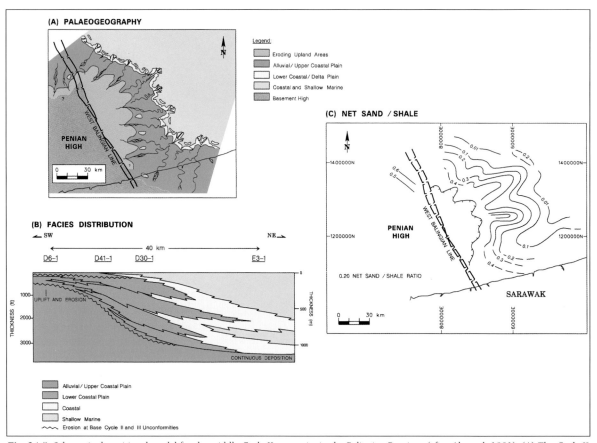

Fig. 14.5. Schematic depositional model for the middle Cycle II reservoirs in the Balingian Province (after Almond, 1991). (A) The Cycle II depositional system represents a northeastward progradation of coastal-deltaic systems that derived sediment from the uplifted Penian High. (B) Cycles of coastal progradation and retreat resulted in intercalated sand-prone and shale-prone facies, with reservoir quality generally deteriorating away from the coastline, i.e., to the northeast. (C) The distribution of net sand/shale across the area closely follows the main depositional trend.

Although the present-day coastline trends approximately NE-SW, previous studies (e.g. Doust, 1981; Ismail Che Mat Zin and Swarbrick, 1997) have shown that the Early Miocene palaeocoastline might have trended NW-SE, almost perpendicular to that of today. Figure 14.5 illustrates schematically the Early Miocene (Cycle II) palaeogeography of western Balingian, which is interpreted to have been controlled by fault movements along the West Balingian Line, marking the eastern boundary of the Penian High. A transect parallel to this palaeocoastline (present NW-SE) would show that, over short distances, source facies could vary between estuarine, marsh, peat swamp, distributary streams, fluvial plains, and even possibly lacustrine. In contrast, a transect perpendicular to the palaeocoastline (present NE-SW) would show a gradual change from a continental setting, through coastal, and into a marine depositional setting. Significant variation in the quality and quantity of the preserved organic matter may occur over fairly small distances. In addition, frequent marine transgressions and regressions may result in complex spatial and temporal facies relationships. A very complex system of source rock depositional environments will build up over time, giving rise to interfingering wedges of source rock facies, many of which are not fully understood in terms of the quality and quantity of the organic matter that they preserve. Since the accumulation and preservation of source rocks and reservoir rocks in a rapidly varying environment is strongly influenced by palaeogeography, a good understanding of the palaeogeographic evolution is crucial to exploration success in the offshore Sarawak Basin, particularly in Balingian.

Source Rock Quality

Swinburn et al. (1994) and Wan Hasiah et al. (1995) presented results of geochemical analyses of source-rock samples and oils which indicate strongly that oils in the Balingian Province were generated from coals and coaly shales in the Oligocene-Lower Miocene coastal plain sequences. These coals and carbonaceous shales, which are found mainly in Cycles I and II, were deposited in coastal and inland peat swamps, as overbank muds deposited on floodplains and delta tops, as well as transgressive marine muds

Balingian Province

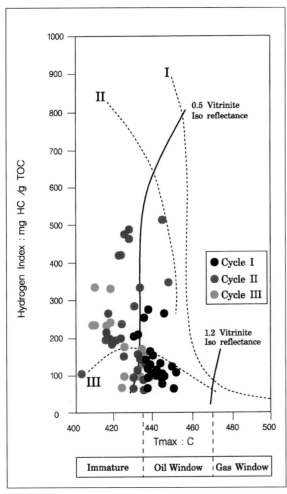

Fig. 14.6. Plot of Hydrogen Index against Tmax for Cycles I, II, and III source rocks in the Balingian Province.

deposited during sea-level highstands. A large part of East Balingian may have been covered by shallow sea during the Early Miocene. Hence, the potential source rocks here are likely to be marine shales.

Source rock studies show that coals and carbonaceous shales are, by definition, not only organically the richest source rocks but also tend to contain the most oil-prone organic matter. Figure 14.6 shows a plot of Hydrogen Index against Tmax data from a D35 and a C2 well that have good Cycle I and Cycle II penetrations, respectively. The data show that lean or mediocre shales, with total organic carbon (TOC) values of < 1.00 wt%, generally have hydrogen indices of between 100 and 200 mgHC/gTOC. This suggests that the leaner shales are generally only gas prone, not only because of low organic richness but also because of the poor source quality of Type III kerogen.

Coals and carbonaceous shales, on the other hand, have hydrogen indices of between 200 and 300 mgHC/gTOC. They contain a mixed assemblage of Types II/III kerogen and, therefore, have a greater potential as an oil source. Organic petrographic studies of Cycles I/II equivalent coals from onshore Balingian have revealed a highly oil-prone nature, as suggested by a high abundance of liptinitic material and the presence of microscopic oil-generative features (Wan Hasiah, 1997). These features, combined with the good organic richness,

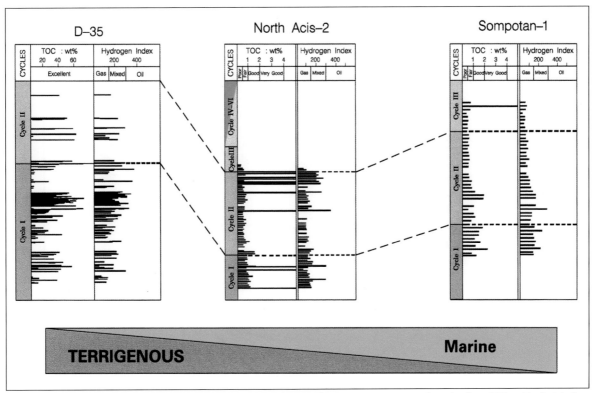

Fig. 14.7. Geochemical summary logs for Cycle I and Cycle II, showing the variation in source rock quality from D35 to North Acis-2 to Sompotan-1.

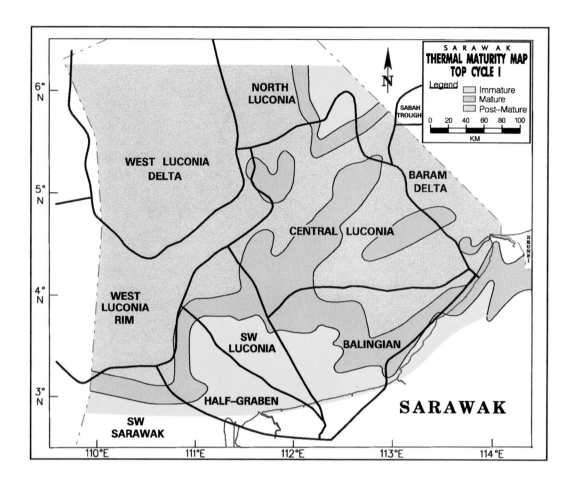
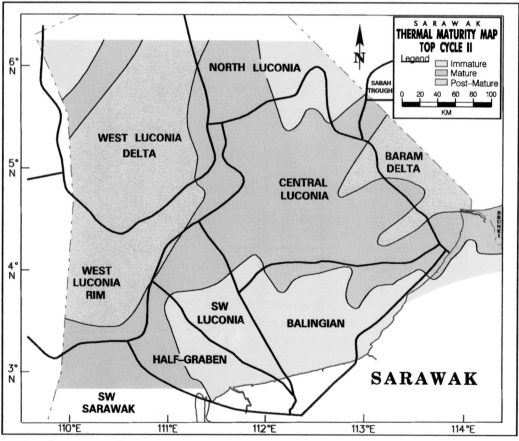

Fig. 14.8. Thermal maturity maps for the Sarawak shelf for top Cycle I (Upper) and top Cycle II (Lower).

contribute to the good quality of the source rocks in the Cycles I and II sequences.

Coal abundance generally decreases basinwards in the Balingian Province which suggests that the non-marine source rock potential also decreases basinwards, ie. eastwards. D35 and C2 wells, for example, contain numerous coal seams of up to 1m thick. Eastwards, however, although numerous, the coals are generally poorly developed. For example, coals in North Acis-2 and Yu-1 tend to be thin and discontinuous. Further eastwards, in Sompotan and Serunai, no coals have been reported (Fig. 14.7).

The difference in the kerogen quality between lean/mediocre shales and coals/carbonaceous shales becomes even greater in East Balingian. The hydrogen indices of East Balingian coals, which are thinner and less abundant than in West Balingian, imply a good source potential for oil, with values of between 200 and 300 mgHC/gTOC, as found in wells such as North Acis-2 and Yu-1. In such wells, the hydrogen indices of lean shales are less than 100 mgHC/gTOC, which suggests that the source rocks are most probably gas prone.

Generally, Cycle I source rocks tend to have higher organic richness compared to those in Cycle II. Furthermore, comparative examples of Cycles I and II geochemical summary logs (Fig. 14.7) clearly demonstrate the gradual easterly decrease in source rock quality within the Balingian Province. This general trend suggests that in eastern Balingian, marine sequences may not be important hydrocarbon source rocks. However, although good marine source rocks have not been found, condensates with marine influence do occur, for example, in Cycle II in Sompotan-1. The Sompotan condensate therefore suggests that good quality marine source rocks may be present, at least locally, but are yet to be penetrated. Clearly, additional work is required to investigate the possible presence of marine source rocks in east Balingian.

Maturity and Timing of Generation

In addition to source rock quality, the maturity is equally important. If the source rock never reaches oil window maturity, it will not generate hydrocarbons, regardless of source rock quality. Figure 14.8 is a set of simplified maturity maps for the Balingian Province. In most parts of the province, Cycle I and lower Cycle II source rocks have been buried deeply enough to have generated oil, whereas upper Cycle II and Cycles III/IV are mature only in the deepest parts of the basin. At any given location, Cycle II sediments are less mature than the underlying Cycle I sequence. Over much of western Balingian, Cycle II source rocks are generally within the oil window (e.g. C2) or are immature to early mature (e.g. D35). In eastern Balingian, Cycle II tends to mature rapidly away from the present-day coastline, so that only in the more coastal area is Cycle II immature. On the whole, Cycle III and younger sediments are thermally immature for oil and gas generation over much of the Balingian Province.

Because of the relatively high geothermal gradients in the Balingian Province (Chapter 4), source rocks become thermally mature for oil/gas generation at relatively shallow depths; in general, about 1800 m for oil and 2750 m for gas (Fig. 14.9). Swinburn et al. (1994) suggested that most of the oils in the Balingian Province were derived from Oligocene-Early Miocene (Cycles I and II) coastal plain deposits in the maturity range of about 0.8-1.0% VRo. Figure 14.9 shows that, in the western Balingian, only Cycles I and II are mature enough to have generated oil and gas. In eastern Balingian, the lower part of Cycle III, in addition to Cycles I and II, has been buried to sufficient depths and may have also reached the oil window.

Cycle I sediments are presently past the oil generation window over most of E Balingian and may be generating gas. In western Balingian, however, the upper part of Cycle I is probably still within the oil window although older, deeper sections of Cycle I may pass progressively into the gas window. Cycle I sediments are therefore expected to generate oil and/or gas depending on the location. For example, over the C2 structure, much of Cycle I is within the gas window. But to the north, over D35, Cycle I is generally of oil window maturity.

The presence of major hydrocarbon kitchens has provided the necessary charge to fill the structures in the Balingian Province. Several Miocene depocentres developed in the province and caused the deep burial of Cycles I/II sediments. These depocentres, namely the Acis, South Acis, and Balingian sub-basins (Fig. 14.2), subsequently became the kitchen areas for generation of oil and gas from Cycles I and II source rocks. Most of the oil/gas accumulations in NW and SW Balingian are thought to have been charged from these sub-basins. In the East Balingian Sub-province, deeper burial of Cycles I/II sediments may have contributed to the hydrocarbon charge in structures that are far from the kitchen areas.

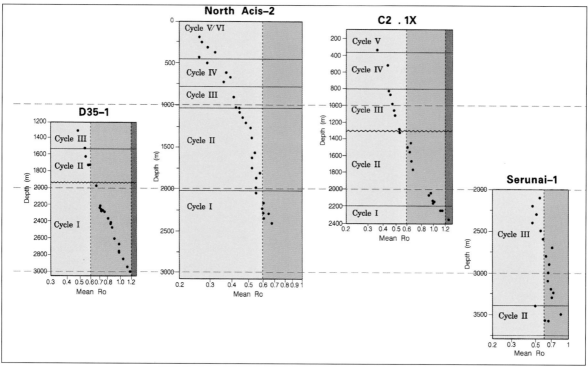

Fig. 14.9. Thermal maturity (VRo) profiles for selected Balingian wells.

Another factor to consider is the relative timing of hydrocarbon generation compared to structure formation. Because of the structural complexity of the Balingian Province, the relative timing of structuration and migration is believed to have been a major controlling factor in determining prospectivity. Tectonic and structural studies have revealed that Middle Miocene folding in western Balingian predated the main phase of oil generation (Middle to Late Miocene) and so the traps were closed when the fields were being charged with oil. In eastern Balingian, however, structuration occurred much later (Late Miocene to Pliocene), either synchronous with, or post-dating, the main phase of oil generation (Middle to Late Miocene). Hence, most traps had not been formed in time to receive the oil migrating into the area. By the time that the traps were forming, the kitchen areas were generating mainly gas and condensate.

TRAPS

Most of the hydrocarbons found in Balingian are associated with structures that formed as a result of multiphase convergent and divergent wrench deformation. In the NW and SW Balingian sub-provinces, west of the Acis and Balingian sub-basins, Oligocene to Early Miocene deformation resulted in major WNW-trending anticlinal structures which form some of the larger oil fields in the province (Fig. 14.1). There is a major play of *en echelon* anticlines formed by dextral motion of the West Balingian Line fault zone (Fig. 14.2). Some anticlines may form traps up to 100 km long, as in the D35-D18 trend. East of the Acis and Balingian sub-basins, the major structures trend ENE-WSW and seem to be related to compressional deformation across the Anau-Nyalau overthrust zone, with a left-lateral wrench component associated with oblique thrusting. Middle Miocene deformation resulted in N-S trending faults that cut across many of the fold culminations on both sides of the Acis and Balingian depocentres.

The hydrocarbon traps in West Balingian are commonly associated with NW-trending anticlines that are cut by NE-trending extensional normal faults. The main reservoir targets in these structures are the Cycles I, II, and III stacked fluvial sandstones. In many structures, especially on the flanks of the Balingian Sub-basin, structural closure is achieved partly by the presence of NE-trending growth faults, such as in the C2.1 and C2.2 structures. Structures such as C8, N Bayan, D18, and D26, are the result of the juxtaposition of NE-trending normal faults and NW-trending anticlines.

Trap styles in East Balingian are dominated by NE-trending anticlinal fault traps (Fig. 14.4). Examples include the S Acis, W Patricia, Patricia, and E Patricia structures. These structures characterise the so-called "Balingian Thrust Belt" which was affected by Late Miocene compressional deformation that resulted in ENE-

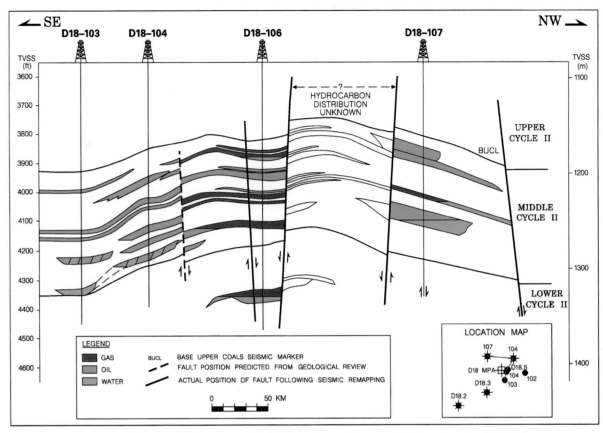

Fig. 14.10. Cross section of the D18 Field showing complex faulting and reservoir connectivity (after Almond et al., 1990a).

trending thrusts and anticlinal-fault traps. The currently producing Temana Field is one of the earliest discoveries made in this type of structure. The E3/North Acis structure is the largest of such anticlines, which are typically cut by major almost N-trending normal faults that form many fault blocks/compartments. The folds are commonly elongate, slightly asymmetric, verge to the southeast, and are bounded to the NW and SE by reverse faults. The intensity of deformation decreases from land to offshore. The stratigraphic relationships suggest that folding started mainly after the deposition of Cycle II.

The main target horizons in the East Balingian thrust/anticlinal structures are the Cycle I carbonates and Cycles I to V siliciclastics. While Cycles I and II reservoirs form the main drilling objectives in West Balingian, Cycles III and IV sequences are also prospective targets in East Balingian. An example of this play type is Cochrane, where non-associated gas was encountered in Cycle III coastal sands. Other hydrocarbon discoveries in Cycle III include E2.1X, NE Cochrane, and S Acis. Cycle V deltaic sediments, similar to the West Baram Delta play, are also being pursued in East Balingian. The Cycle V sequence consists of mainly coastal and fluviomarine deposits, passing into holomarine sediments northwards. NE Cochrane-1 tested some fluviomarine Cycle V sediments and found immature humic source rock and good sand development.

The presence of large hydrocarbon columns in the Balingian Province attests to the presence of good seals throughout the Cycles I to III succession. Of these, intraformational Cycles I/II and basal Cycle III transgressive shales are the most important. It has been suggested by Shell geologists that a "safety valve" mechanism functions in crestal areas of fields whereby seals are temporarily breached to release gas. This mechanism is thought to prevent the oil from being flushed from the trap. The presence of gas chimneys is good evidence for this mechanism.

RESERVOIRS

The most important reservoirs in the Balingian Province are Cycles I/II/III (Oligocene to Lower Miocene) stacked sandstones in the anticlinal-fault traps described above. The Cycles I and II reservoirs are coastal plain and nearshore sandstones, and are the primary target, while the basal Cycle III coastal and shallow marine sandstones constitute secondary reservoir objectives. The currently producing fields in the Balingian province (Bayan, Temana, D18, D35) are all producing from these Cycles I, II, and III reservoirs. Cycle II lower coastal plain sandstones

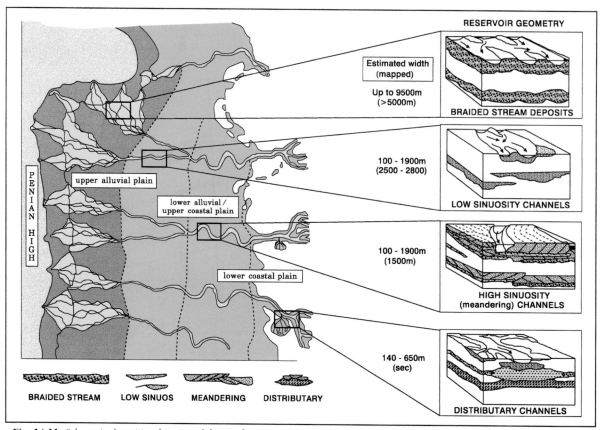

Fig. 14.11. Schematic depositional setting of the North Bayan reservoir sandstones. Reservoir geometry and connectivity depend on the sandbody type and depositional setting.

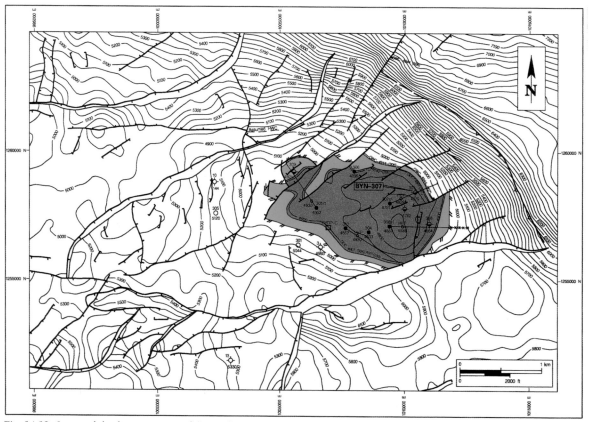

Fig. 14.12. Structural depth contour map of the North Bayan area at the top of the S7.7 sand. Note the complex fault pattern. Contour interval in feet.

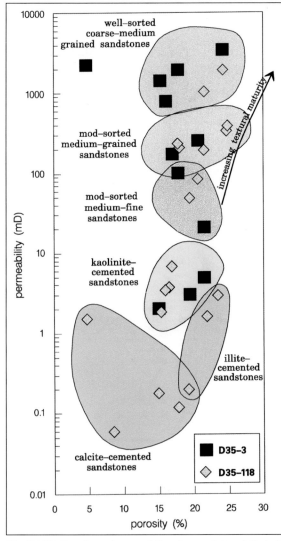

Fig. 14.13. Porosity and permeability characteristics of Cycle II reservoir sandstones in the D35 Field. Low permeabilities (<10mD) in some reservoirs are caused by pore-filling authigenic cements such as calcite, kaolinite, and illite.

are volumetrically the most important, with the exception of Temana which produces oil mainly from Cycle I reservoirs. Intraformational Cycles I/II and basal Cycle III shales provide the seal for these structures. Almond et al. (1990a, b) described the reservoirs in the D18 Field, NW Balingian. Figure 14.10 shows a cross-section of the D18 Field. The reservoirs in this field consist of stacks of lower coastal plain sandstones interbedded with shales and coals that were deposited in an overall transgressive setting. The sandstones were deposited in braided and meandering streams, channels, crevasse splays and channel mouth bars, and locally in nearshore marine sands.

The distribution and geometry of Cycle II sands are generally controlled by three factors: (1) regional E/NE shale-out away from the palaeo-coastline (Fig. 14.5), (2) localised onlap and drape over Cycle I palaeotopography, and (3) truncation by the base Cycle III unconformity-related regional and/or local uplift. The Penian High, whose western margin is marked by the West Balingian Line, was the main source of sediment for the reservoirs in the western Balingian fields (Fig. 14.5). Cycle II strata are completely truncated by the base-Cycle III unconformity towards the Penian High. Intraformational Cycle II shales provide the top seals for the Cycle II play, but Cycle III transgressive shales form a thick regional top seal for several Cycle II fields in the province.

Several genetic sand body types can be expected to be present within a particular field, each with its own reservoir characteristics. Figure 14.11 shows a schematic depositional model for the North Bayan Field. The three main types of sand body that have been recognised are channel sands (of which several types are present), overbank sands, and shallow marine sands. Although channel sands are good quality reservoirs, they tend to be discontinuous, typically being up to 1500 m wide and 4000 m long and are therefore not correlateable over a large distance. Similarly, overbank sand bodies are generally not correlateable for more than about 2 km. Marine sands tend to be correlateable over a larger distance but are generally lower quality reservoirs.

A complicating factor in this play type is the high degree of faulting that can lead to compartmentalisation of the field (Fig. 14.12). Detailed structural studies are, therefore, required for optimal field development. Another type of compartmentalisation that can occur in this play type is stratigraphic compartmentalisation, caused by the pinching out of sandbodies. In the absence of dense stacking, stratigraphic compartmentalisation is most likely for channel sands and overbank sands. Although it is common for an element of stratigraphic trapping to contribute to the reserves of Balingian fields, the contribution tends to be small compared to the total reserves of the fields.

Cycles I and II coastal plain sandstones have porosities ranging between 15% and 32%. Porosity declines rather rapidly in the Balingian Province; the economic porosity floor of about 15% is generally at depths of less than 2100 m, compared with 3350 m in the Baram Delta. This feature is probably related to the higher geothermal gradient in Balingian (36-50° Ckm^{-1}) than in the Baram Delta (25-30°Ckm^{-1}). Apart from burial-related compaction, diagenetic factors, such as compaction and quartz cementation, are important controlling factors on reservoir properties in some fields. In the D35 field, for instance, the poor performance of some

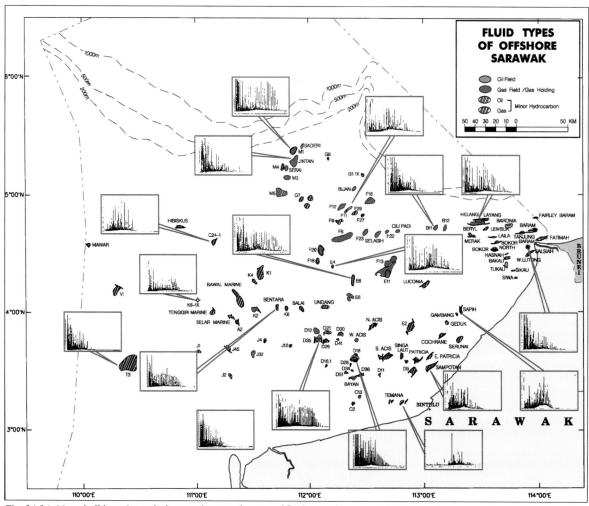

Fig. 14.14. Map of offshore Sarawak showing the great diversity of fluid types that occur. Fluids range from very light condensates, such as T3 in the far west, to very waxy oils, such as at M4.1 in the far north. This range of oil types occurs as a consequence of the source facies and maturity variations that occur throughout the different geological provinces of offshore Sarawak.

Cycle II reservoirs is attributed to tight sandstones that have been adversely affected by compaction and cementation during burial diagenesis. Cementation by calcite, illite, and kaolinite have resulted in very low permeabilities in some reservoirs (Fig. 14.13).

HYDROCARBON TYPES

The hydrocarbon distribution map (Fig. 14.1) shows that many western Balingian structures are oil bearing, including the producing oil fields in the Balingian Province (Bayan, D18, and D35). In East Balingian only minor oil and gas discoveries have been made, such as N Acis, S Acis, Singa Laut, Patricia, and the producing field, Temana. These accumulations tend to occur close to the kitchen areas (Acis and Balingian sub-basins), although minor gas discoveries have been made further eastwards in East Balingian, such as at Sompotan, Serunai, and Cochrane.

Figure 14.14 is a map of the main hydrocarbon types in the Sarawak Basin. The large variation in hydrocarbon types suggest that the kitchen areas probably contain a complex system of source facies, ranging from predominantly lower coastal plain in the west to increasingly marine to the east. It is suggested that the observed hydrocarbon distribution in the Balingian Province is determined by two important factors. One factor is source rock type: leaner, poorer quality marine source rocks, with fewer coals, becoming more dominant to the east. The different source rock facies is clearly demonstrated in Fig. 14.15, which shows the varying degree of higher plant influence in the oils and condensates throughout offshore Sarawak. The other factor is the relative timing of hydrocarbon generation and trapping: early structuration (Early-Middle Miocene) in West Balingian enabled oil to be trapped, whereas late structuration (Late Miocene-Pliocene) in East Balingian meant that most traps were formed when the source rocks were already within the gas generation zone.

The variation in source rock facies in

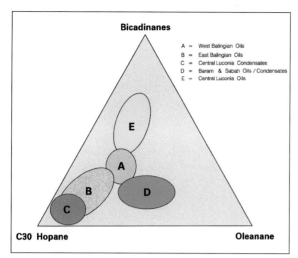

Fig. 14.15. The oils and condensates of the different geological provinces of offshore Sarawak can be distinguished by variation in their biomarker compositions. Such variation reflects different organic matter input to the source depositional environment. This figure demonstrates such a distinction based upon the relative amount of biomarkers with a higher plant origin (oleanane and bicadinanes) and biomarkers with a microbial origin (C_{30} hopane).

Other common geochemical features of the West Balingian oils are a moderate to high 18a-oleanane content, a moderate to high bicadinane content, a C_{29} sterane preference, a high hopane/sterane ratio, a low concentration of diasteranes, and a moderate to high Tm/Ts ratio. Figure 14.16 shows the typical biomarker distribution for the West Balingian oil family.

The presence of oleanane and bicadinanes and the predominance of C_{29} steranes

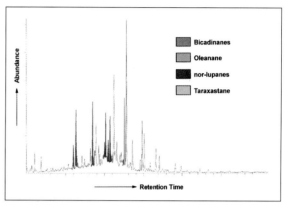

Fig. 14.16. Biomarker characteristics of the West Balingian oil family. The data show typical higher plant markers (good nonmarine indicators) and typical coal indicators (see text).

Balingian manifests itself in the oils and gases generated, particularly in the detailed biomarker distributions. West Balingian oils and gases vary compositionally to reflect a complex interplay of source facies. The main variation, however, tends to occur at the molecular level rather than at the bulk compositional level, as discussed below. The variation in East Balingian oils is greater, representing both a more diverse source facies variation and a greater variation in source rock maturity.

West Balingian Oils

As outlined above, oils tend to occur more commonly in the western part of the province. West Balingian oils share many common features; they tend to be moderately waxy, low-sulphur crudes with API gravities of between 30° and 40°. Severe biodegradation is rarely encountered although moderate levels of biodegradation have been encountered in the Temana Field and at Patricia (Awang Sapawi Awang Jamil et al., 1991). Elsewhere in western Balingian, biodegradation, if present, is very mild. Oil quality is therefore considered to be good with few undesired characteristics. Table 14.2 summarises typical characteristics of some Balingian crudes.

At the molecular level, all the West Balingian oils possess high Pr/Ph ratios, commonly exceeding 5.0, indicative of oxic depositional environments for the source rock. Pr/nC_{17} ratios commonly exceeds 1.0, and are often greater than 1.5. Such features are typical of lower coastal plain depositional environments.

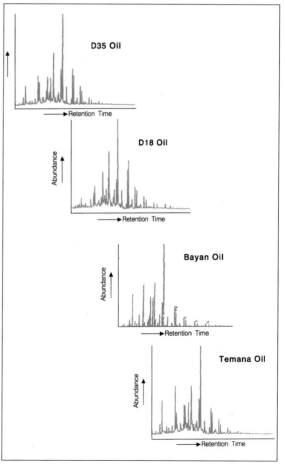

Fig. 14.17. A comparison of the triterpane distribution in Balingian oils. Examples from D35, D18, Bayan, and Temana, which are arranged in a NW-SE trend from top to bottom, almost parallel to the palaeocoastline, and therefore indicates a wide variation in triterpane characteristics.

confirm that the oils were generated from source rocks rich in higher plant material. Hopanes can outweigh steranes by an order of magnitude in West Balingian oils, a typical feature of non-marine oils. The low quantities of diasteranes and a moderate to high Tm/Ts ratio strongly suggest that coals and carbonaceous shales have contributed significantly to these oils. Such characteristics therefore clearly show that West Balingian oils were generated from source rocks deposited in a lower coastal plain environment, as predicted by the source facies model described above. Figure 14.17 shows some examples of the variation in triterpane distributions of oils in selected wells. The wells were specifically chosen to show the variation in biomarker distribution parallel to the palaeocoastline. The figure clearly demonstrates the low degree of variation in oil composition parallel to the coastline; the main variation being in the relative abundance of major higher plant biomarkers, although this variation is still within a limited range.

Despite the general similarities of West Balingian oils, variation still exists. This is not surprising, considering the complex facies relationships outlined above. The relative amount of bicadinanes, in particular, can vary from very high, such as in D30, to moderate in D35. Similarly, absolute values of Tm/Ts and hopane/sterane can vary significantly. Although the reasons for these variations are not well understood, they are probably related to the main source component contributing to the charge in a specific area (e.g., peat swamp versus mangrove forest). The implication of high quantities of bicadinanes on source facies is unclear and requires further work.

East Balingian Oils

The occurrence of oils in the Balingian Province decreases towards the east, with the majority of wells drilled encountering gas and condensate. This is possibly an artefact of the low number of wells drilled. Oils are found in East Balingian at North Acis, South Acis, Sompotan, and Patricia.

Well	Cycle	API gravity (°)	Sulphur (Wt%)	Wax (Wt%)
North Acis-2	II	36.1	<0.05	1.29
Yu-1	II (2070 m)	36.9	<0.05	1.61
	II (2187 m)	35.9	<0.05	0.55
Patricia-1	III (827 m)	37.8	0.10	-
D35	II	34.0	0.10	-
D18	III	34.7	0.10	-

Table 14.2. Summary of typical crude oil characteristics of some Balingian crudes.

Oil and gas have also been encountered onshore at Suai. East Balingian oils (Fig. 14.4) tend to be lighter than west Balingian oils. They also show a larger variation in physical and geochemical properties, as documented below, which is to be expected when considering the source rock depositional model described earlier.

East Balingian oils include lower coastal plain and marine-influenced oils. The lower coastal plain oils have features that are very similar to those of the West Balingian oils described above; they are characterised by high relative abundances of higher plant derived biomarkers. In contrast, the marine-influenced oils and condensates are very different from the West Balingian oils with regard to some important geochemical characteristics, including low quantities of higher plant markers such as oleanane and bicadinanes, high relative abundances of C_{28} and C_{29} tricyclic terpanes, enhanced development of C_{34} and C_{35} extended hopanes, low hopane/sterane ratios (compared to

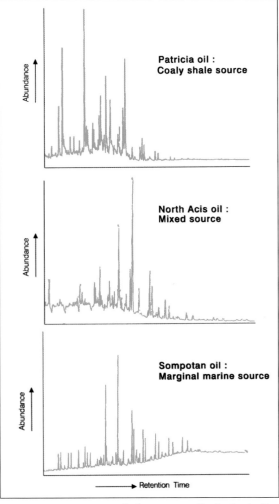

Fig. 14.18. A comparison of the triterpane distribution in East Balingian oils, displaying a transition in source rock facies (depositional environments), from a coaly source to a marginal marine source.

west Balingian oils), abundant diasteranes, and a C_{27} sterane preference over C_{29}. All these features tend to suggest an increased marine contribution compared to West Balingian oils.

Examples of lower coastal plain oils include those in Patricia and Yu-1, whereas examples of marine-influenced condensates occur in Sompotan. Figure 14.18 shows the biomarker distributions for three East Balingian crudes, which clearly indicate that the marine influence increases eastwards. Figure 14.15 shows clearly that the East Balingian oils have generally less higher plant input compared to West Balingian oils.

An important feature of East Balingian oils/condensates is the elevated C_{34} and C_{35} extended hopanes, which is commonly associated with good quality source rocks. The rarity of Sompotan-type oils in East Balingian suggests, however, that the marine source rock is unlikely to have been widespread. Nevertheless, it does suggest that marine incursions took place during the Early Miocene and that good quality marine source rocks were deposited, at least locally. The Sompotan-1 well also tested oils that are very similar to those of West Balingian, indicating that a lower coastal plain source rock may have also been effective in the area. The lack of higher plant markers in the marine oil type suggests that no mixing had taken place and, therefore, two distinct pods of active source rock may be present. It is very likely that the lower coastal plain oils had been generated from the Acis or South Acis Sub-basin to the west whereas the marine oils were generated from an active source pod to the east where marine conditions had prevailed.

MIGRATION PATHWAYS

The geographical distribution of oil and gas in Balingian clearly demonstrates the importance of the three main depocentres as kitchen areas for the accumulated hydrocarbons. Major accumulations occur in closures on structural highs peripheral to these kitchen areas. This, and the absence of highly mature condensates, at least in West Balingian, suggest that up-dip migration has been the main migration mechanism for the Balingian oil fields. Vertical drainage through deep-seated faults does not appear to have been a major mechanism for oil migration, but may have been the cause of some gas flushing that has been recognised in some fields.

Figure 14.19 shows a schematic model for hydrocarbon migration in the Balingian Province developed by Shell (Dolivo, 1991). The high percentage of sand in Cycles I and II, and its close stratigraphic association with potential

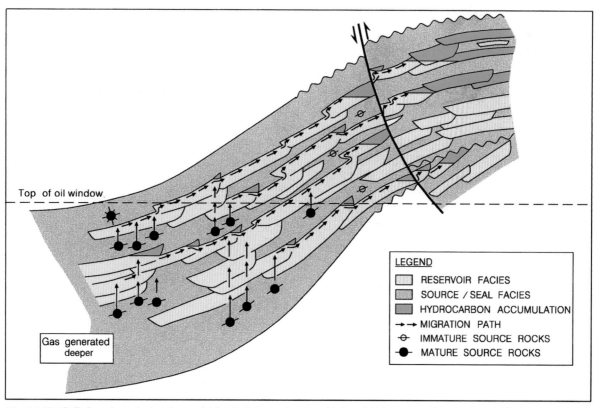

Fig. 14.19. Shell's hypothetical migration model for the Balingian Province (Dolivo, 1991). Diagram shows migration of oil generated from mature source rocks that lie within the oil window through interconnected sandbodies to be trapped updip, including in the thermally immature section, by sealing faults and/or structurally-closed reservoir sandbodies.

source rocks, will have facilitated primary migration from source to carrier bed. Buoyancy is the main driving force for migration, with migration pathways established preferentially across the highest structural dips. Although many of the individual sandstone carrier-beds are laterally discontinuous, the high degree of vertical stacking of sandbodies in Cycles I and II often results, over geological time, in an unbroken pathway from source to trap. The migration model in Fig. 14.19 predicts the presence of many down-dip stratigraphic traps but such traps, although numerous, are likely to be small. It is postulated that such down dip stratigraphic traps may act as sinks for gas flushing, effectively protecting the main oil accumulation from later gas charges.

Detailed migration routes have been established only in limited areas of Balingian. Such routes are controlled, not by current structural styles, but by the structural styles prevalent at the time of migration. Because the Balingian Province has suffered several structural phases, determining the migration paths requires complex structural restoration and backstripping.

PLAY TYPES AND PROSPECTIVITY

The different hydrocarbon play types in East and West Balingian have an impact on the prospectivity of these areas. The main hydrocarbon plays in West Balingian are Oligocene-Early Miocene Cycles I and II coastal plain sands, sealed by intraformational shales and trapped in wrench-related anticlines. Basal Cycle III and Cycle V shallow marine sands constitute secondary objectives in the western sub-province. Carbonates in the upper Cycle III, Cycle IV, and Cycle V are important plays in the northern and northwestern parts of this area.

In East Balingian, the main play types are the siliciclastics in Cycles III to V, as have been penetrated in places such as Cochrane, Sapih, and Gambang. There are also potential traps in the Cycle I limestones, as have been tested in the Rebab-1, A1.1X, and Serunai-1. Cycles I, II, III clastics are also potential plays in the Patricia and Sompotan areas in eastern Balingian.

Cycle II sandstones form the main reservoirs in the existing oil fields of the Balingian Province. These sediments were deposited in an overall transgressive stack of lower coastal plain interbedded sands and shales, and coals. Sand depositional environments included braided/meandering streams, channels, crevasse splays, and channel mouth bars, and locally, nearshore marine sands. The distribution and geometry of these sandbodies are controlled by the position relative to the palaeocoastline. Intraformational Cycle II shales provide top seals, whereas Cycle III transgressive shales form a thick regional ultimate top seal for several of the Cycle II accumulations.

Hydrocarbon accumulations in the Balingian Province were derived from source rocks within Cycles I and II, which are primarily coals and coaly shales deposited in lower coastal plain environments. There are abundant coals in Cycles I and II that are rich in liptinite and amorphous organic matter which have both oil and gas potential. The coals are presently mature on the flanks of the Acis and Balingian sub-basins. Oil generation is thought to have occurred since Cycle III times, whereas expulsion and migration began in Cycle IV times and continues today.

In West Balingian, the peak oil generation in the Cycle II kitchens occurred sometime during the Late Miocene to Pliocene and post-dates the main period of Cycle II trap formation during the Middle Miocene. The timing of generation for Cycle II in East Balingian is less favourable. Here, peak oil generation in the Cycle II kitchens occurred around the Middle to Late Miocene and pre-dates the main trap formation which occurred during the Late Miocene-Pliocene. The late trap formation has caused much of the generated oil to be lost. Cycle II kitchens in East Balingian are probably at full peak gas generation window now, whereas the West Balingian kitchen areas are less mature for gas.

The greater abundance of gas than oil in East Balingian may be attributed to two main factors, as discussed earlier. One is that there are mainly humic and/or gas-prone marine source rocks in the Cycles I and II kitchen areas. Another factor is the contrasting maturation and trap-timing histories: the late timing of structuration in East Balingian caused the trapping of only the later products of hydrocarbon generation. Most of the structures in East Balingian that are associated with the Anau-Nyalau Fault Zone, such as the A1, Cochrane D, Suai offshore, D14-A, D14-B, and D8 structures, were formed during Cycle VI times. Structuration seems to have occurred even later away from the fault zone. Hence, unlike at Cochrane-D and A1, Cycle VI is almost undeformed at Temana. There is also the possibility of gas coming from the more deeply buried source rocks, as at Cochrane-

2, where gas in the Cycle III reservoirs may have migrated from a Cycle II source.

Apart from siliciclastic reservoirs, carbonates are also potential targets in some parts of the Balingian Province. Cycle I carbonates have been drilled into in East Balingian, namely at A1.1X, Suai, Subis, Serunai, and Rebab. Gas was found in Suai-3, -4, and -5, while oil has been tested at Suai-3 at a rate of 28 BOPD. Suai-4 encountered limestones with 22%-27% porosity. Asphalt has been observed in fractures at A1.1X. The source rock for these hydrocarbons may have been from the deeply buried interbedded shales. The seal is provided by Cycles I/II shales. The upper Cycle III, Cycle IV, and Cycle V carbonates, the main targets in the Central Luconia Province, may also form potential plays in the NW Balingian subprovince, and in the northern part of East Balingian. An important gas play in the Cycle IV carbonates and Cycle V sands in the northwestern part of the province was confirmed by the J32.1 well, although this accumulation has significant CO_2 and N_2 contamination.

CONCLUSION

The Balingian Province of offshore Sarawak is a proven petroliferous region, with 4 oil fields currently in production and several other fields slated for future development. The province can be subdivided into an oil-prone western sub-province (West Balingian) and a more gas-prone eastern sub-province (East Balingian). Current production comes from 3 fields in West Balingian (D35, D18, and Bayan) and 1 field in East Balingian (Temana).

Hydrocarbon accumulations in the province occur mostly in structural traps formed during several phases of tectonic deformation that affected the area since Oligocene times. In West Balingian, the traps are formed by NW-trending folds, thought to be related to dextral wrench movements associated with the West Balingian Line during the Oligocene to Pliocene. The D35 and D18 fields lie along one of the major NW-SE fold trends in the western subprovince. Structures in East Balingian have a NE-SW trend, and are thought to have resulted from sinistral wrench deformation during the Late Miocene to Pliocene. Major anticlinal trends, some with reverse faults, occur close to the Anau-Nyalau thrust fault zone. These include the producing Temana oil field. The West and East Balingian sub-provinces are separated by a series of major depocentres that are aligned in a N-S trend (the Acis and Balingian sub-basins). These depocentres are thought to be the kitchen areas that generated the hydrocarbons in the surrounding structural traps.

The main reservoirs are Cycles I, II, and III (Upper Oligocene to Lower Miocene) fluvial sands of variable reservoir quality. The source rock type changes gradually from west to east, in response to the changing palaeogeography. In West Balingian, the main source rocks are oil-prone coals and organic-rich shales deposited in a lower coastal plain setting. Source rocks become increasingly marine-influenced and increasingly gas-prone towards the east, with coals and organic-rich shales becoming less abundant. There is evidence to suggest, however, that oil-prone marine shales occur locally in the east.

The variation in source rock quality and distribution in the Balingian Province is reflected in the different types of reservoired hydrocarbons. The hydrocarbons in West Balingian tend to be moderately waxy oils, with a prominent land-plant biomarker distribution. In contrast, the light oils and condensates from East Balingian have biomarker distributions that indicate only low contributions from higher plants.

Factors other than source rock quality are also responsible for the differences in hydrocarbon quality and distribution between East and West Balingian. The timing of hydrocarbon generation relative to structural development is significant. Structuration in the west pre-dates hydrocarbon generation and, therefore, traps were already formed when the early generated oils were being expelled. To the east, however, traps were charged by late-generated hydrocarbons, such as light oils and condensates, because structural development post-dates the main oil generation phase.

REFERENCES

Agostinelli, E., Mohamed Raisuddin Ahmad Tajuddin, Antonielli, E. and Mohamad Mohd Aris, 1990. Miocene-Pliocene palaeogeographic evolution of a tract of Sarawak offshore between Bintulu and Miri. Bulletin of the Geological Society of Malaysia, 27, 117-135.

Ahmad Said, 1982. Overview of exploration for petroleum in Malaysia under the Production Sharing Contracts. Proceedings of the Offshore South East Asia Conference, 9-12 February 1982, Singapore. (preprint).

Almond, J., 1991. Reservoir geological/sedimentological evaluation of the Cycle II sediments in the Balingian Province. Part One. Sarawak Shell Berhad, Unpublished internal report ER:SSB:3:91-10.

Almond, J., Vincent, P. and Williams, L.R., 1990a. The application of detailed reservoir geological studies in the D18 Field, Balingian Province, offshore Sarawak. Bulletin of the Geological Society of Malaysia, 27, 137-159.

Almond, J., Mohd Reza Lasman, Vincent, P. and Williams, L., 1990b. The application of integrated 3D seismic and reservoir geological studies in a complex oilfield, D18 Field, Sarawak, Malaysia. Proceedings of the 8th Offshore South East Asia Conference, 4-7 December 1990, Singapore, 47-60. (OSEA 90185)

Awang Sapawi Awang Jamil, Mona Liza Anwar and Seah, E.P.K., 1991. Geochemistry of selected crude oils from Sabah and Sarawak. Bulletin of the Geological Society of Malaysia, 28, 123-149.

Chiu, S. K. and Mohd Khair Abd Kadir, 1990. The use of SAR imagery for hydrocarbon exploration in Sarawak. Bulletin of the Geological Society of Malaysia, 27, 161-182.

Dolivo, E., 1991. Hydrocarbon habitat of the offshore Balingian and SW Luconia provinces, Sarawak, East Malaysia. Sarawak Shell Berhad, Unpublished internal report. ER: SSB:3:91-04.

Doust, H., 1981. Geology and exploration history of offshore central Sarawak. In: Halbouty, M.T., ed., Energy Resources of the Pacific Region. American Association of Petroleum Geologists, Studies in Geology, 12, 117-132.

Ismail Che Mat Zin and Swarbrick, R.E., 1997. The tectonic evolution and associated sedimentation history of Sarawak Basin, eastern Malaysia: a guide for future hydrocarbon exploration. In: Fraser, A.J., Matthews, S.J. and Murphy, R.W., eds., "Petroleum Geology of Southeast Asia". Geological Society of London Special Publication, 126, 237-245.

OXY, 1989. North Borneo Regional Study: Offshore Sarawak Occidental Malaysia Inc. Unpublished Report, ER: OXY: 3:89-01

Shell, 1994. SK5 Exploration Data Book 1995. Sarawak Shell Berhad, Unpublished Report.

Scherer, F.C., 1980. Exploration in East Malaysia over the past decade. In: Halbouty, M.T., ed., Giant Oil and Gas Fields of the Decade 1968 - 1978. American Association of Petroleum Geologists Memoir, 30, 423 - 440.

Swinburn, P., 1994. Structural styles in the Balingian Province, Offshore Sarawak. Abstracts of American Association of Petroleum Geologists International Conference & Exhibition, Kuala Lumpur, Malaysia, 21-24 August 1994, American Association of Petroleum Geologists Bulletin, 78, 62.

Swinburn, P., Burgisser, H. and Jamlus Yassin, 1994. Hydrocarbon charge modeling, Balingian Province, Sarawak, Malaysia. Abstracts of American Association of Petroleum Geologists International Conference & Exhibition, Kuala Lumpur, Malaysia, 21-24 August 1994, American Association of Petroleum Geologists Bulletin, 78, 62.

Wan Hasiah, A., 1997. Common liptinitic constituents of Tertiary coals from the Bintulu and Merit Pila coalfields, Sarawak, and their relation to oil generation from coal. Bulletin of the Geological Society of Malaysia, 41, 85-94

Wan Hasiah, A., Mohamad Jamaal, H. and Abolins, P. 1995. Aspects of oil generation from coals: A Sarawak case study. The importance of exsudatinite and variations in organic facies characteristics. Abstracts or the Geological Society or Malaysia Petroleum Geology Conference 1995, Kuala Lumpur, Malaysia, 11-12 December, 1995.

Chapter 15

Central Luconia Province

*Mohammad Yamin B. Ali and
Peter Abolins*

INTRODUCTION

The Central Luconia Province, offshore Sarawak, is bounded to the east and west by the Baram and West Luconia deltas, respectively, and to the south by the compressive Balingian Province (Fig. 15.1). The Central Luconia Province is a broad and stable continental shelf platform, characterised by extensive development of Late Miocene carbonates. More than 200 carbonate buildups have been seismically mapped (Fig. 15.2) and some 65 buildups have been tested. To date, about 20 carbonate buildups are proven to contain commercial quantities of non-associated gas. With a total of more than 40 TSCF of gas initially in place and over 30 TSCF ultimate recovery, the Central Luconia Province constitutes about 40% of the total non-associated gas reserves in Malaysia.

The regional geology of onshore Sarawak was earlier studied by Liechti et al (1960) and the geology and stratigraphy of the offshore area were first published by Doust (1981) and Ho (1978). These were followed by Epting (1980, 1988) and Aigner et al. (1989) who worked particularly on carbonates. This chapter is a review of the geology and hydrocarbon potential of the province based on previous works and the results of recent exploration efforts.

EXPLORATION HISTORY

Sarawak Shell Berhad, the first oil company operating in Sarawak, has been engaged in hydrocarbon exploration since 1908. Their exploration activities started on land on the Baram Delta area. Early success was achieved with the discovery of the Miri Field in 1910. The focus of exploration activities remained onshore until offshore technology was introduced in the late 1950s. Between the two World Wars (1914 and 1945), land surveys were carried out in the coastal areas of Sarawak. In 1938, a number of wells, such as Mukah and Teres, were drilled on anticlinal structures where surface oil seeps occurred. After the 2nd World War, the entire area was re-mapped with the aid of aerial photographs. Since then, it was realised that many of the structures are highly faulted. Although many oil shows were recorded, none of them proved to contain commercial quantities of hydrocarbons. Exploration in the onshore area was halted in 1956 after the drilling of the Balingian-3 well.

With the introduction of marine seismic surveys in the late 1950s, exploration activity moved offshore onto the Sarawak shelf. The first regional grid seismic survey was limited to the shallow water area up to 100 km offshore. This survey revealed several anticlinal structures equivalent to those that were tested in eastern Balingian. In 1961, the Cochrane-1 well was drilled to test the first offshore structure. In 1962, many other prospects were tested, and oil was found in coastal plain sands in the Temana, Patricia and South Acis structures. In 1965 and 1966, reconnaissance digital seismic surveys were extended out to the shelf edge and it became possible to study the structure and stratigraphy of the entire Sarawak shelf. It was at this time that the offshore province of Central Luconia, dominated by carbonate buildups, was recognised. The first attempt to test the carbonate potential was made on the F1 structure, which was drilled down to Cycle III in 1968, followed by several other carbonate structures. Between 1968 to 1981, Sarawak Shell drilled 46 carbonate structures and encountered 27 gas reserves bigger than 0.5 TSCF. A number of major gas discoveries, such as E8, E11, F6 and F23, and several indications of oil rims, such as E1, E6 and M1, were recorded. Most discoveries have been made in platform-type buildups in the central part of the province. Some gas columns, occasionally reaching down to the structural spill-point, have also been found in pinnacle-type buildups. Hydrocarbons were also found in clastic sequences below the carbonates,

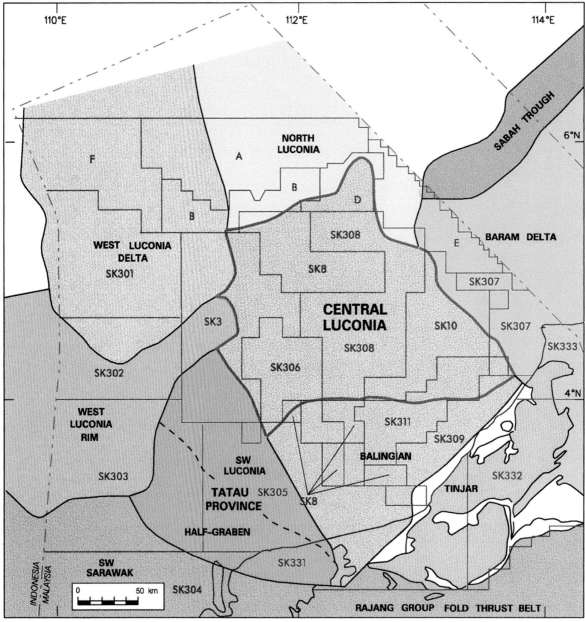

Fig. 15.1 Location of Central Luconia Province relative to other major provinces of northern Borneo.

particularly in the J area, SW Luconia Province.

After testing the J9 buildup in 1982, Shell ceased exploration activities in the Central Luconia Province. This was partially due to a low demand for gas. Shell relinquished most of the blocks within the Central Luconia Province and retained only the discovered fields. When the demand for gas increased in the late 1980s, PETRONAS launched several Liquefied Natural Gas (MLNG) projects based in Bintulu to cater for the industrial needs nationwide. To fulfill this demand, several new acreages were opened under new Production Sharing Contracts.

In 1992 Occidental Petroleum Malaysia Limited (OXY) was awarded the SK8 block in NW Central Luconia. After shooting thousands of line kilometres of seismic, OXY made several gas discoveries in the carbonates, such as in the Bijan, Jintan, and Serai structures (Fig. 15.2). These carbonate structures are proven to contain commercial quantities of gas, which will be supplied to the MLNG3 project which is due to operate in the year 2003. To further evaluate the pre-carbonate potential, Jintan Deep-1 was spudded in the middle of 1998.

STRUCTURAL FRAMEWORK AND STRATIGRAPHY

The Central Luconia Province is located between an extensional area in the north and a compressive realm in the south. This area is

Fig. 15.2 The major carbonate buildups of the Central Luconia Province.

characterised by southwest to northeast trending submarine plateaus and elongated troughs. Sea floor spreading in the South China Sea Basin during the Oligocene to Middle Miocene affected the continental crust to the south. The deepening of the South China Sea Basin, and its opening to the southwest, allowed marine currents to supply large amounts of nutrient-rich water to the Sarawak shelf and enabled the extensive growth of Middle to Late Miocene carbonate buildups. Contemporaneous crustal extension in the Central Luconia Province resulted in the development of a horst-graben pattern, which controls the size and distribution of these buildups. Large platform-type buildups developed on highs, whereas pinnacle-type buildups formed in adjacent elevated blocks within the basinal areas where the subsidence is stronger, and the distance from the source of clastic materials is closer. The development of some of the buildups was also controlled by the re-activation of thrust faults that occur during carbonate deposition (eg. the E8 buildup). The southwest-northeast alignment of the buildups, especially in the central and eastern part, probably reflects rift-induced structural trends. A geological profile across the Central Luconia Province is shown in Fig. 15.3.

The Central Luconia Province has undergone several episodes of structural deformation. The dominant structural feature in the southern part of Central Luconia is the central ridge, which trends NNE-SSW and plunges gently to the NNE as a series of tilted fault blocks. This ridge is bounded to the west by extensional faults whereas to the east and southeast, the ridge is relatively unfaulted. The northern part of the province is dominated by the Jintan and G2 highs, separated by a heavily faulted graben that extends to the south to form the F19 graben.

The Central Luconia Province has also undergone several episodes of sedimentation. A number of individual structural elements had a long-lived effect on sedimentation on the platform, and have directly affected the geometry of the pre-carbonate siliciclastics, the carbonate buildups, and the overlying siliciclastics. During Cycle I times, the area underwent an early synrift graben fill whereby deepwater argillaceous and shallow marine siliciclastic succession were deposited. This was followed by a late phase of synrift sedimentation throughout Cycles II and III times during the opening of the South China Sea. Continuous subsidence and formation of half-grabens resulted in widespread Middle to Upper Miocene carbonate deposition during Cycles IV and V times. This was eventually terminated by the influx of siliciclastic sediments derived from the uplifted Rajang Fold-Thrust Belt during Cycles V to VIII times. Figures 15.3 and 15.4 show the general stratigraphy and the stratigraphic cross-section of the Central Luconia Province.

Carbonate deposition in the Central Luconia Province started during the Early Miocene (Late Cycle III times) and continued until the present day in deeper water areas. Most of the Cycle III carbonates were deposited as localised discontinuous banks that became tight argillaceous limestone stringers encased in siliciclastic sediments. During Cycles IV and V times, carbonate production exceeded siliciclastic deposition within the Central Luconia Province, resulting in the deposition of extensive carbonate buildups.

The ages of the carbonate buildups were inferred from the dating of the under- and overlying clastic sequences, as the biostratigraphic marker species are not well preserved within the carbonate sequences. Because conventional biostratigraphy, based on foraminifers and nannoplanktons, is not feasible within the carbonate interval, strontium isotope stratigraphy has been used to accurately date the carbonates. The Sr-derived ages indicate that the carbonates were deposited as early as 20.5 Ma (Cycle III) and have continuously developed until the present day in deeper water areas. The majority of the Cycle V carbonates were deposited during the Messinian and Tortonian stages, between 10 and 5 Ma ago.

EVOLUTION OF THE CENTRAL LUCONIA PLATFORM

The continental shelf of Sarawak is the easternmost segment of the Sunda Shelf. The shelf is very broad, exceeding 300 km from coast to shelf edge, and exhibits a relatively smooth and gentle topography. Landwards, the Sarawak shelf is bounded by the Rajang Fold-Thrust Belt (Chapter 5), which was caused by the South China plate subducting beneath Borneo from the Late Cretaceous to the Late Eocene in Sarawak. This subduction is thought to have continued in Sabah through the Late Oligocene to Early Miocene. The southern and eastern parts of this subduction complex became the source of sediments for the basin. Major uplift and deformation took place during the Eocene, and a series of episodic uplift and erosion, continuing through to the Pliocene, resulted in the deposition of a thick succession of siliciclastic sediments in the Sarawak Basin.

The tectonics and sedimentation in offshore Sarawak are largely controlled by the continuous opening of the South China Sea Basin from the Middle Oligocene to the Middle Miocene. The Luconia Shoal (platform) is a micro-continental block bounded by transform faults, which separate it from the Baram Delta Province to the east and the Rajang (W Luconia) Delta Province to the west. The area remained stable throughout the Tertiary, resulting in carbonate deposition and reef growth, contemporaneous with clastic deposition in the Baram and Rajang deltas. The basin fill consists of several kilometres of sediments of Oligocene to Recent age, ranging from coastal plain to deeper marine sequences, representing 8 regressive depositional cycles.

Middle Miocene regional subsidence and extensional tectonics affected the lower part of the sedimentary sequence. In the southern part of the Sarawak shelf and onshore areas structural features indicate uplift and compression; the intensity of deformation decreases away from the coast. Central Luconia was not affected by compressional tectonics and major structuring was restricted only to block-faulting. The structural elements within and adjacent to the Central Luconia Province are shown in Fig. 15.3.

HYDROCARBON OCCURRENCES AND PLAY TYPES

The proven oil and gas reserves of the Central Luconia Province testify to the existence of at least one effective petroleum system. Although

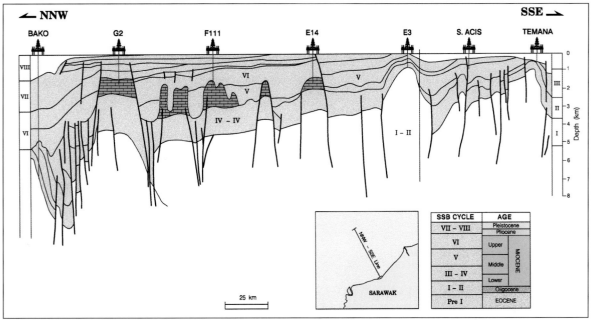

Fig. 15.3 A schematic NNW - SSE cross section across the continental shelf of offshore northwest Sarawak.

the major hydrocarbon accumulations are trapped within Cycles IV/V/VI carbonates, several other habitats occur. Four main proven play types are known in the Central Luconia Province.

1) Cycles IV/V/VI Carbonates.
2) Cycles I/II Pre-Carbonate Siliciclastics.
3) Post-Carbonate Cycles V/VI Siliciclastics
4) Cycle II Carbonates.

Cycles IV/V/VI Carbonates

Cycles IV/V/VI carbonates possess total gas reserves of 24 TSCF in 17 buildups, including 8 that contain oil-rims. This carbonate play contributes to about 40% of the total gas reserves in the country. The carbonate reefs developed on top of a transgressive carbonate ramp (transgressive systems tract) that were either terminated by drowning during a sea level rise or subsequent influx of regressive clastic sediments (highstand systems tract). The growth rate of the buildups was controlled by the rate of basin subsidence. Flat-top, platform-like, buildups developed over pre-carbonate highs, while pinnacle buildups formed in the basinal areas overlying the deep half grabens. Mid-cycle shales commonly form excellent top and lateral seal for the reef plays, while regressive clastic wedges and basal trangressive sands constitute potential seal rocks.

Pre-carbonate closures act as focal points for hydrocarbon migration, and the overpressure and faulting in Cycle III cap rocks trigger re-migration into the overlying reefs.

The Cycles IV and V carbonates (both platform and pinnacle-types) are the main reservoir and represent about 80% of total reservoirs in the Central Luconia Province. This type of reservoir mostly contains trapped non-associated gas, as in the F6, F23, E8 and E11 buildups. Some gases are associated with oil rims, as found in the E1, E6, M1 and M3 buildups. The hydrocarbons can be trapped in various conditions; normal to near hydrostatic pressure (eg. E8, E11, F13), low to mild overpressure of 140 to 615 psi (eg. B11, F23, G8, G1, K11, F6, F28, Bijan), or high overpressure of 1300 to 1376 psi (eg. G7, Jintan, K5).

Cycles I/II Pre-Carbonate Siliciclastics

Cycles I and II sediments represent the early synrift graben fill and late-phase syn-rift sedimentation, respectively. The depositional environments and distribution of these sediments are quite complex and highly variable as a result of syn-sedimentary tectonics. Although this has an impact in terms of reservoir, seal and source rock distribution, the potential for this type of play is still high within the area.

The possible types of traps within the Cycles I/II pre-carbonate are pre-carbonate palaeohighs, graben depocentres, and tilted fault blocks. In addition, the potential for stratigraphic and turbidite plays is high, turbidite plays being particularly likely to the north.

The pre-carbonate sequences can be potential reservoirs within the Central Luconia Province. The occurrence of oils within the pre-carbonate reservoirs of Cycles II and III is limited. Stratigraphically trapped hydrocarbons are seen in Bunga Pelaga, Lada Hitam, Bijan, Selar Marine and J4 structures, while structurally trapped hydrocarbons occur in Balai and Bentara. The hydrocarbons are trapped at different pressure regimes, ranging from normal hydrostatic to highly overpressured conditions. The pre-carbonate sequence therefore has a potential, but remains under explored.

Post-Carbonate Cycles V/VI Siliciclastics

This play type is relatively unexplored. Wells that have tested it include F13, E11, E18, E14, E8, F19 and Undang-1. The play consists of a combination of highstand deltaic and shelf sandstones and lowstand turbidites, which can be structurally or stratigraphically closed. The Cycle V deltaic deposits in F13, E11 and E18 areas are structurally closed and contain some 2 TSCF of gas. Reservoirs appear silty but porosity is good (up to 34%). The E6 field has approximately 75 m of Cycle V deltaic sandstones with about 20% porosity and Undang has 290 m of Cycles V/VI clastic sediments with similar porosity. The post-carbonate, dip-closed clastics over F19 and E8 were found to have poor porosity.

Seals for the post-carbonate clastics and turbidite sandstone are provided by transformational shales. These plays are expected to be charged by re-migrated hydrocarbon that leaked out through the vertical faults from nearby Cycles IV/V carbonate reefs.

Substantial quantities of oils and gas are believed to have been flushed out of reefs into adjacent post-carbonate clastics through thief horizons located on reef flanks. The discovery of 2 TSCF of gas in clastics above the E11, F13 and E18 reefs confirmed the prospectivity of the post-carbonate clastics as good reservoirs.

Cycle II Carbonates

Transgressive Cycle II shelfal limestone and patch reefs provide an additional play within the Central Luconia Province. The play is still unexplored and only penetrated by the G2.1, G10.1, Rebab-1 and Jintan Deep-1 wells. They constitute potentially attractive targets when sealed by Cycles II/III shales. The porosity has not been well established due to lost recovery during drilling. The limestones overlying Cycle I clastics are the best potential reservoirs. In the G10.1 area, Cycles II/III are represented by reefal limestones but they are not very well sealed.

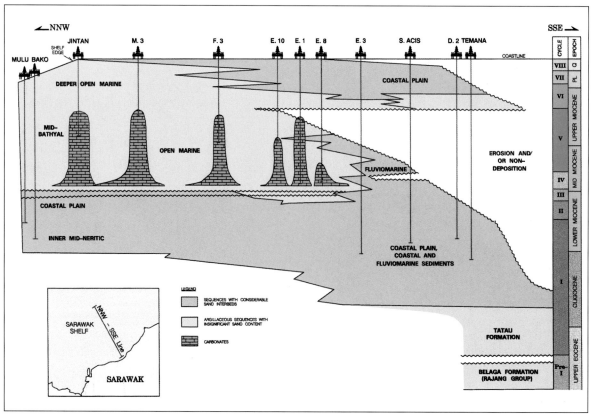

Fig. 15.4 A stratigraphic cross section over the Central Luconia Province.

SPECIAL TOPIC ON CARBONATE RESERVOIRS

Buildup Geometry and Morphology

The growth pattern, size and distribution of the Central Luconia carbonates were mainly controlled by a combination of relative topographic uplift of the substrate and monsoonal conditions during the Miocene. The horst-graben pattern dissecting Central Luconia determined the shape of the carbonate tops. Large platform-type buildups developed on highs, whereas the pinnacle-type buildups formed in basinal areas where subsidence is strong. The monsoonal weather patterns produced an elongated and asymmetrical fringing reef morphology with a steep slope at one end (windward) and gentle slope at the other (leeward). These layers form different flank morphologies and produce strong seismic reflections. The size of the buildups vary between about 3-8 km wide and 10-20 km long.

The morphology of the Central Luconia carbonates can be classified into two major types (Fig. 15.5):

(a) *Pinnacle-type buildups* - Pinnacles are high relief, conical shape buildups located at the margin of the regional highs or in the basinal areas. They are characterised by multiple crests and very steeply dipping flanks associated with carbonate stringer development. The carbonate tops are generally poorly defined seismically, and the flanks are often complicated by the steepness and development of stringers. E6 and E11 are examples of pinnacle-type buildups.

(b) *Platform-type buildups* - Platforms are large, elongated and fairly flat-topped buildups located on fault-bounded regional highs. They are characterised by a single crest and asymmetrically dipping flanks. The windward flanks of these buildups are steeper than the leeward flanks. The morphology of the buildups is well expressed by modern seismic. The F23 and F6 are examples of platform-type buildups.

Carbonate Architecture and Depositional Cyclicity

An integrated approach, using seismic, wireline log, and core data, reveals possible eustatic signals in the Central Luconia carbonates. Several scales of cyclicity and depositional patterns can be recognised for both types of buildups (Caline and Mohammad Yamin Ali, 1991; Caline et al., 1992).

(a) *Layer-cake features in seismic* : Layer-cake seismic character is particularly apparent in platform-type buildups (Fig. 15.5). In some seismic lines karst topography can be clearly seen, indicating a regional subaerial exposure of the buildup. Similar layer-cake architecture extends into the inter-build-up areas. The buildup is made up of a series of porous and non-porous carbonates. The deposition of sediments at the margin was enhanced during sea level lowstands. This would form different types of flank morphology at the margin of the buildups.

(b) *Elementary sequence and parasequence set in core* : Examination of continuous cores indicates that the reservoir interval consists of a vertical stacking of shallowing upward sequences separated by sharp lithological contacts. A typical sequence consists of a 0.3-1.0 m thick argillaceous algal limestone that is overlain by a 1-10 m thick coral or foraminiferal limestone. The lithological variations between these two units are further enhanced by a contrast in reservoir properties. The bluish algal limestone forms a tight bed whereas the overlying whitish/yellowish limestone is porous. Fossil assemblages indicate that the algal beds were deposited in an open marine environment. The foraminiferal and coral beds represent deposition in a reefal environment (reef-front, back-reef, lagoonal). Each elementary sequence results from a gradual shallowing of the depositional environment; the deeper algal limestone being overlain by shallower coral limestone. The elementary sequences are correlateable between the cored wells in each buildup and form the building block of the larger scale depositional cycles. These periodic sequences are believed to be caused by climatic fluctuations associated with Milankovitch-scale orbital cycles (period <500,000 years). They correspond to the fourth- and fifth-order cycles defined in sequence stratigraphy. The relationship between eustatic sea-level changes and depositional cycles and subaerial exposures are given in Figs. 15.6 and 15.7. The details on how the depositional environment interpretation was done are illustrated in Fig. 15.8.

(c) *Depositional sequences* : The stacking of elementary sequences displays a cyclic depositional pattern, separated by subaerial exposure surfaces (sequence boundaries) and maximum flooding surfaces (mfs). These two surfaces are best identified in core. Subaerial exposure surfaces, which are caused by sea-level lowstands, have been identified in all the cored wells. These horizons are characterised by colour change due to soil development, intraformational brecciation, extensive patchy cementation, and vertical solution channels. The exposed horizons can be correlated with the uncored development wells, as these surfaces and the underlying vadose zones are marked by high gamma ray (GR) readings caused by high uranium content. Evidence of subaerial exposure can be found in some seismic lines where weak-developed terraces can be traced at the margin of the buildups.

Each depositional cycle consists of transgressive and regressive sequences and records the succession of 4 basic phases of carbonate growth (Fig. 15. 6). The buildup phase develops during moderate rates of relative sea-level rise, the build-in phase corresponds to the maximum rate of relative sea-level rise, the build-out phase is related to a slow rate of relative sea-level rise and fall, and finally, subaerial exposure occurs during relative sea level fall. The maximum flooding surfaces can be identified by the faunal content of the carbonate lithofacies. A faunal assemblage revealing a relatively deep open marine environment is a good indicator of a maximum flooding surface in a transgessive systems tract.

An integrated correlation between wells within the buildup and the depositional cyclicity is given in Fig. 15.9.

Carbonate Rock Types

The Central Luconia carbonates can be subdivided into 6 basic types based on their diagenetic features.

i) Chalkified limestone
ii) Mouldic limestone
iii) Tight argillaceous limestone
iv) Sucrosic dolomite
v) Mouldic sucrosic dolomite, and
vi) Overdolomite.

Although these rock types are different from the original (deposited) rock types, the distribution of porous and tight carbonate within the reef complex is mainly controlled by the primary environment of deposition.

Diagenesis and Reservoir Quality

The Central Luconia carbonates have undergone extensive early diagenesis under surface-related physico-chemical conditions. Burial diagenesis is less important and is restricted to argillaceous intervals only. A leaching phase, that may relate to subaerial exposure, occurred later than the formation of stylolites. The fresh water leaching process also caused the chalky appearance.

Although all porosity encountered in the Central Luconia carbonates is secondary, the distribution of porous and tight carbonates within each buildup is still controlled by the primary depositional environments (Epting, 1980; Mohammad Yamin Ali, 1994). Sediments deposited in protected and reefoid environments were preferentially converted by dolomitization to chalky and mouldic limestones, and sucrosic and mouldic-sucrosic dolomites. These fresh

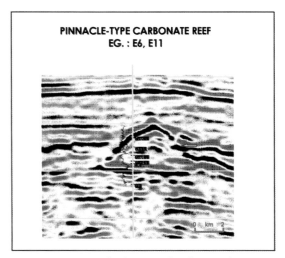
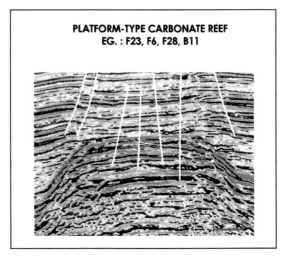

Figure 15.5 Types of carbonate reefs in the Central Luconia Province.

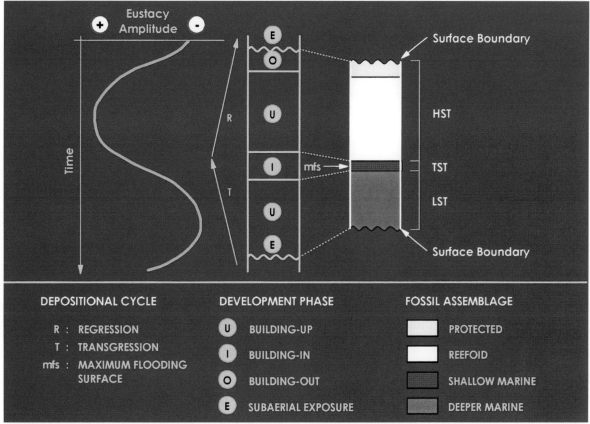

Fig. 15.6 An idealised eustatic cycle and its relationship with depositional cycle and subaerial exposure.

water stabilization and dolomitization processes are extensive at the centre of the buildups and possibly decrease towards the margin of the buildups.

The sediments deposited in open marine off-reef and bank (argillaceous limestone), were deposited in deeper water environments with an abundance of impurities. They were then subjected to mechanical compaction and pressure dissolution (that eventually developed into stylolites) that destroyed the porosity. The mechanical compaction that occurred at the very early stage turned them into low permeability barriers to subsequent flow of fresh water and dolomitizing brines. In contrast, the sediments deposited in protected and reefoid environments were preferentially transformed by dolomitization and fresh-water leaching into mouldic and chalkified limestones, and mouldic- and mouldic-sucrosic limestones and dolomites. The dolomitization and fresh water leaching enhanced porosity within the buildups. As a result, there is a marked alternation of tight and porous rock layers that are distinctly observed in the seismics,

Detailed core studies indicate that each buildup in Central Luconia has experienced at least two distinct regional episodes of subaerial exposure related to sea-level lowstands. Subaerial exposures are related to the existence of layered dolomites that occur simultaneously within the buildups. Figures 15.9 and 15.10 show that dolomitization during periods of sea-level lowstand enhances the reservoir quality of the carbonates. The present lithofacies with good reservoir potential developed during the building-up phase, while the lithofacies with poor reservoir potential were deposited in the building-in phase (end of transgression) and building-out phase (end of regression) of each depositional sequence.

The depositional pattern of small-scale sequences (<0.3 m) to reservoir unit scale (>6 m) is dominantly controlled by the cyclic variation of sea level during the Late Miocene. The resulting stacking of alternately good and poor reservoir facies has been recognised in the core intervals and agrees with the predicted pattern derived from the eustatic signals.

Diagenetic Modelling

The diagenetic model indicates that all the porosity encountered in the buildups is secondary in origin. The changes in porosity distribution occurred at an early stage under surface-related physical-chemical conditions. Generally, diagenetic alteration that affected the carbonate during the burial stage are less important and are entirely restricted to the argillaceous intervals. The diagenetic history of the carbonates is summarised below where surface and burial diagenetic domains have been clearly differentiated. Similar observations were also made by Epting (1980) and Caline and Mohammad Yamin Ali (1991).

Surface Diagenesis

The simultaneous creation of secondary porosity and cementation of the primary pore spaces occurred during freshwater stabilization. Little or no cement was adddded during this process of calcium carbonate redistribution. The loss of metastable carbonate minerals at an early stage explains why the bulk of the carbonates resisted porosity destruction during burial and retained their overall good reservoir properties at present burial depths. Freshwater stabilization occurred during repeated subaerial exposure where lenses of rain water could penetrate deeply into the buildups. Most of the macroscopic porosity in the limestones is formed by moulds of dissolved aragonitic skeletal grains. In muddy sediments, microscopic dissolution occurred at the surface of the crystals leading to a chalkified texture. Simultaneously, stable rhombic calcite crystals were precipitated within primary and secondary pore spaces. Chalkification by freshwater fluids is supported by stable isotope analysis (average $\delta^{18}O$ (PDB) = -6 per mil, $\delta^{13}C$ = -0.5 per mil)

Dolomitization occurred intermittently during buildup growth and predominantly affected the protected and reefoid environments. Lagoonal mudstones and wackestones were often transformed into sucrosic dolomite with excellent reservoir characteristics. Two dolomitization models have been envisaged : a meteoric/marine mixing model related to subaerial exposure, and an evaporitic model during low sea levels and dry seasons. Stable isotope measurements do not conclusively favour either of the two models (average $\delta^{18}O$ (PDB)=-0.5 per mil, $\delta^{13}C$ (PDB) = +1.2 per mil). Tight dolomite intervals are present in the Cycle IV carbonates. These "overdolomites" resulted from the cementation of the intercrystalline pore space by a second generation of dolomite rhombs. Carbon isotope results ($\delta^{13}C$ (PDB) = -12 per mil) suggest that bacterial degradation of organic matter incorporated in the muddy sediments may supply additional CO_2 for dolomite precipitation.

Burial Diagenesis

Burial diagenesis comprises the porosity-destroying processes of mechanical/chemical compaction and the porosity-creating, late leaching process. Mechanical compaction started early in the burial history and has exclusively affected argillaceous facies that were rich in ductile clay. During deeper burial, compaction was eventually replaced by pressure solution, which resulted in nodular bedding and horse-tail texture. As surface diagenesis has had little effect upon these argillaceous intervals, these eventually became tight limestones.

Pressure dissolution also occurs in non-argillaceous carbonates and results in the formation of stylolites that commonly accumulate insoluble residues such as clay seams and structureless organic matter. Although stylolite frequency increases with depth, it only marginally affects the reservoir quality of the non-argillaceous intervals.

The concentration of porosity in the vicinity of stylolites is a common feature in the buildups. A leaching phase apparently occurred later than the formation of the stylolites. The chalky appearance of many intervals is caused by the same leaching process. Scanning Electron Microscope (SEM) images show that many stable calcite and, to a lesser extent, dolomite crystals have been corroded. Two fluid flow mechanisms have been proposed to explain the late leaching event :

1). Penetration of normal meteoric water into the buildups from the emergent surface. This would imply the occurrence of freshwater lenses that penetrated the stabilised carbonates to depths where pressure dissolution had already occurred.

2). Generation of CO_2-rich fluids during the thermal maturation of coals and organic-rich shales of the underlying Cycles I-III clastics. Intervals with good permeability were flushed preferentially, whilst in tighter rocks, porosity-creation is restricted to the vicinity of the stylolites, where apparent fluid circulation was concentrated.

Modelling of the Carbonates

The reconstruction of the buildups in Central Luconia has been successfully achieved using the BASMOD basin modelling programme (Aigner et. al., 1989). Quantitative modelling demonstrated that buildup type, reefal growth, period of subaerial exposure and final drowning can be satisfactorily reproduced using three critical parameters. Fluctuation of sea level during the Miocene appears to be the prime parameter that controls the buildup development. Secondary parameters include a constant tectonic subsidence, rate per buildup, and a variable carbonate growth potential to account for the reduction of carbonate production due to suffocation by clastics. Moreover, distribution and timing of subaerial exposure surfaces within the buildups could also be derived from this simulation. This should allow prediction of the early diagenetic overprinting on the carbonates and therefore a better understanding of the secondary porosity distribution which is responsible for the reservoir quality of these gas-bearing carbonates.

A good match between eustasy and platform evolution has been observed in F23, Jintan and other buildups in the Central Luconia Province. Application of sequence stratigraphy resulted in the recognition of three complete depositional sequences in the Miocene (as opposed to 6 sequences in Natuna (Rudolph and Lehmann, 1989). Seismic modelling has been used to map and predict reservoir distribution within the Luconia carbonates. It appears that the architecture of the buildups in Central Luconia is also related to the sea-level fluctuations during the late Miocene as shown in Fig. 15.7, and the internal reservoir properties can be partly predicted and quantitatively modelled.

Implications on Porosity Distribution

Although all porosity is diagenetic in origin, the distribution of porous and tight carbonates in the buildups is still controlled mainly by the primary environment of deposition. The sediments have undergone a complete reversal in terms of porosity development. Sediments deposited in protected and reefoid environments were preferentially converted to chalky and mouldic limestones and to sucrosic and mouldic-sucrosic dolomites with very good porosity. Both dolomitization and freshwater stabilization processes operated in the central part of the buildup and decreased in intensity towards the flanks. The open-marine off-reef and bank carbonates, originally deposited in deeper water and enriched with non-carbonate impurities, were subjected to mainly porosity-destroying compactional processes and produced bad quality reservoirs.

Multiphase dolomitization has affected most of the the Central Luconia buildups. The isotopic results indicate that the dolomitization of the F23 buildup occurred at around 5.5 Ma and between 5.5 and 8.0 Ma, respectively. Older phases of dolomitization have probably affected the E8 and E6 buildups at around 10 Ma and the F13 buildup at around 13 Ma. These dolomitization events seem to be related to subaerial exposures.

Factors Controlling Hydrocarbon Distribution

Undrilled carbonate buildups within the hydrocarbon-proven Central Luconia Province represent obvious targets for further exploration. Numerous major discoveries have been made in carbonates, firstly by Shell (eg. B11, E11, F23, F6) and more recently by OXY (eg. Jintan, Serai) demonstrating the large rewards possible. However, the success ratio in drilling the buildups has not been great, and sometimes the discovered volumes have not met expectations. An important issue that needs to be addressed therefore is what geological factors control whether a particular carbonate buildup is hydrocarbon bearing, and to what degree it is filled.

There are several risk factors to consider. These are focusing, crest-flank faults, thief beds and overpressure. (Nippon Oil, 1996)

Focusing

The presence of a four-way dip closure in the pre-carbonate sequence underneath a prospective buildup will act as a focal structure and improve the likelihood of the buildup being hydrocarbon-bearing. For example, F13 and E11 are large gas fields which are underlain by large focal areas. In contrast, buildups such as Pudina and Kunyit are dry due to the lack of focusing below the carbonates.

Crest and Flank Faults

Faulting has long been recognised as a major cause of leakage from hydrocarbon traps. Carbonate buildups in the Central Luconia Province are prone to faulting in crestal and flankal positions. Such faults represent a major risk factor controlling the amount of hydrocarbon within a particular buildup by providing a structurally high spill point, limiting the thickness of any hydrocarbon column present. The presence of a crestal fault in buildups such as

G2, G5, and Kantan probably explains why these buildups are dry. The presence of flanking faults in buildups such as F11 and F29, limits the thickness of the hydrocarbon column, despite the presence of focusing in the pre-carbonate sequence.

Thief Beds

Upper Cycle V sequences tend to be composed of sand-rich sediments and are widespread in the Central Luconia Province. These sandy intervals may act as thief beds to hydrocarbons. Whether or not these sandy intervals will act as thief beds, thereby draining carbonate buildups, depends upon the sedimentological/stratigraphic relationship between the sandy beds and the carbonate buildups. Several scenarios exist and explain why buildups such as E14 and E18 are dry, and buildups such as Selasih, F23, and E1 are only partly filled due to the presence and location of sandy thief beds.

Overpressure

The sealing capacity of a cap rock is controlled by a relationship between, among others, overburden, aquifer, and gas pressures. The presence of overpressure in a carbonate buildup will impact negatively on the maximum sustainable height of hydrocarbon column. In areas where overburden is thin, there is an increased likelihood of seal breach. A possible example of such seal breach is suspected in the Lang Merah buildup resulting in it being a water wet structure.

The presence of source and reservoir is obvious in the Central Luconia Province, as shown by the presence of several large hydrocarbon-bearing structures. As the source and reservoir are proven elements of the petroleum system, the prospectivity of a particular carbonate buildup in Central Luconia is influenced by the presence or absence of the risk factors outlined above.

SOURCE ROCKS

Source Rock Quality

Considering that the Central Luconia Province essentially co-evolved (at least depositionally, if not tectonically) with the Balingian Province, the Central Luconia Province represents the northward extension of the source rock belts seen in the Balingian Province. Chapter 14 provides a detailed account of the source rock development in the Balingian Province. The reader is therefore referred to Chapter 14 for a detailed account of the depositional setting of Central Luconia source rocks.

In brief, the most important source rocks are coals and carbonaceous shales deposited in a lower coastal plain setting during Cycle I and Cycle II times (Late Oligocene to early Miocene). In Balingian, Cycle III sources were generally considered unimportant, based in part upon their thermally immature nature. Considering the greater depths of burial that the Central Luconia sediments have experienced, compared to those in the Balingian Province, the potential for Cycle III to be an effective source rock is there, kerogen quality and organic richness provided.

As with the Balingian Province, in Central Luconia the lower coastal plain influence decreases eastwards, the depositional setting becoming increasingly marine. In addition, in the Central Luconia Province marine conditions probably also became increasingly more important to the north. In view of this, the prospectivity of large tracts of Central Luconia will depend upon either (1) the existence of good quality marine source rocks, or (2) intermittent progradation of lower coastal into an otherwise marine setting. The traditional view has been that all the hydrocarbons of Central Luconia have been sourced from lower coastal plain sediments and therefore the latter of the two source rock options has applied. As outlined later in this chapter, and also discussed in Chapter 14, the influence of marine source rocks is now thought likely.

Less is known of the true quality of Central Luconia Cycle I - Cycle III due to the lower number of well penetrations, compared to Balingian. This is because the primary reservoir target in Central Luconia are the Cycles IV/V carbonates; pre-carbonate penetration have therefore been few. A further problematic point is that the distinction of what is Cycle I and Cycle II (and often Cycle III) is not easy to make and they are therefore often considered together.

A striking feature of the penetrated Cycles I/II sequence in Central Luconia is the scarcity of coals. This is in contrast to the Balingian Province where coals are numerous and probably represent the main source rock. To date, this has not been the case in Central Luconia. Up to now, coals have only been observed in the very south of the province, such

Central Luconia Province

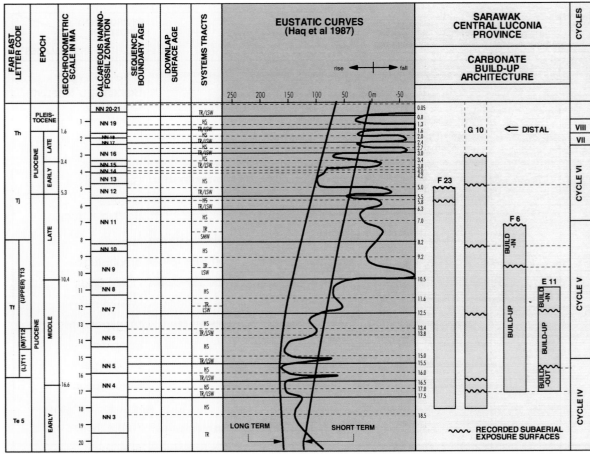

Fig. 15.7 Relationship between eustatic sea-level changes and subaerial exposures in Luconia Province during the Tertiary.

as at Bentara and Balai. The Bako-1 deepwater well drilled in 1996 by Mobil, however, penetrated Cycle II coal in the North Luconia Province, providing indirect evidence that they may also exist in Central Luconia but have not yet been penetrated. However, the Jintan Deep-1 well, drilled by OXY, penetrated no coals in the pre-carbonate sequence.

The source rock quality in Cycles I - III are summarised in Fig. 15.11 which shows plots of hydrogen index (HI, a kerogen quality parameter) versus T_{max} (a maturity parameter). These plots indicate, as expected, the progressive maturity increase from Cycle III through to Cycle I. They also reveal that, for any given Cycle, the Central Luconia sequence is generally of higher maturity that the corresponding Balingian sequence (Figs. 14.6 and 15.11). Figure 15.11 also demonstrates the lower quantity of coal in Central Luconia compared to Balingian, resulting in a greater dependence on shales as the main oil source in Central Luconia. If coals are present in Central Luconia they are deeply buried and would therefore be expected to be of high maturity (see next section) and would now only be generating gas.

Figure 15.11 shows that the hydrogen indices for Cycles I/II source rocks generally range from 100 to 200, occasionally rising to 250. This suggests therefore that much of Cycle I and Cycle II is generally gas prone with a limited potential for oil. Kerogen assemblages are predominantly Type III vitrinitic with only subordinate quantities of Type II, oil-prone, material. Perhaps the most interesting feature of Fig. 15.11 is the Cycle III plot. Cycle III is predominantly marine and yet, as shown, source rock quality does not deteriorate. This is contrary to traditional thought on Central Luconia source rocks. This Cycle III source rock quality represents the first argument for potential marine source rocks in Central Luconia. Other evidence is given in the section on hydrocarbon characterisation, which discusses how biomarker distributions in some condensates show a marine influence.

Source Rock Maturity

The HI-Tmax plots in Fig. 15.11 indicate, as expected, the progressive maturity increase from Cycle III through to Cycle I in the Central Luconia Province. They also reveal that, for any given cycle, the Central Luconia sequence is generally of higher maturity than the

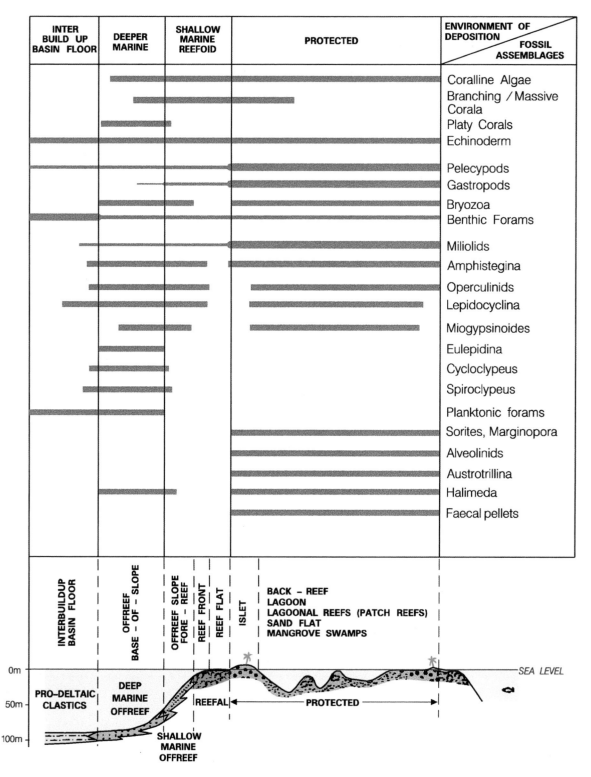

Fig. 15.8 Physiographic zones and depositional environments within the carbonate buildup, based on fossil assemblages.

corresponding Balingian sequence. The plots suggest that, on structure, Cycle I and Cycle II are presently within the oil window. This is confirmed by the vitrinite reflectance profiles for typical wells in Central Luconia (Fig. 15.12). Cycle III tends to be only marginally mature, at least on structure. It is therefore expected that at greater burial depth in the main depocentres, such as in the Central Graben, Cycle III will be of main oil window maturity, whereas Cycle I and Cycle II will be post mature. A regional impression of thermal maturity in the Central Luconia province can be seen in the maturity maps shown for the Balingian Province (Chapter 14), which support this interpretation.

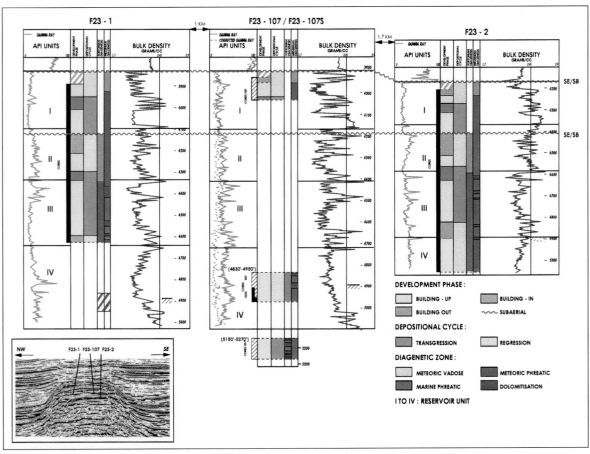

Fig. 15.9 Depositional and diagenetic cyclicity as observed in the F23 buildup.

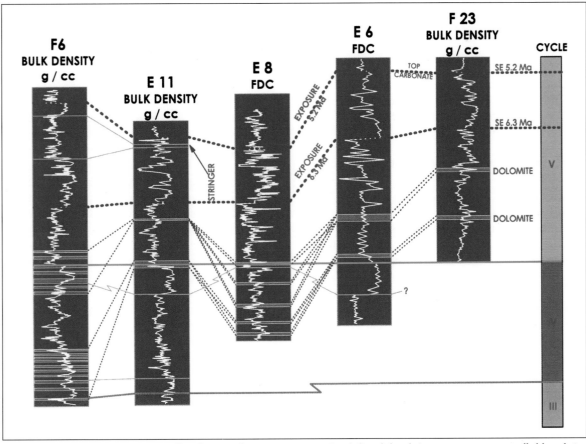

Fig. 15.10 Dolomites occur in every well in the Central Luconia Province. It is believed that their occurrences are controlled by relative sea-level falls.

HYDROCARBON TYPES

Hydrocarbons have been tested in several habitats in the Central Luconia Province as mentioned earlier (Hydrocarbon Occurrences and Play Types). The geochemical characteristics and mutual relationships of these hydrocarbon types are discussed by Abolins and Mona Liza Anwar (1996) and Abolins (1998) and are outlined below. As only little geochemical data exists on post-carbonate trapped hydrocarbons, they are

A variable Pr/Ph ratio, ranging from high (>4.0) in the west (suggesting an oxidising depositional environment), to low (<1.0) in the east indicative of anoxic conditions. The relatively low abundance of common higher plant biomarkers, such as oleanane and bicadinanes, is an important distinguishing feature, as many other Sarawak oils contain considerable quantities of these compounds. This is demonstrated in Fig. 15.14 which graphically demonstrates the terpane distributions of the oils studied here.

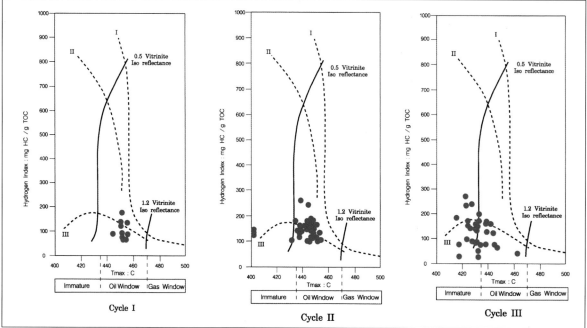

Figure 15.11 A cycle by cycle summary of Central Luconia source rock quality. Data collected from numerous unpublished reports.

not discussed. Furthermore, a good gas dataset is presently not available, and this section therefore concentrates on oils and condensates.

Carbonate–Reservoired Condensates

Condensates occur as gas accumulations in Middle Miocene–Pliocene reefs throughout much of Central Luconia. Examples of such condensates include Jintan-1, F23, Bijan-1, and B12. A typical example of biomarker distributions is shown in Fig. 15.13 for the Jintan condensate. Characteristic features include C_{20+} components in low abundance, high API gravities, bicadinanes and oleanane in low abundance, tricyclics present low to moderate abundance, high $C_{29}H/C_{30}H$ (occasionally exceeding 1.1, e.g. F23), low to high Pr/Ph ratio (decreasing eastwards), Tm/Ts <2.0, sulphur contents > 0.20wt%, and C_{27} sterane preference over C_{28} and C_{29}.

The low abundance of C_{20+} components and the high API gravity are in good agreement with one another and are typical of condensates.

The distribution of regular steranes, shown graphically in Fig. 15.15, reveals a slight preference of C_{27} steranes over C_{28} and C_{29} steranes; a feature commonly seen in marine oils. The high $C_{29}H/C_{30}H$ ratio and moderate development of C_{33} - C_{35} hopanes would be in line with such a suggestion being characteristic of marine-sourced oils. Such oils would normally be expected to possess quite high sulphur content. Although the condensates do possess the highest sulphur contents of the Sarawak oils/condensates, they are not particularly high, although this could be a consequence of the suspected high maturity of the condensates.

It is considered that these condensates display many features typical of marine-sourced oils, and that these features are stronger in the more easterly fields. Marine incursions across Central Luconia are thought to have begun in the Early Miocene. Cycle II could therefore have reached sufficient maturity to generate light oils and condensates. These biomarker distributions are the second, and perhaps most important, argument (the first being described earlier under source rock quality) that an effective marine source rock is present in the Central Luconia Province.

Carbonates – Reservoired Waxy Oil Rims

A number of carbonate reefs in Central Luconia contain a thin rim of waxy oil below the main gas/condensate reservoir. Available data suggest that such rims are geographically restricted to the western edge of SK8 and SK308. Examples of such waxy oil rims include M4.2 and E6.2. It is interesting to note the extremely similar biomarker distributions possessed by these two oils, despite considerable geographic separation. Typical biomarker distributions for these waxy oils are shown in Fig. 15.16. Characteristic geochemical features include a waxy gas chromatogram (GC) trace, a biomarker distribution dominated by bicadinanes, oleanane and other higher plant terpanes, tricyclics present in very low abundance or absent, $C_{29}H/C_{30}H = 0.5 - 0.7$, high Pr/Ph ratio (> 5), Tm/Ts > 2.0, sulphur contents < 0.20wt%, and slight C_{29} sterane preference over C_{27} and C_{28}.

The general waxiness would attest to a non-marine source for these oils. The high Pr/Ph ratio suggests the source of the oil was deposited in an oxidising depositional environment. The high abundance of land plant-derived biomarkers, such as bicadinanes and oleanane, is a clear indication of a major terrigenous input to the source of these oils. In fact, the bicadinane concentration is so high that they tend to swamp the regular steranes, although there is an indication that the regular steranes show a C_{29} preference over C_{27} and C_{28}, further supporting a major land plant input to the source. The presence of possible taraxastane, eluting just before the C_{31} hopanes, is also indicative of higher plant derived oils.

Overall, the biomarker distribution observed in these waxy oils is typical of a source deposited in a lower coastal plain setting.

Pre-carbonate Stratigraphically – Trapped Oils

Pre-carbonate, stratigraphically-trapped oils have been discovered by OXY in a small number of wells, such as Lada Hitam-1 and Bunga Pelaga-1, both of which contain moderately waxy oils. Figure 15.17 shows representative biomarker distributions of the Lada Hitam-1 oil, features of which include a moderately waxy GC trace,

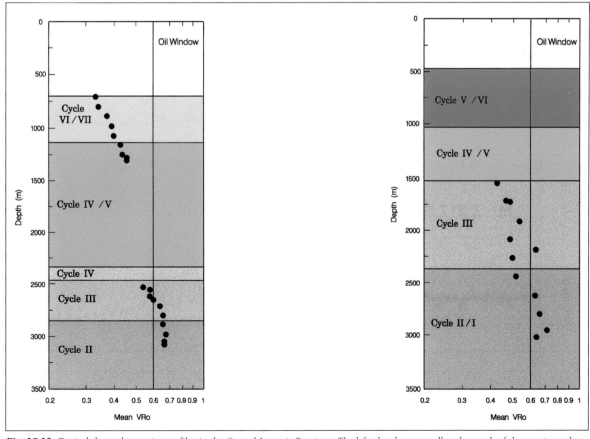

Fig. 15.12 Typical thermal maturity profiles in the Central Luconia Province. The left plot shows a well to the north of the province; the right plot shows a well to the south.

moderate to high concentration of bicadinanes, oleanane and other higher plant terpanes in fair abundance, tricyclics present in moderate abundance, $C_{29}H/C_{30}H = 0.7$-0.8, moderately high Pr / Ph ratio (3 - 5), Tm/Ts << 2.0, sulphur contents < 0.20wt%, slight C_{27} sterane preference over C_{28} and C_{29}.

A noteworthy feature of these oils is the presence of tricyclic terpanes, such compounds often being associated with marine environments of deposition. Such an interpretation is supported by the regular steranes showing a slight C_{27} preference. Interestingly, however, the moderately waxy nature, and the presence of moderate quantities of higher plant biomarkers, such as oleanane and bicadinanes, also suggest a considerable terrigenous input.

It is therefore considered that the stratigraphically-trapped pre-carbonate oils, at least at Lada Hitam, are probably derived from a source rock receiving a mixed marine/terrigenous input, a marginal marine setting being likely.

Pre-carbonate Structurally Trapped Oils

Structurally trapped oils exist in pre-carbonate Lower Miocene sequences, and have been tested by OXY, examples being Bentara-1 and Balai-1. Two main oil types have been recognised, both types being present in well Balai-1. The first oil type is waxy and the second is a light oil or condensate. Stratigraphically, the two oil types can be quite closely associated. Characteristic features of the light oil include bicadinanes in low abundance, oleanane and other higher plant terpanes in low abundance, tricyclics present in fair abundance, high $C_{29}H/C_{30}H$, high Pr/Ph ratio, Tm/Ts < 2.0, sulphur contents < 0.20 wt%, and a C_{27} sterane preference over C_{28} and C_{29}.

The low abundance of higher plant markers, and the C_{27} regular sterane preference, as mentioned earlier when discussing the condensates, are features tending to suggest a marine-influenced source rock for these oils.

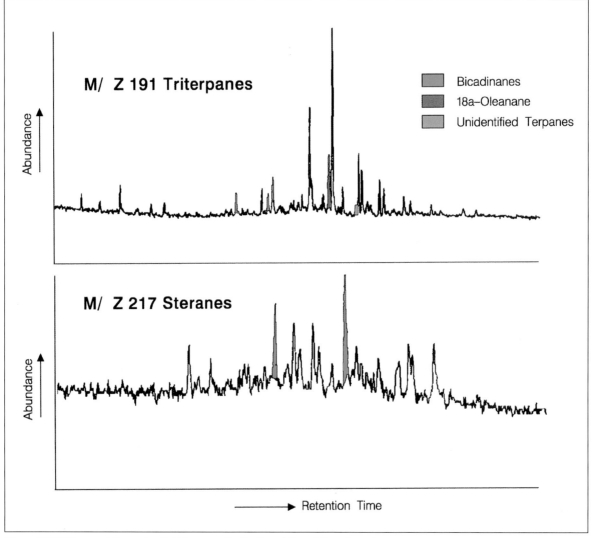

Figure 15.13 Characteristic geochemical biomarker distributions of Central Luconia condensates.

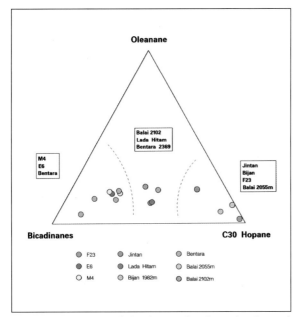

Fig. 15.14 A geochemical plot showing the relative abundance of land-plant biomarkers (oleanane and bicadinanes) relative to microbial biomarkers (C_{30} hopane). This figure demonstrates the compositional variation of Central Luconia oils and condensates.

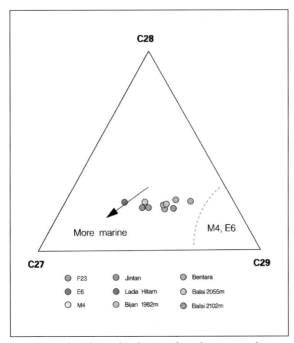

Fig. 15.15 The relative distribution of regular steranes showing an increasing marine influence for the more easterly wells.

Characteristic features of the waxy oils include bicadinanes in high abundance, oleanane and other higher plant terpanes in moderate abundance, tricyclics present in low abundance, $C_{29}H/C_{30}H = 0.5$-0.7, high Pr/Ph ratio (> 5), Tm/Ts > 2.0, sulphur contents << 0.20 wt%, and a slight C_{29} sterane preference over C_{27} and C_{28}.

In contrast to the light oils, the waxy oils display features indicating a major terrigenous input, examples being the high abundance of biomarkers such as bicadinanes and oleanane.

The presence of tricyclic terpanes may also reveal some marine influence thereby suggesting a mixed source input.

Oil Families

An oil family can be defined as a group of oils genetically related by derivation from the same source rock unit, regardless of the maturity of the source rock at the time of generation.

The preceding section has outlined and described the oils and condensates from the four proven play types recognised in the Central Luconia Province. Such play types are, of course, a consequence of structural geology, sedimentology, and the timing of hydrocarbon generation, and so the hydrocarbons they contain do not necessarily represent different oil families. For example, the pre-carbonate oils may represent one family of oils, whether they accumulate in structural or stratigraphic traps, the trap being controlled by geology rather than the geochemistry of the source.

An important issue to be addressed is the relationship between the gases/condensates trapped in the carbonate reefs and the waxy oil rims that occur in a number of these reefs. It is customary to assume that the waxy oil rims are the early generated equivalents of the gases and condensates, the gases and waxy oils therefore being genetically related. However, the characteristics of these two hydrocarbon types discussed in the previous section indicate some clear differences in their respective biomarker distributions. This is clearly seen by comparing Fig. 15.13 with Fig. 15.16 of the waxy oils (exemplified by M4.2) and the condensates (exemplified by Jintan-1). Such a comparison shows a clear difference between the waxy oils and the condensates. Such differences cannot be attributed to maturity differences between the waxy oils and the condensates. It is more likely that they are a result of different sources for the waxy oils and the condensates which therefore represent different oil families.

A similar situation exists in the pre-carbonate structurally trapped oils in the southern part of Central Luconia (e.g. Balai-1), in which a waxy oil and a condensate are intimately juxtaposed stratigraphically. There are considerable geochemical differences between the two oil types, differences which can only be attributed to two distinct source rock units, and not to differences in maturity of the same source rock. So, again, two distinct oil families are present.

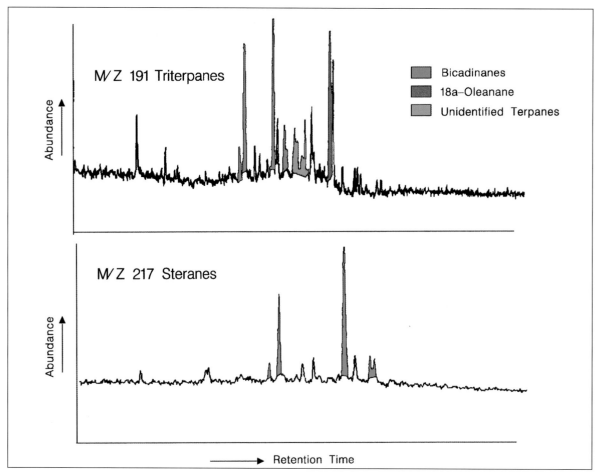

Fig. 15.16 Characteristic geochemical biomarker distributions of Central Luconia oil rims.

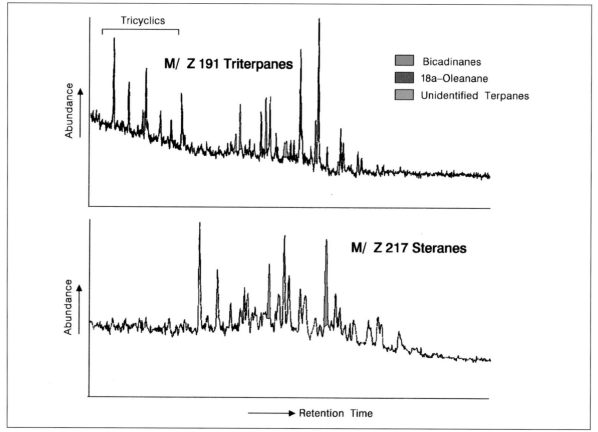

Fig. 15.17 Characteristic geochemical biomarker distributions of Central Luconia stratigraphically-trapped oils.

The geochemical relationship of some Central Luconia oils and condensates are shown in Fig. 15.14 in which three main groupings can be seen. It is considered, as discussed in the source rock discussion, that a major controlling factor on this grouping is the Late Oligocene/Early Miocene palaeogeography. If we consider a palaeocoastline running approximately northwest from Bintulu, with sea to the east and land to the west, then the above groupings agree well with the depositional environment interpretations made earlier for the oils in each play type; the West group oils consisting mainly of higher plant-derived, lower coastal plain-sourced condensates; the East group consisting predominantly of marine influenced condensates; and the Central group being transitional between the two other groups. The occurrence of Bentara oils in the West and the Central groups, and of Balai oils in the Central and East groups, is a consequence of a fluctuating coastline transgressing and regressing; depositional environments locally changing through time.

It is therefore feasible that the three oil families discussed above may represent a continuous spectrum of oils, each family representing a different source facies in the same sedimentary sequence as we move laterally offshore.

FUTURE EXPLORATION OUTLOOK

Because many of the structures within and above the carbonates have been tested, the future exploration outlook will focus more on stratigraphic traps and deeper reservoirs within the pre-carbonates. Although the pre-carbonate clastic deposition is highly variable, the distribution of reservoirs, seals, source rocks and migration pathway is primarily controlled by syn-depositional tectonics. Siliciclastic lithofacies seems to show a continuity from the more proximal in the south to more distal in the north.

Therefore, the future target for exploration will focus on turbidite plays, stratigraphic traps, and pre-carbonate palaeohighs.

REFERENCES

Abolins, P. 1998. A hydrocarbon characterisation study of oils and condensates from the West Baram Delta area. Unpublished PRSS Report No: PRSS TCS 08-98/25.

Abolins, P. and Mona Liza Anwar, 1996. Geochemical characterisation of oils in SK6 and SK8, offshore Sarawak. Unpublished PRSS Report No: PRSS TCS 08-96/12.

Aigner, T., Doyle, M., Lawrence, D., Epting, M., and Van Vliet, A., 1989. Quantitative modelling of carbonate platforms : some examples. In: Crevello, P.D., Wilson, J.L., Sarg, J.F., and Read, J.F., eds., Controls on carbonate platform and basin development, Society of Economic Paleontoligists and Mineralogists Special Publication No. 44, 27-37.

Caline, B. and Mohammad Yamin Ali, 1991. Depositional and diagenetic cyclicity in a Miocene carbonate build-up (F23) Central Luconia gas province, offshore Sarawak. A contribution to Carbonate Research Workshop, 28 Oct. to 2 Nov. 1991, 1-16.

Caline, B., Ton Ten Have, Updesh Singh, Saiful Bahri Zainal, Mohd Reza Lesman and Mohammad Yamin Ali, 1992. Geological architecture of the Miocene carbonate buildups from the Central Luconia Gas Province offshore Sarawak, Malaysia, Abstracts of the Symposium on Tectonic Framework and Energy Resources of the Western Margin of the Pacific Basin, Kuala Lumpur, Malaysia, 29 Nov.-2 Dec. 1992, 99.

Doust, H., 1981. Geology and exploration history of offshore central Sarawak. In: Halbouty, M.T., ed., Energy Resources of the Pacific Region. American Association of Petroleum Geologists, Studies in Geology Series, 12, 117-132

Epting, M., 1980. Sedimentology of Miocene carbonate buildups, Central Luconia, offshore Sarawak, Bulletin of the Geological Society of Malaysia, 12, 17-30.

Epting, M., 1988. The Miocene carbonate buildups of Central Luconia, offshore Sarawak. In: Bally, A.W.B., ed., Atlas of Seismic Stratigraphy: American Association of Petroleum Geologists, Studies in Geology Series, 27, 168-173.

Ho, K.F., 1978. Stratigraphic framework for oil exploration in Sarawak. Bulletin of the Geological Society of Malaysia, 10, 1-13.

Liechti, P., Roe, F. W. and Haile, N.S., 1960. Geology of Sarawak, Brunei and western North Borneo. British Borneo Geological Survey, Bulletin 3, Vols. 1 and 2.

Mohammad Yamin Ali, 1994. Reservoir development in the Miocene Carbonate, Central Luconia Province, Offshore Sarawak. American Association of Petroleum Geologists International Conference (Abstract), Kuala Lumpur, 21-24 August, 1994.

Nippon Oil, 1996. Remaining potential of Miocene carbonate play. Unpublished Nippon Oil report, PETRONAS report no. ER : Nippon 3:96-01.

Rudolph, K.W. and Lehmann, P.J., 1989. Platform evolution and sequence stratigraphy of the Natuna Platform, South China Sea. In: Crevello, P.D., Wilson, J.L., Sarg, J.F., and Read, J.F., eds., Controls on carbonate platform and basin development. Society of Economic Paleontologists and Mineralogists Special Publication No. 44, 353-364.

Chapter 16

Tinjar Province

*Mohd Idrus B. Ismail
and Redzuan B. Abu Hassan*

INTRODUCTION

The onshore Tinjar Province is a wrench-related foldbelt in the Sarawak Basin which covers largely the offshore Balingian, Central Luconia, West Baram Delta, SW Luconia, Tatau and W. Luconia provinces (see Fig. 16.1). The Tinjar Province, which is uplifted relative to the offshore provinces, is separated from the offshore Balingian by the NE-SW to ENE-WSW Anau Nyalau fault. The Upper Eocene to Upper Miocene sedimentary sequences in the Tinjar Province were deposited after the uplift of the Upper Cretaceous to Eocene indurated sediments (Liechti et al., 1960). Major structures formed at the end of the Early Miocene and Late Miocene are dominantly NE-SW trending, with some NW-SE anticlinal structures and faults, including the Tatau Horst, the Arip Pelagau Anticline, and the Dulit Range. The regional framework resulted from the southeasterly plate movements due to rifting and opening of the South China Sea Basin from the Late Cretaceous to the Middle Miocene (e.g. Doust, 1981; Scherer, 1981; James, 1984). Furthermore, part of its tectonic history was also due to subduction of the South China Sea Plate under Borneo.

Except in its northern part, the Tinjar Province is not well explored, and recent reviews of the onshore areas have excluded the Tinjar Province. Further north are the Balingian, Central

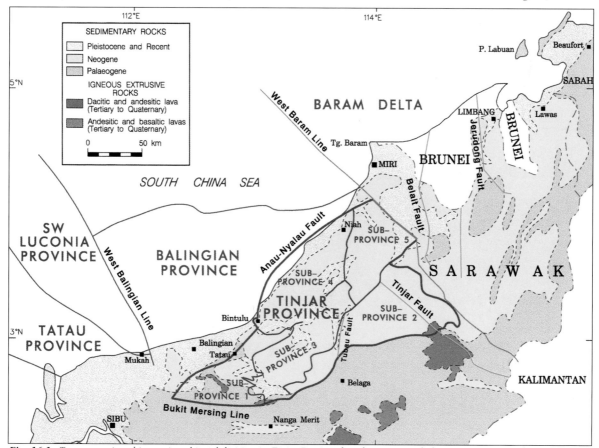

Fig. 16.1 Tectonostratigraphic stratigraphic subdivision in Tinjar Province based on structural characteristics arising from end- Early Miocene and Late Miocene deformations.

Luconia and West Baram Delta provinces. These contain sedimentary sequences of post- Late Eocene ages, and their producing horizons are the Oligocene- Lower Miocene (Cycles II/I) clastics, Middle Miocene carbonates (Cycles IV/V) and Upper Miocene deltaic sandstones respectively. In the Tinjar Province there are wrench-related large complexly faulted anticlinal traps. The structural closures are associated with NW-SE dextral faults and NNE-SSW sinistral faults.

The most comprehensive structural analysis of onshore Sarawak was based on air photo and synthetic aperture radar (SAR) acquired in 1990 by PETRONAS. The regional data were interpreted in 1993 (Intera, 1993). Furthermore the SAR interpretation was calibrated with field checks (Tjia et al.,1997). The data shows many NE-SW anticlinal and synclinal structures, NE-SW regional transform faults (e.g. Tinjar Fault) and NNE-SSW faults (e.g. Tubau Fault). Domes (e.g. the Ulu Suai Dome) and NW-SE-trending anticlines were also mapped. Wrench-related structures, formed during the Early Miocene, were reactivated in the Late Miocene in the northeastern part of the province. Early Miocene wrench faults dominate the western part of the Balingian Province (Swinburn,1993), whereas Late Miocene deformation occurred in the eastern Balingian Province (Doust, 1981). However, the regional deformation was mainly attributed to the end of the Miocene phase (Liechti et al., 1960). This chapter describes the structural setting of petroleum accumulations in the Tinjar Province with reference to the structural development in both East and West Balingian provinces.

EXPLORATION HISTORY

The Anglo-Saxon Petroleum Company acquired the rights under the Sarawak Oil Mining Lease to explore the whole of onshore Sarawak in 1909. The land exploration targeted several areas having oil seepages around the Baram Delta. By 1910 the first oil field was discovered when the Miri-1 well penetrated hydrocarbon bearing intervals within Upper Miocene deltaic sands. Between the

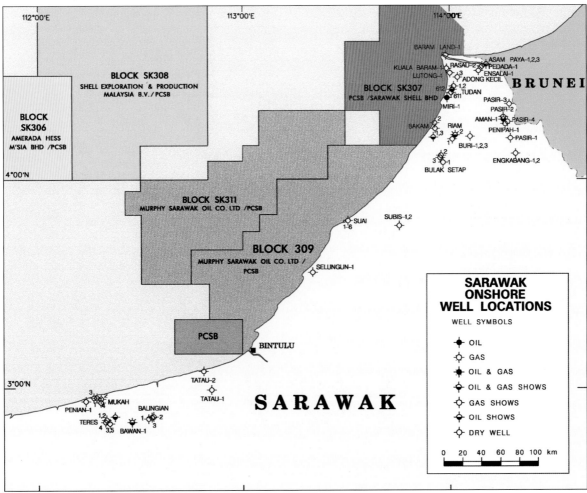

Fig. 16.2 Map of onshore well locations in Tinjar Province and Baram Delta, including 11 wells in Tinjar Province. None of the wells in the Tinjar Province to date was drilled in the southwestern part which falls in the oil/gas trend of offshore Balingian Province.

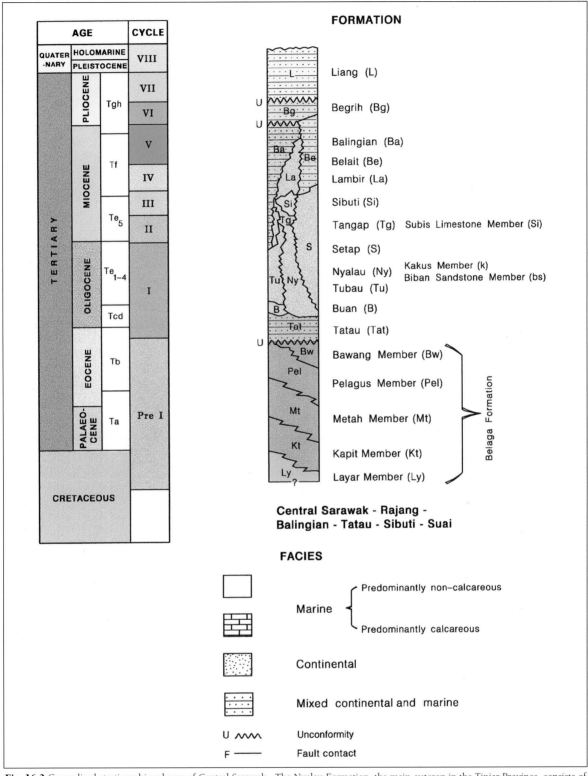

Fig. 16.3 Generalised stratigraphic scheme of Central Sarawak. The Nyalau Formation, the main outcrop in the Tinjar Province, consists of shallow marine to paralic successions grading into marine clastics from SW to NE. It interfingers with the Setap Shale Formation. Towards the NE, the Lambir and Belait formations are posssible reservoirs.

1920-1950's, geological mapping, acquisition of land gravity and seismic were conducted around the coastal areas. The belt between Suai and Bulak was covered with 483 line-km of single-fold land seismic in 1952/53. Several wells were drilled in the onshore part of the Tertiary basin around this coastal belt. The Suai 1-3 core holes drilled in 1952-53 (see Fig. 16.2) reached Middle Miocene (Cycles I/II) interbedded shales and sandstones (see Fig. 16.3). The Suai-3 well produced oil at a rate of 27 BOPD from an isolated channel sandstone. This well was abandoned at total depth (TD) of 1286 feet within the Lower Miocene after a gas blow out.

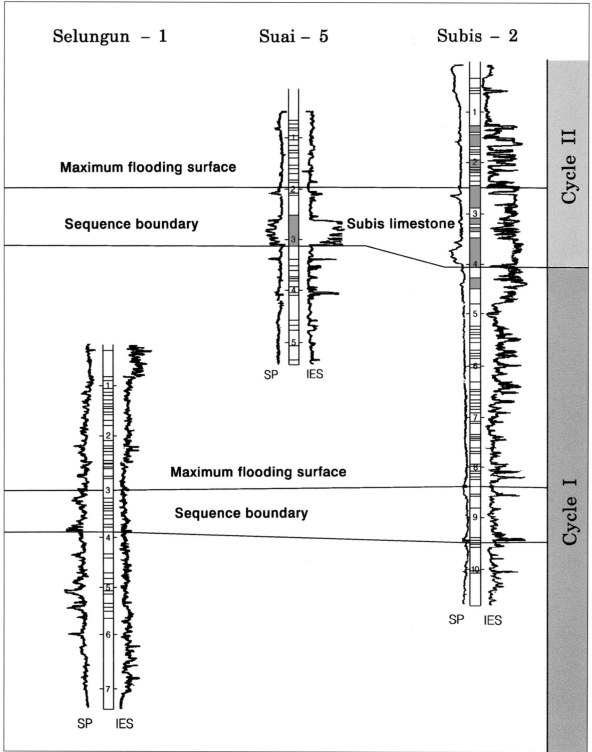

Fig. 16.4 Well correlation from Selungun-1, Suai-1 and Subis-2 showing the Subis Limestone. Due to erosion after strong uplift of the structure, the Cycle II objective in Selungun -1 was never encountered.

In 1955 three more wells (Suai-4, 5 and 6) were drilled nearby. These penetrated deeper into Lower Miocene (Cycles I/II) intervals comprising the Subis Limestone, and interbedded sandstones and shales (see Fig. 16.4). The limestone represent deep-marine and off-reef facies, whilst the higher section are mixed clastics and carbonates comprising thinner, 30 foot limestone, interbedded with a mixture of shale, sandstone and calcareous sands. At well locations the calcareous and argillaceous sandstone intervals show only marginal reservoir properties (Porosity, 8-20%; permeability up to 2.8 mD). A drill stem test (DST) within the reservoir intervals recovered only traces of gas and water.

Located about 25 km northeast from the

Suai wells, the other onshore well in the near coastal areas in the Tinjar Province, Subis-2, was spudded in 1956. The Subis well encountered several Lower Miocene intervals of thinner limestone in a mixed carbonate- clastic sequence and two thick carbonate units of between 1200 to 4000 feet thickness. DST's show that the amount of hydrocarbons at Subis-2 was insignificant.

The final well, Selungun-1, was drilled in 1956 in the area more towards Bintulu. The Selungun-1 well drilled into an anticlinal feature and spudded into Oligocene (Cycle I) shales, silts and thin sands and 150-400 feet thick shallow marine and coastal plain sandstones. The porosity of the sandstones varies between 3 to 10%, but improves near the surface to between 9 to 24%.

The former company, renamed Sarawak Shell Berhad (SSB) in 1958, resumed exploration activities around the coastal belt. In 1967 6-12 fold CDP stack seismic data in the area around Suai were acquired in 1974 - in the area around Bungai. An aeromagnetic survey of all offshore Sarawak was extended landwards into a small belt in the Tinjar Province in 1965.

Since 1957, the exploration company relinquished several packages of its onshore Sarawak concession. As the deformed and least accessible hinterland areas were considered unprospective, these were the first to be relinquished, including the Rajang Accretionary Prism and the Dulit Range. Exploration activities continued in some of the the inland belt around Pandan Kakus, Ulu Suai Dome and Kelabu Anticline (Fig. 16.5) but in 1971 this belt, including the Grabit-Mentebai Syncline, was released under a second relinquishment exercise. Since 1960's, exploration moved offshore with several successes, including the discovery of the Temana Field. Consequently onshore exploration became limited. In 1976 Shell signed the Production Sharing Contract, relinquished the remaining areas around the coastal belt of the Tinjar Province, but retained those around the Miri Field. Several wells along the Miri-Tudan trend in onshore Baram Delta were drilled, but subsequently the remaining contract area was relinquished.

STRATIGRAPHY AND STRUCTURE

Stratigraphic scheme

In 1960 a review of the stratigraphy of onshore Sarawak, Sabah and Brunei was conducted by Liechti et al. (1960) and subsequently the stratigraphic nomenclature based on both rock stratigraphy and time stratigraphy (based on foraminiferal parameters), was formalised. The offshore Tertiary succession in the Sarawak shelf is classified into eight regressive cycles (Ho, 1978). Limited subsurface onshore stratigraphic penetrations and poor age controls make recognition of sedimentary cycles difficult, and hence the formation terminology cannot be dispensed with. The Sarawak stratigraphic scheme and its approximate age correlation to the cycles is shown in Figure 16.3.

Formations

The Tatau Formation (Upper Eocene to Oligocene) unconformably overlies the steeply folded Bawang Member of the Belaga Formation at the Arip Pelagau Anticline. At its base, the Tatau Formation consists of mainly medium to fine grained sandstones and siltstones with some intervals of marls, thick shales and intercalated limestone lenses. The formation is separated by the Arip acid volcanics into a lower (Eocene or Tb) and upper (Oligocene or Tcd) part. The depositional environment for the formation is littoral to neritic.

Near Tatau and the Arip-Pelagau Anticline the Tatau Formation passes gradually up into the laminated marine shales of the Buan Formation (Oligocene). Elsewhere, as at Bukit Mersing, Buan Formation shales overstep the Tatau Formation to unconformably overlie the Belaga Formation. Tubau Formation shelfal shales and marls occur in the Tubau Valley, ranging in age from Oligocene to Lower Miocene (Te_{1-4} to early Te_5). Along the Jelalong river, and towards the coast, this formation merges into the Oligocene-Miocene Setap Shale Formation. Similar rocks occur in several of the onshore wells (Subis-2, Selungun-1 and Bulak Setap-3). These intervals had been assigned to the Setap Shale Formation. It has a conformable contact with overlying Nyalau Formation.

Most of the outcrops in the Tinjar Province belong to the Oligocene-Miocene Nyalau Formation (Cycles I/II). In the Subis-Bintulu area the Nyalau Formation occurs as interbedded offshore - subtidal - estuarine sandstones, sandy shales and shales with dispersed lignite bands and marls. In the Selungun well the interval consists of silty sandstones, partly calcareous, grading into sandy limestone. In the Tatau-Bintulu area, the lower part of the Nyalau Formation is distinguished as the Biban Sandstone Member. This is of Oligocene- Miocene age (Tcd-Te) and is

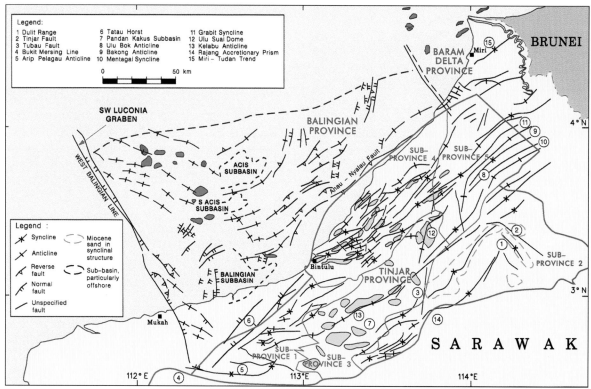

Fig. 16.5 Structural map of the Tinjar Province and sub-provinces (modified after Leichti, et. al., 1960) and the offshore Balingian Province. Offshore oil and gas fields are shown.

predominantly fine to medium grained sandstones, with some siltstones interbeds and contains calcareous nodules. Limestone beds and calcareous sandstones in the Nyalau Formation are present as the Oligocene (Tcd) Sarong Limestone and Bekuyat Limestone. The 250 to 500 feet thick limestones are included within the Biban Sandstone Member. In the Pandan Kakus and Ulu Suai areas several intervals of massive sandstone, laminated clays and brackish shales and lignites, attributed to interdistributary environments, are designated the Kakus Member (Lower-Middle Miocene, Te_5 to Tf_1).

The Lower to Middle Miocene (Te_{1-4} to Te_5, and possibly Tf_1), Setap Shale Formation consists of littoral to inner neritic clay- shales and silty clays. These intervals are occasionally interbedded with sandstones, calcareous sandstones and thin to moderately thick limestones (5-400 feet). This interval laterally interfingers with the Nyalau Formation. The boundary of this formation with the overlying Belait and Lambir formations is mostly conformable and diachronous, although an occasional unconformable contact, described as insignificant, has been observed in some places, as in the Paroh anticline.

The Nyalau Formation grades into the Tangap Formation (Oligocene-Lower Miocene) to the north of Subis. The inner to outer neritic of marls and shales include the Subis Limestone Member. The limestones comprise coral-algal reefoid growth and foraminiferal open shoal limestone. The Nyalau Formation grades up into the Sibuti Formation (Lower-Middle Miocene). This consists of inner neritic clay-shales with minor siltstones and limestone.

The Sibuti Formation grades up into the Lambir Formation (Middle Miocene). This consists of alternating inner neritic to littoral sandstones and shales, succeeded by clay shales, marls, thin limestone lenses and calcareous sandstones. Sandstones with some impure lignite layers occur in the uppermost intervals in the Grabit Syncline. The Lambir Formation grades up into the Belait Formation towards the Dulit Range.

The Belait Formation (Middle to Upper Miocene, Tf_1) consists of intervals of thick bedded, barren, massive, sometimes pure white, and medium to coarse grained sandstones interbedded with claystones. The formation is the time stratigraphic equivalent of the Lambir, Miri and Tukau formations.

Structure

Five sub-provinces have been recognised in the Tinjar Province based on SAR data (Tjia et al., 1997) (see Fig. 16.5). Structural styles in the Tinjar Province are believed to result from two phases of deformation in wrench-fault systems. Major structures were formed at the end of the

Early Miocene and in the Late Miocene (Tjia et al., 1997). Previously the deformation history of the Tinjar Province was mainly attributed to a Late Miocene regional event (e.g Liechti et al., 1960). The Early Miocene wrenching is related to southeasterly directed movements as a result of Late Cretaceous-Miocene spreading and opening of the South China Sea Basin. The Late Miocene wrenching is related to the southwards movement of the South China Sea Basin, which resulted in several N-S transform faults such as the Morris Fault, Jerudong Line and Tubau Fault.

Sub-provinces 1 and 3 are related to Early Miocene deformation. Sub-provinces 2 and 4 show deformation during the Early and Late Miocene. Several NE-SW trending transpressional structures in East Balingian Sub-province have large amplitude anticlines with cutting cross faults and flanking reverse faults (Doust, 1981). Sub-province 5 is related to Late Miocene deformation. However, the structural patterns are different from the rest of the Tinjar Province, and are interpreted as a conjugate fault system.

The sedimentary section in the Tinjar Province is mostly Oligocene to Middle Miocene in age (Nyalau, Setap Shale formations). Outcrops in the Sub-province 1 are older, and include Upper Eocene-Oligocene outcrops of the Tatau and Buan formations.

Sub-province 1

Sub-province 1 forms an area of convergent wrenching in the western part of the Tinjar Province. Late Eocene and Late Miocene movements along the northward-opening fan of dextral wrench faults produced a zone of strong transpression and uplift of the Upper Cretaceous-Lower Miocene sediments. This sub-province contains a number of structural elements, the Arip Pelagau Anticline (see Fig. 16.5), the Tatau Horst, NE-SW trending anticlines and NW-SE dextral wrench faults. The NW-SE aligned Arip-Pelagau Anticline is wedged tightly between NW SE faults in Tinjar Province and the thrust foldbelts in the south along the Bukit Mersing Line.

Sub-province 2

Sub-province 2, mostly made up of the Dulit Range (see Fig. 16.5), shows two distinctive structural trends that are at right angles to each other. The Dulit Range, which is located in the southeastern part of the Tinjar Province, is bounded on the east by Tinjar Fault, a major transcurrent fault in Sarawak. Movement along the Tinjar Fault, a dextral NE-SW trending wrench fault, was active in the Early Miocene, and was reactivated in the Late Miocene. The structures consists of two NE-SW trending tight shale-cored folds, sharply bending to a NW-SE trend, the NW-SE Tinjar Fault and NNE-SSW-trending Tubau Fault. This is a sinistral wrench fault that was active in the Late Miocene in the opposite part of the Dulit Range.

The sense of movement of the dextral Tinjar Fault was deduced from a number of tension fractures that are left stepping around several folds, near Tinjar Fault. NE-SW trending faults are left lateral offsetting one anticline near the sinistral Tubau Fault.

Tjia (1996) proposed that the Dulit Syncline became refolded during the post-Belait deformation (Late Miocene-Early Pliocene) and developed into the right angle or dog-leg pattern of the Sekiwa-Dulit ridges. The deformation operated by detachment tectonics that was facilitated by strike-slip motion along the Tinjar, Dengan, Tubau and another NE-SW, hitherto unknown zone of weakness, just north of Bukit Selika. The term "frame-folding" was used for this phenomenon. The detachment of the Belait from older sequences was lubricated by shales of the Setap Shale Formation.

Sub-province 3

Sub-province 3 is a large wrench fold zone that was deformed and faulted along a few NW-SE dextral wrench faults in the southern part of the Tinjar Province (see Fig. 16.5). The movement along the NW-SE wrench faults, which are occasionally WNW-ESE and almost sub-parallel to the Bukit Mersing Line, was active during the Early Miocene. There are NW-SE wrench faults and NE-SW transpressional structures. The large NW-SE trending anticlines (see Figs. 16.6a, 16.6b), which are often cut by several NW-SE faults showing right lateral offsets, are up to 15 km long and 5 km wide. The smaller anticline which lies in the orientation of NW-SE major wrench faults is considered to be a drag fold (see Fig. 16.6c). The sub-province, which is less uplifted than both sub-provinces in the east and west, also forms an E-W trending, Pandan-Kakus Sub-basin containing a relatively thick Oligocene - Middle Miocene sedimentary section (see Fig.

16.6b). The sub-basin is located in the south and measures about 18 km long and 8 km wide. Most of the areas exhibit deformation mainly in Lower Miocene sediments. Only a small northeastern part of this sub-province, within which is the Ulu Suai dome (see Fig. 16.5) which has N-S long axis and E-W short axis may have been reactivated by Late Miocene wrench movement. Strong uplift in the southern part of the fold zone exposed the Bekuyat and Sarong limestones (Oligocene-Lower Miocene).

Sub-province 4

Sub-province 4 contains a wrench foldbelt in the northeastern part of Tinjar Province which developed due to Early Miocene NW-SE dextral wrench faulting (Fig. 16.5), and to Late Miocene reactivation associated with NNE-SSW sinistral wrench faulting. The structures include NE-SW folds, NW-SE dextral wrench faults, NNE-SSW sinistral wrench faults and possible rhombohedral basins. Early Miocene deformation is more prominent in the southwest. The NE-SW wrench anticlines are cut by NW-SE dextral faults with displacements of up to 5 km, measure about 14-28 km long with wavelengths of about 6 km (Fig. 16.6d) The Late Miocene structural compression, which is stronger towards the northeastern part of the sub-province, resulted in NE-SW striking structures cut by NNE-SSW sinistral faults. These usually have flat crestal areas and steep flanks. Tight folds and compression with low-angle reverse faults were observed in outcrops (Tjia et al., 1997). Furthermore certain parts of Sub-province 4 underwent both Early Miocene and the Late Miocene deformations which resulted in superposed structures. In the Balingian Province the Temana structure (Doust, 1981) underwent deformation during both times. Fault-bounded rhombohedral-shaped basins in the southwestern part of the onshore measure 24 km long by 12 km wide. In the Balingian Province are several depressions (Fig. 16.5).

Sub-province 5

Sub-province 5 consists of a wrench fold zone in the eastern part of Tinjar Province that was deformed in the Late Miocene (Fig. 16.5). This sub-province of Oligocene to Upper Miocene sediments is different from the rest of the Tinjar Province since the Late Miocene deformation is associated with a NW-SE dextral wrench fault (Fig. 16.7a). The distribution of the folds is controlled by the shear zone. These structures include a major NW-SE dextral wrench fault NNE-SSW minor faults, elongated synclines (e.g. the Mentebai, Ulu Bok and Grabit synclines), transpressional anticlines (e.g. the Ulu Bok and Bakong anticlines), and broad NE-SW synclines (e.g. the Jelalong Syncline) in the south. The orientation of the fractured sandstones in the shear zone is 290° (Fig. 16.7b). Transpressive NE-SW synclines and narrow zones of anticlinal uplift are situated north of the shear zone. Their arrangement is *en echelon* with a dextral sense. The cores of the anticlines are composed of the Nyalau and Setap Shale formations (Oligocene-Lower Miocene). To the south of the NW-SE shear zone is the broad Jelalong syncline (Fig. 16.7b). It is 45 km long and 25 km wide. The NE-SW trend of the syncline bends NNE-SSW near another possible shear zone which is at its eastern limit that coincides with a narrow N-S trending (010°) zone of steep beds.

It is thought that these structures resulted from shearing that locally developed conjugate faults within a system of a regional N-S Late Miocene left lateral motion (Fig. 16.7b). The maximum principal stress is oriented 335°. The conjugate faults developed close to basement inhomogeneities across the Tinjar Fault at Dulit Range. Furthermore, Tjia (1996) concluded that the Setap Shale Formation acted as a zone of detachment near the Dulit Range.

HYDROCARBON OCCURRENCES AND PLAY TYPES

Hydrocarbon Occurrences

The principal reservoir plays in the Tinjar area are the Oligocene-Lower Miocene (Cycles I/II) clastics. These occur in the southwestern part of the Tinjar, and are only partially tested by the Selungun-1 well (Fig. 16.2). Due to the uplift at the core of the anticline, only Cycle I was encountered. Hydrocarbons indicators in the southwestern part include oil stained sandstones/residual oil traces (Rahdon, 1974; Tjia et al.,1997). Other occurrences support the presence of hydrocarbon expulsion in the southwestern part of the province, where carbonaceous matter, believed to be heavy hydrocarbons, fills cracks which are up to 2 metres in length and at right angles to coal beds (Rahdon, 1974). The stains represent immature to mature oils based on geochemical signatures (see Table 16.1).

Oligocene -Lower Miocene carbonates

TOC (wt. %)	Source Parameters	2.18
S_1 (mg HC/g rock)		0.46
S_2 (mg HC/g rock)		4.65
PI		0.09
HI		213
Tmax (°C)		430
VRo (%)	Maturity Parameters	0.54
CPI		0.99
Tm/Ts		16.70
C_{32} Hopane 22S/(22S + 22R)		0.59
C_{29} Hopane 20S/(20S + 20R)		0.47
MPI 1		0.61
% Rc		0.77

Table 16.1 Selected geochemical data of the oil-stained sandstone in the Tinjar Province.

(Cycles I/II) are secondary objectives occurring mainly in the northern part of the province. Minor gas shows were encountered in the Cycle II limestones in several of the Suai and Subis wells.

The Lambir Formation clastics are secondary plays in the northeastern part of the province. Oil in the shallow clastics of the Lambir Formation were encountered in the Suai 3 wells (Fig. 16.2) in the northern parts.

TRAP STYLES, RESERVOIR AND SEAL

Trap Styles

Over 40 NE-SW oriented structures have been mapped in Sub-provinces 3 and 4 (see Fig. 16.5). Due to structural complexities and erosion of the reservoir, Sub-provinces 1 and 2 are the least prospective. There are also Early Miocene (end Cycle II), Late Miocene (end Cycle V) and superposed Late Miocene/ Early Miocene traps.

Early Miocene (end Cycle II) traps are formed by NE-SW elongated anticlines, occasional small NW-SE anticlines, and large NE-SW structures. NE-SW to E-W anticlines are often structurally compartmentalised by NW-SE dextral faulting and oblique faulting (see Figs. 16.6a and 16.5). They occur in various parts of the Pandan Kakus Sub-province and Kelabu anticline. The NW-SE drag folds (see Fig. 16.6b) are occasionally found along NW-SE transcurrent faults. Large NW-SE structural complexes (see Fig. 16.6c) in the Pandan Kakus Sub-basin are compartmentalised by NW-SE faults into several fault blocks cutting the central part and the western flank. Individual fault blocks may be prospective, although shallow sandy intervals and structural faulting reaching the surface are the major risks. Figure 16.8 shows the trapping model for hydrocarbons in reservoirs within basinal parts and over areas of structural highs, based on similar examples in the Balingian Province.

Late Miocene (end Cycle V) traps are formed by elongated NE-SW anticlines. These are compartmentalised by NE-SW faults, and are characterised by flat to gently dipping crests and steep flanks (see Fig. 16.6d) (Tjia, 1996). These traps have been drilled in Selungun, Suai and Subis. The Tiban structure is as yet undrilled. Similar traps in the East Balingian Sub-province have been described (Doust, 1981).

Superposed Late Miocene/ Early Miocene traps are formed by elongated anticlines cut by two sets of wrench faults - one right lateral NW-SE fault and the other left lateral NE-SW fault (see Fig. 16.6e). Several of these anticlines occur around Bintulu, where the faults are commonly developed on the eastern noses, and the zones of complex fault intersections.

Reservoirs

Cycles I and II clastics and limestones

The Nyalau Formation (Cycles II/I) is the main potential reservoir in the southwestern part of Sub-provinces 3 and 4. The reservoirs consist of offshore marine-subtidal to estuarine clastics. The upper part is preserved in the Pandan Kakus area. Near Bintulu the reservoir grades into distributary channel and interdistributary sandstones, silty sandstones and shales/coals interbeds. Porosities and permeabilities are between 12-30% and <1 to 1210 mD. The Kakus Member (Cycle II) has higher porosity and permeability. Porosity reduction is caused by calcareous cementation, compaction and authigenic quartz.

Due to thicker overlying shales

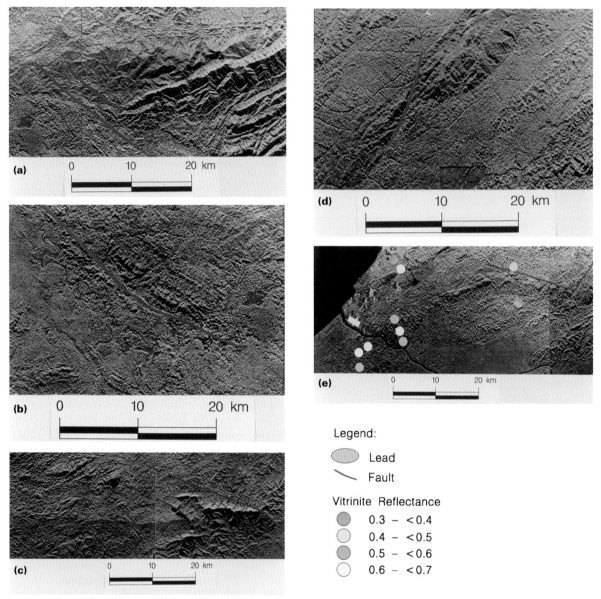

Fig. 16.6 SAR images of the Tinjar Province. a) End Cycle II anticlinal structures having major NW-SE faults on its western flank and interpreted as wrench related, is also faulted oblique to axis and plunges to the SE. b) Rhombohedral basin containing end- Cycle II large anticlines, which are moderately folded. These anticlines are bounded and cut by NW-SE strike-slip faults and NE-SW extensional faults. Fault compartmentalisation could be a positive factor in trap/ seal capacity for this trap type, c) Symmetrical, small anticline which are associated with drag along major NW-SE transcurrent fault and has NW-SE trend. Such trap types are also found in other tectonostratigraphic provinces such as in West Balingian, d) End- Cycle II anticlinal structures are symmetrical to asymmetrical, elongated structures with flat to gentle core and steep flanks. These are often broken by large wrench related NE-SW faults and shows SW plunges. Other areas with such structures include East Balingian Province, e) Mixed end- Cycle II/Cycle V anticlines are cut by two sets of wrench related fault systems- NW-SE (right lateral) and NE-SW (left lateral). In the superposed structures, both ENE and SW structural plunges are seen, and a rhombohedral basin is present to the south. Also plot of vitrinite reflectance data on the SAR shows that mature shales and organic matter have been sampled close to the rhombohedral basin.

preserving the objective section, the Nyalau Formation in the rhombohedral basins could have better potential. Major accumulations have been found in several major depressions close to the boundary between the West and East Balingian sub-provinces, comprising the Central Balingian Sub-basin, Acis and North Acis sub-basins (see Fig. 16.5).

Lower Miocene (Cycles II/I) carbonates are another play type in the northeastern part of the province. The Suai and Subis wells tested off-reef carbonate facies.

Cycles III and IV clastics.

The Lambir Formation (Cycle III) is regarded as a secondary reservoirs play in the eastern part of the province. The formation consists of shallow marine to paralic sandstones. Porosities and permeabilities range between 5-25% and <1 to 450mD. Porosity reduction is by calcite cementation, and quartz cementation. The Lambir Formation probably underlies the Belait Formation in the Mentegal, Ulu Bok and Grabit Synclines (see Fig. 16.7).

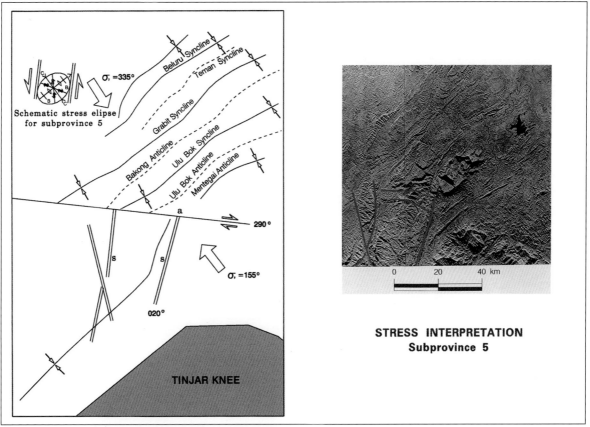

Fig. 16.7 SAR image (right) and interpretation (left). Structural patterns in Sub-province 5 comprise a) major NW-SE faults interpreted as dextral wrench fault, and NNE-SSW minor faults, b) elongated synclines (in yellow), c) transpressional anticlines and d) in the south, broad NE-SW synclines. Model for the area north of the Tinjar Knee involves N-S lateral motion, local stresses and shearing with development of conjugate faults and compression. The maximum principal stress is oriented 335°.

Seal

Seals in the Tinjar Sub-provinces 3 and 4 are the intraformational shales in Nyalau and Lambir formations. Regional shales are largely absent. In the Balingian Province Lower Miocene (Cycle II) intraformational shales are important in the Bayan Field, but generally Middle Miocene (Cycle III) shales form the regional seal.

SOURCE ROCKS

Oligocene-Lower Miocene (Cycles II/I) sediments are the main source rocks in the southwestern part of the Sub-provinces 3 and 4. These include coals and shales that were both deposited in either interdistributaries, or in the floodplains of the estuaries. Their equivalents are Cycles II and I respectively.

Outcrop samples from coal and shale intervals of the Nyalau Formation within the Tinjar Province have been analysed. The coals show excellent organic carbon richness (TOC: up to 78 wt.%) and excellent hydrocarbon-generating potential. The coals also contain oil-prone (Type II) and gas prone organic matter (Type III), as suggested by the HI values of 219 - 475. A plot of Hydrogen Index against Tmax in Figure 16.9 demonstrates the quality of selected coals and shales from the Nyalau Formation. The percentage of oil prone kerogens in the coals ranges from a trace to 33%. Source facies are only found in certain intervals of the Nyalau Formation. The occasional shale samples from the shale source facies deposited mainly in the southwestern part of the Sub-provinces 3 and 4 show excellent hydrocarbon generating potential (S_2 up to 7.8 mg HC/g rock). Most of the shales of the Nyalau Formation have fair to good organic carbon richness (TOC up to 6.6 wt. %) but generally poor hydrocarbon generating potential (S_2 < 2 mg HC/g rock, HI < 100). .

The kitchen area is below the local depression in the southwestern part of the province. Figure 16.6d provides thermal maturity data for several sampling points, showing the regional distribution of maturity for the Nyalau Formation in the southwestern part of the province, including these local depressions. In some of the depressions the specimens are in the early mature state with VRo: 0.62-0.65%.

The Setap Shale Formation contains generally poor source rock. The shales were deposited in the north in an open marine setting. Source rock studies have been carried out on both outcrop and subsurface samples of the Setap Shale Formation. The data show that the shales possess fair organic carbon richness and poor hydrocarbon generating potential. The samples contain Type III organic matter with potential to generate predominantly gas. All the Setap samples are thermally immature to early mature.

Biomarker assemblages are characterized by the presence of moderately high Tm/Ts, moderate C_{29} $\alpha\beta$ hopane, low oleanane, high bicadinanes, and C_{29} preference (Fig. 16.11). Generally, these features suggest that the source rocks were deposited within a lower coastal plain setting with considerable terrestrial organic matter input/higher land-plant (resins).

The Setap Shale and the Nyalau

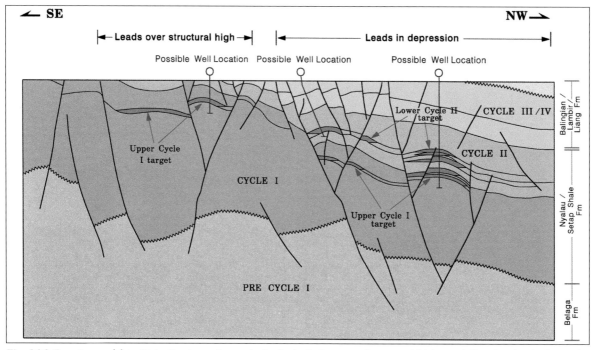

Fig. 16.8 Trapping model in Tinjar Province. Reservoirs in the basinal parts and structural highs are believed to be similar to the hydrocarbon accumulations in offshore tectonostratigraphic provinces such as West Balingian Sub-province.

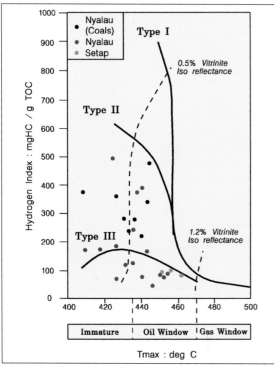

Fig. 16.9 Hydrogen Index against Tmax plot showing the quality of selected Cycles I/II coals and shales from the Nyalau and Setap Shale formations.

formations in Cycle I at Selungun-1 have poor to good organic carbon richness (TOC: 0.40-1.41 wt.%). Since the analyzed samples from 769m to 2212m are thermally mature to post mature (VRo: 0.76-1.50 %, Fig. 16.12), the S_2 values (< 2 mg HC/g rock) merely indicate their residual hydrocarbon potential. The HI's are generally less than 100, indicating the ability to generate predominantly gas. The kerogen types are mainly Type III/IV of terrigenous origin.

More detailed biomarker analyses show that the Nyalau Formation coals and shales display certain similarities in their biomarker assemblages : moderate to very high Tm/Ts, moderate to extremely high C_{29} $\alpha\beta$ hopane, trace to low oleanane, moderate to high bicadinanes, and C_{29} preference (Figures 16.13 and 16.14). These features are generally characterized as coastal plain (lower and upper) depositional setting with substantial terrestrial organic matter input and also higher land-plant inputs (resins). The degree of the input (especially oleanane and bicadinanes) varies from location to location. The biomarker assemblages observed in the rock extracts in the Tinjar Province also show a

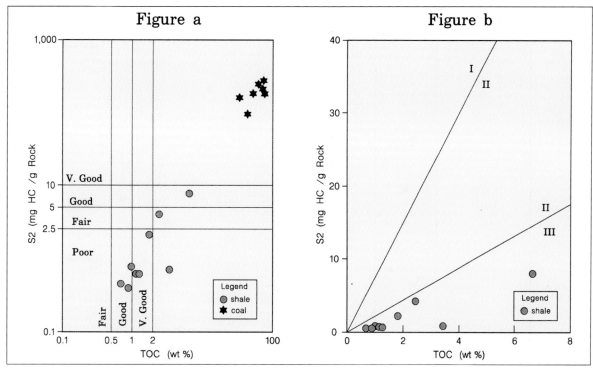

Fig. 16.10 Source potential summary diagram (S₂ vs TOC) for the selected Cycles I/II coals and shales from the Nyalau Formation.

positive correlation with the biomarker characters displayed by the Balingian oils (see Balingian Province Chapter 14).

In the Balingian Province, Cycles II/I coals have oil and gas potential and shales have poor oil-prone potential. All the best source rocks occur within lower coastal plain deposits. The kerogens are Type III with some Types III/IV gas prone. Some intervals of Cycle I in the D35 field is liptinitic.

Coals from the Merit-Pila coalfield of central Sarawak was studied by Wan Hasiah (1997) who focused on the petrographic details of the early generation of oil-like substances from the precursor suberinite (coal). The Merit-Pila coalfield forms an east-west elongated, block-faulted, erosional outlier of the Miocene Nyalau Formation which rests unconformably on the Eocene Belaga Formation. The study of these coals clearly demonstrates the early generation of liquid hydrocarbon from the maceral suberinite, and establishes a model for a possible pathway for which liquid hydrocarbon could be naturally expelled from coal source rocks.

PLAY TYPES

The main play types are a) Oligocene- Lower Miocene (Cycles II/I) clastics, b) Oligocene - Lower Miocene limestones in the northeastern part, and c) Lambir Formation in the eastern part of the province.

The dominant play in the Tinjar Province is the Oligocene- Lower Miocene Nyalau (Cycles II/I). The play occurs in the southwest of the Tinjar Province where the traps include Lower Miocene structural traps and superposed Upper Miocene/Lower Miocene structural traps.

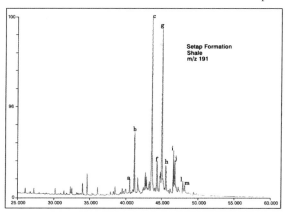

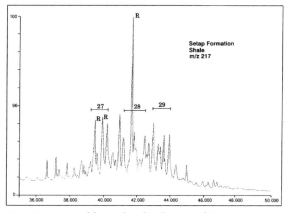

Fig. 16.11 Typical biomarker distributions of the shale extract from the Setap Shale Formation.

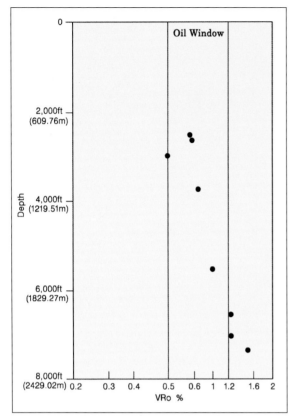

Fig. 16.12 Depth-vitrinite reflectance relationship showing the maturity profile of the analyzed samples from the Selungun-1 well.

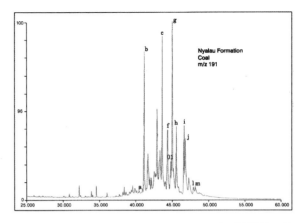

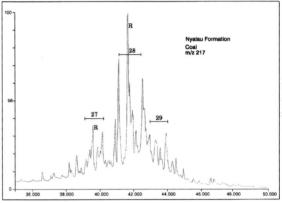

Fig. 16.13 Typical biomarker distributions of the coal extract from the Nyalau Formation.

Numerous structures are found in Sub-provinces 3 and 4 (see Fig. 16.6). Half of the 40 untested leads are found in the southwestern part of the Tinjar Province where possible rhombohedral basins may be present. These structures are generally shallow, but Nyalau Formation is thicker in the local depressions due to subsidence. The factors that determine the success of this type of prospect include a) thickness of intraformational shales, b) amount of fault throw and position of fault with respect to structure and c) preservation of Oligocene -Lower Miocene (Cycles II/I) target intervals (see Fig. 16.8). Porosities and permeabilities are good. The depositional pattern and structural trend seen in West Balingian are believed to extend to the southwestern part of the Tinjar Province. In the 1970-80's, drilling in East and West Balingian led to the discoveries of several fields (Temana, Bayan, D21, D18, D30 and D35). Oil migrated laterally into structures on the flanks of the local sub-basins from kitchens such as the Central Balingian Sub-basin, and the Acis and North Acis sub-basins (Fig. 16.5).

The clastic and carbonate plays in the Tinjar Province are mutually exclusive. In the northeast, the play is in the Oligocene-Lower Miocene carbonates, where Late Miocene structural traps are present. Drilling at Suai and

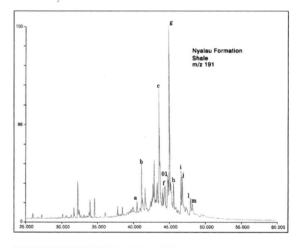

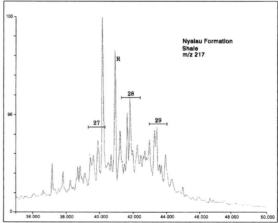

Fig. 16.14 Typical biomarker distributions of the shale extract from the Nyalau Formation.

Subis found sub-commercial gas shows.

Hydrocarbon traps in the Lambir Formation, a secondary play in the eastern part of the Tinjar Province, is difficult to define without seismic data. The traps are possibly present underneath the Upper Miocene sandy Belait Formation in synclines (see Fig. 16.7a). The presence of effective intraformational shales separating the reservoir from the extensive and porous Belait sandstones is critical to the success of the Lambir Formation play. In the Balingian Province oil and gas have been found in the Lower Cycle III reservoirs at Temana, D12 and Patricia. Coals and coaly shales are present in the northern and eastern part of Tinjar Province and could provide internal hydrocarbon sourcing for the play.

Extensive land seismic work is necessary to explore for these three main play types in the future. Seismic acquisition selectively over some structures is critical to upgrade leads into prospects for drilling. In addition, aerogravity in the Tinjar Province was acquired in September 1998. Sequence stratigraphy may also be useful in predicting the continuity of reservoirs in the subsurface.

SUMMARY

The Tinjar Province has been studied geologically to find petroleum. Three main play types in the province have been identified. There are nearly 40 leads in Sub-Provinces 3 and 4 within the Oligocene- Lower Miocene (Cycles I/II) main clastic and limestone plays. The Lambir Formation in the eastern part of the province is a secondary play. The reservoirs in the main Nyalau clastic play are the subtidal to estuarine and interdistributary intervals. These are the Cycles II/I equivalents of the hydrocarbon prolific West Balingian Sub-Province. The interbedded shales are generally thick enough to act as effective seals. Oils could migrate laterally from kitchens in local depressions into structures on the flanks of adjacent sub-basins. In the Balingian Province such local depressions include the Central Balingian, Acis and North Acis sub-basins.

Current seismic coverage is too sparse to develop leads into prospects. Further exploration efforts should certainly include seismic and potential field data acquisition in order to upgrade some of the leads.

REFERENCES

Doust, H., 1981. Geology and exploration history of offshore central Sarawak. In: Energy resources of the Pacific region. (Ed. Halbouty M.T.), American Association of Petroleum Geologists Studies in Geology 12, 117-132.

Ho Kiam Fui, 1978. Stratigraphic framework for oil exploration in Sarawak. Bulletin of the Geological Society of Malaysia, 10, 1-13.

Intera, 1993. Radar Geologic Intepretation of Sarawak, Malaysia (Unpublished report: PETRONAS).

James, D.M.D., (Ed.), 1984. "The geology and hydrocarbon potential of Negara Brunei Darussalam". Muzium Brunei Special Publication, 169 pp.

Leichti, P., Roe, F.N., Haile, N.S. and Kirk, H.J.C., 1960. The Geology of Sarawak, Brunei and western part of North Borneo. British Borneo Geological Survey, Bulletin 3.

Rahdon, A.E., 1974. Geological field investigations-NW Borneo (Nov 1969 - July 1974)- Summary of data (Unpublished report: Sarawak Shell Berhad).

Sarawak Shell Berhad, 1955. Resume of deep coreholes Suai # 1,2 and 3. (Unpublished report : Sarawak Shell Berhad).

Sarawak Shell Berhad, 1971. Sarawak Basin Study (Unpublished report : Sarawak Shell Berhad).

Scherer, F.C., 1981. Exploration in East Malaysia over the past decade. In: Giant oil and gas fields of the decade 1968-1978 (Ed. by Halbouty M.T.), American Association of Petroleum Geologists Bulletin, 30, 423-440.

Swinburn, P.M., 1993. Tectonic styles of the Balingian Province. Geological Society of Malaysia Petroleum Geology Seminar 1993, Programmes and Abstracts, 15 pp.

Tjia, H.D., 1996. Tectonics of the Sarawak Basin. Warta Geologi, 22, 6.

Tjia, H.D., Mohd Idrus Ismail and Othman Ali Mahmud, 1997. Bintulu-Ulu Suai and Lambir Geological Fieldtrip (3-8 Nov, 1997). (Unpublished report: PETRONAS).

Wan Hasiah A., 1997. Evidence of early generation of liquid hydrocarbon from suberinite as visible under the microscope. Organic Geochemistry, 27, 591-596.

Chapter 17

Tatau Province

*Mazlan B. Hj. Madon and
Redzuan B. Abu Hassan*

INTRODUCTION

The Tatau Province lies between the Balingian and SW Sarawak provinces. It is separated from the Balingian Province by the NW-trending West Balingian Line and from the SW Sarawak Province by the western extension of the Tatau-Mersing Line (Fig. 17.1). The Tatau Province is subdivided into the SW Luconia and Half-Graben sub-provinces. Its onshore extension underlies the area between Mukah and Tg Sirik. Although the Tatau Province is relatively less explored compared to the adjacent Balingian and Central Luconia provinces, there has been a number of discoveries particularly in the SW Luconia Sub-province. Substantial reserves of oil and gas have been found in the J4, J32, and K3 structures, in addition to a number of oil and gas shows nearby.

EXPLORATION HISTORY

The Tatau Province includes parts of the exploration blocks SK303 (formerly SK3 and SK4) and SK305 (SK5) in offshore and onshore central Sarawak (Fig. 17.1). Exploration in the Tatau Province began with gravity surveys and surface mapping in onshore Tatau, which subsequently resulted in the drilling of Rajang-1 by Shell in 1951. The Rajang-1 well encountered 461 m of Cycles V and VI sediments overlying pre-Tertiary basement (Belaga Formation). Shell relinquished their exploration areas in 1976, by which time they had acquired 1400 km of land seismic between 1947 and 1967, and a further 560 km in 1975. Subsequently, there had been little exploration activity until 1989 when a consortium led by Overseas Petroleum and Investment Corporation (OPIC) signed a PSC

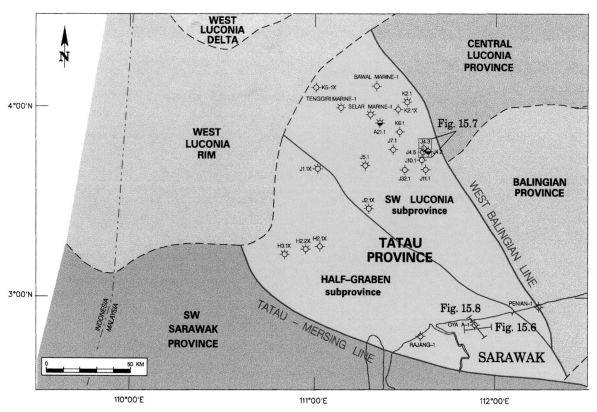

Fig. 17.1. Location of the Tatau Province in relation to other geological provinces, offshore Sarawak. Also shown are the location of the key exploration wells.

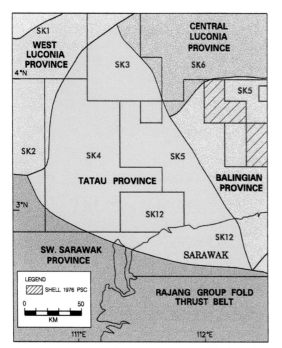 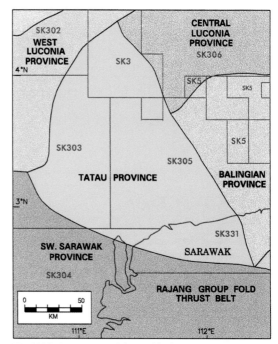

Fig. 17.2. Index maps of the exploration (PSC) blocks in the Tatau Province. (A) Old blocking system (B) New blocking system.

for Block SK12, onshore/offshore Sarawak, which covers parts of the Tatau and Balingian provinces (Fig. 17.2). Extensive exploration activities during the initial phase involved gravity, magnetic, and Synthetic Aperture Radar (SAR) surveys (Chiu and Mohd Khair Abd Kadir, 1990). After re-evaluating the area, OPIC refocused its efforts on the Igan-Oya Graben, in the western onshore part of the province.

Between 1989 and 1991 OPIC acquired more seismic data and carried out aeromagnetic and aerogravity surveys, soil gas analysis, and biostratigraphic analysis. These studies led to the drilling, in 1993, of the Oya-A1 well in the Igan-Oya Graben, a small outcropping extension of the Half-Graben Sub-province. The Oya-A1 well found dry gas in Cycle V sandstones.

The offshore part of the Tatau Province

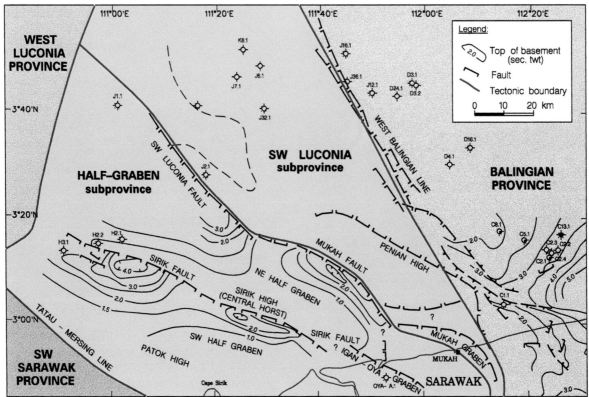

Fig. 17.3. Major structural elements of the Tatau Province and adjacent areas. Simplified from Shell (1973).

was first explored by Shell between 1955 and 1979. Shell acquired 540 km of seismic and drilled six wells between 1970 and 1978, resulting in five gas discoveries: J1.1X, J2.1X, H2.2X, H3.1X, and J5.1X (Fig. 17.1). The bulk of the gas was found in Cycle IV carbonates, with smaller gas accumulations in Cycles V-VI clastics and the pre-carbonate Cycles I-III clastics (Table 17.1). The J wells in the SW Luconia Sub-province encountered significant gas accumulations in Cycle IV carbonates. In addition, some oil and gas-bearing sands were also encountered in the sub-carbonate (Cycles I–III clastic) sequences. The most significant of these discoveries is the J4 oil accumulation which was discovered by Shell in 1978. The H wells in the Half-Graben Sub-province were drilled to test a Cycle IV carbonate pinchout on a tilted fault block. The second well, H2.2X, found 15 m of gas in carbonate reservoirs.

The next exploration phase in the offshore area was by Elf Aquitaine Malaysia in the northwestern part of the SW Luconia Sub-province (formerly Block SK3). This exploration campaign resulted in the drilling of the Tenggiri Marine-1, Selar Marine-1, Bawal Marine-1, and Jerung Marine-1 wells. These encountered several gas and condensate accumulations. Tenggiri Marine-1 and Jerong Marine-1 discovered condensate and associated gas in the Upper Oligocene–Lower Miocene Cycles I to III reservoirs. Selar Marine-1 found condensate in the Lower Miocene Cycles II and III. A small gas accumulation was found in carbonate reservoirs at the top of Cycle IV in the Jerong Marine-1 well. Gas was also encountered in the K5.1 well, within the Middle Miocene Cycles IV and V.

During late 1994, OPIC conducted a 2D seismic survey over the offshore extension of the Igan-Oya Graben (Fig. 17.3). No further drilling was carried out by OPIC until they subsequently relinquished the block.

STRUCTURE AND STRATIGRAPHY

The Tatau Province is characterised by extensional tectonics in the form of NNW-trending normal faults, most of which downthrow to the southwest (Fig. 17.3). Some of these faults have throws of up to 6 km. The Half-Graben Sub-province has two major NW-trending sub-parallel half-grabens — the NE Half-Graben and the SW Half-Graben. These half-grabens contain as much as 5 s (two-way time) of Cycles III to V siliciclastic sediments. The Mukah and the Igan-Oya Graben are regarded as the outcropping extensions of the NE Half-Graben and the SW Half-Graben, respectively. They are separated from one another by the Sirik High. The Sirik Fault bounds the northeastern edge of the Igan-Oya Graben. The southwestern margin of the Igan-Oya Graben is formed by the Patok High, which is part of the Rajang Fold-Thrust Belt. The Mukah Graben is flanked by the Penian High to the northeast. The West Balingian Line, a complex zone of high-relief reverse faults, forms the northeastern margin of the Penian High.

Figure 17.4 shows schematically the stratigraphy of the Tatau Province. Extension of the Tatau half-grabens took place during the Early Miocene (Cycles II and III times). This extensional phase resulted the formation of grabens. Erosion of previously deposited Cycles I/II sediments resulted in base Cycle II and base Cycle III unconformities. The Tatau Province remained a relatively high area that provided sediment to the adjacent West Luconia and Balingian provinces. Cycles I and II are therefore absent or very thin in most parts of the Tatau Province and are preserved only locally (Fig. 17.5). In many areas, Cycle III or Cycle IV rests directly on Palaeocene–Eocene basement, which has been penetrated by the Rajang-1 and J5.1X wells.

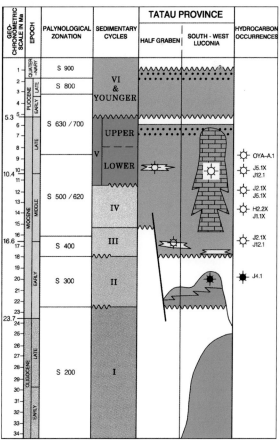

Fig. 17.4. Stratigraphic summary of the Tatau Province. Based on Shell (1994).

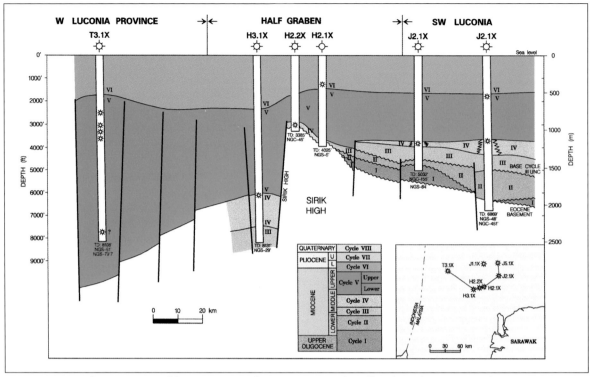

Fig. 17.5. Stratigraphic cross-section of the Tatau Province, from SW Luconia to the Half-Graben Sub-provinces, to the West Luconia Rim.

Fault movement during Cycles III to IV times resulted in the development of deep half grabens filled with thick Cycles III, IV and V sediments. At the end of Cycle III times, the area underwent compressional deformation, followed by the development of carbonate buildups on the uplifted horsts. Earlier phases of deformation have been strongly overprinted by major synsedimentary block faulting and basement rotation during Cycles III to IV times. A regional tectonic event at the end of Cycle V is represented by the base-Cycle VI unconformity, which is also recognised in the West Luconia Province. Northward progradation of siliciclastic sediments occurred during Cycles IV and V times. Generally, the Cycles V to VI strata thicken to the north in response to regional tilting of the Sarawak continental margin.

Figure 17.6 shows a model proposed by OPIC (1993) for the development of the Igan-Oya Graben, in the onshore southeastern part of the Tatau Province. The Igan-Oya Graben is envisaged to have developed by crustal extension associated a listric normal fault system, which is the precursor of the Sirik Fault. Ismail Che Mat Zin and Swarbrick (1997) interpreted this fault, which they referred to as the "Igan-Oya Line", as a major dextral strike-slip fault system that controlled the development of the Sarawak Basin. The syn-extensional phase along that fault system resulted in deposition of clastic wedges belonging to Cycles I and II, which was then followed by tectonic movements that resulted in a major base-Cycle III unconformity. During Cycles III and IV, another phase of extension occurred along a synthetic normal fault that was linked to the sole fault, resulting in the formation of a half-graben close to the Sirik Fault (Fig. 17.6). Another period of tectonic movement during the post-extensional phase resulted in a base-Cycle V unconformity.

Depositional environments in the Tatau Province range from alluvial through coastal plain to open marine shelf and slope, which reflects an overall deepening with time, i.e., up the stratigraphic column (Fig. 17.4). Because of the rift topography, Cycles I to III are very thin or absent except in deeper half-grabens in the northeastern part of the Province. Ganesan (1997) carried out a detailed palaeoenvironment analysis of the western Sarawak shelf, which includes the Tatau Province. The study indicates that the Cycles I to IV (Oligocene to Middle Miocene) were deposited in a nonmarine setting, with Cycle IV marking the beginning of a regional marine transgression that enabled bank carbonates to develop on the Tatau horst blocks. Cycle IV is therefore present throughout the Tatau Province. An example of the Cycle IV carbonate play is the H2.1 structure, which rests directly on the basement (Sirik High). While carbonate deposition occurred on the depositional highs, the Cycle IV sequence may have been predominantly siliciclastic within the intervening half-grabens (Ganesan,

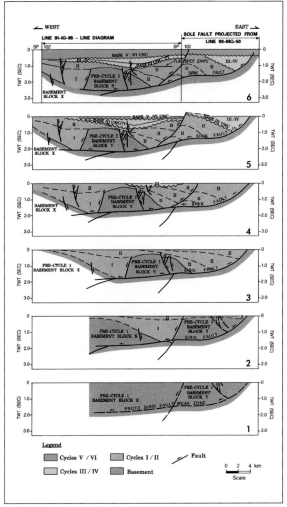

Fig. 17.6. Structural development of the Igan-Oya Graben based on interpretation of onshore seismic (after OPIC, 1993).

1997). Deltaic and shallow marine deposits typify the Cycle V and younger sequences

HYDROCARBON HABITAT

Many oil and gas discoveries have been made in the Tatau Province, although only the J4 accumulation has been declared commercial and is currently under development. Other significant discoveries include J1, J2, J5, H2, and H3 by Shell; Tenggiri Marine, Selar Marine, and Jerong Marine by Elf Aquitaine; and Oya-A by OPIC. Hydrocarbons occur in reservoirs that range in age from Cycles I to V (Oligocene to Miocene age).

Play Types

Two main hydrocarbon play types occur in the Tatau Province. The most common type is formed by Cycle IV carbonate buildups that developed on extensional horst blocks. Cycle IV carbonates are the most common reservoir rocks in the wells drilled on the horsts in the northern part of the Tatau Province. They also form the major gas-bearing reservoirs encountered in the H and J structures. Less commonly, carbonate traps are formed by carbonate pinchouts against tilted fault blocks (e.g. H2.2). The Cycle V neritic shales act as regional seals for these reservoirs. Up to 204 m of net gas pay have been recorded from the carbonate reservoirs in these structures.

The second major play type is the pre-carbonate (Cycles I to III) clastic play which has been proven to be oil-bearing in the Tatau Province and could be an important exploration target in the near future. The Cycles I-III sandstones are better developed in the J structures in the northeastern part of the province. In the southeastern part of the Tatau Province, the Cycles I-III clastic play occurs in anticlinal traps formed by transpressional reactivation of the half-graben bounding normal faults. Hydrocarbons in the pre-carbonate clastic play were probably sourced from Cycles I/II lower coastal plain shales, which are presently mature over most of northeastern part of the Tatau Province. Intraformational, as well as the overlying Cycle IV neritic shales, provide the seals for the pre-carbonate play. The discovery of J4 in 1978 was a significant event because that it has opened a new hydrocarbon play below the Cycles IV/V carbonates in this area.

Besides the Cycle IV carbonates and Cycles I/II/III clastic plays, the upper Cycle IV to Cycle V siliciclastic reservoirs form a minor objective in the Tatau Province. In addition, several leads have been identified, including thrust structures and four-way dip closure below the intra-Cycle V unconformity, and faulted blocks below the Cycle VI unconformity. Potential unconformity traps may also occur as a result of onlapping strata on the intra-half-graben basement highs.

One of the more recent successes in the Tatau Province is the Oya-A gas accumulation discovered by OPIC in late 1993. The Oya-A structure is a small four-way dip closure located on the northeastern edge of the Igan-Oya Graben, onshore Tatau Province, which is the eastern extension of the SW Luconia Graben (Fig. 17.3). Several seismic flat spots were identified in what was interpreted as Cycles I and II sequences. The objective of the Oya-A1 well was to test the potential of Cycles I and II reservoirs. The drilling results, however, showed that the presumed age for the targeted succession was incorrect. The Oya-A1 well found Cycle VI unconformably overlying Cycle V, and was terminated in Cycle

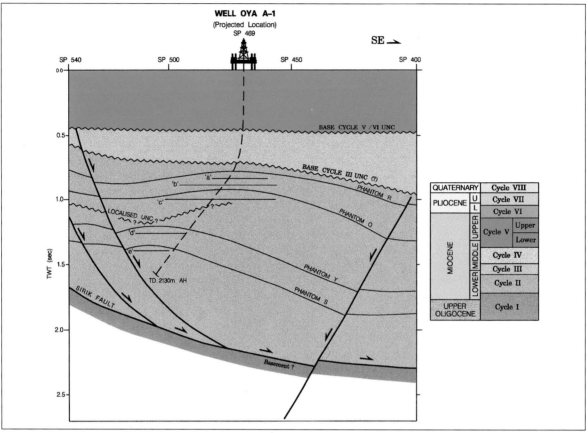

Fig. 17.7. Seismic cross section from the Igan-Oya Graben, showing a hangingwall rollover feature associated with a half-graben bounding normal fault. The Oya-A prospect was tested by the Oya-A1 well. For location see Fig. 17.1. Based on OPIC (1993).

IV (Fig. 17.7). The well had penetrated four gas-bearing sand packages in Cycle V. The deeper two reservoirs tested 7.23 MMSCF per day dry gas and about 48 m NGS was recorded.

Reservoirs

The Cycle IV carbonate reservoirs consist of Lower-Middle Miocene limestone and dolomite with excellent reservoir quality. Their porosities range between 16% and 40%, and average generally between 25% and 35%. Sandstone reservoirs in Cycles I and II were deposited in a lower coastal plain setting and are generally thicker than the shallow marine sandstones in Cycle III. Their porosities generally range between 10% and 32%. The J4 accumulation is an example of a Cycles I/II sandstone reservoir in a N-trending fault-closed anticlinal trap (Fig. 17.8). There is also a stratigraphic component to the trapping mechanism, as part of the reservoir is truncated by the Lower Miocene (I-79) Unconformity. The Cycles I and II reservoirs consist of stacked channel sands deposited in a lower coastal plain environment. There are three main pay zones, each averaging about 28 m thick. They occur at depths of between 1400 m and 1500 m, with a net reservoir thickness of about 80 m. The flow rate in J4.1 was about 3000 BOPD. The reservoirs in J4 has porosities of about 18% to 29%.

Potential hydrocarbon-bearing reservoirs occur also in Cycles IV and V. In H3.1X well, gas has been found in the uppermost Cycle IV calcareous sandstones and sandy limestones with average porosity of 20%. The reservoirs in the Oya-A1 discovery are formed by Cycle V coastal to shallow marine sandstones that are up to 11 m thick and have porosities of up to 27% and permeabilities of about 44 mD.

SOURCE ROCKS

Hydrocarbons in the Cycle IV carbonate play are believed to have been generated from mature Cycles I-III lower coastal plain shales, similar to those in the Balingian Province (Chapter 14). The deposition of source rocks during this period is probably related to the shifting depositional environments associated with a fault-controlled coastline. In general, the Cycles I/II/III shale source rocks, which occur mainly in the SW Luconia Sub-province, have poor to fair organic carbon richness, with poor to good hydrocarbon generating potential. The organic matter is dominated by gas-prone Type III

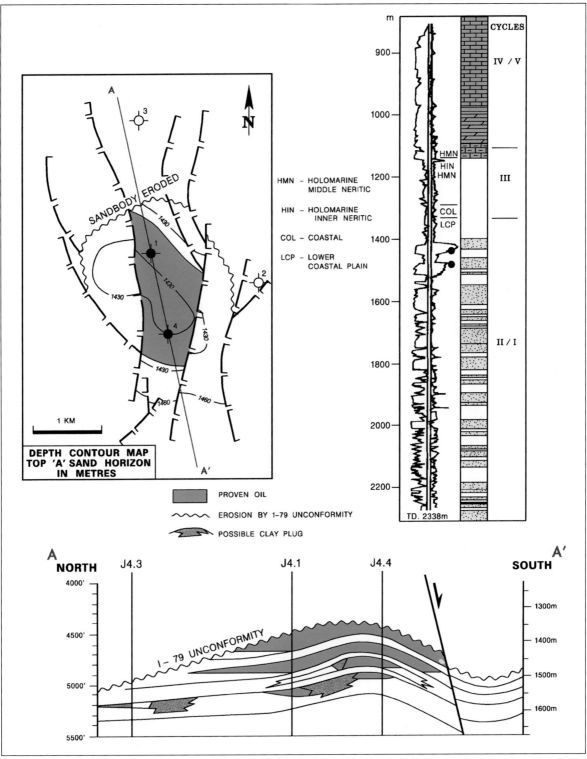

Fig. 17.8. Schematic cross-section of the J4 oil discovery, which is currently being developed. For location see Fig. 17.1. From OXY (1989).

kerogen. Cycles I/III coal samples at Bawal Marine-1, K2.1X and J4.5X have excellent hydrocarbon generating potential. The kerogen types are mainly Type III and Types II/III organic matter with some liquid hydrocarbon potential. The plots of Hydrogen Index against Tmax for selected wells with Cycles I, II and/or III penetrations show the various kerogen types in the Tatau Province (Fig. 17.9).

Cycles IV/V sediments form potential source rocks in the southern and northern margins of the Tatau Province. In Oya A-1 well, the Cycles IV and V sediments contain fair to moderate organic richness with poor hydrocarbon generating potential. The kerogen is predominantly gas prone. The same has been observed in the K5.2/ST1 well, which penetrated Cycle IV to Cycle VII sediments. The shale

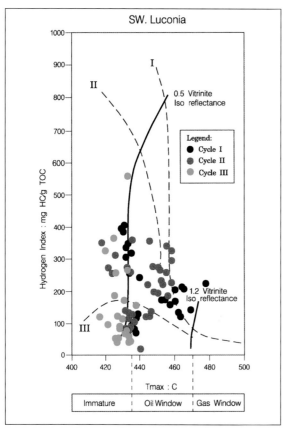

Fig. 17.9. Hydrogen Index against Tmax for selected well penetrations of Cycles I, II, and III in SW Luconia Sub-province.

samples have poor to fair organic carbon richness with poor to fair hydrocarbon generating potential. The kerogen is dominated by Type III gas prone material, with subordinate Type II/III mixed oil-gas prone, mainly terrigenous organic matter.

Vitrinite reflectance data from SW Luconia show that Cycle II and Cycles I/II sequences are presently within the oil generation window (Fig. 17.10). The lower Cycle III is within the early oil-generation window, the upper Cycle III, Cycle IV and younger sequences are thermally immature. However, although the maturity profile generally indicates that Cycles I-III shales/coals are at optimum maturity for generating hydrocarbons, the paucity of lower coastal plain sediments, especially in the southern part of the province, may limit their source potential. The condensate discoveries in K5 and J32.1, however, prove the existence of a viable source rock.

A shift in the top of the oil window from Cycle IV (Half-Graben) to Cycle III (SW Luconia) (Fig. 17.10) suggests that thermal maturity generally increases in a northward direction from the Half-Graben into SW Luconia Sub-province, as the sediments are being buried deeper and deeper to the north. This seems to be the general maturity trend in offshore Sarawak. This trend also implies that Cycle III sediments in the deep grabens are largely overcooked or are in the tail end of the gas window. The Cycle III

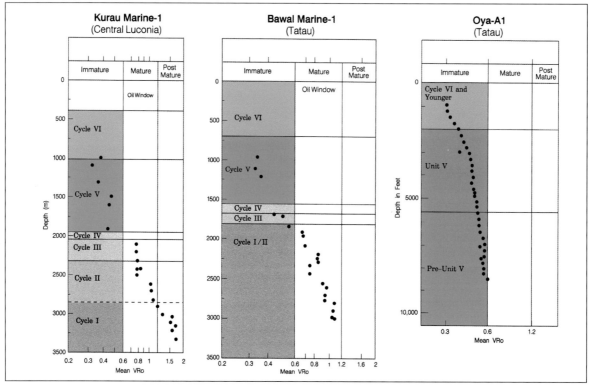

Fig. 17.10. Comparison of maturity profiles from the Kurau Marine-1 (Central Luconia), Bawal Marine-1 (SW Luconia Sub-province), and Oya-A1 wells (onshore Half-Graben Sub-province).

source rocks may have passed through the oil window rather early, probably during early Cycle V times. Hence, by the time the upper Cycle IV carbonates were sealed by the Cycle V shales, the source rock kitchen would have passed through the gas window. This may explain the observed hydrocarbons in H2.2X and Oya A-1. In the Oya A-1 well the oil window is probably near the T.D., which suggests that Cycle V and most of Cycle IV are thermally immature.

The movement of the SW Luconia Fault, running NW–SE across the southeastern part of the province (Fig. 17.3), may play an important role in speeding up the maturation process by acting as a heat conduit. This may be the cause of high heat flow observed in SW Luconia, especially the southern part. The narrow oil window observed across the sub-province is possibly an effect of this high heat flow. Generally, the SW Luconia Sub-province has a thick sedimentary section compared to the Central Luconia and Balingian provinces. Because of the great depth of burial and high geothermal gradient (> 55°C km^{-1}, Ganesan, 1997), the Cycles I/II sequences in SW Luconia may already be in the gas window. The gas discoveries, coupled with the observed lean and gas-prone source rocks, suggest that the SW Luconia sub-province is a gas-prone area. The distribution of oil and gas fields in the Tatau Province (Fig. 17.1) suggests that they may have been charged from the northeastern side (possibly along relatively short migration pathways), where Cycles I-III would have moved through the oil window in late Cycle V times, and had since remained within the gas window until the present day.

HYDROCARBON TYPES

Oil

Oil in the Tatau Province has been found in only two places: in the A21.1 well and in the J4 structure. Hence, there are few data available on the oil characteristics. The J4 oil is light and waxy, with an API gravity of 41° and a pour point of 80°C.

Condensates

Several of the gas discoveries in the Tatau Province have yielded considerable quantities of condensate. Condensates are found in

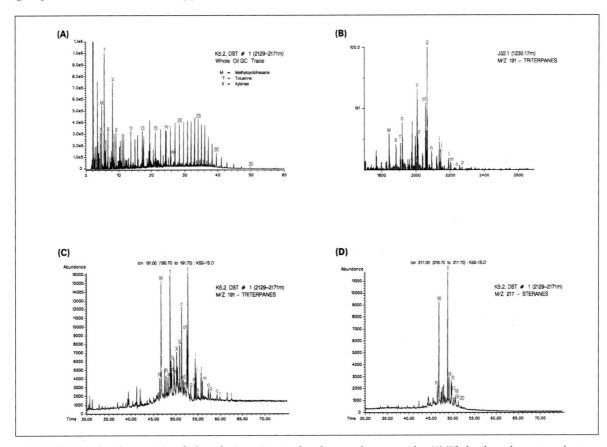

Fig. 17.11. Biomarker characteristics of oils in the Tatau Province based on gas chromatography. (A) Whole-oil gas chromatography trace from K5.2 (2129-2171m), (B) Triterpane fragmentogram from J32.1 (1230.17m), (C) Triterpane fragmentogram from K5.2 (2129-2171m). (D) Sterane fragmentogram from K5.2 (2129-2171m).

the J32.1, Tenggiri Marine-1, Jerong Marine-1, Selar Marine-1 and K5.2/ST1 wells, in various sequences; in Cycle IV (J32.1, K5.2/ST1), in Cycles I to III (Tenggiri Marine-1, Jerong Marine-1) and in Cycles II/III (Selar Marine-1). The J32.1 condensate is non-biodegraded, has high Pr/Ph ratios (> 5.0) indicative of an oxic depositional setting, high Pr/nC$_{17}$ ratios (2.0–2.5), and a highly mature n-alkane envelope. The biomarker characteristics indicate the presence of oleanane, bicadinanes, and dominant C$_{30}$ hopanes (Fig. 17.11). These features suggest that the Tatau condensates were generated from shaly source rocks containing higher land-plant organic matter deposited in a lower coastal plain (swampy) environment subject to significant higher plant input.

In K5.2/ST1 well, which was drilled in 1997, the condensates are enigmatic, being very waxy and solid at room temperature. Compositionally, they contain abundant high molecular waxes, which is not a common feature of condensates (Fig. 17.11). They may represent an oil rim or co-mingling of a stratified condensate and waxy oils. Both situations are known to occur in the Central Luconia Province (Chapter 15). Despite the high wax content, these condensates have similar characteristics as the J32.1 condensates i.e. non-degraded, moderate Pr/Ph ratios (>3.0), high oleanane, high C$_{30}$ hopanes, high hopane/sterane ratios (>6.0) and the dominance of resin-derived bicadinanes (Fig. 17.11). These features indicate that the condensates were derived from a coal or coaly shale source that contains abundant higher land-plant organic matter. The occurrence of J32.1 and K5.2 condensates suggests that the lower coastal plain sediments are effective source rocks in the Tatau Province, in agreement with the source facies depositional model described earlier in this chapter.

Gases

The source and maturity of hydrocarbon gases may be deduced from their chemical and isotopic compositions (Table 17.1). In general, gases generated from Type III kerogen dispersed in shale and concentrated in coal are isotopically heavier and chemically drier than those generated from Type II kerogen at equivalent levels of thermal maturity, especially during the mature stage of hydrocarbon generation (Johnson and Rice, 1990). Gases at K5.2/ST1 are considered as dry because more than 95% of the hydrocarbon gases is methane, although a high proportion of CO_2 (> 50% of the total gases) is present.

Gases generally become isotopically heavier with increasing level of thermal maturity (Johnson and Rice, 1990). Thermogenic gases that are generated during the mature stage have $\delta^{13}C_1$ values ranging between about –55‰ to –4‰, depending on the C_1/C_{1-5} ratios. At higher temperatures, light hydrocarbons are preferentially formed by thermal cracking of pre-existing hydrocarbons. Gases generated during the post-mature stage (dry gas) have $\delta^{13}C_1$ values less than –40 ‰.

Stable carbon isotope data (Table 17.2) indicate that most of the Tatau gases are of thermogenic origin. There appears to be a correlation between CO_2 content and the thermal maturity of gases. Gases that contain over 30% of

WELL	Year Drilled	Reservoir	Net pay (m)	Porosity (%)	%CO$_2$	%N$_2$
J1.1X	1970	Cycle IV carbonate	204	23-28	62	15
		Cycle III sandstones	3	25	18	4
		Cycle VI sandstones	8	32	-	-
H2.1X	1970	Cycle VI	2.4	-	-	-
J2.1X	1970	Cycle IV carbonate	4.7	-	2	31
		pre-carbonate	24	-	6	33
H2.2X	1970	Cycle IV carbonate	13.7	16-28	50	30
H3.1X	1970	Cycle IV calcareous	8.8	20-23	62	11
		sandstone/sandy limestone				
J5.1X	1978	Cycle IV carbonate	137	-	87	4
		Cycle V sandstones	9.4	-	-	-

Table 17.1. Results of wells drilled by Shell in the Tatau Province, showing the reservoir types and CO$_2$/N$_2$ percentages.

CO_2, such as those in K5.2/ST1, have $\delta^{13}C_1$ values of between –38‰ and –33‰, which suggest that they were generated during the post-mature stage. On the other hand, gases with less than about 15% CO_2, such as in Selar Marine-1, have $\delta^{13}C_1$ values of between –46‰ and –42‰, which implies generation during the mature stage. The Oya-A1 gases do not contain CO_2 and N_2 and are extremely dry, with 99% methane. They are probably of biogenic or early thermogenic origin.

CO_2 and N_2

Many of the gas discoveries on the western Sarawak shelf, including those in the Tatau Province, contain high amounts of non-hydrocarbon gases, mainly CO_2 (up to 88%) and N_2 (up to 39%). For example, the J wells drilled by Shell contain high amounts of CO_2 and N_2 (ca. 60% and 20%, respectively). The H2.2X well encountered gas with 50% CO_2 and 30% N_2. Figure 17.12 shows the distribution of these gases

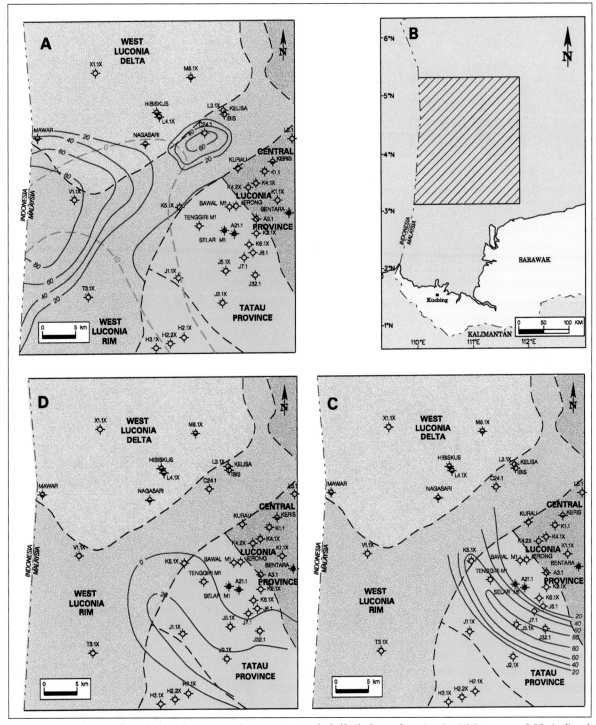

Fig. 17.12. Distribution of non-hydrocarbon gases in the western Sarawak shelf. Clockwise, from A to D: (A) Percentage of CO_2 (red) and N_2 (green) in Cycles IV/V/VI. (B) Index map of area. (C) CO_2 in Cycles IV/V/VI. (D) N_2 in Cycles IV/V/VI. Modified from Triton (1994).

Wells	TGM-1		SLM-1		K5.2/ST1	
Tests	DST2A	DST1	DST3	DST2	DST1	DST2
Depth (m)	2401-2412	2591-2597	2316-2327	2470-2485	2129-2171	1988-2013
N_2	6.65	7.73	0.95	1.30	2.43	2.57
CO_2	11.75	38.52	10.92	13.49	54.33	54.23
C_1	61.38	45.87	59.83	61.34	41.88	41.57
C_2	11.22	5.43	11.06	10.61	0.94	0.93
C_3	5.45	1.45	11.52	9.06	0.16	0.40
iC_4	1.25	0.38	2.39	1.73	0.09	0.09
nC_4	1.13	0.28	2.17	1.61	0.08	0.10
iC_5	0.51	0.16	0.66	0.46	0.04	0.04
nC_5	0.30	0.08	0.35	0.26	0.03	0.06
C_6	0.33	0.09	0.15	0.14	0.02	0.01
C_1/C_{1-5}	0.75	0.85	0.68	0.72	0.97	0.96
$\delta^{13}C\ CO_2$ ‰	-6.9	-3.7	-10.8	-7.5	-2.6	-2.4
$\delta^{13}C\ C_1$	-42.1	-38.0	-45.6	-43.5	-33.3	-33.3
$\delta^{13}C\ C_2$	-31	-27.6	-32.7	-32.0	-24.9	-24.3
$\delta^{13}C\ C_3$	-27.1	-24.5	-30.1	-29.1	-24.0	-23.4

Table 17.2. Gas compositions and carbon isotope data from selected wells in the Tatau Province.

in the western Sarawak shelf. In the Tatau Province, more CO_2 is observed in carbonate than in clastic reservoirs (Table 17.1), unlike in the West Luconia Province, where high CO_2 gases are commonly found in clastic reservoirs. The distribution of CO_2 appears to follow the fault trend in the Tatau Province, suggesting that the extensional normal faults, particularly those in the SW Luconia Sub-province, may have acted as conduits for migration of the gases from deeply buried sources. This explains the high concentrations of CO_2 in SW Luconia and much lower concentrations in the Half-Graben Sub-province (Fig. 17.12C). Besides CO_2, N_2 is also a significant inert component in some parts of the Tatau Province, reaching over 30% in places. High N_2 concentrations (>25%) occur in the H2 and J2 structures in the Half-Graben Sub-province. The rest of the western Sarawak shelf seems to have relatively low N_2 concentrations.

Isotopic analysis of gases provide some insights into the origin of the CO_2. The CO_2 in natural gas can come from either an organic or inorganic source (e.g. Hunt, 1996). Inorganic CO_2 may be produced by thermal decomposition of carbonates that are buried at great depths, by mantle degassing, through volcanic activity, or diagenetic reactions involving carbonate and silicate minerals, whereas organic CO_2 is generated directly from organic matter during diagenesis via processes such as the decarboxylation of organic acids, oxidation of organic matter by minerals, bacterial degradation of organic matter, thermochemical sulphate reduction and hydrolytic disproportionation of petroleum. Figure 17.13 shows the range of stable carbon isotope ($\delta^{13}C$) values for CO_2 of various origins.

In general, the $\delta^{13}C$ values of CO_2 for the offshore Sarawak show a wide range from –11‰ to –2.4‰ suggesting that the CO_2 is predominantly inorganic in origin. High concentrations of CO_2, coupled with heavier isotopic ($\delta^{13}C\ CO_2$) values in the range of –2.4‰ to –3.7‰ in K5.2/ST1 and Tenggiri Marine-1 (Table 17.2) probably resulted from the thermal breakdown of deeply buried, marine carbonate sediments. The correlation between CO_2 content and thermal maturity, mentioned earlier, also implies that the CO_2 may have been derived at higher temperatures from a deeply buried source. In contrast, the $\delta^{13}C$ values of gases containing lower CO_2 concentrations, as in Selar Marine-1, are less (lighter) than –7.5‰. This suggests a mixed source with a minor organic component, probably derived from the

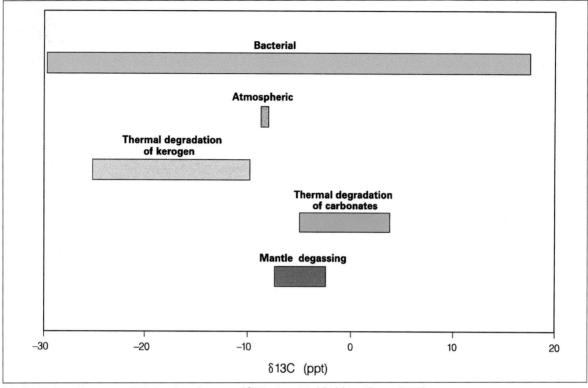

Fig. 17.13. Possible sources of CO_2 based on the range of $\delta^{13}C$ values. Modified from Clayton (1995).

maturation of the organic matter.

When the carbon isotopic compositions of methane, ethane, and propane from Selar Marine-1, Tenggiri Marine-1 and K5.2 wells are plotted on James's (1983) maturity diagram using the ethane to propane separations, the gases seem to cluster in two groups (Fig. 18.8 in Chapter 18). The K5.2 gases appear to have been generated at higher maturity (VRo > 1.5%), whereas the other gases have a source maturity of 0.7 to 1.0%, which incidentally is the peak oil generation window.

A study by Idris (1991) on CO_2 in J32.1 showed that the CO_2 may have been derived from dissolution of carbonates within the sedimentary section by acidic meteoric or formation fluids. Idris (1991) also studied the N_2 contaminants in J32.1. The origin of N_2 is poorly understood but may have been derived from deep crustal sources/basement. Other possible origins are from maturation of humic organic matter (particularly coals), dissolved atmospheric N_2 and thermal degradation of nitrate minerals found in some evaporites (Hunt, 1996).

CONCLUSION

The Tatau Province is characterised by an extensional structural style that dominates the nonmarine Upper Oligocene to Lower Miocene strata (Cycles I to III). Cycle IV times mark the beginning of marine sedimentation in the province that resulted in the development of carbonate buildups on some of the extensional fault-blocks. Meanwhile, the deltaic and shallow marine siliciclastic sedimentation from Cycle IV onwards continued in the intervening half-grabens.

The main hydrocarbon play types comprise the carbonate reservoirs in Cycle IV and the siliciclastic reservoirs in the pre- and post-Cycle IV sections. These reservoirs derived hydrocarbons from source rocks deposited in the lower coastal plain environments during Oligocene–Early Miocene times. The source rocks, mainly from Cycles II and III, are currently in the mature stage of hydrocarbon generation.

The most significant hydrocarbon discovery in the province is the J4 accumulation in the SW Luconia sub-province, discovered by Shell in 1978. J4 is important in that it has opened a new hydrocarbon play below the Cycles IV/V carbonates. In addition, the Oya-A gas discovery has encouraged further exploration activities onshore, not only in the Tatau area but also in the adjacent Tinjar Province to the east.

REFERENCES

Chiu S. K. and Mohd Khair Abd Kadir, 1990. The use of SAR imagery for hydrocarbon exploration in Sarawak. Bulletin of the Geological Society of Malaysia, 27, 161-182.

Clayton, C.J., 1995. Controls on the carbon isotape ratio of carbon dioxide in oil and gas fields. In: Grimatt, J.O. and Dorronsoro, C., eds., Organic Geochemistry: Developments and Applications to Energy, Climate, Environment and Human History. AIGOA, Donastia-San Sebastian, 1073-1074.

Ganesan, B.M.S., 1997. Geology and hydrocarbon potential of the offshore western Sarawak shelfal area. Proceedings of the ASCOPE '97 Conference, Kuala Lumpur, 24-27 November 1997, v. 2, 131-169.

Hunt, J.M., 1996. "Petroleum geochemistry and geology". 2nd Edition: New York, W.H. Freeman, 743pp.

Idris, M.B., 1991. CO_2 and N_2 contamination in J32.1, SW Luconia, offshore Sarawak. Bulletin of the Geological Society of Malaysia, 32, 239-246.

Ismail Che Mat Zin and Swarbrick, R.E., 1997. The tectonic evolution and associated sedimentation history of Sarawak Basin, eastern Malaysia: a guide for future hydrocarbon exploration. In: Fraser, A.J., Matthews, S.J. and Murphy, R.W., eds., Petroleum Geology of Southeast Asia. Geological Society of London Special Publication, 126, 237-245.

James, A.T., 1983. Correlation of natural gas by use of carbon isotopic distribution between hydrocarbon components. American Association of Petroleum Geologists Bulletin, 67, 1176-1191.

Johnson, R.C. and Rice, D.D., 1990. Occurrence and geochemistry of natural gases, Piceane Basin, Northwest Colorado. American Association of Petroleum Geologists Bulletin, 74, 805-829.

OPIC, 1993. Well proposal, OYA A-1, Block SK-12, onshore Sarawak. Overseas Petroleum and Investment Company (Malaysia), Unpublished Report No. 1177.

OXY, 1989. North Borneo Regional Study: Offshore Sarawak. Occidental Malaysia Inc. Unpublished Report, ER: OXY:3:89-01.

Shell, 1973. Geology of the Tatau Province, Sarawak Shelf, northwest Borneo (A case study for exploration of faulted basin areas on continental structures). Sarawak Shell Berhad, Unpublished Report. ER:55B:3:73-04.

Shell, 1994. SK5 Exploration Data Book 1995. Sarawak Shell Berhad, Unpublished Report.

Triton, 1994. Malaysia Offshore Sarawak SK-1 Block Evaluation. Triton Energy Company, Unpublished Report.

Chapter 18

West Luconia Province

*Mazlan B. Hj. Madon and
Redzuan B. Abu Hassan*

INTRODUCTION

The West Luconia Province is located in the western part of the Sarawak continental shelf and appears to continue westwards into the East Natuna Basin (Fig. 18.1). The province covers an area of about 20,000 km^2 between two major Miocene carbonate provinces – the Central Luconia Province to the east and the Natuna Platform to the west. These carbonate platforms are relatively high basement areas that were not affected by rapid subsidence due to loading by deltaic sedimentation. To the south lies the SW Sarawak Province, which is the offshore extension of the Sibu Zone. The SW Sarawak Province is bounded by the Lupar Line to the southwest and the Tatau-Mersing Line to the northeast. It extends westwards into the northeastern margin of the Natuna Arch.

The West Luconia Province represents a large deltaic depocentre, analogous to the Baram Delta complex of east Sarawak and Brunei. This major delta probably evolved as the ancestral Lupar-Rajang delta system which built out northwards across the SW Sarawak Province beginning in the Middle Miocene, during Cycle III to Cycle VIII times. It appears to have prograded beyond the present shelf edge onto the continental slope (Rajang Slope), where the sediment thickness is in excess of 7.5 km. Unlike the Baram Delta, however, the petroleum potential of the West Luconia Province has not been fully explored. Its location between two prolific hydrocarbon provinces, the Central Luconia Province and the Natuna Platform, indicates that there is still potential for major discoveries here. This chapter gives an update of the petroleum geology and exploration potential of this area. Since no studies have been published on the area, this account is based on previously unpublished compilations, mainly by Shell (Tan and Swinburn, 1993) and Triton (1994), incorporating some new geochemical data from PETRONAS.

EXPLORATION HISTORY

The West Luconia Province was part of the greater Sarawak Offshore Concession explored by Shell between 1966 and 1970. During this period a total of 6593 km of 2D seismic with a resolution down to 2.5 x 5 km were acquired. The province encompasses the exploration blocks SK301, SK302, SK303, and part of SK3 (Fig. 18.1), and has been explored only lightly by two operators during the early 1970s and late 1980s. Only 9 wells have been drilled, which resulted in 1 gas discovery (with minor oil) and 4 wells with oil and gas shows. There has been little activity since 1990 when the last operator, Idemitsu, relinquished the area.

Prior to drilling in West Luconia, Shell drilled three wells in the SW Sarawak Province to the south: P1.1X, P1.2X, and P1.3X (Fig. 18.1). All the wells penetrated at relatively shallow depths (about 500 m) the pre-Cycle I basement, which consists of hard lithic/feldspathic sandstones, siltstone, and shale, overlain by a thin succession of Cycles V and VI clastic sediments. No Cycles I to IV sediments were found. Well P1.1X encountered about 23 m of good porosity (30%) Cycle V sandstone, while P1.3X encountered about 60 m of Cycle V sediment, but no hydrocarbons have been found.

In 1970 Shell drilled three wells – L4.1X, V1.1X, and X1.1X – in the central part of the West Luconia Province, in the present SK301 and SK302 blocks (Fig. 18.1). Well L4.1X was drilled on a fault-closed hangingwall rollover of a north-hading growth fault with a direct hydrocarbon indicator (DHI), thought to be related to gas-bearing sands. The well encountered 34 m of net gas sand and some 33-38° API oil in Upper Miocene Cycle V sandstones. Well V1.1X was drilled on a four-way dip closure. Although good reservoir was found in the Upper Miocene primary objectives, V1.1X was drilled off-structure except at the topmost Upper Miocene level, and recovered high-CO_2 (88%) gas. Because of mechanical problems the third well, X1.1X, was prematurely abandoned in Lower Pliocene

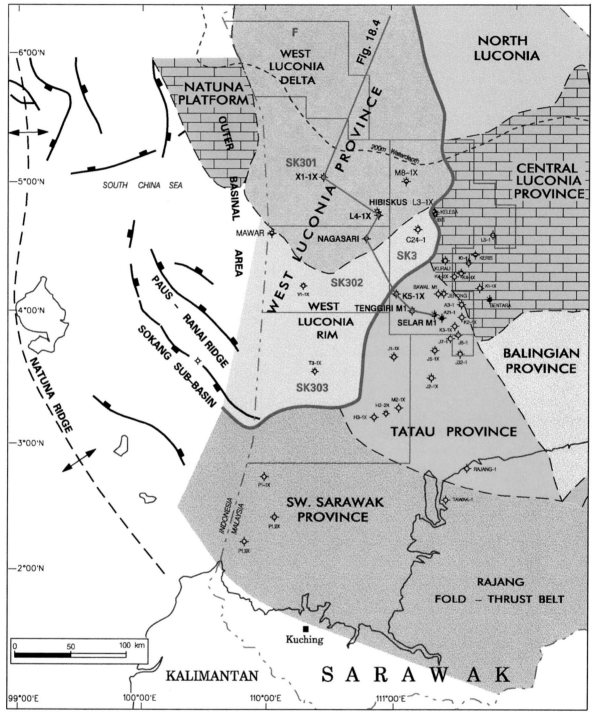

Fig. 18.1. Location of the West Luconia Province, in relation to other provinces in the region. Also shown are the exploration block boundaries and the names of key wells.

marine clays and did not find any hydrocarbons. As a result, the targeted Upper Miocene sandstones in the hangingwall fault block still remain untested. On the West Luconia Rim, south of X1.1X, Shell drilled the T3.1X well to test Cycle V clastic reservoirs. They found gas shows with a high CO_2 content (ca. 60%). An appraisal well, T3.2, was drilled by PETRONAS Carigali in 1997, to confirm the gas in T3.1 and test the gas potential of intra-Cycle V and lower Cycle V sandstones. A test was done at 2019-2020m and, unfortunately, found mainly CO_2.

On 19 May 1987, the concession area originally held by Shell was awarded to a consortium headed by Idemitsu (40%), with Pecten holding another 40% and PETRONAS Carigali the remaining 20%. During that year, Idemitsu acquired 7988 km of seismic on grids down to 1x2 km in the central part of the concession. They also reprocessed 1866 km of Shell's older vintage lines. In 1989 PETRONAS acquired a further 45.5 km of seismic to tie X1.1X to the North Luconia Province. Idemitsu drilled three wells in 1988 in the central part of the province - Hibiskus-1, Mawar-1, and Nagasari-1. All three wells had oil and gas shows in Upper Miocene Cycle V sands (Table 18.1). Hibiskus-1 was drilled on a dip- and fault-closed structure that was previously tested in a crestal position by the L4.1X well. The Hibiskus-1 well was drilled to 2455 m and encountered about 12 m NGS and minor oil shows in Cycle V sands.

The Mawar-1 well tested a faulted anticlinal structure with an areal closure of 24 km^2 and vertical relief of 180 m at the base of Cycle VI. After re-drilling and side-tracking because of mechanical problems, Mawar-1AS encountered minor gas shows at 2360 m. The final well drilled by Idemitsu, Nagasari-1, tested a dip- and fault-closed structure with an areal closure of 67 km^2 and vertical relief of 180m at the base-Cycle VI level. Although the porosity of Cycle V sandstones was high (11-26%) only minor CO_2-bearing gas was found. Idemitsu and its partners relinquished the area in May 1990.

In 1990, as part of their operations in Block SK3 (see Fig. 18.1), Elf Aquitaine drilled the C24.1 well in the northeastern end of the West Luconia Rim. The drilling objectives were Upper Miocene sandstones in an elongated anticlinal trap formed on the upthrown side of a NE-trending antithetic normal fault. The well found good quality reservoir sands with greater

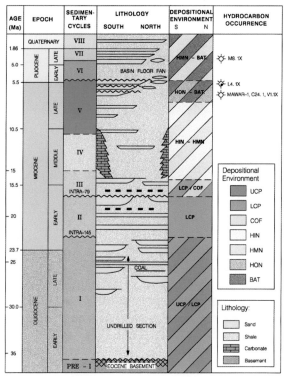

Fig. 18.2. Summary of the stratigraphy and hydrocarbon occurrences in the West Luconia Province. Key for depositional environments: UCP- upper coastal plain, LCP- lower coastal plain, COF - coastal fluviomarine, HIN - holomarine inner neritic, HMN- holomarine middle neritic, HON- holomarine outer neritic, BAT- bathyal.

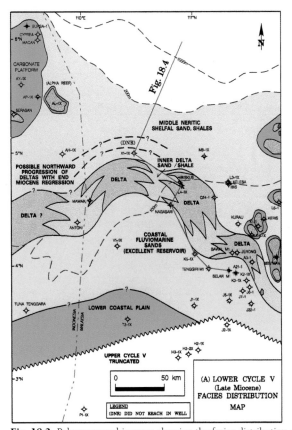
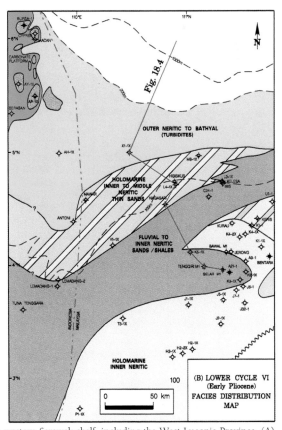

Fig. 18.3. Palaeogeographic maps showing the facies distribution on the western Sarawak shelf, including the West Luconia Province. (A) lower Cycle V (Late Miocene) (B) lower Cycle VI (Early Pliocene). Modified after Triton (1994).

than 30% porosity and 1000mD permeability at depths of around 1500 m, but a test of one zone at 1515-1527 m found mainly CO_2.

STRUCTURE AND STRATIGRAPHY

Like the Baram Delta, which is confined between the West Baram Line and the Jerudong-Morris faults, the West Luconia Delta developed between the carbonate shelf edges of the Central Luconia platform to the east and the Natuna platform to the west. The West Luconia Province consists of two sub-provinces: the West Luconia Delta and the West Luconia Rim (Fig. 18.1). The delta forms the core of the province and is characterised by north-hading, east-west trending growth faults. It was an area of rapid sedimentation that resulted from the northward progradation of the ancestral Rajang/Lupar delta since the Early Miocene. Sedimentation occurred in coastal plain to marginal marine environments during the Oligocene to Early Miocene, and in predominantly shallow marine environments during the Middle Miocene and later (Fig. 18.2).

Generally, Cycles I to III have not been penetrated in the province. Middle Miocene Cycle IV carbonates occur in some places, especially on the fringes of the Central Luconia platform, as at J1.1X (Fig. 18.2). Away from areas of carbonate sedimentation, the main phase of deltaic progradation occurred during Cycles III, IV, and V times (Middle Miocene to Pliocene), resulting in the deposition of more than 7 km of sediment on the shelf edge.

A much thicker post-Cycle V succession was deposited by the northward-progradation of the delta (Fig. 18.3). Further northwards beneath the Rajang Slope, the Cycle V and younger succession rest directly on basement, and are characterised by north-verging deltaic toe-thrusts and associated asymmetrical folds (Fig. 18.4). The Early Pliocene was characterised by a major marine transgression over the shelf, with deposition taking place in environments ranging from inner neritic in the south, to bathyal in the north (Fig. 18.3). Continuous deltaic progradation and subsidence have produced east-west trending, down-to-the-basin, growth faults, with typical rollover features, especially in the central and northern parts of the delta. Rapid sedimentation and

WELL	Year	Total depth (m)/ formation at TD	Results
L4.1X	1970	2672.4 Cycle VI	34.2 m of net gas sand in lower Cycle IV and Cycle V sands. Trace of 38° API oil recovered in one RFT.
X1.1X	1970	2267.4 Cycle VI	Abandoned dry
V1.1X	1970	2706.9 Cycle V	Minor gas shows.
T3-1X 1	1970	2471 Cycle IV(?)	15.5 m net gas sand, 60% CO_2
Hibiskus-1	1988	2454.3 Cycle V	12 m net gas sand in Cycle V. Oil shows in Cycle V sands at 2271-2362m.
Mawar-1A, 1AS	1988	2360.4 Cycle V	net gas-bearing sands: 4 m in Cycle VI and 21 m in Cycle V.
Nagasari-1	1988	2211.3 Cycle V	Trace oil shows 2136-2191m, minor gas in 3 RFTs (12-77% CO_2 and 2.3-3.1% N2).
C24.1	1990	2100 Cycle V	mainly CO_2 tested at 1515-1527 m.
T3.2	1997	2602 Cycle V	oil and gas shows in Cycle V sands at 2019-2020 m.

Table 18.1. Summary of some major drilling results in the West Luconia Province, offshore Sarawak (obtained partly from Tan and Swinburn, 1993).

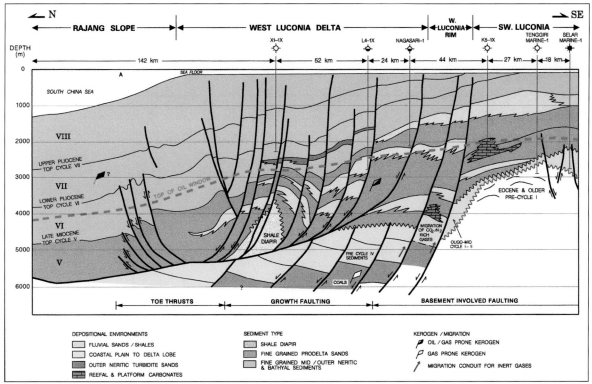

Fig. 18.4. Structural-stratigraphic cross-section of the western Sarawak shelf, showing the structural styles and traps in the West Luconia Province (West Luconia Delta and West Luconia Rim) and adjacent areas (SW Luconia Sub-province and the Rajang Slope). Also shown are the major drilling targets and hydrocarbon plays. Modified after Triton (1994).

loading on the continental margin have resulted also in shale diapirism beneath some growth faults, forming potential hydrocarbon traps.

The West Luconia Rim represents the edge of the West Luconia Delta, flanking the Miocene carbonate platforms of the Central Luconia and Natuna, and the landward margin of the delta bordering with the SW Sarawak and Tatau provinces. It is characterised by a thick succession of Cycles III, IV, and V sediments prograding northwards over a pre-Cycle I basement. Faulting is much less intense in the West Luconia Rim than in the delta core.

HYDROCARBON HABITAT

Table 18.1 shows the well results from the West Luconia Province. Generally, the wells encountered only thin columns of gas in Cycles V/VI sands and with traces of oil. The presence of oil and gas in the Nagasari-1, Hibiskus-1, L4.1X, Mawar-1, and V1.1X wells is proof of an active petroleum system in the West Luconia Province. There is very little data available on the hydrocarbon types in the West Luconia Province. The available data, however, indicate that the majority of the oils are of normal gravity and slightly waxy.

Traps

Figure 18.4 shows a cross section of the western Sarawak shelf, depicting the various structures that are present in the West Luconia and adjacent provinces. The dominant structural trap style in the main West Luconia Delta is the rollover anticline associated with deltaic growth faults, rather like those in the Baram Delta Province. The rollover anticlines usually occur on the downthrown (northern) sides of major growth faults and are formed in Cycles V and VI sediments. Wells X1.1X, L4.1X and Nagasari-1 tested some of the rollover anticlines in the central and northern parts of the province. Some of the larger faulted anticlines, especially in the north, are cored by shale diapirs formed by Cycle V prodelta shales. In the central region, larger dip/fault-closed structures associated with growth faults are formed by Cycle V and some Cycle VI reservoirs, as tested by the Nagasari-1, L4.1X and Hibiskus-1 wells. Combined structural/stratigraphic traps also occur as a result of tilting of Cycle V fault blocks due to loading of Cycles VI/VII sediments.

Figure 18.5 illustrates some of the main trap configurations in the West Luconia Province. Growth fault related structures form the most common trap style. In the northern part of the province, traps occur in middle-upper Cycle V delta-front sand pinch-outs and

Wells	T3.2			C24.1
Tests	DST3	DST2	DST1	DST1
Depth (m)	625-718	1392-1416	721.5-1730.5	1515-1527
N_2	1.69	9.51	12.32	1.29
CO_2	77.76	1.93	11.86	78.89
C_1	20.38	83.58	69.98	19.03
C_2	0.00	3.09	2.75	0.58
C_3	0.12	0.91	1.82	0.15
iC_4	0.00	0.57	0.77	0.00
nC_4	0.05	0.19	0.26	0.06
iC_5	0.00	0.12	0.13	0.00
nC_5	0.00	0.07	0.08	0.00
C_6	0.00	0.03	0.03	n.d.
C_1/C_{1-5}	0.99	0.94	0.92	0.96
$\delta^{13}C\ CO_2$ ‰	-3.3	-7.4	-4.8	-3.3
$\delta^{13}C\ C_1$	-31.5	-36.0	-35.7	-32.0
$\delta^{13}C\ C_2$	-24.0	-26.5	-26.0	-25.1
$\delta^{13}C\ C_3$	-24.2	-25.6	-25.2	-23.5

Table 18.2. Composition of gases in the West Luconia Province, offshore Sarawak. (n.d. - not determined)

truncations associated with shale-cored anticlines (Fig. 18.5A). These traps are capped by bathyal shales in Cycle VII and sealed by shale-smeared growth faults. More commonly, traps occur on the hanging wall rollover anticlines in Cycles V and VI coastal-shelf sands and sealed by Cycle VI prodelta shales and sealing growth faults (Fig. 18.5B). In places, anticlinal closures also occur in the footwall of growth faults (Fig. 18.5C). Three-way faulted traps in the southern part of the province are provided by faulted Cycle V coastal plain sandstones and Cycles IV/V reefal buildups sealed by Cycle V prodelta shales (Fig. 18.5D). Traps are also formed by small low-relief faulted drapes over basement highs, with potential reservoirs developed in the Cycle V sands. Such a structure has been tested at V1.1X. Other structural traps include highly faulted domal structures associated with basement highs and wrench faults, as at Mawar.

Reservoir

The primary reservoir objectives in the West Luconia Province are the Upper Miocene lower coastal plain to coastal fluviomarine deltaic sandstones belonging to the upper Cycle V, especially in the southern and central parts of the delta (Figs. 18.4, 18.5). The gas-bearing reservoirs encountered in Mawar-1, L4.1X, and V1.1X belong to this type of reservoir. Well results have shown that these sandstones have porosities of between 17% and 25% and permeabilities of up to 930 mD. In the C24.1 well, the sandstones have porosities of greater than 30% and permeabilities of over 1000 mD. Sandstone units with individual thickness of 1 to 15 m are commonly stacked to produce 50 to 150 m of gross sand. Lower Pliocene turbidite sandstones in Cycle V form the secondary reservoir objectives. These are extremely fine grained, generally less than 1 to 5 m thick, and have poorer poroperm characteristics. Amplitude anomalies indicate that cleaner sands may occur on the downthrown sides of growth faults. Besides the clastic reservoir objectives, there is also potential development of reefal carbonates in the Cycles IV/V succession (Fig. 18.5D).

Seal

Most of the traps are believed to be sealed by Lower Pliocene Cycles VI/VIII shales (Figs. 18.4, 18.5). Regional seals are found in upper Cycle VI

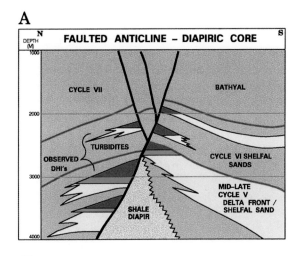
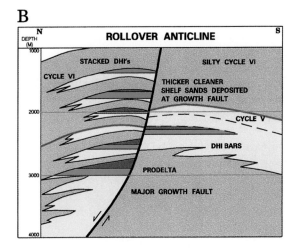
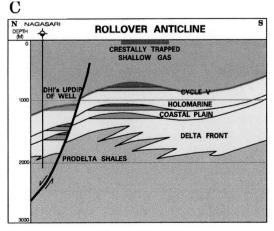
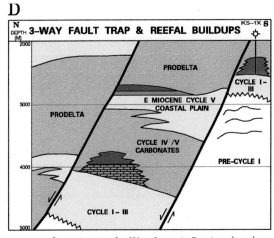

Fig. 18.5. Schematic illustration of the structural styles and potential trapping configurations in the West Luconia Province based on seismic and well results. Some of the structures have been drilled into, while others are yet to be tested. From Triton (1994).

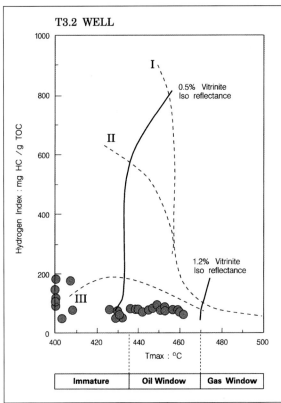

Fig. 18.6. Hydrogen Index vs Tmax plot for the T3.2 well, showing the organic matter type and thermal maturity.

and Cycle VII prodelta and bathyal shales in the central and northern areas, whereas locally, intraformational shales in Cycle V may act as seals when significant upper Cycles VI/VII is absent because of erosion and/or non-deposition. Hydrocarbon seals also occur along faults, as shale smears the interfaces of juxtaposed sandstone and shale on either side of the faults. Hydrocarbons in reefal buildups may be sealed by regional shales (Fig. 18.5D).

SOURCE ROCK

Quality

The main source rock beds in the West Luconia Province probably occur within Cycle V (Ganesan, 1997). As in the Baram Delta, thinly laminated carbonaceous layers within the Upper Miocene section are the probable source for the oils. These rocks have a total organic carbon (TOC) of between 0.2% and 0.8%, and contain predominantly Types II and III kerogen, and only minor Type I. In addition, these layers may also contain up to 40% Types I/II oil-prone kerogen. In the T3.2 well, the analysed samples are generally gas prone, with very low capability for liquid hydrocarbon generation at optimum thermal maturity (Fig. 18.6). Two different source facies have been identified in the T3.2 extracts. The Cycle V is dominated by terrigenous source facies, while the lower Cycle V and most of Cycle IV intervals contain a mixed facies in which the marine component is more dominant than the terrigenous. While most of the shales penetrated contain less than 1% TOC, some kerogen-rich layers containing Types I/II kerogen may have generated oil and charged the L4.1X structure. Generally, the source rocks contain a mixed Types II/III kerogen with increasing marine affinities offshore, towards the northern part of the province.

The hydrocarbon shows encountered in L4.1X, M8.1X, and C24.1 indicate the existence of another potential source rock (possibly within Cycles IV/V/VI) sequence in the northern part of the West Luconia Province. Based on the recent data from T3.2 well, the presence of condensates within Cycles V and VI sands indeed suggests the presence of mature source rocks in the Cycles IV/V or Cycles II/III equivalent. The condensates were probably generated from source rocks similar to those that generated the Luconia gas condensates. The West Luconia oils and condensates, such as those at L4.1X, M8.1X, C24.1, and Mawar-1AS, typically have relatively low pristane/phytane ratios (Pr/Ph < 2.5), suggesting a source environment that was not strongly oxidising. The low Pr/Ph ratios are distinct from the Balingian and Luconia oils, which are generally characterised by high Pr/Ph ratios. The low Pr/Ph ratios suggest either a marine or lacustrine source. Another important feature in these oils is the presence of sulphur compounds (benzothiopenes), which are generally observed in carbonate depositional environments.

Maturity

The approximate depth to the top of the oil window in the West Luconia Province is shown in Fig. 18.4. As expected, the top of the oil window gradually becomes shallower from 4000 m to 2000 m towards the south, as older sediments are at shallower depths in the south compared to the north. Figure 18.7 shows a maturity-depth profile for the T3.2 well which indicates that the lower Cycle V and much of Cycle IV lie within the oil window. The top of the oil window (taken as 0.6% VRo) is reached at about 1900m, while the top of the gas window is at about 3000 m (1.2% VRo). However, although the T3.2 section is within the oil window, no significant quantities of oil would have been generated because of the poor source potential of the source rocks at this location. In most other wells, the vitrinite reflectance data suggest that the top of the oil window occurs between 2500 m and 3500 m. Most of the Cycle V in the West Luconia Province is within the oil window. Cycle VI is also within the oil window in some parts of the province. The hydrocarbons encountered in some of the wells shows that petroleum generation has occurred.

Unlike the Baram Delta, the West Luconia Delta has relatively high geothermal gradients (45°Ckm^{-1}). Ganesan (1997) attributed the low maturity, despite the high heat flow, to a recent heating phase that did not affect the maturity of the sediment. Early to peak maturity encountered in the Idemitsu wells attests to the high palaeogeothermal gradients in the area, which apparently resulted from a late heating phase, and does not affect the maturity of the sedimentary section. Oil migration from Cycle V probably started during the Late Pliocene and is

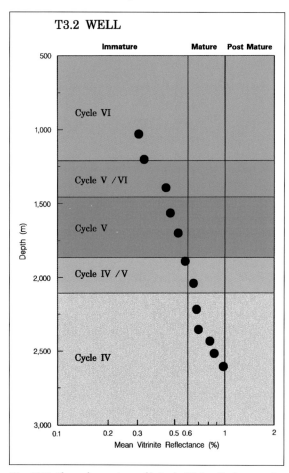

Fig. 18.7. Thermal maturity profile in the T3.2 well, shown as the mean vitrinite reflectance vs depth.

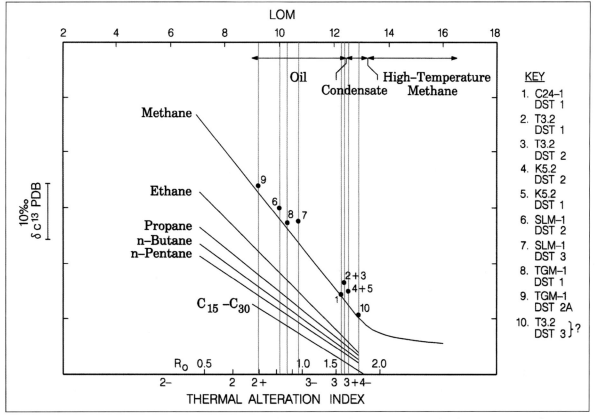

Fig. 18.8. Theoretical maturation diagram, using the method of James (1983), showing the calculated carbon isotope ratios ($\delta^{13}C$) of gas components plotted against the level of organic maturation (LOM) of the source rock. Thermal alteration index (TAI) and vitrinite reflectance (Ro) are also shown at the bottom of the plots.

probably started during the Late Pliocene and is probably progressing at the present day. Hydrocarbons may have migrated from deeper pre-Cycle IV section up the fault conduits. The analyses done on the T3.2 samples showed that some samples, especially interval between Cycle IV - Upper Cycle V, are stained with migrated hydrocarbons. In the Nagasari-1 and Hisbiskus-1 wells, analysed samples contain over 20% migrated oil.

The isotopic composition of gases can be used as a maturity indicator for the source rocks that generated the gases. Figure 18.8 shows the carbon isotopic compositions of methane, ethane, and propane from the C24.1 and T3.2 wells, plotted on James's (1983) maturity diagram using ethane-to-propane separation. The data indicate that all the C24 and T3 gases have a source rock maturity higher than 1.5% VRo, which implies that the hydrocarbon generation occurred during the post-mature stage.

CO_2

The concentration of CO_2 in the gases of the West Luconia Province is high, as indicated by the results from Nagasari-1, Mawar-1, C24.1, T3.2, and L3.1X (the latter is situated at the border with the Central Luconia Province, see Fig. 18.1).

The possible sources of the CO_2 have been discussed in Chapter 17. High concentrations of CO_2 tend to occur at the margins of the carbonate platforms, and where there are numerous basement-involved faults. The deep basement faults may have acted as conduits, bringing CO_2-charged fluids into the reservoirs. It is also possible that the CO_2 was sourced from the Rajang Group turbidites or older carbonate rocks in the sub-Oligocene basement. The CO_2 may have been derived also from overmature source rocks in the sub-Cycle IV section, similar to those that fed the Central Luconia carbonate reservoirs.

Ganesan (1997) suggested that there is a vertical (stratigraphic) variation in the concentrations of inert gases (mainly CO_2 and N_2) in the western Sarawak shelf. The data from T3.2 well (Table 18.2), however, show no vertical trend. The abundance of CO_2, in agreement with the heavier carbon isotope values ($\delta^{13}C < -5‰$), suggests that the CO_2 in T3.2 and C24.1 wells were generated from an inorganic source, possibly deeply buried carbonate sediments. The low concentration of CO_2, as seen in some T3.2 gases, associated with a slightly lighter carbon isotope value ($\delta^{13}C = -7.4‰$), is indicative of a mixed organic/inorganic origin or, perhaps, mantle degassing.

PLAY TYPES AND PROSPECTIVITY

The hydrocarbon occurrences so far discovered in the West Luconia Province provide evidence for a generative hydrocarbon system. The West Luconia Province has only been lightly explored and still has the potential for discoveries to be made. In the past, the primary drilling objectives were predominantly fault-controlled structural traps. The emphasis has been to go for these large structural traps associated with growth faults. The main play type in the province consists of upper Cycle V sands trapped in rollover structures in the hangingwall of growth faults. Secondary objectives include the thin upper Cycle VI sands and deeper plays within the lower Cycle V or older, and some Cycle IV carbonate reefs on the West Luconia Rim.

There are other potential play types that require further investigation. These include the Oligocene-Miocene extensional fault blocks in the basement, and the Upper Miocene deltaic and distal slope fans in the northern part of the province. Stratigraphic traps will be important targets in future exploration, as the remaining undrilled prospects are relatively small. Most of the large amplitude anomalies visible on seismic sections have not been tested during previous exploration phases.

In addition to the conventional plays such as the rollover structures, there is also potential for fault traps within the deeper Cycle V(?) section in the southern and central areas. The Cycle V(?) has apparently better reservoir development in the tilted fault blocks of the northern zone. Drilling into the overpressured zone is also one of the challenges in the West Luconia Province. As an example, the X1.1X well in central West Luconia was drilled to test the Lower Pliocene Cycle VI shallow marine sandstones, but was terminated because of overpressure near the top of Cycle VI, and therefore has not fully tested the potential of the Cycle VI play. As this play type occurs down-dip into the basin, there is potential for traps in the turbidites sourced from palaeohighs to the south.

CONCLUSION

The West Luconia Province is relatively poorly explored, compared to neighbouring provinces. It is characterised by a large deltaic system that prograded northwards since the Middle Miocene, resulting in a total sediment thickness exceeding 7.5 km. The province has been explored previously by Shell, Idemitsu, Elf Aquitaine, and PETRONAS. There has been 1 gas discovery (L4.1X) and a number of significant oil and gas shows since exploration began in the late 1960s. The hydrocarbon accumulations occur in Upper Miocene Cycle V (and minor Cycle VI) deltaic sandstones in growth fault structures (fault-closed rollover anticlines). These sandstones have high porosities (17-25%) and permeabilities (> 900mD). Some reefal carbonates form minor objectives in some parts of the province.

The source rocks for the hydrocarbons in the West Luconia Province are thought to occur within Cycle V. Geochemical data indicate that the oils and condensates recovered from the wells were derived from either a marine or a lacustrine source, possibly within Cycle V, which are presently within the oil generation window. Carbon isotope data suggest that some of the gases may have been generated during the post-mature stage (equivalent VRo > 1.5%).

Although there has been some exploration activity in the West Luconia Province during the last three decades, there has been relatively little economic success. This could be attributed to our poor understanding of the petroleum system in the area. The province has been perceived to have a low potential because the main source rocks are gas-prone, and also because the geothermal gradient is relatively high (average $45°Ckm^{-1}$). There is still the need to improve our understanding of the West Luconia petroleum system, which has been proven to exist. Among the geological risks that need to be addressed are the high CO_2 (expected > 15%), abnormal pressures, and relatively poor reservoir development. In the central and southern parts of the province, the main risks are the inefficient fault seals, insufficient charge, and contamination by CO_2 and N_2. Further work on the petroleum systems and prospectivity of the West Luconia Province is in progress.

REFERENCES

Ganesan, B.M.S., 1997. Geology and hydrocarbon potential of the offshore western Sarawak shelfal area. Proceedings of the ASCOPE '97 Conference, Kuala Lumpur, 24-27 November 1997.

James, A.T., 1983. Correlation of natural gas by use of carbon isotopic distribution between hydrocarbon components. American Association of Petroleum Geologists Bulletin, 67, 1176-1191.

Tan, D. and Swinburn, P.M., 1993. Evaluation of the remaining exploration potential of Block SK-1, West Luconia Province, offshore West Sarawak. Unpublished Sarawak Shell report. EXP.R.50741.

Triton, 1994. Malaysia Offshore Sarawak SK-1 Block Evaluation. Triton Energy Company, Unpublished Report, (PETRONAS BD.1: TR1: 3: 94-01).

Chapter 19

North Luconia Province

Mazlan B. Hj. Madon

INTRODUCTION

The North Luconia Province lies beyond the continental shelf edge in the northernmost part of the Sarawak offshore, in water depths ranging from 200 m to greater than 2000 m (Fig. 19.1). Petroleum exploration in this province began only during the early 1990s. Activities included 2D seismic, gravity, and magnetic surveys and the drilling of two wildcat wells. Although exploration drilling has been unsuccessful so far, more detailed work is being carried out by present operators in the area to further explore the potential of this province. Understanding of the petroleum potential of this new exploration area has come about through the application of sequence stratigraphy in delineating reservoir, source, and seal facies, and the appreciation of regional tectonic controls on the structural and sedimentation history of the Sarawak continental margin. This chapter summarises the results of the past efforts in realising this potential. Little has been published on the North Luconia Province, except for the regional structural studies by Mohd Idrus Ismail et al. (1994). Besides this work, the following account of potential plays in the area is based on unpublished work by Mobil (1996) and Shell (1996).

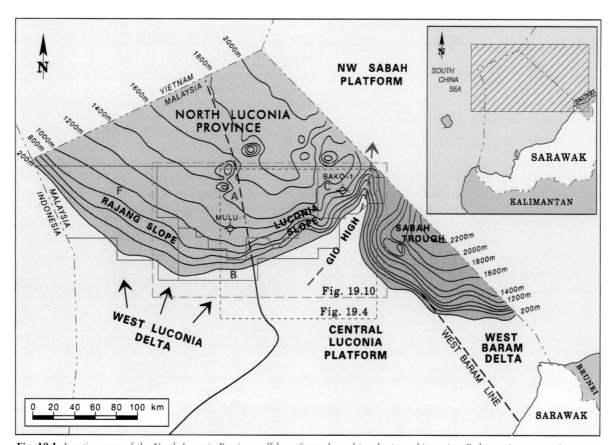

Fig. 19.1. Location map of the North Luconia Province, offshore Sarawak, and its physiographic setting. Bathymetric contours in metres below mean sea level. Only two wells have been drilled: Mulu-1 and Bako-1. Note the G10 High, which separates the North Luconia Province from the Sabah Trough and the Baram Delta Province.

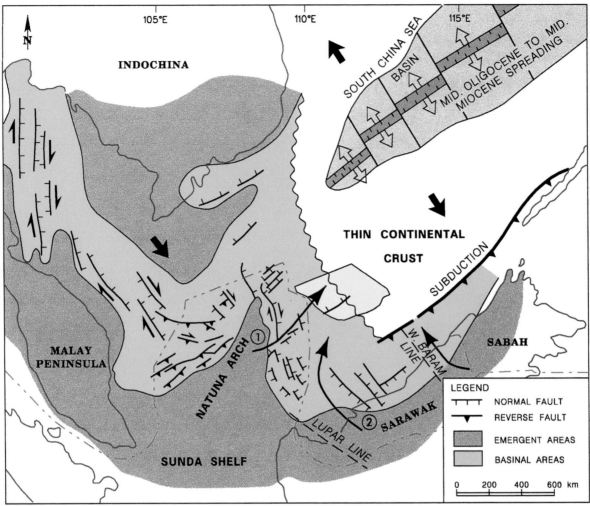

Fig. 19.2. Schematic Early Miocene tectonic setting of the North Luconia Province, located at the northern margin of the Sarawak Basin, which is underlain by thin continental crust that resulted from the sea-floor spreading in the South China Sea Basin. The Sarawak Basin is a peripheral foreland basin formed by continental collision along the Lupar Line, while subduction continues in northwestern Sabah. Arrows show possible source of sediments for the North Luconia Province: 1- Natuna Arch, 2- Rajang Fold-Thrust Belt.

EXPLORATION HISTORY

Before 1989, not much work has been done to explore for petroleum in the North Luconia Province. The western part of the province is generally less explored than the eastern part, with only sparse 10-20 km 2D seismic grid available. In 1989 PETRONAS acquired over 4000 line-km of 20-25 km grid regional 2D data in the eastern part of the province, north of Central Luconia, to identify new structural traps and better define the structures that had been identified by several operating companies using early 1960s vintage seismic. More detailed work began only when Mobil, along with partners Nippon Oil, Norsk Hydro, JAPEX and PETRONAS Carigali, signed the Deepwater PSC in November 1993. During the following year Mobil shot a 2x5 km seismic grid totalling 6345 km in Blocks A and B of the PSC area (Fig. 19.1). A further 3585 km of seismic data were acquired by Mobil in 1996.

During late 1994 to 1995 Mobil drilled the first two deepwater exploration wells in Malaysia, Mulu-1 and Bako-1, in water depths of 1165 m and 1112 m, respectively. Both wells lie within the eastern part of the province. Mulu-1 well was drilled to evaluate the potential of the Oligocene-Lower Miocene Cycles I-III reservoirs in the Mulu prospect, which is a large fault-closed "buried-hill" structure. The well reached a total depth of 5056 m and encountered only minor gas shows, but no indication of liquid hydrocarbons. Bako-1 well tested Cycle I reservoirs in a similar

Partners	BLOCKS A&B	BLOCKS C&D
Esso*	32.50	32.50
Mitsubishi	12.50	12.50
Mobil	17.50	15.29
Nippon Oil	12.05	10.00
JAPEX	5.74	10.00
Norsk Hydro	4.71	4.71
PETRONAS Carigali	15.00	15.00

* — operator

Table 19.1. Current partners in the North Luconia PSCs and their stakes. Location of blocks in Fig. 19.1.

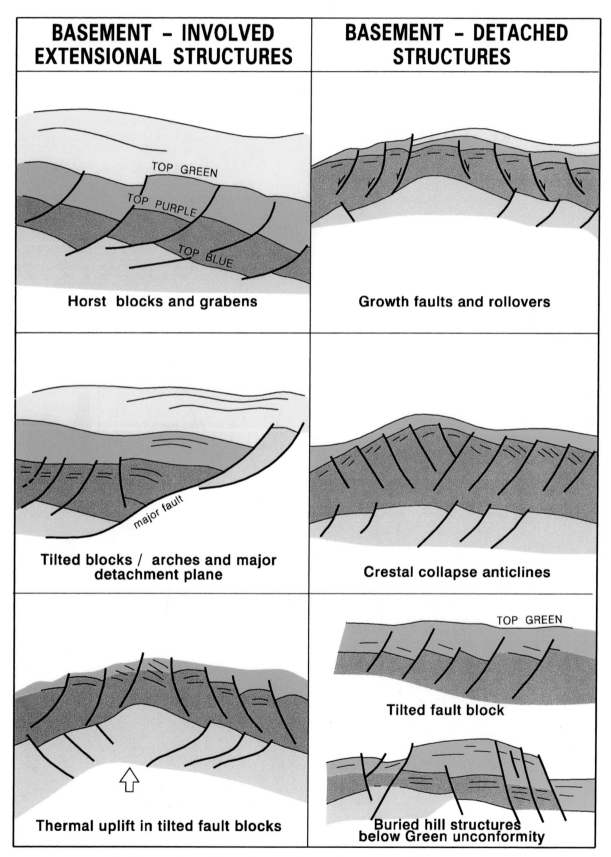

Fig. 19.3. Some examples of structural trap styles on the Luconia Slope, in the eastern part of the North Luconia Province. Structures are categorised into basement-involved and basement-detached structures. After Mohd Idrus Ismail et al. (1994).

structure, the Bako prospect, and reached TD at 3875 m. No hydrocarbons were found. Mobil farmed out the Deepwater PSC in late 1997. The area is now operated by Esso under a new production sharing agreement. The partners in this venture are Mitsubishi, Mobil, Nippon Oil, Norsk Hydro, JAPEX, and PETRONAS Carigali (Table 19.1).

STRUCTURE AND TECTONIC SETTING

The North Luconia Province occupies the southwestern margin of the South China Sea Basin (Fig. 19.2) which began opening and spreading during the Early Oligocene (about 32 Ma) (Taylor and Hayes, 1980). The province is believed to be underlain by thin continental crust, similar to the NW Sabah Platform and the Reed Bank, which are believed to be foundered continental fragments that drifted off southern China when the South China Sea Basin opened (Taylor and Hayes, 1980). The relatively low-relief NE-facing continental slope and rise in the North Luconia Province contrasts with the more rugged bathymetry of the NW Sabah Platform, which is formed by numerous carbonate banks and atolls. A north-trending positive bathymetric feature, the G10 High, separates the North Luconia Province from the NW Sabah Platform and the relatively featureless Sabah Trough to the east (Fig. 19.1). This ridge may have been a positive feature since Oligocene times as it is underlain by the G10 carbonate reef complex that is still growing today.

The North Luconia Province includes two sub-provinces that have markedly different structural styles. Its western part, the Rajang Slope, lies in front of the West Luconia Delta complex and has more than 13 km of sediment. The eastern part, the Luconia Slope, is the continental slope north of the Central Luconia Province (Fig. 19.1). There is a major contrast in structural style between the Rajang (western) and Luconia (eastern) slopes. Mohd Idrus Ismail et al. (1994) described the structural style in the North Luconia Province as comprising thick-skinned

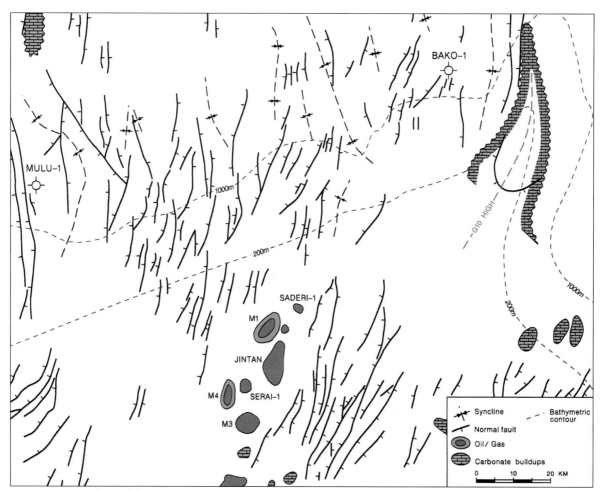

Fig. 19.4. Structural map of the eastern part of the North Luconia Province (Luconia Slope). Note the north-trending normal fault pattern extending from the Central Luconia Province where major oil/gas fields are located. For location see Fig. 19.1.

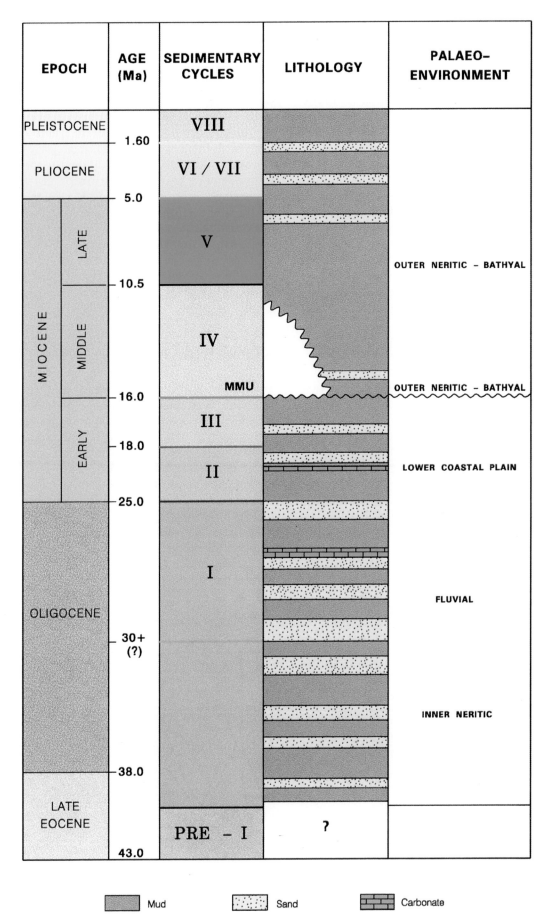

Fig. 19.5. Stratigraphy of the North Luconia Province, depicted in terms of the sedimentary cycles in the inboard areas (after Mobil, 1996). Note the major hiatus associated with the Middle Miocene Unconformity (MMU), based on the results from Mulu-1 and Bako-1. The MMU marks a sudden change from lower coastal plain to outer neritic and bathyal environments.

(basement-involved) extensional style and thin-skinned (basement-detached) structural style (Fig. 19.3).

The Luconia Slope is characterised by basement-involved normal faults trending NE-SW, similar to those in the Central Luconia Province. These extensional faults mainly cut the sub-Middle Miocene strata. Most of these are west-dipping listric normal faults that bound *en echelon* half-grabens and tilted fault-blocks formed by Cycles I and II sequences. Pinnacle reefs have been identified on some of the horsts created by the extensional faulting. Some faults extend from basement up to the regional Middle Miocene Unconformity and beyond. The N-S to NE-SW trends in the extensional structures seem to be a continuation of those in the Central Luconia Province.

The Rajang Slope, on the other hand, is dominated by basement-detached, deltaic growth faults associated with sedimentation in the West Luconia Delta. The structures here are dominated by east-trending normal faults and gravity-induced toe thrusts formed by deltaic progradation, and appear to be confined to the sedimentary cover (basement-detached) (Fig. 19.3). In the eastern part of the province the structural grain of the Luconia Slope is generally N- to NE-trending, roughly parallel to the G10 High (Fig. 19.4).

Two major deformation events have been recognised from detailed seismic interpretation in the eastern part of North Luconia. A pre-Miocene extension, probably related to the initiation of sea-floor spreading in the South China Sea, resulted in half-grabens into which thick packages of Miocene sediments were later deposited. The structural styles are dominated by basement-involved, high-angle normal faults. A second deformation occurred during the early Middle Miocene. This event resulted in the major Middle Miocene (or Green) Unconformity (MMU, Fig. 19.5), and is thought to be related to the cessation of sea-floor spreading in the South China Sea Basin. The MMU is therefore correlated with the Deep Regional Unconformity (DRU) of Levell (1987) in the Sabah Basin. The deformation style during this event is mainly thin-skinned fault reactivation, detached normal faulting, and shale diapirism.

STRATIGRAPHY

Seismic sequence stratigraphy, as opposed to cycle stratigraphy (Chapter 6), has been used by PETRONAS and Mobil to subdivide the stratigraphy of the North Luconia Province. In addition to the transgressive or marine flooding surfaces used in cycle stratigraphy in the inboard areas, key stratal onlap, downlap and erosional surfaces were identified and interpreted as major sequence boundaries within the succession. These surfaces and sequence boundaries have been correlated with the cycle boundaries in the Balingian and Central Luconia provinces (Fig. 19.5).

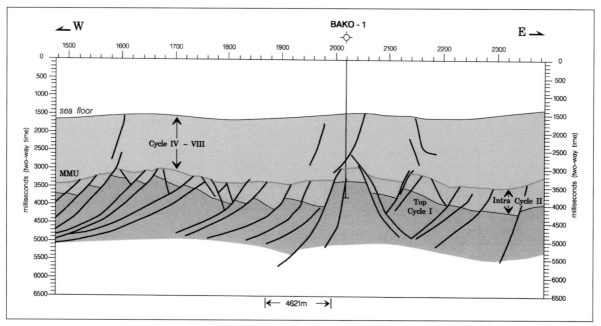

Fig. 19.6. E-W seismic section across the Bako structure, showing the extensional structural style in the Cycle I to III succession, overlain unconformably by undeformed Cycle V and younger sediments. The Middle Miocene Unconformity (green horizon) separates the two successions. For location see Fig. 19.9.

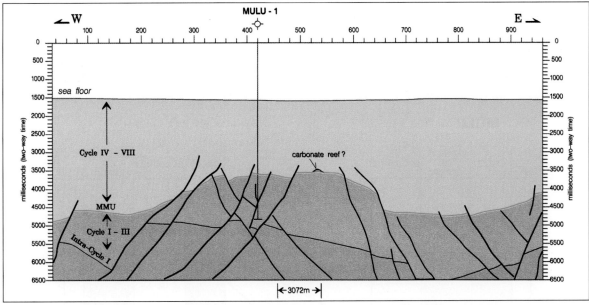

Fig. 19.7. W-E seismic section across the Mulu structure, showing the horst and graben morphology as a result of extensional tectonics prior to the Middle Miocene event that caused the major unconformity (green horizon). This "buried hill" structure is bounded by major normal faults and half-grabens on both sides. For location see Fig. 19.9.

The Middle Miocene Unconformity (MMU) or "Green Unconformity" is the most prominent seismically identified sequence boundary in the North Luconia Province, and perhaps also in the entire Sarawak shelf. The unconformity is characterised by erosion, truncation, and onlap/downlap relationships. The age of the MMU is probably about 17-15 Ma, and seems to be coincident with the cessation of sea-floor spreading in the South China Sea Basin. This event is also believed to have resulted in the Deep Regional Unconformity (DRU) in offshore NW Sabah (Levell, 1987; Hazebroek and Tan, 1993). Seismic reflection characteristics differ markedly across the MMU-strong high-amplitude reflections in the sub-MMU section, and high frequency, slightly transparent, continuous reflections above the MMU. The MMU is recognisable in the Luconia Slope but is covered by 8-9 km of sediments in the distal parts of the encroaching West Luconia Delta.

Sequence stratigraphic mapping has enabled several prospects to be delineated. The major ones were tested by Mobil's Bako-1 and Mulu-1 wells, respectively (Figs. 19.6, 19.7). At the time of drilling, both the structures were expected to have been charged from coal source beds in the sub-MMU section (Cycles I/II) and sealed by post-Middle Miocene deep marine shales. The Mulu structures is a N-S extensional horst bounded to the west by a major SW-hading fault that intersects with a conjugate NE- and SE-hading fault pair. Many smaller post-Middle Miocene faults divide the structure into smaller compartments. The Mulu horst appears to have remained stable during post-Middle Miocene times, while the flanking areas on the downthrown sides of the bounding faults continued to subside. Possible carbonate reefs have also been identified on the structure.

The Bako structure (Fig. 19.6) is a NE-trending horst between extensional half-grabens, and is bounded by steep, east- and west-dipping basement normal faults. As in Mulu, the structure is draped by relatively undeformed Cycles V to VIII sediments. Note that the extensional faults, which appear to sole out near the base of Cycle I (Oligocene), generally terminate at the MMU, and rarely extend beyond into the Middle Miocene and younger strata. This suggests that the extensional tectonics ended before the Middle Miocene. The Cycles IV-VIII strata above the MMU are generally undeformed and drape sub-MMU structures. A few normal faults in the strata above the major horsts were probably caused by the compactional drape, similar to the "drape-slip" fault of Bertram and Milton (1989).

Micropalaeontological data from Mulu-1 indicate that Cycles I to II sediments were deposited in bathyal environments and are overlain by Cycle III inner neritic and coastal sediments (Fig. 19.8). In contrast, bathyal environments were not established at Bako-1 until Cycle II times, when outer neritic to bathyal environments prevailed until today. In both wells, the entire Cycle IV section is missing at the MMU. Cycle V therefore rests directly on Cycle III, representing a depositional and/or erosional hiatus of 5.5 Ma duration.

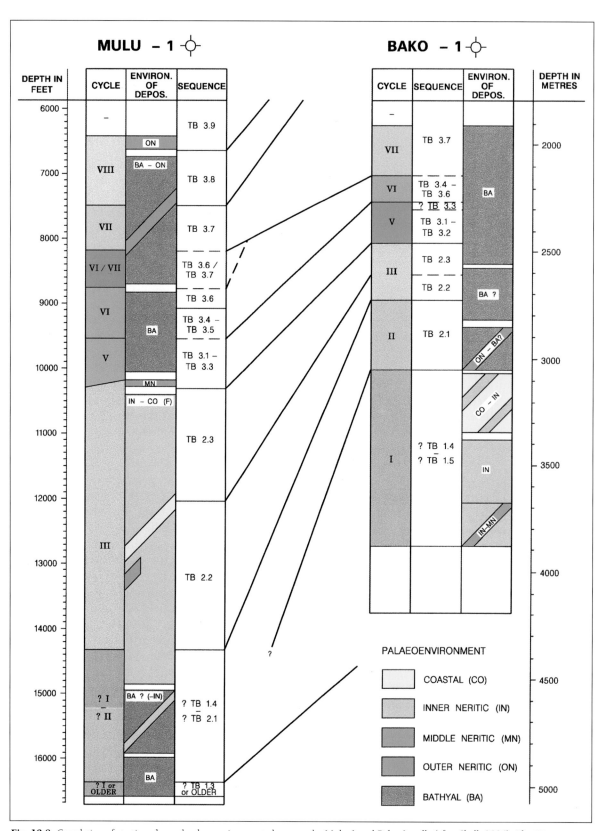

Fig. 19.8. Correlation of stratigraphy and palaeoenvironments between the Mulu-1 and Bako-1 wells (after Shell, 1996). The TB sequences correspond to the 3rd order cycles of Haq et al. (1987).

POTENTIAL HYDROCARBON PLAYS

The contrasting structural styles between the Rajang and Luconia Slopes (Fig. 19.3) have resulted in different hydrocarbon plays. On the Rajang Slope, the potential traps are synsedimentary thin-skinned deformational features in the Upper Miocene-Pliocene Cycle VI and younger succession. The Rajang Slope is the slope-to-basin extension of the West Luconia Delta. It is dominated by anticlinal structures formed by thin-skinned, gravitationally-induced deltaic toe-thrust in the Upper Miocene-Pliocene Cycle VI and younger succession. These large, north-verging, arcuate thrust structures are spaced at 5-10 km intervals, and appear to die out upwards and sole out below about 5 s two-way time. The faults are detached from the basement via a decollement surface formed by shale layers above the MMU. Thrusting via gravity sliding seems to have developed during Cycles VI and VII times, i.e. during the Late Miocene to Early Pliocene (Mohd Idrus Ismail et al., 1994). At least 20 such structures have been identified from seismic mapping. Potential reservoirs in these structures could be formed by turbidites enclosed in bathyal shales, sand pinch-outs on the flanks of anticlines, ponded sands in synclines, and the up-dip pinch-outs of basin floor fans.

Potential drilling objectives on the Luconia Slope are the Oligocene-Lower Miocene Cycles I, II, and III reservoirs below the MMU. The most common trap style is represented by the dip and/or fault-closed gentle anticlinal features on upthrown horsts bounded by basement-involved normal faults. The primary reservoirs are probably lower coastal plain and shallow marine deposits within the Lower Miocene Cycles I, II, and III intervals below the MMU, particularly between the green and purple sequence boundaries (see Fig. 19.5). Lithological and micropalaeontological data from the Bako-1 and Mulu-1 wells indicate that the sandstones were deposited in coastal to inner neritic settings, similar to the oil and gas reservoirs in the Balingian Province.

Seismic facies mapping has identified a Cycle II/III deltaic lobe encroaching from the south and southwest (Fig. 19.9). The delta system would have supplied sand into the basin to form potential reservoirs in the province. Source rocks are expected within the same sequences and may also act as seals. The traps developed before the maturation of the source rocks in Cycles I and II and the deposition of the post-MMU deepwater strata. Seals are formed by overburden bathyal shales above the MMU, which attain a thickness of over 1350 m. Lateral shallow marine equivalents of these shales on the shelfal areas have proven to be effective seals for the carbonate plays in the Central Luconia Province.

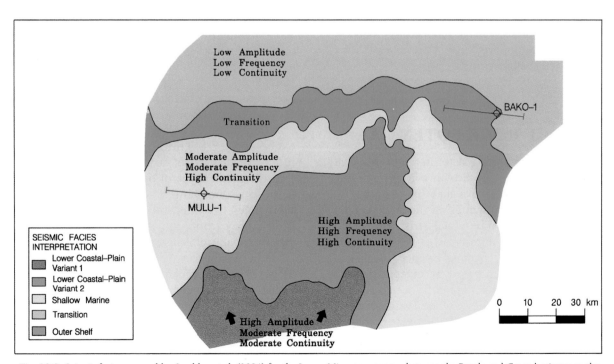

Fig. 19.9. Seismic facies mapped by Snedden et al. (1994) for the Lower Miocene sequence between the Purple and Green horizons on the Luconia Slope around the Mulu-1 and Bako-1 wells. Note the outline of a delta lobe prograding into the basin from the southwest. For location see Fig. 19.1.

RESERVOIR ROCKS

Direct evidence for the presence of reservoirs in the N Luconia Province are provided by wells drilled on the N Luconia Slope to test the Mulu and Bako structures. The main exploration uncertainty before Mulu-1 and Bako-1 were drilled was reservoir development in the sub-MMU section i.e. Cycles III and older. The depositional model in Fig. 19.9 envisages a delta prograding northwards onto the shelf during Cycles II and III times. Results from the Mulu-1 and Bako-1 wells show that good to excellent reservoir sandstones are present. Most sandstones are fine to medium grained lithic and quartz arenites (Shell, 1996; Fig. 19.10). The lithic component is mainly derived from sedimentary and metasedimentary rocks (PRSS, 1994; Shell, 1996). Although porosity deteriorates with increasing depth in both wells, significant porosity (11%) is still preserved at TD in Bako-1 which is about 3800 m below sea level. Generally, though, the sand is better developed in Bako-1 than in Mulu-1.

Sedimentary facies in Cycles I and III sandstones have been studied from cores taken from the Mulu-1 and Bako-1 wells (Shell, 1996). Cores taken from Cycle III (Mulu-1) indicate an inner neritic shelf environment, with wave/tidal influence, whereas those from the Cycle I in Bako-1 suggest deposition in a bay to foreshore environment with a tidal shoal/bar unit (Fig. 19.11). The dominant facies are ripple and low-angle cross-laminated sandstones containing holomarine body fossils and trace fossils. Facies associations are indicative of the distal part of a deltaic depositional system, in an overall wave-dominated shelfal environment.

SOURCE ROCK POTENTIAL

One of the determining factor in the prospectivity of the North Luconia Province is the presence of source kitchen(s) that may have charged the large extensional structures on the Luconia Slope and the toe-thrust anticlinal structures on the Rajang Slope. It is uncertain that there are source rocks in the North Luconia Province and they are unlikely to be penetrated by current drillable depths. Source rocks are difficult to identify even in mature prolific deltaic basins such the Baram Delta. It is expected, however, that the source-to-reservoir system in the North Luconia Province to be analogous with that in the shelfal areas of the West and Central Luconia provinces. On the Luconia Slope, the likely candidates for

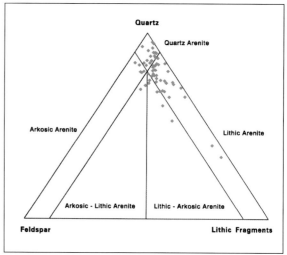

Fig. 19.10. Ternary compositional diagram (Q-F-L) for 62 sandstone samples from Mulu-1 and Bako-1 cores (data from Shell, 1996).

the source rocks would be the Oligocene shales and coals within the coastal plain to shallow marine Cycles I/II/III section below the MMU. The same rocks are thought to have generated the hydrocarbons found in the Central Luconia carbonate reservoirs. The sequence stratigraphic depositional model shown in Fig. 19.9 predicts the presence of a large deltaic lobe prograding to the north in the North Luconia Province. Coals and coaly shales are expected to have been deposited within this lobe, especially in the upper Cycle II to Cycle III interval.

At Mulu-1, the sediments at TD are within the late oil-generation window, while at Bako-1 the sediments at TD are in the early stages of oil generation. Since bathyal sediments generally do not contain sufficient organic matter to be of significant potential, source rock intervals are expected to be present in the inner neritic to coastal deposits of Cycles I to III. At Bako-1 there are numerous coal layers preserved in the coastal plain deposits which have good to excellent source rock quality (TOC generally exceeding 2%). The upper Cycle I interval, in which these coals occur, is in the immature to early mature stage of oil generation. Type III kerogen is dominant. Unfortunately, in Mulu-1 the Cycles II/III coastal plain shales are poor in organic carbon content (TOC < 0.5%) and may be capable of generating only gas.

The source rock potential of the Rajang Slope is more speculative, for no well has been drilled there. Nevertheless, the potential source rocks will have to be from the deeper section of the West Luconia Delta succession, probably within Cycles III and older. The source rocks are probably coastal plain to shallow marine shales that were laid down during the early stages of delta outbuilding.

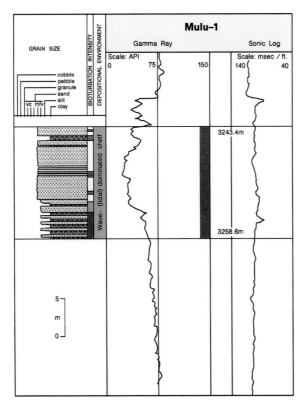
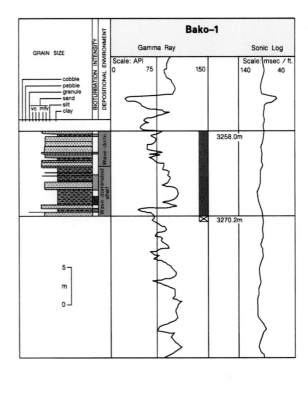

Fig. 19.11. Sedimentary facies in cores and well logs from the Mulu-1 and Bako-1 wells (from Shell, 1996). Core depths in metres below derrick floor.

CONCLUSION

Numerous structures exist in the North Luconia Province. Two of these have been drilled in early 1995. Although the results have been disappointing, the area still remains attractive for oil companies to invest in exploration work. As a result of past exploration efforts, including seismic surveys by PETRONAS and Mobil, which led to the drilling of the two exploratory wells in 1995, a large database now exists in the province which will enable more detailed geological studies to be undertaken. The area is actively being explored, and more data will become available to improve our understanding of this exciting new exploration area. Currently, further studies are under way by Esso and partners to improve the understanding of the geology and the assessment of its hydrocarbon potential.

Because the North Luconia Province is still a new exploration area, the existing hydrocarbon play concepts are based mainly on 2D seismic and data from the Bako-1 and Mulu-1 wells. The ongoing studies may help to refine the existing models regarding the prospectivity of the area and to identify potential drilling prospects. The tested plays so far are the "buried hill" type of structures, as exemplified by Mulu and Bako. There are several other large untested prospects, including fault-bounded anticlinal closures on horsts, fault and dip-closed structures in tilted fault blocks, and faulted anticlinal structures. The potential of these play types is dependent upon major issues such as the presence of reservoirs, availability of source rocks to charge the structures, and the integrity of the seal rocks. Ongoing work is being carried out to address these issues.

REFERENCES

Bertram, G.T. and Milton, N.J., 1989. Reconstructing basin evolution from sedimentary thickness; the importance of palaeobathymetric control with reference from the North Sea. Basin Research, 1, 247-257.

Haq, B.U., Hardenbol, J. and Vail, P.R., 1987. Chronology of fluctuating sea levels since the Triassic. Science, 235, 1156-1167.

Hazebroek, H.P. and Tan, D.N.K., 1993. Tertiary tectonic evolution of the NW Sabah continental margin. In: Teh, G.H., ed., Proceedings of the Symposium on Tectonic Framework and Energy Resources of the Western Margin of Pacific Basin. Bulletin of the Geological Society of Malaysia, 33, 195-210.

Levell, B.K., 1987. The nature and significance of regional unconformities in the hydrocarbon-bearing Neogene sequence offshore West Sabah. Bulletin of the Geological Society of Malaysia, 21, 55- 90.

Mobil, 1996. Mobil Blocks A, B, C, D, offshore Sarawak, Malaysia: Benchmark Report. Mobil New Exploration & Producing Ventures (Asia/Pacific), Unpublished Report, ER:MOBIL:96-01.

Mohd Idrus Ismail, Abdul Manaf Mohamad, Sahalan Abd Aziz, Abdul Rahman Eusoff, and Barney Mahendran, 1994. Geology of Sarawak deepwater and surroundings. American Association of Petroleum Geologists International Conference & Exhibition, Kuala Lumpur, Malaysia, 21-24 August 1994, Bulletin of the Geological Society of Malaysia, 37, 165-178.

PRSS, 1994. The petroleum geology of offshore Sarawak: a provenance, heat flow and geochemistry study including implications for deepwater Blocks A, B, C and D. Unpublished Report, December 1994. PETRONAS Research & Scientific Services Sdn Bhd

Shell, 1996. Geological study of Mulu-1 and Bako-1 wells, offshore Sarawak. Sarawak Shell Berhad, Geological Review Department. Unpublished Report, EXP.R.50784

Snedden, J.W., Sarg, J.F. and Md Yazid Mansor, 1994. Pre-drill interpretation of sequence stratigraphy, deepwater blocks A, B, C, & D, Sarawak (Malaysia). Mobil, Unpublished Report, ER:MOBIL:3:94-09.

Taylor, B. and Hayes, D.E., 1980. The tectonic evolution of the South China Basin. In: Hayes, D.E., ed., The Tectonic and Geologic Evolution of Southeast Asian Seas and Islands. American Geophysical Union, Geophysical Monograph, 23, 89-104.

Chapter 20

Petroleum Resources, Sarawak

*Othman Ali B. Mahmud and
Salahuddin B. Saleh*

INTRODUCTION

The petroleum resources of Sarawak is described with reference to the intensity of exploration drilling and play types in the different geological provinces. The first discovery well Miri-1 was drilled in 1910 in the onshore part of Sarawak. The Miri Field produced 80 million barrels of oil until it was shut down in 1972. In the early and middle 1960s major discoveries were made in the offshore part of Sarawak, in particular the West Baram Delta. The distribution and density of exploration wells and oil and gas fields in the Sarawak Basin are shown in Figures 20.1 and 20.2.

There are currently 12 Production Sharing Contracts (PSCs) in Sarawak (Fig. 20.3). Sarawak Shell Berhad (SSB) operates the 76 PSC', Block SK5, Deep Water Block E, Block SK308 and Block SK 8; PETRONAS Carigali Sdn Bhd (Carigali) operates Baram Delta fields (BDO) and Block SK307 under the Revenue over Cost (R/C) PSC; Amerada Hess, Block SK306 also under R/C PSC; Esso, Deep Water Blocks A,B,C and D under the 1993 Deep Water PSC; YPF (M) Ltd. operates Block SK 301 under R/C PSC and Murphy Sarawak Oil Co. operates Blocks SK 309 and SK 311 also under R/C PSC.

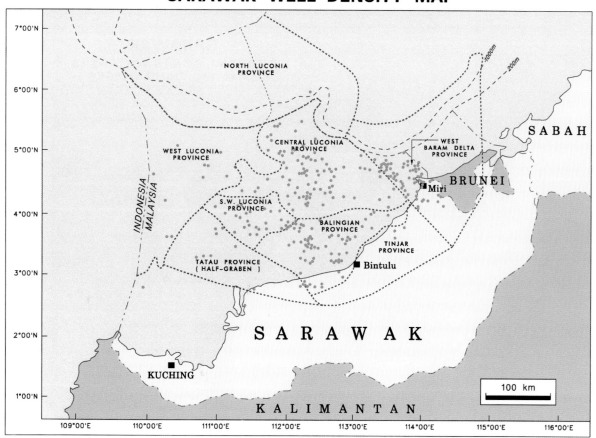

Fig. 20.1 As of January 1998, a total of 237 wildcats and 202 appraisal wells had been drilled in Sarawak Basin alone. New exploration ideas are being tested in the mature area, while attention is once more being given to the West Luconia and Tatau provinces.

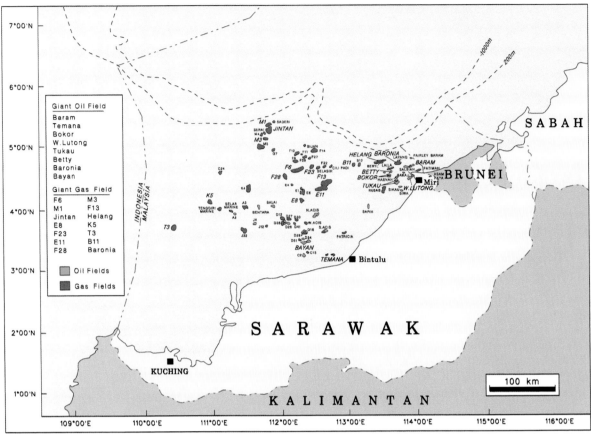

Fig. 20.2 The West Baram Delta and Balingian provinces contribute most producing oil fields while Central Luconia has the most producing gas fields.

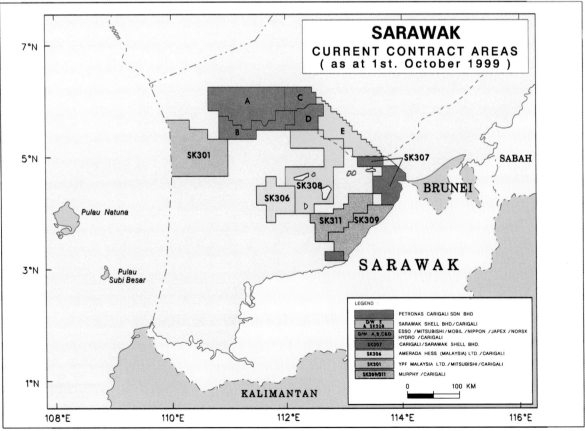

Fig. 20.3 As of October 1999, 14 exploration blocks out of 24 being offered were awarded with more anticipated to be signed by end of 1999.

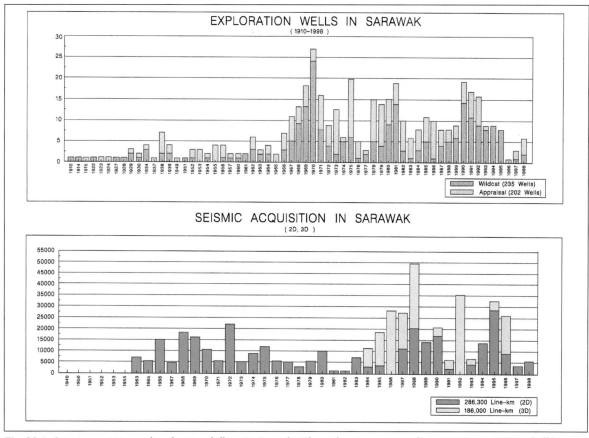

Fig. 20.4 Seismic acquisition and exploration drillings in Sarawak. The mid sixties saw a steady increase in acquisition of offshore 2D seismic in Sarawak. With the signing of the 1985 PSC's seismic acquisition in offshore Sarawak started to pick up as new contractors entered Sarawak exploration blocks. 1984 saw the introduction of 3D seismic acquisition in Sarawak. In some years acquisition was mainly of 3D seismic over mature fields. Exploration drilling in Sarawak shows a tri modal distribution tied to the three contracting rounds; a) Pre PSC b) 1976 PSC, and 1985 PSC.

As at 1.1.1998, about 280,000 line-km of 2D seismic and 186,000 line-km of 3D seismic data have been acquired in the Sarawak region, and a total of 431 exploration wells, 233 wildcats and 198 appraisals, have been drilled (Figs. 20.4 and 20.5). These resulted in the discovery of 2.1 BSTB estimated recoverable oil and 52.3 TSCF estimated recoverable gas in 40 oil fields and 85 gas fields (excluding small oil and gas fields with EUR less than 8 MMSTB and 50 BSCF respectively). Out of these fields, 8 are 'giant' oil fields and 14 'giant' gas fields each with recoverable reserves of more than 100 MMSTB and 1 TSCF, respectively. Currently, there are 13 producing oil and 18 gas producing fields. Thirteen of the gas producing fields consist of associated gas reservoirs, where part of the produced gas are reinjected for the pressure maintenance for the oil fields, while the rest is vented out and sold to local consumers. The 5 main producing gas fields are dedicated for MLNG I, II projects which supply LNG to Japan, Korea and Taiwan. Sarawak's remaining oil reserves stand at 0.9 BSTB and total gas reserves at 44.3 TSCF (40.6 TSCF non-associated gas). Total production of oil and gas are 1.2 BSTB and 8.0 TSCF (6.3 TSCF non-associated gas) respectively.

It should be noted that the resource figures relate to activities as at 1.1.98. During 1998, a total of 6 exploration wells were drilled and 6305 km of 2D seismic acquired.

OIL AND GAS RESOURCES BY PROVINCES, SARAWAK BASIN

Sarawak Basin Overview

The formation of the Sarawak Basin and the petroleum geology of the tectonostratigraphic province have been described earlier in Chapters 12-19.

Seven geological provinces have been identified in the Sarawak Basin, namely the West Baram Delta, Balingian, Central Luconia, Tinjar, Tatau, West Luconia and North Luconia. It should be noted that the Tatau Province incorporates the previously named SW Luconia and the Half-Graben provinces (Chapter 17).

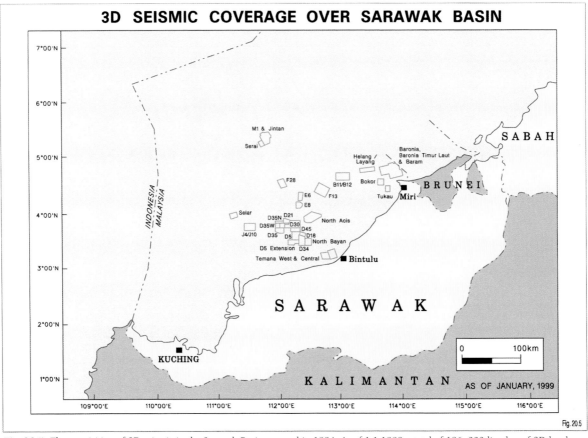

Fig. 20.5 The acquisition of 3D seismic in the Sarawak Region started in 1984. As of 1.1.1998 a total of 186, 000 line km of 3D has been acquired. Most of the acquisitions were concentrated in the major oil producing fields in the Balingian and the West Baram Delta.

However, the drilling statistics and discoveries in SW Luconia and the Half-Graben, sub-provinces of the Tatau Province, are shown separately.

All the seven provinces have been drilled. Extensive exploration activity was carried out in the Balingian, West Baram Delta and Central Luconia provinces, whereas the other four provinces were tested with a limited number of wells (Figs. 20.6, 20.7 and 20.8). Commercial hydrocarbons have been found in the West Baram Delta, Balingian and Central Luconia provinces.

West Baram Delta Province

The West Baram Delta Province (WBD) straddles the border between Brunei and the northeast corner of Sarawak . The western part of the province known as the West Baram Delta lies in the northeastern portion of Sarawak waters.

The WBD is the most prolific in terms of hydrocarbon resources. As at 1.1.1998 a total of 132 exploration wells, 46 wildcats and 86 appraisals, have been drilled, representing about 30% of the total wells drilled in Sarawak (Fig. 20.6). The main reservoirs are the Cycles V/VI prograding deltaic to shallow marine sandstones. Extensive synsedimentary growth faulting with very large throws form the principal petroleum traps in this area (Rijks, 1981).

The WBD has about 53% of the total oil-in-place, 69% of the estimated recoverable oil and has contributed to 82% of oil production in Sarawak. The province still has the largest remaining oil reserves, about 51% of total. As for gas, the WBD has about 13% of the total recoverable gas, contributing to 19% of total gas production. It has about 12% of the remaining gas reserves (Fig. 20.9, 20.10 and 20.11).

The field sizes in the WBD range from 3 to 346 MMSTB estimated recoverable oil. Six of the 8 giant oil fields and 9 of the 13 producing oil fields in Sarawak come from the WBD. Among the major oil producing fields are Baram, Baronia, Bokor, W. Lutong and Tukau The crude oil is piped (between 15 km to 45 km) to the onshore facilities. In spite of extensive exploration activities carried out in the WBD, the potential of discovering more oil and gas in this province is still considered high.

The onshore field Asam Paya, with EUR less than 8 MMSTB, is scheduled to come on stream in 1999. The emerging play of Cycles V/VI HPHT (High Pressure High Temperature)

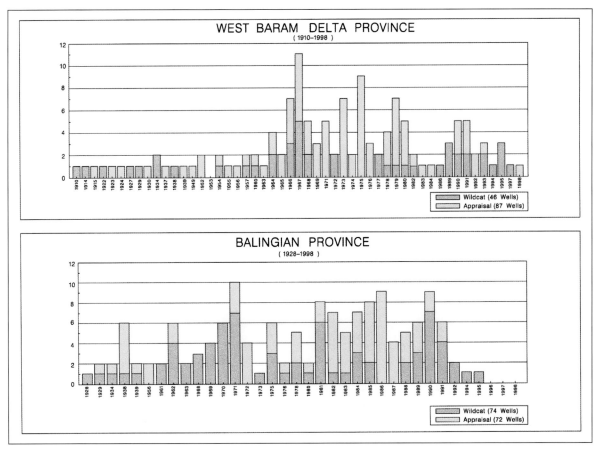

Fig. 20.6 Exploration drillings in the West Baram Delta and Balingian provinces.

could contribute to significant new discoveries. The drilling results of the latest development well, Bakau-5, and exploration well Bakau Deep-1 are very encouraging for deeper reservoir objectives.

Balingian Province

The Balingian Province is the most extensively explored region in Sarawak with a total 146 exploration wells, 74 wildcat and 72 appraisal, representing about 35% of the total wells drilled in Sarawak (Fig. 20.6). Exploration activity started in 1928 with the drilling of onshore Balingian-1 well by Shell (Sarawak Shell Berhad). This was followed by the drilling of 16 more wells between 1939 to 1956 with no commercial success. In 1962, Shell made major discoveries in the offshore part of the Balingian Province with the drilling of Temana-1 and Patricia-1 wells. From then onwards, further success has been achieved with the discovery of more major oil fields like Bayan, D18 and D35 in the 1970's and 1980's. Hydrocarbons are found in the structurally and stratigraphically complex Cycles I/II/III clastics play. The Miocene compressional deformation events have resulted in complexly faulted structures (Fig. 20.12).

The Balingian Province has about 32% of total oil in-place, 28% of estimated recoverable oil and has contributed to 18% of oil production. The province still has about 41% of remaining oil reserves in Sarawak. The average field sizes ranges from 2 MMSTB to 148 MMSTB estimated recoverable oil. The lower volume recoverable oil is reflective of the lower recovery factor and generally small field sizes found in the province. As for gas the Balingian Province also has about 6% of estimated recoverable gas and has contributed to 2% of gas production in Sarawak (Figs. 20.9, 20.10 and 20.11).

The main producing reservoirs in the Balingian Province are of Cycle II and Cycle III lower coastal plain sands including channel, crevasse splay and shoreface sands. In fields like Bayan and D18 the main reservoirs are stacked channel/crevasse splay sands with complex fluid contacts and compartmentalisation of the hydrocarbon accumulations.

In spite of intense exploration activities and more oil discoveries in the Balingian Province than in the WBD, the latter has more than twice the volume of recoverable oil (Fig. 20.9). The oil field sizes in the WBD are large, whereas in the Balingian Province, the fields are smaller with the hydrocarbons trapped in more complex structures, resulting in lower recoverable oil and

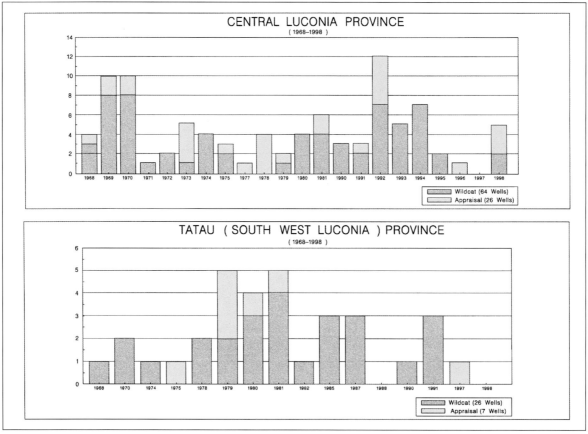

Fig. 20.7 Exploration drillings in the Central Luconia and SW Luconia provinces.

lower production rate. The channel reservoir sands developed in most of the oil fields in the Balingian Province is another factor contributing to the poor reservoir continuity.

The emerging play of Cycles II/III stratigraphic traps in the east Balingian area has yet to be tested. The distal turbidite and possibly lowstand wedge sands provide potential for the prospective exploration objectives.

Central Luconia Province

The Central Luconia Province is a stable carbonate platform area, known as the gas province in Sarawak.

Exploration activities in the area started in the 1960's. As at 1.1.1998 a total of 85 exploration wells, 62 wildcats and 23 appraisals have been drilled in the province (Fig. 20.7). About 100 carbonate buildups have been identified in the Central Luconia Province, half of them have been drilled and resulted in the discovery of 33 fields (60 % success rate). Most of the gas is found in the Cycles IV/V Miocene carbonate play. The trap sizes range from very large carbonate platforms to small transgressive buildups. Subareal exposure and secondary diagenetic processes experienced by the carbonates provide good reservoir quality, while a thick Cycle V shale overlying most of the carbonate platform acted as an effective caprock for hydrocarbon entrapment.

The Central Luconia Province has about 67% of total recoverable gas and about 75% of the recoverable non-associated gas in Sarawak respectively. The field sizes in the Central Luconia Province range from 5 BSCF to as large as 5 TSCF estimated recoverable gas. Ten fields exceed 1 TSCF recoverable gas. The major producing gas fields are E11, F6, F23, M3 and M1. Major gas fields to be developed in the near future are B11, Jintan, E8 and F28. As the main gas province, the Central Luconia Province has contributed about 78 % of gas production in Sarawak, all from non-associated gas fields. In terms of remaining gas reserves, the Central Luconia Province has the largest, about 65% of total gas, and 71% of total non-associated gas in Sarawak respectively (Figs. 20.10 and 20.11).

Beside gas resources, the Central Luconia Province also has about 14% of total oil in-place and 3% of estimated recoverable oil in Sarawak (Fig. 20.9). The higher volume of oil in-place and small volume of recoverable oil reflect the low recovery factor of oil in this province. Most of

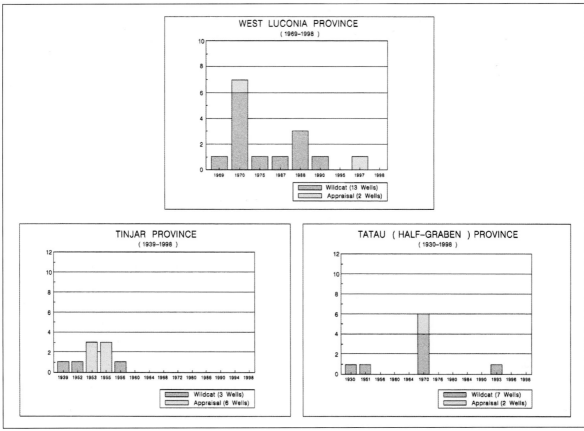

Fig. 20.8 Exploration drillings in the West Luconia, Tinjar and Tatau provinces. Hydrocarbon prospectivity in the West Luconia province remains high despite the low exploration drilling activity in the area. The entrance of new players in the area will see a revival of activity as new play concepts are tested.

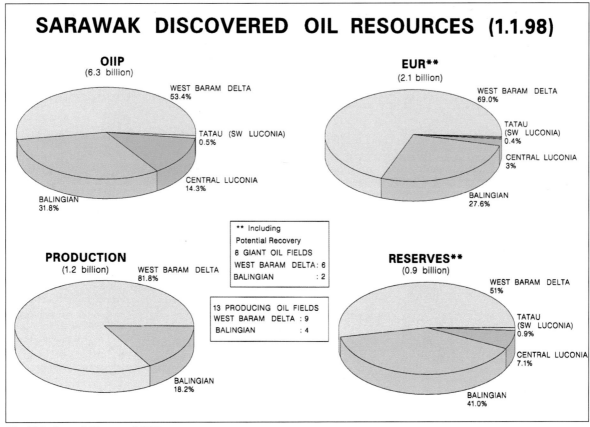

Fig. 20.9 Sarawak Discovered Oil Resources.

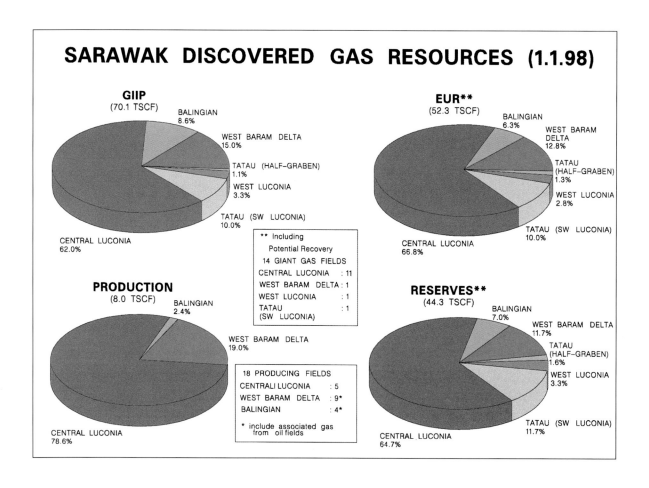

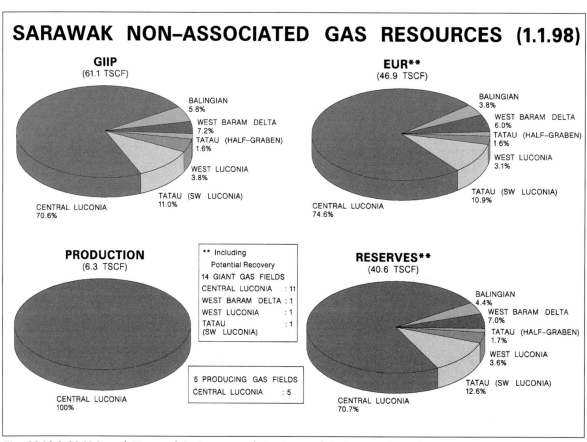

Figs. 20.10 & 20.11 Sarawak Discovered Gas Resources and Non-Associated Gas Resources (1.1.98).

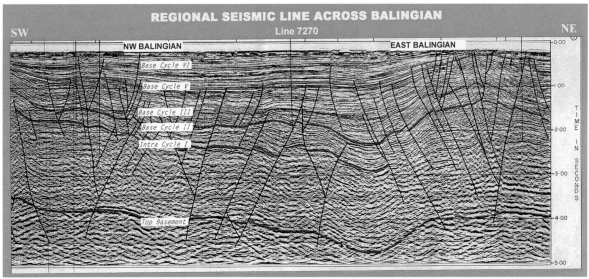

Fig. 20.12 Regional Seismic lines across the Balingian province. Complex structures and reservoir discontinuity provide exploration and development challenges in the Balingian province. innovative techniques and the latest technology are needed to fully realise and exploit the hydrocarbons in the area. (Modified after Sarawak Shell unpublished report)

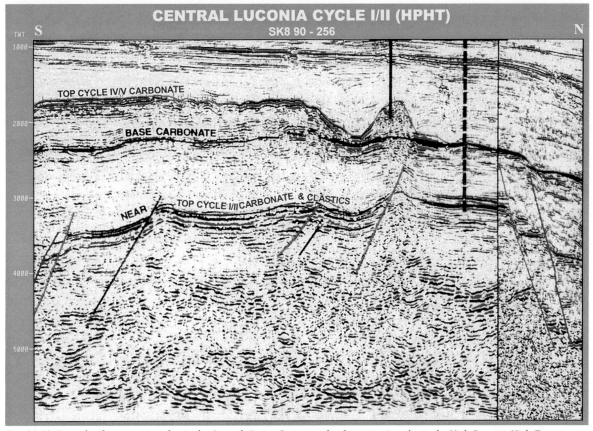

Fig. 20.13 Example of an emerging play in the Sarawak Basin. One example of an emerging play is the High Pressure High Temperature (HPHT) play. The figure shows a HPHT prospect with deeper Pre-carbonate objectives (Cycles I/II).

the oil is found in small amounts in poor quality reservoirs of Cycles I/II/III pre-carbonate clastics, as in the Bentara, Balai, Lada Hitam and Bunga Pelaga fields. Oil also occurs as a thin rim beneath the thick gas columns, as in the M1, M4, E6 and F28 carbonate gas fields. Production tests of the recently drilled F28-2 well showed the oil is non-recoverable or residual. With high water saturation in the oil zone, and strong aquifer beneath the carbonate platform, the recovery factor for the oil is considered negligible or nil.

Even though a significant amount of hydrocarbons have yet to be found in the Cycles I/II/III pre-carbonate clastic play, the potential of this play in oil and gas exploration is still promising. Previously drilled wells have proved the existence of matured source rock, hydrocarbon migration and effective sealing in this area. The future challenge is to locate larger pools and better quality reservoirs. The Cycle I/II/III in the western and southern part of the Central Luconia Province are dominated by coastal plain facies. Further to the north in the G2-1 & Jintan Deep-1 wells' area, the sediments are shallow marine siliciclastics and carbonates. In the Jintan Deep-1 well the Cycle II carbonate reservoir exhibited a reasonably good porosity of about 12% at depth of 3500 m. Accurate structural definition is difficult due to poor seismic data quality below the carbonate. Special seismic processing, such as `Pre-stack Depth Migration', which could enhance the data quality below the carbonate, should be used in future exploration programme (Fig. 20.13).

Another potential hydrocarbon play in the Central Luconia Province is the Cycles V/VI post-carbonate turbidite play (Fig. 20.14). This new play of toe-of-slope and basin floor fans was identified by seismic stratigraphic interpretation. The nearest analogue for the play is the Lower Stage IV-D toe-of-slope fans from the Tembungo Field in the Sabah Basin (Esso, 1997).

Other Provinces

The other non-producing provinces, Tatau (incorporating SW Luconia and Half-Graben sub-provinces), Tinjar, West Luconia and North Luconia contain about 16% of the total remaining recoverable gas resource of 44.3 TSCF, and negligible oil resource. Most of the gas contain considerable amounts of CO_2 and N_2. More than 60 exploration wells have been drilled in these provinces, mostly in the SW Luconia sub-province (33 wells), which constitutes about 15% of the total exploration wells drilled in Sarawak region (Figs. 20.7 and 20.8).

Recent drilling activity occurred in the Half-Graben Sub-province of Tatau (Oya-A1 in 1994) and West Luconia (T3.2 in 1997; K5.2 in 1997). Oya-A1 is a minor gas discovery, indicating that the Cycles III/IV/V clastics within grabens in the Tatau are potential exploration targets for gas (OPIC, 1990; Wee et al., 1995). In West Luconia, the gas is moderately to highly contaminated with inert gases (CO_2 and N_2). Except for the T3 gas discovery, which also encountered gas bearing reservoirs with very low CO_2 in the Intracycle V the other fields such as C24 and Mawar are considered minor gas

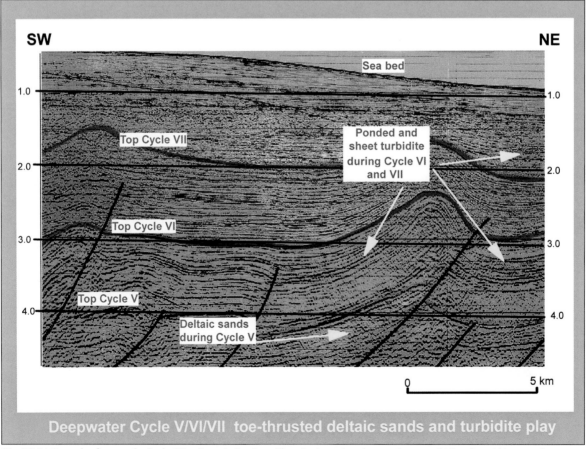

Fig. 20.14 Example of a new play in the West Luconia Province. The seismic section shows a deeper turbidite play within a toe thrust area in Deepwater Sarawak. Exploration of this type of play requires thorough geological and geophysical studies. New models for source and deposition of reservoirs are needed.

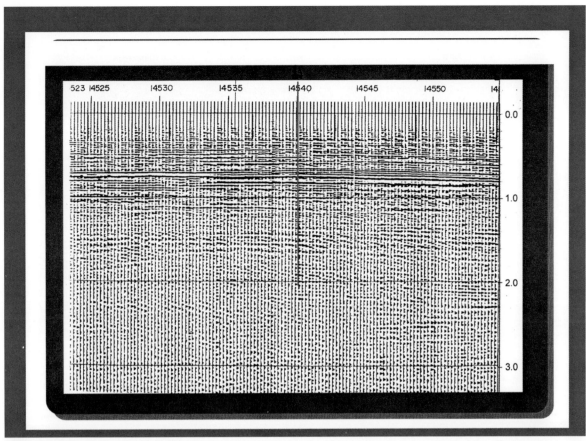

Fig. 20.15 An example of 6 fold 1969 vintage seismic section in the West Luconia Province. Much of the effort in assessing the hydrocarbon prospectivity in the West Luconia Province was hampered by poor quality seismic data. Exploration drilling in the early seventies relied on this kind of data.

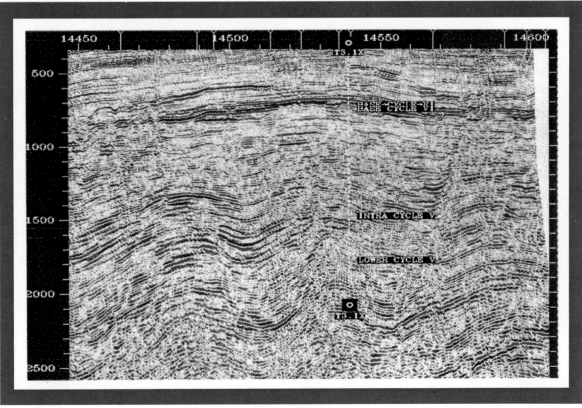

Fig. 20.16 An example of a reprocessed 1969 seismic line. With the advent of new and more powerful reprocessing techniques and workstation based interpretation, most of the old vintage data was reprocessed and used together with newly acquired data to reassess the area's prospectivity.

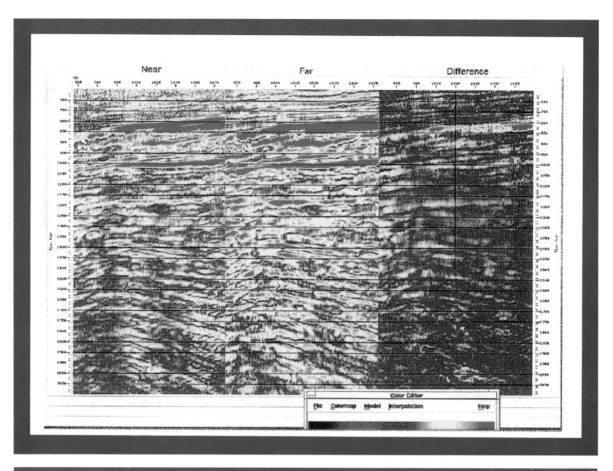

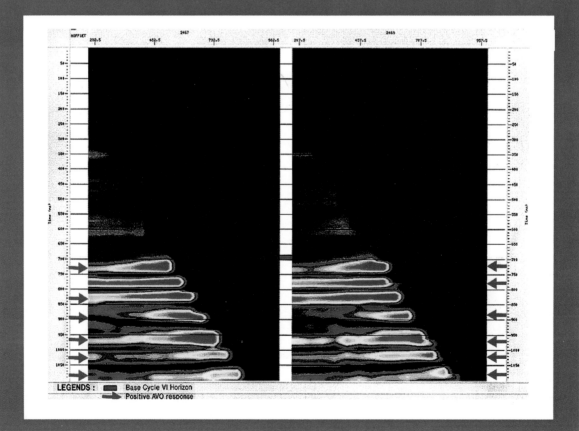

Figs. 20.17a & 20.17b AVO anomalies over a gas field in Block SK-303 in the West Luconia Province. AVO analysis has been proven effective as an exploration tool in the area and is routinely used together with other techniques in exploration in the West Luconia Province. Figure 20.17a show a reflection strength angle limited plot of a multiple gas reservoir. Figure 20.17b show a AVO difference plot of near and stack.

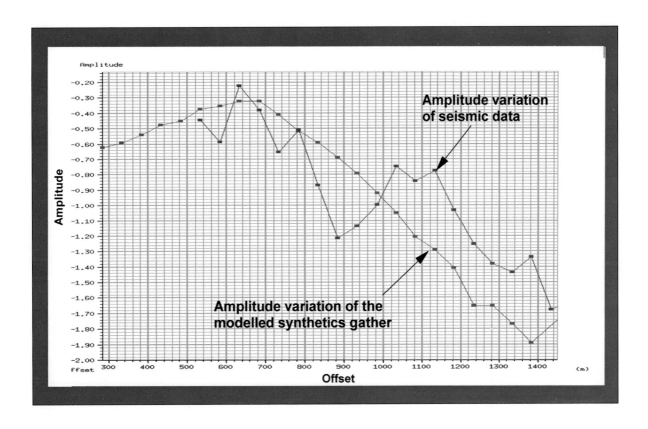

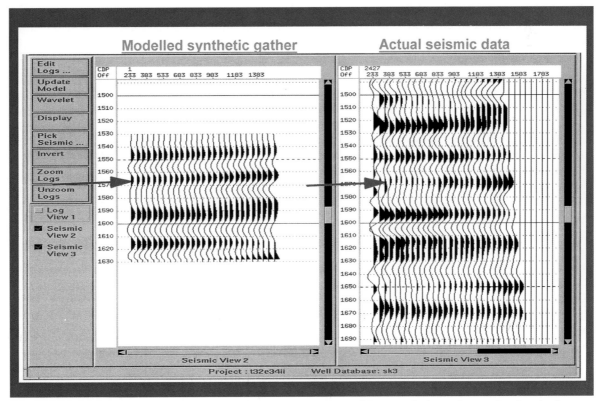

Figs. 20.18a & 20.18b Amplitude versus offset plot of a gas reservoir at T3.2 well location. The figures shows a plot of amplitude versus offset and CDP gathers of a modelled synthetic gather and actual seismic data which both shows agreement in increase of amplitude with offset.

discoveries with average field sizes of about 30 BSCF estimated recoverable gas. Exploration effort in the area is being revitalised with new activities being carried out. Figures 20.15 and 20.16 show 1969 seismic lines reprocessed in 1997. The use of workstations has enhanced the interpretation of exploration potential in the area. Furthermore the acquisition of new seismic lines in 1995 permits a more thorough assessment of the area. AVO analysis was also carried out on the newer lines. Figure 20.17a shows a line across a gas field with positive AVO response. The T3.2 well results provide good information on the potential of the Intra Cycle V play. Follow-up post-drilled AVO modelling confirms the validity of using AVO in the area. Figure 20.17b shows a modelled gas reservoir at T3.2. Figure 20.18a and 20.18b shows a reconnaissance AVO for a prospect in the area.

In the SW Luconia Sub-province, there are several gas discoveries in Cycles II/III clastics and Cycles IV/V carbonates. The gas is reservoired in Cycles II/III clastics and Cycles IV/V carbonates. The major discovery is the K5.1 well drilled in 1970, which penetrated about 310 m of gas column in the Cycles IV/V carbonate. Recent drilling of K5.2 well to the north of K5.1 well encountered thicker gas column of about 350 m. However, two production tests conducted in the well show a contamination of about 54% CO_2 and 2% of N_2. A number of studies have been conducted to investigate the origin of CO_2 and N_2 in the hydrocarbon bearing-reservoirs in this area. Idris (1992), based on the vertical distribution profile of CO_2 and N_2 contamination in the J32-1 well's reservoirs, believed that the CO_2 is most likely due to late in situ inputs while the N_2 was derived from the basement.

Other discoveries in the province are J5, A3, Tenggiri Marine, Selar Marine, A21 and J32. In the A21.1 well drilled in 1991 about 12 m net oil sand was encountered in the lower Cycle III clastics.

As shown in Figures 20.10 and 20.11, the SW Luconia Sub-province contributes about 12 % of the estimated recoverable of gas. Except for K5 gas discovery, the other discoveries in this province are considered minor with average field sizes (without inerts) ranging from 20 BSCF to 700 BSCF. The oil potential of the Cycles II/III clastics play which was tested by A21 well should be followed up in a future exploration programme.

All 9 wells drilled in the Tinjar Province are onshore, and the drilling locations were selected from surface geological information in the late 1930's and 1950's. Although oil shows are present in most of the wells, no significant hydrocarbons were encountered in the wells. However, recently aerogravity and aeromagnetic surveys were conducted for improved definition of structural traps and re-estimates of sediment thicknesses, as the Cycles I/II (Nyalau) clastics and carbonates are still considered prospective. The data are being processed and interpreted. To date, 2 wells have been drilled in the deepwater area. Exploration activities are continuing.

FUTURE OUTLOOK

New and Emerging Plays

Two categories of play types, i.e new and emerging plays, have been delineated and identified. A new play is defined as a geological concept based on experience or on a geological analogue derived from comparative data and information. More work has to be done in order to mature the concept and to reduce the uncertainties and related risk. An emerging play is defined as a geological concept and model which has attained a state of maturity and confidence to be tested with drilling, or that has been proven to be valid by limited drilling. Each play is described in the following format : Location/Reservoir Type/Play Type.

Ten hydrocarbon plays comprising 5 emerging and 5 new plays have been identified in Sarawak. The 5 emerging play types identified are as follows:

i) West Baram Cycles V/VI High Pressure High Temperature/ Stratigraphic Play
ii) Central Luconia Cycles I/II High Pressure High Temperature Play
iii) Balingian East Cycles II/III Stratigraphic Play
iv) West Luconia Cycles V/VI/VII Stratigraphic Play
v) S.W. Shelfal Cycle V Structural Play

An example of an emerging play is illustrated in Figure 20.13. This is a pre-carbonate palaeohigh beneath the carbonate platforms in the Central Luconia Province. This prospect has a large areal extent, up to 300 km2, and deep culmination at 3,800 m (Taylor et al., 1997). To date 7 wells (Jintan Deep-1, Keris-1, Bunga Pelaga-1, F6.6, Lada Hitam-1, Bijan-1 and G2-1X RDR) have been drilled to test this play. They have made some minor oil and gas discoveries. Some of the wells were based on 2D seismic to test the Cycle IV carbonates as the primary objective and the pre-carbonate clastics as a secondary objective. However, the Jintan Deep-1 well drilled in July 1998 targeted Cycles I/II carbonates and clastics as the primary and secondary objectives respectively. The well was drilled to depth of -3510 mss, and terminated in the Cycle II carbonates with gas shows/residual gas. Minor late movement to the sealing fault is believed to cause the gas accumulation to leak upward to the overlying Cycles IV/V carbonates and ultimately to the seabed.

The Jintan Deep-1 well results proved the presence of hydrocarbon migration and charge in the Cycles I/II carbonates and clastics, as well as finding reasonably good porosity at a deeper level. The thick Cycle III shale could provide a good sealing capacity for the hydrocarbon. The results upgrade the hydrocarbon prospectivity of the Cycles I/II pre-carbonate play in the Central Luconia Province.

The main risk associated with this play are high pressure and temperature and inaccurate structural definition due to poor seismic data quality below the carbonate.

As for the new play types identified which need further work to be carried out to reduce the risk prior to forming up the prospects includes:

i) West Baram Turbidite Play
ii) Tinjar Cycles I/II Structural Play
iii) Tatau Province Cycle V Stratigraphic Play
iv) Deepwater Cycles V/VI Stratigraphic Play
v) Central Luconia Lowstand Carbonate Play

An example on the new play is displayed in Figure 20.14. The seismic line shows a possible trap in Cycles V/VI/VII toe-thrusted deltaic sands and deep water ponded/sheet turbidites. This turbidite play has yet to be tested by any well. The main risks associated with this play are reservoir quality, sealing efficiency and the sufficiency of oil expulsion/migration.

CONCLUSION

To date the West Baram Delta, Balingian and Central Luconia provinces contain more than 95% of the oil and 85% of the gas resources and remaining reserves in Sarawak. They have also contributed to the entire oil and gas production for Sarawak. Despite extensive exploration activities in these 3 provinces, further discoveries are expected with the identification of emerging and new plays.

For the other provinces, Tatau, Tinjar, West Luconia and North Luconia, some significant gas discoveries have been made. Ongoing exploration activities in the West Luconia and North Luconia provinces indicate that these 2 provinces hold promise for new discoveries. Efforts are already underway to identify prospective areas in the Tatau and Tinjar provinces.

REFERENCES

Esso, 1997. Technical report for Block SK8/308/10. Central Luconia Province, Sarawak. Internal Report.

Idris, M.B., 1992. CO_2 and N_2 contamination in J32-1, SW Luconia, Offshore Sarawak. Geological Society of Malaysia Bulletin, 32, p. 239-246.

OPIC, 1990. Geological Completion Report Oya-A1 Block SK12 onshore Sarawak, Malaysia. Internal Report No. EXPL 1198/94.

Taylor, G., Powel, C., Newall, M., Ngau, A., Hermans, L., Wiemer, M., Nasaruddin Ahmad, 1997. PETRONAS-SSB Joint Regional Study of the Pre-Carbonate Clastics, Central Luconia Province, Offshore Sarawak. Internal Report No. EXP. R. 50793.

Wee, E.S., Lai, S.K. and Salim Sahed, 1995. Geology and Hydrocarbon Potential of Igan-Oya Half Graben Block SK12, Sarawak, Internal Report No. EXPL 1242/95.

Part 5

SABAH SEDIMENTARY BASINS AND PETROLEUM RESOURCES

Chapter 21

Geological Setting of Sabah

Leong Khee Meng

INTRODUCTION

Sabah, situated in the northern part of the island of Borneo, and lying adjacent to actively moving plates in the Southeast Asian region, has a complex geological history. The well developed compressive and extensional structures are the manifestation of several regional tectonic events in Sabah since the early Tertiary (Figs. 21.1 and 21.2). Geomorphologically, a distinct mountain range comprising Palaeogene Crocker Formation and older rock units occurs near the west coast, and rises to between 3,000 to 4,000 m above sea level, culminating in the Late Neogene Kinabalu granodiorite intrusion, which peaks at 4101m. Offshore NW Sabah, the water depth increases rapidly to 2800-3000m towards the NE-SW elongated Sabah Trough, in the South China Sea. Similarly in offshore SE Sabah, the water depth rapidly increases towards the Celebes (Sulawesi) Sea.

The early NW Borneo geosynclinal model of Haile (1969) has been superceded by several plate tectonic models especially for the NW Sabah continental margin (Noor Azim Ibrahim, 1994; Tongkul, 1990, 1994; Hazebroek and Tan, 1993; Xia and Zhou, 1993; Hutchison, 1996, 1992a, 1988; Rice-Oxley, 1991; Tan and Lamy, 1990; Rangin et al., 1990; Benard et al., 1990; Tjia et al., 1990; Tjia, 1988; Levell, 1987; Hinz et al., 1985, 1989; Bol and van Hoorn, 1980; Scherer, 1980). Tectonic models for eastern Sabah are given by Bell and Jessop (1974), Tjia (1988), Tjia et al. (1990), Hutchison (1992a), Clennell (1991, 1996), Walker (1993), and Rangin et al. (1990). Tongkul (1991a), however, gave the first comprehensive and coherent account on the tectonic evolution for the whole of Sabah, supplemented later by further research work in northern and central Sabah (Tongkul, 1994, 1997).

The main geological elements of Sabah (see Figs. 21.3, 21.4 and 21.5) may be categorized as follows :

- Pre-Tertiary Core, metamorphic and igneous complex;
- Cretaceous-Early Palaeogene, ophiolites and indurated deep marine sediments;
- Palaeogene Basins, mainly N-S elongated basins containing deep marine sediments to the east and west of the Cretaceous-Early Palaeogene deep marine sediments.
- Neogene Basins - the petroleum-producing Sabah Basin to the northwest; the Northeast (NE) Sabah Basin and Southeast (SE) Sabah Basin, flanking the Palaeogene Basins (Leong, 1978a; Figs. 21.2 and 21.7). A total of some 190 exploration wells have been drilled in both the onshore and offshore Sabah portion of the Sabah Basin, resulting in the discovery of some 1.2 billion barrels of oil in 8 commercial accumulations and over 5.0 TSCF of gas in yet to be developed gas fields. Several oil and gas seeps occur onshore (Fig. 21.6).

The distinctive tectonic/geomorphic elements in offshore NW Sabah deepwaters are the Sabah Trough and the NW Sabah Platform, covering the Dangerous Grounds area in the South China Sea (Figs. 21.1 and 21.2). The Sabah Trough is a NE-SW linear bathymetric feature with water depths of 2.8 to about 3.0 km, and extending over 300 km; its average width is 80km. The Sabah Trough has been interpreted by earlier writers as a Palaeogene subduction trench, underlain by southward dipping oceanic crust (Haile, 1973; Hamilton, 1979; Ludwig et al., 1979; Taylor and Hayes, 1980, 1983). This has not been substantiated by recent deep seismic reflection data (Hinz and Schluter, 1985; Hinz et al., 1989, 1991), geological data (Hazebroek and Tan, 1993), and gravity data (Milsom et al., 1997). The Sabah Trough (together with Palawan Trough) has also been recently interpreted as a failed rift system developed on proto-China continental margin in Late Cretaceous-Early Palaeogene times (Schluter et al., 1996).

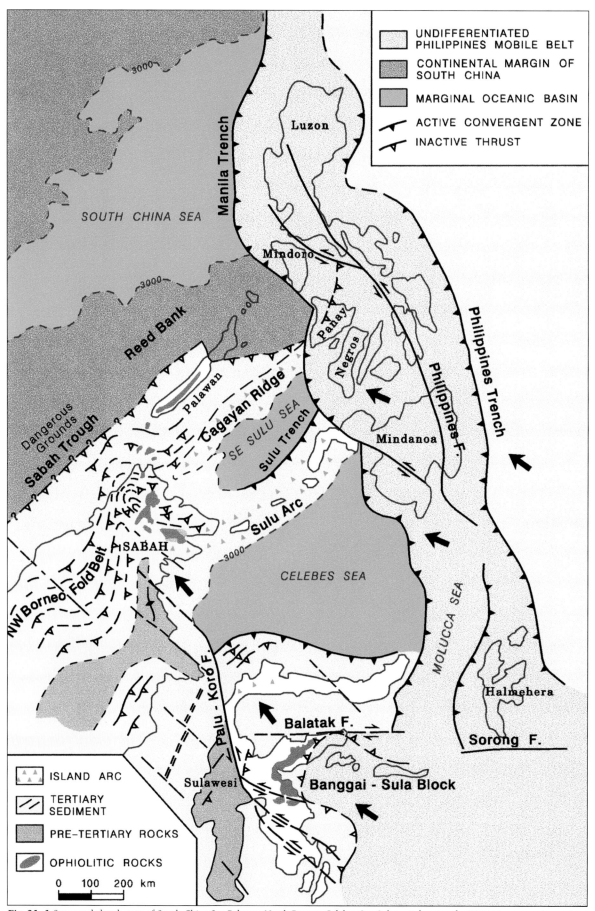

Fig. 21. 1 Structural sketch map of South China Sea-Palawan-North Borneo-Celebes Sea-Sulawesi showing the NW-SE compressive regime in Sabah as a result of the opening of the South China Sea and the collision of the Banggi-Sulu Platform with Sulawesi since early to late Tertiary (after Tongkul, 1997). Reprinted with permission from Elsevier Science.

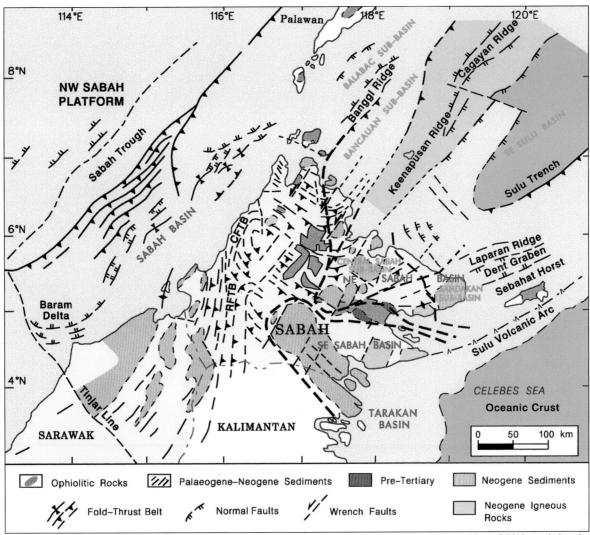

Fig. 21.2 Structural trends of Sabah and surrounding areas showing outline of the 3 Neogene basins. The 2 fold-thrust belts, the Palaeogene Rajang Fold-Thrust Belt (RFTB) and the early Neogene Crocker Fold-Thrust Belt (CFTB) are also indicated. (modified after Tongkul, 1993).

PRE-TERTIARY TO EARLY PALAEOGENE

Pre-Tertiary Core

The pre-Tertiary Core of onshore Sabah is represented by :

- *Crystalline Basement* : a metamorphic and igneous complex (amphibolites, hornblende schists and gneisses; granodiorites, tonalites, granites), radiometrically dated in the range Cretaceous to Triassic (Kirk, 1962; Leong, 1974, 1998; Hutchison, 1997).

Cretaceous-Early Palaeogene

- *Chert-Spilite Formation* : complex of extrusive igneous rocks (basalts, spilites, volcanic breccias, agglomerates) associated with radiolarian cherts, limestones and clastics. The radiolarian cherts are of Early Cretaceous age; the limestones, Late Cretaceous (Leong, 1974,1977; Basir Jasin and Sanudin Tahir, 1988; Basir Jasin, 1992). Associated with ultrabasic and basic igneous intrusives.

- *Madai-Baturong Limestone* : algal, oolitic and massive limestone of Cretaceous age (Adams and Kirk, 1962; Leong, 1974; Fontaine and Ho, 1989).

- *Sapulut, Trusmadi and East Crocker formations* : thick turbiditic clastics; partly metamorphosed to slates and phyllites.

The above rock units occur mostly in central and eastern Sabah. The Chert-Spilite Formation and the associated large bodies of ultrabasic and basic intrusive igneous rocks occur in an arcuate belt extending from eastern Sabah through central Sabah (Labuk Valley) to the Mt. Kinabalu area and on to

Geological Setting of Sabah

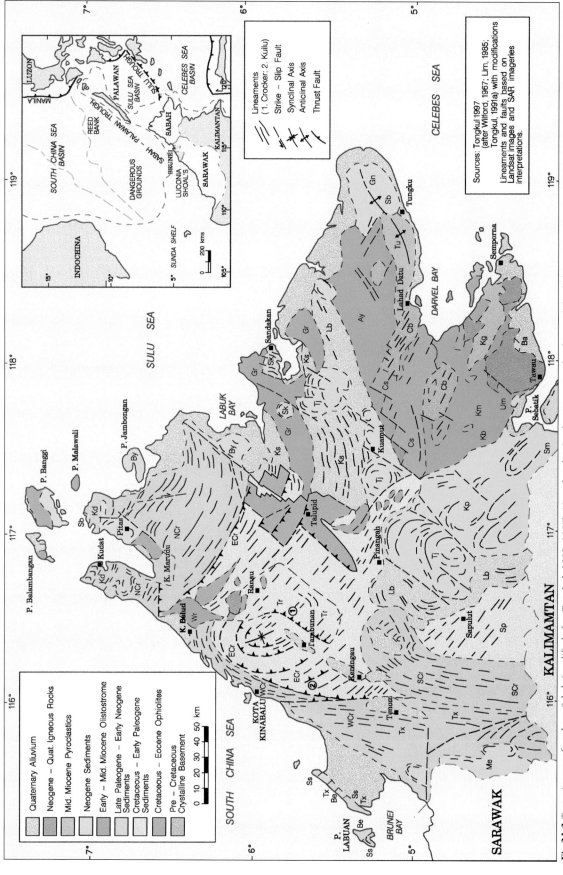

Fig. 21.3 Tectonic map of onshore Sabah (modified after Tongkul, 1993; 1997). Note that the Crocker Formation has been subdivided into West Crocker (WCr), South Crocker (SCr) and North Crocker (NCr). Reprinted with permission from Elsevier Science.

the Kudat Peninsula area and the northern islands of Banggi and Malawali.

Upper Cretaceous limestones of the Chert-Spilite Formation unconformably overlie the Crystalline Basement (Wong and Leong, 1968) and Upper Cretaceous mudstones fill fissures of the Madai-Baturong Limestone (Adams and Kirk, 1962; Leong, 1974). The Cretaceous shallow marine Madai-Baturong Limestone appears to be the lateral equivalent of the deeper marine limestones and radiolarian cherts of the Chert-Spilite Formation.

The Cretaceous to Early Palaeogene Sapulut, Trusmadi and East Crocker formations consist of thick rhythmic alternations of sandstones and shales with some limestone and tuff, and are partly metamorphosed to phyllite and slates (Figs. 21.3 and 21.4). The East Crocker is probably Late Cretaceous to Middle Eocene (Collenette, 1958; Tongkul, 1997) and this formation together with the Sapulut and Trusmadi formations comprises the 'older' Crocker Fold-Thrust Belt i.e. Rajang Group Fold - Thrust Belt (RFTB). They are believed to have been deposited in the centre of a deep marine trough, partly contemporaneous with the emplacement of the ophiolitic sequences, of the Chert-Spilite Formation and associated ultramafic and mafic rocks (Liechti et al., 1960; Collenette, 1958,1965; Stauffer, 1968; Lee, 1980; Tongkul, 1991a,1994,1997; Tjia, 1974; Bernard et al., 1990; Geological Survey of Malaysia, 1995; Wilford, 1967; Lim, 1985a).

The oldest rocks penetrated offshore northern Sabah consisted of Eocene shales and mudstones, partly conglomeratic, containing smooth, rounded to angular fragments of sandstone, siltstone, chert and serpentinite in the wells Kalutan-1 and Balambangan-1. The pre-Tertiary to Lower Palaeogene sedimentary rock units are correlated with the Stage I and II in the NW Sabah offshore lithostratigraphy of Bol and van Hoorn (1980). The occurrences of Palaeocene to Eocene limestones in all parts of Sabah (Tongkul 1991a) indicate that the Early Palaeogene and older units were uplifted during this time.

The Crystalline Basement and Chert-Spilite Formation and the associated ultramafic and mafic bodies have been considered together as an ophiolite series representing remnants of oceanic crust (Hutchison, 1975). Tjia (1988) and Tjia et al. (1990), however, interpret the Crystalline Basement as fragments of an allochthonous pre-Cretaceous consolidated continental margin crust. Tongkul (1991a) refers to the Crystalline Basement as 'older metamorphosed oceanic basement with rare acidic intrusive rocks', distinct from the 'new unmetamorphosed oceanic basement' (Chert-Spilite). Recent palaeogeographic reconstruction of the tectonic evolution of the Sulu Sea area by Schluter et al. (1996) indirectly indicates that the Crystalline Basement in eastern Sabah was part of the drifted continental fragment that now underlies the Cagayan and Sulu ridges. In summary, the acid igneous rocks of the Crystalline Basement probably represent a continental basement and perhaps have a different origin and tectonic history to the amphibolites and metagabbros, which are interpreted as metamorphosed oceanic or ophiolitic basement (see discussions in Hutchison, 1997 and Leong, 1998).

The pre-Tertiary crust in offshore Sabah is largely unknown. However, in the deepwater area offshore NW Sabah in water depths between 1.2 to 2.5km, the platform area known as the Dangerous Grounds is characterized by a complex system of horsts, tilted fault blocks and synrift half-grabens largely orientated NE-SW (Hinz et al., 1989). The tilted fault blocks comprise probably Mesozoic sediments/ metasediments (Triassic to Cretaceous) resting on continental basement (Kudrass et al., 1986). The sedimentary fill of the half-grabens is probably Lower Palaeogene paralic to bathyal clastics. The same structural style of tilted fault blocks and half-grabens is observed below the Dangerous Grounds area and in the Sabah Trough area (Hinz et al., 1985,1989; Hazebroek and Tan, 1993; Noor Azim Ibrahim, 1994; Schluter et al., 1996).

SEDIMENTARY BASINS AND STRATIGRAPHY

Palaeogene Basins

A thick succession of marine outer shelf to bathyal sediments, as young as Early Miocene, fills two large, deep Palaeogene troughs on the western and eastern sides of the Early Palaeogene and older core (Fig. 21.4 and 21.5).

Western Palaeogene Basin

Onshore northwestern and western Sabah is characterized by the mountainous Crocker Range which consists of highly deformed, imbricated and thrusted deepwater sediments of the Crocker Formation (e.g.: Lee, 1980; Tongkul, 1991a;

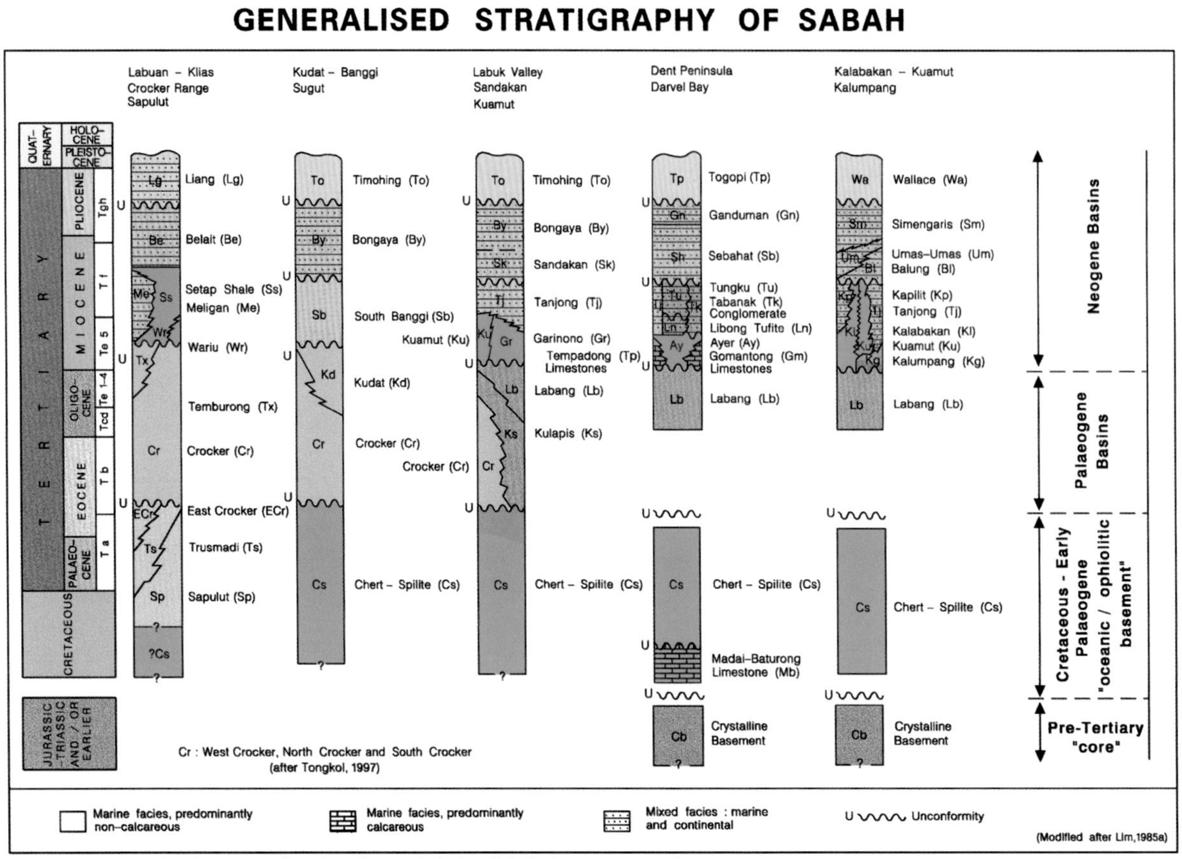

Geological Setting of Sabah

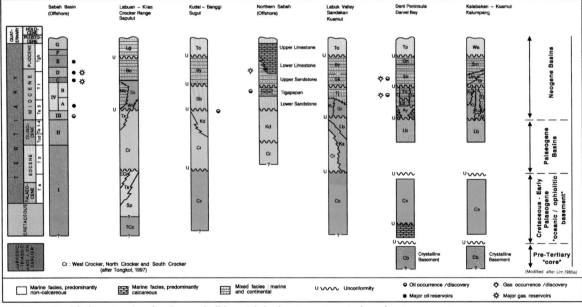

Fig. 21.4 Generalized stratigraphy of onshore Sabah (modified after Lim, 1985a).

Fig. 21.5 Simplified stratigraphy of onshore and offshore Sabah, showing major petroleum-bearing reservoirs.

Bernard et al., 1990; Stauffer, 1968). The entire Crocker Formation has been referred to as the Crocker accretionary prism, Crocker fold belt, a northern continuation of the Rajang Group accretionary prism, or Rajang Group Fold-Thrust Belt (RFTB), composite Rajang-Crocker Belt, Kinabalu fold-thrust belt and others. The Crocker Formation can be further sub-divided into East and West Crocker, and/or undifferentiated Crocker (Liechti et al., 1960; Wilford, 1967; Lee, 1980).

Wilford (1967) adopted this sub-division in his compilation of the Geological Map of Sabah. The West Crocker (Crocker) Formation consists of non-metamorphic sandstone with minor shale, whereas the East Crocker Formation consists of sub-metamorphic predominantly argillaceous strata. West Crocker (also undifferentiated Crocker) is considered to be younger (Liechti et al., 1960) and from recent studies by Tongkul (1994), a late Middle Eocene to Early Miocene age is indicated. The boundary between West and East Crocker is marked by a prominent lineament 'Kiulu Lineament', probably a thrust zone (Lee, 1980). Tongkul (1997) recently introduced North Crocker and South Crocker formations in place of Crocker undifferentiated in the northern and southern area (Fig. 21.3).

Further south, the West Crocker (and South Crocker) becomes more argillaceous (Temburong Formation). To the north in the Kudat Peninsula, the rhythmic sandstone and shale alternations pass into massive grain flow sandstone fan deposits, known as the Kudat Formation (Oligocene-Early Miocene). The Crocker Formation in general is more deformed than the Kudat Formation. These differences are consistent with the Crocker Formation being generally older than the Kudat Formation. The Kudat Formation is described in more detail, as oil seeps occur in the area.

The Kudat Formation, represented by the Tajau Sandstone and the Sikuati Member, has broad lithological similarities to the Crocker Formation. The Tajau Sandstone consists of massive sandstones, and the Sikuati Member, carbonaceous sandstone/shale alternations. The Kudat Formation has been interpreted as deep water fan deposits, the Tajau Sandstones as grain flow deposits on the inner/middle fan (Frank, 1981). The common occurrences of plant remains and resins in the Sikuati Member sandstones suggest a nearby deltaic source which has since been eroded. The transition from the Tajau Sandstone to the Sikuati Member may represent a shallowing phase, just prior to the late Early Miocene/early Middle Miocene uplift and deformation.

This shallowing phase is also supported by occurrences of Lower Miocene limestone and limestone breccia occurrences in the Kudat Peninsula (see Tongkul, 1994). The Kudat Formation has also been interpreted as entirely of shallow water depositional environment by other workers (e.g. Hazebroek and Tan, 1993; Tongkul, 1994).

Palaeocurrent patterns show that the Crocker sediments appear to have come from the southwest in western Sabah and were then deflected to the east in northern Sabah, while the Kudat sediments came mainly from the north and northeast (Tongkul, 1994). Support for the provenance of Kudat Formation sediments is indicated by the occurrences of Lower Miocene quartz-rich sands to the north in Balabac Island (Irving, 1949), which from lithologic descriptions are similar to the Tajau Sandstone.

Hutchison (1996) has re-emphasized the distinction which should be drawn between the older Rajang Group, to which the Sabah East Crocker Formation may be equivalent, and the 'younger' Crocker group (ie. West Crocker and equivalent formations). The Middle to Late Eocene 'Sarawak Orogeny' has affected the Rajang Group and East Crocker (and older) Formation (Hutchison, 1996; Tongkul, 1997). Following this strong tectonic uplift, deep water sediments continued to be deposited in Sabah i.e. West Crocker, Temburong, Setap Shale and Kudat formations. This younger Crocker group (i.e. 'younger' Crocker Fold-Thrust Belt [CFTB]) was deformed in Early Miocene to early Middle Miocene in the so-called 'Sabah Orogeny' whose uplift and erosion gave rise to the Deep Regional Unconformity (DRU), and the onset of the clastic shelf/slope deposition in the Neogene Sabah Basin.

Eastern Palaeogene Basin

To the east, the trough was filled with grey coloured turbidites and semi-pelagic shales (Labang Formation), and predominantly red-coloured shales and sandstones and thick mudstones (Kulapis Formation), which Clennell (1996) interpreted as probably grain-flow deposits of bathyal and unstable outer-shelf environment.

A sudden change in, or the shallowing of, the environment, due to uplift in the Late Oligocene-Early Miocene, led to the deposition of shallow water coralline limestones throughout Sabah. Examples of these limestones are in the Kudat Peninsula area, in eastern Sabah e.g. Gomantong and Tempadong (Mensuli) limestones (Haile and Wong, 1965; Leong, 1974) and in south and southeastern Sabah (Collenette, 1965).

In the Labuk Bay area, eastern Sabah, seismic data indicate a strong reflector with mounded features. These features are interpreted as carbonate buildups similar in age to the onshore Gomantong Limestone. They overlie clastics of probable Labang or Crocker formation (Wong and Salahuddin Saleh Karimi, 1995).

Offshore NW Sabah

In the deepwater offshore NW Sabah, the change in depositional environment was also recognised by the widespread occurrence of

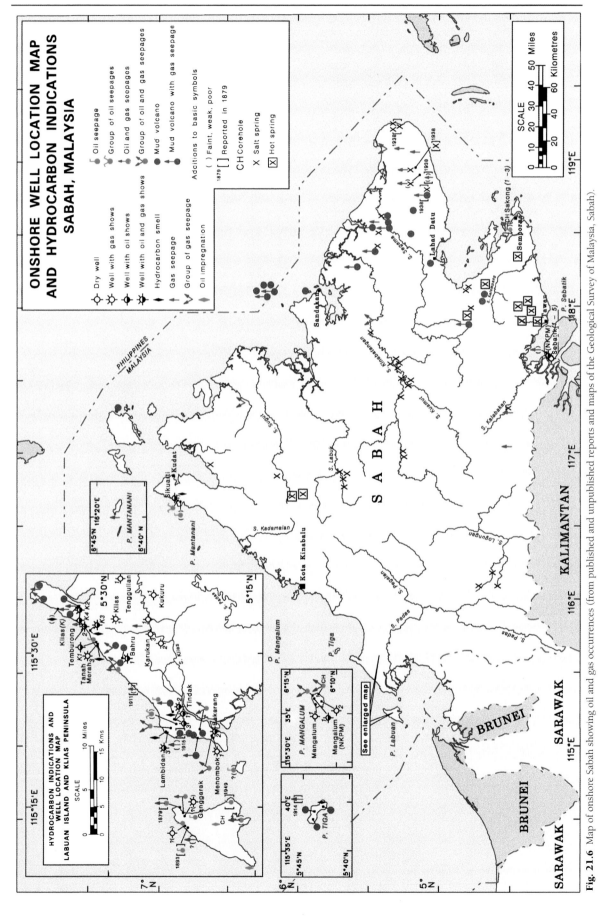

Fig. 21.6 Map of onshore Sabah showing oil and gas occurrences (from published and unpublished reports and maps of the Geological Survey of Malaysia, Sabah).

platform carbonates of predominantly Oligocene to Early Miocene age (Hinz and Schluter, 1985; Fulthorpe and Schlanger, 1989).

The carbonate platform is characterized by a distinct seismic pattern unconformably overlying the syn-rift sediments of the half-grabens (Hinz et al., 1985). The seismic reflector is mappable and is found at 2.5-3.0 sec TWT in the northwestern part of the deepwater area. It plunges southeastwards below the Sabah Trough at depths varying between 4.5 and 6.0 sec TWT; and extends below the Miocene-Pliocene sediments of the NW Sabah continental shelf basins (Hazebroek and Tan, 1993) to disappear at about 8.0 sec TWT.

It is unclear whether the southeastern limit of the 'Oligo-Miocene' carbonate reflector is the actual edge of the carbonate platform or its 'disappearance' is due to poor seismic penetration below 8.0 sec. However, the occurrences of similar age limestones in onshore Sabah indicate the depositional environment change was widespread and regional.

Neogene Basins

The Neogene Basins flank the Palaeogene Basins and the Early Palaeogene and older core (Fig. 21.7). Post-carbonate deposition took place during regional extension due to the opening of the Southeast Sulu Sea, and associated volcanism in eastern Sabah. This was followed by the structural rearrangement/collision of northwestern Sabah with continental terranes linked to the opening of the South China Sea.

Eastern Neogene Basin

The formation of Neogene Basins, the Northeast (NE) and Southeast (SE) Sabah basins, began in the Early Miocene by extensional tectonism in eastern Sabah. The NE Sabah and SE Sabah basins were probably contiguous during this period. The earlier carbonate depositional phase was followed in eastern Sabah by chaotically disrupted rock units that now crop out over some 12,000 sq km, making this area one of the largest melange (olistostrome) terrane in the world (Leong, 1976; Clennell, 1991). The chaotic rocks are mainly mud-matrix olistostromes and broken formations, and include tuff, and tuffaceous layers. They are known variously as the Kuamut, Garinono, Ayer and Kalabakan formations. The olistostromes of eastern Sabah were formed by slumping into a deep marine basin, mainly detached from the Tertiary sediments, but cutting down locally to the ophiolitic substrate. They were generated during regional extension at the time of early rifting of the Southeast Sulu Sea (Clennell, 1991,1996; Hutchison, 1992a). Explosive volcanism accompanied olistostrome deposition, as indicated by large amounts of tuffaceous material in the Kuamut, Garinono and, especially, the Ayer Formation in the Dent Peninsula area.

The Tanjong Formation and its equivalent shallow marine clastics may have been initially deposited in the Early Miocene on the flanks of the NE and SE Sabah basins contemporaneously with the olistostrome deposition in the deeper parts of the basins.

In eastern Sabah, the olistostrome deposits and older rock units were uplifted in the early Middle Miocene. The uplifted Crystalline Basement and ophiolitic Chert-Spilite Formation probably acted as a barrier or ridge separating the NE Sabah Basin from the SE Sabah Basin. The continued opening of both the SE Sulu Sea (Hutchison, 1992a; Clennell, 1996), and the Makassar Trough in Kalimantan (Lentini and Darman, 1996) probably led to the formation of two prominent northeasterly and southeasterly oriented 'failed rifts' - named the 'Sandakan Rift' and 'Tarakan Rift' by Tjia et al. (1990). These 'failed rifts' were the sites of thick and extensive deposition of coastal to shallow marine clastics and coal layers. They are named here as the Central Sabah Sub-basin of the NE Sabah Basin and the Northern Tarakan Sub-basin of the SE Sabah Basin. The lithostratigraphic units are the Tanjong Formation, and Tanjong and Kapilit formations respectively. These rock units are now exposed in circular, sub-circular and synclinal basins in two main depositional trends, northeasterly and southeasterly.

While these mainly deltaic-shallow marine clastics were being deposited in several sites along the 'failed rifts', volcanic activity continued in the eastern platform areas of the NE Sabah and SE Sabah basins. These are indicated by the occurrences of tuffaceous and volcaniclastics of the Segama Group (Tungku, Libong Tuffite, Tabanak Conglomerate) in the Dent Peninsula and the Kalumpang Formation in the Semporna area respectively.

A tectonic event in the late Middle Miocene to early Late Miocene uplifted and eroded the Tanjong and equivalent units as well as older units, and resulted in another cycle of deltaic to shallow marine sediment filling in new depocentres. The larger depocentre on the eastern Sabah side, and extending to the Philippines, is the Sandakan Sub-basin, which

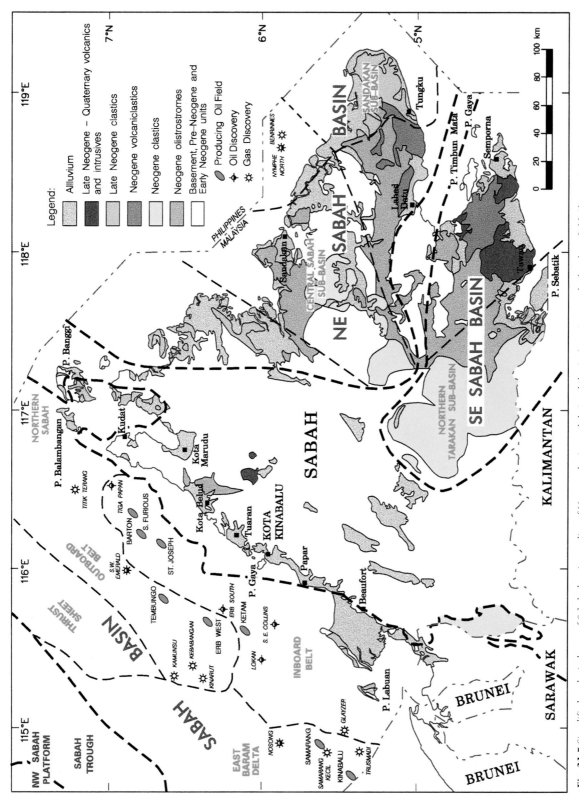

Fig. 21.7 Simplified geological map of Sabah showing outline of Neogene basins and their sub-basins/ provinces and oil and gas fields/discoveries.

contains the Dent Group of sediments. These are mainly clastics with minor carbonates - the Sebahat, Ganduman and Togopi formations (Haile and Wong, 1965; Bell and Jessop, 1974; Walker, 1993; Wong, 1993; Ismail Che Mat Zin, 1994; Graves and Swauger, 1997). Elsewhere, remnants of the younger cycle of sedimentation in both the NE Sabah and SE Sabah basins include the Sandakan and Bongaya formations, and the Simengaris, Umas Umas and Balung formations, respectively. The latter thicken towards the offshore Tarakan Basin in Indonesia. Gas and condensate have been discovered in the pre-Dent and Dent Group clastic reservoirs of the Sandakan Sub-basin (Bell and Jessop, 1974; Walker, 1993).

Plio-Pleistocene sediments, mostly limestones, calcareous sandstones, conglomerates and carbonaceous shales occur in the coastal areas around eastern Sabah (Togopi and Wallace formations).

Western Neogene Basin

While eastern Sabah was in the initial stage of basin formation during the Early Miocene, in western Sabah, deep marine shales and turbiditic sands continued to be deposited (Setap Shale and Kudat formations). In western Sabah, some chaotic deposits (Wariu Formation) are present. The shallow marine Meligan Formation may have been initially deposited on the flanks of the deep marine trough in southern Sabah. The deposition of the deep marine shales and turbiditic sands in Early Miocene extended to the NW Sabah continental margin area. The unit is designated Stage III in Shell's classification (Bol and van Hoorn, 1980). Minor oil and gas are reservoired in Stage III sands (e.g. Pondu-1 well); and oil seeps occur in the onshore Kudat Formation (Stephens, 1956).

In the offshore area of Balabac-Southern Palawan, the Oligo-Miocene carbonate layer is overlain by thick Lower Miocene deep marine shales and/or allochthonous thrust wedge of older Crocker/Chert-Spilite-equivalent unit (Hinz et al., 1985). The uplift and tilting to the northwest of the Crocker Fold-Thrust Belt resulted in the drowning of the Oligo-Miocene carbonates in the present South China Sea, especially in the deep Sabah Trough area, where hemipelagic shales overlie the carbonates (Hinz et al., 1989).

This was followed in northwestern Sabah by collision tectonism with the 'Palawan Continental Terrane' resulting in the formation of imbricate thrust slices and wedges of the pre-Early Miocene units, including the ophiolitic Chert-Spilite Formation and associated ultramafic and mafic bodies, and the Crocker Formation (Hinz et al., 1985, 1989).

The subsequent uplift of the folded and thrust units resulted in a regional erosional surface known as the Deep Regional Unconformity (DRU) of late Early Miocene-early Middle Miocene age (Bol and van Hoorn, 1980; Levell, 1987; Tan and Lamy, 1990; Hazebroek\ and Tan, 1993). A review of well results and seismic data in the South China Sea, offshore SW Palawan and Sabah and western Sulu Sea, shows a distinct unconformity designated as Unconformity 'C' (equivalent to the Deep Regional Unconformity), which appears to represent the end of a major regional tectonic event up to late Early Miocene time. Unconformity 'C' is correlated with the end of seafloor spreading in the South China Sea, between 17 to 15 Ma, i.e. early Middle Miocene (Hinz et al., 1985,1989; Schluter et al., 1996; Clennell, 1996). The DRU marks the base of the Sabah Basin (also known as the NW Sabah Basin).

The sedimentary units deposited during the Middle to Late Miocene are of primary importance in providing hydrocarbon source and reservoirs. The search for commercial quantities of petroleum are focused in the mainly clastic units of Middle to Late Miocene in the offshore areas (Scherer, 1980; Whittle and Short, 1978). There have been several pulses of major deposition during the Middle to Late Miocene.

Onshore outcrops are limited to a few localities in the south - represented by Belait and Liang formations (Mazlan Madon, 1997), and in the north by South Banggi, Bongaya and Timohing formations (Tongkul, 1994). A carbonate unit, the Mantanani Limestone, exposed on Mantanani Islands, is correlatable to the Balambangan Limestone Member of the Bongaya Formation and is at least of Middle Miocene age (Idris and Kok, 1990).

Extensive exploration efforts in the offshore western Sabah in the Sabah Basin have revealed at least 2 major cycles of sedimentation. The post-DRU period shows the initial phase of widespread clastic shelf/slope deposition, followed by the westward and northwestward progradation of thick sediment wedges. These Middle Miocene and younger sediments are designated as Stage IV (Scherer, 1980; Bol and van Hoorn, 1980; Tan and Lamy, 1990; Rice-Oxley, 1991; Hazebroek and Tan, 1993). The major cycles of deposition are separated by tectonic events in the late Middle/early Late Miocene, Late Miocene and Early Pliocene. In the

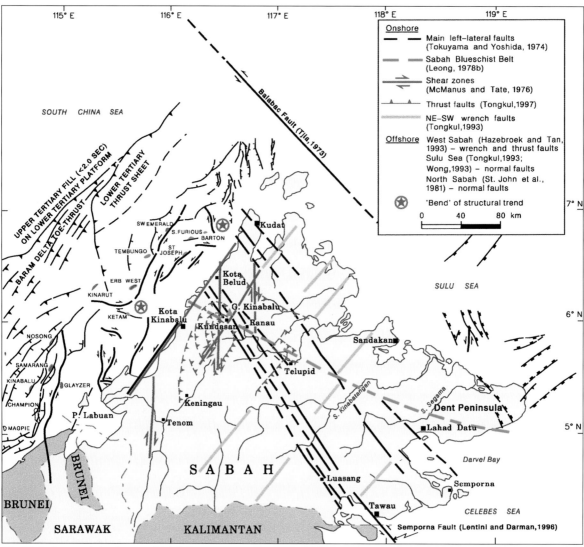

Fig. 21.8 Structural sketch map of Sabah from various sources Note (i) the 3 main sets of faults : N-S trending wrench faults; NW-SE trending wrench faults; both NE-SW trending extension and wrench faults; and (ii) the 'bends' of the fold-thrust belts (RFTB and CFTB), and similar pattern offshore.

southern offshore NW Sabah, the major depositional phases, Stage IVC/older and Stage IVD/younger, are separated by a pronounced unconformity known as the Shallow Regional Unconformity (SRU). Noor Azim Ibrahim (1994) studied the post-early Middle Miocene sediments of the Sabah Basin, and concluded that there were 2 distinct phases of subsidence history since the early Middle Miocene - i) rapid tectonic subsidence (>500m/Ma) between 14.7 and 9 Ma followed by ii) a generally slow subsidence accompanied by pronounced uplift along the eastern margin. The inflection point between these two phases is the major regional unconformity known as the Shallow Regional Unconformity (SRU), separating the Stage IVA-C and Stage IVD-G sedimentary packages.

During post-SRU times, 2 new large depocentres, the East Baram Delta and the Outboard Belt, were formed. The depositional environment varied from coastal plain to shallow marine and deepsea fans and turbidites, in a westerly and northwesterly trend.

In northern offshore NW Sabah, including the offshore Malawali Sub-basin, the main cycle of deposition commenced in the late Middle Miocene or early Late Miocene. These sediments are post-South Banggi Formation equivalent and are separated by an unconformity (Wong et al., 1997), which is probably the equivalent of the late Middle/early Late Miocene Upper Intermediate Unconformity.

The Plio-Pleistocene sediments, mostly limestones, calcareous sandstones, conglomerates and carbonaceous shales, occur in the coastal areas around western Sabah (Liang, Timohing).

In offshore western and northern Sabah, the sediments continued to prograde westerly and northwesterly. The Pliocene sediment packages, known as Stages IVF/G in the offshore western portion of the Sabah Basin, consist largely of deepsea fan turbidites

(Salahuddin Saleh Karimi et al., 1997; Marzuki Mohamad and Labao, 1997).

In northeastern Sabah, just west of Balambangan Island, thick reefal limestone occur in a platform area, known as the Kudat Platform, a high since the Late Miocene. The limestones are known as the Lower and Upper Limestone Units of the Kalutan Group (Frank, 1981).

Igneous Activities

Igneous activities commenced in the Late Miocene and extend into the Quaternary. They are represented by the acid to intermediate intrusive rocks of Mount Kinabalu in onshore western Sabah, and the extensive extrusive volcanic rocks with minor intrusives in the Semporna and Tawau areas, eastern Sabah (Kirk, 1968; Jacobson, 1970; Lim, 1981; Lee, 1988).

REGIONAL STRUCTURES

Several regional tectonic events have occurred in Sabah since the early Tertiary, thus, contributing to the diverse structural and depositional framework of Sabah (Fig. 21.8). Mud volcanoes occur at several localities in eastern, western and northern Sabah. Occasionally, they are activated through earth movements (Tjia, 1978; Lee, 1990; Tongkul, 1988). According to Lim (1981), shallow earthquakes in Sabah indicate renewed fault activities mainly along the north to northeast structural trend. Lim (1986) described 3 concentrations of seismic activities (see also the Seismo-Tectonic Map of Malaysia, Geological Survey of Malaysia 1994), offshore Sulu Sea, the shallow-depth earthquakes in the Dent-Semporna area, closely related to major northeast and northwest trending faults, and the central and north Sabah zones, related to deep-seated northeast-trending faults (Ranau area). Seismic activities in the Labuk Bay area appear to be related to renewed movements along major north to northeast trending faults.

Western Basinal Areas

Onshore

The most striking structural feature of Sabah is the large 'bend' around Mt. Kinabalu, from the prominent north to northeasterly trend, the NW Borneo Trend, to a southeasterly Sulu Trend in northeastern Sabah. These N-S, NE-SW and NW-SE trends represent bedding strikes and two sets of major faults - a NNE-SSW set and a NW-SE set (Wilford, 1967; Lee, 1980; Tongkul, 1991a,1994,1997; Benard et al., 1990).

Another major feature is the arcuate belt of ophiolites (including the Crystalline Basement in eastern Sabah) extending from the Darvel Bay area, trending E-W, to the northern Kudat area through central Sabah, trending N-S.

Tokuyama and Yoshida (1974) proposed that a wide zone of strike-slip faults cuts across Sabah from the southeast to northwest. The most prominent fault is the NW-SE trending 'Kinabalu Fault', a sinistral strike-slip fault, extending from the southeastern coast between Cowie Harbour, near Tawau, to the northwestern coast, through Labuk Valley and Mt. Kinabalu. Its topographic expression is generally obscured, although inferred from onshore geology. The 'Kinabalu Fault' zone appears to extend southeastwards to the 'Semporna Fault', demarcated by Lentini and Darman (1996).

NW Sabah Continental Margin

Bol and van Hoorn (1980), Tan and Lamy (1990) and Hazebroek and Tan (1993) have described the structural styles and tectonic evolution of the NW Sabah continental margin. In essence, the area is dominated by N-S and NNE-SSW structural grain, ranging from faulted, tight, anticlinal features separated by wide, deep synclines to major extensional systems and large down-to-basin normal faults. Piercement structures by shale diapirism also produced anticlinal features.

The dominant trend of major deep-seated strike-slip faults appear to be N-S, with a sinistral sense. Major N-S trending sinistral deep-seated strike-slip faults are shown to be present in the reconstruction of the tectonics of the South China Sea region (Tongkul, 1994). These N-S trending deep-seated wrench faulting have created independently moving basement blocks, resulting in separate deformation at different times (see Levell, 1987). Tjia (1997) has also described the widespread occurrences of N-S striking regional faults (meridian-parallel faults) in the Tertiary basins of 'Sundaland'.

However, the N-S trend deviates to northeast, east and southeast in the north, similar to the 'bend' observed onshore. The deviation from N-S to E-W occurs in two areas. The first 'bend' is in the Mangalum area and the second 'bend' is the Batumandi area extending to onshore Kudat Peninsula. The second 'bend' extends eastwards to the Bonanza area (see Levell, 1987).

These 'bends' probably represent important structural breaks. The first 'bend' is marked by the dextral Kinarut-Mangalum fault. The fault is one of several major NW-SE and NNW-ESE trending 'transtensional' faults, the others being the SW Luconia-Mukah Line, West Baram Line, Strait Balabac Fault, which are probably related to deep-seated N-S sinistral strike-slip faults in the South China Sea region. Tan and Lamy (1990), James (1984) and Ismail Che Mat Zin and Swarbrick (1997) indicate that these NW-SE trending 'transtensional' faults (especially in the Sarawak area) have considerable dextral strike-slip movements. The offshore dextral Kinarut-Mangalum fault probably continues onshore as the 'Kinabalu Fault', described by Tokuyama and Yoshida (1974), as a sinistral strike- slip fault zone. A NW-SE trending belt, known as the 'Sabah Blueschist Belt' has also been proposed by Leong (1978b).

Sabah Deepwater areas

In the Sabah deepwater areas, Hinz et al., (1989) mapped several tectonostratigraphic terranes, including the Sabah Trough, all trending N-S to NE-SW, and bounded by thrusts or normal down-to-the-basin faults. The terranes are cut by a system of NW-SE trending strike slip faults (Schluter et al., 1996).

Eastern Basinal Areas

Onshore

The prominent features in onshore eastern Sabah are a series of circular, subcircular and synclinal basins trending northeasterly and southeasterly (Fig. 21.3). They are believed to represent 2 rift zones (Tjia et al., 1990). Tongkul (1991b; 1993) proposed that 2 main structural trends, namely NE-SW and NW-SE, have controlled the development, distribution and shape of these Neogene basins.

Offshore NE Sabah and Sulu Sea

Offshore NE Sabah extending to the Sulu Sea is characterised by several NE-SW trending tectonic elements e.g. the NW Sulu Basin, Cagayan Ridge, SE Sulu Basin etc., as described by Schluter et al., (1996). Off the Dent Peninsula, a series of N-S to NE-SW trending horst and graben features including normal/growth faults are present (Bell and Jessop, 1974; Wong, 1993). The splitting or rifting of the Cagayan-Sulu Ridge in the Early Miocene extended into Sabah (Hinz et al., 1991; Hutchison, 1992a; Clennell, 1996; Schluter et al., 1996) and has also created NE-trending extensional structural features. These NE-SW trending terranes are cut by a series of NW-SE trending strike-slip faults. A prominent NW-SE strike-slip fault separating the Cagayan Ridge and the NE Sabah area, including the Sandakan Sub basin (Fig. 9 in Schluter et al., 1996) probably continues northwesterly to join another major fault just south of Balabac island, known as the 'Strait Balabac Fault' (Fitch, 1963), interpreted as a sinistral strike-slip fault (Bosum et al., 1972; Tjia, 1973; Beddoes, 1976; Wood, 1985), but dextral by James (1984).

Summary

According to Tongkul (1991a), the diverse and complex structures in Sabah resulted from at least 5 episodes of deformation, that began in the Early Cretaceous. At least 3 major episodes were linked to NW-SE compressions coinciding with the intermittent opening of the South China Sea sub-basins during the Middle Eocene, early Middle Miocene and Late Miocene (Tongkul, 1994, 1997). Based on the study of central Sabah area, Tongkul (1997) concluded that the Middle Eocene deformation episode is characterized by folding and thrusting of basement rock and older Palaeogene sediments trending N70E with associated N-S left lateral horizontal faults i.e. the Rajang or older Crocker Fold-Thrust Belt (RFTB). The early Middle Miocene deformation is characterised by imbrication of the basement rocks and overlying sediments to the NE, with associated NE-SW left lateral horizontal faults. The Late Miocene deformation is characterised by thrusting of the earlier deformed basement rock and overlying sediments to the NW. The origins of the polyphase deformations are discussed in the next section.

The complexity of the structural patterns are further enhanced by extensional tectonics in eastern Sabah in Early Miocene to early Middle Miocene. The continued extension in eastern Sabah probably initiated subduction processes in the southeast to produce the extensive volcanic activity in Late Miocene to Quaternary.

TECTONIC EVOLUTION

Tongkul (1991a) has given a fairly comprehensive account of the tectonic evolution of Sabah. The account given here is a slightly revised model (see Figs. 21.9 and 21.10).

Pre-Cretaceous

Tongkul (1991a) referred to the Sabah Crystalline Basement as 'older metamorphosed oceanic basement with rare acid intrusive rocks'. However, as discussed earlier, the acid igneous rocks and the metamorphic complex of the Crystalline Basement probably represent different origins and tectonic history.

Early Cretaceous-Early Eocene

A complex environment of deposition, perhaps in several unstable deep troughs, in a block-faulted emerging 'older basement' occurred during the Early Cretaceous. Shallow marine limestones (Madai-Baturong) were deposited on highs. The deep trough or basin, variously referred to as an ocean basin (Tjia, 1988), proto-South China Sea, (e.g. Hinz et al., 1991), a major marginal sea/gulf of Pacific Ocean into Sundaland (Hutchison, 1992b) or 'Rajang Sea' (Chapter 4) consisted of relatively complete ophiolite complex and its cover pelagic sediments, including radiolarian cherts and *Globotruncana*-bearing limestones (Chert-Spilite), which constitute the 'new oceanic basement' (Tongkul, 1991a). The ophiolite complex i.e. the 'new oceanic basement', probably forms the basement to the thick sedimentary systems of Sabah (Hutchison, 1989).

In the centre of the deep marine troughs are thick successions of predominantly argillaceous sediments, contemporaneously and coeval with the Chert-Spilite Formation. These are represented by the Sapulut, Trusmadi and East Crocker formations. The ophiolite complex and these agillaceous sedimentary units were all later deformed and uplifted in the Eocene tectonic episode, which was strong in Sarawak (Geological Survey of Malaysia, 1995; Benard et al., 1990; Hutchison, 1992b; Tongkul, 1991a, 1994, 1997).

The Eocene deformation has been described as related to the southeastward subduction of the proto-South China Sea oceanic lithosphere under the NW margin of Borneo. This resulted from the NW-SE extension in the southern continental margin of China, and enhanced by the anticlockwise rotation of the continental basement of Borneo (Kalimantan), the West Borneo Basement (Tongkul, 1997; Taylor and Hayes, 1983; Hinz et al., 1991; Ru and Piggot, 1986; Haile et al., 1977; Schmidtke et al., 1990). The former subduction trench zone, probably trending NE-SW, would have been located in central Sabah, west of Labuk Valley (see Figure 11 of Hazebroek and Tan, 1993).

The closing of the deep trough of the hypothetical proto-South China Sea Basin or 'Rajang Sea' could also have been the result of opposing movements of two continental masses from the NW (the southern continental margin of China) and the SE (the Borneo microcontinental plate). Xia and Zhou (1993) have proposed that the 'Nansha Block' - a microcontinental block of transitional crust in the Dangerous Grounds-Reed Bank area - collided with 'palaeo Borneo' in the Late Eocene, before the initial opening of South China Sea. This Late Eocene collision folded the Palaeogene and older units in Sabah.

Oligocene-Early Miocene (Figs. 21.9a, 21.9b and 21.10)

Following the Late Eocene deformation, uplift and erosion of the Rajang Fold-Thrust Belt (RFTB) fed new depocentres to the west and east (Palaeogene Basins).

Southward directed movement of the NW Sabah continental plate/platform linked to the opening of the SCS resulted in the initial uplift of the Temburong (Te) and West Crocker (WCr) formations. To the south, shallow marine sediments were deposited (Meligan, Me). However to the north, deep marine clastic deposition continued (Stage III and Kudat). Prior to impending intense deformation, widespread carbonates were deposited manifesting initial uplift of Crocker Fold-Thrust Belt (CFTB). In eastern Sabah, two rift zones (the NE trending Sandakan Rift and SE trending Tarakan Rift) began to form, accompanied by chaotic deposits (olistostromes) and explosive volcanism.

Early Miocene-Middle Miocene (Figs. 21. 9b, 21.9c and 21.10)

In the Early Miocene to early Middle Miocene, the eastern Sabah rift zones widened and were infilled by synrift deposits, the shallow marine Tanjong (Tj) and Kapilit (Kp) formations. Contemporaneous with the extension tectonism in eastern Sabah, intense compressional tectonism occurred with the collision of southward directed movement of the NW Sabah continental plate and the northwest margin of Borneo. According to Tongkul (1997), the deformation was due to collision of microcontinents against older fold belts in the NW Borneo region. To the north the Palawan and Reed Bank terrane collided with the fold belt of NW Borneo as a result of the N-S opening of the South China Sea. To the south, the Celebes Sea

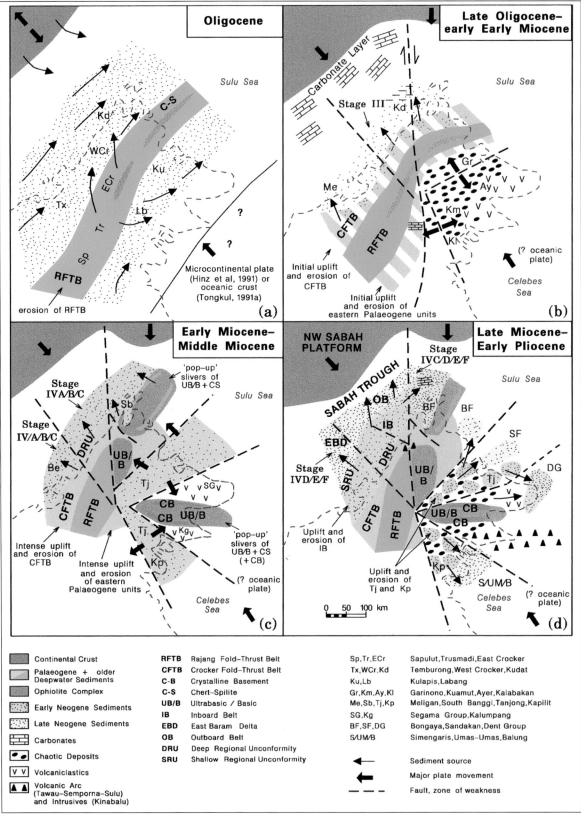

Fig. 21.9 Palaeogeographical and tectonic sketches of Sabah (modified after Tongkul, 1991a). Reprinted with permission from Elsevier Science.

oceanic lithosphere was subducted northwestward (Rangin et al., 1990) as a result of the NW collision of the Banggai-Sula Platform with older fold belt of eastern Sulawesi (Simandjuntak, 1993). The collision from both ends probably caused the bending of the CFTB.

Uplift and erosion of the CFTB (and the older RFTB) fed new depocentres to the north and northwest (Inboard Belt, IB). Thrust sheet emplacement also occurred (Hazebroek and Tan, 1993). During this period, the eastern and central Sabah ophiolitic complex (CS, UB/B) were

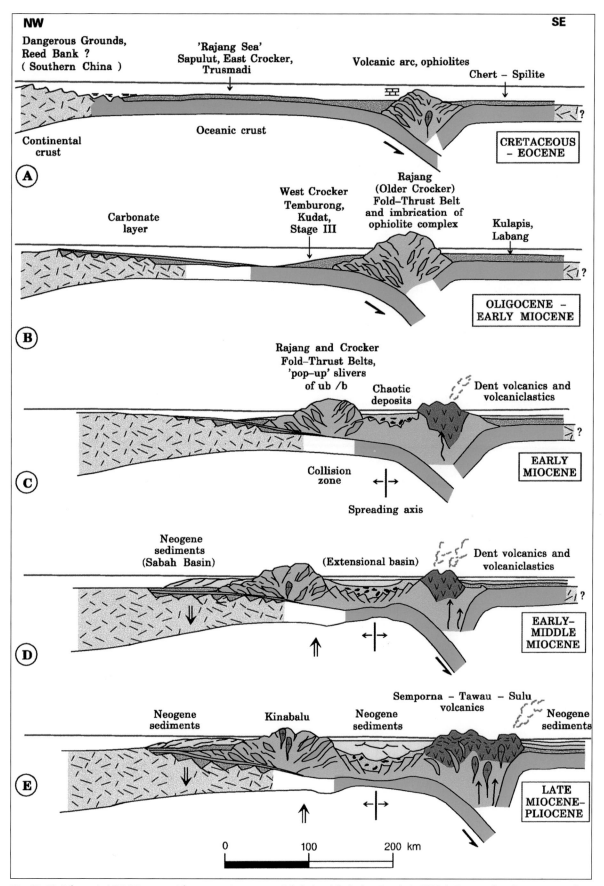

Fig. 21.10 Schematic NW-SE sequential cross-sections across Sabah (modified after Tongkul, 1991a). Reprinted with permission from Elsevier Science.

exhumed, probably as 'pop-up' slivers. In eastern Sabah, volcanic activity continued.

The contemporaneous regional tectonic deformations in Sabah and neighbouring areas, compressional in west and northwest Sabah and Palawan areas, and extensional in eastern Sabah and Southeast Sulu Sea area, have caused a major period of uplift and erosion, which produced the so-called Deep Regional Unconformity (DRU) or Unconformity 'C'.

Late Miocene-Early Pliocene (Figs. 21.9d and 21.10)

In western offshore Sabah intense deformation with large scale N-S strike-slip movements occurred in Late Miocene with the continuing plate movements. The Inboard Belt (IB) were uplifted and eroded to feed new depocentres, East Baram Delta (EBD) and Outboard Belt (OB) (Hazebroek and Tan, 1993). The major erosional surface is the Shallow Regional Unconformity (SRU). In eastern Sabah the Miocene synrift deposits were not only uplifted and eroded, but formed into circular and sub-circular shapes, probably caused by both shale diapiric movement and wrench faulting. The erosion of the synrift deposits fed new depocentres, to the north, east and southeast. The eastern Sandakan Sub-basin has considerable sediment thicknesses. A new phase of volcanic activity (the Tawau-Semporna-Sulu volcanics) and igneous intrusions (Kinabalu) occurred from Late Miocene to Quaternary.

According to Tongkul (1997), the Late Miocene deformation, recognized throughout Borneo, was probably related to northwestward shortening and uplift of the Sabah fold belt as a result of the collision of the Banda Arc with northern Australia (Simandjuntak, 1993; Richardson, 1993) and the NW movement of the Philippine Mobile Belt against the southeastern margin of Asia (Rangin, 1991).

Pliocene-Recent (Figs. 21.9d and 21.10)

A Pliocene, probably extending into Pleistocene, compressional event affected most of Sabah in various degrees. The Pliocene compressive event was probably a continuation of the Late Miocene event described above. The volcanic arc in eastern Sabah appeared to have migrated from the Dent Peninsula southward to the Tawau area and Semporna Peninsula with a phase of basaltic, dacitic and andesitic volcanism and minor intrusives. Subduction of the Celebes (Sulawesi) Sea towards the north may have produced the extensive volcanic activity.

In offshore NW Sabah, the Pliocene tectonic phase, also referred to as Horizon II (Levell, 1987), primarily affected a northern area between Mangalum and Kudat where upthrust anticlines were separated basinwards by large normal faults from gentle anticlines with crestal faults (Bol and van Hoorn, 1980). In eastern Sabah, the Upper Miocene and younger Dent Group sediments were deformed into a series of *en echelon* folds with ENE trending axes, which according to Walker (1993) 'reactivated the deltaic growth faults, creating large faulted anticlines'. These folds are high relief on the Sabah side and low relief on the Philippines side (Graves and Swauger, 1997). Recent offshore well results indicate that the Upper Miocene sections towards the northwest have almost entirely been eroded. In southeastern Sabah, the Miocene sediments were also affected by the Pliocene compression with uplift occurring in response to continued northeast-southwest compression. For the nearby Tarakan Basin to the south, Lentini and Darman (1996) attributed the Late Pliocene tectonic phase to the ongoing spreading of the Makassar Strait, resulting in several wrench fault zones, including the Semporna Fault.

The recent seismic activities in Sabah and the presence of active mud volcanoes indicate continuing tectonic activity in Sabah (Lim, 1985b, 1986).

REFERENCES

Adams, C.G. and Kirk, H.J.C., 1962. The Madai-Binturong Limestone Member of the Chert-Spilite Formations, North Borneo. Geological Magazine, 44, 289-303.

Basir Jasin and Sanudin Tahir, 1988. Barremian Radiolaria from Chert-Spilite Formation, Kudat, Sabah. Sains Malaysiana, 17, 67-79.

Basir Jasin, 1992. Significance of radiolarian chert from the Chert-Spilite Formation, Telupid, Sabah. Bulletin of the Geological Society of Malaysia, 31, 67-84.

Beddoes, L.R., 1976. The Balabac Sub-Basin Southwestern Sulu Sea, Philippines. SEAPEX Program, Offshore South East Asia Conference, Paper 13.

Bell, R.M. and Jessop, R.G.C., 1974. Exploration and Geology of the West Sulu Basin, Philippines. Australian Petroleum Exploration Association Journal, 1, 21-28.

Benard, F., Muller, C., Letouzey, J., Rangin, C. and Yahir, S., 1990. Evidence of multiphase deformation in the Rajang-Crocker Range (northern Borneo) from Landsat imagery interpretation : geodynamic implications. In : Angerlier, J. (ed.) Geodynamic evolution of the eastern Eurasian margin.Tectonophysics, 183, 321-339.

Bol, A.J. and van Hoorn, B., 1980. Structural styles in western Sabah offshore. Bulletin of the Geological Society of Malaysia, 12, 1-16.

Bosum W., Fernandez, J.C., Kind, E.G. and Teodoro, C.F., 1972. Aeromagnetic survey of the Palawan-Sulu Offshore Area of the Philippines. UN ECAFE, CCOP Technical Bulletin, 6, 141-160.

Clennell, B., 1991. The origin and tectonic significance of melanges of Eastern Sabah, Malaysia. Journal of the Southeast Asian Earth Sciences, 6, 407-425.

Clennell, B., 1996. Far-field and gravity tectonics in Miocene Basins of Sabah, Malaysia. In: Hall, R. and Blundell, D.J. (eds.). Tectonic Evolution of Southeast Asia, Geological Society of London Special Publication, 106, 307-320.

Collenette, P., 1958. The geology and mineral resources of the Jesselton-Kinabalu area. British Borneo Geological Survey Memoir 6.

Collenette, P., 1965. The geology and mineral resources of the Pensiangan and Upper Kinabatangan area, Sabah, Malaysia. Geological Survey Borneo Region, Malaysia, Memoir 12.

Fitch, F.H., 1963. Possible role of continental core movement in the geological evolution of British Borneo. Proceedings of British Borneo Geological Conference, 1961.

Fontaine, H. and Ho, W.K., 1989. Note on the Madai-Baturong Limestone, Sabah, East Malaysia; Discovery of Caprinidae (Rudists). CCOP Newsletter, 14, 27-32.

Frank, P.L., 1981. The onshore geology of the Kudat Peninsula and northern islands, Sabah. Carigali-BP Sdn. Bhd., Report 2021.

Fulthorpe, C.S. and Schlanger, S.O., 1989. Paleo-Oceanographic and tectonic settings of early Miocene reefs and associated carbonates of offshore South East Asia. American Association of Petroleum Geologists Bulletin, 71, 281-289.

Geological Survey of Malaysia, 1994. Seismo-Tectonic Map of Malaysia, scale 1:500,000.

Geological Survey of Malaysia, 1995. Annual Report 1995.

Graves, J.E. and Swauger, D.A., 1997. Petroleum systems of the Sandakan Basin, Philippines. In J.V.C. Howes and R.A. Noble (Eds.), Proceedings of the Indonesian Petroleum Association, Petroleum Systems of SE Asia and Australasia Conference, Jakarta, Indonesia, May 1997, Indonesian Petroleum Association, 799-813.

Haile, N.S. and Wong, N.P.Y., 1965. The geology and mineral resources of the Sandakan area and parts of the Kinabatangan and Labuk Valleys, Sabah. Geological Survey Borneo Region, Malaysia, Memoir 16.

Haile, N.S., 1969. Geosynclinal theory and the organizational pattern of the Northwest Borneo Geosyncline. Quarterly Journal Geological Society of London, 124, 171-194.

Haile, N.S., 1973. The recognition of former subduction zones in Southeast Asia. In : Tarling, D.H. and Runcorn, S.K., eds., "Implications of Continental Drift to the Earth Sciences", 2. Academic Press, London, 885-892.

Haile, N.S., McElhinny, M.W. and McDougall, I., 1977. Palaeomagnetic data and radiometric ages from the Cretaceous of West Kalimantan (Borneo), and their significance in interpreting regional structures. Quarterly Journal of Geological Society of London, 133, 133-144.

Hamilton, W., 1979. Tectonics of the Indonesian region. U.S. Geological Survey Professional Paper, 1078.

Hazebroek, H.P. and Tan, D.N.K., 1993. Tertiary tectonic evolution of the NW Sabah continental margin. In : Teh, G.H., ed., Proc. Symp. Tectonic Framework and Energy Resources of the Western Margin of Pacific Basin. Bulletin of the Geological Society of Malaysia, 33, 195-210.

Hinz, K. and Schluter, H.V., 1985. Geology of the Dangerous Ground, South China Sea and Continental Margin off SW Palawan: Results of SONNE Cruise SO-23 and SO-27, Energy 10, 297-315.

Hinz, K., Kempter, E.H.K. and Schluter, H.U., 1985. The Southern Palawan-Balabac area: an accreted or non-accreted terrane? ASCOPE Proceedings, 2, 48-72.

Hinz, K., Fritsch, J., Kempter, E.H.K., Mohammad, A.M. Meyer, J., Mohamed, D., Vosberg, H., Weber, J. and Benavidez, J., 1989. Thrust tectonics along the North-western Continental Margin of Sabah/Borneo. Geologische Rundschau, Band 78, Heft 3.,

Hinz, K., Block, M., Kudrass, H.R. and Meyer, H., 1991. Structural elements of the Sulu Sea, Philippines. Geol. Jahrb. Reihe A., 127, 483-506.

Hutchison, C.S., 1975. Ophiolites in Southeast Asia. Bulletin of the Geological Society of America, 86, 797-806.

Hutchison, C.S., 1988. Stratigraphic-tectonic model for eastern Borneo. Bulletin of the Geological Society of Malaysia, 22, 135-151.

Hutchison, C.S., 1989. "Geological evolution of South-East Asia". Clarendon Press, Oxford, 368pp.

Hutchison, C.S., 1992a. The Southeast Sulu Sea, a Neogene marginal basin with outcropping extensions in Sabah. Bulletin of the Geological Society Malaysia, 32, 89-108.

Hutchison, C.S., 1992b. The Eocene unconformity in Southeast Asia and East Sundaland. Bulletin of the Geological Society of Malaysia, 32, 69-88.

Hutchison, C.S., 1996. The 'Rajang accretionary prism' and 'Lupar Line' problem of Borneo. In: Hall R. and Blundell, D. (eds.). Tectonic Evolution of Southeast Asia. Geological Society of London Special Publication, 106, 247-261.

Hutchison, C.S., 1997. Tectonic framework of the Neogene basins of Sabah. Abstracts Geological Society of Malaysia Petroleum Geology Conference 1997.

Idris, M.B., and Kok, K.H., 1990. Stratigraphy of the Mantanani Islands, Sabah. Bulletin of the Geological Society of Malaysia, 26, 35-46.

Irving, E.M., 1949. Notes on the geology of Balabac Island, Palawan, Philippines Island. Philippines Geologist, 3, 14-21.

Ismail Che Mat Zin, 1994. Dent Group and its equivalent in the offshore Kinabatangan area, East Sabah. Bulletin of the Geological Society of Malaysia, 36, 127-143.

Ismail Che Mat Zin and Swarbrick, R.E., 1997. The tectonic evolution and associated sedimentation history of Sarawak Basin, eastern Malaysia, a guide for future hydrocarbon exploration. In : Fraser, A.J., Matthews, S.J. and Murphy, R.W. (eds.) Petroleum Geology of Southeast Asia, Geological Society of London Special Publication, 126, 237-245.

Jacobson, G., 1970. Gunong Kinabalu area, Sabah, Malaysia. Geological Survey of Malaysia. Report 8.

James, D.M.D., 1984. Regional geological setting. In : James, D.M.D. (ed). "The geology and hydrocarbon resources of Negara Brunei Darussalam". Muzium Brunei Special Publication, 169pp.

Kirk, H.J.C., 1962. The geology and mineral resources of the Semporna Peninsula, North Borneo. British Borneo Geological Survey. Memoir 14.

Kirk, H.J.C., 1968. The igneous rocks of Sarawak and Sabah. Geological Survey of Malaysia. Bulletin 5.

Kudrass, H.R., Wiedecke, M., Cepeck, P., Kreuser, H. and Muller, P., 1986. Pre-Quaternary rocks dredged from the South China Sea (Reed Bank area) and Sulu Sea, during Sonne Cruises in 1982-1983. Marine and Petroleum Geology 3, 19-30.

Lee, D.C.T., 1980. Application of Landsat to regional geologic studies with reference to the geology of central and west coast Sabah and adjacent areas. Geological Survey of Malaysia, Geology Papers, 3, 126-133.

Lee, D.C.T., 1988. Gunung Pock area, Sabah, Malaysia. Geological Survey of Malaysia, Report 9.

Lee, D.C.T., 1990. Formation of Pulau Batu Harian and other islands around Pulau Banggi, Northern Sabah. Bulletin of the Geological Society of Malaysia, 26, 71-76.

Lentini, M.R. and Darman, H., 1996. Aspects of the Neogene tectonic history and hydrocarbon geology of the Tarakan Basin. Proceedings of the Indonesian Petroleum Association, 1996, 241-251.

Leong, K.M., 1974. The geology and mineral resources of the Upper Segama Valley and Darvel Bay area, Sabah, East Malaysia. Geological Survey of Malaysia Memoir 4 (Revised), 354pp.

Leong K.M., 1976. Miocene chaotic deposits in eastern Sabah : characteristics, origin and petroleum prospects. Geological Survey of Malaysia Annual Report for 1975, 238.

Leong, K.M., 1977. New ages from radiolarian cherts of the Chert-Spilite Formation, Sabah. Bulletin of the Geological Society of Malaysia, 8, 109-111.

Leong, K.M., 1978a. Sabah. In : ESCAP Atlas of Stratigraphy I. Burma, Malaysia, Thailand, Indonesia, Philippines. United Nations, Mineral Resources Development Series, 44, 26-31.

Leong, K.M., 1978b. The 'Sabah Blueschist Belt' - a preliminary note. Warta Geologi, 4, 45-51.

Leong, K.M., 1998. Sabah Crystalline Basement : 'Spurious radiometric ages? Continental?'. Warta Geologi, 24, 5-8.

Levell, B.K., 1987. The nature and significance of regional unconformities in the hydrocarbon-bearing Neogene sequence offshore West Sabah. Bulletin of the Geological Society of Malaysia, 21, 55-90.

Liechti, P., Roe, F.W. And Haile, N.S., 1960. The geology of Sarawak, Brunei and the western part of North Borneo. Geological Survey Department for the British Territories in Borneo, Kuching, Bulletin 3.

Lim, P.S., 1981. Wullersdorf area, Sabah, Malaysia. Geological Survey of Malaysia, Report 15.

Lim, P.S., 1985a. Geological Map of Sabah, 1:500,000 (3rd Edition). Geological Survey of Malaysia.

Lim, P.S., 1985b. History of earthquake activities in Sabah, 1897-1983. Geological Survey of Malaysia Annual Report for 1983, 350-357.

Lim, P.S., 1986. Seismic activities in Sabah and their relationship to regional tectonics. Geological Survey of Malaysia Annual Report for 1985, 465-480.

Ludwig, W.J., Kumar, N. and Houltz, R.E., 1979. Profiler-Sonobuoy measurements in the South China Sea Basin. Journal of Geophysical Research, 84, 3505-3518.

Marzuki Mohamad and Labao, J.J., 1997. The Lingan Fan : Late Miocene/Early Pliocene Turbidite Fan Complex, North-West Sabah. In : Howes, J.V.C. and Noble, R.A., eds., Proceedings of the IPA Petroleum Systems of SE Asia and Australasia, 21-23 May 1997, Indonesian Petroleum Association, Jakarta, 787-798.

Mazlan B. Hj. Madon, 1997. Sedimentological aspects of the Temburong and Belait Formations, Labuan (offshore West Sabah, Malaysia). Bulletin of the Geological Society of Malaysia, 41, 61-84.

McManus, J. and Tate, R.B., 1976. Volcanic control of Structures in north and west Borneo. SEAPEX Program, Offshore South East Asia Conference, Paper 5.

Milsom, J., Holt, R., Dzazali bin Ayub and Smail, R., 1997. Gravity anomalies and deep structural controls at the Sabah-Palawan margin, South China Sea. In : Fraser, A.J., Matthews, S.J.and Murphy, R.W. (eds.). Petroleum Geology of Southeast Asia. Geological Society of London Special Publication, 126, 417-427.

Noor Azim Ibrahim, 1994. Major Controls on the Development of Sedimentary Sequence, Sabah Basin, Northwest Borneo. Unpublished PhD Thesis, University of Cambridge, 254 pp.

Rangin, C., Bellon, H., Benard, F., Letouzey, J., Muller, C. and Sanudin, T., 1990. Neogene arc-continent collision in Sabah, North Borneo (Malaysia). Tectonophysics, 183, 305-319.

Rangin, C., 1991. The Philippine Mobile Belt : a complex plate boundary. Journal of Southeast Asian Earth Sciences, 6, 209-220.

Rice-Oxley, E.D., 1991. Palaeoenvironments of the Lower Miocene to Pliocene sediments in offshore NW Sabah area. Bulletin of the Geological Society of Malaysia, 28, 165-194.

Richardson, A., 1993. Lithosphere structure and dynamics of the Banda Arc collision zone, Eastern Indonesia. Bulletin of the Geological Society of Malaysia, 33, 105-118.

Ru, K. and Piggot, J.D., 1986. Episode rifting and subsidence in South China Sea. American Association of Petroleum Geologists Bulletin, 70, 1136-1155.

Salahuddin Saleh Karimi, Labao, J.J. and Wannier, M.M., 1997. Seismic identification of depositional processes in a turbidite fan environment, Deepwater Block SB-G, NW Sabah. Bulletin of the Geological Society of Malaysia, 41, 13-29.

Scherer, F.C., 1980. Exploration in East Malaysia over the past decade. In : Halbouty, M.T., ed., Giant Oil and Gas fields of the Decade 1968-1978. American Association of Petroleum Geologists, Memoir 30, 423-440.

Schluter, H.U., Hinz, K. and Block, M., 1996. Tectono-stratigraphic terranes and detachment faulting of the South China Sea and Sulu-Sea. Marine Geology, 130, 39-78.

Schmidtke, E.A., Fuller, M. and Haston, R., 1990. Paleomagnetic data from Sarawak, Malaysia Borneo, and the Late Mesozoic and Cenozoic tectonics of Sundaland. Tectonics, 9, 123-140.

Simandjuntak, T.O., 1993. Neogene Tectonics and orogenesis of Indonesia. Bulletin of the Geological Society of Malaysia, 33, 43-64.

Stauffer, P.H., 1968. Studies in the Crocker Formation, Sabah. Geological Survey of Malaysia Bulletin 8, 1-13.

Stephens, E.A., 1956. The geology and mineral resources of the Kudat and Kota Belud area, North Borneo, with an account of the Taritipan manganese deposits. British Borneo Geological Survey Memoir 5.

St. John V.P., Leong Khee Meng and Frank, P., 1981. Regional tectonics and the Neogene geology of Northwest Sabah. Abstract Petroleum Geology Seminar, Geological Society of Malaysia.

Tan, D.N.K. and Lamy, J.M., 1990. Tectonic evolution of the NW Sabah continental margin since the Late Eocene. Bulletin of the Geological Society of Malaysia, 27, 241-260.

Taylor, B. and Hayes, D.E., 1980. The tectonic evolution of the South China Sea Basin. In: The Tectonic and Geologic Evolution of Southeast Asian Seas and Islands. (Geophysical Monograph 23) American Geophysical Union, Washington, 89-104.

Taylor, B. and Hayes, D.E., 1983. Origin and History of the South China Sea Basin. In: The Tectonic and Geologic Evolution of Southeast Asian Seas and Islands. (Geophysical Monograph 27) American Geophysical Union, Washington, 23-55.

Tjia, H.D., 1973. Displacement patterns of strike-slip faults in Malaysia-Indonesia-Philippines. Geol. Mijnbauw, 52, 21-30.

Tjia, H.D., 1974. Sense of tectonic transport in intensely deformed Trusmadi and Crocker sediments, Ranau-Tenompok area, Sabah. Sains Malaysiana, 3, 129-161.

Tjia, H.D., 1978. The Lahad Datu (Sabah) earthquakes of 1976 : Surface deformation in the epicentral region. Sains Malaysiana, 7, 33-64.

Tjia, H.D., 1988. Accretion tectonics in Sabah : Kinabalu Suture and East Sabah accreted terrane. Bulletin of the Geological Society of Malaysia, 22, 237-251.

Tjia, H.D., Ibrahim Komoo, Lim, P.S. and Tungah Surat, 1990. The Maliau Basin, Sabah : geology and tectonic setting. Bulletin of the Geological Society of Malaysia, 27, 261-292.

Tjia, H.D., 1997. Meridian-parallel faults and Tertiary basins of Sundaland, Warta Geologi, 23, 3.

Tokuyama, A. and Yoshida, S., 1974. Kinabalu Fault, a large strikeslip fault in Sabah, East Malaysia. Geology and Palaeontology of Southeast Asia, 14, 171-188.

Tongkul, F., 1988. Geological control on the birth of Pulau Batu Harian mud volcano, Kudat, Sabah. Warta Geologi, 14, 153-165.

Tongkul, F., 1990. Structural styles and tectonics of Western and Northern Sabah. Bulletin of the Geological Society of Malaysia, 27, 227-239.

Tongkul, F., 1991a. Tectonic evolution of Sabah, Malaysia. In : Nichols, G. and Hall, R.(eds.) Proceedings of the Orogenesis in Action Conference London 1990. Journal of Southeast Asian Earth Sciences, 6, 395-405.

Tongkul, F., 1991b. Basin development and deposition of the Bongaya Formation in the Pitas region, northern Sabah. Bulletin of the Geological Society of Malaysia, 29, 183-193.

Tongkul, F., 1993. Tectonic control on the development of the Neogene basins in Sabah, East Malaysia. Bulletin of the Geological Society of Malaysia, 33, 95-103.

Tongkul, F., 1994. The geology of Northern Sabah, Malaysia : its relationship to the opening of the South China Sea Basin. Tectonophysics, 235, 131-147.

Tongkul, F., 1997. Polyphase development in the Telupid area, Sabah, Malaysia. In: Geosea 1995 Proceedings, Journal of Asian Earth Sciences, 15, 175-184.

Walker, T.R., 1993. Sandakan Basin prospects rise following modern reappraisal. Oil and Gas Journal, May 10, 1993, 43-47.

Whittle, A.P. and Short, G.A., 1978. The petroleum geology of the Tembungo Field, East Malaysia. Southeast Asia Petroleum Exploration (SEAPEX) Conference Paper.

Wilford, G.E., 1967. Geological Map of Sabah 1:500,000 (2nd Edition), Geological Survey of Malaysia.

Wong, N.P.Y. and Leong, K.M., 1968. Unconformity between the Chert-Spilite Formation and Crystalline Basement around Sungai Agob and Sungai Dabalan, East Malaysia. Geological Survey of Malaysia Bulletin, 9, 32-33.

Wong, R.H.F., 1993. Sequence stratigraphy of the Middle Miocene Pliocene, southern offshore Sandakan Basin, East Sabah. Bulletin of the Geological Society of Malaysia, 33, 129-142.

Wong, R.H.F., and Salahuddin Saleh Karimi, 1995. Regional interpretation of SBP93 seismic lines, Block SB5. PETRONAS Internal Report.

Wong, R.H.F., and Salahuddin Saleh Karimi, and Mohd Idrus Suhud, 1997. Hydrocarbon potential of the Malawali Basin. Abstracts Geological Society of Malaysia Petroleum Geology Conference 1997.

Wood, B.G.M., 1985. The mechanics of progressive deformation in crustal plates - a working model for SE Asia. Bulletin of the Geological Society of Malaysia, 18, 55-99.

Xia, K.Y. and Zhou, D., 1993. The geophysical characteristics and evolution of northern and southern margins of the South China Sea. Bulletin of the Geological Society of Malaysia, 33, 223-240.

Chapter 22

Sabah Basin

Mazlan B. Hj. Madon, Leong Khee Meng and Azlina Anuar

INTRODUCTION

The Sabah Basin (also known as the NW Sabah Basin) is a predominantly offshore Middle Miocene sedimentary basin that underlies the continental margin off western Sabah (Fig. 22.1). It represents a major segment of the greater Sarawak-Brunei-Sabah continental margin that evolved since Late Cretaceous times (Chapters 4 and 5). The 'basement' of the Sabah Basin is represented in western Sabah by the relatively unmetamorphosed succession of turbidites belonging to the Oligocene-Lower Miocene West Crocker, Temburong, and Kudat formations. These rocks represent the younger Rajang Group rocks that defines the 'Crocker Fold-Thrust Belt' (Liechti et al., 1960), which is underlain by older Rajang Group rocks (East Crocker, Trusmadi, and Sapulut formations) that are equivalent to the Belaga and Lupar formations in Sarawak (Chapter 5). The older and younger Rajang Groups are separated by a major Late Eocene unconformity. The Sabah Basin includes the Baram Delta complex which lies mainly in Sarawak and Brunei, and extends into the Philippines to the northeast. Only the eastern part of the Baram Delta lies in Sabah waters. This chapter reviews the petroleum geology of the Sabah Basin based on numerous studies by production-sharing contract (PSC) operators, primarily Shell, as well as those conducted in-house by PETRONAS.

There have been many studies describing the geology of the Sabah Basin in general, and the northwestern Sabah continental margin in particular. The first detailed account of the structural framework and stratigraphy of the Sabah Basin (slope and shelf areas) was published by Bol and van Hoorn (1980). The tectonic evolution of the Sabah Basin has been discussed by many workers (e.g. Whittle and Short, 1978; Levell, 1987; Tan and Lamy, 1990; Hazebroek and Tan, 1993; Tongkul, 1994). A major unpublished compilation of the petroleum geology of the Sabah Basin was done by Occidental (Abbott et al., 1990). More specifically, Levell

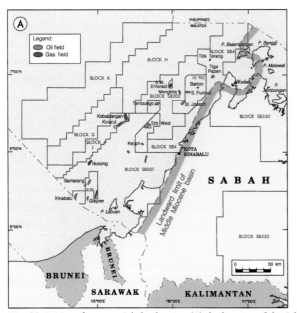

Fig. 22.1. Map of western Sabah, showing (A) the location of the Sabah Basin, its oil and gas fields and the current PSC blocks. The brown-shaded line represents the approximate landward limit of the Middle Miocene Sabah Basin, as defined in Chapter 5. The embayment in this line near Kudat includes the Malawali Sub-basin. (B) The old PSC block nomenclature, referred to in this chapter.

(1987) described the tectonic controls on the stratigraphic architecture and structural evolution of the NW Sabah margin based on regional seismic interpretation. Noor Azim Ibrahim (1994) studied the sequence stratigraphy and subsidence history of the basin, while Azlina Anuar (1994) studied the geochemistry of source rocks in the area. The petroleum geology of Esso's Tembungo Field was presented by Whittle and Short (1978). Shell's Samarang Field was described by Scherer (1980). Other field studies published by Shell include those on the Erb West (Johnson et al., 1987) and St. Joseph (Johnson et al., 1989) fields.

This chapter also includes a detailed discussion of the source rock potential and quality based on geochemical studies of both onshore and offshore samples by PETRONAS Research (PRSS). New geochemical data on the source rocks and hydrocarbons of both offshore and onshore Sabah were also reviewed. This provided some insights into the nature of the hydrocarbon habitats in the basin. In addition to the tectonostratigraphic provinces described previously by Tan and Lamy (1990) and Hazebroek and Tan (1993), this chapter also describes the Northern Sabah Province, which includes the Kudat Platform and surrounding depocentres such as the Malawali Sub-basin and Siagut Syncline, located at the northern tip of Sabah. This new province is additional to the tectonostratigraphic provinces described previously by Tan and Lamy (1990) and Hazebroek and Tan (1993). The existing literature on the Sabah Basin rarely includes mention of the Northern Sabah Province, but the occurrence of onshore oil seeps at Sikuati (Stephens 1956, p. 115-117) has generated interest among oil companies. The discovery of oil at Tiga Papan, near its southern boundary, and gas at Titik Terang have enhanced the prospectivity of this area.

There are currently 7 producing fields in the Sabah Basin (Ketam ceased production recently) and, except for Kinabalu, all the fields were discovered before 1980. There have been encouraging discoveries also in the former PSC blocks SB1 and SB2, such as Mengkira, Kinarut, and Kebabangan (Fig. 22.1B). Most of the exploration activities have traditionally been concentrated on the inboard areas of the continental shelf. It is only recently that exploration moved towards deeper waters in the outer shelf and slopes of the East Baram Delta (including the EBD Toe-Thrust Zone), the distal Outboard Belt (Thrust Sheet Zone), and the Sabah Trough. Murphy Oil recently signed a PSC for Block K which includes part of the lower continental slope and Sabah Trough.

EXPLORATION HISTORY

Western and northern Sabah have been explored for more than 100 years, since the first oil seeps were reported from the Klias and Kudat peninsulas. Oil seeps were also reported from a coal mine on Labuan Island in the late 1800s. Exploration for oil started in 1897 with the drilling of the Menombok-1 well on the Klias Peninsula by the Bombay-Burma Company. Three more wells were drilled in the following year, but all were unsuccessful. Among the other pioneering companies that explored in onshore western and northern Sabah were the British Borneo Exploration Company, Burma Petroleum Syndicate, N.K.P.M. (STANVAC), Kuhara Mining Company, and the Singapore Oil Syndicate. Between 1897 and 1954, a total of 21 exploration wells were drilled in onshore Sabah. By the end of 1998, a total of 113 exploration wildcat and 79 appraisal wells have been drilled, including 23 onshore wells.

The first offshore well in the Sabah Basin was Hankin-1, drilled by Sabah Shell/Pecten (SSPC) in 1958 from a fixed platform. Shell drilled several more offshore structures during the 1960s, including the Barton structure. In the mid and late 1960s, Esso and Oceanic Exploration began exploring in offshore western Sabah. Between 1958 and late 1976 (when the Concession Period ended), a total of 53 offshore wells were drilled (37 by SSPC, 15 by Esso, 1 by Oceanic), with a peak of 18 wells drilled in 1972. These efforts resulted in several oil discoveries which are currently on production: the Tembungo Field by Esso, and the Barton, Erb West, South Furious, Samarang, and St Joseph fields by SSPC. Other oil and gas discoveries were Erb South (SSPC) and Kinarut (Esso) respectively.

Exploration in the Sabah Basin under Production Sharing Contracts (PSCs) during the 5-year period from 1977 to 1981 resulted in 4 oil discoveries by Shell: Ketam, Lokan, SE Collins, and SW Emerald. There were also 2 gas discoveries: Samarang Kecil (Esso) and Glayzer (SSPC). A total of 46 wells were drilled during that period. When the PSC expired, exploration activity continued in the retained development and production areas.

The Tembungo Field was relinquished prematurely by Esso in 1986, and is now operated by Carigali. The Samarang Field PSC was originally held by SSPC, Pecten, and PETRONAS Carigali from 1991 to 1995. A new PSC arrangement was awarded in 1995 to Carigali as the sole contractor and operator. This PSC also includes the Erb South, SE Collins,

Lokan, SW Emerald, Glayzer, Kinarut, Samarang Kecil fields in offshore Sabah, and the Asam Paya Field in onshore Sarawak. The Erb West Field was relinquished by SSPC in 1996 and is now included in the new Samarang Field PSC. The St Joseph, South Furious, Barton, and Ketam fields now fall under a new North Sabah PSC, which was awarded to SSPC (as operator), Sabah Shell Selatan (SSS) and Carigali in 1996.

The original Concession Area in northern Sabah which was held by Oceanic Exploration was expanded to incorporate areas relinquished by SSPC and Esso. In 1980 a new PSC was awarded to British Petroleum (BP), Carigali, and Oceanic Exploration. The joint operators were Carigali and BP, then known as Carigali-BP (CBP). A total of 2998 line-km 2D seismic data were acquired, and 8 wells were drilled, resulting in an oil discovery at Tiga Papan. In 1984 Overseas Petroleum Investment Company (OPIC) assumed operatorship of the area, and drilled 2 wells without success. Some 557 line-km 2D seismic data were acquired, before the area was relinquished in 1986.

Following the promulgation of the 1985 PSC terms, the open exploration areas in offshore western Sabah were re-blocked as SB1, SB2, SB3, and SB4 (Fig. 22.1A). All these blocks are in water depths of 200 m or less, except for SB1, which contains a strip of deepwater area. Block SB1 was awarded to Sabah Shell, Pecten (later to Sabah Shell Selatan or SSS), and Carigali in 1987. A total of 14,604 line-km 2D and 82,662 line-km 3D seismic data were acquired, and 13 exploration wells were drilled. This resulted in the discovery of the Kinabalu oil field and the Kebabangan gas accumulation. Block SB1 was relinquished in 1997 but the Kinabalu Field was retained as a production area, and is now producing.

Block SB2 was awarded to Tenneco, Occidental, and Carigali in 1988. British Gas became the operator of this block after acquiring Occidental's and Tenneco's interests. In 1992 British Gas assumed 100% equity in the area. Approximately 7000 line-km of 2D seismic data were acquired. Five wells were drilled, but only 2 found minor oil and gas shows. The acreage was relinquished in 1997. In 1983 Block SB3, which includes parts of the Klias Peninsula, was awarded to a consortium that comprised Phoenix, Enron, Kufpec, Monument Resources, and Carigali. A total of 4575 line-km of 2D seismic data were acquired, and 2 onshore wells were drilled without success. The operatorship of the block changed from Phoenix to Enron and to Monument Resources, before it was relinquished in 1995. Enron drilled South Lokan-1 and Gava-1 wells. Block SB4 is an expanded Carigali-BP-Oceanic-OPIC acreage that includes a narrow strip of the offshore area around the Kudat Peninsula. The block was awarded to Hall-Houston (as operator), Ensearch, Global, and Carigali in 1991. A total of 1594 line-km 2D and 4276 line-km 3D seismic data were acquired. Three wells were drilled, with a gas discovery at Titik Terang-1. The PSCs for blocks G and J were awarded under the 1993 Deepwater Terms to SSPC (operator), SSS, and Carigali in 1995 and 1997, respectively. Block H was awarded to Esso (as operator) and Carigali in 1997, while Block K was awarded to Murphy Oil and Carigali in early 1999.

With the introduction of the 1997 PSC Revenue-over-Cost (R/C) terms, the open acreage was once again re-demarcated and re-designated. Figure 22.1A shows the block nomenclature as at 1.1.99. The blocks are now designated as SB301 and SB302, while Block SB4 remains. Block SB301 is now the combined Blocks SB1 and SB3, while the deep-water portion of SB1 was re-designated as Deepwater Block J. Block SB302 is formerly Block SB2. In 1997 Block SB301 was awarded to SSPC (operator), SSS and Carigali, and Block SB302 to Esso (operator) and Carigali, under the R/C terms. Exploration activities are currently under way.

TECTONIC SETTING

The Sabah Basin contains a post-early Middle Miocene succession overlying Oligocene-Early Miocene West Crocker Formation and related rocks (the younger Rajang Group) in western and northern Sabah (see Chapter 5). The Crocker Fold-Thrust Belt consists of deep marine turbidites that crop out along strike of the Rajang Fold-Thrust Belt of Sarawak (Fig. 22.2). This belt of deep marine turbidites forms the shore-parallel Crocker Range (Stauffer, 1967), and is slightly younger than the Rajang Fold-Thrust Belt in Sarawak, which is Late Cretaceous-Late Eocene (Hutchison, 1996). The equivalents of the older Rajang Group in Sabah include the East Crocker, Trusmadi, and Sapulut formations (see Chapter 21).

A large part of the Sabah Basin lies offshore (Fig. 22.1A). The basin is bounded to the west by the West Baram Line, which is believed to be the offshore extension of the NW-trending Tinjar strike-slip fault (James, 1984; Hutchison, 1989; Tan and Lamy, 1990; Hazebroek and Tan, 1993), and to the east by the transcurrent Balabac Strait Fault, also known as

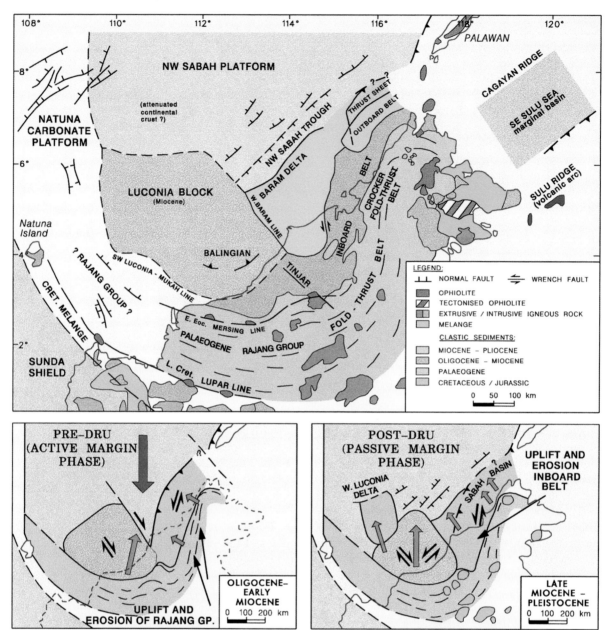

Fig. 22.2. Tectonic framework and regional setting of the Sabah Basin. (Top) The major tectonic elements of Sabah and surrounding areas. The basin is situated between continental platform (NW Sabah and Luconia platforms) and the Rajang Fold-Thrust Belt. (Bottom, left). Basin initiation during the Oligocene to Early Miocene when the Rajang Sea crust was subducted beneath Sabah, which ended with uplift of the Crocker accretionary prism. (Bottom, right). The Middle Miocene-Recent Sabah Basin resulted from progradational sedimentation across the once-active continental margin, deriving the sediment from the uplifted Rajang Group. (Diagrams modified from Hazebroek et al., 1994).

Strait Balabac Fault, the Balabac Line or Sabah Shear (Beddoes, 1976; Wood, 1985). The basin extends northwestwards beyond the continental shelf, into the Sabah Trough, which is a linear bathymetric feature at the foot of the continental slope. Further northwestwards is the NW Sabah Platform, which is a shoal area that is geologically similar to the adjacent Reed Bank and Dangerous Grounds.

The Sabah Basin was interpreted by earlier workers as a trench-associated (fore-arc) basin (e.g. Kingston et al., 1983; Levell, 1987). Based on recent studies that integrate gravity, subsidence history, and sequence stratigraphy, it is now believed that the basin evolved as a foreland basin following the collision of a South China Sea continental fragment (now forming the NW Sabah Platform) with western Sabah during the early Middle Miocene (Hazebroek et al., 1994; Noor Azim Ibrahim, 1994; Milsom et al., 1997) (Fig. 22.2). The Sabah Trough, formerly thought of as a subduction zone (Hamilton, 1979) is probably a sediment-starved foreland trough formed by thrust-front loading of the NW Sabah Platform (Noor Azim Ibrahim, 1994; Milsom et al., 1997).

The Sabah Basin was formed following the uplift and exhumation of the Crocker Fold-Thrust Belt since the early Middle Miocene (Fig. 22.2). The pre-early Middle Miocene West

Crocker and related formations are therefore considered as the basement to the Sabah Basin. Although no evidence of a crystalline basement has been reported in the offshore part of the basin, wells Kalutan-1 and Balambangan-1 in offshore northern Sabah encountered Eocene deep-marine shales and mudstones containing fragments of chert and serpentinite.

At least 12 km of Neogene sediments occur on the northwestern Sabah margin, including 6 to 8 km of post-lower Middle Miocene sediments that fill the Sabah Basin. Post-lower Middle Miocene rocks crop out in a few localities in northern and southern Sabah. The pre- and post-lower Middle Miocene successions are separated by a major regional unconformity, known as the Deep Regional Unconformity (DRU) (Bol and van Hoorn, 1980; Levell, 1987; Tan and Lamy, 1990; Rice-Oxley, 1991; Hazebroek and Tan, 1993). The DRU marks a major contrast in the siliciclastic sedimentation style on the NW Sabah margin: a pre-early Middle Miocene phase of generally deep-marine sedimentation (the pre-Sabah Basin) and a post-lower Middle Miocene phase of progradational shelf/slope deposition which constitutes the Sabah Basin proper. The DRU therefore represents the base of the Sabah Basin. Almost the entire pre-DRU succession (basement) consists of deep marine siliciclastic sediments. An older part of the succession consists of the East Crocker, Trusmadi, and Sapulut formations, which are coeval with the Chert-Spilite Formation of Cretaceous-Early Eocene age. This is overlain by a younger package comprising the Middle Eocene to Lower Miocene West Crocker, Temburong, and Kudat formations (see Chapter 21). In the offshore Sabah Basin, these pre-DRU formations are represented by Stage III sequences (see below).

Figure 22.2 shows a reconstruction of the pre- and post-Middle Miocene history of the NW Sabah margin based on Hazebroek et al. (1994). Prior to the formation of the Sabah Basin, Sabah was part of the southern margin of the Rajang Sea, which is represented by the older Rajang Group rocks (East Crocker, Trusmadi and Sapulut formations). The Crocker and Rajang fold-thrust belts represent an accretionary prism related to the diachronous closure of the Rajang Sea crust from the Late Cretaceous to the Early Miocene (see Chapter 4). Closure of the Rajang Sea was concomitant with rifting in the South China Sea Basin to the north. The Sabah Basin was formed after the cessation of sea-floor spreading in the South China Sea Basin sometime at the end of Early Miocene (Taylor and Hayes, 1980) or early Middle Miocene (Bol and van Hoorn, 1980), when the Reed Bank/Dangerous Ground continental fragment, of which the NW Sabah Platform is part, collided with western Sabah (Hinz and Schluter, 1985; Hinz et al., 1989). This folded and uplifted the deep-marine turbiditic sediments that now crop out in western Sabah in the Crocker Fold-Thrust Belt. Bol and van Hoorn (1980) described the Early Miocene deformation preceding the formation of the Sabah Basin, which is attributed to the collision. This Early Miocene collision was contemporaneous with the rifting of the Sulu Sea (Clennell, 1996), which may have intensified the deformation in western Sabah.

STRATIGRAPHY AND PALAEOGEOGRAPHY

The post-DRU sedimentation history of the Sabah Basin involved the northwestward progradation of a siliciclastic shelf and slope system. Basin filling was punctuated by 6 deformation phases which resulted in major localised to semi-regional, angular unconformities (Levell, 1987). These are the Lower and Upper Intermediate Unconformities (LIU and UIU, respectively) of Middle Miocene age, the Shallow Regional Unconformity (SRU) and Horizon III in the Late Miocene, Horizon II in the Pliocene, and Horizon I in the Plio-Pleistocene. Each deformation phase is typically marked by uplift of the hinterland, basinward tilting, deltaic progradation, marine erosion and overlap, and angular unconformities which become conformable basinwards.

Figure 22.3 schematically shows the stratigraphy of the Sabah Basin, from south to north. The regional unconformities are used to subdivide the basin succession into unconformity-bounded sequences, called stages (Bol and van Hoorn, 1980; see also Chapter 6). Stages I to III belong to the pre-DRU "basement", and consist of strongly deformed pre-early Middle Miocene deep-marine siliciclastic sediments. The Sabah Basin begins with Stage IV, which comprises an overall regressive unit of alluvial plain to bathyal sediments. Stage IV is subdivided into seven unconformity-bounded sequences: Stages IVA to IVG, based on planktonic foraminifera, palynomorph, and nannofossil biostratigraphy. These sub-stages can be recognised and mapped relatively easily on the inner shelf, but are more difficult to pick in the outer shelf and slope areas where their boundaries are conformable.

This stage nomenclature is used in most

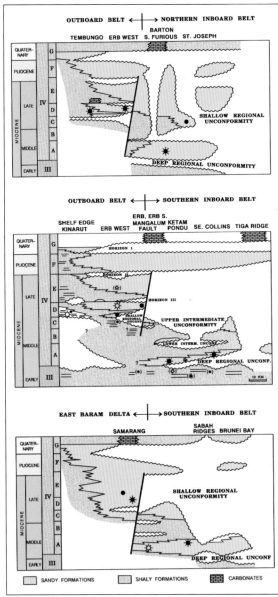

Fig. 22.3. Schematic illustration of the stratigraphy of several east-west tracts of the Sabah Basin. Based on Shell (1993).

parts of the Sabah Basin. In the northern part of the basin, however, different nomenclatures based on either chrono- or litho-stratigraphies have been used. The lithostratigraphy of the northern region is shown in Fig. 22.4. The onshore geology of N and NE Sabah is dominated by a thick succession of deepwater sediments (Crocker and Kudat formations) of Late Eocene Early Miocene age. These are fringed by a relatively narrow strip of Middle-Upper Miocene to Pliocene deltaic-shallow marine sediments (South Banggi, Bongaya, and Timohing formations). Thicker sequences of the latter are preserved in the offshore areas, i.e., in the Kindu-Mangayau Sub-basin, the Malawali Sub-basin in the Marudu Bay, the Siagut Syncline to the north, and in several isolated, "ponded" depocentres offshore. The pre-Crocker section is represented by the ophiolitic Chert-Spilite Formation (Cretaceous-Lower Eocene) and associated ultrabasic/basic igneous bodies, which represent the oceanic basement in northern Sabah (Chapter 5).

In the Northern Sabah Province, Carigali-British Petroleum (CBP) subdivides the Neogene into 4 lithostratigraphic units or formations (Frank, 1981). Figure 22.4 shows a summary of the stratigraphy of this province, approximately correlated with Shell's stages (cf. Bol and van Hoorn, 1980; Levell, 1987). The Eocene shales penetrated in the Kalutan-1 well, on the Kudat Platform in the northern part of the province, are correlated with the onshore Crocker Formation (not shown in Fig. 22.4), while the Lower Miocene sequences penetrated by Dudar-1 in the Kindu-Mangayau Sub-basin to the southwest are considered equivalent to the Kudat Formation (Stage III) exposed on the Kudat Peninsula. The Middle Miocene to Lower Pliocene siliciclastic sequence penetrated in Bangau-1 towards the east is subdivided into the Lower Sandstone Unit, Mudstone Unit, and Upper Sandstone Unit (Fig. 22.4). The Lower Sandstone Unit is correlated with Stage IVA, Mudstone Unit with Stage IVB, and the Upper Sandstone Unit with Stages IVC, IVD, and IVE. The Upper Sandstone Unit is also correlated with the onshore Bongaya Formation. The Upper Miocene to Pleistocene carbonate-siliciclastic sequences penetrated by Kalutan-1 is referred to as the Kalutan Group (Stages IVE, IVF, and IVG), and is correlated onshore to the Timohing Formation. The reefal limestones in Kalutan-1 are referred to as the Lower Limestone and Upper Limestone units of the Kalutan Group. The mixed siliciclastic-carbonate oil-bearing unit in Tiga Papan-1, known as the Tiga Papan Unit (Stage IVB), is correlated with the onshore South Banggi Formation.

Palaeogeographic maps of the Sabah Basin from the Early Miocene (Stage III) to the Late Pliocene (Stage IVF) were compiled by Rice-Oxley (1991). Biostratigraphic data from over 100 wells and seismostratigraphic data mainly in the southern area were integrated to produce these maps. Figure 22.5 shows the maps which have been slightly modified to include data from the northernmost part of the basin. During pre-DRU times western Sabah was underlain by a deep ocean basin (Chapter 5), where deep-marine muds and turbidite sands (Stage III, Kudat Formation, Setap Shale) were deposited. During the post-DRU period, there were at least 3 major basinward shifts in facies related to tectono-eustatic events during the middle Middle Miocene (Stage IVA), Late Miocene (Stages IVC and IVD), and latest Miocene-Early Pliocene (Stage IVF). The regressive siliciclastic sediments were deposited mainly in coastal plain and

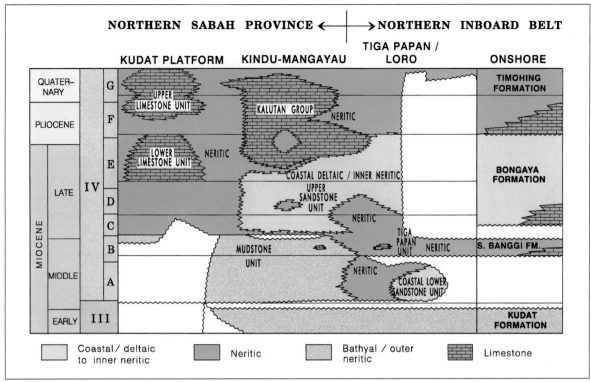

Fig. 22.4. Stratigraphic summary of the Northern Sabah Province. Compiled from various well data.

fluviomarine depositional environments. Some deep-marine sequences are also present. Stage IVB (late Middle Miocene) represents the regional transgression which resulted in thick mudstone sequences. Stage IVE (Late Miocene) is generally characterised by aggrading siliciclastic sedimentation, with sands on inner shelf and muds on the outer shelf.

The uplift related to the Upper Intermediate Unconformity (UIU) and the consequent relative sea-level fall triggered a regressive episode during the early Late Miocene, i.e. Stage IVC. Stage IVD is also a regressive phase during a time of relative tectonic quiescence after the tectonic uplift that gave rise to the SRU. A transgression with marine erosion and onlap is evident on the Kinabalu Culmination by the transgressive sands fringing the high, and forming the oil reservoirs in the Erb West Field. The ensuing regression gave rise to Stage IVD delta and shelf systems that prograded for a considerable distance beyond the Morris Fault (in the East Baram Delta) and the Emerald Fault Zone (in the Outboard Belt).

In the northernmost part of the Sabah Basin, Stage IVD (Upper Sandstone Unit) sedimentation continued northwest of the Batumandi Ridge and in smaller depocentres such as the Kindu-Mangayau Sub-basin, Siagut Syncline, and Malawali Sub-basin. On the Kudat Platform, reefal carbonates (Lower limestone Unit of the Kalutan Group) grew on pre-existing highs that were reactivated along N-trending faults.

During the latest Miocene to Early Pliocene (Stage IVF), towards the west and northwest, a series of deep-water submarine fans were deposited. An example is the Lingan Fan Complex, described by Marzuki Mohamad and Lobao (1997) and Salahuddin Saleh Karimi et al. (1997). In the outboard area of the East Baram Delta, deep-water fans were also deposited during Stage IVF times. Meanwhile, carbonate deposition of the Upper Limestone Unit continued in the northern area.

STRUCTURAL AND SUBSIDENCE HISTORIES

The complex structural history of the Sabah Basin is the result of multiple phases of structural development that may be local or regional in nature. Bol and van Hoorn (1980) recognised three major phases of structural deformation in the basin:

(i) Late Miocene phase in the central and southern parts of the basin, which corresponds with the SRU.
(ii) Early Pliocene phase folding in the northeast (represented by Horizon II)
(iii) Late Pliocene phase along the westernmost margin of Sabah Basin (represented by Horizon I).

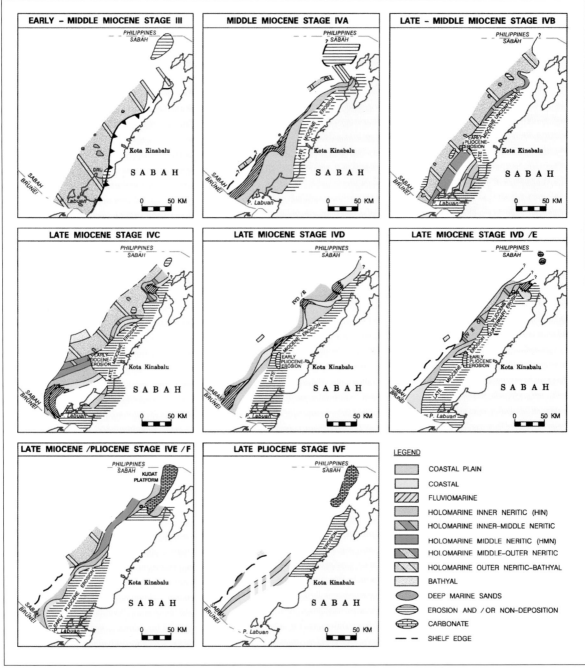

Fig. 22.5. Palaeogeographic reconstructions. Based on Shell (1993), with additions for the Northern Sabah Province.

Levell (1987) also noted the progressive regional tilting of the basin northwestwards, which is evident from the migration of depocentres, and the basinward migration of the hinge lines that separate unconformities from their correlative conformities. These indicate a gentle flexure of the basement in response to isostatic uplift of the landward margin and the consequent exhumation of the Crocker Fold-Thrust Belt. Superimposed upon the basement flexure is the independent movement of crustal blocks, which resulted in contrasting structural styles in the different provinces, particularly between the northwestern and southern areas (Levell, 1987; Tan and Lamy, 1990; Hazebroek and Tan, 1993). The overprinting of several structural events has also produced a complex fault pattern. Furthermore, the obliquity of the compressive forces, and the presence of mobile/diapiric shale within the deltaic depositional system, have produced very complex transpressive structural trends, commonly associated with wrench faulting.

Uplift and deformation preceding the middle Late Miocene SRU was severe in the southern and central nearshore areas of Sabah, known as the Inboard Belt. Here, compressive structures are sometimes referred to as the "Sabah Ridges" (see Fig. 3 of Bol and van Hoorn, 1980).

This Late Miocene tectonic activity has been attributed to the resumption of convergent forces/collision between the NW Sabah Platform and Borneo (Hazebroek and Tan, 1993) and was also accompanied by the intrusion of the Kinabalu granodiorite. The complexity of structures in the northern Sabah Basin may be due to its proximity to the intersection of the NW Borneo Trend and the Sulu Trend. The SRU deformation, strong in the southern portion of the Sabah Basin, is negligible in the northern area. The more significant unconformity is the equivalent of UIU, which can be correlated in most wells and the seismic sections.

Noor Azim Ibrahim (1994) analysed the subsidence history of the Sabah Basin and found 2 distinct phases of subsidence: a rapid tectonic subsidence (> 500m/Ma) between 14.7 and 9 Ma, followed by a generally slow subsidence accompanied by pronounced uplift along the eastern margin of the basin (Fig. 22.6). The boundary between the two phases is marked by the SRU, which separates Stages IVA–IVC and Stages IVD–IVG sedimentary packages. The change from rapid to slow subsidence occurred earlier in the south and progressively moved northwards. Rapid subsidence is still continuing today in the north. This was attributed to the delicate interplay between tectonic uplift/subsidence and eustatic sea-level changes (Noor Azim Ibrahim, 1994).

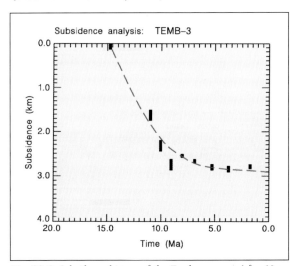

Fig. 22.6. Subsidence history of the Tembungo area (after Noor Azim Ibrahim, 1994).

The subsidence pattern cannot be explained by lithospheric extension, for there is no evidence to suggest a Middle Miocene rifting in the basin. Subsidence in the Sabah Basin has also been attributed to wrenching of the West Crocker Formation during the Oligocene (Whittle and Short, 1978). Some authors have also suggested that the structural evolution of the basin was controlled by basement-involved strike-slip faults (e.g., Bol and van Hoorn, 1980; Levell, 1987; Tan and Lamy, 1990; Hazebroek and Tan, 1993; Hazebroek et al., 1994). Noor Azim Ibrahim (1994), however, pointed out that the regional nature of rapid subsidence in the basin does not necessitate strike-slip motion as the primary subsidence mechanism. In central Sabah Basin, strike-slip deformation has been associated with younger deformation, and yet there is no evidence of rapid subsidence.

Some studies (e.g. Hinz and Schluter, 1985; Hinz et al., 1989) have shown that the NW Sabah continental margin is part of an overthrust sheet, loaded onto a downwarped extended margin of the South China Sea that drowned the Oligocene to upper Lower Miocene carbonate buildups in the Sabah Trough. It is speculated that the thick uplifted wedge of strongly deformed Crocker Formation and its lateral equivalents formed a sizeable portion of the necessary load. Furthermore, the negative free-air gravity anomaly in the Sabah Trough may be attributed to the depression of the lithosphere into the mantle through thrust loading (Noor Azim Ibrahim, 1994; Milsom et al., 1997).

TECTONOSTRATIGRAPHIC PROVINCES

Bol and van Hoorn (1980) noted a variation in the structural styles across the continental margin off NW Sabah. Generally, from southeast to northwest, the structural styles show (i) dominantly compression with minor early extension, (ii) dominantly extension with compression in some areas, and (iii) dominantly compresssion, with thrust belts that could be the result of gravity sliding or subduction or a combination of both. Subsequent workers (Tan and Lamy, 1990; Hazebroek and Tan, 1993) recognised 7 structural provinces on the basis of the different structural styles and sedimentation histories. These are the Crocker Fold-Thrust Belt, Inboard Belt, Outboard Belt, East Baram Delta, Thrust Sheet Zone, Sabah Trough, and the NW Sabah Platform (Fig. 22.7). Figure 22.8 shows the structural styles across the Sabah Basin. Chapter 5 has given a summary of the structural styles of these provinces. Here, a more detailed account of the structures in the main provinces is given. The latter 3 provinces are relatively unexplored, and thus will not be described further.

Sabah Basin

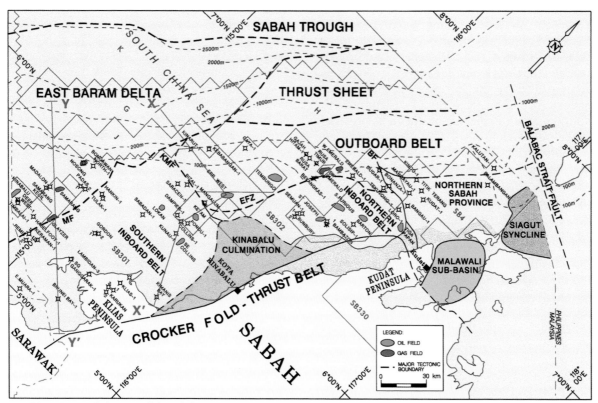

Fig. 22.7. Structural elements of the Sabah Basin, showing the basin boundaries, tectonic elements, and tectonostratigraphic provinces (modified from Abbott et al., 1990). Major fault zones are labelled: MF - Morris Fault, KMF - Kinarut-Mangalum Fault, EFZ - Emerald Fault Zone, BF - Bonanza Fault. XX' and YY' refer to locations of cross-sections in Fig. 22.8

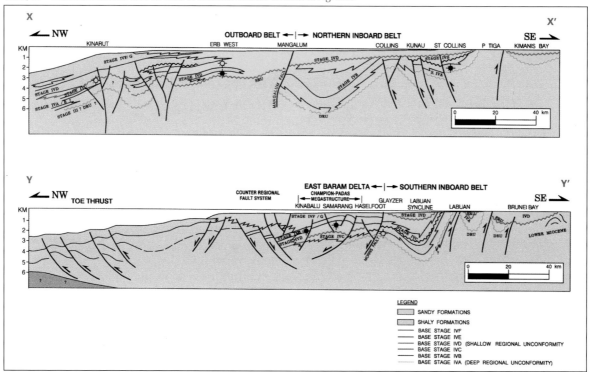

Fig. 22.8. Regional cross-sections of the Sabah Basin (modified from Tan and Lamy, 1990). Location of sections in Fig. 22.7.

Inboard Belt

The Inboard Belt (Tan and Lamy, 1990; Hazebroek and Tan, 1993), sometimes referred to as the "Ridge and Syncline" Province, is the innermost structural belt basinward of the Crocker Fold-Thrust Belt. It stretches from the onshore inner Baram area of Brunei (possibly represented by the Meligan Formation) northeastwards into offshore Kudat Peninsula. The Inboard Belt does not appear to continue onshore in northern Sabah although there are isolated outcrops of Middle Miocene sediments

(South Banggi, Bongaya and Timohing formations), which are probably equivalent to the sequences in the belt.

The Inboard Belt consists of northern and southern segments (Fig. 22.7), both having a characteristic structural style, of sharp anticlines and wide, deep synclines. The southern belt consists of the large N-trending Labuan-Paisley Syncline (Fig. 22.9) which is bounded to the north by the Morris-Padas-Saracen fault line, and numerous folds to the east and south of it. Many of the drillable structures are located in this area. The Morris Fault, which marks the southern segment's western boundary, appears to be a major wrench fault system that continues into Brunei as the Jerudong Fault. The southern and northern segments of the Inboard Belt are separated by the northward-dipping Kinarut-Mangalum Fault and the Kinabalu Culmination, which is a basement high area underlain by a relatively shallow subcrop of the Crocker Fold-Thrust Belt. The E-trending Kinarut-Mangalum Fault (Tan and Lamy, 1990) has been interpreted as a right-lateral wrench fault (see Fig. 4. of Hazebroek and Tan, 1993).

The post-DRU sedimentation history of the Inboard Belt can be simplified into 3 phases: an early Middle Miocene regression (Stage IVA), a late Middle Miocene transgression (Stage IVB), and a Late Miocene to Pliocene regression (Stages IVC, F/G; Stages IVD and IVE are thin or absent).

Southern Inboard Belt

The Southern Inboard Belt is characterised by N-S to NNE-SSW trending anticlines with steep flanks and strongly faulted crests, spaced generally between 5 and 20 km apart. Apart from the Labuan-Paisley Syncline, other major synclines include the Kimanis and Winchester synclines (Fig. 22.9). The Labuan-Paisley Syncline contains up to 4000 m of Stage IVC deep-marine sands and shelfal sediments formed by deltaic progradation (van Vliet and Schwander, 1987; Wong, 1997). The synclines are believed to be the kitchen source areas for hydrocarbons in the surrounding structures. The cores of the anticlines commonly consist of uplifted deep-marine Stage III shales. The

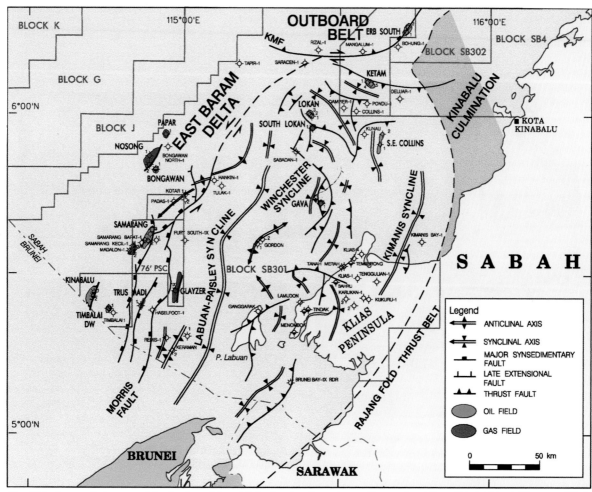

Fig. 22.9. Structural elements of the Southern Inboard Belt, showing the major synclines and anticlines (Sabah Ridges). Figure modified from Tan and Lamy (1990), and Hazebroek and Tan (1993). Note that only the Kinabalu and Samarang fields are currently (1999) producing. KMF - Kinarut-Mangalum Fault.

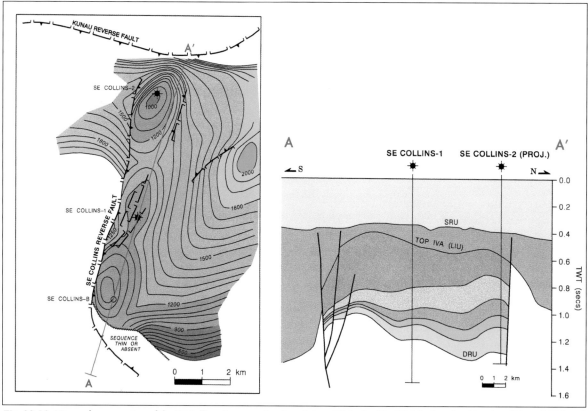

Fig. 22.10. Map and cross section of the SE Collins discovery, Southern Inboard Belt. See Fig. 22.9 for location. Based on Wong (1996) and Phoenix (1988).

Fig. 22.11. Some field photographs of the Belait Formation (equivalent to basal Stage IVA) sediments on Labuan Island, Southern Inboard Belt. Top: Photomosaic of cliff section of fluvial deposits comprising trough cross-bedded channel sandstones. Cliff is about 10 m high. Bottom left: Close-up of sandstone with trough cross-bedding. Hammer is 33 cm long. Bottom right: Hummocky cross-stratified sandstone with a solitary *Ophiomorpha* burrow. Pencil is 15 cm long. All photographs by Mazlan Madon.

anticlinal ridges exhibit flower structures, which suggest they are transpressional structures. Large-scale sinistral strike-slip movements with cumulative horizontal displacements of up to 100 km have been estimated in the Southern Inboard Belt (Hazebroek and Tan, 1993).

A prominent feature of onshore western Sabah is the change in the structural grain from NE-SW NW Borneo Trend to the NW-SE Sulu Trend (Tongkul, 1993; 1994). This feature is also observed offshore. Between the Sabadan and Mangalum wells (Fig. 22.9) is a structurally complex zone situated at the intersection of WNW- and N- trending lineaments. The result is a domal, faulted fold interference pattern. Unlike in the south, where the structural grain is generally NNE-SSW, the main structural trend here is E-W. Compressional wrench movements along the lineaments occurred in 3 main phases during the Middle to Late Miocene, and resulted in the major unconformities DRU, LIU/UIU, and SRU. Several structures in this complex zone have been drilled and have resulted in several oil discoveries, e.g. Ketam, Lokan, and SE Collins. The SE Collins Field (Fig. 22.10) was described by Wong (1996). Hydrocarbon shows were also reported from the Pondu-1 and Dampier-1 wells. Separating the Southern Inboard from the Outboard belt is the Kinarut-Mangalum Fault. This fault zone appears to be on trend with the NW-trending Sabah Blueschist Belt of Leong (1978) or the Kinabalu strike-slip fault of Tokuyama and Yoshida (1974). It is interesting to note that the Kinabalu Culmination and the Kinarut-Mangalum Fault also separate two main stress domains identified from well-bore breakouts (Tjia and Mohd Idrus Ismail, 1994).

The initial deltaic progradation in the Southern Inboard Belt can be traced as far as the Labuan-Paisley Syncline, and was followed by a rapid northwestward progradation of a major delta towards the Samarang area. This subsequent outbuilding is maintained by uplift of the hinterland and erosion of older foresets (Rice-Oxley, 1991). Stage IVA represents the first significant deposition of alluvial, coastal plain, and deltaic sediments in the Inboard Belt. Part of this basal succession is exposed on the island of Labuan as the Belait Formation (Fig. 22.11). The sedimentological aspects of the Belait Formation on Labuan were described by Mazlan Madon (1994, 1997).

The overlying Stage IVB is a relatively thin transgressive marine sequence, which is absent over some of the syn-depositional highs. Exploration wells in this area have encountered Stage IVB mudstones, but much of the sand-rich upper portion has been eroded. Intense deformation during the Late Miocene (post-Stage IVC, associated with the SRU) and subsequent tectonic stability is characteristic of the Southern Inboard Belt. This event resulted in the tightening of the earlier-formed structures and the inversion of the depositional troughs to form the complex pattern of ridges and synclines.

The SRU is essentially a peneplain with only a thin cover of younger sediments. Towards the east, much of Stage IVC, Stage IVB, and Stage IVA have been eroded. Consequently, the lower Stage IVA sediments and the Stage III turbiditic sands occur at shallower depth, just below the SRU. They form prospective exploration targets. To the west, however, these potential reservoirs are much deeper.

Northern Inboard Belt

This area lies to the north of the Kinabalu Culmination, east of the Emerald Fault Zone (Fig. 22.12). It is characterised by an intersecting network of N-, E- and SE-trending compressional ridges and synclines. The synclines often contain a full Neogene succession. Figure 22.13 shows some of the structural styles in the area. Between the major structural ridges are synclinal areas. Among the deeper synclines are the Pritchard and Furious synclines (Fig. 22.14). The synclines and ridges terminate abruptly against the NE-trending Emerald Fault Zone to the west and the Crocker Fold-Thrust Belt to the east. Whereas the SRU deformation was intense in the Southern Inboard Belt, it was less so in the Northern Inboard Belt, where Stages IVA, IVB, and IVC sequences, including carbonates, are better preserved in the synclinal areas.

As in the Southern Inboard Belt, the structural grain in the Northern Inboard Belt also changes sharply from NE-striking to E-striking between the Bunbury-St. Joseph trend to the South Furious-Barton trend and the Batumandi Tiga Papan structures (Fig. 22.14). The E trending structures continue onshore into the Kudat Peninsula. The Tiga Papan oil accumulation is on trend with the onshore Sikuati oil seep. The major E- and NE-trending mainly offshore ridges, namely Bunbury-St Joseph, Mantanani, S.Furious-Barton, and Batumandi-Kudat Peninsula, were subjected to severe Pliocene shown by the Horizon II unconformity. The resulting anticlinal structures resemble the typical "pop-up" or flower structures (Fig. 22.13). Significant oil reserves have been found in the anticlines, e.g. in the St. Joseph (Stages IVA, IVC), South Furious (Stage IVA), and Barton (Stage IVA) fields.

Sedimentation in the Northern Inboard

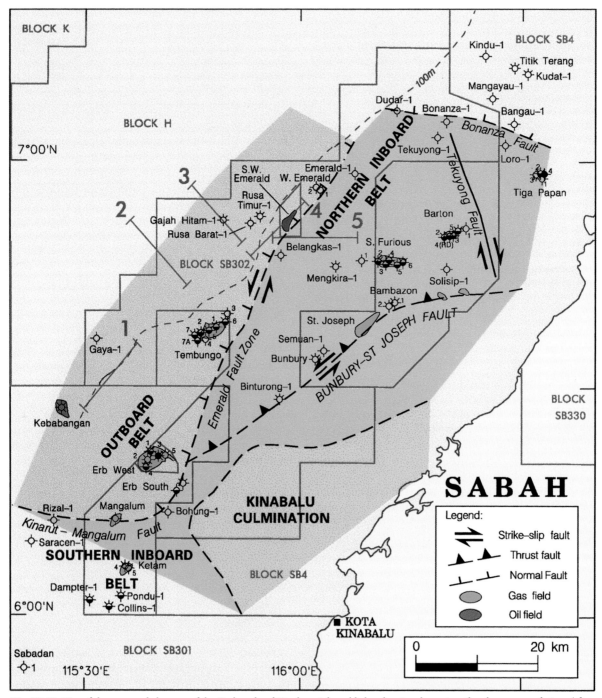

Fig. 22.12. Map of the structural elements of the Outboard and Northern Inboard belts, showing the major oil and gas accumulations (after British Gas, 1991). Lines labelled 1 to 5 refer to cross-sections in Fig. 22.13.

Belt is characterised by a prolonged Late Miocene to Pliocene regression marked by a well-developed Stage IVC succession, and the presence of Stages IVD and IVE sediments. Although Stage IVB generally consists mainly of mudstones, some Stage IVB sandy coastal plain sediments have been encountered at Binturong-1 (Carr and Wakefield, 1995). Slump scars occur in the Bunbury-St. Joseph and Mantanani Ridges (Levell and Kasumajaya, 1985). Thick Stage IVC turbidites were deposited basinward from these ridges to the northwest (Fig. 22.12). The Belangkas-1 well on the northwestern flank of the Pritchard Syncline encountered some 500 m of Stage IVC turbidite sands.

In the northeastern end of the Inboard Belt, south of the Bonanza Fault, lies the Tiga Papan Sub-basin (Fig. 22.14). This E-trending triangular depocentre is bounded by major wrench and thrust faults, and occurs in a transition zone between the Northern Inboard Belt and the Northern Sabah Province (see below). The transition zone characterised by a sharp bend in structural grain from N-S to E-W. The bend appears to have also affected older rock units onshore, such as the Kudat Formation,

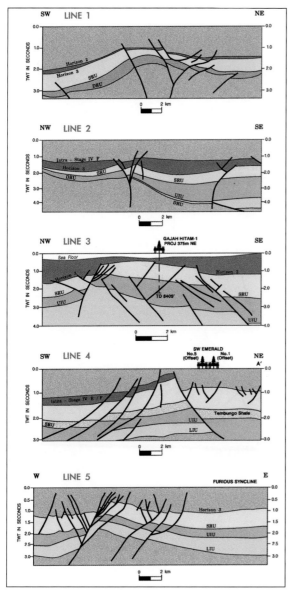

Fig. 22.13. Structural styles in the Outboard Belt (see Fig. 22.12 for location). Two main trap styles are present: compressional wrench-related flower structures (Lines 1, 2, 5) and extensional structures (Lines 3, 4).

which shows a predominantly E-W structural trend. Middle Miocene units have been encountered in the Batumandi-1, Tiga Papan, and Loro-1 wells, and indicate a rapid deepening to the northwest. They are known as the Lower Sandstone Unit (Stage IVA), and the Mudstone Unit and the oil bearing Tiga Papan Unit (Stage IVB). The Tiga Papan Unit lithologically resembles the calcareous South Banggi Formation onshore. A detailed description of the mixed siliciclastic-carbonate Tiga Papan Unit is given by Mohammad Yamin Ali (1992, 1995).

Northern Sabah Province

The Northern Sabah Province is defined here as the area westward of Banggi Island and the Bengkoka Peninsula extending towards the present-day shelf edge (Fig. 22.14). The province is bounded to the northwest by the Outboard Belt, to the south by the Bonanza Fault, and to the northeast by the Balabac Strait Fault (Fig. 22.7). It is characterised by several depocentres and sub-basins of Upper Miocene-Pliocene deltaic to shallow marine sediments. North of the Bonanza Fault is the Kindu-Mangayau Sub-basin, where the Titik Terang-1 gas discovery is located. To the northeast of this sub-basin is the Kudat Platform, which is characterised by the development of Upper Miocene carbonates above an Eocene basement. Eastwards, there are 2 major N-S trending synclinal sub-basins. To the east of the Kudat Platform, north of Balambangan Island, is the Siagut Syncline, which extends into Philippine waters. South of Balambangan, in Marudu Bay is the Malawali Sub-basin. These sub-basins are described below.

Siagut Syncline

Towards the Sabah-Philippines boundary, east of the Kudat Platform, a thick succession of Upper Miocene and younger sediments occurs in a synclinal structure or depocentre, known as the Siagut Syncline (Figs. 22.14, 22.15A). Some 4500 m of Upper Miocene and younger sediments have been mapped from seismic. These sediments rest on Eocene economic basement. The area is characterised by intersecting faults that generally trend NNE. The western edge of the sub-basin with the Kudat Platform is marked by a major west-hading rotational normal fault, with as much as 400 m of offset at the top Eocene level (Wiebe, 1996). Seismic data indicate syn-sedimentary fault growth associated with the westward progradation of a major delta. The Siagut Syncline has been affected by Pliocene to Recent E-W folding.

Kindu-Mangayau Sub-basin

Much of the Northern Sabah Province is underlain by Upper Miocene deltas with N- to NE-trending growth faults (Fig. 22.14). The Kindu-Mangayau Sub-basin is one such delta which has prograded westwards from the Bangau area, and is apparently bounded to the south by the Bonanza Fault. The sub-basin is characterised by several NE-trending growth faults. Two wells, Bangau-1 and Kudat-1, have been drilled on the footwall of a major growth

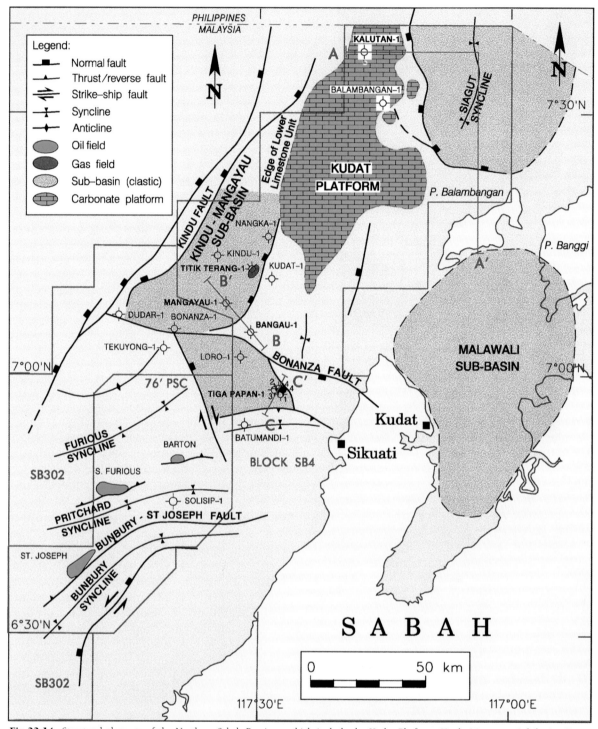

Fig 22.14 Structural elements of the Northern Sabah Province, which include the Kudat Platform, Kindu-Mangayau Sub-basin, Siagut Syncline, and the Malawali Sub-basin. Cross sections are shown in Fig. 22.15. For a section across the Malawali Sub-basin, see Fig. 5.18 of Chapter 5. Map based on Hall-Houston (Wiebe, 1996) and Johnson et al. (1989).

fault, whereas 3 wells, Mangayau-1, Titik Terang-1, and Nangka-1, were drilled on the downthrown side of the fault. The Titik Terang-1 is a gas discovery, drilled on a faulted 4-way dip closure developed in the hanging wall rollover of a major NE-trending growth fault. The fault has been active since the Middle Miocene (Stage IVA) and has at least 1000 m of throw. The structural style is similar to the Samarang Field, with minor synthetic growth faults on the eastern flank, and a relatively unfaulted western flank.

Malawali Sub-basin

Southwards, and on trend with the Siagut Syncline, is the Malawali Sub-basin, which is also a relatively small synclinal area characterised by deltaic sedimentation. A cross section of the sub-

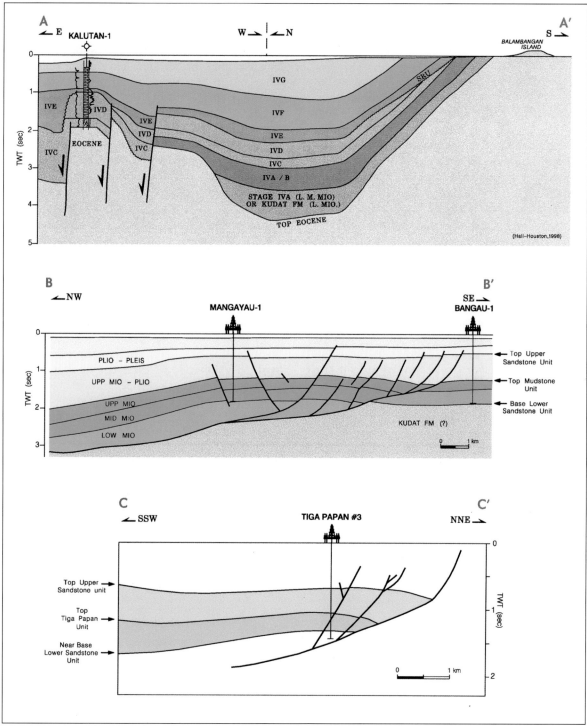

Fig. 22.15. Structural styles in the northern part of the Sabah Basin. (A) Schematic cross section of the Siagut Syncline, Northern Sabah Province, from the Kalutan-1 well on the Kudat Platform to Balambangan Island. (B) NW-SE section across the Kindu-Mangayau Sub-basin (Northern Sabah Province) through the Mangayau-1 and Bangau-1 wells, showing a hanging wall rollover structure associated with a listric growth fault system. (C) Hanging wall rollover structure in the Tiga Papan area, Northern Inboard Belt. See Fig. 22.14 for location. (From AIPC, 1990).

basin is shown in Fig. 5.18 of Chapter 5. Although relatively unexplored, seismic studies have shown possible hydrocarbon indicators. Onshore, similar sediments compose the Upper Miocene Bongaya Formation, which consists of thick interbeds of calcareous sandstones and shales with some coals, limestones, and conglomerates. These Upper Miocene deltaic-shallow marine sediments are similar to those occurring on the western side of the Kudat-Balambangan High, which is underlain by the Ophiolite Complex/Crocker Formation. This suggests that the Malawali Sub-basin may have been part of the greater Sabah Basin before the Pliocene uplift of the Kudat-Balambangan High.

Kudat Platform

The Kudat Platform is a N-S trending area characterised by the development of Upper Miocene-Pliocene carbonate reefs, known as the Lower and Upper Limestone units of the Kalutan Group (Fig. 22.14). Large N-S trending growth faults on the western edge of the Kudat Platform extend up from basement and displace the Miocene and Pliocene sections by as much as 600 m (Wiebe, 1996). The shallow marine carbonates probably grew on intrabasinal highs produced by the re-activation of the N-S trending faults. Sedimentation of the carbonates appeared to have been contemporaneous with deltaic deposition in the flanking basinal/ synclinal areas.

The Kudat Platform is thought to be underlain by an allochthonous thrust sheet emplaced during the early Middle Miocene closure of the Rajang Sea in Sabah, and is analogous to the larger Lower Tertiary Thrust Sheet in the Thrust Zone further to the west. Eocene shales have been penetrated by the Kalutan-1 and Balambangan-1 wells on the Kudat Platform, as well as in the Likas-1 well to the north in Philippines. These shales are probably equivalent to the deepwater sediments of the Crocker Formation. Data from Kalutan-1 and Balambangan-1 (Fig. 22.15A) indicate that the Oligocene through to much of the upper Middle Miocene is missing. This suggests that the Kudat Platform was probably an area of non-deposition before Late Miocene times. It is possible that the Kudat Platform may have acted as a barrier that restricted the westward progradation of deltas in the Northern Sabah Province.

OutBoard Belt

The Outboard Belt is an elongated Late Miocene-Pliocene depocentre that lies roughly along strike of the East Baram Delta (Tan and Lamy, 1990). It is differentiated from the latter by a thinner cover of deltaic and shelf (topset) sediments. Its complex structural features indicate both extensional and compressional deformation. In the Stages IVD-IVG deltaic and shelf (topset) sequence, the deformation is essentially extensional, represented by large NE-trending NW-hading normal/growth faults. In the underlying shale-prone, deeper marine succession, however, the structural style is compressional; the deformation is indicated by wrench-induced features with characteristic high-relief anticlines cored by shale diapirs.

The Outboard Belt is an area of Stage IVD and younger depocentres, comprising Stages IVD, IVE, IVF, and IVG shallow to deep marine sediments prograding northwestwards over Stage IVC and older deep-marine sediments. The slumping of the shelf edge along the once-active Bunbury-St Joseph Ridge (Levell and Kasumajaya, 1985) provided large volumes of sediments, eroded from Stages IVA and IVB, into the Outboard Belt. These Stage IVC turbiditic sandy sediments form submarine fans and provide the reservoirs at Tembungo (a producing oil field) and Gajah Hitam (Fig. 22.12). Upper Miocene slump scars occur west of the Kinarut-Mangalum Fault and the Ketam Field in the Saracen Bank area (Levell and Kasumajaya, 1985). Downflank to the northwest are the Stage IVC turbiditic sands of the Kinarut and Kebabangan gas fields (Outboard Belt).

Following the SRU deformation, Stage IVD was a period of relative tectonic stability and the Outboard Belt was predominantly shelfal. Stage IVD shallow-marine sandstones provide the reservoirs at Erb West (producing oil and gas field), SW Emerald, West Emerald, Rusa Timur and Gajah Hitam.

East Baram Delta

The East Baram Delta (EBD) is a major depocentre for Stage IVD sediments, although the whole succession consists of late Middle Miocene and younger (Stages IVC-IVG) deltaic, shelf, and slope sediments deposited in an overall regressive setting. The EBD may be subdivided from SE to NW into a proximal zone dominated by growth faults, a transitional zone formed by the delta slope, and a distal zone dominated by NW-verging overthrust anticlines, which represent the toe-thrust zone of the delta (Hazebroek and Tan, 1993). The synthetic (down-to-the-basin) growth faults and corresponding counter-regional (antithetic) faults in the proximal zone trend NNE-SSW. Large rollover anticlines occur in the hanging walls of many of these growth faults and form major hydrocarbon traps. The major structural trends include the prolific Champion-Padas trend (Fig. 22.16), which comprises the Champion (in offshore Brunei), Timbalai, Madalon, Samarang Kecil, Samarang, and Padas structures. The Kinabalu Field is on the same trend, but in a more basinward position. All the growth-fault structures have experienced some degree of basement-controlled wrench faulting. Mild compressional tectonics during the Late Pliocene resulted in numerous crestal faults on some of the anticlines, such as at Samarang. Some of the growth faults appear to have been

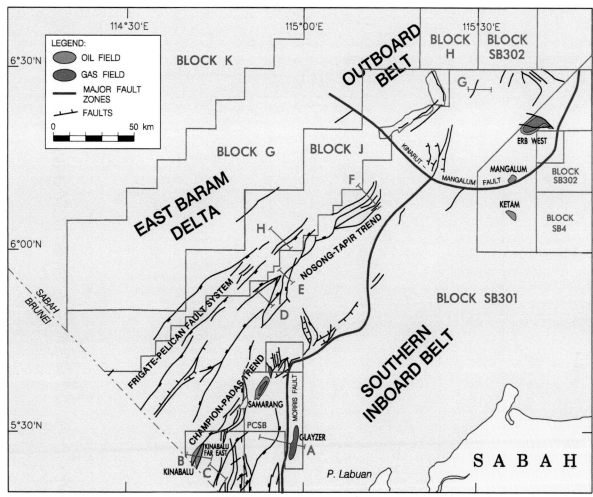

Fig. 22.16. Map of the East Baram Delta Province, showing the major structures and hydrocarbon occurrences.

reactivated as reverse faults during the deformation (Bol and van Hoorn, 1980).

The transition zone, stretching basinwards from the Timbalai-Samarang fault zone, underlies much of the outer shelf area (Fig. 22.16). It is bounded on the seaward side by a major counter-regional fault system, the Frigate-Pelican fault system. The zone is dominated by extensional syn-sedimentary faults which also produce major anticlinal trends associated with rollover structures. An example is the Nosong-Tapir trend (Fig. 22.16).

The distal zone of the EBD is in water depths of >200 m, and includes the outer shelf, delta-slope, and the toe-thrust zone. It is 50–80 km wide, but narrows towards the northern fringe of the delta. Up to 6 NE-trending ridges, formed by elongated, broadly parallel, asymmetrical overthrust anticlines, with intervening mini-basins, occur in the toe-thrust zone. The ridges young in a NW direction. The mini-basins are filled completely nearshore, but are progressively under-filled further down the slope. The toe-thrust features are believed to result from gravitational delta tectonics (Hazebroek and Tan, 1993).

HYDROCARBON OCCURRENCES, TRAPS AND RESERVOIRS

Hydrocarbons in the Sabah Basin are found mainly in the Stages IVA, IVC, and IVD (Table 22.1). There are minor occurrences in other stratigraphic units, including Stage III (e.g. at SE Collins). Most of the hydrocarbons occur in complex wrench-induced faulted anticlines, rollover anticlines associated with deltaic growth faults, and other fault-related closures. There are no major seal horizons in the Sabah Basin, but intraformational shale and mudstone units throughout the stratigraphic column provide effective top and flank seals in many known accumulations (Table 22.1). In some places, shale-filled slump scars and shale diapirs also act as seals.

The timing of structural trap formation is related to the 3 main deformation events that are manifested in the unconformities SRU (Late Miocene), Horizon II (Early Pliocene), and Horizon I (Late Pliocene). There are two types of structures of different ages:

Field/ discovery	Year of discovery	Reservoir	Trap	Seal
Erb West	1972	Stage IV D shallow marine sands. 6 main reservoirs at around 2286 m. ϕ=15-26%, κ=10-2000 mD.	E-W anticline cut by normal faults, oil rim with gas cap.	Intraformational shales.
Ketam	1977	Stage IVA coastal/barrier and fluviomarine sands at 1615 m. 48 m net reservoir thickness. ϕ=18-25%, κ=200-300 mD.	ESE-trending overthrust anticline, cut by ESE-trending normal and north-verging reverse faults. Accumulation fault-dip closed.	Intraformational shales and fault seals.
Kinabalu	1988	Stage IVD, IVE, IVF coastal plain and shallow marine sands ϕ=12-27%, κ=69-630 mD. 262 m NOS, 36 m NGS, up to 8000 BOPD	NNE-trending closure on a downthrown side of W-hading growth fault	Intraformational shales.
Samarang	1972	Stage IVD, IVE, IVF stacked reservoirs at 1400-2280 m. Shallow marine/deltaic sandstones, ϕ=17-32%, κ=100-1000mD	Rollover anticline on a NW-hading growth fault, with collapsed crest dissected by faults. Sealing by E-hading antithetic faults.	Fault seals
South Furious	1974	Stage IVA lower coastal plain, distributary and tidal channel sands at 540 m. ϕ=20-30%, κ=100-500 mD.	Elongate E-W upthrust anticline dissected by E-W high-angle reverse faults and N-S, E-W normal faults.	Intraformational shales.
St. Joseph	1975	Stage IVC (minor IVA) shallow marine sands at 610-1220 m, 60 m net oil sand, ϕ=18-25%, κ=200-300 mD.	Fault-dip closure in steep-sided anticline controlled by deep-seated wrench faults.	Intraformational shales.
SW Emerald	1980	Stage IVD coastal sandstones at 1066 m. 17 m NGS, 20 m NOS. ϕ=28-34%	Combined structral and stratigraphic trap: NE-trending growth-faulted anticline formed by Late Pliocene deformation. Oil and gas trapped beneath Horizon III unconformity.	Stage IVE shales above Horizon III unconformity.
Tembungo	1971	Stage IVC, IVD marine turbidites, 3-18 m thick. ϕ=18-25%, κ=200-500 mD.	ENE-trending anticline cut by NNW-trending normal faults.	Fault seals.
Tiga Papan	1982	Tiga Papan-1 well encountered 1 gas and 3 oil zones in Stage IVB/C shallow marine mixed carbonate-siliciclastic sediments at 1520 m. Porosity about 10%.	NW-trending rollover anticline bounded by SW-hading growth fault.	Stage IVD shales above SRU.

Table 22.1. Summary of fields in the offshore Sabah Basin.

- Middle Miocene structures affected by Late Miocene (SRU) and/or Early Pliocene (Horizon II) deformations.
- Late Miocene/Early Pliocene structures affected by Late Pliocene (Horizon I) deformation.

The Middle Miocene structures (containing Stage IVA reservoirs) occur mainly in the Southern Inboard Belt and were severely deformed during the Late Miocene (SRU) deformation. Relatively small accumulations in this area probably resulted from the loss of hydrocarbons due to the destruction of earlier-formed traps. In the Northern Inboard Belt, the SRU deformation appears to have been less severe. Hence, larger accumulations are preserved. The Northern Inboard Belt, together with the Northern Sabah Province, appear to have been affected by the strong Early Pliocene deformation (Horizon II). In the St Joseph Field, this deformation may have caused the leakage of hydrocarbons from the Stage IVA reservoirs into the overlying Stage IVC reservoirs. The Late Miocene/Early Pliocene structures, containing mainly Stages IVC/IVD reservoirs, are mostly related to extensional tectonics and growth faulting in the Outboard Belt (Erb West, Tembungo) and East Baram Delta (Samarang, Kinabalu, Nosong). They were variably affected by the Late Pliocene deformation (Horizon I). Shale diapirism has possibly contributed to the structural modification of some of these structures.

Besides the structural traps, stratigraphic traps are also recognised on seismic. These traps may be formed by Stage IVA sands onlapping on to the Deep Regional Unconformity (DRU) and, especially in the Northern Inboard Belt, by Stage IVC sands truncated by slump scars. Similar traps of younger turbidite sands against shale-filled slump scars may also be present in the distal areas of the East Baram Delta. Stratigraphic traps formed by Stages IVC/IVD and/or older turbidite sands in the Labuan-Paisley and South Furious synclines are potential plays.

The hydrocarbon reservoirs in the Sabah Basin are predominantly siliciclastic. Good quality reservoirs are formed by coastal to fluviomarine and stacked shallow-marine sandstones in Stage IVA and Stages IVC/IVD. Reservoir thickness can reach up to 300 m in places. In general, lower coastal plain facies have better quality (higher permeability) and tend to form thicker reservoir than the fluviomarine facies. Reservoirs of fluviomarine facies tend to have better lateral continuity (Scherer, 1980). In addition to fluviomarine and shelf deposits, deep-marine turbidites in Stages IVC/IVD also form thick, high-quality reservoirs. Carbonate reservoirs, although a minor component, have fair to excellent reservoir qualities. These carbonate mounds and reefs occur on the Kudat Platform in the Northern Sabah Province. They have been assigned to Stage IVC and to Stages IVE-IVG (Lower and Upper Limestone Units). In the Tiga Papan oil discovery, Northern Inboard Belt, mixed carbonate-siliciclastic reservoirs (most probably Stage IVB) are thought to be equivalent to the onshore South Banggi Formation. A detailed diagenetic study of these rocks was carried out by Mohammad Yamin Ali (1992, 1995).

The following section describes the hydrocarbon plays and trap styles in the different tectonostratigraphic provinces.

East Baram Delta

The East Baram Delta Province is one of the most explored areas in the Sabah Basin, apart from the Inboard and Outboard belts. There are 2 producing giant oil fields, Samarang and Kinabalu, which also contain considerable volumes of gas. Other discoveries in the province include Nosong (oil and gas), Samarang Kecil (gas), and Padas (gas). More details of the Samarang Field are given in Scherer (1980) and Lee (1993).

The EBD is characterised by Middle Miocene and younger Stages IVC to IVG deltaic, shelf, and slope sediments deposited in an overall progradational system. Seismic facies analysis is the key to mapping the distribution of reservoirs in this depositional setting. The progradational nature of the sediments means that potential reservoirs are mainly to be found in coastal fluviomarine to open marine sands of the delta-top or topset facies. These sand-prone deposits are identified on seismic as the topsets of sigmoidal reflection packages, commonly forming traps associated with growth faults. Figure 22.17 shows some of the structural and stratigraphic trap styles that have been identified through seismic mapping in the East Baram Delta.

The topset facies have been pursued vigorously, e.g. in the wells drilled within the proximal zone of the province. There is also potential development of deep marine turbidites in the toe-set facies in the deeper, distal portion of the delta/shelf. Here, the main targets are deep marine turbiditic sands in stratigraphic traps, dip-fault closures, and rollover anticlinal traps. The hydrocarbon charge is thought to be from source rocks rich in landplant-derived organic matter. Both coastal plain and deep-marine shales are potential source rocks.

Major growth fault systems associated with deltaic progradation have resulted in several NE-trending megastructural trends, made up of a series of hanging wall rollover anticlines. The Champion-Padas Trend contains the larger and more prospective structures in the proximal part of the province. It includes the giant Champion (Brunei, cf. James, 1984; Sandal, 1996), Samarang, and Kinabalu fields. The reservoirs are Stages IVC to IVE coastal to fluviomarine sands in growth-fault related traps, which have been affected by the Late Pliocene deformation (Horizon I).

The Samarang Field (Fig. 22.18A) was described briefly by Scherer (1980). The field was discovered by Shell in 1972, but has been operated by Carigali since 1995. The Samarang structure is a highly faulted, NNE-trending, rollover anticline bounded to the east and southeast by several major west-hading growth faults, and is cut by NNW-trending antithetic and synthetic normal faults. Production of the field started in 1975, with a peak in the production rate in 1979 at 73,000 BOPD. The present production rate is 10,000 BOPD. The reservoirs are shallow marine subtidal sheet sandstones above the SRU, in the Upper Miocene Stages IVD to IVF, and 90% of these are in the depth range

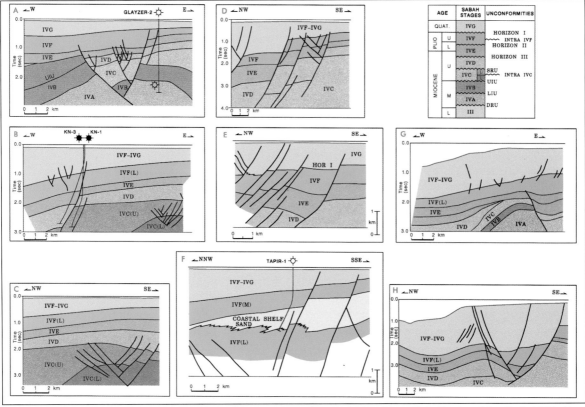

Fig. 22.17. Some examples of the structural styles and trapping mechanisms in the East Baram Delta Province (after Shell, 1993). Most structures are associated with deltaic growth faults and related deformation. See Fig. 22.16 for location of the line drawings. (A) Possible traps formed by normal faults in the Trusmadi North area, basinward of the Glayzer gas discovery (Southern Inboard Belt). Traps are formed by tilted Stages IVA and IVC sands beneath the SRU. (B) The Kinabalu Field occurs in fault-dip closures in both the hanging wall and footwall of the main Kinabalu Fault. (C) Complex flower structures, with potential closures in the Stage IVC level. (D) Fault-dip closed structures in the hanging wall of growth faults. (E) Tilted Stage IVD strata between two growth faults. (F) Combined stratigraphic-structural traps, as drilled at Tapir-1, where topset sands abut against a major growth fault. (G) Shelf-edge play, which has a closure at Stages IVA-IVC levels beneath the Intra-IVE (Horizon III) Unconformity. A tilted flat event is seen on seismic. (H) Another shelf-edge play formed by structural closure at Intra-IVE and older stratigraphic level.

between 1370 and 2285 m subsea. The oil gravity is about 30° API. The porosity and permeability of the reservoirs are 20-35% and 600-2000 mD, respectively.

The Kinabalu Field is located slightly northwestwards, somewhat off the Champion-Padas Trend. Figure 22.17B shows a cross section of the Kinabalu structure, which is a NNE-trending rollover anticline developed on a NNE-trending growth fault. The main phase of growth on this fault appears to be during Stage IVD times, becoming negligible by late Stage IVF times. The reservoirs occur in Stages IVD to IVE, and comprise stacked upward-coarsening deltaic to shallow-marine sandbodies. Unlike the Samarang structure, the Kinabalu structure is less faulted. The discovery well Kinabalu-1 was drilled in 1988 to test the flank of the structure, and encountered hydrocarbons in Stages IVD, IVE, and IVF sands between about 1500 and 4000 m subsea. Three intervals were tested, which flowed at a combined rate of about 16,000 BOPD.

The Kinabalu discovery proves the existence of a play fairway between Kinabalu and Samarang, in which there are several other prospects of similar structural/stratigraphic configuration (e.g. Madalon and Ambun). These structures are favourably located on the flank of a major kitchen source area to the north. Many wells have been drilled on some of these structures, and some encountered minor oil and gas, e.g., Timbalai Deep West-1. The Samarang Kecil wells found mostly gas. The Padas structure, at the distal end of this trend, has also been tested unsuccessfully because of poor fault seal and a high sand content in the Stages IVD and IVC.

Basinwards from the Champion-Padas Trend is the Nosong-Tapir Trend, which contains the minor Nosong gas discovery. This structure is also a hanging wall rollover anticline on the northwestern side of the Nosong Fault. Hydrocarbons were detected from seismic amplitude anomalies in Stage IVC sediments. Along the same trend are the Tapir and Papar structures, which were tested by Tapir-1 and Papar-1 wells, respectively. In these structures, minor amounts of gas were found in the topset beds within the hanging wall

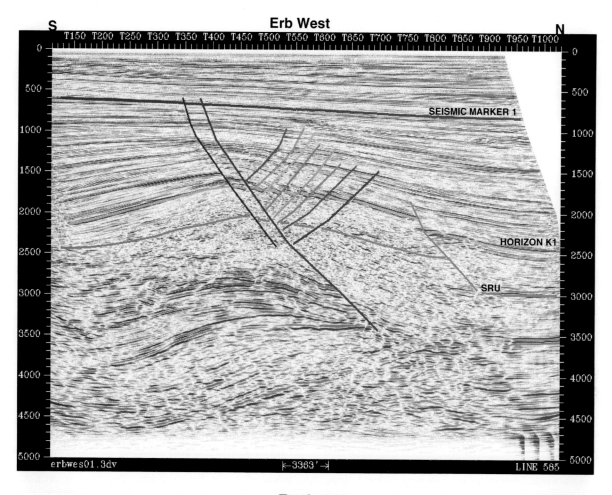
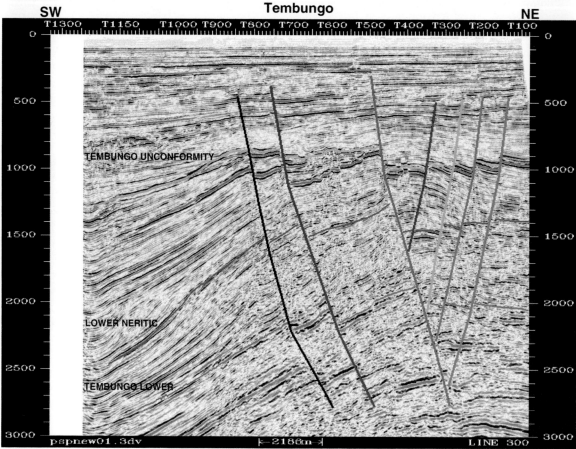

Fig. 22.18. Examples of seismic section across the (top) Erb West and (bottom) Tembungo fields, vertical scale in milliseconds.

fault closures associated with major synthetic growth faults.

Further basinwards, there is the deepwater trend of Stage IVF and older turbiditic sands. Here, faulted and anticlinal traps occur in the footwall of the counter-regional (antithetic) growth faults, which are probably associated with shale diapirs or delta-toe thrusts. This play concept has not been tested here in Sabah, but in Brunei has been proven to be charged with hydrocarbons. Work on the turbidite depositional systems is currently on-going to understand better reservoir distribution and quality, such as in the Stage IVF Upper Lingan Fan Unit, as described by Salahuddin Saleh Karimi et al. (1997).

Inboard Belt

There are many oil and gas occurrences in the Southern Inboard Belt, including its onshore part. Many of the wells drilled onshore in Labuan and on the Klias Peninsula have oil and/or gas shows (Fig. 22.9). Except for Ketam, however, there are no commercial discoveries. More details of the discoveries are given in Ahmad Said (1982) and Wong (1996). Besides the formerly producing Ketam Field, the other oil discoveries are Lokan, SE Collins, and Erb South, while the only major gas discovery is Glayzer. With the exception of the Erb South discovery which is in Stage IVD shallow marine sands, most of the reservoirs in this part of the province are in Stage IVA. They occur in very steeply dipping, complexly faulted anticlines, many of which have a core of diapiric shale. The Ketam structure, for example is an upthrust anticline bounded by a series of reverse faults. The reservoirs occur in the Stage IVA interval in the footwall of the thrust fault (Ahmad Said, 1982).

On the northwestern flank of the Labuan-Paisley Syncline, there is a major ENE-trending wrench-induced anticlinal feature extending from Hankin to Saracen (Fig. 22.9). The main prospective structures include Kawag and Lakutan. At Hankin, fault-related closures are developed due to crestal collapse of the rollover structures. The Hankin-Saracen trend is atypical of other wrench features of the Inboard Belt in that there has been little uplift and erosion during the Late Miocene (SRU) (Shell, 1993). As a result, Stage IVC is still preserved and may form an exploration objective in this area. In other areas Stage IVA commonly subcrops beneath the SRU at shallower depths due to intense deformation, and hence forms the main reservoir targets. In contrast, Stage IVA is probably not prospective in the Hankin-Saracen area. The Late Miocene (SRU) deformation was severe in this province and has contributed to the uplift and erosion of the reservoirs, and probably resulted in secondary migration and leakage of hydrocarbons.

In addition to Stage IVA, minor amounts of oil have been found in Stage III at SE Collins and Brunei Bay-1 well. At SE Collins-1, where Stage III sediments have been uplifted to shallower depths, a 3 m net oil sand was encountered in Stage III at a depth of about 1770 m. The Brunei Bay-1 well was drilled on an upthrust anticline where, in addition to a 8 m NGS in Stages IVE/IVF between 240-990 m depth, a 2 m net gas sand in Stage III at a depth of about 1220 m was also found.

There are currently (1999) 3 producing oil fields in the Northern Inboard Belt: South Furious, St. Joseph, and Barton. All the fields are operated by Shell. The main reservoirs are the sands of Stage IVA, except at St. Joseph where the main reservoirs are in Stage IVC, occurring in complex faulted anticlines. Some details on these fields are given in Bol and van Hoorn (1980), Guest and Deckers (1982), and Johnson et al. (1989). The South Furious Field is located at the northern tip of the Pritchard Syncline, and is on trend with the Barton Field. The field is an E-W trending anticline bounded by a major thrust fault to the south. Discovered by Shell in 1974, the South Furious Field came on stream in 1979. It had a peak production rate of 9000 BOPD in 1981-1982, but is currently producing at about 6000 BOPD. The South Furious crude is classified as light (24-36° API) and waxy (5-23 wt% wax).

The Barton Field is structurally similar to the South Furious Field, an E-W trending asymmetrical anticline bounded to the south by a major thrust fault. The reservoirs are Stage IVA lower coastal plain/deltaic sandstones. The seals are formed by interbedded lagoonal/estuarine shales. The field was discovered by Shell in 1976 but came on stream only in 1982. Peak oil production was at 15,000 BOPD in 1989. The current rate is about 7000-8000 BOPD using gas lift.

The St Joseph Field (Fig. 22.18B) has been described by Johnson et al. (1989). It is one of the giant oil fields in Sabah. The field was discovered by Shell in 1975. It is a structurally and stratigraphically complex field formed by early Pliocene wrench-fault related deformation. It is situated along the Bunbury-St Joseph-Bambazon trend, which is a major NE-trending anticlinal structure bounded to the south by a major thrust fault. This fault is believed to have had significant left-lateral movement, and separates the Bunbury Syncline in the south from the Pritchard Syncline to the north. The latter is thought to be the main kitchen area that provided the hydrocarbons to the St Joseph Field and other accumulations nearby. The main hydrocarbon-bearing zones are in Stage IVC, which directly overlies Stage IVA at the Upper Intermediate Unconformity (UIU) as a result of submarine erosion and slumping at the Late Miocene shelf-edge (Levell and Kasumajaya, 1985). The structures were affected by the Late Miocene (SRU) deformational event which resulted in secondary migration of hydrocarbons, probably from Stage IVA. The reservoirs are shallow marine storm-wave influenced environment, with slight fluviomarine influence (Johnson et al., 1989). The reservoirs are part of a prograding shelf-slope system that built out over the tectonically active shelf-margin.

The Tiga Papan oil accumulation, in the northern end of the Inboard Belt, was discovered by Carigali-BP in 1982. The structure is a hanging wall rollover anticline associated with a major growth fault system. The discovery well, Tiga Papan-1, encountered oil and gas at about 1600 m depth, in mixed carbonate-siliciclastic sediments, later known as the Tiga Papan Unit (Foo and McDonald, 1983). The Tiga Papan Unit consists of bioclast-rich, medium- to fine-grained calcareous sandstones and sandy limestones interbedded with calcareous siltstones. Abundant calcareous nannofossils and planktonic foraminifera in these rocks indicate an open shallow marine environment, probably in the middle to outer shelf. The sandstones exhibit a complex history of dolomitization and calcitization, including the occurrence of clay mineral-associated dolomite formed during burial diagenesis.

Outboard Belt

The Outboard Belt is a significant petroleum producing province with a variety of reservoir facies from shallow-marine sandstones to deep-marine turbidites. There are 2 producing oil fields in the Outboard Belt, Tembungo and Erb West, both of which are operated by Carigali. The Erb West Field also produces gas. More details of the fields, in particular Tembungo, are given in Whittle and Short (1978) and Ismail Che Mat Zin (1992), and Erb West in Johnson et al. (1985) and van der Harst (1991). Several other significant oil discoveries include SW Emerald and Rusa Timur, while significant gas discoveries include Kinarut, Kebabangan, and Kamunsu.

Hydrocarbon shows have also been reported from Gajah Hitam and West Emerald. The main types of traps in this province are extensional and growth-fault related faulted anticlinal traps which have been affected by the Late Pliocene deformation (associated with Horizon I). The main prospective reservoirs in the Outboard Belt are the Stage IVC and/or older deep-marine turbiditic sandstones, such as at Tembungo, Kinarut, and Kebabangan and the Stage IVD and younger coastal to shallow marine sandstones, as at Erb West, SW Emerald, and Rusa Timur.

The Erb West Field was discovered by Shell in 1971, but has been operated by Carigali since 1996. The field is in a NW-trending anticline cut by several E-W normal faults. Development of the field started in 1981, with peak production rate of about 34,000 BOPD in 1989. The current rate is about 10,000 BOPD. The reservoirs in the Erb West Field consist of Upper Miocene Stage IVD shallow marine storm/wave influenced transgressive shelf sands (Johnson et al., 1987). These sand-bodies onlap the SRU to the southeast against the Erb High, which also served as the sediment source. The field has a large gas cap above a 39 m-thick oil column. The reservoirs are at depth between 1829-2112 m. The porosity and permeability are 15-25% and 25-1000 mD, respectively.

The Tembungo Field, Sabah's first developed offshore field, was discovered by Esso in 1971, but has been operated by Carigali since 1986. Production started in 1974, peaking in 1977 at a rate of 14,000 BOPD. The current production rate is about 5680 BOPD. The Tembungo structure is a ENE-trending anticline cut by numerous N-S normal faults which divide the field into many compartments. The reservoirs, occurring at depths of 1045-2560 m, are turbidites deposited in palaeo-water depths of about 1000 m by slumping at the palaeo-shelf edge due to movement of the Bunbury-St Joseph Ridge during Stage IVC times. The oil gravity is about 37° API. Most of the reservoirs at Tembungo are overpressured by about 500 psia above hydrostatic pressure. The porosity ranges between 16 and 33%, while permeability ranges from 30 to 3000 mD.

Shallow marine sands and deepwater turbidites are the main reservoir objectives in the Outboard Belt. In the SB301 part of the Outboard Belt, Shell has identified several potential play types, including the Stages IVA, IVC, IVD, and IVF turbidites. Stages IVA to IVD turbidites have been tested at Kinarut, Kebabangan, and Tembungo, whereas the Stage IVF turbidites are still untested. The Lingan Fan complex (Marzuki Mohamad and Lobao, 1997) seems promising.

Stage IVC turbidites are proven reservoirs in the Outboard Belt, west of the Emerald Fault Zone, such as in the Tembungo Field. Stage IVB is mostly absent because of erosion, while Stage IVD proximal turbidites have been identified as potential reservoirs. Turbidites in the upper Stage IVC and lower Stage IVD form the reservoirs at Kinarut-1 and Kebabangan-1. At SW Emerald, Erb South, Rusa Timur, Gajah Hitam, and W Emerald, the main reservoirs are the upper Stage IVD shelf sandstones.

Figure 22.13 shows some of the trap styles in the Outboard Belt. The main trap types in this area include normal-fault bounded horst blocks, such as at Gajah Hitam, where closure is achieved by dip and fault seals. The Gajah Hitam structure is bounded to NW, SW, and NE by major normal faults, and is gently dipping to southeast. There are also structures formed by fault-dip closures associated with wrench faults (Fig. 22.13, Lines 1, 2, and 5). The Belangkas-1 well proved the presence of Stage IVC reservoirs. The Ungka structure is a 4-way dip-closed anticline dissected by major reverse faults, which appear to be splays from deep-seated basement wrench faults. Reservoir sandstones are expected in Stages IVC/IVD. The Rusa Timur-1 well was drilled by British Gas in 1990 to test the shallow marine sandstones of Stage IVD, equivalent to the SW Emerald sandstone unit. The structure is a dip/fault- closed structure between the south-dipping Belangkas and Emerald faults. The Tembungo sandstone unit was a secondary objective of the well. The source for hydrocarbons in the Outboard Belt is thought to be plant-derived organic material in Stage IVA offshore shales. There is also some indication of bathyal shales as a potential source in Rusa Timur-1. In the synclinal areas, Stages IVC and IVD may provide mature source rocks.

Northern Sabah Province

The occurrence of onshore oil seepages in the Sikuati Member of the Lower Miocene Kudat Formation is evidence for a viable petroleum system in the vicinity of the Kudat Peninsula. However, the Northern Sabah Province is still relatively poorly explored. Hydrocarbons have been found only at Titik Terang-1 in 1992 by Hall-Houston. The Titik Terang structure is a 4-way dip closure on the hanging wall of a major N-S trending growth fault that forms the eastern boundary of the Kindu-Mangayau Sub-basin. The prospective horizons were associated with bright spots on seismic. The Titik Terang-1 well was drilled to a total depth of 2750 m and found several 25-30 m thick gas-bearing shallow marine sandstones in the Upper Sandstone Unit of the Bangau Group (Stages IVC and IVD), at depths of between 950 and 2500 m subsea.

There are at least 2 potential plays in the Northern Sabah Province. One is the siliciclastic reservoirs in growth fault structures mainly in the Kindu-Mangayau sub-basin. The growth faults in the Northern Sabah Province appear to be low-angle and have a listric geometry (Fig. 22.15). Some of the structures have been tested at Mangayau-1, Kindu-1, Nangka-1, and Titik Terang-1. Kindu-1 well was drilled by Esso in 1972, and found an outer shelf sequence of claystones, sandstones, and a few limestone intervals. Kudat-1 was drilled in 1981 on the footwall of the major growth fault and appears to have been a poorly placed well. Mangayau-1 was drilled in 1985 on a rollover structure but failed to find hydrocarbons.

Another potential play in the Northern Sabah Province is the carbonate reef limestone play on the Kudat Platform. Two carbonate reef prospects have been tested: Esso's Kalutan-1 (1972) and Carigali-BP's Balambangan-1 (1981). Both wells were dry. The Kalutan-1 well was spudded in a modern reef that extends down to a depth of about 760 m. The primary drilling objective was the Upper Miocene reef, which was encountered at about 960 m down to 1670 m depth. The limestone has a porosity of about 15 to 28%, but no hydrocarbons were found. The Kalutan structure is similar to Balambangan, and was equally unsuccessful.

HYDROCARBON CHARACTERISTICS

Crude oils in the Sabah Basin are generally light, slightly waxy, and have low sulphur contents (Table 22.2). API gravity ranges between 25° to 50°, and averages 35° for normal, non-degraded oils, and 25° for degraded oils. The variation in wax content and API gravity may be attributed to the amount of landplant-derived components in the source rock and the extent of biodegradation of the oils. The Sikuati oil seep is a low-sulphur, low-asphaltene, and high-aromatic crude. The very heavy (low) API gravity (13.9°) probably resulted from the biodegradation of a light condensate due to subaerial chemical and biological weathering.

Variations in the wax content are thought to be related to the depositional environment of

Sample	Gravity (°API)	Wax (wt%)	Sulphur (wt%)	$\delta^{13}C_{oil}$	$\delta^{13}C_{sat}$	$\delta^{13}C_{aro}$
Bambazon	35	0.7	0.30	-28.0	-28.5	-26.9
Barton	32	0.7	0.36	-27.9	-28.9	-26.8
Bongawan	32	0.4	0.07	-27.7	-28.5	-26.8
Dampier	36	0.3	0.10	-27.8	-28.4	-26.3
Erb South	20	-	0.20	-26.7	-26.8	-26.4
Erb West	28	0.6	0.30	-27.5	-28.0	-26.4
Glayzer	31	0.1	0.10	-28.0	-28.4	-26.8
Kebabangan	29	-	0.10	-27.0	-27.6	-26.5
Ketam	37	0.4	0.05	-27.2	-27.8	-26.3
Kinabalu	37	0.6	0.10	-	-28.7	-29.9
Nosong	47	-	0.10	-	-27.6	-26.6
Lokan	33	0.3	0.30	-27.60	-28.00	-26.60
Pondu	24	-	0.20	-27.5	-28.1	-26.4
Rusa Timur	32	1.0	0.10	-	-28.7	-26.5
Samarang	34	0.4	0.20	-27.6	-28.1	-26.8
Sikuati	14	Low	0.50	-27.1	-	-
South Furious	31	0.8	0.10	-27.6	-28.0	-26.6
St. Joseph	34	0.8	0.20	-28.0	-28.9	-26.7
SE Collins	34	0.4	0.10	-27.9	-28.4	-26.7
Tembungo	40.0	0.3	0.30	-27.4	-27.9	-26.6
Tiga Papan	26-36	4-10	0.10	-27.6	-	-
Titik Terang	50	-	0.05	-	-	-
W. Emerald	20	-	0.2	-26.7	-26.7	-26.4

Table 22.2. Bulk properties of hydrocarbons recovered from the offshore Sabah Basin.

the source rock. Generally, deltaic source rocks generate oils with higher wax contents compared to source rocks deposited in marine shelf/slope environments. It is observed that most of the slightly waxy oils are found in the northeastern part of the basin, while some occur in the southwest. Low-wax Stage IVA accumulations at Ketam, SE Collins, Lokan, and Dampier are all underlain by Stage III turbiditic sediments. The oils and condensates commonly have a low sulphur content (Table 22.2), which suggests that the source rocks are most likely non-marine, coastal plain deposits.

In general, stable carbon isotope ratios of whole oils and saturated hydrocarbons become progressively heavier with decreasing amounts of saturated hydrocarbons, while the carbon isotope ratios of aromatics remain relatively constant (Fig. 22.19). As such, the aromatics are probably more reliable for correlation, because they are not affected by gross composition. The narrow band of stable carbon isotope values for the aromatics indicates that they were generated from very similar types of organic matter, i.e. from land plants.

SOURCE ROCKS

Hydrocarbons in the Sabah Basin are essentially very similar in composition and have originated from source rocks rich in mainly terrigenous organic matter (Scherer 1980; Awang Sapawi Awang Jamil et al., 1991; Abdul Jalil Muhamad and Mohd Jamaal Hoesni, 1992; Azlina Anuar, 1994; Azlina Anuar and Abdul Jalil Muhamad, 1997). The source rocks are most likely within the post-DRU (Stage IV) sequences, as the pre-DRU (basement) deep marine shales are generally lean and thermally over-mature. Widespread erosion of the NW Sabah margin during the early Middle Miocene, and the extensive outbuilding of the Stage IV siliciclastic wedge, resulted in the deposition of source beds that are rich in terrigenous organic matter, interbedded with sand-prone reservoir facies.

The palaeogeographic controls on the distribution and preservation and the quality and maturity of source rocks are discussed, based on a compilation of data from onshore and offshore areas of the Sabah Basin. Numerous studies on both onshore and offshore samples have identified the source rock horizons in the Sabah

Basin. The major source lithology is lower coastal plain shales containing disseminated organic matter derived mainly from land plants, with coals occurring only as transported fragments. These studies concluded invariably that the majority of potential source rock samples have poor to good organic matter contents (total organic carbon, TOC < 2 wt%) with correspondingly poor to fair hydrocarbon generating potentials (S_2 < 5 mgHC/gTOC). Good quality source rocks (S_2 > 5 mgHC/g TOC) are rare. They are commonly associated with coaly shales.

Palaeogeographic Controls

The deposition and preservation of the source beds occurred in 3 types of palaeoenvironment: (i) coastal/lower coastal plain environments, which include interchannel peat swamps, back-mangrove swamps, lagoons, and interdistributary areas, (ii) fluviomarine environments, and (iii) continental shelf to deep marine environments (Azlina Anuar and Abdul Jalil Muhamad, 1997). A hot, humid and ever-wet tropical climate throughout the Neogene, rather similar to the present day, ensured a high terrigenous productivity during the deposition of the Middle to Upper Miocene source-rock sequences. Based on the present-day vegetation, the type of organic matter preserved would be dominated by mangrove and Nypa palm products, along with those transported by rivers draining the hinterland. Since the spores and pollen preserved in the sediments are comparable to modern genera, reasonable deductions on their palaeo-ecology can be made.

Favourable conditions for source bed deposition can occur in different environments and are not limited to specific time period. Land plant material is more likely to accumulate in swampy lower coastal plains and delta tops, in front of large estuaries/river mouths, and in deep-marine turbidite and submarine fan deposits. In these environments, high organic matter productivity during periods of rapid sedimentation (300-1000 m/Ma, Noor Azim Ibrahim, 1994) results in sub-oxic conditions that inhibit the degradation of organic matter. Furthermore, the presence of tannin in the waters of the mangrove, back-mangrove, and peat swamps and lagoons creates bacteria-free water columns that effectively impede organic matter degradation by bacteria. The successful preservation of source rocks in the Sabah Basin is the result of the high input of terrigenous organic matter and high sedimentation rates, and seemingly not due to the establishment of anoxia (Azlina Anuar, 1994; Azlina Anuar and Abdul Jalil Muhamad, 1997). Coaly and carbonaceous shales are the most prolific source rocks because they occur in large volumes (>2000 m thick in some areas) and are closely interbedded with sandstones which act as migration conduits for hydrocarbon migration once the saturation threshold is reached. This efficiency in expulsion allows liquid hydrocarbons to leave the source rocks, as opposed to being retained and converted to gas.

Onshore Source Rock Studies

Detailed geochemical studies on the source rock potential of onshore rock formations were carried out by Azlina Anuar and Abdul Jalil Muhamad (1995, 1996a, 1996b). Outcrop samples were collected along the northwestern Sabah coastline, from the Balambangan and Banggi islands in the northeast to the Klias Peninsula and Limbang areas in the southwest. Shales from all the major formations were analysed for organic matter content, thermal maturity, and hydrocarbon generating potential. The results are summarised in Tables 22.3 and 22.4.

In northern Sabah, the shales analysed are from the Timohing, Bongaya, South Banggi, Kudat and Crocker formations. In the south, samples were taken from the Miri, Belait, Setap Shale, Meligan and Temburong formations. The majority of samples, regardless of formation, contain essentially terrigenous organic matter and have poor to fair hydrocarbon-generating potentials. Furthermore, samples of the Eocene and Oligocene Crocker, Temburong, and Kudat formations (equivalent to Stages II and III) are generally mature to late mature. There are, however, coaly/carbonaceous shale, coal, and resin in some formations, e.g. Kudat, Temburong, and Bongaya, which show fair to excellent source rock qualities. Some samples from the Sikuati and Tajau Sandstone members of the Kudat Formation, for example, contain abundant carbonaceous material and laminae. Generally, coaly and carbonaceous materials are more common in the Sikuati Member than in the Tajau Sandstone Member. TOC contents range from poor (0.28-0.56 wt%) in the Tajau Sandstone Member to fair (1.33 wt%) in the Sikuati Member. A coaly shale sample from the Sikuati Member contains up to 27% organic carbon. Kerogen studies indicate the presence of Type III organic matter with little or no marine influence. The Sikuati Member sample comprising coaly

Age	Formation	Offshore equivalent	Organic Geochemical Evaluation Results : Summary
Pliocene	Timohing	Stage IVF	Sample from Balambangan Island. TOC 0.16wt%, S_2 < 0.2mg/g; non-source. Immature.
Miocene	Bongaya	Stage IVD	Samples from the southern tips of the Banggi and Balambangan islands, Berungus and Pitas (West Bengkoka). The shales all show non to poor source potential (TOC 0.33-2.35 wt%, S_2 < 0.8 mg/g). High TOC samples, however, may have had higher S_2 values; their unweathered counterparts may possess fair hydrocarbon generating potentials. Coal samples from West Bengkoka show good source rock potentials (TOC 33-49wt%, S_2 32-78mg/g). Immature
	South Banggi	Stage IVA	Samples from the southern tip of the Banggi Island show non-source characteristics (TOC 0.34wt%, S_2 < 0.10mg/g). Immature.
Oligocene	Kudat	Stage III	Samples from the eastern parts of the Bengkoka Peninsula and on the Kudat Peninsula. The shales have poor to fair source rock potentials (TOC 0.33-1.98wt%, S_2 < 2.0mg/g); however, the majority are poor with S_2< 1.0mg/g. Coals from Kudat and Bengkoka peninsulas and coaly shale from Bengkoka Peninsula show fair to very good hydrocarbon generating potentials (TOC 4-58wt%, S_2 16-29mg/g). Resin nodules from West Bengkoka show excellent source characteristics with TOC values of 85wt% and S_2 values as high as 650mg/g. These sediments are immature to early mature.

Table 22.3. Summary of results of the organic goechemical analyses of outcrop samples from the Balambangan and Banggi islands, and Kudat and Bengkoka peninsulas, northern Sabah.

Age	Formation	Offshore equivalent	Organic Geochemical Evaluation Results : Summary
Pliocene	Miri	Stage IVF	Samples from Jerudong, Brunei, show non-source characteristics (TOC 0.49-1.25wt%, S_2 < 0.4mg/g). Immature.
Miocene	Belait	Stage IVD	Shale samples from the Klias Peninsula, Limbang and Brunei have poor source-rock potential (TOC 0.53-5.27wt%, S_2 < 1.3mg/g), except 1 sample from Limbang which has fair hydrocarbon-generating potential (TOC 1.93wt%, S_2 2.84mg/g). Coals from this formation generally have good to very good source potential (TOC 26-68wt%, S_2 9-136mg/g); the best was from Limbang. Immature to early mature.
	Setap Shale	Stage IVA	Samples from the Klias Peninsula and Brunei show negligible source-rock potential (TOC 0.55-0.85wt%, S_2 < 0.6mg/g). Early mature - mature.
	Meligan		Samples from Sipitang and Merapok-Lawas (SW of Sipitang) areas. The Sipitang shales have poor to fair hydrocarbon-generating potential (TOC 0.23-3.2wt%, S_2 < 2mg/g with the majority being < 0.5mg/g). Coaly shales and carbonaceous siltstones show good source potentials (TOC 2.8-6wt%, S_2 5-11mg/g). The Merapok-Lawas shales show non-source potentials (TOC 0.23-0.5wt%, S_2 0.1-0.5mg/g). Early mature to mature.
Oligocene	Temburong	Stage III	Samples from the Klias Peninsula, Trusan-Lawas and Brunei. All the samples have none to poor hydrocarbon-source potentials (TOC 0.31-1.87wt%, S_2 < 2mg/g with a majority having values < 0.7) except for a carbonaceous shale sample collected from the Klias Peninsula (TOC 0.64-4.3wt%, S_2 0.7-9mg/g). Mature.
Eocene	Crocker	Stage II	Samples from Kg. Melinsung are mature to late mature but show non-source potential with TOC 0.41-1.64wt% and S_2 < 0.1mg/g.

Table 22.4. Summary of results of the organic geochemical analyses of outcrop samples collected from the Sipitang-Klias-Limbang areas, southwestern Sabah.

material with thin sand horizons is rated as having an excellent potential for both oil and gas, while the Tajau Sandstone Member has moderate to good gas potential. The South Banggi and Timohing formations lack potential source rocks and are thermally immature. Only some of the Kudat Formation samples are immature to early mature.

In the southern onshore area, coals and shales from the Belait Formation (time equivalent of the Bongaya Formation in north Sabah and to Stage IVD offshore) in the Limbang area have fair to very good source rock potential (Table 22.4). Shales of the Meligan Formation in Sipitang have poor to fair hydrocarbon generating potential, but are non-source in the Merapok-Lawas area. There are, however, coaly shales and carbonaceous siltstones in the Meligan Formation that show good source rock potential. The Miri and Setap Shale formations appear to have negligible potential for hydrocarbon generation. The samples exhibit a range of thermal maturity. The Belait is immature to early mature, while the Setap Shale and Meligan formations are in the early mature to mature stages.

Offshore Source Rock Quality

Figure 22.20 summarises the source rock potential of 3 offshore provinces in the Sabah Basin: (A) Southern Inboard Belt (data from Glayzer, Hankin, Ketam), (B) East Baram Delta (Kinabalu, Samarang, Bongawan, Papar), (C) Northern Inboard Belt (Binturong, St. Joseph, Mengkira, South Furious, Barton), and (D) Outboard Belt (Tembungo, Kinarut, Malutut, Gajah Hitam, Rusa Timur, SW Emerald). The figure shows that fair to good quality source rocks of varying ages are present in some areas. Coaly shales of Stages IVC and IVD penetrated in the

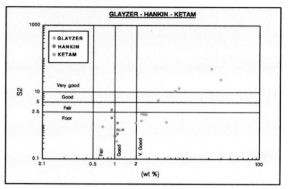

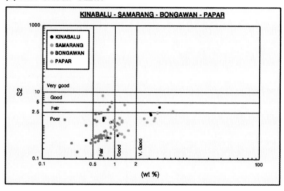

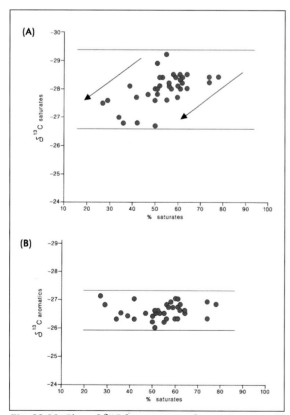

Fig. 22.19 Plots of $\delta^{13}C$ for saturates and aromatics versus % saturates. (A) Stable carbon isotopic values of the saturate fractions become heavier (more positive) with decreasing % saturate. This is not observed in the aromatics (B), which makes them more reliable for correlation. The narrow band of the aromatics also tends to suggest that they had a similar origin; in this case, from terrigenous organic matter.

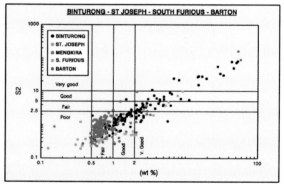

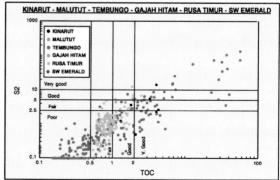

Fig. 22.20. Source-rock potential summary diagrams (S_2 vs TOC) for the (A) Southern Inboard Belt, (B) East Baram Delta, (C) Northern Inboard Belt, and (D) Outboard Belt.

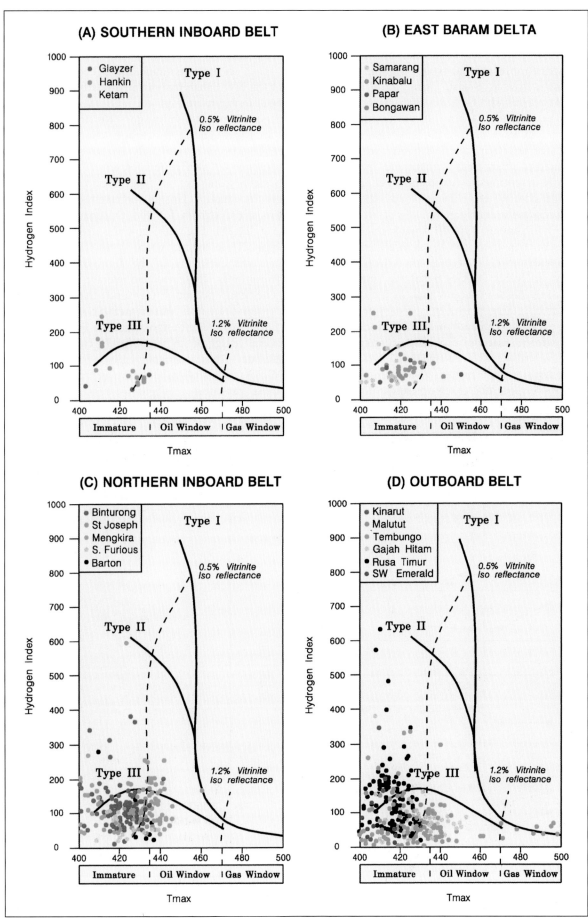

Fig. 22.21. Hydrogen index vs pyrolysis Tmax cross-plots for the (A) Southern Inboard Belt, (B) East Baram Delta, (C) Northern Inboard Belt, and (D) Outboard Belt.

Binturong area, Northern Inboard Belt, show the best hydrocarbon generating potential. In contrast, Stage IVD carbonaceous shales at Mengkira represent fair to good quality source rocks, while at South Furious, Stage IVA carbonaceous shales have the desired source rock properties. A similar variation is also observed in other parts of the Sabah Basin, which indicates that source rock occurrence is not restricted to a particular stratigraphic interval, but is partly determined by depositional environment and the selective enrichment of liptinitic organic matter during sedimentation.

In general, Type III organic matter is the main constituent of the kerogen assemblage in the Sabah Basin, as seen in the corresponding Hydrogen Index (HI) vs pyrolysis Tmax cross-plots (Fig. 22.21). Maceral analysis indicates that vitrinite is the primary constituent while structureless organic matter is secondary. Structureless material probably originated from mechanically and chemically degraded terrigenous organic matter. Varying quantities of liptinitic organic matter, comprising cutinites and resinites, are also present. These may enhance the hydrocarbon-generating potential of some of the sediments.

In the outer shelf and deeper water areas of the NW Sabah margin, hydrocarbons have also been encountered in turbiditic sequences, e.g. Rusa Timur, Tembungo, Kebabangan, and Kamunsu. The hydrocarbon charge for these reservoirs is most likely to have come from transported terrigenous organic matter, with possibly some minor contribution from marine organic matter. This is indicated by the HI vs Tmax cross-plots for the Outboard Belt wells (Fig. 22.21). The data show the predominance of Type III kerogen, with Types II/III kerogen mixtures occurring in the Tembungo and Rusa Timur areas. The calculated average hydrocarbon-generating potential (S_2) and the average HI increase in the outer neritic and bathyal sediments at Tembungo and Rusa Timur, with an overall decrease in the organic carbon content (Azlina Anuar et al., 1993; Fig. 22.22). Microscopic analysis of samples from these two areas revealed a slight increase in exinitic

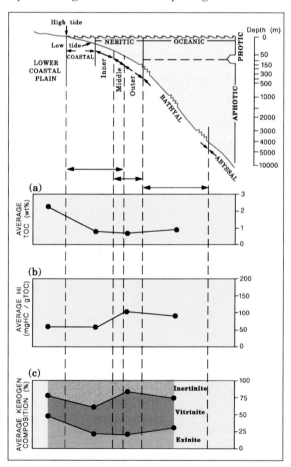

Fig. 22.22. Lateral variation across the continental margin of the average (a) total organic carbon content, (b) hydrogen index (HI), and (c) kerogen composition. The apparent increase in quality of the deep marine sediments despite having a low corresponding total organic carbon content is attributed to the selective segregation and transportation of light, hydrogen-rich, land-derived organic matter (from Azlina Anuar et al., 1993).

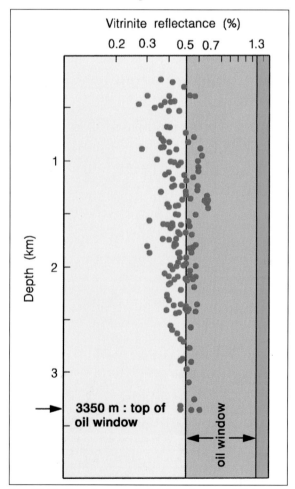

Fig. 22.23. Plot of measured vitrinite reflectance (VRo) versus depth for offshore Sabah Basin. The data show that the sediments penetrated by wells in the basin are immature. The slope of the data set indicates that early hydrocarbon generation may occur at approximately 3350m depth (where VRo = 0.5%).

material in the outer neritic and bathyal sediments. This may explain the increase in HI values despite the lower organic carbon contents. Probably in most of the deeper marine hydrocarbon plays, comprising basin slope and basin floor fans, the hydrocarbon source would still be land-derived organic matter despite being in a marine environment, with selectively segregated hydrogen-rich liptinitic materials being the main hydrocarbon-generating components.

In the offshore Northern Sabah Province (off Kudat Peninsula) the effective hydrocarbon source rocks have not been penetrated, although the presence of oil proves their existence. Several coal horizons in nearby Tiga Papan within the Kalutan Group, and the underlying Upper Sandstone Unit of the Bangau Group, suggest that there were periods of high-organic matter deposition and preservation. Kerogen typing revealed the presence of mainly vitrinite and amorphous vascular plant material, with some resin particles and inertinites. Dinoflagellate cysts, acritarchs, and foraminiferal linings are present only in trace amounts. The depositional environment for these organic matter is likely to be a nearshore/paralic setting with a strong terrigenous influence. Coals in the Kalutan Group equivalent and Upper Sandstone Unit show excellent pyrolysis yields of 47–123 mgHC/gRock, but pyrolysis-gas chromatograph results indicate that they are gas prone with minor condensate or light oil potential.

Maturity Modelling

Figure 22.23 shows a plot of vitrinite reflectance (VRo) against depth from wells in the offshore Sabah Basin. The VRo measurements were taken from coal and isolated kerogen samples. The general VRo-depth trend suggests that the top of the oil window (assuming an equivalent VRo of 0.5%) is at about 3350 m. This indicates that the sediments penetrated by the offshore wells are thermally immature and, at their location and burial depths, are not likely to have been the effective source rocks for the discovered hydrocarbon accumulations. The same sediments, however, may be mature if buried to sufficient depths, for example in the kitchen areas, such as the Pritchard and Furious synclines (Fig. 22.24). One-dimensional (1D) thermal maturity modelling of the Southern Inboard Belt and the East Baram Delta indicates that the top of the oil window is shallower at Malutut (2600 m) and Kinabalu (3500 m) than in the Pritchard, Furious, and Rusa Timur synclines (4000, 4100, 4480 m respectively). Assuming that temperature gradients are the same, this implies that Stages IVC and IVD sediments in the south were more deeply buried than those in the north, and were

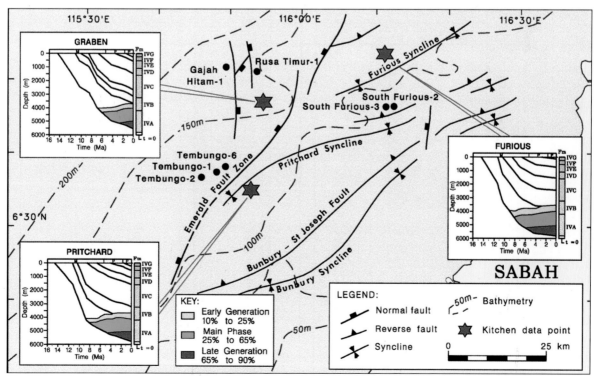

Fig. 22.24. 1D modelling of 3 major kitchen areas in the northern part of the Sabah Basin (from Azlina Anuar, 1994). The results indicate that hydrocarbons were generated from the Stage IVA source rocks in the Furious and Pritchard synclines between 8 and 9 Ma ago, and around 5.6 Ma ago in the Rusa Timur graben. Stage IVB sediments are in the early to middle mature phases in both synclines, but are still immature in the graben. Details of the modelling methodology are discussed by Hunt and Hennet (1992) and Azlina Anuar (1994). Sediment thicknesses were estimated based on sequence stratigraphic correlation by Noor Azim Ibrahim (1994).

subsequently uplifted.

1D modelling was also carried out to determine the depth and timing of hydrocarbon generation in the Pritchard and Furious synclines, and the Rusa Timur graben (Fig. 22.24). In the Pritchard and Furious synclines, hydrocarbons were generated from Stage IVA source rocks around 8 to 9 Ma ago, but much later, at 5.6 Ma, in the Rusa Timur syncline. The lower part of Stage IVA ceased generating at around 4 to 6 Ma in the synclines and at around 1 to 3 Ma in the graben. Stage IVB sediments are in the early to main oil generation window in the two synclines, but are immature in the graben. Thus, although source rocks in the Sabah Basin are of marginal quality, they have matured through "fast cooking" of the organic matter such that enough hydrocarbons were generated for expulsion. Rapid maturation of the source rocks resulted in large volumes of hydrocarbons generated, thus overcoming the saturation threshold for expulsion and migration to occur. This is considered to be an important contributing factor to the presence of hydrocarbon accumulations in the Sabah Basin.

In offshore Kudat Peninsula, the most likely section to contain good quality, mature source rocks are Stage III (Kudat Formation equivalent) and Stages IVA-IVB (equivalent to the lower part of the Bangau Group). Vitrinite reflectance data from offshore wells, however, indicate the sediments penetrated are generally immature. Onshore, the Kudat Formation is also thermally immature, with VRo value ranging between 0.38-0.53%. In the Tiga Papan-1 well, the entire penetrated section is immature; the top of the oil window occurs at about 2800-3100 m (Fig. 22.25). The effective source rocks for the Tiga Papan oil accumulation are probably in the Tiga Papan Sub-basin. The VRo data from Titik Terang-1 also indicate an immature to early mature state of the penetrated section (Fig. 22.25). The main hydrocarbon generation window (VRo 0.7%-1.3%) occurs just below the total depth at about 2900-3050 m. The effective source rocks for the Titik Terang-1 accumulation are believed to be the lower coastal plain and prodelta clays of the Upper Sandstone Unit and the middle to outer neritic clays of the Mudstone Unit in the Kindu-Mangayau Sub-basin.

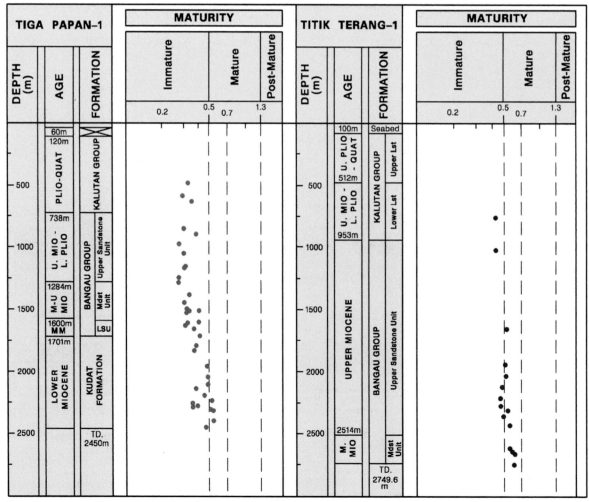

Fig. 22.25. Measured vitrinite reflectance profiles for the Titik Terang-1 and Tiga Papan-1 wells, offshore Sabah Basin.

OIL-SOURCE CORRELATION

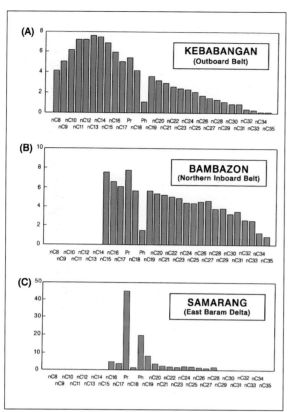

Fig. 22.26. Typical gas chromatographic envelopes of oils (normalised gas chromatographic distribution of the saturate fractions) in the Sabah Basin. The oils vary from (A) normal crudes, as found in Kebabangan, (B) slightly waxy crudes, as in Bambazon, and (C) biodegraded crudes, as found in the top intervals at Samarang.

Figure 22.26 shows some gas-chromatographic (GC) trace envelopes normally observed in the Sabah Basin oils. The oils range from normal (as shown by the Kebabangan oil) to slightly waxy (Bambazon) to biodegraded (Samarang). The Pr/Ph ratios, ranging from 4.5 to 6.5 (average 5.8), indicate that the source rocks for these oils were deposited in oxidising environments, as discussed in the earlier section on source rock deposition. Oil biomarker studies have shown that the source for the organic matter is predominantly terrigenous. This is indicated by high triterpane to sterane ratios, the presence of diagnostic landplant-derived compounds such as oleanane and the resins W, T and R, the absence of C_{34} and C_{35} hopanes, and the general dominance of C_{29} regular steranes over its C_{27} and C_{28} counterparts (Fig. 22.27). Based on these features, the Sabah Basin oils can be grouped into one family. The consistently comparable mass-fragmentograms of both m/z 191 and m/z 217 suggest that the oils originated either from the same source rock, or from different source rocks with similar chemical compositions. The biomarker peaks shown by the oil samples are

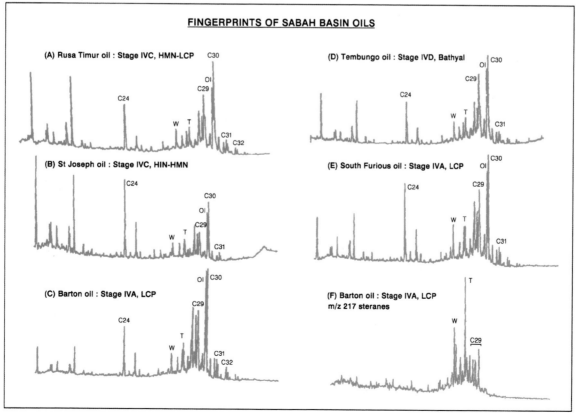

Fig. 22.27. Triterpane m/z 191 and sterane m/z 217 traces of selected Sabah Basin oils. These have similar biomarker distributions, suggesting that they were all generated from source rocks with comparable chemical compositions. The presence of diagnostic landplant-derived biomarkers, along with other supporting analytical evidence, points to a terrigenous, Type III source kerogen for the oils. An example of an m/z 217 sterane from Barton (see F) shows the swamping effect by bicadinanes and the lack of C_{27} and C_{28} regular steranes.

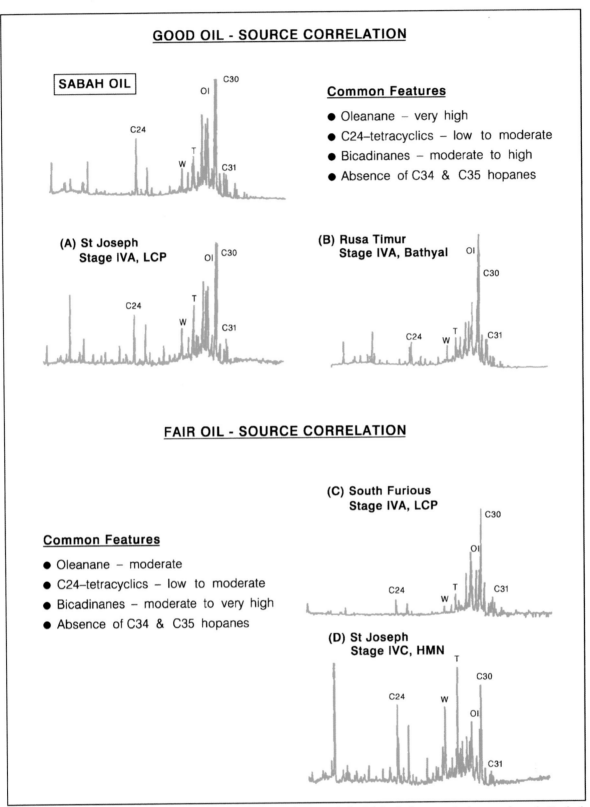

Fig. 22.28. Some examples of source-rock extracts that display similarities with the Sabah oils (an example shown at the top). Good oil-source correlation is observed for (A) Stage IVA lower coastal plain (LCP) sediment extracts from St Joseph, and (B) Stage IVA bathyal sediment extracts from Rusa Timur. Fair to good correlation is observed in (C) Stage IVA LCP sediment extracts from South Furious, and (D) Stage IVC holomarine-inner neritic (HIN) sediment extracts from St. Joseph.

similar even though the oils are from different palaeo-environmental settings. This suggests that the detrital plant components were derived from similar types of higher land plants and were redistributed in different environments by sediment transport processes.

It is difficult to correlate the oils with their potential source sediment extracts, as the sediments are generally immature. Furthermore, the organic matter assemblages in the shales

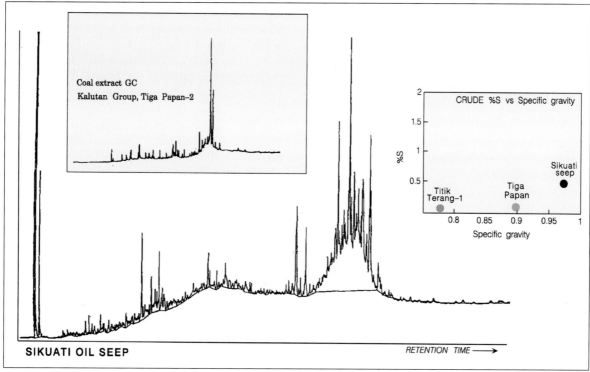

Fig. 22.29. Gas chromatograph of a sample from the Sikuati oil seep, showing the effects of microbial degradation. The light ends and most of the saturated hydrocarbons have been removed, resulting in an apparent enrichment of sulphur, relative to the samples from the Titik Terang-1 (Hall-Houston, 1993) and Tiga Papan-2 (Ward, 1983) wells (right inset). An extract from a Kalutan Group coal sample taken from the Tiga Papan-1 (left inset, data from Mason and O'Reilly, 1983) shows some similarities with the Sikuati oil seep.

penetrated in the well sections may be different from those in the kitchen areas where effective source rocks are more likely to occur but have not been drilled (Azlina Anuar and Abdul Jalil Muhamad, 1997). Nevertheless, there are a few source-rock extracts that show some resemblance to the oils. For example, the Stage IVA lower coastal plain sediment extracts from St Joseph, and bathyal sediment extracts from Rusa Timur correlate well with the oils (Fig. 22.28). Fair to good correlation with the oils were also observed for the Stage IVA lower coastal plain sediment extracts from South Furious and the Stage IVC holomarine/middle neritic sediment extracts from St Joseph. The variations observed between the source extracts and the Sabah oil are mainly due to differences in oleanane and bicadinane concentrations (Azlina Anuar and Abdul Jalil Muhamad, 1997).

Despite being subjected to intense weathering, the Sikuati oil seep has a similar stable carbon isotope ($\delta^{13}C$) composition as the Tiga Papan whole oil sample (Table 22.2). This suggests that both oils were derived from similar source rocks. Coals in the Kalutan Group at Tiga Papan are potential source rocks for these oils. Figure 22.29 shows the saturate fraction of a coal extract from the Kalutan Group at Tiga Papan-2. The coal extract is geochemically similar to the Sikuati oil seep, both dominated by sterane and triterpane compounds, although the Kalutan Group is immature at that location. It may be concluded that the effective source rock for the Sikuati and Tiga Papan oils has chemical properties that are similar to those of the Kalutan Group coals, and probably occurs deeper in the basin.

CONCLUSION

The Sabah Basin is a structurally complex basin that was formed on the southern margin of a foreland basin that resulted from the collision between the NW Sabah Platform and western Sabah during the early Middle Miocene. Its complex syn-tectonic sedimentary history resulted in the recognition of major unconformity-bounded sedimentary packages Stages IVA to IVF. Major hydrocarbon accumulations have been found and produced from siliciclastic reservoirs in mainly Stages IVA, IVC, and IVD. The reservoirs range from coastal/shallow-marine sediments to deep-marine turbidites, and are found in structural/stratigraphic traps including extensional growth fault structures and wrench-related compressive anticlines. Excellent data from the basin provide a good insight into the hydrocarbon habitats in the basin, which can be used for future exploration. The recent PSCs awarded by PETRONAS in the

Sabah Basin indicate the confidence of major oil companies, such as Shell and Esso, in the exploration prospectivity of the basin. Esso, in partnership with Carigali, was awarded 2 PSCs covering the Outboard Belt, the Thrust Sheet Zone, and the deepwater area. Shell, also in partnership with Carigali, was awarded 3 PSCs to explore the Southern Inboard Belt, the East Baram Delta, and the deepwater area. The potential of further discoveries in these and other areas are outlined below.

Stage IVA coastal fluviomarine and coastal plain sands and Stage IVC shallow-marine sands are the main exploration objectives in the Inboard Belt. Because of the multiphase deformation history of the Stage IVA sediments, however, future discoveries are likely to be in subtle stratigraphic traps, such as sand onlapping the DRU, and truncation traps formed by Stage IVA sands subcropping beneath the LIU or UIU. Dense 2D, or even 3D, seismic coverage is required to delineate the closures. Stage IVC shallow-marine sands, which form the main reservoirs in the St. Joseph Field, provide new exploration targets in the sub-unconformity stratigraphic traps beneath massive shale-filled slump scars. Stage IVC turbiditic sands in the Labuan-Paisley Syncline are also potential plays. Other play types in the Inboard Belt include Stage IVD shallow-marine sands (which have been tested successfully in the Erb South discovery), Stage IVB turbidite sands (untested), and Stage III turbidite sands (tested in several wells, but thought to be invalid).

In the Outboard Belt, structural traps formed by Stage IVC and/or older turbidite sands have been successfully tested in the Tembungo Field and in the Kinarut and Kebabangan gas discoveries. These form an important play in the Outboard Belt, especially in the deep-water areas. Stratigraphic traps formed by Stages IVC-IVD and/or older turbidites are also a potential play, but uncertainties in reservoir development, distribution, and continuity need to be addressed.

The main play in the Outboard Belt, however, is the basal Stage IVD shallow-marine sands, which form the reservoirs in Shell's Erb West Field. Shelf sands in the upper Stage IVD have also been successfully tested in Block SB2 under the operatorship of British Gas, e.g. Rusa Timur and Gajah Hitam. Earlier discoveries in this block by Shell, SW Emerald and West Emerald, were also in Stage IVD sands. There is also a gas/oil discovery in turbidite sands at Kebabangan-1 to the southwest of SB302. In addition, there are potential new plays in the deeper waters of the belt, including turbidite sands in the upper Stage IVD and Stage IVF. Many drillable prospects have also been identified in the Outboard Belt.

Stages IVD-IVE coastal fluviomarine and shallow-marine sands in the proximal part of the East Baram Delta are proven to be hydrocarbon-bearing (Samarang, Kinabalu). The Stage IVC play was tested by the Timbalai, Timbalai Deep West-1, and Trusmadi-1 wells, and remains a viable, albeit gas-prone, play. Other potential targets are the deeper Stage IVA and Stage IVF outer shelf to slope/basin turbidite sands. Although situated in the deeper waters of both the Outboard Belt and the EBD, the Stage IVF and/or older turbidite sands are potential exploration targets. Source rocks are expected to consist primarily of landplant-derived organic matter, that was deposited on the shelf, and later reworked and transported into the deep-marine environment by turbidity currents. Such a depositional process may have operated in the case of the Kebabangan gas discovery and the Tembungo oil field in the Outboard Belt.

The results of the recent studies on the multi-sequence Lingan Fan turbidite unit by Marzuki Mohamad and Lobao (1997) and Salahuddin Saleh Karimi et al. (1997) provided a better insight into the deepwater turbidite fan systems in the Sabah Basin, which will help future exploration in these areas. The deepwater areas west of the East Baram Delta and the Outboard Belt are still relatively unexplored. The exploration campaign to be conducted by Shell, Esso, Murphy Oil, and Carigali will enable more data and information to be gathered, particularly in the Sabah Trough and the Thrust Sheet Zone.

REFERENCES

Abbott, W.O., Alder, K.E., Boodoo, W., Coker, P., Stevens, A.H., Toland, M.D. and Vlierboom, F.W., 1990. North Borneo regional study, offshore Sabah. Unpublished Report by Occidental Malaysia, Inc., PETRONAS Exploration Dept. Report No. ER:OXY:2:90-02.

Abdul Jalil Muhamad and Mohd Jamaal Hoesni, 1992. Possible source for the Tembungo oils: evidences from biomarker fingerprints. Bulletin of the Geological Society of Malaysia, 32, 213-232.

Ahmad Said, 1982. Overview of exploration for petroleum in Malaysia under the Production Sharing Contracts. Proceedings of the Offshore South East Asia Conference 1982, 9-12 February 1982, Singapore.

AIPC, 1990. Review of Malaysian Petroleum Potential, Sabah Block SB-4. Unpublished report, American International Petroleum Corporation.

Awang Sapawi Awang Jamil, Mona Liza Anwar and Eric Seah, P.K., 1991. Geochemistry of selected crude oils from Sabah and Sarawak. Bulletin of the Geological Society of Malaysia, 28, 123-149.

Azlina Anuar, 1994. Source rock evaluation of Middle-Late Miocene sequences, North Sabah Basin, Malaysia. Unpublished PhD thesis, Imperial College of Science, Technology and Medicine, University of London, 296 pp.

Azlina Anuar and Abdul Jalil Muhamad, 1995. Geochemical evaluation of the Sandakan-Lahad Datu-Dent Peninsula areas, East Sabah, Malaysia. PETRONAS Research & Scientific Services Sdn. Bhd., Report No. PRSS-RP5-95-13.

Azlina Anuar and Abdul Jalil Muhamad, 1996a. Geochemical evaluation of the Klias-Limbang-Brunei areas. PETRONAS Research & Scientific Services Sdn. Bhd., Report No. PRSS-R01-96/01(B).

Azlina Anuar and Abdul Jalil Muhamad, 1996b. Geochemical evaluation of the Kudat-Banggi Island-Balambangan Island-Bengkoka-Lawas areas, Sabah, Malaysia. PETRONAS Research & Scientific Services Sdn. Bhd., Report No. PRSS-RP5-96-02.

Azlina Anuar and Abdul Jalil Muhamad, 1997. A comparison of source rock facies and hydrocarbon types of the Middle Miocene sequence, offshore NW Sabah Basin, Malaysia. In: Howes, J.V.C and Noble, R.A, eds., Proceedings of the IPA Petroleum Systems of SE Asia and Australasia, 21-23 May 1997, Indonesian Petroleum Association, Jakarta, 773-786.

Azlina Anuar, Kinghorn, R.R.F., Cocksedge, M.J. and Carr, A.D., 1993. Source rock evaluation and thermal maturity of Middle-Late Miocene sequence in the northern part of the Sabah Basin, Malaysia. In: Oygard, K., ed., 16th International Meeting on Organic Geochemistry, Stavanger, European Association of Organic Geochemists, Extended Abstract, 67-72.

Beddoes, L.R., Jr., 1976. The Balabac Sub-Basin, southwestern Sulu Sea, Philippines. Offshore South East Asia Conference, SEAPEX Program, Feb 1976, Paper 15, 22pp.

Bol, A.J. and van Hoorn, B., 1980. Structural styles in western Sabah offshore. Bulletin of the Geological Society of Malaysia, 12, 1-16.

British Gas (Malaysia) S.A., 1994. Block SB-2 Offshore Sabah, East Malaysia: Unpublished Technical Brochure.

Carr, A.D. and Wakefield, M.I., 1995. A regional geochemical evaluation of the SB-2 licence area, offshore Sabah, Malaysia, British Gas Report No. GRC R 0990, PETRONAS Research & Scientific Services/British Gas Project 122/91 Petroleum Geochemistry, Sabah.

Clennell, B., 1996. Far-field and gravity tectonics in Miocene basins of Sabah, Malaysia. In: Hall, R. and Blundell, D.J., eds., Tectonic Evolution of Southeast Asia. Geological Society of London Special Publication, 106, 307-320.

Foo, W.Y. and McDonald, D.I.M., 1983. Proposal of a new lithostratigraphic term – the Tigapapan Unit. Unpublished Carigali-BP Company Letter, Kuala Lumpur.

Frank, P.L., 1981. The onshore geology of the Kudat Peninsula and northern islands, Sabah. Unpublished Carigali-BP Report, No. 2021.

Guest, J.W. and Deckers, F., 1982. The role of pressure, dipmeter, and seismic data in planning the development of the structurally complex South Furious Field. Proceedings of the Southeast Asia Petroleum Exploration Society (SEAPEX) Conference, 1982.

Hall-Houston, 1993. Titik Terang-1, Block SB4, Offshore Sabah, Geological Final Well Report.

Hamilton, W., 1979. Tectonics of the Indonesian region. United States Geological Survey Professional Paper, No. 1078.

Hazebroek, H.P. and Tan, D.N.K., 1993. Tertiary tectonic evolution of the NW Sabah continental margin. In: Teh, G.H., ed., Proceedings of the Symposium on Tectonic Framework and Energy Resources of the Western Margin of Pacific Basin. Bulletin of the Geological Society of Malaysia, 33, 195-210.

Hazebroek, H.P., Tan, D.N.K. and Swinburn, P., 1994. Tertiary evolution of the offshore Sarawak and Sabah basins, NW Borneo. Abstracts of American Association of Petroleum Geologists International Conference & Exhibition, Kuala Lumpur, Malaysia, 21-24 August 1994, American Association of Petroleum Geologists Bulletin, 78, 1144-1145.

Hinz, K., Fritsch, J., Kempter, E.H.K., Mohammad, A.M., Meyer, J., Mohamed, D., Vosberg, H., Weber, J. and Benavidez, J. 1989. Thrust tectonics along the northwestern continental margin of Sabah/Borneo. Geologische Rundschau, 78, 705-730.

Hinz, K. and Schluter, H. U., 1985. Geology of the Dangerous Grounds, South China Sea, and the continental margin off southwest Palawan: results of Sonne Cruises SO 23 and SO 27. Energy, 10, 3/4, 297-315.

Hunt, J.M. and Hennet, R.J.-C., 1992. Modelling petroleum generation in sedimentary basins. In: Whelan, J.K. and Farrington, J.W., eds., Organic matter: Productivity, Accumulation and Preservation in Recent and Ancient Sediments, 20-52.

Hutchison, C.S., 1989. "Geological Evolution of South-East Asia". Oxford monographs on Geology and Geophysics no. 13, Clarendon Press, Oxford.

Hutchison, C.S., 1996. The "Rajang accretionary prism" and "Lupar Line" problem of Borneo. In: Hall, R. and Blundell, D.J., eds., Tectonic Evolution of Southeast Asia. Geological Society of London Special Publication, 106, 247-261.

Ismail Che Mat Zin, 1992. Regional seismostratigraphic study of the Tembungo area, offshore West Sabah. Bulletin of the Geological Society of Malaysia, 32, 109-134.

James, D.M.D. (ed.), 1984. "The geology and hydrocarbon resources of Negara Brunei Darussalam". Muzium Brunei, 169pp.

Johnson, H.D., Chapman, J.W. and Ranggon, J., 1989. Structural and stratigraphic configuration of the Late Miocene Stage IVC reservoirs in the St. Joseph Field, offshore Sabah, NW Borneo. Bulletin of the Geological Society of Malaysia, 25, 79-118.

Johnson, H.D., Levell, S. and Mohamad, A.H., 1987. Depositional controls on reservoir thickness and quality distribution in Upper Miocene shallow marine sandstone (Stage IVD) of the Erb West Field, offshore Sabah, NW Borneo. Bulletin of the Geological Society of Malaysia, 21, 195-230.

Kingston, D.R., Dishroon, C.P. and Williams, P.A., 1983. Global basin classification system. American Association of Petroleum Geologists Bulletin, 67, 2175-2193.

Lee, B.S., 1993. Samarang K-5/7 Reservoir Simulation Study. Society of Petroleum Engineers, Asia Pacific Oil & Gas Conference and Exhibition, Singapore, 8-10 February 1993, SPE Paper 25351.

Leong, K.M., 1978. The 'Sabah Blueschist Belt' - a preliminary note. Warta Geologi, 4, 45-51.

Levell, B.K., 1987. The nature and significance of regional unconformities in the hydrocarbon-bearing Neogene sequence offshore West Sabah. Bulletin of the Geological Society of Malaysia, 21, 55-90.

Levell, B.K. and Kasumajaya, A., 1985. Slumping at the late Miocene shelf-edge offshore West Sabah: a view of a turbidite basin margin. Bulletin of the Geological Society of Malaysia, 18, 1-29.

Liechti, P., Roe, R.W and Haile, N.S., 1960. Geology of Sarawak, Brunei and western North Borneo. British Borneo Geological Survey Bulletin, 3.

Marzuki Mohamad and Lobao, J.F., 1997. The Lingan Fan: Late Miocene/Early Pliocene turbidite fan complex, North-west Sabah. In: Howes, J.V.C and Noble, R.A, eds., Proceedings of the IPA Petroleum Systems of SE Asia and Australasia, 21-23 May 1997, Indonesian Petroleum Association, Jakarta, 787-798.

Mason, P.C. and O'Reilly, S., 1983. Petroleum geochemistry of the well Tiga Papan-1, offshore Northwest Sabah, Malaysia. CBP Report No. GCB/151/82, PETRONAS Carigali Report No. 4109/1.

Mazlan B. Hj. Madon, 1994. The stratigraphy of northern Labuan, NW Sabah Basin, East Malaysia. Bulletin of the Geological Society of Malaysia, 36, 19-30.

Mazlan B. Hj. Madon, 1997. Sedimentological aspects of the Temburong and Belait Formations, Labuan (offshore west Sabah). Bulletin of the Geological Society of Malaysia, 41, 61-84.

Milsom, J., Holt, R., Dzazali Bin Ayub and Ross Smail, 1997. Gravity anomalies and deep structural controls at the Sabah-Palawan margin, South China Sea. In: Fraser, A.J., Matthews, S.J. and Murphy, R.W., eds., Petroleum Geology of Southeast Asia. Geological Society of London Special Publication, 126, 417-427.

Mohammad Yamin Ali, 1992. Carbonate cement stratigraphy and timing of hydrocarbon migration: an example from Tigapapan Unit, offshore Sabah. Bulletin of the Geological Society of Malaysia, 32, 185-211.

Mohammad Yamin Ali, 1995. Carbonate cement stratigraphy and timing of diagenesis in a Miocene mixed carbonate-clastic sequence, offshore Sabah, Malaysia: constraints from cathodoluminescence, geochemistry, and isotope studies. Sedimentary Geology, 99, 191-214.

Noor Azim Ibrahim, 1994. Major controls on the development of sedimentary sequences Sabah Basin, Northwest Borneo, Malaysia. Unpublished PhD Thesis, University of Cambridge, 254pp.

Phoenix Resources Company, 1988. Offshore Malaysia, Sabah Block SB3 Acreage Evaluation. Unpublished report, PETRONAS, BD.1:Phoenix:2: 88-02.

Rice-Oxley, E.D., 1991. Palaeoenvironments of the Lower Miocene to Pliocene sediments in offshore NW Sabah. Bulletin of the Geological Society of Malaysia, 28, 165-194.

Salahuddin Saleh Karimi, Lobao, J.J., and Wannier, M.M, 1997. Seismic identification of depositional processes in a turbidite fan environment, Deepwater Block SB-G, NW Sabah. Bulletin of the Geological Society of Malaysia, 41, 13-29.

Sandal, S.T. (ed.), 1996. "The geology and hydrocarbon resources of Negara Brunei Darussalam". 2nd Ed., Syabas, Bandar Seri Begawan, 243pp.

Scherer, F.C., 1980. Exploration in East Malaysia over the past decade. In: Halbouty, M.T., ed., Giant Oil and Gas Fields of the Decade 1968-1978. American Association of Petroleum Geologists Memoir, 30, 423-440.

Shell, 1993. Block SB1 Exploration Data Book 1994. Sabah Shell Petroleum Company.

Stauffer, P.H., 1967. Studies in the Crocker Formation, Sabah. Geological Survey Borneo Region Bulletin, 8, 1-13.

Stephens, E.A., 1956. The geology and mineral resources of the Kota Belud and Kudat area, North Borneo. British Borneo Geological Survey Memoir, 5, 137pp.

Tan, D.N.K. and Lamy, J.M., 1990. Tectonic evolution of the NW Sabah continental margin since the Late Eocene. Bulletin of the Geological Society of Malaysia, 27, 241-260.

Taylor, B. and Hayes, D.E., 1980. The tectonic evolution of the South China Basin. In: Hayes, D.E., ed., The Tectonic and Geologic Evolution of Southeast Asian Seas and Islands. American Geophysical Union, Geophysical Monograph, 23, 89-104.

Tjia, H.D. and Mohd Idrus Ismail, 1994. Tectonic implications of well-bore breakouts in Malaysian basins. Bulletin of the Geological Society of Malaysia, 36, 175-186.

Tokuyama, A. and Yoshida, S., 1974. Kinabalu Fault, a large strike-slip fault in Sabah, East Malaysia. Geology and Palaeontology of Southeast Asia, 14, 171-188.

Tongkul, F., 1993. Tectonic control on the development of the Neogene basins in Sabah, East Malaysia. In: Teh, G.H., ed., Proceedings of the Symposium on the Tectonic Framework and Energy Resources of the Western Margin of Pacific Basin. Bulletin of the Geological Society of Malaysia, 33, 95-103.

Tongkul, F., 1994. The geology of northern Sabah, Malaysia: its relationship to the opening of the South China Sea Basin. Tectonophysics, 235, 131-147.

van der Harst, A.C., 1991. Erb West: An Oil Rim Development With Horizontal Wells. Society of Petroleum Engineers, Asia Pacific Oil & Gas Conference, 4-7 November 1991, Perth, Western Australia, SPE Paper 22994.

van Vliet, A. and Schwander, M.M., 1987. Stratigraphic interpretation of a regional seismic section across the Labuan Syncline and its flank structures, Sabah, North Borneo. In: Bally, A.W., ed., Atlas of Seismic Stratigraphy. American Association of Petroleum Geologists, Studies in Geology, No. 27, 163-167.

Ward, H.E., 1983. The geochemistry of sediments and an oil sample from the well Tiga Papan-2, offshore northwest Sabah, Malaysia. Carigali-BP Report No. GCB/42/83, PETRONAS Carigali Report No. 4118.

Whittle, A.P. and Short, G.A., 1978. The petroleum geology of the Tembungo Field, East Malaysia. SEAPEX Offshore South East Asia Conference, 21-24 February 1978, Singapore, 18p.

Wiebe, M., 1996. Siagut Basin Geochemistry Study, Block SB4, Offshore Borneo. Hall-Houston Malaysia Ltd., PETRONAS Exploration Dept. Report No. ER:Hall-Houston:2:96-01 (02674).

Wood, B.G.M., 1985. The mechanics of progressive deformation in crustal plates- a working model for Southeast Asia. Bulletin of the Geological Society of Malaysia, 18, 55-99.

Wong, R., 1996. Seismic sequence stratigraphic interpretation enhances remaining hydrocarbon potential of the SE Collins Field. Nada PETMOS, November 1996, 12-23.

Wong, R.H.F., 1997. Sequence stratigraphy of the Upper Miocene Stage IVC in the Labuan-Paisley Syncline, Northwest Sabah Basin. Bulletin of the Geological Society of Malaysia, 41, 53-60.

Chapter 23

Northeast Sabah Basin

Leong Khee Meng and Azlina Anuar

INTRODUCTION

The Northeast (NE) Sabah Basin is defined here to include the central and eastern Sabah basins, which were considered by various authors as part of the West Sulu Basin (Bell and Jessop, 1974; Tamesis, 1990), the Southwest Sulu Basin (ASCOPE 1984) and the Central Sabah Basin (Hutchison, 1992; Clennell, 1996). Previously, the major depocentres commonly mentioned in petroleum exploration literature are the Labuk Bay and Sandakan sub-basins. The NE Sabah Basin is of Miocene-Pliocene age and overlies the Labang and/or older formations (Fig. 23.1).

The NE Sabah Basin can be divided into two deep sedimentation areas and a platform area. The deep sedimentation areas comprise (i) the Central Sabah Sub-basin which occupies the 'Sandakan Rift' of Tjia et al. (1990); and (ii) the Sandakan Sub-basin (Bell and Jessop, 1974). The former sub-basin is represented by a northeast trending series of circular to subcircular depocentres occurring both onshore (Bangan, Tangkong, Latangan-Malua) and offshore (Labuk Bay). The latter consists the Dent Group sediments which occur onshore and offshore (Bell and Jessop, 1974; Haile and Wong, 1965). The

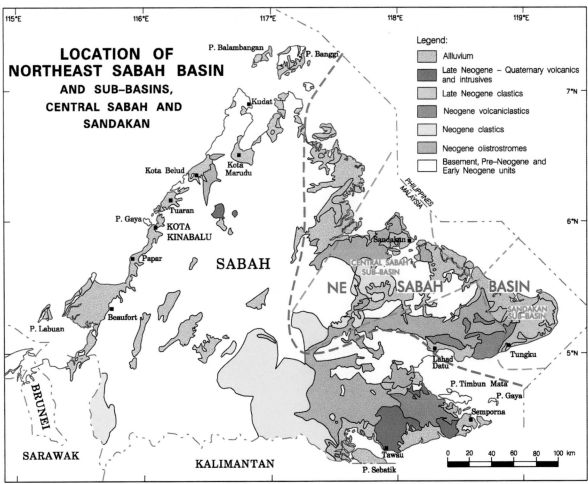

Fig. 23.1 Location of Northeast Sabah Basin and simplified onshore geology.

platform area spans the Kinabatangan area and the Dent Peninsula, and consists of olistostromes and volcaniclastic units of the Segama Group.

The western and southern margins of the NE Sabah Basin consist of an arcuate Crystalline Basement - Ophiolite belt, extending from the northern Banggi-Bengkoka Peninsula, through the Central Sabah Labuk Valley and the Upper Segama Valley-Darvel Bay area. The northeastern and eastern boundaries are probably offshore in the Philippines.

EXPLORATION HISTORY

Several gas seepages in the Dent Peninsula gave impetus to petroleum exploration in the area (Haile and Wong, 1965). Following Shell's relinquishment in 1979, active exploration was focused on the offshore areas of the Dent Peninsula in the Sandakan Sub-basin. The main operators were Elf Aquitaine Petroleum Co., Sabah Teiseki Oil Co. and Western Mining Corporation (WMC) during the periods 1965-1976, 1968-1977 and 1990-1995, respectively. Eleven (11) wells were drilled, most with hydrocarbon shows (Fig. 23.2); the well Nymphe North-1 flowed oil and wells Nymphe-1 and Mutiara Hitam-1 flowed gas. Between 1965 and 1977 Esso carried out exploration activities in north and northeastern Sabah which resulted in the drilling of the Labuk Bay-1 well in the Labuk Bay.

Except for the Sandakan Sub-basin offshore and the Labuk Bay area, the NE Sabah Basin is practically unexplored. To encourage exploration in this area PETRONAS acquired a total of 1176 km of good quality seismic lines in 1993 in the Malawali and Labuk Bay areas, where thicker Miocene sediments are present. In addition, regional tie lines were also acquired over the Tigapapan-1 oil discovery located to the west of the Kudat Peninsula and the Nymphe North-1 oil/condensate and gas discovery. The results of the interpretation of the seismic data are given in Wong and Salehuddin Saleh Karimi (1995).

REGIONAL GEOLOGY

With the exception of the eastern end of the Dent Peninsula, the NE Sabah Basin lies in a complex zone variously termed 'Melange' or 'Suture Belt' (Beddoes, 1976), 'Sabah-Palawan Orogenic Belt' (Rangin et al., 1989, Rangin et al., 1990), 'North Borneo-Sulu Collision Belt' (Hinz et al., 1991) and in part the "Rajang Group Orogene" (Hutchison, 1992). Tjia (1988) and Tjia et al. (1990) proposed that the whole of northeast and east Sabah was an allochthonous fragment of continental crust which collided and accreted onto Borneo (see Chapter 26).

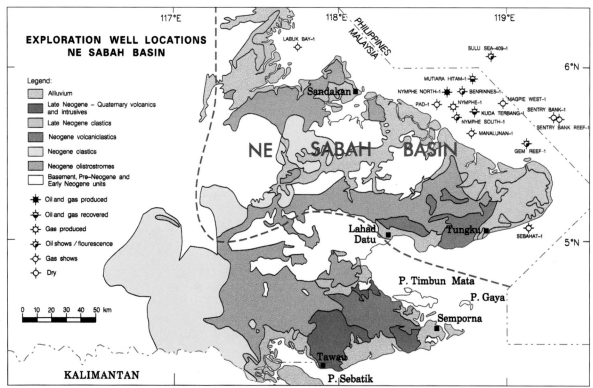

Fig. 23.2 Exploration well locations in offshore Northeast Sabah Basin. Gas, condensate and light oil were recovered in some wells.

Pre-Basin and Basin Units

The NE Sabah Basin comprises several pre-basin (i.e. pre-Neogene) and basin (i.e. Neogene) units (Fig. 23.3). The pre-basin units are (i) Crystalline Basement and ophiolite complex, and (ii) Pre-Neogene sediments. The basin units are (i) Neogene olistostromes, fluviodeltaic to marginal marine sediments and volcaniclastics (ii) Late Neogene fluviodeltaic to marginal marine sediments and (iii) Late Neogene to Quaternary sediments and igneous rocks.

Limestones of various ages are also present. In north and northeast Sabah, Eocene nummilitic and Lower Miocene coralline and reefal limestones (e.g. Tempadong and Gomantong Limestones - Leong, 1974; Haile and Wong, 1965) have been reported. Recent seismic data acquired north of Labuk Bay show distinct mounded features which are interpreted as carbonate buildups overlying the Crocker or Labang Formation (Wong and Salehuddin Saleh Karimi, 1995). These carbonates grew on uplifted areas and are contemporaneous with the

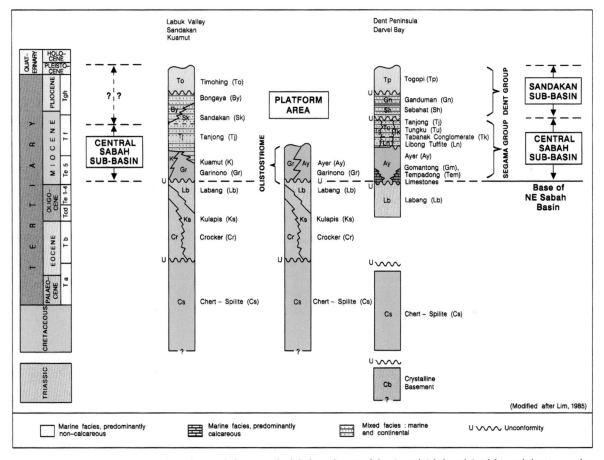

Fig.23.3 Generalized stratigraphy of Northeast Sabah Basin. The lithological units of the Central Sabah and Sandakan sub-basins are also indicated. (modified after Lim, 1985).

The pre-basin deposits comprise mainly deep marine Early Neogene/Palaeogene and older clastics (Crocker, Kulapis and Labang formations), Cretaceous-Early Tertiary ophiolitic complex and deep marine sediments in central and south Sabah (Chert-Spilite, Sapulut, Trusmadi and East Crocker formations), Cretaceous recrystallized massive, and oolitic mollusc-rich and algal limestone, the Madai-Baturong Limestone (see Fontaine and Ho, 1989), and pre-Cretaceous igneous and metamorphic complex (Crystalline Basement). Blocks of these rock-units, especially the ophiolite complex, Labang and Kulapis formations are abundant in the olistostrome.

olistostrome deposits. They may also be equivalent to the reported Oligocene-Lower Miocene carbonate platform in the South China Sea extending to the western shelf of Palawan Island and beyond, and probably underlie Balabac Island and the shallow parts of NW Sulu Basin (Hinz et al., 1985, 1989). These limestones may be potential petroleum exploration targets in offshore northeast Sabah. The Early Neogene olistostrome deposits are extensive in the NE Sabah Basin. They have also been described as 'slump breccias' and 'melange' (Haile and Wong, 1965; Collenette, 1965; Lee, 1970; Leong, 1974; 1976; Clennell, 1991) and are considered by Hutchison (1992) as early rift-fill material.

The Central Sabah Sub-basin is a northeasterly trending circular to subcircular basinal area that overlies the pre-Neogene Labang and/or Kulapis Formation. The sub-basin contains a thick, continuous sequence of Miocene fluviodeltaic and marginal marine sediments. These deposits represent at least 2 cycles of sedimentation, with the earlier deposition of the Tanjong Formation in the southeast and the later deposition of the Sandakan and Bongaya formations in the northeast. The Sandakan Formation in Sandakan town has been mapped in detail, and palaeontological studies indicate that formation is younger than the Tanjong Formation (Lee, 1970; Clennell, 1996). Clennell (1996) proposed that the Tanjong and Sandakan formations were once contiguous, but were later segmented during subsidence into discrete circular to sub-circular basins. Contemporaneous with the filling of the Central Sabah Sub-basin by the Tanjong Formation sediments, thick volcaniclastics continued to be deposited in the Dent Peninsula area. These volcaniclastic units, namely the Libong Tuffite, Tungku and Tabanak Conglomerate formations, are collectively known as the Segama Group. Curiously, on the 2nd edition of the Geological Map of Sabah (Wilford, 1967), these onshore volcaniclastics units also appear as elongate, sub-circular to circular outcrops.

The other prominent deep sedimentation area of the NE Sabah Basin is the Sandakan Sub-basin, occupying the easternmost end of the Dent Peninsula. It is primarily a Late Neogene depocentre, in contrast to the Early Neogene Central Sabah Sub-basin. The prograding fluviodeltaic sediments in the Sandakan Sub-basin (i.e. the Dent Group comprising the late Middle Miocene and younger Sebahat, Ganduman and Togopi formations) are equivalent to the younger cycle of sedimentation in the Central Sabah Sub-basin represented by the Sandakan and Bongaya formations. In the late Middle Miocene, are the uplifted and eroded pre-Neogene Crocker, Kulapis and Labang formations, as well as the synrift Tanjong Formation in the western interior of Sabah, provided the source of the quartz-rich Sebahat and Ganduman formations. The eroded sediments were fed into the depocentre by the ancestral Kinabatangan River (Walker, 1993). The younger cycle of fluviodeltaic to marginal marine deposits commonly show progradational features and is considered as post-rift fill deposition. After a major Late Miocene-Pliocene deformational phase of uplift and erosion, the shaley Togopi Formation was deposited.

Regional Structures

Onshore northern, central and eastern Sabah is characterised by the 'Sulu Trend' which is a linear pattern of parallel ridges trending WNW-ESE to W-E (see Chapter 21). These ridges are delineated by bedding strikes, fold axis or faults. In northern Sabah especially, these ridges are cut by several strike-slip faults, trending N-S, NE-SW and NW-SE. In central Sabah, the NE-SW ridges represent elongate bodies of ophiolitic rocks (Tongkul, 1993,1997). According to Tongkul (1993), it is often difficult to differentiate between older structures and younger structures once they are reactivated.

The dominant structures found in the pre-Neogene units are WNW-ESE trending folds and faults, following the Sulu Trend. A zone of NW-SE trending sinistral strike-slip faults cut across central Sabah (Tokoyama and Yoshida, 1974). In the Neogene units, NE-SW horsts and grabens are common. These NE-SW oriented faults also have a wrench component in the Sandakan Sub-basin (Wong, 1993).

The NE Sabah Basin is dominated by two structural trends, the apparently older WNW-ESE Sulu Trend, and the younger NE-SW trend, which parallels the orientation of the grabens formed by the opening of the southeastern Sulu Sea (Hutchison, 1992; Clennell, 1991,1996). However, the WNW-ESE Sulu Trend appears to have been reactivated in Late Miocene or Early Pliocene times as evidenced by similarly oriented faults, probably with strike slip components, displacing the Middle Miocene and older sediments.

Basin Evolution

Following the Late Eocene uplift of the ophiolite-deepwater sediments complex (Chert-Spilite, Sapulut, Trusmadi and East Crocker formations), depressions were formed and filled by deepwater clastics (Labang, Kulapis formations) during the Oligocene. Renewed uplift of these pre-Neogene units, with the consequent deposition of limestones on local highs heralded the Early Miocene opening of the southeastern Sulu Sea. Rift basins extend southwest into the interior of Sabah (Clennell, 1991,1996; Hutchison, 1992; Walker, 1993). The early rift deposits are represented by the widespread olistostromes, intermingled with volcanic material. Following movements along NE trending faults, the 'Sandakan Rift' depocentre (i.e. the Central Sabah Sub-basin) was formed (Tjia et al., 1990). Other sub-basins in the Sulu Sea, e.g. Balabac and

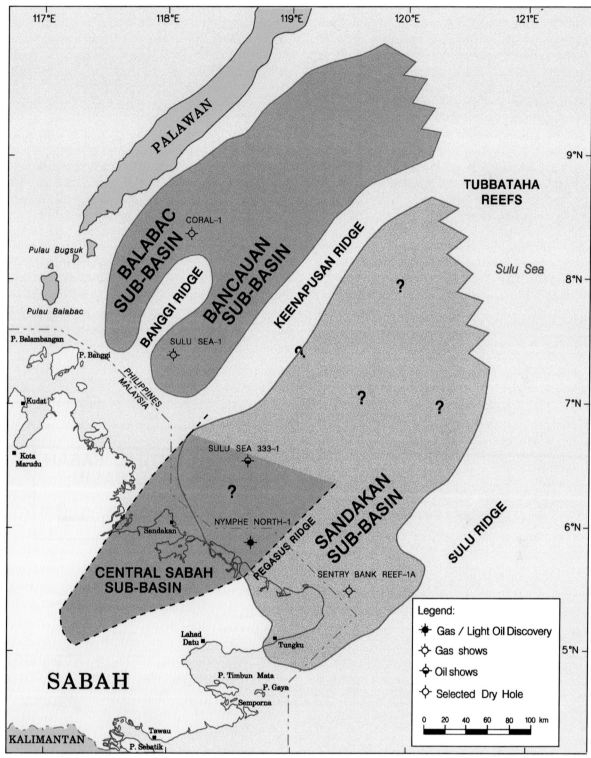

Fig 23.4 Sub-basins in the SW Sulu Sea area. The Central Sabah Sub-basin is on trend with the Balabac and Bancauan sub-basins; the Sandakan Sub-basin is younger and partly overlies the Central Sabah Sub-basin. (modified after Durkee, 1993).

Bancauan, also generally trend in a NE-SW direction (Fig. 23.4). The structure in at least one of them, (the Balabac Sub-basin) suggests that it may have originally been a rift basin which was subsequently inverted by NW-SE compression during the Late Miocene (Clennell, 1996). The NE-SW trending 'Sandakan Rift', interpreted as an aulocogen, was filled by Lower-Middle Miocene fluviodeltaic to marginal marine deposits of the Tanjong Formation. The rate of subsidence approximately kept pace with the rate of sediment supply and deposition, the provenance of which is believed to be the pre-Neogene units of Sapulut, Labang, Kulapis and East Crocker. It appears that the deposition of the syn-rift Tanjong Formation extends offshore towards the north of the Dent Peninsula, where seismic data reveal several circular features. Contemporaneous with the deposition of the Tanjong Formation, volcanic activity continued in the east resulting in the deposition of volcaniclastic units of the Segama Group. The continued volcanism commencing in Early Miocene may be due to a volcanic arc related to subduction of the Celebes (Sulawesi) Sea (Walker, 1993).

A major shift in sedimentary provenance occurred in the late Middle Miocene. From then on erosion of the uplifted Crocker, Kulapis, Labang and Tanjong formations transported quartz-rich sediment by the ancestral Kinabatangan River that prograded into the Sandakan Sub-basin depocentre to deposit the Dent Group. The sequence stratigraphy of these progradational sequences in the Sandakan Sub-basin have been studied by Wong (1993). The shift in the sedimentary provenance from previously volcaniclastic island arc geology (as represented by the Segama Group) and associated rift basins (Tanjong) to classic deltaic deposition from late Middle Miocene onwards, followed the cessation of the Sulu Sea spreading. This was due to collision and subsequent uplift of exotic micro continents related to the Australian and Philippines blocks (Walker, 1993).

A Late Miocene compressional phase with inversion and wrenching is evident in the Sandakan Sub-basin. Volcanic activity is indicated by the presence of basalt with K/Ar dating of 6.3 Ma (Late Miocene) in a core sample from the well Magpie West-1 (Western Mining Corporation, 1991a). The Pliocene/Late Miocene unconformity is clearly seen in several wells in the Sandakan Sub-basin. Compression closed the newly opened Sulu Sea and resulted in the formation of several NE trending arches which reactivated the deltaic growth faults and created large faulted anticlines. In the distal areas, reefs on pre-existing highs were later buried by prodelta shales (Walker, 1993).

The NE Sabah Basin is still seismically active. There are active mud volcanoes, testifying to the continuing mud/shale diapirism (Tongkul, 1988).

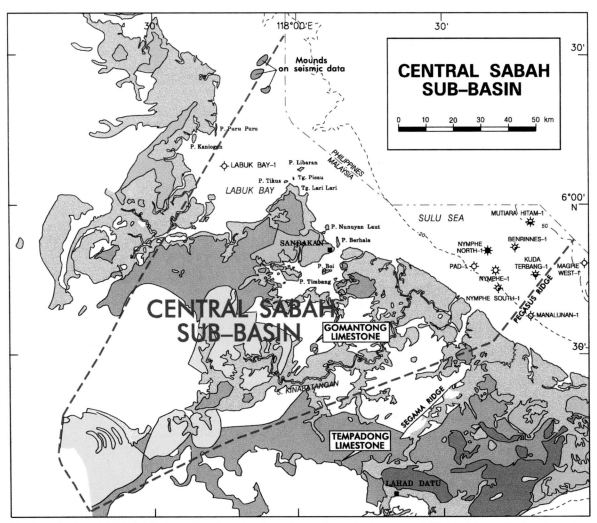

Fig. 23.5 The Central Sabah Sub-basin consists of a series of circular to sub-circular depocentres infilled by Miocene Tanjong Formation, and appears rift-related. Mounded features on seismic data in Labuk Bay area are probably Miocene reefal carbonates similar to the onshore Gomantong and Tempadong limestones.

THE CENTRAL SABAH SUB-BASIN

The Central Sabah Sub-basin is characterized by a series of circular to sub-circular basins situated in the northeasterly trend, from the onshore Bangan basin in the southwest to the offshore area in the Sulu Sea (Fig. 23.5). These basins or depocentres, each some 20km across, are infilled by fluviodeltaic to marginal sediments of the Tanjong Formation. For discussion of the origin of these enigmatic circular basins, readers are referred to Lee and Tham (1989), Tongkul (1993) and Clennell (1996). The Tanjong sediments extend offshore in sub-circular basins that can be mapped from seismic data. The northern and southern boundaries of the 'rift' are probably complex fault zones trending NNE to ENE . The northern boundary may extend offshore to the Keenapusan Ridge which probably continues northeast to the Cagayan Ridge in the Sulu Sea. The southern boundary merges with a major tectonic dislocation in the Dent Peninsula, the onshore Segama Ridge and offshore Pegasus Ridge (Fig. 23.4, 23.5).

Exploration Results

The Central Sabah Sub-basin is little explored with only one dry well drilled in the Labuk Bay (Labuk Bay-1) by Esso in 1977. The offshore well Labuk Bay-1 objective target was the younger Sandakan/Bongaya Formation fluviodeltaic sequence. What has been interpreted in the Labuk Bay prospect as a faulted anticlinal feature turned out to be a shale diapir. The underlying Garinono shale mass appeared to have pierced through the prognosed reservoir sand units which were subsequently uplifted and eroded during the Pliocene tectonism. The numerous active mud volcanoes in the surrounding areas are indicative of continuing diapiric activity.

Traces of hydrocarbon residue have been reported in an onshore sandstone sample of the Tanjong Formation (see Leong, 1974, p.186). However in the overlapping offshore area of the Central Sabah and Sandakan sub-basins (Fig. 23.4), Lower and Middle Miocene units contain several hydrocarbon-bearing reservoirs. These units, Segama Group and/or Tanjong Formation equivalent, belong to the Central Sabah Sub-basin, but are described in more detail in the Sandakan Sub-basin section.

Stratigraphic and Structural Setting

Hutchison (1992) has described the stratigraphic and structural setting of the Central Sabah Sub-basin in some detail. The early stage of the opening of the Sulu Sea was characterized by explosive volcanic activity, and the rifting resulted in extensive olistostrome deposits. The Lower Miocene olistostrome deposits interbedded with volcaniclastics are known onshore as the Kuamut, Garinono and Ayer formations (see Clennell, 1991 for detailed descriptions). The chaotic deposits contain blocks from the underlying Labang, Kulapis and Chert-Spilite formations, including eclogites. Very soon after the limited rifting and olistostrome formation, the pre-olistostrome rock units, e.g. Crocker, Labang and Kulapis, were uplifted and eroded to provide the provenance for the thick quartz and detrital chert sands of the quartz-rich, Lower-Middle Miocene Tanjong Formation. The subsidence along the NE trend kept pace with the supply of sediments.

The tectonic map of Sabah (Tongkul, 1991) shows several major NE-SW trending faults, mostly normal, although several may have wrench components. The faults appear to cut across the circular basins, and are later than the prominent WNW-ESE and E-W Sulu Trend. According to Tongkul (1993), differential movements of re-activated NE-SW and NW-SE wrench faults, due to the extension in the Sulu Sea, may have initiated the formation of several horst and graben of various sizes in the Central Sabah Sub-basin.

It is widely presumed that the Tanjong sediments were funnelled into the rift basin(s) associated with extension and spreading of the southeastern Sulu Sea in a NE-SW trend (e.g. Hutchison, 1988, 1992; Tjia et al., 1990). This appears to be the most likely explanation of the sedimentation process, but direct evidence of extensional structures over much of onshore eastern Sabah is lacking (Clennell, 1996).

Meanwhile, to the east in the Dent and Semporna Peninsula areas, the volcanic and volcaniclastic units of the Segama Group (Libong Tuffite, Tungku) and Kalumpang Formation, respectively were deposited. Coeval reefal and coralline limestones were deposited on structural highs. Subsequent uplift in the mid-late Middle Miocene exposed the Tanjong sediments and Segama Group volcaniclastics to erosion which provided source material for a new phase of Miocene deltaic deposition, represented by the Sandakan/Bongaya Formation to the north and northwest, and the Dent Group (Sebahat and Ganduman formations) to the east and northeast.

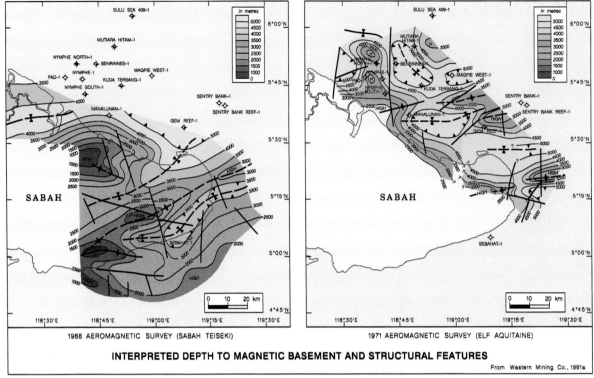

Fig. 23.6 The interpreted depth to magnetic basement maps from aeromagnetic data over the Dent Peninsula area show a series of troughs and anticlinal highs. The magnetic basement is most likely the Chert-Spilite Formation.

These post-rift Middle Miocene and younger units show strong progradational features on seismic.

The later Pliocene/Late Miocene uplift of the rift-fill and post-rift rock units caused deep erosion of crestal areas. The Tanjong Formation sandstones in circular to sub-circular basins were elevated up to 1600m above sea level. This phase of tectonism probably remobilized the mud-rich matrix of the underlying olistostromes and gave rise to shale diapirism. The post-tectonic units comprise undeformed shallow marine deposits of the Timohing Formation to the north and northwest and of the Togopi Formation in the east and northeast.

The Central Sabah Sub-basin is currently seismically active. Mud volcanoes are still active and new ones are being formed (Lee, 1990; Tongkul, 1988).

Recent seismic data acquired over offshore Labuk Bay indicate horst and graben features along NE-SW trending faults. The seismic data also confirmed the major deformation episodes interpreted from onshore geology which are as follows (Wong and Salahuddin Saleh Karimi, 1995) :

- Early Miocene : opening of Sulu Sea in a NE-SW trend; rift basin formation; olistostromes and volcaniclastic deposition;.

- Late Middle Miocene : uplift and erosion of rift-filled Tanjong Formation and Segama Group volcaniclastic units.

- Late Miocene/Pliocene : Inversion of the Tanjong Formation; and uplift and erosion of post-rift Sandakan/Bongaya Formation; strong shale diapirism.

THE SANDAKAN SUB-BASIN

The Sandakan Sub-basin is located in the southern portion of the Sulu Sea off eastern Sabah, with a small part of the basin lying onshore Sabah in the Dent Peninsula (Fig. 23.4, 23.7). It is filled up to 8km of sediments, including the pre-Dent Group (CCOP, 1991). The sub-basin is bounded to the northwest by the Keenapusan Ridge extending northwesterly to the Cagayan Ridge, and to the east by the Sulu Ridge. To the northeast, in the Sulu Sea, the sediments are deformed by toe-of-slope compressional folds. Northeast of these folds the sediments thin to 2.5 km and downlap onto the SE Sulu Sea oceanic crust, marking the northeastern boundary of the sub-basin (Graves and Swauger, 1997). Earlier papers on the Sandakan Sub-Basin include Bell and Jessop (1974), ASCOPE (1984), Tamesis (1990), Gopakumar (1993), Walker (1993) and Wong (1993). Unpublished reports by Western Mining Corporation (1991a, 1991b, 1995a, 1995b) have also been used in this account of the Sandakan Sub-basin.

The interpreted results of two aeromagnetic surveys by Sabah Teiseki and

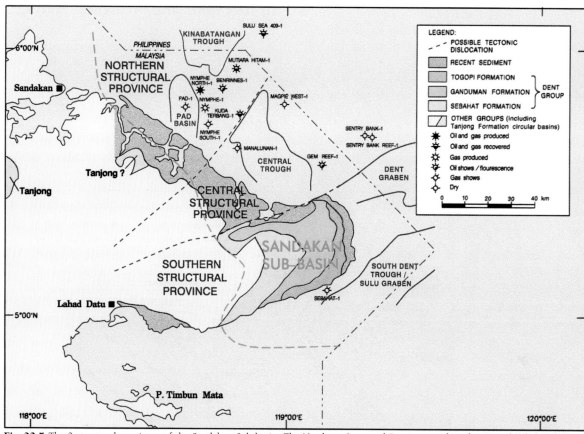

Fig. 23.7 The 3 structural provinces of the Sandakan Sub-basin. The Northern Structural Province overlaps the Central Sabah Sub-basin and the troughs in this province are interpreted as extension of the Tanjong-filled onshore circular depocentres.

Elf Aquitaine in 1968 and 1971 respectively, over the Sandakan Sub-basin are shown in Figure 23.6. The depth to magnetic basement maps (Chert-Spilite Formation and older) show a series of ridges or anticlines and graben areas trending N-S, NE-SW and E-W. The series of magnetic highs and troughs approximately match the ridges and troughs interpreted from seismic data (Fig. 23.7, 23.8) except for the Benrinnes area. The significant troughs to the south and southeast are the Dent Graben and South Dent Trough or Sulu Graben, in the centre is the Central Trough and to the north and northeast, the Pad Basin and the Kinabatangan Trough, which are situated in the overlapping areas of the Central Sabah and Sandakan sub-basins (Fig. 23.7).

Exploration Results

Prior to the drilling of Mutiara Hitam-1 and Kuda Terbang-1, Walker (1993) summarized the list of previous valid and invalid structural or stratigraphic tests. In Sabah, Walker (1993) listed the following wells as invalid tests :

On the Philippines side, the following wells were listed as invalid - Clotilde-1 (as for Magpie West-1), Sentry Bank-1, Sulu Sea 389-1, and Sulu Sea 409-1 (as for Sebahat-1). The only valid test considered by Walker (1993) were :- Nymphe-1, Nymphe North-1 and Benrinnes-1 on the Sabah side; and Sentry Bank Reef-1 and probably Sulu Sea 333-1 and Dokan-1 on the Philippines side.

WELL	RESULT	REASON
Manalunan-1	invalid stratigraphic test	unclosed deltaic topsets
Magpie West-1	invalid structural test	apparent "anticlinal structure" was actually volcanic complex
Gem Reef-1	invalid structural test	apparent "anticlinal structure" was actually shallow reefs
Nymphe South-1	invalid structural test	outside closure
Sebahat-1	invalid structural test	unclosed closure or below closure
Pad-1	invalid stratigraphic test	too far downdip from pinchout play

The anticlines tested by wells Mutiara Hitam-1 and Kuda Terbang-1, although valid tests, indicate a discrepancy between pre-drilled seismic correlation and post-drilled biostratigraphic results. The entire predicted

Upper Miocene sandy Ganduman Formation and most of the Middle Miocene Sebahat Formation are missing and are presumably eroded off.

Hydrocarbon shows have been noted in thin sand in most of the wells and these range from minor gas shows to recordable volumes flowing to the surface during drill stem tests (DSTs). The best results were obtained in Nymphe North-1, Nymphe-1 and Mutiara Hitam-1. Nymphe North-1 tested 504 BOPD oil

Stratigraphic and Structural Setting

The Sandakan Sub-basin is a post-rift successor basin, or a passive margin basin, unconformably overlying the syn-rift Tanjong Formation and/or the volcaniclastic Segama Group (i.e. Central Sabah Sub-basin). The fluviodeltaic sediments show strong progradation from generally west to east, and comprise the Dent Group (Sebahat, Ganduman and Togopi formations) ranging from Middle Miocene-Pleistocene age.

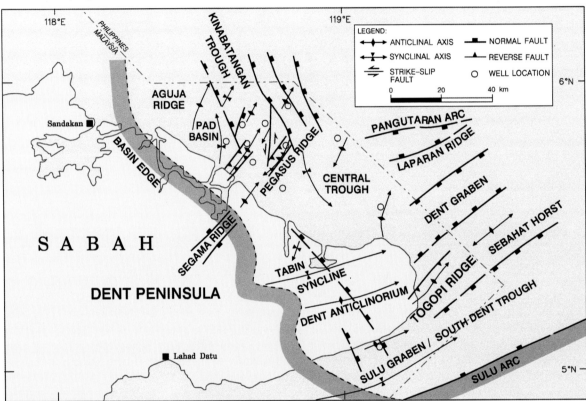

Fig. 23.8 The structural elements in the Sandakan Sub-basin, interpreted from mainly seismic data for the offshore area. There is a broad correlation with the interpretation from aeromagnetic data in fig.23.6. Note the online trend of the Pegasus and Segama ridges. The former marks the boundary of the Central Sabah Sub-basin. Also note the NS to NNW trend of the extensional faults in the overlapping sub-basinal area.

(46° API) and 3.88 MMSCF/D on a 1/4" choke from 1658m-1662m within the Middle Miocene. Nymphe-1 tested 2 good flows of gas (about 15 MMSCF/D each) with 94 BOPD condensate from two zones between 2047m-2184m near the Middle and Lower Miocene boundary. Mutiara Hitam-1 flowed gas at a rate of 4.15 MMSCF/D. According to Western Mining Corporation (1995b), the Kuda Terbang-1 and Benrinnes-1 wells would have also flowed hydrocarbons to the surface if they had been drill stem tested. The majority of the hydrocarbon-bearing sands have been dated as Early to Middle Miocene in age, which corresponds to the Segama Group (most likely Tungku Formation) sequence. However, they could also be assigned to the equivalent of Tanjong Formation.

Based on the study of onshore outcrops (Ismail Che Mat Zin, 1994), deltaic deposition began in the late Middle Miocene with the aggrading shale-rich Sebahat Formation, accompanied by distal reef growth sometimes on subsiding volcanic high (Fig. 23.9). This was followed by the deposition of the sandy southeast prograding Ganduman Formation. Following a phase of major uplift and erosion in the Early Pliocene, the marine Togopi Formation was deposited. Ismail Che Mat Zin (1994) concluded that the Sebahat and Ganduman formations were deposited as a delta unit. The Togopi Formation is a marl with allochthonous reef detritus. The provenance for both the Sebahat and Ganduman is believed to be the 'Older Sebahat' or Tanjong Formation equivalent which is interpreted to fill the offshore Pad Basin.

Stratigraphic and Correlation Schemes

Various stratigraphic schemes have been devised for lithologic units in the sub-basin. However, difficulty remains in assigning lithogical units to these schemes for all but the youngest formation, due to lateral variation of units and poor biostratigraphic control. Thus, the age of the Dent Group appears unresolved. Walker (1993) placed the underlying Segama Group, including the Tanjong Formation, as Early to Middle Miocene, and the Dent Group, late Middle Miocene to Late Miocene for the Sebahat and Ganduman formations, and Plio-Pliestocene for the Togopi Formation. Wong (1993), however, extends the Ganduman Formation from Late Miocene to Late Pliocene based on sequence stratigraphic studies. Ismail Che Mat Zin (1994) assigns a Pliocene age to both the Sebahat and Ganduman formations. On the 3rd edition of Geological Map of Sabah (Lim, 1985), the onshore Sebahat and Ganduman formations are in the Middle Miocene-Pliocene age range and the Togopi, Plio-Pleistocene. In this chapter, the age ranges used are those of Walker (1993).

Seismic data show a pronounced unconformity at the Pliocene/Late Miocene or base Pliocene boundary, overlain by the Togopi Formation (Fig. 23.10). The geological section beneath this unconformity is still not well understood due to structural complexity (including shale diapirism), poor biostratigraphic control and seismic record quality. In the pre-drilled prognosis for wells Mutiara Hitam-1 and Kuda Terbang-1, the Upper Miocene sandstones of the Ganduman and Sebahat formations beneath the pronounced base Pliocene unconformity were the primary targets. However, a major portion of the predominantly sandy Upper Miocene sequence has been eroded in Mutiara Hitam-1, and completely eroded off in Kuda Terbang-1. The biostratigraphic analysis results, therefore, do not support the seismic correlations. Palynology was the main discipline used to interpret the ages of rocks. However, it should be noted that the age of the sequences encountered beneath the unconformity, ranging from the Early Miocene to early Middle Miocene, could be due to reworking of older fossils.

The Lower and Lower to Middle Miocene sections, which have proven to be gas-bearing (with condensate) in Nymphe North-1, Nymphe-1, Benrinnes-1 and Mutiara Hitam-1 as well as in other wells, have been mainly assigned to the Segama Group and, to a lesser extent, the Sebahat Formation by Western Mining Corporation (1995b). However, what has been assigned to the Segama Group (in particular the Tungku Formation) could well be the synrift Tanjong Formation, which occurs onshore in the circular to sub-circular basins described earlier.

Several phases of structural growth are evident within the Dent Group. Prior to the deposition of the Dent Group, the Segama Group and the Tanjong Formation were deformed and uplifted in the mid-late Middle Miocene. The most important tectonic event, however, was uplift and erosion at the end of the Miocene which created the base Pliocene unconformity. This erosion removed the upper portion of the Sebahat Formation and all of the Ganduman Formation in the northwestern part and over the Benrinnes structure, and considerably thinned the Ganduman Formation over the Pegasus Ridge, which is on trend with the onshore Segama Ridge (Fig. 23.8). The Togopi Formation was deposited on the base Pliocene unconformity, commonly with strong angular truncation of the underlying beds around ridges striking NNW through NNE. Offshore, particularly in synclinal areas, the unconformity loses most of its angularity, but in the Sandakan onshore area, widespread peneplanation occurs.

Structural Provinces

The Sandakan Sub-basin may be subdivided into the Northern, Central and Southern structural provinces, based mainly on the distinct lithologies and facies and structural styles of both the Sandakan and underlying Central Sabah sub-basins (Western Mining Corporation, 1995b; Fig. 23.7).

The Northern Structural Province, overlying rift basins, is structurally complex with large NE trending anticlines and extensional and compressional faults. Large N-S trending grabens include the Kinabatangan Trough, the Pad Basin and a synclinal area west of the Kuda Terbang-1 well. Various phases of deformation from the Middle Miocene to Pleistocene have affected this Province. The western area contains NE-SW trending wrench-related 'flower' structures, while the eastern area is associated with growth faults. During the Pliocene, a period of strong transpression wrenching rejuvenated many existing faults. The anticlinal structures (flower structures) were probably modified by cores of mobile shale of the Ayer Formation (Segama Group). Growth faults were locally reversed, and the wrenching is well developed over zones of basement faults and basin hinge-lines. The Benrinnes structures are examples of such compressive wrench anticlines. The Nymphe and Nymphe North structures are examples of growth

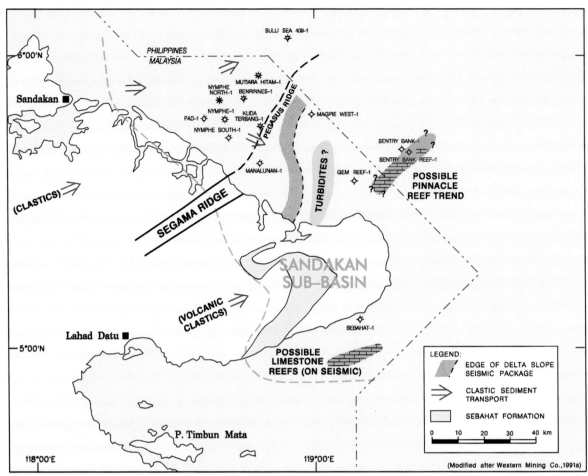

Fig. 23.9 The interpreted palaeoenvironment of the Sebahat Formation from generally deltaic deposition to probably turbidites in the Central Trough. The provenance of the Sebahat ranged from clastics of the Tanjong and older formations in the north to volcaniclastics of the Segama Group and the older Chert-Spilite Formation in the south. (modified after Western Mining Corp. 1991a).

fault rollovers. The growth faults in the Benrinnes and Nymphe areas appear to have been reactivated around the NE-SW compressive trend. The province is also characterised by the absence of the Upper Miocene Ganduman Formation and the upper section of the Sebahat Formation, as a result of pronounced Early Pliocene uplift and erosion. The western and northwestern boundary of the Northern Structural Province is defined by the Aguja Ridge and the eastern and southeast boundary, also known as the Pegasus Ridge, is a major tectonic dislocation, with slices of older Chert-Spilite Formation exposed onshore (Fig. 23.8).

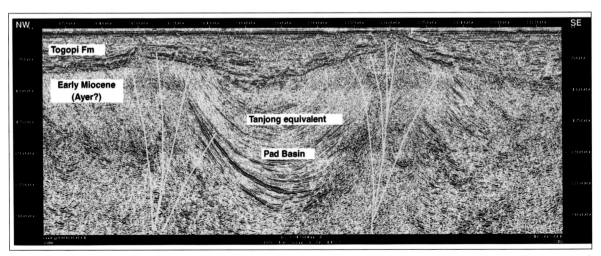

Fig. 23.10 NW-SE seismic section across the Pad Basin showing its circular feature. Note also the Upper Miocene/Lower Pliocene erosional unconformity and indications of shale diapiric movement on the edges of the basin.

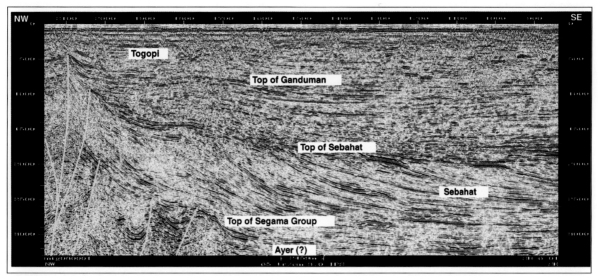

Fig. 23.11 NW-SE seismic section across the Central Trough showing progradational features of the Sebahat Formation.

Incidentally, this boundary also merges with the approximate southeastern boundary of the Central Sabah Sub-basin rift basin margin. In this province, the sandstones of the pre-Dent Group (Tanjong Formation equivalent) as well as the Dent Group have proved hydrocarbon-bearing.

The Central Structural Province is underlain by volcanics and volcaniclastics of the Segama Group and is relatively unstructured. The Pegasus Ridge represents the northern margin, and probably marks the shelf edge of the easterly prograding post-Tanjong sediments in the later Miocene. The southern margin of the Central Structural Province is marked by the Tabin Fault, a transpressive fault first identified on SAR and further verified by field work (Ismail Che Mat Zin, 1994). The Central Structural Province, also referred to as the Central Trough or Dent Graben, was starved of sediments until the later Miocene. The Miocene section within the Central Trough offshore is marked by an eastward prograding sequence of the Sebahat Formation, as seen on several seismic lines (Wong, 1993; Ismail Che Mat Zin, 1994) (Fig. 23.11). The central part of the Central Trough marks the toe end of the prograding sequence and is predicted to consist of basin floor and basin slope turbidite sands (Wong, 1993).

Three wells have been drilled in this province. Magpie West-1 and Gem Reef-1 were drilled near the crest of two highs, probably underlain by volcaniclastics of the Segama Group. Manalunan-1, drilled through the prograding delta slope zone encountered poor reservoir development. However, Gem Reef-1 encountered gas-bearing sands in the Sebahat Formation. The evidence for the presence of reefal carbonates updip from Gem Reef-1 is poor, but is not ruled out.

The Southern Structural Province is the least explored area. The Segama Group of volcaniclastic units occur extensively. Unconformably overlying the Segama Group with marked angularity is the Middle Miocene to Pliocene Dent Group. The Dent Group has been gently folded into an ENE plunging anticlinorium forming the Sebahat Arch. Smaller sub-parallel anticlines and synclines, the largest being the NE-SW trending Dent Graben and the Sulu Graben/South Dent Trough, have been mapped at the eastern end of the Dent Peninsula. The provenance for the Middle Miocene and younger sediments in this province is likely to be the volcaniclastic Segama Group and the pre-Tertiary igneous/metamorphic complex to the west. Sebahat-1, the only well drilled in this province, encountered poor reservoir rocks. However, carbonate buildups have been identified on seismic data to the south and east of Sebahat-1, and close to the Philippines-Sabah boundary. These build-ups are probably of Early-Middle Miocene age, and are correlatable with the onshore Tempadong and Gomantong limestones.

HYDROCARBON OCCURRENCES AND PLAY TYPES

Numerous gas seeps reported in the onshore areas of the Dent Peninsula are commonly associated with mud volcanoes and salt springs (see Haile and Wong, 1965). The existence of a generative petroleum system in the Sandakan Sub-basin is proven by the occurrences of oil and gas reservoired in the Sebahat Formation (Dent Group) and the Segama Group/Tanjong Formation equivalent sandstones (pre-Dent Group), with most being in the latter. The schematic correlation chart for the wells drilled in the offshore Sandakan and Central Sabah sub-basins is shown in Figure 23.12. A compilation on the results of earlier wells, including geochemistry, is given in Western Mining Corporation (1991a, 1991b).

The play types tested are the fault and/or dip closed structures along the anticlinal ridges and stratigraphic traps. The former play type has yielded hydrocarbons (Table 23.1) in mainly the pre-Dent Group units (e.g. Tanjong Formation equivalent) and, to a lesser extent, the Dent Group Sebahat Formation. The latter play was tested by Manalunan-1 and Pad-1, but these were considered invalid stratigraphic tests by Walker (1993).

sands within the volcaniclastic Segama Group, have porosities in the range 20-25% and permeabilities 10-300 md. The sands which have produced hydrocarbons in Nymphe North-1, Nymphe-1 and Mutiara Hitam-1 are all assigned to the Segama Group (Western Mining Corporation, 1995b); however, the Lower-Middle Miocene sands are more likely to be equivalent to the Tanjong Formation which occurs onshore in circular to subcircular basins. Offshore, near the above mentioned wells, similar circular basins e.g. the Pad Basin, are present (Fig.23.10). Intraformational claystones form the seals to the reservoirs.

On the Philippines side, reservoirs occur as stacked, good quality, relatively thick (7-25m), distributary channel sands and delta front/channel mouth bar (Graves & Swauger, 1997). These reservoir units are correlatable to the Tanjong Formation, rather than the Dent Group in Sabah. Regional seals are absent and as such, intraformational mudstones play the role of

Well Name	Reservoir Formation	Hydrocarbon Indications
Gem Reef-1	Sebahat	gas/from FIT
Benrinnes-1	Segama/Tanjong Eq./Sebahat	gas/condensate from FIT
Magpie West-1	Dent Group	no shows
Manalunan-1	Sebahat/Dent Group	gas shows in cuttings
Nymphe North-1	Segama/Tanjong Eq./Sebahat	gas/light oil from DST
Nymphe-1	Segama/Tanjong Eq.	gas/condensate from DST
Pad-1	Segama /Tanjong	gas shows in cuttings
Nymphe South-1	Segama/Tanjong Eq.	gas shows and fluoresence in cuttings
Sebahat-1	Dent Group	gas shows in cuttings
Mutiara Hitam-1	Segama/Tanjong Eq.	oil from RFT; gas from DST
Kuda Terbang-1	Segama/Tanjong Eq.	oil/gas from RFT

RFT: Repeat formation test;   FIT: Formation interval test;   DST: Drill stem test

Table 23.1 Well results, offshore east Sabah.

Reservoir, Trap and Seal

Reservoirs are present in the pre-Dent rock units - Tanjong Formation and Segama Group; and in the Dent Group - Sebahat, Ganduman and Togopi formations. The Sebahat, where encountered, has high (>25%) porosities and permeabilities (1 darcy range). However, in the Northern Structural Province intensely drilled area, it is absent or deeply eroded. The underlying Tanjong Formation equivalent, or thin

seals. The incompetent nature of shales, as shown by the occurrence of numerous diapiric structures in the area, may enhance the sealing potential of faults by infilling and moving along the fault planes.

The Pliocene/Late Miocene compressional anticlines in the offshore Northern Structural Province have proven to be hydrocarbon-bearing (Nymphe, Benrinnes, Mutiara Hitam and Kuda Terbang structures). These structural traps lie at the intersection of a NE fold trend and pre-existing N-S deltaic growth faults, commonly reactivated as reverse faults.

Other significant traps, possibly stratigraphic, may occur adjacent to the Pad Basin where closure of the anticlinal flank or sealing of the traps is provided by a diapiric shale mass and a normal fault. The Pad Basin is sub-circular, similar to the onshore basins, and is infilled with the Lower to Middle Miocene Tanjong Formation. Almost all hydrocarbon-bearing stratigraphic sections encountered in the well Nymphe-1, Nymphe South-1, Benrinnes-1, Mutiara Hitam-1 and Kuda Terbang-1 are also of Early to Middle Miocene age.

For the Central Structural Province, buildups on basement highs are also potential targets (e.g. near Gem Reef-1). Stratigraphic traps include low stand fans. These basin floor, basin slope and turbiditic sands have been interpreted from sequence stratigraphic studies in the offshore Central Structural Province (Wong, 1993). The potential of the Central Structural Province is recently enhanced by the identification of DHIs (Direct Hydrocarbon Indications) on seismic on the Philippines side, and by an indication of a distal marine-influenced source kitchen located seaward of the major normal faults (Graves and

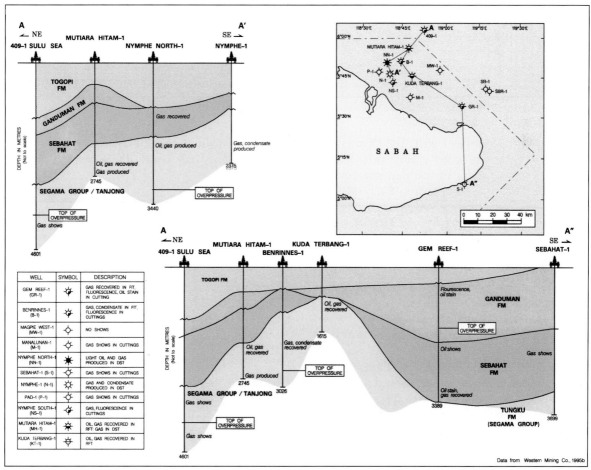

Fig. 23.12 Schematic sections showing the formations pentrated by the exploration wells, the occurrences of oil and gas, the onset of overpressures. Note that most of hydrocarbon occurrences were in the Segama Group/Tanjong equivalent, ie in the underlying Central Sabah Sub-basin. However, there is still need to firm up the age-dating from microfossils. (modified from Western Mining Corp. 1995b).

It should be noted that the pre-Dent Group reservoirs (e.g. Tanjong Formation) have been affected by at least two episodes of deformation in late Middle Miocene and Late Miocene/Pliocene times. These multiple periods of deformation may have created fault-bounded compartments on the structure especially in the Northern Structural Province. Shale diapirism may have either complicated the trapping mechanism or enhanced the traps by providing effective seals.

Swauger, 1997).

The Pliocene-Pleistocene carbonates of the Togopi Formation are regarded as a secondary reservoir objective. Limestone reefs with vuggy porosity were intersected in the Benrinnes and Magpie West areas. However, the lack of seals poses the main problem with this type of reservoir objective.

In the Southern Structural Province, the Sebahat Formation becomes more argillaceous, as evidenced by both onshore investigations and

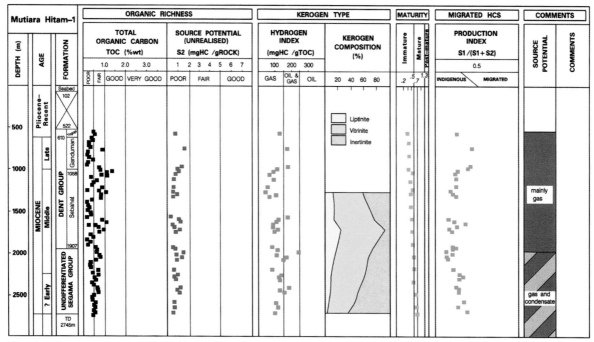

Fig. 23.13 Geochemical summary log of Mutiara Hitam-1.

results from the Sebahat-1 well. The Dent Group is likely to be shaley with volcaniclastic detritus derived from erosion of the Segama Group. Probable Carbonate buildups, Early-Middle Miocene age, identified on seismic data near the Phillipines-Sabah boundary may be considered potential exploration targets (Comexco, 1993). However, the quality of the carbonate reservoirs is unknown. Wrench-related anticlinal traps with large relief are also present in this province positive linear 'flower' structures.

In the Central Sabah Sub-basin, reservoirs include, in addition to the Tanjong, the Sandakan/Bongaya formations or their equivalents, and the Gomantong Limestone or its equivalent. However, it should be noted that the reservoir quality varies from very tight to friable. The less crestally faulted anticlinal traps of the growth faulted Tanjong (and Sandakan/Bongaya sediments) are attractive targets, especially where the reservoirs are in lateral continuity with the possible mature source rocks in the synclinal or basinal areas, e.g. Pad Basin.

Source Rocks

Conditions for hydrocarbon generation have been attained in parts of the basins, as proven by the presence of hydrocarbons. According to Graves and Swauger (1997) the understanding of the source rock quality, its volume and distribution poses a primary challenge for continuing exploration activity in this area. As such, the palaeogeographic development throughout the Miocene plays an important part in attempting to determine the above factors.

Offshore

In the Central Sabah and Sandakan sub-basins, poor to marginal hydrocarbon source rock potentials are possessed by the Tanjong Formation equivalent and Segama Group and the Dent Group, namely the Ganduman and Sebahat formations (Bates, 1995). Geochemical review of earlier wells are given in Connan and Porthault (1971) and Western Mining Corporation (1991b).

The Segama Group sequence, in particular the Tungku Formation, was deposited during an overall regressive stage with coal and plant-rich clays overlying tuffaceous/carbonaceous sandstones and conglomerates (Libong Tuffite) and highly disturbed tuffs, slump breccias, chert and pebbly mudstones (Ayer Formation). The Tanjong Formation appears to extend into the offshore area as indicated by circular and sub-circular features on seismic data. Total organic carbon contents of the Tanjong Formation equivalent are generally poor to good (range : 0.10 - 2.23 wt%, average =0.64 wt%). As described earlier, the Tanjong-filled onshore circular basins appear to extend offshore based on seismic data. One such depression is the Pad Basin (Fig. 23.10), off the Dent Peninsula, at the mouth of Kinabatangan River, which is interpreted to be filled by 'Older Sebahat' or Tanjong Formation equivalent sediments (Ismail Che Mat Zin, 1994). Pad-1 well penetrated shales with TOC values as high as 15% at the interval 1275m to 1510m, i.e. in the interpreted Tanjong Formation equivalent section. Although, it is not known if rocks of similar high organic carbon content occur at deeper levels in the Pad Basin, it is sufficiently deep for the potential source rocks to be mature. Interestingly, the majority of the exploration wells drilled in the Sandakan Sub-basin encountered hydrocarbons in the Lower to Middle Miocene units, which have been correlated to the Segama Group (most likely the Tungku Formation) or the equivalent of the Tanjong Formation.

Corresponding Hydrogen Index (HI) values of the Tanjong Formation equivalent in Mutiara Hitam-1 are particularly promising, ranging between 86-243 with an average of 134. This formation is thus expected to generate mainly gas and possibly condensate upon maturation.

The Dent Group formations contain only poor to fair organic carbon contents with values ranging from 0.11 to 1.38 wt% (average=0.64wt%). The generally low corresponding HI values (range : 45-177, average = 108) indicate these formations would generate mainly gas. The Sebahat Formation comprises dark grey to black massive claystone with carbonaceous limestone, argillaceous limestone concretions, sandstone and conglomerate beds. Reducing conditions were thought to have prevailed based on the abundance of pyrite. The Ganduman Formation is primarily a sequence of sandstone, clay and lignite deposited within a range of coastal plain-deltaic-neritic environmental settings. Leaves and resin are reported to have been preserved.

In both groups, kerogen is predominantely vitrinite and inertinite with minor oil-prone liptinite. This suggests that the sequence is mainly gas-prone with relatively minor oil or condensate generating potentials.

Graves & Swauger (1997) reported an Early to Middle Miocene source interval which corresponds to the Segama Group/Tanjong Formation equivalent in the Sandakan Sub-basin, in the Clotilde-1 and 333-1 wells in the Philippines waters. This source interval displays some influence from marine organic matter, has correspondingly high HI values (275-350) and mixed oil and gas potential. Indeed, sediments of similar quality have yet to be penetrated within the Malaysian waters but, as will be discussed later, the oil from such a source rock may well have been encountered in Mutiara Hitam-1.

Onshore

Recent geochemical evaluation of onshore outcrops samples of the Tanjong and Sandakan formations, as well as the Garinono olistostrome and the Gomantong Limestone indicate the following conclusions (Abdul Jalil Muhamad, 1993; Azlina Anuar and Abdul Jalil Muhamad, 1995) :

- Tanjong Formation : Fair to good TOC, mainly terrigenous organic matter input; poor hydrocarbon generating potential; immature to early mature.

- Sandakan Formation : Very low to very high TOC, mainly terrigenous organic matter input. The coals have excellent hydrocarbon generating potential for both liquid and gaseous hydrocarbons; immature.

- Garinono shales : Low TOC, poor hydrocarbon generating potential, mature.

- Gomantong Limestone : Low TOC, mixed marine and terrigenous organic matter source with higher land plant input being more abundant; may generate small amounts of gas.

It should, however, be noted that coals are also present in the Tanjong Formation and they may have good to excellent hydrocarbon generating potential for both liquid and gaseous hydrocarbons.

Outcrop samples from the Dent Group (Ganduman and Sebahat formations) and the Segama Group/Tanjong Formation equivalent all exhibited poor to negligible hydrocarbon source rock potentials with Types III/IV kerogen predominating. They are all immature (VRo < 0.4%) and as such, the source potentials indicated can be regarded as their true hydrocarbon generating abilities. However, it is also possible that extensive weathering may have affected the organic matter by reducing their hydrogen contents.

Maturity

For an average geothermal gradient of 35°C/km, the potential hydrocarbon source rocks, namely the Tanjong Formation equivalent, and the Sebahat and Ganduman formations of the Dent Group, should be buried to at least 3000m before becoming effective source rocks. A lower

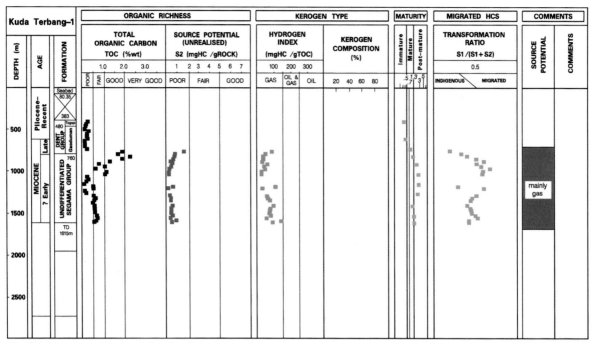

Fig. 23.14 Geochemical summary log of Kuda Terbang-1.

geothermal gradient of 21°C/km was observed in Sebahat-1 and is thought to be the result of blanketing by the thick Sebahat Formation shale sequence. This effectively moves the top of the hydrocarbon generation window in the southern Dent Peninsula area to a relatively deeper level.

The measured vitrinite reflectance profile of Mutiara Hitam-1 shows that the Segama Group/Tanjong Formation equivalent is presently early mature, with VRo values ranging between 0.52 - 0.69%; the Dent Group is still immature (Fig. 23.13). In Kuda Terbang-1, however, the Sebahat Formation is absent and the Ganduman/Togopi Formation directly overlies the Segama Group/Tanjong Formation equivalent. A large difference in measured VRo values is observed at this boundary, namely 0.40% in the overlying Ganduman/Togopi Formation and 1.11% in the underlying Segama Group/Tanjong Formation equivalent (Fig. 23.14). The latter is basically mature to overmature, indicating that it must have undergone burial to depths greater than 3000m prior to being uplifted to its present position. Any hydrocarbon generating potential that it may have had would have been spent.

A number of depressed pyrolysis Tmax values were recorded. This phenomenon is common in samples containing migrated hydrocarbons (Production Index (PI) values greater than 0.2), and also in the coal samples of NW Sabah. These Tmax values thus cannot be used as an indicator of sediment maturity. High PI values observed within the overpressured zones are deemed to be due to low expulsion efficiencies which, in turn, result in higher than normal hydrocarbon accumulations.

Kitchen areas

Hydrocarbon kitchens occur locally within individual normal fault blocks, for example in the Nymphe complex and the Mutiara Hitam-1 area (Fig. 23.7). A large NE-SW trending syncline, the Central Trough, occurs to the east of the Manalunan-1 and Kuda Terbang-1 wells. Based on the presence of hydrocarbons in Kuda Terbang-1 and its absence in Manalunan-1, hydrocarbon migration from the Central Trough was northwards towards the Pegasus Ridge. Another possible hydrocarbon kitchen that has not been tested, is the Pad Basin. The lack of hydrocarbons in Pad-1, that tested a stratigraphic trap (a sand body encased in shale) to the east, is insufficient evidence to assess the syncline as non-generative.

Overpressures

Overpressures were seen in a few wells (Fig. 23.12) with the tops varying between 1597m in the Gem Reef-1 to 3840m in Sulu Sea 409-1. Only Gem Reef-1 drilled through the overpressure zone, where there was an indication of overpressured conditions preventing effective migration. Additionally, maturation was reported to be suppressed within this overpressured zone, indicated by static maturation indices throughout the zone and their rapid increase below the zone. As such, it is expected that the oil and gas are being generated at greater depths and higher temperatures. This apparent retardation of hydrocarbon generation basically agrees with results and observations of various workers (Osborne & Swarbrick, 1997 and references therein). However, there are exceptions to this, for example in the Baram Delta where overpressured zones had elevated maturity levels, suggesting that overpressure actually enhanced the maturation process. It is clear that the effects

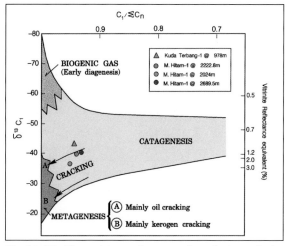

Fig.23.15 The origin and state of evolution of the Sandakan/Central Sabah sub-basins pooled gases.

of overpressure on organic matter maturation, at least in this region, is not yet well understood.

Hydrocarbon Characterisation

Bulk Properties

Table 23.2 lists the bulk properties of the hydrocarbons encountered in the Sandakan Sub-basin. The oils are generally medium to light gravity crudes (32° and 51° API), Their low to very low pour points are consistent with the correspondingly low wax contents. Based on the low ph contents, a non-marine organic source facies for the oils is envisaged. Characterization of the Mutiara Hitam-1 gas samples indicates that dry methane gas constitutes between 91-93% of the bulk composition, with minor non-hydrocarbon gases and wet gases. Stable carbon isotope values obtained from the methane component (-36.4 to -39.8 ppt) indicate that these gases were generated from a late mature source rock (Table 23.2). Similarly, the Kuda Terbang-1 gas sample also had a high dry methane gas content (95% of total gases) with a CH_4 isotope value of -40.0 ppt, thus indicating a thermogenic origin. These thermal maturity levels are equivalent to vitrinite reflectance values of approximately 1.0% up to greater than 2.5%, between the late stages of kerogen catagenesis and the early cracking of oil into gaseous products (Fig. 23.15).

Two distinct hydrocarbon types can be identified in the overlapping Sandakan/Central Sabah sub-basins through organic geochemical methods; one generated from a marginal fluvial-coastal marine source rock, and the other from a more terrigenous-influenced environment with evidence of resins and higher land plant input, possibly in an estuarine or a delta front setting.

Whole oil gas chromatographic traces of the Mutiara Hitam-1 samples displayed two different envelope shapes; shallower samples had relatively mature, unimodal n-alkane envelopes with the n-alkane distribution ranging between nC_9 up to nC_{31}, while the deeper sample exhibited a trace typical of a condensate-like oil with the n-alkane distribution ranging from nC_7 to nC_{27} (Fig. 23.16). Pr/Ph ratios between 2.26 and 7.33 suggests that the depositional conditions were moderately oxic to sub-oxic, possibly in a fluvio-deltaic setting. The two oils in the Kuda Terbang-1 well had similar gas chromatographic fingerprints, that are typical of biodegraded oils, with the significant depletion of n-alkanes and unresolved cyclo-alkane humps (Fig. 23.16). Again, Pr/Ph ratios between 3.0 and 3.3 suggest a moderately oxic depositional environment.

The shallow Mutiara Hitam-1 samples correlate well with each other and they were expelled from a marginal fluvial marine source rock. Positive oil-oil correlations were also established for the deep Mutiara Hitam-1 condensate samples with those encountered in Kuda Terbang-1, and these were generated by a more terrestrially-influenced source horizon.

More insight as to the origins of these oils was gained from gas chromatograph-mass spectrometry analysis. The hydrocarbons found in Mutiara Hitam-1 were confirmed to have been generated by at least two different source rocks. These source rocks could either be discrete horizons or they could also be a single source horizon with lateral variations in its organic facies. The shallower RFT sample was generated by a source rock with generally marine characteristics within the early to middle oil window thermal maturity level (Figs. 23.13 and 23.16). Common features include the relatively low concentrations of higher plant derived resins and oleanane, the extension of the homohopanes up to nC_{35} and the comparable amounts of the C_{27} and C_{29} regular steranes. As such, a marginal-

2(a) Oils

Well	HC	Group Formation	Gravity (°API)	Wax (wt%)	Sulphur (wt%)	Pour pt. (°C)
Nymphe North-1	Oil (DST)	Segama Group Tanjong Formation Eq.	46	-	-	-
Nymphe-1	Condensate (DST)		51	-	-	-
Mutiara Hitam-1	Oil (RFT) 2089.2m		38.6	< 0.5	< 0.05	9
	Oil (RFT) 2222.8m		34.7	< 0.5	< 0.05	0
	Condensate (DST-1A) @ 2689.5m		39.6	< 0.5	< 0.05	< -30
Kuda Terbang-1	Oil (RFT) 953.1m		31.7	< 0.5	0.05	< -30
	Oil (RFT) 978.0m		32.3	< 0.5	0.05	< -30

2(b) Gases

Well	Hydrocarbon type	Strat. Group	Specific Gravity	CO_2 (%)	C_1 (%)	> C_2 (%)	C_1 / total C_n	$\delta^{13}C_1$ (ppt)
Mutiara Hitam-1	Gas (RFT) @ 2024.0m	Undiff. Segama Group	0.605	< 2.0	92	6	0.94	-38.8
	Gas (RFT) @ 2089.2m		0.607	<2.0	91	-	-	-
	Gas (RFT) @ 2222.8m		0.612	< 2.0	91	7	0.93	-39.8
	Gas (RFT) @ 2687.1m		0.613	< 2.0	92	-	-	-
	Gas (DST-1A) @ 2689.5m		0.616	1.6	93	5.4	0.95	-36.4
Kuda Terbang-1	Gas (RFT) @ 978m		0.601	2.28	95	2.72	0.9700	-40

Table 23.2 Bulk properties of hydrocarbons recovered from the overlapping Sandakan and Central Sabah sub-basins, offshore east Sabah.

coastal marine depositional environment situated laterally away from the main river flow of an estuarine or a delta front is envisaged for its source rock. The presence of a marine-influenced source rock was recently proven by the Clotilde-1 and Sulu Sea 333-1 wells (Graves & Swauger, 1997) drilled in the Philippines waters.

By contrast, the deeper condensate-like oil sample recovered from the drillstem test had more higher land plant indicators, generally low amounts of algal-derived biomarkers and the relatively higher abundance of C_{29} to C_{27} regular steranes (Figs. 23.13 and 23.16). Evidence for a possible slight marine influence can be seen in the presence of the homohopane series up to nC_{35}. The source rock of this oil would have been deposited in a lower coastal plain-delta front type of setting with higher land plant debris brought in by the river system. The maturity level of this oil is also higher than its shallower counterparts, as shown by the equilibrated C_{32} S/(S+R) homohopane ratio. Consequently, multi-phase migration is thought to have occurred, with an earlier phase involving the older Tanjong Formation equivalent marginal fluvial-coastal marine source rock and a later phase which involved the younger, more terrigenous-influenced Dent Group source rock. The Kuda Terbang-1 RFT oil samples resemble the deeper DST sample of Mutiara Hitam-1 (Fig. 23.14 and 23.16). From the m/z 412 triterpane traces, high quantities of C_{30} resins and oleanane, which are

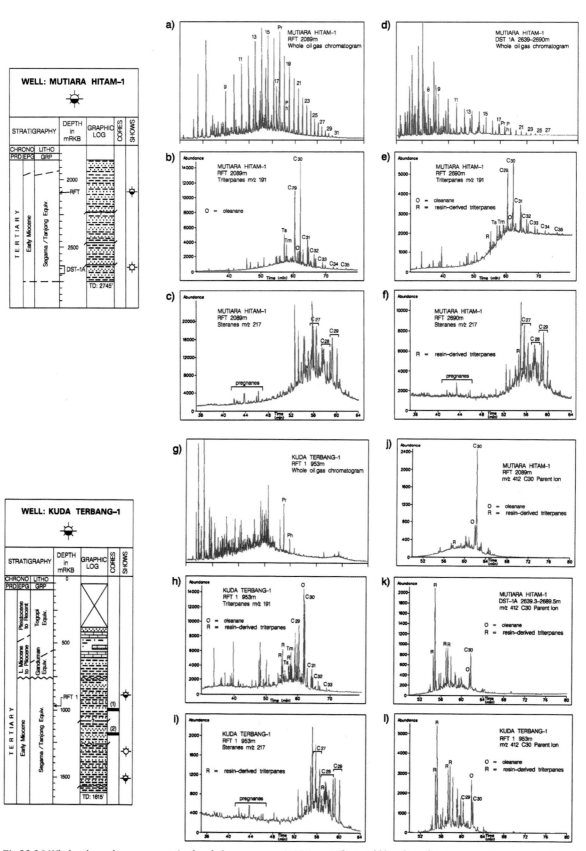

Fig.23.16 Whole oil gas chromatograms (a, d and g), triterpane m/z 191 traces (b, e and h) and regular sterane m/z 217 traces (c, f and i) for Mutiara Hitam-1 shallow sample (RFT 2089m), Mutiara Hitam-1 deep sample (DST 2690m) and Kuda Terbang-1 sample (RFT 953m), respectively. From the m/z 412 traces (j, k and l), it is clearly observed that sample Mutiara Hitam-1 RFT 2089m (j) lacks higher landplant input compared to the other two samples (k and l) which show abundant resin-derived compounds.

indicative of significant higher land plant input to the source facies, are clearly seen in the two Kuda Terbang-1 RFT samples and the deeper Mutiara Hitam-1 DST sample, but are relatively low in the shallower RFT samples of Mutiara Hitam-1 (Fig. 23.16). Both marginal -coastal marine and lower coastal plain-delta front source facies are present within the Tanjong Formation equivalent (or Segama Group-probably Tungku Formation) and the Ganduman and Sebahat formations (Dent Group). However, direct oil to source rock correlations are currently unavailable.

SUMMARY

Hydrocarbon discoveries have been made in the overlapping areas of Central Sabah and Sandakan sub-basins mainly in thin sands of Middle and Early Miocene age, where NS to NNE-SSW trending anticlinal structures with marked vertical relief are developed. These sands are correlatable to the Central Sabah Sub-basin Tanjong Formation exposed onshore in the circular and subcircular basins. On seismic sections, similar features (e.g. Pad Basin) are also present in the overlapping areas i.e. Northern Structural Province. The late Middle to Upper Miocene sequences (Sebahat and Ganduman formations of the Dent Group) of the Sandakan Sub-basin also contain hydrocarbons but are poor reservoirs compared to the pre-Dent Group formations. Moreover, they are also deeply eroded.

Hydrocarbon potential still exists in the undrilled fault blocks in the vicinity of the following discoveries : Nymphe North-1, Nymphe-1 and Benrinnes-1. Undrilled faulted anticlinal and dip-closed structures in the Northern Structural Province offer the best prospects, particularly where nearby synclinal basins (e.g. Pad Basin, Kinabatangan Trough) containing the Tanjong Formation equivalent shales as possible mature source rocks are present. Thicker sands would be attractive targets.

Untested plays include the predicted basin flow and basin slope sandstone reservoirs encased in shales, west of the prograding sequence, in the Central Trough of the Sandakan Sub-basin. The predicted turbiditic sands are probably time-equivalent to the Middle-Upper Miocene Sebahat Formation. Another play is the predicted carbonate build-ups, updip from Gem Reef-1. This play is similar to the Sentry Bank Reef-1 in the Philippines. However, it should be noted that similar carbonate plays have been tested unsuccessfully in the Philippines.

In the Southern Structural Province of the Sandakan Sub-basin, the least explored part of the sub-basin, anticlinal structures trending ENE-WSW are present. However, the presence of reservoirs and source are speculative. Towards the south, near the Philippines boundary, probable carbonate buildups of Early-Middle Miocene age are potential exploration targets. It appears that similar carbonate build-ups equivalent to the Lower-Middle Miocene Gomantong Limestone, are also potential exploration targets in the offshore Labuk Bay area in the Central Sabah Sub-basin. Reservoir quality is, however, unknown. The Tanjong and Sandakan/Bongaya sands are also exploration targets in the Central Sabah Sub-basin.

In summary, 3D seismic surveys need to be acquired over the discoveries to resolve the fault patterns in the overlapping Central Sabah and Sandakan sub-basins. Seismic modelling to define reservoirs, and refinement of geological model are also required prior to a new exploration round in both the Sandakan and Central Sabah sub-basins. Identification and mapping of seismic closures over the pre-Dent Group Tanjong Formation equivalent in the overlapping Central Sabah and Sandakan sub-basins need to be better defined.

REFERENCES

Abdul Jalil Muhammad, 1993. Geochemical evaluation of Tertiary Sediments. PETRONAS Report No. PRSS-RP5-93-09.

ASCOPE, 1984. Tertiary sedimentary basins of the Southwest Sulu Sea, Makassar Strait and Java Sea : stratigraphy, structure and hydrocarbon occurrences. Technical Paper TP/3 Asean Council on Petroleum (ASCOPE).

Azlina Anuar and Abdul Jalil Muhammad, 1995. Geochemical evaluation of the Sandakan-Lahad Datu-Dent Peninsula areas, East Sabah, Malaysia. PETRONAS Report No. PRSS-RP5-95-03.

Bates, C.R., 1995. Geologic well evaluation reports : Kuda Terbang1; Mutiara Hitam-1. WMC Petroleum (Malaysia) Sdn. Bhd.

Beddoes, L.R., 1976. The Balabac Sub-Basin Southwestern Sulu Sea, Philippines. SEAPEX Program, Offshore South East Asia Conference, Paper 13.

Bell, R.M. and Jessop R.G.C., 1974. Exploration and Geology of the West Sulu Basin, Phillipines. Australian Petroleum Exploration Association Journal. 1, 21-28.

CCOP, 1991. Total sedimentary isopach map, offshore East Asia. Sheet 4, 1:2,000,000 scale.

Clennell, B., 1991. The origin and tectonic significance of melanges of Eastern Sabah, Malaysia. In : Nichols, G. and Hall, R. (eds.) Proceedings of the Orogenesis in Action conference. London 1990. Journal of Southeast Asian Earth Sciences, 6, 407-425.

Clennell, B., 1996. Far-field and gravity tectonics in Miocene Basins of Sabah, Malaysia. In : Hall, R. and Blundell, D.J. (eds.). Tectonic Evolution of Southeast Asia, Geological Society of London Special Publication, 106, 307-320.

Collenette, P., 1965. The geology and mineral resources of the Pensiangan and Upper Kinabatangan area, Sabah, Malaysia. Geological Survey, Borneo Region, Malaysia, Memoir 12.

Comexco, Inc., 1993. Hydrocarbon potential of the Sibutu block (GSEC-70). The Philodrill Corporation, Report No. BD1.:Philodrill:5:93-01.

Connan, J. and Porthault, B., 1971. Alkanes analysis of some samples of Gem Reef-1 (GER 1), Benrinnes-1 (BER 1) and Magpie West-1 (MPW 1), Sabah, Malaysia. Elf Aquitaine Report No. 183/71, PETRONAS Exploration Department Report No. WD.1:Aqui:2:Gem Reef-1/3.

Durkee, E.F., 1993. Oil, geology and changing concepts in the Southwest Philippines (Palawan and the Sulu Sea). Bulletin of the Geological Society of Malaysia, 33, 241-262.

Fontaine, H. and Ho, W.K., 1989. Note on the Madai-Baturong Limestone, Sabah, East Malaysia; Discovery of Caprinidae (Rudists). CCOP Newsletter, 14, 27-32.

Gopakumar, V., 1993. Wells haven't tested Sandakan Basin potential. Asian Oil and Gas, 20-26.

Graves, J.E. and Swauger, D.A., 1997. Petroleum systems of the Sandakan Basin, Philippines. In : J.V.C. Howes and R.A. Noble (Eds.), Proceedings of the Indonesian Petroleum Association, Petroleum Systems of SE Asia and Australasia Conference, Jakarta, Indonesia, May 1997, Indonesian Petroleum Association, 799-813.

Haile, N.S. And Wong, N.P.Y., 1965. The geology and mineral resources of the Dent Peninsula, Sabah. Geological Survey, Borneo Region, Malaysia, Memoir 16.

Hinz, K., Kempter, E.H.K. and Schluter, H.U., 1985. The Southern Palawan-Balabac area : an accreted or non-accreted terrane? ASCOPE Proceedings, 2, 48-72.

Hinz, K., Fritsch, J., Kempter, E.H.K., Manaf Mohammad, A., Meyer, J., Mohamed, D., Vosberg, H., Weber, J., and Benavidez, J., 1989. Thrust tectonics along the northwest continental margin of Sabah, Borneo. Geologische Rundschau, Band 78, Heft 3.

Hinz, K., Block, M., Kudrass, H.R., and Meyer, H., 1991. Structural elements of the Sulu Sea, Philippines. Geol. Jahrb. Reihe A., v. 127, 483-506.

Hutchison, C.S., 1988. Stratigraphic-tectonic model for eastern Borneo. Bulletin of the Geological Society of Malaysia, 22, 135-151.

Hutchison, C.S., 1992. The Southeast Sulu Sea, a Neogene marginal basin with outcropping extensions in Sabah. Bulletin of the Geological Society of Malaysia, 32, 89-108.

Ismail Che Mat Zin, 1994. Dent Group and its equivalent in the offshore Kinabatangan area, East Sabah. Bulletin of the Geological Society of Malaysia, 36, 127-143.

Lee, D.C.T., 1970. Sandakan Peninsula, Eastern Sabah, Eastern Malaysia. Geological Survey of Malaysia, Report 6.

Lee, D.C.T., 1990. Formation of Pulau Batu Harian and other islands around Pulau Banggi, northern Sabah. Bulletin of the Geological Society of Malaysia 26, 71-76.

Lee, C.P. and Tham, K.C., 1989. Circular basins of Sabah. Proceedings of the Geological Society of Malaysia Petroleum Seminar, Kuala Lumpur, 1989 [abstracts]. Bulletin of the Geological Society of Malaysia, 29, 54.

Leong, K.M., 1974. The geology and mineral resources of the Upper Segama Valley and Darvel Bay area, Sabah, East Malaysia. Geological Survey of Malaysia Memoir 4 (Revised), 354pp.

Leong, K.M., 1976. Miocene chaotic deposits in eastern Sabah: characteristics, origin and petroleum prospects. Geological Survey of Malaysia, Annual Report for 1975, 238.

Lim, P.S., 1985. Geological Map of Sabah (3rd Edition), Geological Survey of Malaysia.

Osborne, M.J. And Swarbrick, R.E., 1997. Mechanisms for generating overpressure in sedimentary basins : A reevaluation. American Association of Petroleum Geologists Bulletin, 81, no.6, 1023-1041.

Rangin, C. 1989. The Sulu Sea, a marginal basin setting with a collision zone. Tectonophysics, 161, 119-141.

Rangin, C., Bellon, H., Benard, F., Letouzy, J., Muller, C. and Sanudin, T., 1990. Neogene arc-continent collision Sabah, North Borneo (Malaysia). Tectonophysics, 183, 305-319.

Tamesis, E.V., 1990. Petroleum geology of the Sulu Sea Basin, Philippines. Offshore South East Asia Conference 1990, 109-118.

Tjia, H.D., 1988. Accretion tectonics in Sabah : Kinabalu Suture and East Sabah accreted terrane. Bulletin of the Geological Society of Malaysia, 22, 237-251.

Tjia, H.D., Ibrahim Komoo, Lim, P.S. and Tungah Surat, 1990. The Maliau Basin, Sabah : geology and tectonic setting. Bulletin of the Geological Society of Malaysia, 27, 261-292.

Tokuyama, A. and Yoshida, S., 1974. Kinabalu Fault, a large strikeslip fault in Sabah, East Malaysia. Geology and Palaeontology of Southeast Asia, 14, 171-188.

Tongkul, F., 1988. Geologic control on the birth of Pulau Batu Harian mud volcano, Sabah. Warta Geologi, Geological Society of Malaysia, 14, 153-165.

Tongkul, F., 1991. Tectonic evolution of Sabah, Malaysia. In : Nichols, G. N. and Hall, R. (eds.) Proceeding of the Orogenesis in Action Conference London, 1990. Journal of Southeast Asian Earth Sciences, 6, 395-405.

Tongkul, F., 1993. Tectonic control on the development of the Neogene basins in Sabah, East Malaysia. Bulletin of the Geological Society of Malaysia, 33, 95-103.

Tongkul, F., 1997. Polyphase development in the Telupid area, Sabah, Malaysia. In : Geosea 1995 Proceedings, Journal of Asian Earth Sciences, 15, 175-184.

Walker, T.R., 1993. Sandakan Basin prospects rise following modern reappraisal. Oil and Gas Journal May 10, 1993, 43-47.

Western Mining Corporation, 1991a. SB6 Farm-Out Brochure (unpublished).

Western Mining Corporation, 1991b. A review of the petroleum geochemistry, Block SB6. (unpublished).

Western Mining Corporation, 1995a. Regional Geological/Geophysical Cross-Sections, Block SB6, Sabah (unpublished).

Western Mining Corporation, 1995b. Prospectivity Review of Remaining Leads, SB6, Sabah (unpublished).

Wilford, G.E., 1967. Geological Map of Sabah (2nd Edition), Geological Survey, Borneo Region, Malaysia.

Wong, R.H.F., 1993. Sequence stratigraphy of the Middle Miocene Pliocene, southern offshore Sandakan Basin, East Sabah. Bulletin of the Geological Society of Malaysia, 33, 129-142.

Wong, R.H.F. and Salahuddin Saleh Karimi, 1995. Regional interpretation of SBP93 seismic lines, Block SB5. PETRONAS Internal Report.

Chapter 24

Southeast Sabah Basin

Leong Khee Meng and Azlina Anuar

INTRODUCTION

The term Southeast (SE) Sabah Basin was first introduced by Leong (1978) and can be broadly divided into (i) an eastern platform area and (ii) a western deep area (Fig. 24.1). The latter was known as the Tidung Sub-basin and is recently referred to as the Northern Tarakan Sub-basin (PETRONAS, 1994). It comprises uplifted and eroded Palaeogene to Early Neogene sediments and is considered as the onshore equivalent of the petroliferous Late Neogene Tarakan Basin in Kalimantan, Indonesia (Wight et al., 1993).

EXPLORATION HISTORY

The area was earlier known for its coal mining activities between 1904-1932. The coal measures occur in the Lower Miocene strata in the Silimpopon Syncline area (Collenette, 1954). Following the discovery of oil in Pulau Tarakan in 1905, similar antiformal structures in eastern Borneo were explored for hydrocarbons. Shell drilled a well in Pulau Sebatik in 1914, adjacent to a gas seepage. However, only insignificant oil shows were encountered and the well was abandoned at 645 m.

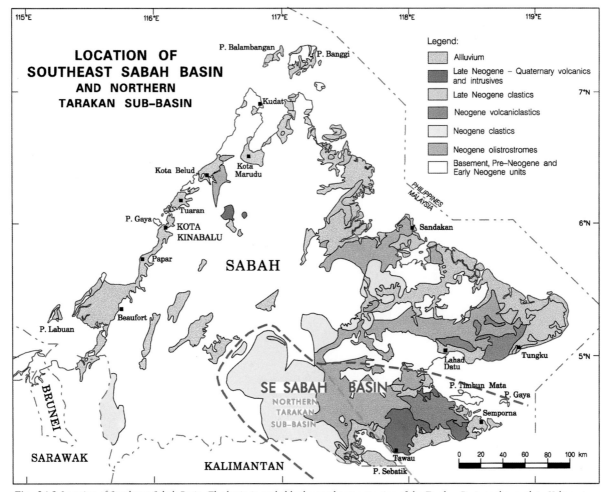

Fig. 24.1 Location of Southeast Sabah Basin. The basin is probably the northeast extension of the Tarakan Basin to the south in Kalimantan Indonesia.

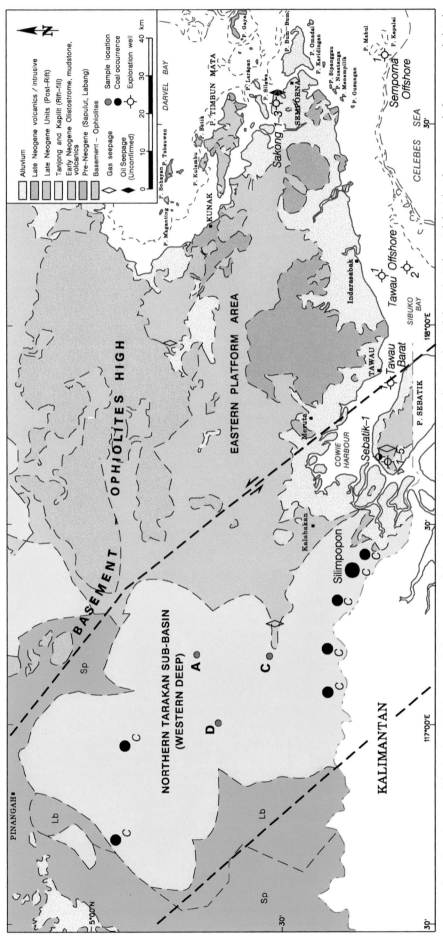

Fig. 24.2 The geological setting of the Southeast Sabah Basin. The basin consists of the western deep area known as the Northern Tarakan Sub-basin (or Tidung Sub-basin) and an eastern platform area, underlain by volcaniclastics. Geology simplified from Lim (1985).

Prior to 1989, exploration was concentrated mainly on the platform area of the SE Sabah Basin. The onshore SE Sabah Basin acreage, including the Semporna Peninsula, was acquired by a Shell company for oil exploration in 1934. Oil seepage had been reported at Kuala Sakong on the northeastern tip of the Semporna Peninsula in 1948 (see Kirk, 1962). Subsequently in 1950, 3 shallow coreholes, Sakong 1-3, were drilled near the reported seep. The coreholes encountered a Quaternary sequence of marls, shale and volcanic rocks with a maximum depth of 188m. Two of the coreholes were dry. Sakong-2 encountered methane gas at 57m. The acreage was relinquished in 1951 because of unfavourable results and geology. The first comprehensive geological surveys were those conducted by Geological Survey Department. Their work (Kirk, 1962; Collenette, 1965) provided a sound stratigraphic framework and a preliminary structural model for southeast and east Sabah. Recent work focused more on the Late Neogene igneous rocks (Lim, 1981; Lee, 1988), including onshore Balung Formation by Lim (1981).

Modern oil exploration began in the mid-1960s when Sabah Teiseki commenced exploration in the platform area of the SE Sabah Basin. The work included aeromagnetic and gravity surveys, reflection seismic on Cowie Harbour and field geological mapping. The most striking gravity anomaly is the positive feature associated with the Sebatik Anticline. From 1966-1967, 5 shallow wells (Sebatik ST 1-5) were drilled in Pulau Sebatik, near the site of the 1912 well, without any success. The 5 wells each penetrated a thin Pleistocene sequence (Wallace Formation) unconformably overlying an undated but presumed Miocene sequence of hard, dark mudstone with thin sandstone interbeds (Kalabakan or Kalumpang Formation equivalent). Up to 800 m of the Miocene section was penetrated in each of the 5 wells; none of the wells intersected a complete section.

Following seismic and broad aeromagnetic surveys in 1969 to 1970, 3 offshore wells were drilled in 1970. Wells Tawau Offshore-1 and Tawau Offshore-2 penetrated a Recent and Late Miocene section and also encountered a 200-300m thick limestone within the predominantly Upper Miocene clastic sequence (Balung Formation). Gas (almost entirely methane) was tested, but the flow rate was too small to measure, from a bioclastic limestone in Tawau Offshore-1. Another well, Semporna Offshore-1, encountered igneous rocks at shallow depth (912m). Subsequently the area was relinquished in stages between 1973 - 1977. More details of the exploration activities conducted by Sabah Teiseki are given in Chung et al. (1977).

The more recent exploration activities by Sun Malaysia Petroleum Company in 1989-1991 (Sun, 1991) focused specifically on the onshore areas containing the circular, subcircular and synclinal basins, extending to Cowie Harbour, ie. the Northern Tarakan Sub-basin or the deep portion of SE Sabah Basin (Figs. 24.1 and 24.2). The activities included extensive field work, petrographic and mineralogical studies, acquisition of synthetic aperture data (SAR), gravity and magnetic data onshore and offshore seismic data, amounting to 358 and 182 km respectively, and identification of several onshore and offshore leads/prospects. The exploration activities culminated in the 1992 drilling of an offshore exploration well, Tawau Barat-1, in Cowie Harbour, without success. Tawau Barat-1 was drilled to a total depth of 6700ft (2042 m) and encountered a clastic sequences dominated by shales and mudstones from Late Pliocene and younger to Late Miocene or earlier (Fig. 24.5). Geochemical evaluations indicate the clastic sections possess good organic richness, although of high thermal maturity.

REGIONAL GEOLOGY

The SE Sabah Basin comprises 5 units - (i) Crystalline Basement - Ophiolite Complex and associated sediments (ii) Pre-Neogene sediments (iii) Olistostrome-chaotic and mudstone units (iv) Neogene sediments and (v) Late Neogene to Quaternary volcanic and minor intrusives. In addition to published reports, the following unpublished reports were referred - Technical Brochure Block SB7 by PETRONAS (1990), Integrated Interpretation of Data from Exploration Surveys, 1989-1991, SB8, Sabah, Malaysia by Sun (1991), Biostratigraphy, Geochemistry and Sedimentology by Robertson Research for Sun (1992) and Core Laboratories Report of Tawau Barat-1 (Abolins, 1993).

Stratigraphy

The onshore stratigraphy (Fig. 24.3) can be divided into pre-olistostrome and post-olistostrome associations. In the SE Sabah Basin, the olistostromes, or chaotic deposits, include the Kuamut and Kalumpang formations and the predominantly mudstone Kalabakan Formation.

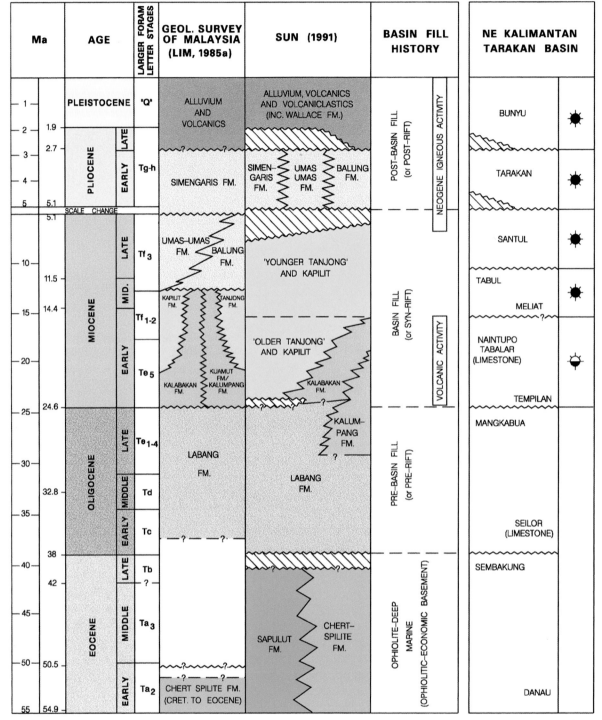

Fig. 24.3 Stratigraphy of the Southeast Sabah Basin with basin fill history and correlation of equivalent units in the Tarakan Basin.

The pre-olistostrome associations are represented by the following units:

(i) Crystalline Basement:
igneous and metamorphic complex of Triassic and/or earlier age

(ii) Chert-Spilite Formation (and associated ultrabasic and basic igneous rocks): ophiolite sequence (Cretaceous-Eocene)

(iii) Sapulut Formation:
deep water clastics (Cretaceous-Eocene)

(iv) Labang Formation:
deep water clastic with shallow limestone towards the top (Oligocene)

The Chert-Spilite and the Sapulut formations are probably co-eval.

The eastern platform area of the SE Sabah Basin consists mainly of olistostromes and predominantly mudstone units with tuffaceous layers (Kalumpang, Kalabakan and Kuamut formations), Late Neogene to Quaternary volcanics and minor intrusives, and small outliers

Southeast Sabah Basin

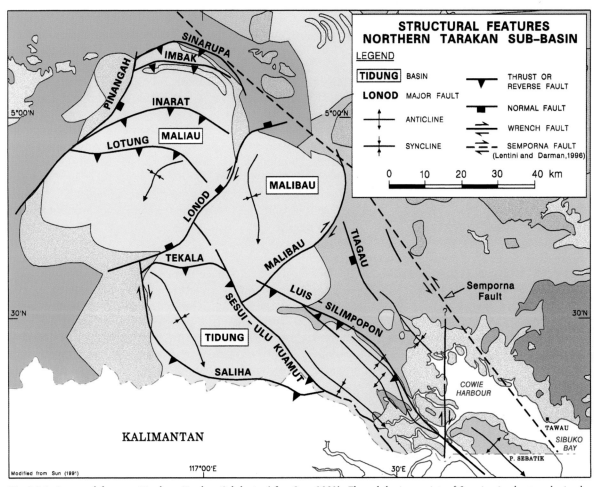

Fig. 24.4 Structural features, Northern Tarakan Sub-basin (after Sun, 1991). The sub-basin consists of 3 main circular to sub-circular basins or depocentres, Maliau, Malibau and Tidung, and several NW-SE trending synclines. Thrust faults bound the circular basins.

of Late Neogene sediments (Balung Formation). The area is bounded to the north by the Crystalline Basement- Ophiolite Complex. As indicated by the results of the offshore wells, Tawau Offshore-1 and Tawau Offshore-2, the (?)Pliocene Balung Formation thickens towards the south. Volcanics were encountered at shallow depth in Semporna Offshore-1. To the west, the Kalumpang and Kuamut formations appear to grade into the more argillaceous Kalabakan Formation. A bounding regional strike-slip fault to the west has been postulated by Lentini and Darman (1996).

The deep area of the SE Sabah Basin, i.e. the Northern Tarakan Sub-basin, consists of an elongated basinal area with a pronounced northwest-southeast structural trend overlying the Labang Formation. Overlying the olistostrome and mudstone associations (Kuamut, Kalumpang and Kalabakan) are the fluvio-deltaic sequences known as the Tanjong and Kapilit formations occurring in circular, or sub-circular basins and synclinal features. The Northern Tarakan Sub-basin consists of three basins (or sub-basins), about 30km in diameter, Maliau, Malibau and Tidung basins ranging from near circular, saucer shaped to sub-circular and elongated synclines (Fig. 24.4) consisting of thick Miocene sequences (Tanjong and Kapilit formations) of almost continuous fluvio-deltaic and marginal marine sedimentation. SAR data geological interpretation and field work have shown that these thick Miocene sequences represent two successive upward-coarsening, prograding fluvio-deltaic sequences (Sun, 1991).

The original total thickness of these Miocene fluvio-deltaic sediments is estimated by Sun (1991) to be 7000m, considerably less than the 12,200m cited by Collenette (1965). The two sequences are also informally referred to as 'Older Tanjong' and 'Younger Tanjong' (Sun, 1991). The former - 'Older Tanjong' - consists of (i) mudstones and siltstones interbedded with thin sandstone beds and minor coals. Some limestones and unconsolidated fine-grained quartzose sandstones are also present; and (ii) predominantly sandstone beds forming escarpments in certain areas. The sandstone beds are up to 80m thick, interbedded with mudstone, siltstone and coal. Conglomerates with chert fragments are locally present. Low rank coal occurs in beds up to 2m thick, particularly towards the top of the unit.

The 'Younger Tanjong' consists of similar lithologies, except for the occurrence of the limestones and unconsolidated sands.

There are scattered small outcrops of post-Tanjong and Kapilit formations, variously known as the Simengaris, Umas-Umas and Balung formations, deposited in shallow marine and swamp environments. They have been assigned a probable Early Pliocene age. The Simengaris Formation occurs as interbedded mudstones and sandstones with a locally developed conglomerate. The sandstones are friable with moderate porosity.

The Umas-Umas Formation crops out as an isolated cuesta 30km northwest of Tawau (Kirk, 1962). The formation consists of sandstone and carbonaceous mudstone. The sandstones are highly compacted, possibly due to tectonism.

The Balung Formation consists predominantly of grey volcanic ash, mudstone, tuff shale and rare thin coral beds. Abundant organic matter, particularly plant remains, is present. The Early Pliocene age of the formation cannot be confirmed. According to Lim (1981), the formation is probably of Middle-Upper Miocene age, based on the presence of sporomorphs and its stratigraphic position. The Balung Formation has been encountered in Tawau Offshore-1 and Tawau Offshore-2 wells, with bioclastic limestones in the upper part of the formation. The Balung Formation is 1488m thick in Tawau Offshore-1 and appears to thicken offshore towards the Tarakan Basin.

The Late Pliocene to Quaternary Units consist of the Wallace Formation - a claystone and tuffaceous deposit with minor sandstone, conglomerate and lignite and Pleistocene terrace deposits, with alluvium; and extensive dacitic, andesitic and basaltic volcanics and minor intrusives in the Tawau and Semporna peninsulas.

Structures in the Northern Tarakan Sub-basin

The Northern Tarakan Sub-basin (Fig. 24.2, 24.4) is bounded to the west by pre-Neogene indurated and deformed sediments (Sapulut and Labang formations), to the north by pre-Neogene sediments and early Neogene chaotic deposits, and to the east by Early Neogene chaotic and mudstone deposits (Kuamut, Kalabakan, Kalumpang). The western boundary of the Maliau basin and the eastern boundary of the Malibau basin are structurally complex zones.

The eastern boundaries of the Northern Tarakan Sub-basin probably represent intensely faulted zones, the manifestation of the NW-SE trending postulated strike-slip fault zones described earlier as 'Kinabalu Fault' (Tokuyama and Yoshida, 1974) or 'Semporna Fault' (Lentini and Darman, 1996).

The Maliau basin is a sub-circular outcrop of Miocene sandstone-mudstone units with slices of Oligocene (Labang) clastics. It is completely bounded by faults - the Pinangah to the west, the curvilinear Lonod to the south and east and by a series of N-verging imbricated thrust faults to the north. The Maliau basin comprise two provinces - the northern zone of imbricated thrusts and the southern near circular, saucer shaped Maliau basin *senso stricto* - designated a Nature Conservation Area by the Sabah State Government. In the northern zone, the cross-cutting relationship of the folds to the thrust faulting indicate that folding may have occurred during initial compression and was subsequently broken by emergent thrusts. The southern zone (Maliau basin *senso stricto*) is a broad, gently deformed synclinal feature with significant volumes of good quality coking coals (Tjia et al., 1990).

The southwest Tidung basin, contains 3 *en-echelon* oval-shaped southeast plunging synclines, and is confined and imbricated by northeast and north-verging thrust faults. The thrust sheets are ramped into the Malibau basin. Major NW-SE folds are parallel or sub-parallel to the thrust fault trace. The eastern boundary of the Tidung basin probably coincides with the regional NW-SE strike-slip fault the 'Kinabalu Fault' or the 'Semporna Fault'.

The least tectonized of the three sub-basins is the Malibau basin. It is a broad NE-SW graben, bounded to the west by the Lonod Fault and to the east by the Malibau Fault. The bounding faults are probably growth faults with strike-slip components.

The origin of the circular features of the Miocene sediments in the Northern Tarakan Sub-basin, as well as in the NE Sabah Basin "Sandakan Rift" - the Bangan, Tangkong, Latangan-Malua, Bukit Garam circular basins trending NE-SE - is still debated. Interested readers can refer to the discussions in Tongkul (1991, 1993) and Clennell (1996). Views proposed to explain the geomorphological shape of the 'basins', i.e. whether post-depositional, combination of syn- and post-depositional and less probable pre-depositional, must take into consideration the great thicknesses of the basin-fill consisting of almost entirely similar depositional environment, the curvilinear fault

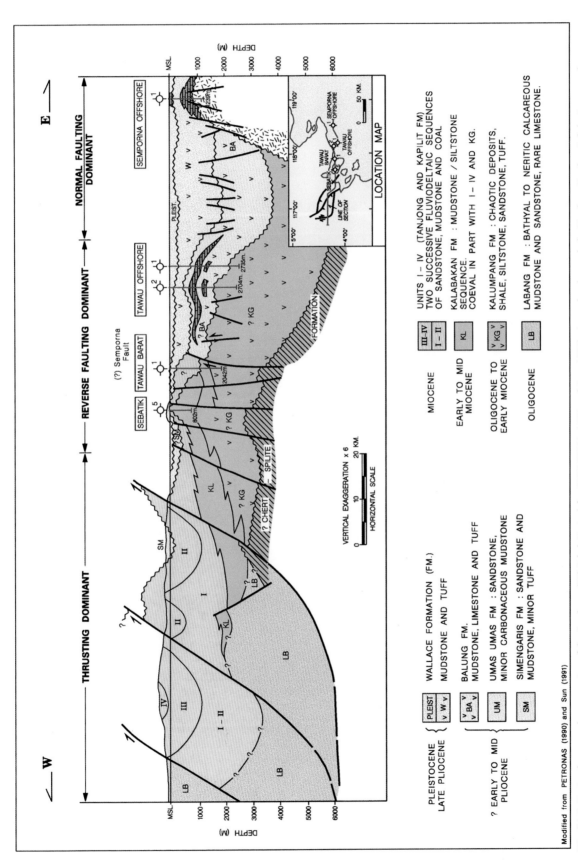

Fig. 24.5 Regional structural cross-section of the Southeast (SE) Sabah Basin (after PETRONAS, 1990 and Sun, 1991). As can be observed, the Northern Tarakan Sub-basin proper has not been explored by drilling.

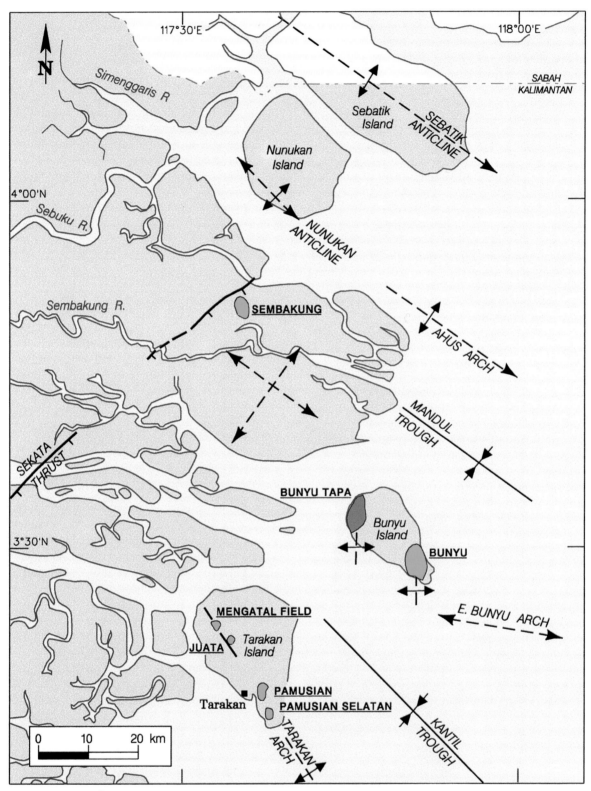

Fig. 24.6 Locations of established oil production in Tarakan Basin, Kalimantan Indonesia (after Wight et al., 1993).

boundaries and the underlying strata of mainly thick mudstone and chaotic assemblages, which are diapiric in places.

Field mapping and SAR interpretation by Sun (1991) support the interpretation of Collenette (1965) that the three circular basins (Maliau, Malibau and Tidung) were originally one major proto-basin. Further evidence is provided by similarities in palynology for the coal zone of the inner Maliau basin (Gunung Lotung Escarpment) and that of the Malibau basin from Broken Hill Propriety (BHP) coal investigations. The palynological similarities show that the coal layers were deposited in a laterally extensive and continuous sequence. The segmentation of the Miocene proto-basin is likely to have occurred in the Late Pliocene compressive event, following the deposition of Early Pliocene transgressive units, Simengaris, Umas-Umas and Balung formations, now being restricted to small outliers. The Miocene fluvio-deltaic sediments had been buried to considerable depths for coalification to occur.

The SAR and photogeological interpretations indicate that almost all of the boundary faults, originally interpreted as listric normal or growth faults, appear to have strike-slip components.

These intense faultings, including imbricate thrusting to the north and northeast, probably occurred in the Late Pliocene compressional tectonic phase, although initial deformation (folding and uplift) of the Miocene-Pliocene sediments could have commenced in the Late Miocene/Early Pliocene. Volcanic activity (with minor intrusives) in the Tawau-Semporna area also reached its peak in the Pliocene and continued through its waning stages to Quaternary.

BASIN EVOLUTION

Following the Late Eocene uplift of the ophiolite-deepwater sediments complex (Chert-Spilite, Sapulut-Trusmadi-East Crocker), depressions were formed and infilled with deep water clastics (e.g. the Labang Formation) during the Oligocene. Renewed uplift of these pre-Neogene units, with deposition of limestones on structural highs, heralded the opening or rifting of the southeastern Sulu Sea and the separation of north and west Sulawesi from east Kalimantan (Lentini and Darman, 1996). The rifts were infilled with chaotic assemblages with mudstone matrix known as olistostromes, intermingled with volcanic material.

The SE Sabah Basin, including the thick Miocene sediment-filled Northern Tarakan Sub-Basin, is believed to be rift-related (Tjia et al., 1990; Wight et al., 1993; Lentini and Darman, 1996). However, the margins of the rift basins are poorly defined. The early basin deposits include the olistrostrome, mudstone and volcaniclastic units i.e. Kalumpang, Kuamut and Kalabakan formations. To the north, in the NE Sabah Basin, similar olistostromes with volcaniclastic material are considered as early rift deposits (Hutchison, 1992).

Following the olistostrome, mudstone and volcaniclastic deposition, the pre-Neogene pre-rift units to the north and west and the early rift olistostromes were uplifted and eroded. This tectonism marked a change of environment from generally deeper water to shallow marine and fluvial. Much of the eroded material were probably funnelled as thick synrift basin-fill sediments into a newly formed palaeo-depression trending NW-SE, whose rate of subsidence approximately kept pace with the rate of sediment supply. The NW-SE orientation of the palaeo-depression, known as the Northern Tarakan sub-basin, may be related to the pre-existing pre-Neogene structural grain which was re-activated as faults with wrench components in the Late Pliocene.

The thick Miocene synrift basin fill sediments (Tanjong and Kapilit formations) occurred in two phases of fluvio-deltaic sedimentation, informally referred to as 'Younger Tanjong' and 'Older Tanjong'. A relative sea-level drop in Late Miocene/Early Pliocene exposed the Miocene synrift clastics to erosion. This was soon followed by the deposition of the transgressive units, (Simengaris, Umas-Umas and Balung formations). These transgressive units (predominantly argillaceous) are believed to be Pliocene in age, through lithological correlation with similar units in the Tarakan Basin to the south in Indonesian Kalimantan.

Late Pliocene compression, probably related to westward movements of the Pacific Plate and renewed plate movement/spreading in the Sulu Sea, folded the mid-Pliocene units and strongly imbricated the Miocene units with thrust faults verging to the north and northeast. The thick Miocene sediments were also segmented into discrete circular to subcircular basins described earlier. The sediments must have been buried to considerable depth for coalification to occur before uplift and subsequent erosion. The boundaries of the Northern Tarakan Sub-basin were also reactivated as wrench faults. Rapid erosion of the Miocene and Pliocene units transported the material to the southeast, where they are present offshore as the Bunyu and Wallace formations.

Andesitic and dacitic volcanism in the Semporna and Tawau areas occurred between the Late Miocene and the Quaternary, with intense volcanic activities taking place in the Pliocene (Kirk, 1962, 1968; Lim, 1981; Lee, 1988; Rangin et al., 1990). There were also Quaternary basalt lava flows. These volcanic phases coincide with Late Miocene and Late Pliocene tectonism and were probably due to persistent plate movements in the Sulu and Sulawesi seas. Recent earthquakes in the Tawau and Semporna areas attest to the continuing instability (Lim, 1986).

HYDROCARBON OCCURRENCES AND PLAY TYPES

Oil and gas seeps have been reported at various onshore locations (Fig. 24.2). Analyses of the gas samples show most are biogenic in origin, based on the 100% methane composition and the accompanying strong fetid odour typical of marsh gases.

In the Semporna Peninsula, a coastal oil seepage was discovered in 1948 (Kirk, 1962) in a mangrove swamp fringing the coast at Sakong. Subsequent shallow drilling (Sakong 1-3) in 1950 did not encounter hydrocarbons.

Sebatik Island represents the most explored area in the south-western part of Sabah where 5 shallow wells, Sebatik-1 to 5, were drilled. Offshore, 4 wells had been drilled, namely the Tawau Offshore-1 and Tawau Offshore-2, Semporna Offshore-1 and the latest well, Tawau Barat-1 (Fig. 24.2, 24.5). All these wells, however, proved to be dry and, to date, no production has been established in south-eastern Sabah.

The 4 offshore exploration wells have only marginally tested the Neogene reservoirs and structures in the Northern Tarakan Sub-basin (Fig. 24.5). None of the onshore structures in the Sub-basin has been tested. In terms of source, reservoir and trap, the 'Early Anticlinal Play' (Tanjong and Kapilit formations) in early formed traps (pre-Late Pliocene) rank high. However, most of the traps were formed later, thus the Late Pliocene anticlinal structures or Late Pliocene Anticlinal Play would most likely be charged by remigrated oil, assuming that initial migration commenced in Early-Middle Miocene. The Sub-Unconformity Trap Play, Carbonate Play and Reactivated Fault Trap Play can be considered as secondary plays. Intraformational mudstones would provide adequate trap seal, but may be broken by faulting.

To the south, oil and gas production have been established on the Indonesian Tarakan (since 1905) and the Bunyu Islands (since 1922), and the Sembakung area from Middle Miocene to Late Pliocene clastic reservoirs (Fig. 24.6).

RESERVOIR, TRAP AND SEAL

Reservoir

There is negligible reservoir potential in the pre-Neogene units (Pre-Tertiary Crystalline Basement to Early Cretaceous-Palaeogene ophiolites and deep water clastics of the Sapulut and Crocker formations) and the sandstones within the early Neogene olistostromes and mudstones. The sandstones of the Oligocene Labang Formation consist of lithic fragments, comprising mainly chert, cemented by quartz, and, thus, have poor porosity. Associated limestones also have poor reservoir characteristics.

The Neogene Tanjong and Kapilit formations comprise two coarsening-upwards fluvio-deltaic sequences consisting of 10-15% and 25-30% sandstone in the lower and upper cycles respectively. Reservoir characteristics range from poor to fair with total porosity averaging 17.5%. Effective porosity is, however, much lower in sandstones with high amounts of chert or clay matrix. The sandstones range from feldspar-rich to feldspar-poor, with chert and monocrystalline or polycrystalline quartz lithic fragments. The depositional environments range from lower delta plain to coastal plain to near shore environments. The sandstones are mostly crevasse splays, fan or channel deposits, and also stream mouth bar and shoreface deposits (Sun, 1991). Coals indicate swamp environments.

The probable Pliocene shallow marine units (Simengaris, Umas-Umas and Balung formations) include highly porous sands (Simengaris) and compacted or argillaceous sandstones (Umas Umas and Balung). Lim (1981) described the Balung Formation as consisting predominantly of volcanic ash, mudstone, tuff, shale and rare thin coal beds, with abundant plant remains. Offshore exploration wells Tawau Offshore-1 and Tawau Offshore-2 encountered bioclastic limestone up to 60m thick assigned to the Balung Formation. The Late Pliocene-Pleistocene volcaniclastic Wallace Formation has poor reservoir potential, although isolated channel sandstones correlative to the hydrocarbon-bearing Bunyu Formation in the Tarakan Basin may be present.

Diagenetic studies, including hot stage cathodoluminescence-SEM (CL-SEM), fluid inclusion analysis (FIA) and apatite fission track analysis (AFTA) techniques reported in Sun (1991) were carried out for several sandstone samples, mostly from the Tanjong and Kapilit formations.

The results of the analyses are summarized as follows.

- Quartz cementation and brittle fracturing occurred contemporaneously.

- The broad range of FIA temperatures for secondary inclusions (50-120°C and >200°C) indicates deep burial or localized hydrothermal activity (e.g. Plio-Pleistocene?

volcanic activity) in Tawau-Semporna areas. Lim (1981) also reported changes of spore colour in the sediments of the Balung Formation correlateable to the distance from the volcanic activity.

- The presence of migrated hydrocarbons in 8 samples may have inhibited quartz cementation and preserved porosity.

- AFTA on one sample from Labang Formation indicates burial depths of up to 1800m.

The conclusions that can be drawn on the reservoir potential are :
- If the Late Pliocene tectonic event is the dominant factor for porosity destruction (i.e. grain fracturing, suturing and pressure solution) and hydrothermal activity (mobilizing silica for cementation), then there is sufficient time for the onset of oil migration to charge early formed traps and inhibit diagenesis before the Late Pliocene tectonic event, in particular the sandstone reservoirs in the 'Younger Tanjong'.

- If compaction has greater influence, then porosity destruction of the sandstone reservoirs would probably have predated oil migration.

- In general, early formed traps must therefore be favoured as an exploration target. However, it should be noted that Late Pliocene tectonism may have breached traps and further compacted the reservoirs.

Exploration wells Tawau Offshore-1, Tawau Offshore-2 and Semporna Offshore-1 were drilled outside the boundaries of the Northern Tarakan Sub-basin. The sediments encountered by Tawau Offshore-1 and Tawau Offshore-2 are most likely Early Neogene or earlier mudstone and volcaniclastics of the Kalumpang Formation. Semporna Offshore-1 well encountered igneous rocks at total depth, which may represent a Plio-Pleistocene volcanic plug or older ophiolitic Chert-Spilite Formation. The reservoir targets of Tawau Barat-1 also lie at the margin of the Northern Tarakan Sub-basin and the lithologies encountered are assigned to the Tanjong Formation equivalent.

Trap and Seal

The Late Pliocene tectonism event has inverted the rift sequences in the Northern Tarakan Sub-basin and formed anticlinal structures categorised as :
i) Late Pliocene Anticlines
ii) Late Pliocene domal Anticlines
iii) Pre-Late Pliocene Anticlines

As modelling studies indicate that primary generation and migration of oil could have begun in the Early-Middle Miocene, the pre-Late Pliocene anticlinal traps would be attractive targets.

The Tawau Barat-1 well was drilled into a large roll-over structure which is totally fault-dependent, and has been interpreted as a Late Pliocene inverted half-graben.

The Late Pliocene anticlines are the dominant structural style and include most of the major structures observed at the surface. These anticlines are commonly bounded by faults, some of which have a wrench component.

Domal anticlines may be basement-cored (e.g. Sebatik Anticline) or associated with Plio-Pleistocene igneous bodies (e.g. volcanic plugs). Other trap styles include sub-unconformity, rollover, subthrust, as well as isolated carbonate pinch-out and other stratigraphic types.

Intraformational seals are expected for all reservoirs, as gross sandstone content of any unit rarely exceeds 30%. Bounding fault closures or seals are also important for effective trapping.

SOURCE ROCKS

The Northern Tarakan Sub-basin has been divided into the 3 Maliau, Malibau and Tidung basins (see Fig. 24.4). The presence of minor oil and gas seeps and the evidence from pyrolysis of staining by migrant oil in the onshore outcrop samples indicate the existence of a source rock, which could still be generating hydrocarbons at the present day, or which may have generated hydrocarbons in the past. In the producing neighbouring Tarakan Basin, the best source rocks are in the Middle Miocene Tabul Formation coals which were deposited within a coastal plain to deltaic environment.

In the Northern Tarakan Sub-basin, the Miocene Tanjong and Kapilit formations occupying the large sub-circular Maliau, Malibau and Tidung basins have been identified as potential source rocks. A dull, yellow gold fluorescence was detected in the Tanjong Formation equivalent sediments penetrated by the Tawau Barat-1 well drilled in the Cowie Bay (Abolins, 1993). The formation consists of progradational and regressive sequences of mudstone, sandstone, siltstone, conglomerate,

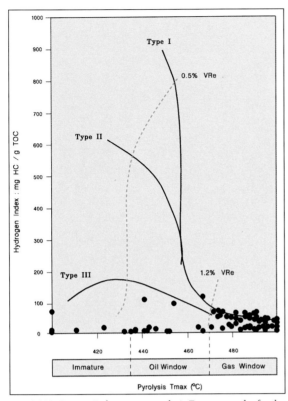

Fig. 24.7 Hyrogen Index versus pyrolysis Tmax cross-plot for the Tawau Barat-1 well indicating the preservation of Types III/IV kerogen in the Tanjong Formation equivalent. Maturation of the organic matter resulted in the observed depletion of hydrogen contents. The penetrated Tanjong Formation equivalent in Tawau Barat-1 is within the mature to post mature stage (refer to Fig 24.8 for the vitrinite reflectance profile). This cross-plot therefore mainly shows the residual hydrocarbon generating potentials of the Tanjong Formation equivalent in the Tawau Barat-1 well.

minor limestone and lignite. Total organic carbon contents of the shales are fair to very good, ranging between 0.86 to 2.21wt%. However, the corresponding hydrocarbon generating potentials and Hydrogen Index (HI) values are rather poor, suggesting only gas generation upon maturation. The HI versus Tmax plot (Fig. 24.7) indicates that the Tanjong Formation equivalent shales contain predominantly Types III/IV organic matter. However, sapropel-rich saline peats are deemed to be potential oil source rocks.

The vitrinite-rich coals in the Northern Tarakan Sub-basin are also thought to represent good quality source rocks. The main coal accumulation is in the Maliau basin (Fig. 24.2); coals also occur to a lesser extent in the southern and south-eastern parts of the Tidung basin, and the northern and central parts of the Malibau basin. These Tertiary coals range from lignite to anthracite with predominating bituminous coals. More than 20 and 130 coal outcrops were mapped by the Geological Survey of Malaysia in the Maliau and Malibau basins, respectively. The Maliau coals have excellent quality, with low ash, low to moderate sulphur contents (1.19-4.70%) and high specific energy, and are classified as a high volatile bituminous coal. Its total carbon content ranges from 77 to 83% (Chen, 1993).

Maturity

Vitrinite reflectance and spore colour analyses indicated alternations of slow and rapid subsidence throughout the history of southeastern Sabah. The maturity levels of the outcropping Miocene sediments range from early mature to late mature. Geochemical modelling indicates a palaeogeothermal gradient of 45°C per km.

Hydrocarbon generation is estimated to have commenced from the oldest source rocks at approximately 20Ma (Early Miocene) and continued successively into younger sources until the Pliocene-Pleistocene uplift/ thrusting event.

The offshore Tawau Barat-1 well penetrated a generally mature to post-mature sequence with measured vitrinite reflectance values ranging between 1.22 to 2.37% (Fig. 24.8).

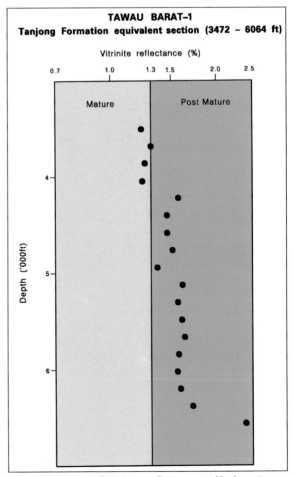

Fig. 24.8 Measured vitrinite reflectance profile from Tawau Barat-1 indicating the mature to post-mature nature of the Tanjong Formation equivalent.

Southeast Sabah Basin

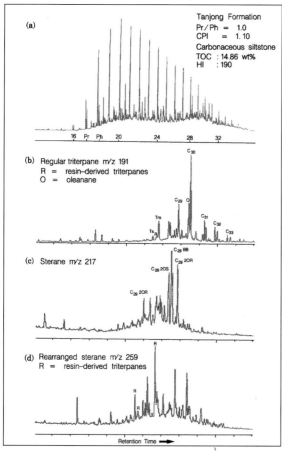

Fig. 24.9 Gas chromatography (a) and gas chromatography-mass spectrometry (b, c, d) traces of an oil stained Tanjong Formation sample taken from Location A (Fig 24.2).

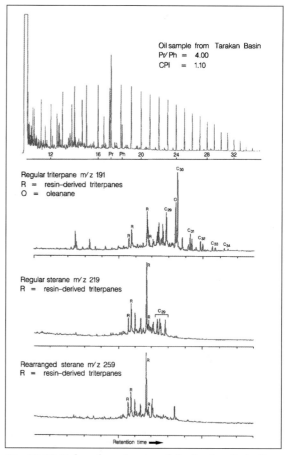

Fig. 24.10 Hydrocarbon character of the Tarakan Basin (Sun, 1991) oil exhibiting all the criteria of being generated by a terrigenously-influenced source rock.

About 400m of section was predicted to have been eroded off during the Pliocene-Pleistocene uplift, removing most of the lower maturity sediments. The Pliocene-Pleistocene extrusives occurring mainly in the vicinity of Sibuko Bay and Cowie Harbour may also have contributed to the accelerated maturation of the sediments.

HYDROCARBON CHARACTERISATION

Indigenous and migrant hydrocarbons in the onshore sediment samples attest to the existence of a hydrocarbon source interval. Figure 24.9 shows the gas chromatographic trace of an extract from an oil-stained Tanjong Formation sample.

The nearest oil generating region is the neighbouring Tarakan Basin and as such, the characteristics of these oils are compared with the oil stains and source extracts from the south eastern Sabah area. Oils from the Bunyu, Tarakan and Pamusian/Tarakan fields were also derived from source rocks containing predominantly land-derived organic matter. The evidence for this is given by the high pristane/phytane ratios with accompanying abundance of sesquiterpanes, oleanane and resin-derived triterpanes (Fig. 24.10).

Oil-Source and Oil-Oil correlations

In general, the potential source rock extracts from the Tanjong and Kapilit formations and their equivalents exhibit an abundance of high molecular weight n-alkanes maximising in the nC_{27} - nC_{31} region (Fig. 24.11), implying the preservation of predominantly waxy land plant-derived organic matter. Pristane/phytane ratios (Pr/Ph) exceeding 2.0 suggest that source deposition took place in moderately oxidising environments. These extract samples are considered to be immature based on their Carbon Preference Index values of 1.13 to 2.11.

The terrigenous nature of the source rock is confirmed by the presence of resin-derived triterpanes, abundant oleanane, and abundant C_{29} regular and rearranged steranes observed in the GC-MS traces. Interestingly, one Tanjong Formation extract taken from Location C (see Fig. 24.2) exhibited some marine influence with the

Southeast Sabah Basin

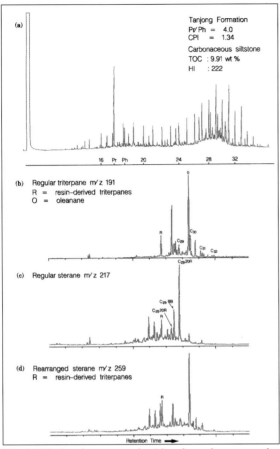

Fig. 24.11 Gas chromatography (a) and gas chromatography spectrometry (b, c, d,) traces of an extracted unstained potential source rock sample from the Tanjong Formation, exhibiting characteristics typical of source rock containing land-derived organic matter.

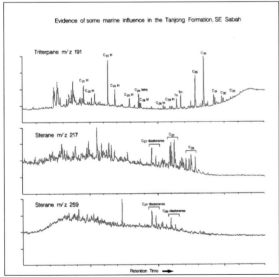

Fig. 24.12 Gas chromatography-mass spectrometry traces of an extract sample from the Tanjong Formation (Location C, Fig 24.2). The abundant occurrence of tricyclic terpanes and the predominance of C_{27} regular steranes over the C_{28} and C_{29} counterparts highlight the presence of some marine influence in parts of the Tanjong Formation. Note the absence of oleanane and resin-derived triterpane peaks.

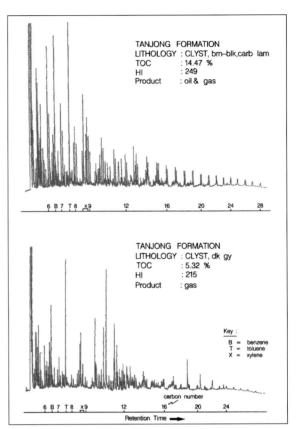

Fig. 24.13 Resulting pyrolysis-gas chromatographic traces of immature Tanjong Formation source rocks indicating their potentials to generate (a) waxy oil and gas, (Location C, Fig 24.2) and (b) gas (Location D, Fig 24.2) upon maturation.

presence of tricydic terpanes in the M/Z 191 trace (Figure 24.12). Kerogen typing studies also illustrated a strongly terrigenous character of the source rocks whereby terrestrially-derived waxy sapropel, vitrinitic and inertinitic organic matter predominate, with the relatively minor input from marine organic matter. These Tanjong Formation source rocks have the potential to generate both gas and waxy oil as indicated by pyrolysis-gas chromatographic analysis (Fig. 24.13).

As was mentioned previously, the Tanjong Formation equivalent sediments penetrated by the offshore Tawau Barat-1 well exhibited some yellow-gold fluorescence. Therefore that particular interval was extracted and chromatographically analysed. The resulting saturate fraction trace of this oil showed a bimodal envelope, maximising at both nC_{18} and between nC_{23} - nC_{24}, and a slight odd-carbon number preference in the nC_{27} - nC_{35} region (Fig. 24.14), suggesting a mixed marine and terrigenous source influence. Pr/Ph ratio of 1.00 indicates that the source rock was most likely deposited in a sub-oxic environment. Based on the post-mature nature of the Tanjong Formation (VRo range : 1.2 - 2.37%) and the CPI value of 1.14, it is concluded that the extracted

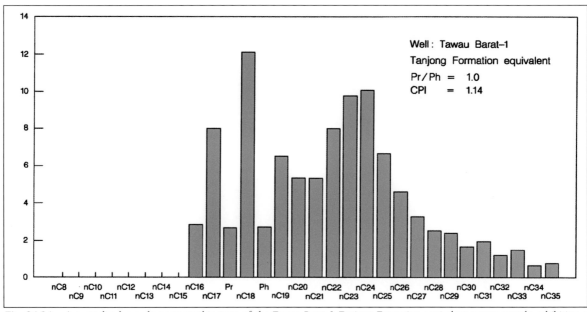

Fig. 24.14 A normalised gas chromatographic trace of the Tawau Barat-1 Tanjong Formation equivalent extract sample exhibiting a bimodal envelope which can attributed to contribution from both marine and terrigenous organic matter.

hydrocarbon was not generated *in situ* and therefore, must have migrated from elsewhere to its present location within the formation. Direct oil to source rock correlations are rather difficult to establish in the southeastern Sabah region due to the generally similar characteristics of the source intervals identified. They all show very similar triterpane and sterane distributions, whereby the triterpane traces are commonly dominated by oleanane and resin-derived triterpanes, and the steranes by the C_{29} compounds both in the regular and rearranged forms (Fig. 24.12). However, as a general guide, most of the source extracts from the Tanjong and Kapilit formations and their equivalents do possess characteristics similar to the discovered hydrocarbons in the neighbouring Tarakan Basin, and as such, they would be the most likely source rocks in the southeastern Sabah area.

It is probable that the oil stained sample from the Tawau Barat-1 well, which exhibits some degree of marine organic matter input, was generated by parts of the Tanjong and Kapilit formations that were deposited in relatively more distal locations.

SUMMARY

There are several encouraging factors with regard to source rocks and hydrocarbon generation in the SE Sabah Basin, in particular the Northern Tarakan Sub-basin. First, derived pyrolysis-gas chromatography shows that the Tanjong Formation equivalent carbonaceous claystones and siltstones are able to generate both waxy oil and/or gas upon maturation. The immature (VRo < 0.5%) and early mature (VRo 0.5 - 0.7%) outcrop samples generally have HI values greater than 100 with an average of 214. These values are comparable to those in the productive Sabah Basin. The source rock depositional environments in these two regions are also very similar, comprising lower coastal plain mangrove swamps and paralic conditions, with enormous input of terrigenous organic matter. Therefore, the original source and hydrocarbon generating potential of the Tanjong Formation equivalent is promising. In addition to the organic rich and carbonaceous shales and siltstones, the Tanjong Formation coals can also be considered as potential sources. Their equivalent is the Tabul Formation in Indonesian Kalimantan which is considered to contain the best source rock in the Tarakan Basin.

Secondly, the Northern Tarakan Sub-basin has undergone advanced thermal maturation and parts of the region are already in the postmature stage, as seen in the Tawau Barat-1 well section. This would imply that source rocks would have already generated both oil and gas. Thus, it is important to locate structures where these hydrocarbons may be trapped, especially in pre-Late Pliocene structures. However, the destruction of existing traps and the formation of new ones following the Late Pliocene deformation, should be taken into consideration in future exploration activities.

REFERENCES

Abolins, P., 1993. A geochemical evaluation of well Tawau Barat-1, SB8, offshore Sabah. Core Laboratories Report No. GSM 92069, PETRONAS Exploration Department Report No. WD1:SUN:2:Tawau Barat-1/9.

Chen, S.P., 1993. Coal as an energy resource in Malaysia. Bulletin of the Geological Society of Malaysia, 33, 399-410.

Chung, S.K., Gan, A.S., Leong, K.M. and Kho, C.H., 1977. Ten years of petroleum exploration in Malaysia. United Nations ESCAP, CCOP Technical Bulletin, II, 111-142.

Clennell, M.B., 1996. Far-field and gravity tectonics in Miocene Basins of Sabah, Malaysia. In : Hall, R. and Blundell, D.J.(eds.). Tectonic Evolution of Southeast Asia, Geological Society of London Special Publication, no.106, 307-320.

Collenette, P., 1954. The coal deposits and summary of the geology of the Silimpopon Area, Tawau District, Colony of North Borneo. British Borneo Geological Survey, Memoir 2.

Collenette, P.,1965. The geology and mineral resources of the Pensiangan and Upper Kinabatangan area, Sabah, Malaysia. Geological Survey of Borneo Region, Malaysia, Memoir 12.

Hutchison, C.S., 1992. The Southeast Sulu Sea, a Neogene marginal basin with outcropping extensions in Sabah. Bulletin of the Geological Society of Malaysia, 32, 89-108.

Kirk, H.J.C., 1962. The geology and mineral resources of the Semporna Peninsula, North Borneo. British Borneo Geological Survey, Memoir 14, 178p.

Kirk, H.J.C., 1968. The igneous rocks of Sarawak and Sabah. Geological Survey of Malaysia, Bulletin 5.

Lee, D.T.C., 1988. Gunung Pock area, Semporna Peninsula, Sabah, Malaysia. Geological Survey of Malaysia, Report 9.

Lentini, M.R.and Darman, H., 1996. Aspects of the Neogene tectonic history and hydrocarbon geology of the Tarakan Basin. Proceedings of the Indonesian Petroleum Association 1996, 25th Silver Anniversary Convention, October 1996, 241-251.

Leong, K.M., 1978. Sabah. In : ESCAP Atlas of Stratigraphy I. Burma, Malaysia, Thailand, Indonesia, Phillipines - United Nations, Mineral Resources Development Series, No.44, 26-31.

Lim, P.S., 1981. Wullersdorf area, Sabah, Malaysia. Geological Survey of Malaysia, Report 15.

Lim, P.S., 1985. Geological Map of Sabah 1:500,00 (3rd Edition) Geological Survey of Malaysia.

Lim, P.S., 1986. Seismic activities in Sabah and their relationship to regional tectonics. Geological Survey of Malaysia Annual Report for 1985, 465-480.

PETRONAS, 1990. Technical Brochure Block SB7 (Unpublished).

PETRONAS, 1994. Petroleum Exploration Potential in Malaysia (Unpublished).

Rangin, C., Bellon, H., Benard, F., Letouzey, J., Muller, C. and Sanudin, T., 1990. Neogene arc-continent collision in Sabah, North Borneo (Malaysia). Tectonophysics, 183, 305-319.

Sun 1991. Integrated Interpretation of Data from exploration surveys, 1989-1991 SB8, Sabah, Malaysia (Unpublished report by Sun Malaysia Petroleum Company).

Sun, 1992. SB8 Block, Sabah, Malaysia : Biostratigraphy, Geochemistry and Sedimentology (Unpublished report by Sun Malaysia Petroleum Company).

Tjia, H.D., Ibrahim Komoo, Lim, P.S. and Tungah Surat, 1990. The Maliau Basin, Sabah : geology and tectonic setting. Bulletin of the Geological Society of Malaysia, 27, 261-292.

Tokoyama, A. and Yoshida, S., 1974. Kinabalu Fault, a large strikeslip fault in Sabah, East Malaysia. Geology and Palaeontology of Southeast Asia, 14, 171-188.

Tongkul, F., 1991. Tectonic evolution of Sabah, Malaysia. In: Nichols, G. and Hall, R. (editors) Proceedings of Orogenesis in Action Conference, London 1990. Journal of Southeast Asian Earth Sciences, 6, 395-405.

Tongkul, F., 1993. Tectonic control on the development of the Neogene basins in Sabah, East Malaysia. Bulletin of the Geological Society of Malaysia, 33, 95-103.

Wight, A.W.R., Hare, L.H. and Reynolds, J.R., 1993. Tarakan Basin, NE Kalimantan, Indonesia : a century of exploration and future potential. Bulletin of the Geological Society of Malaysia, 33, 263-288.

Chapter 25

Petroleum Resources, Sabah

Mohd Idrus B. Ismail

INTRODUCTION

The petroleum resources of Sabah are described with reference to the intensity of exploration activities and the most significant play types in the 3 Tertiary Neogene basins, especially the Sabah Basin. Most exploration activity has been conducted in the Sabah Basin as can be seen from the density and distribution of the exploration wells in Figure 25.1. The first onshore and offshore exploration wells were drilled in 1897 and 1958 respectively. By the end of 1997 a total of 188 wells were drilled, 111 wildcat and 77 appraisal wells (Figs. 25.1, 25.4). Major oil and gas discoveries were made in the 70s and more recently in the late 80s and 90s. In addition, a total of 279,634 line km 2D of seismic has been acquired since 1956, and since the first 3D seismic survey covered the Samarang field in 1988, 13 other major fields have been covered with 3D seismic data, totalling 185,768 line km (Figs. 25.4, 25.5).

These exploration efforts have resulted in the discovery of 3.2 BSTB OIIP with EUR of 1.24 BSTB and 10.7 TSCF GIIP with EUR of 7.0 TSCF in 23 oil fields and 28 gas fields (Fig. 25.2). The resource figures exclude small oil and gas fields with EUR less than 8 MMSTB and 50 BSCF respectively. Amongst these, 4 'giant' oil fields and 2 'giant' gas fields have been found in the

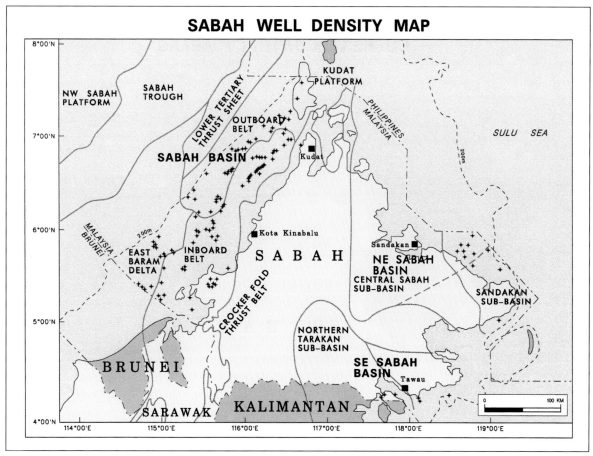

Fig. 25.1 Sabah well density map. About 90% of the wells were drilled in the Sabah Basin and the remainder in the Sandakan Sub-basin and Southeast Sabah Basin.

Petroleum Resources, Sabah

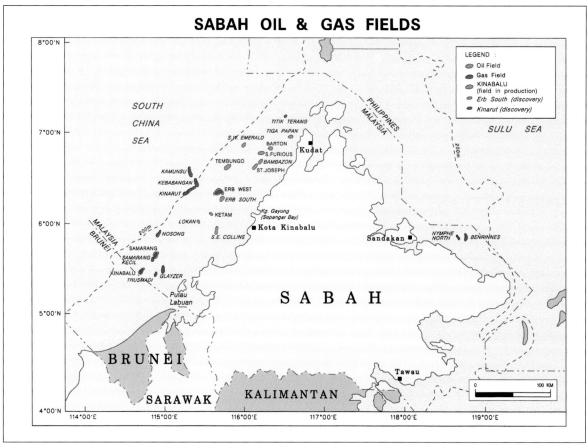

Fig. 25.2 Sabah Oil and Gas Field Map (oil green, gas red). Ketam ceased production recently.

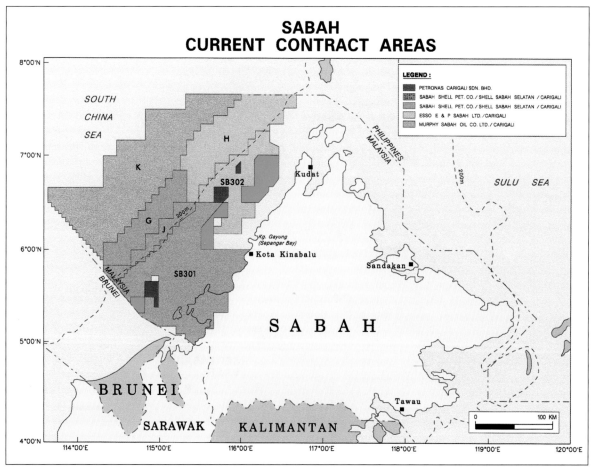

Fig. 25.3 Sabah Current Contract Area Map.

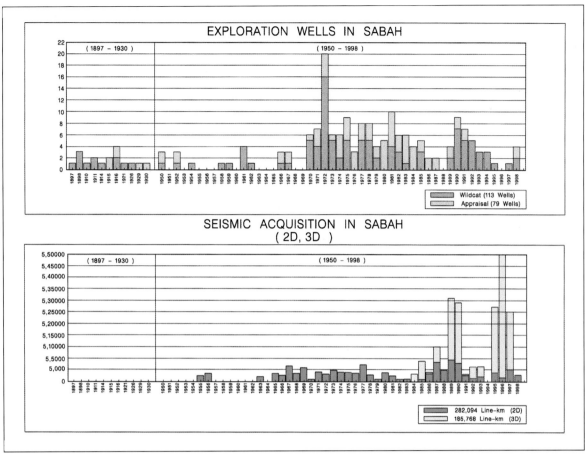

Fig. 25.4 i) Exploration wells in Sabah and ii) Seismic acquisition in Sabah.

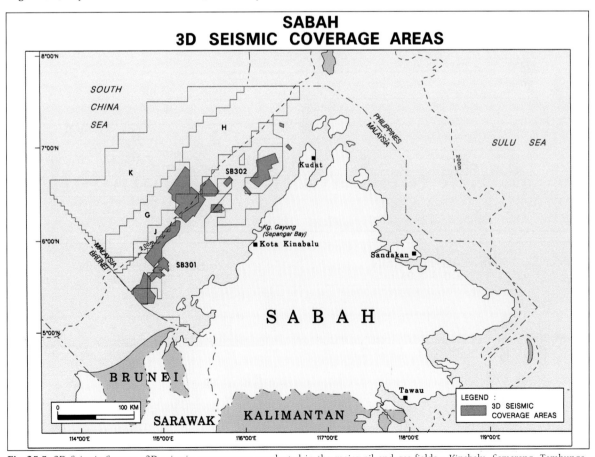

Fig. 25.5 3D Seismic Surveys. 3D seismic surveys were conducted in the major oil and gas fields - Kinabalu, Samarang, Tembungo, St Joseph, Barton, South Furious and the gas fields Kinarut, Kebabangan and Kamunsu.

western Sabah Basin, each with recoverable reserves of more than 100 MMSTB oil and 1 TSCF gas. As at 1.1.1998 there are 8 producing oil fields. It should be noted that the resource figures relate to activities as at 1.1.98. During 1998, 4 exploration wells were drilled and about 3,000 km of 2D seismic acquired. A major gas discovery, Kamunsu, was made.

There are 6 active PSCs in Sabah (Fig. 25.3). The 'Samarang-Asam Paya PSC', which includes the Samarang, Tembungo and Erb West fields, is operated by PETRONAS Carigali (Carigali); whilst the North Sabah 1995 PSC, which includes the St. Joseph, South Furious, Barton and Ketam fields, is operated by Shell. Shell's Kinabalu sub-block is a designated development area. In addition, Shell is operator for Deepwater J and G and exploration block SB301, which includes the onshore Klias Peninsula. Esso is operator for exploration block SB302 and Deepwater H. Blocks 302 and 301 come under the Revenue over Cost (R/C) PSC petroleum arrangement. Block SB4, under the 1985 PSC terms, and held by Hall-Houston/ Ensearch /Global/Carigali expired recently (January 1999), Deepwater K was awarded to Murphy Oil as operator. The recent gas discoveries in Kebabangan under Block SB1 1985 PSC, and Kamunsu in Deepwater G, augur well for the Late Miocene turbidite play in the Sabah Basin continental slope.

OIL AND GAS RESOURCES BY BASIN

The Sabah region consists of 3 major Neogene sedimentary basins, namely Sabah, the Northeast Sabah and Southeast Sabah basins (Fig. 25.1). The Sabah Basin consists of several tectonostratigraphic provinces, and includes those off the continental shelf. These are the Inboard Belt, the Outboard Belt, Kudat Platform, the East Baram Delta and the Lower Tertiary Thrust Sheet, bounded to the NW and W by the Sabah Trough and NW Sabah Platform. The well statistics for the Outboard Belt here include the Northern Sabah Province, but exclude the Kudat Platform (see Chapter 22 for description of the Northern Sabah Province). Towards the east, the basin is bounded by the Crocker Fold-Thrust Belt. Commercial oil fields in the Sabah Basin occur in the Inboard and Outboard Belts and in the East Baram Delta. The petroleum geology of the Sabah Basin is given in Chapter 22.

The Northeast Sabah Basin consists of

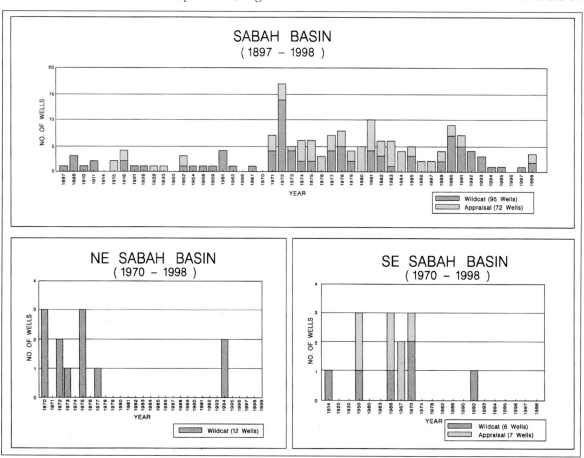

Fig. 25.6 Exploration wells in Sabah Basin (including 1 well in Crocker Fold-Thrust Belt), Northeast Sabah and Southeast Sabah basins.

the Sandakan and Central Sabah sub-basins. Twelve offshore wells have been drilled, 11 in the Sandakan Sub-basin and one in the Labuk Bay area of the Central Sabah Sub-basin (Fig. 25.6). Gas and condensate have been encountered in the Lower to Middle Miocene units in the offshore overlapping area of the Sandakan and Central Sabah sub-basins (see Chapter 23).

The Southeast Sabah Basin consists of the Northern Tarakan Sub-basin and a platform area which is partly underlain by Quaternary volcanics (see Chapter 24). Geological data show that the onshore Northern Tarakan Sub-basin is a cluster of sub-circular basins (e.g. Tidung, Maliau and Malibau basins). A total of 9 wells/core holes were drilled onshore in the Semporna Peninsula and on Sebatik Island between 1914 and 1967, but no commercial hydrocarbons were encountered apart from traces of shallow methane gas. Similarly, results obtained from Tawau Offshore-1 and -2, Semporna-1 and Tawau Barat-1 offshore wells were not encouraging (Fig. 25.6).

Sabah Basin

The Sabah Basin has been explored since 1897 and to date over 90 wildcats have been drilled. In the Sabah Basin the Inboard Belt has been most intensely explored, with some 60% of the drilling activities compared to the Outboard Belt and East Baram Delta (Figs. 25.6, 25.7). In the 70s, Shell made discoveries of commercial petroleum in the northern Inboard Belt (the Barton, St. Joseph and South Furious fields) and in the southern Inboard Belt (the Ketam field). Several smaller oil discoveries, which are not yet commercial, include Lokan, Collins, SE Collins and Erb South. Ketam field has EUR less than 8 MMSTB and ceased production recently.

The Inboard Belt has about 40% of total oil in-place, about 31% the recoverable oil. From the time of its first oil production in 1977, it has contributed about 28% to the oil production in Sabah (Fig. 25.8). The Inboard Belt still has 36% of total recoverable oil reserves. As for gas, the Inboard Belt has a number of gas fields e.g. Glayzer. The main gas production is, however, associated gas from the oil fields. The main play in the Inboard Belt is the Middle Miocene Stage IVA and Stage IVC coastal plain - shallow marine sands (Stage IVC sands are also producers in St. Joseph).

The Outboard Belt has also been explored actively. The Outboard Belt is characterized by deltaic, shelfal and slope deposits (Stages IVD-E) overlying Stage IVC

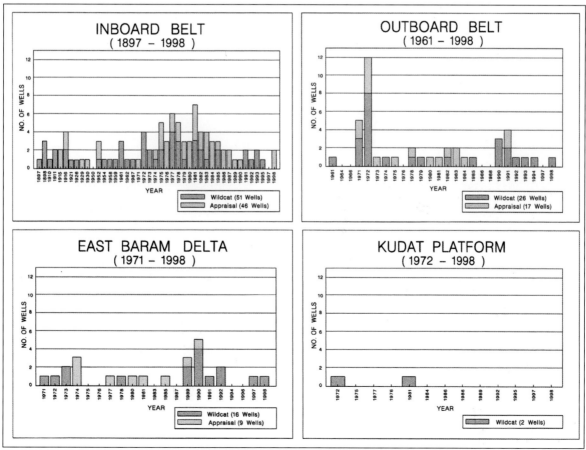

Fig. 25.7 Exploration wells in Inboard and Outboard Belt, East Baram Delta and in Kudat Platform.

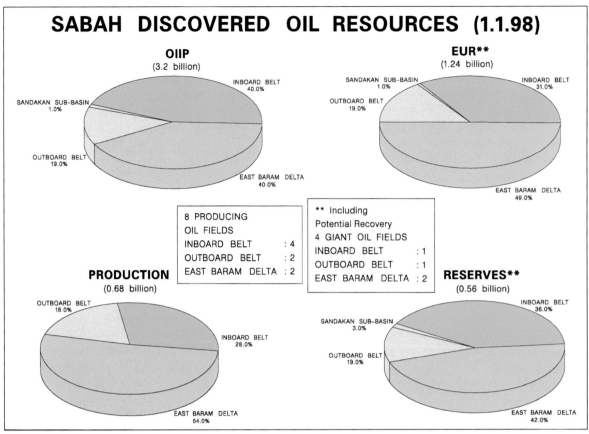

Fig. 25.8 Sabah discovered oil reserves.

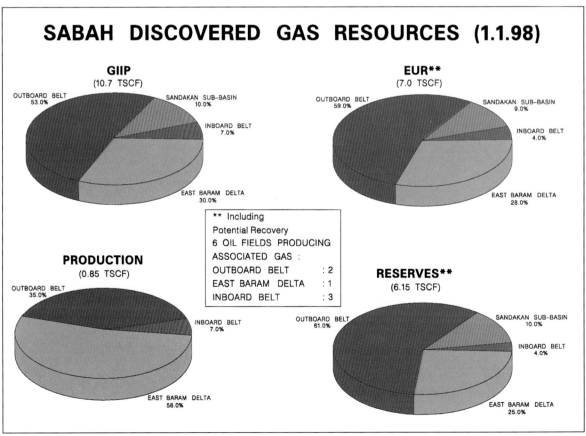

Fig. 25.9 Sabah discovered total gas reserves.

and/or older turbidite. The Outboard Belt is a proven hydrocarbon province with 2 major commercial discoveries, Tembungo and Erb West, and several yet-to-be developed gas discoveries, Kinarut, Kebabangan and Kamunsu. Other smaller oil discoveries include SW and W Emerald, and Tiga Papan, and the Titik Terang gas discovery. The 2 main plays are Stage IVC and/or older turbidites in anticlinal and upthrusted structures (Tembungo and the gas discoveries) and Stage IVD shallow marine sands (Erb West). The Outboard Belt has about 19% of the total oil in-place, about 19% of the recoverable oil, and has contributed about 18% of the oil production in Sabah (Fig. 25.8).

The Outboard Belt is the largest gas province in Sabah, with several recent drilling successes (Kebabangan, Kamunsu). To date it has 53% of the total gas in-place, 59% of the estimated recoverable gas, and 61% of the gas reserves in Sabah (Figs. 25.9, 25.10). Whilst the associated gas is currently being produced from the 2 oilfields, Tembungo and Erb West, the non-associated gas fields are yet to be developed.

The understanding of the distribution of deepwater turbidite sands in the Kebabangan prospect led the way to successes in deepwater sand exploration. Extensive investigation of DHIs was taken before the Kebabangan structure was drilled. Shell subsequently drilled the Kebabangan-1 gas discovery well in 1995, and confirmed the presence of a major turbidite fairway in the Outboard Belt. Currently, several of the turbidite fairways are the focus of exploration in Blocks SB302 and in several deepwater blocks. The recent Kamunsu gas discovery by Shell confirms the presence of a major hydrocarbon pool in the deepwater sands.

In the northern Outboard Belt, the carbonate Kudat Platform extends north to the Philippines. To date, 2 wells have been drilled to test the Upper Miocene carbonates on the Sabah side and one on the Philippines side, all without success (Fig. 25.7).

The East Baram Delta (EBD) province, with the fewest wells drilled compared to the Inboard and Outboard Belts, has 2 'giant' oil field - Samarang and Kinabalu, which have considerable gas sands, and several other smaller discoveries e.g. Nosong, Bongawan. The main play in this province is the Stage IVD and Stage IVE deltaic sands. In spite of having similar volume of total oil in-place as the Inboard Belt (i.e. about 40% each), the recoverable oil in EBD is estimated at 40% of the total, compared to about 31% for the Inboard Belt. The EBD has also contributed to over 54% of the total oil production in Sabah compared to 28% for the Inboard Belt. It has also the largest remaining oil reserves (Fig. 25.8). It also contributes 58% of

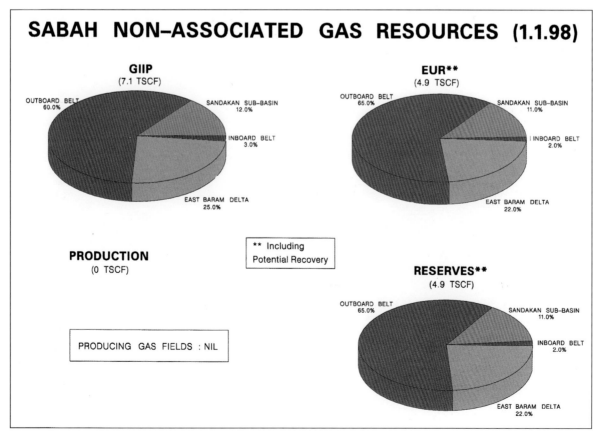

Fig. 25.10 Sabah discovered non-associated gas resources

the total gas production in Sabah (Figs. 25.9, 25.10). Active exploration in this 'giant' field province is continuing and is extending to the delta toe-thrust area to the northwest and west in deeper water.

NW Sabah Platform/Sabah Trough

The NW Sabah Platform/Sabah Trough areas lie in 1500-3000 m water depth. Thick Palaeogene sediments-infilled graben and horst are overlain by Oligo-Lower Miocene carbonates. Apart from reconnaissance seismic, no exploration activity has been conducted in the area. The deepwater block K, which contains some of these leads in the Sabah Trough will soon be explored, as the block has recently been awarded. Several sizeable clastic and carbonate leads have been identified based on reconnaissance seismic.

The Crocker Fold-Thrust Belt is probably the economic basement. It consists of indurated turbidite sandstones and shales that underwent deformation since the Early Miocene. However, oil seeps occur onshore in the younger Kudat Formation in Kudat Peninsula, and oil shows have been encountered in the Lower Miocene Stage III turbidite sands in offshore wells.

Northeast Sabah Basin

In several of the 11 wells drilled in the Sandakan Sub-basin, gas and condensate and minor amounts of light oil in faulted and compartmentalised structures have been encountered. Regional understanding of the hydrocarbon source within the grabens need to be pursued to improve understanding of the pre-Dent Group (i.e. Tanjong equivalent/ Segama Group) for further exploration success.

Southeast Sabah Basin

To date, 4 offshore wells and several shallow onshore wells/coreholes drilled in this basin have not encountered significant hydrocarbons.

The Lower to Middle Miocene Tanjong and Kapilit formations sands remain the primary targets, with the younger sediments, e.g. the Simengaris sands, as secondary target.

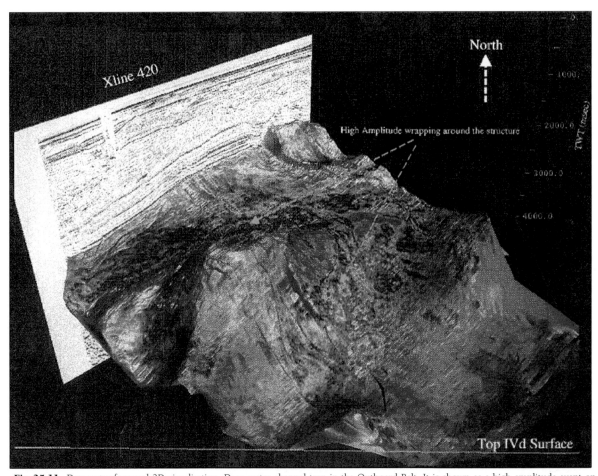

Fig. 25.11 Deepwater fans and 3D visualisation. Deep water channel trap in the Outboard Belt. It is shown as a high amplitude event on the 3D visualisation.

However, the wells drilled to date were either located on the margin of, or outside, the Northern Tarakan Sub-basin. Thus, the sub basin, not fully tested on the Sabah side, still warrants further studies as its extension offshore in the Indonesian side has been proven hydrocarbon-bearing.

FUTURE OUTLOOK

Near field Potential

The Stage IVA play in the complexly faulted flank closures near the main oilfields (Barton, South Furious) in the Inboard Belt have to be delineated using dense 2D and 3D seismic coverage. The understanding of the complex hydrocarbon fluid distribution in these fields is also important. In addition, the fault closures and untested traps in the complexly faulted areas of the East Baram Delta need to be delineated by new 3D seismic. Sandy turbidites (Stage IVC) in the Outboard Belt are major plays targets (Fig. 25.11). The Stage IVD coastal plain to shallow marine sandstones (e.g. Erb West) are also significant.

New and Emerging Play

Four emerging and 5 new petroleum plays have been outlined in the Sabah Region.

The 4 emerging plays consist of three plays in the Sabah Basin and one play in the Northeast Sabah Basin. These are :
1) Inboard Belt/Stage IVC
2) Outboard Belt/Stages IVC-IVD
3) East Baram Distal Delta/Stages IVC-E/ turbidite and deeper play
4) Sandakan Sub-basin/Sebahat Formation and Pre-Dent Group

Stage IVC slump scar and channel stratigraphic traps have been delineated in north Sabah, where Stage IVC has been proven oil bearing in St. Joseph and in Bambazon (carbonates).

Stages IVD and IVC turbidite stratigraphic traps in the flank of several sub-basins are potential plays. In order to explore for these new traps 3D visualisation (Fig. 25.11) and optical stacking techniques are being used (Fig. 25.12).

The 5 new plays are :
1) Outboard Belt and East Baram Distal Delta/ Deepwater Turbidite and Underthrust Play

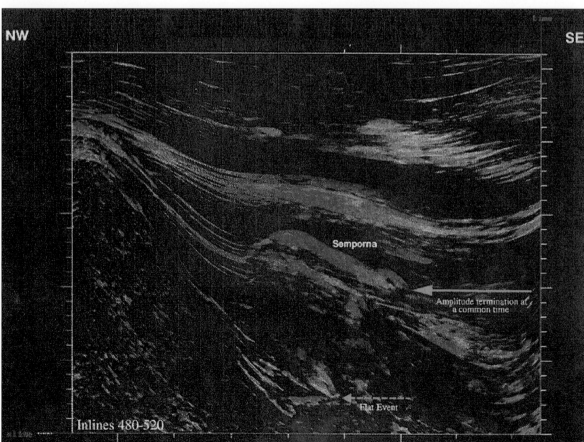

Fig. 25.12 State of the art geophysical visualisation tool. Optical stacking is increasingly being used as a standard technique for visualising 3D seismic data in Sabah. This allows a general overview of structure and pinpoints the prospective area. In this example shallow and deeper amplitude termination that is common or flat is shown in a turbidite play in offshore Sabah.

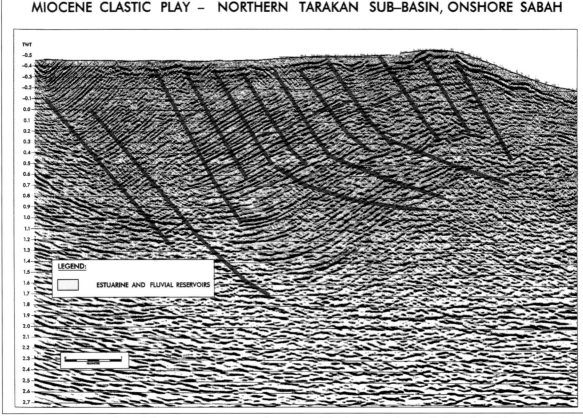

Fig. 25.13 Northern Tarakan Sub-basin-Tanjong Formation Play. Structural development in the so-called 'circular sub-basin' is conducive for trap development.

2) Deepwater and NW Sabah Platform/Carbonate and Synrift Play
3) Inboard Belt/Stages III-IVA/B
4) Central Sabah Sub-basin/Sandakan and Tanjong formations
5) Northern Tarakan Sub-basin/Tanjong and Kapilit formations

Stratigraphic traps formed by a) onlapping of Stage IVA sands and truncation traps, known in the inboard belt around Ketam area, have been observed, b) Stage IVB turbidite play and c) Stage III deep marine sediments are known to contain hydrocarbon bearing turbidites but wells were probably not valid due to mapping difficulties from poor seismic.

The Lower Tertiary Thrust Sheet, Sabah Trough and NW Sabah Platform are untested provinces. The recently released (early 1999) Sabah deepwater block K includes part of these provinces. Thick Palaeogene sediment in horsts and graben, and the overlying carbonate in the NW Sabah Platform and Sabah Trough are potential targets.

The Tanjong and Kapilit formations in the Northern Tarakan Sub-basin, within the so-called circular sub-basins, is an attractive play (Fig. 25.13) due to a) near complete Miocene section b) presence of mature source rocks, probably oil prone, and c) fair to good reservoirs in the Tanjong and Kapilit formations.

CONCLUSION

Exploration efforts have resulted in EUR of 1.24 BSTB oil and 7.0 TSCF gas in 23 oil fields and 28 gas fields. Amongst these, 4 oil fields and 1 gas field are considered as giant fields. The Inboard Belt has benefited from improved structural definition and imaging petroleum distribution due to better seismic quality and new techniques in 2D and 3D seismic. The Outboard Belt underwent 2 important phases of exploration in the 1970s and from mid 1990s. The gas and oil discovered to date suggest that it is poised to be an important gas province. The East Baram Delta remains an important oil province with several opportunities for further discoveries. New and emerging plays include the untested onshore and frontier carbonates, and underlying synrift clastics in the Sabah continental margin.

Part 6

MALAYSIA PRE-TERTIARY

Chapter 26

Pre-Tertiary Hydrocarbon Potential

H.D. Tjia

INTRODUCTION

Malaysia comprises several geological terranes, each having evolved differently from its neighbours (Fig. 26.1). Recent findings indicate that five geological terranes can be distinguished: (1) the core, (2) Western Belt of Peninsular Malaysia, (3) Sarawak east of the Lupar valley and western Sabah, (4) East Sabah Terrane, and (5) the Kinabalu suture zone that divides Sabah.

The core terrane (Terrane 1) comprises Central and Eastern Peninsular Malaysia, westernmost Sarawak and the Sunda Shelf. The oldest rocks are Carboniferous, about 320 Ma (million years) old, that originated as sediments deposited in marine and continental settings. Mainly marine conditions prevailed during the Permian and Early Triassic, and in western Sarawak continued into the Jurassic. The peninsular part of the core terrane was already land in the Jurassic- Cretaceous when large, probably intramontane lakes dominated the landscape. A similar situation existed in the Cretaceous and Early Cenozoic in western-most Sarawak. By the end of the Mesozoic, or at the beginning of the Cenozoic, the core terrane had become tectonically stabilised, and through erosion and weathering it had become a vast peneplain.

The Western Belt of Peninsular Malaysia (Terrane 2) contains the proven oldest rocks consisting of Middle Cambrian (about 520 Ma) deltaic deposits similar to fossil-bearing beds in adjacent southern Thailand. On Langkawi island, small outcrops of quartzitic sandstone below the

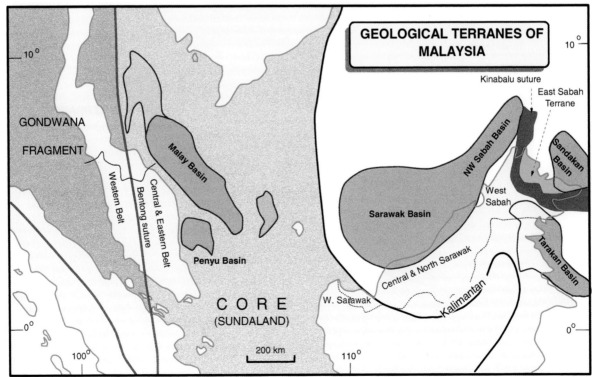

Fig. 26.1 Geological terranes of Malaysia. The core region has been part of Asia since at least the Late Palaeozoic. The Western Belt of the Malaysia-Thailand peninsula is a crustal fragment derived from the Gondwana continent and became attached to the core region by the Early Triassic. The East Sabah Terrane probably originated from eastern Asia and through opening of the South China Sea basin drifted to arrive at its present position by the Early Miocene. Important sutures of the crustal plates are the Kinabalu suture zone, Lupar Suture, and Bentong Suture.

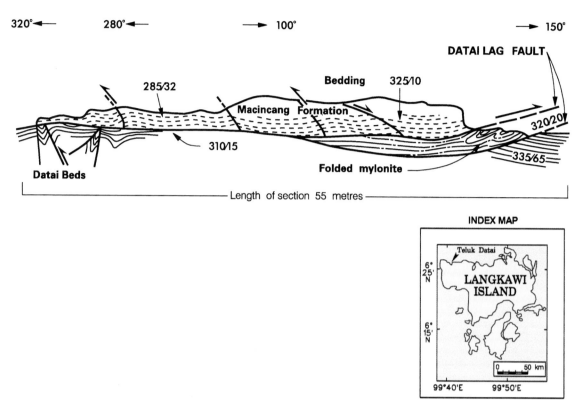

Fig. 26.2 The Datai lag-fault zone transported Middle Cambrian Macincang Formation towards SW and separated the formation from stratified quartzitic Datai Beds (Precambrian?). Eastern shore of Datai Bay, Langkawi island. From Tjia (1989b).

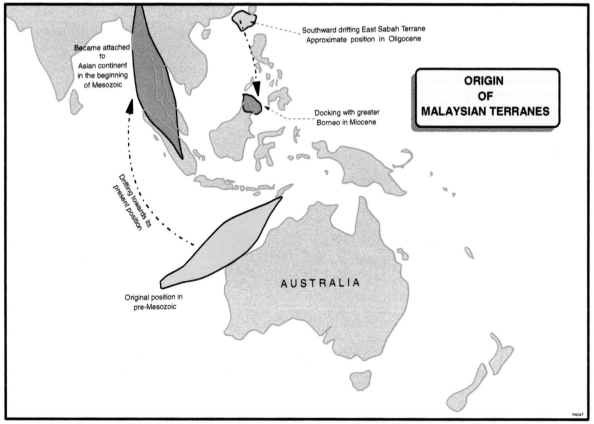

Fig. 26.3 Provenance of the two allochtonous Malaysian terranes: Western Belt (including) Straits of Melaka and East Sabah Terrane.

Middle Cambrian sequence may be Precambrian (Fig. 26.2) Shallow to deep-marine deposition alternated throughout the Palaeozoic into Early Triassic. Fossil evidence, and the presence of pebbly mudstone in Upper Carboniferous-Lower Permian strata, and their absence outside the western belt, strongly suggest that it belonged to the Gondwana continent (Fig. 26.3). This Gondwana fragment broke off, and through sea-floor spreading drifted north to eventually become attached along a well defined tectonic zone, the so called Bentong suture, to the rest of the peninsula. The docking of the western belt was accomplished by the Early Triassic. The main igneous intrusion took place in the Late Triassic-Early Jurassic all along the eastern margin of the Gondwana fragment and resulted in the Main Range (or Titiwangsa) granitoid complex. Subsequently, the entire peninsula evolved in similar fashion.

Central-North Sarawak and western Sabah (Terrane 3) evolved as a subduction-related accretionary prism in Cretaceous-Palaeogene time. Structural style in the older Tertiary rocks consists of asymmetrical to overturned folds and reverse fault zones. Synthetic aperture radar (SAR) images show distinct curving, regional lineaments subparallel to strike. These curvilinear lines probably separate discrete thrust packets in the fold-thrust zone known as the Rajang Group (Fig. 26.4).

In the East Sabah Terrane (4) mainly Cenozoic sedimentary formations crop out with some inliers of the Cretaceous-Eocene, the so called Chert-Spilite Formation in the Dent Peninsula. Structural grain is East-West which is distinctly different from the NE trend of western Sabah. It has been proposed that eastern Sabah is an exotic terrane originating from the continental margin of eastern Asia (Fig. 26.3). Through sea-floor spreading of a proto-South China Sea Basin (possibly at the beginning of the Cenozoic), this continental fragment broke off eastern Asia and drifted in a southerly direction to ultimately become attached onto Borneo by the Early Miocene (Tjia, 1988). Rifting of the Sulu Sea basin in the Middle Miocene (Hutchison, 1988) also affected the East Sabah Terrane by creating the NE-striking Sandakan rift and the [Greater] Tarakan rift (Fig. 26.23b). Cretaceous to Middle Eocene ophiolites became exhumed in the latter and constitute Terrane 5 or the Kinabalu suture zone. This Middle Miocene rifting also resulted in widespread chaotic deposits (Ayer, Kuamut,

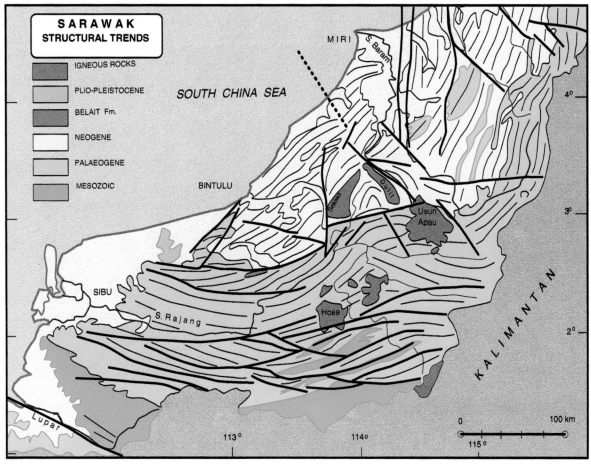

Fig 26.4 Structural trends in Sarawak. The heavy lines in the Cretaceous-Eocene Rajang fold-thrust belt (or accretionary prism) represent major reverse faults.

Pre-Tertiary Hydrocarbon Potential

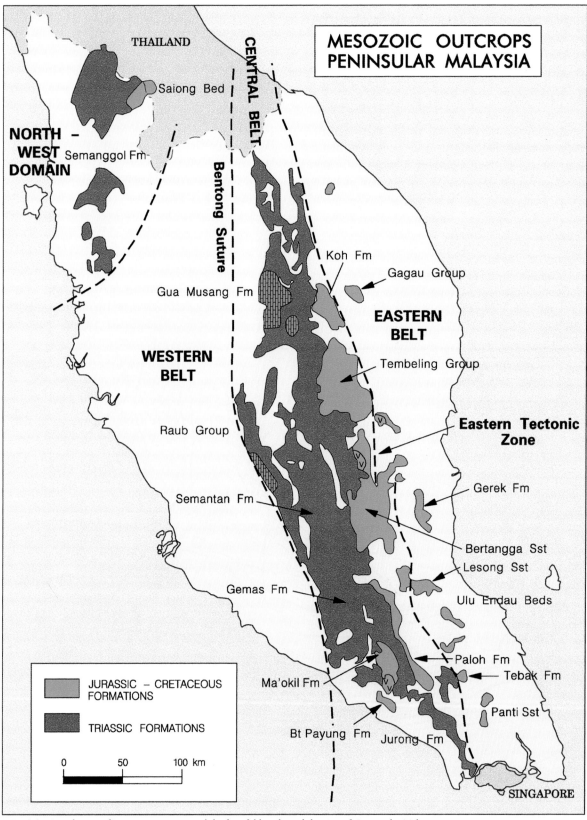

Fig. 26.5 Distribution of Mesozoic outcrops and the four-fold geological division of Peninsular Malaysia.

Garinono, Kalabakan formations); most formed by submarine slumping according to Clennell (1991), and further controlled the location of circular Neogene basins. Locally, the chaotic deposits have been remobilized by mud diapirism. Active mud cones are common.

The Kinabalu suture zone (Terrane 5) contains almost all the banded chert, spilite and ultrabasic bodies of Sabah. Other rock types are Mesozoic and perhaps also older crystalline basement rocks comprising metamorphics and acid-intermediate igneous bodies. The suture reaches 80-km in width and has been interpreted as the remnant of a Cretaceous-Middle Eocene ocean basin that became closed and was possibly almost completely over-ridden by the East Sabah Terrane. This is suggested by the occurrence of Chert-Spilite inliers along the Lower Segama river and elsewhere in the Dent Peninsula (see Lim 1985). Within the suture olistostromes are olistoliths of folded rocks of the Palaeogene Trusmadi Formation, and also pervasively sheared argillaceous matrix of this formation and the younger Palaeogene Crocker Formation at widely different localities in western Sabah. Other evidence of compressional tectonics in the suture zone is the presence of a blueschist belt (Leong, 1978).

PENINSULAR MALAYSIA

Onshore

A four-fold geological division is recognised for the outcropping pre-Tertiary formations of Peninsular Malaysia (Fig. 26.5). These geological domains are distinguished on the basis of a combination of differences and/or variations in structural trend and styles, types of mineralisation, dominant lithology and facies, and palaeogeography (Foo, 1983; Tjia and Zaiton Harun, 1985). The Central Belt has especially well-defined boundaries. In the west is the Lower Triassic (?) Bentong suture (Hutchison, 1975; Tjia, 1989a), while in the east is the newly recognised Eastern Tectonic Zone (ETZ). The ETZ comprises the Lebir fault zone while its southern continuation is marked by N-S faults and a row of low-gravity anomalies (Tjia in Liew et al., 1996).

The tectonic history of Peninsular Malaysia is in Figure 26.6. Middle Cambrian to Recent formations are present and major breaks separate Lower from Upper Palaeozoic rocks, Triassic from later Mesozoic, and these from Tertiary formations. The Triassic and older sedimentary formations are mainly marine; those younger than the Triassic are almost completely non-marine. Four major episodes of granitic emplacements occurred in the (1) Late Carboniferous, (2) Early Triassic, (3) Late Triassic- Early Jurassic, and (4) Late Cretaceous. The latter comprises small bodies, but the older igneous intrusives form large mountain ranges, such as the Titiwangsa (Main Range) granitoid complex and the mountain ranges near the eastern shoreline. Regional metamorphism is widespread; contact aureoles, however, are generally narrow. Multiple deformations characterise Lower Triassic, and especially pre-Permian and older formations (Tjia, 1986). Therefore, the pre-Permian formations exhibiting two or more episodes of superimposed deformations possess very low, if any, hydrocarbon prospectivity. For this reason, the brief description on stratigraphy in this chapter comprises only Permian and younger formations. Details of the stratigraphy are published by the Geological Survey of Malaysia in memoirs and various types of reports and maps.

Permian

Permian sediments are dominantly calcareous (often reefal), and except for the Lower Permian, are widely distributed with thicknesses up into the 600 m to over 1000 m range. The Permian ages have been determined by many diagnostic fossils. In the Central and Eastern belts the limestones are often associated with volcanic tuffs. In the Gua Musang area, however, the rocks have been intricately deformed. Marker beds are absent. Recent detailed studies identified Triassic conodonts, ammonites, algae and foraminifers in several of the limestone occurrences, that on lithological similarities, were once considered of Late Palaeozoic age (Fontaine and Ibrahim Amnan, 1994). Elsewhere, Permian limestones are usually grey to black and occur as smaller outcrops.

Triassic

Except in a few areas, there was a depositional break between the Permian and Triassic periods. Figure 26.7a is the Mesozoic stratigraphy of Peninsular Malaysia. By the Middle Triassic, flysch sedimentation became widespread. Reliable subdivisions were made using abundant ammonites. Sandy shallow marine environments prevailed along the east side of the Titiwangsa,

Pre-Tertiary Hydrocarbon Potential

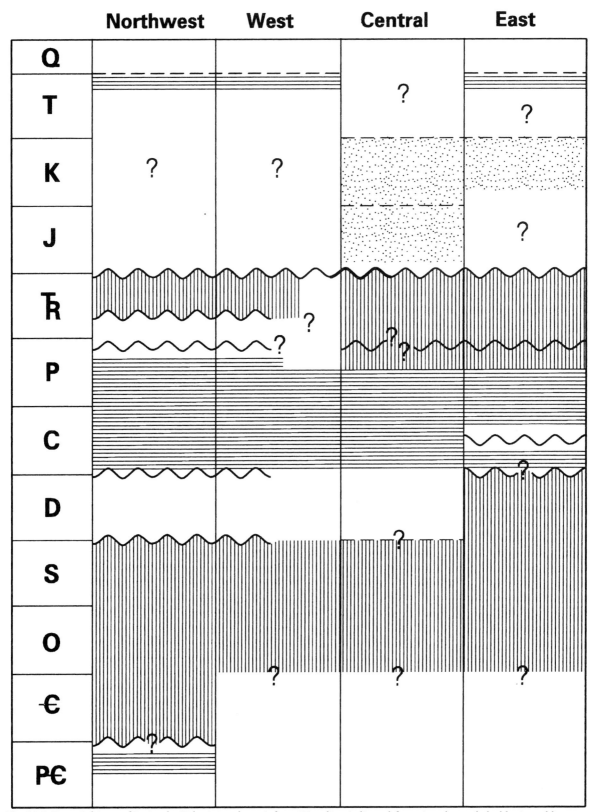

Fig. 26.6 Tectonic history of Peninsular Malaysia in schematic form. Wavy lines indicate deformation phases; dashed horizontal lines are stratigraphic boundaries of uncertain nature; filled-in spaces represent deposition: horizontal lines = coastal (Precambrian), or shallow-marine (Carbo-Perm), or lacustrine (Tertiary); vertical lines = holomarine; dots = continental deposits, some are redbeds. Columns are the geological domains.

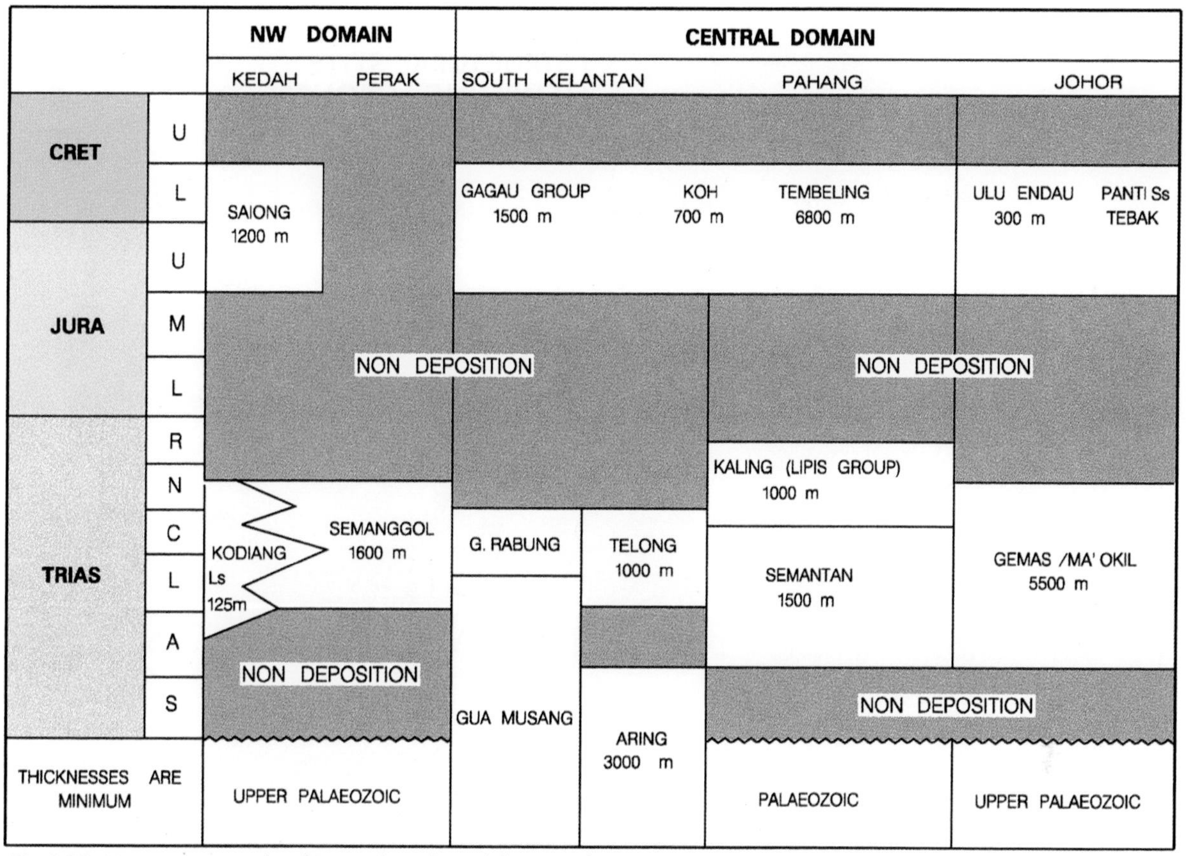

		NW DOMAIN		CENTRAL DOMAIN			
		KEDAH	PERAK	SOUTH KELANTAN	PAHANG		JOHOR
CRET	U						
	L	SAIONG 1200 m		GAGAU GROUP 1500 m	KOH 700 m	TEMBELING 6800 m	ULU ENDAU 300 m / PANTI Ss TEBAK
JURA	U						
	M			NON DEPOSITION		NON DEPOSITION	
	L						
TRIAS	R				KALING (LIPIS GROUP) 1000 m		
	N						
	C	KODIANG Ls 125m	SEMANGGOL 1600 m	G. RABUNG	TELONG 1000 m	SEMANTAN 1500 m	GEMAS /MA' OKIL 5500 m
	L						
	A						
	S	NON DEPOSITION		GUA MUSANG	ARING 3000 m	NON DEPOSITION	
THICKNESSES ARE MINIMUM		UPPER PALAEOZOIC			PALAEOZOIC		UPPER PALAEOZOIC

Fig. 26.7a Mesozoic stratigraphy of Peninsular Malaysia (Khoo, 1983).

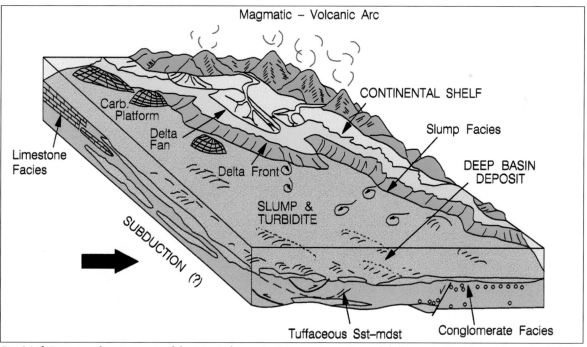

Fig. 26.7b Depositional environments of the Triassic deep-water Semantan Formation and shallow-water Gua Musang Formation (Sahalan Abd Aziz in Liew et al., 1996).

while elsewhere deeper-water conditions are indicated by *Daonella* in association with *Halobia* and *Posidonia* biofacies. The clastics intercalate with limestone and pyroclastics, that in the Central Belt, consist of felsic tuffs. Towards the end of the Triassic, or the beginning of the Jurassic, conditions became continental as indicated by the appearance of the plant fossil *Sagenopteris* sp. (Geological Survey of Malaysia, 1995). Figures 26.5 and 26.7b show the distribution of Mesozoic sedimentary rocks and a facies model of the Middle-Upper Triassic Semantan Formation by Sahalan Abdul Aziz (in Liew et al., 1996).

Pre-Tertiary Hydrocarbon Potential

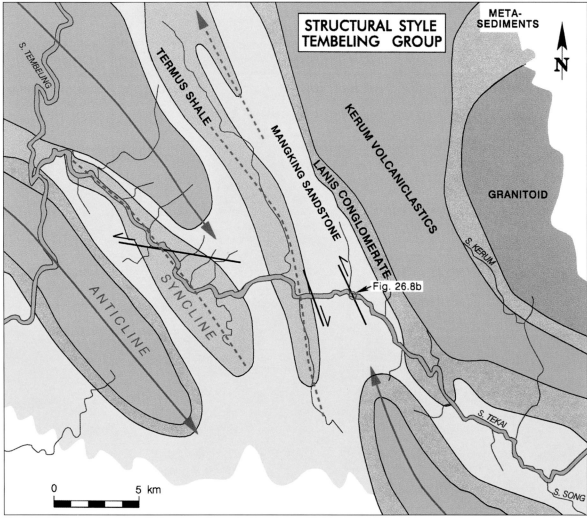

Fig. 26.8a Structural style of the Jurassic-Cretaceous Tembeling Group. Base map is by Khoo (1983) with additions from recent PRSS fieldwork.

Jurassic-Cretaceous (JK)

Post-Triassic sedimentation occurred under continental conditions, although foraminifers in some calcareous rock in the Tekai river area, Pahang, where the JK Tembeling beds crop out extensively, suggest that part may have been deposited in marine surroundings. Figure 26.5 shows that JK-sediments occur predominantly in the Central Belt. The 6800 m thick Tembeling Group is representative (Khoo, 1983). In ascending order the group comprises the Kerum Volcaniclastics, Lanis Conglomerate, Mangking Sandstone and Termus Shale. Reddish colours are dominant. Well-preserved plant fossils mark certain horizons. Early Cretaceous forms occur in the Upper parts of the group (Smiley, 1970). An open forest of few plant taxa is interpreted; while the climate has distinctly dry seasons. Abundant gymnosperms and pteridophytes pollen support the earlier interpretation (Shamsudin Jirin and Morley, 1994). However, Uyop Said (1997) doubts that the pollen assemblage he collected from the Tembeling Group along the Tembeling river can be as old as Early Cretaceous, as *Cicatricosporites* sp. and *Classapollis* sp. are missing. He also believes the climate was warm and humid from the absence of bisaccate pollen while fungal spores are abundant. Recently a study was completed on the hydrocarbon potential of the JK sequences onshore Peninsular Malaysia (Liew et al., 1996). Palaeoslopes and palaeocurrents combined with structural information suggest that moderately large intermontane basins existed during Jurassic-Cretaceous time. Most of these basins were grabens or half-grabens associated with major north-south, dextral strike-slip faults. Prolonged activity along the wrench faults caused strong deformation in some of the basins, while elsewhere the Jurassic-Cretaceous beds were essentially undisturbed (Figs. 26.8a & 26.8b). Figure 26.9a explains the occurrence of deformed as well as undeformed JK sediments. On Figure 26.9b is an example of compressive deformation in central Pahang state, where probable Permian

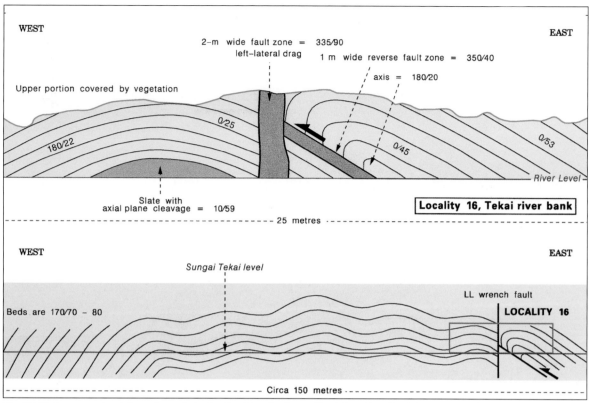

Fig. 26.8b Representative deformation style in the Mangking Sandstone (Tembeling Group) exhibiting an anticlinorium and west-verging reverse fault. Middle course of the Tekai river just to the east of the sinistral NW-striking wrench fault (location indicated in Figure 26.8a).

strata were thrusted over, and in the process, overturned the underlying JK beds

Hydrocarbon Prospectivity

Mohd Jamaal Hoesni (in Liew et al., 1996) studied the geochemistry of outcropping Mesozoic sedimentary rocks, mainly the JK Tembeling Group and equivalent sequences, and the Triassic Semantan and Gemas formations. He concluded from 25 Triassic samples that the total organic carbon (TOC) content ranged from poor to moderate (0.24 to 4.31 wt per cent). High thermal maturity was indicated by reflectance exceeding 2.7 %. In spite of this, the relatively high residual carbon (average 1 wt per cent) suggested that it could have been much higher, especially in the thick black shales. On the other hand, the Jurassic-Cretaceous sedimentary rocks (57 samples) generally showed poor TOC of less than 0.5 wt per cent. Occasional high TOC values correlated with coaly and carbonaceous beds which, however, were thin. Vitrinite reflectance ranged from late mature to post-mature, with a few exceptions showing low and very high maturity. The latter could be attributed to proximity to intrusive bodies.

Evidence of a possible third oil family in the Malay Basin, in addition to the common biomarkers for non-marine coaly and lacustrine source rocks, was recently discovered by Mohd Jamaal Hoesni and Abolins (1996). They suggested, among other possibilities, that its source could have been pre-Tertiary, that is, Triassic carbonates similar to those occurring in the Gua Musang area.

Quartzose, volcanic and volcaniclastic JK sandstones experienced four diagenetic events: quartz overgrowth, iron-oxide cementation, feldspar dissolution, and minor clay authigenesis and replacement (Mohd Fauzi Abdul Kadir in Liew et al., 1996). Visible sandstone porosity is negligible to poor (< 10 %) and this is mainly in the form of secondary pores. However, one Lanis Conglomerate sample shows up to 28% bulk-volume porosity. The reservoir quality of the samples studied is poor, but it cannot be ruled out that originally the quartzose sandstone was clean, possessing moderate sorting and, therefore, was of good quality.

To date, no undisputed surface hydrocarbon indications are known from the onshore Mesozoic areas. However, based on the results of fieldwork, Liew (in Liew et al., 1996) postulated the following possible pre-Tertiary petroleum systems. The Jurassic-Cretaceous strata were deposited in separate pull apart depressions. The larger ones were controlled by repeated dextral strike-slip movements along major north-

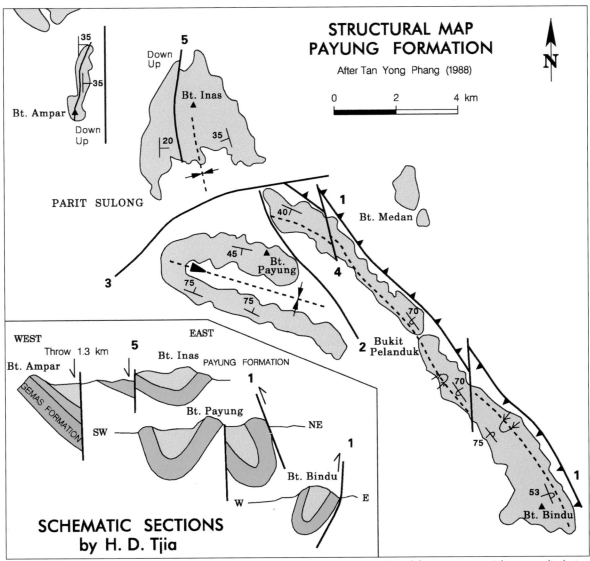

Fig. 26.8c Major synclines and faults in the JK Payung Formation and M.-U. Triassic Ma' Okil Formation, in Johor state, displaying contrasting structural styles. Bukit Payung is indicated on Figure 26.5.

south fault zones that occupy the 25 to 30 km wide Eastern Tectonic Zone (ETZ). Deposition occurred in a semi-arid, distinctly seasonal climate. The stacked, thick channel sandstones are the most likely reservoirs. Structural deformation was stronger at the rims of the depocentres, while open folds prevailed away from those rims. Most JK strata outside the ETZ are undeformed. Vertical drainage through fractures and faults is the most likely migration style. Entrapment could have been by shale interbeds, while the Termus Shale probably acted as a regional seal. Preliminary geochemical results show relatively low TOC, and maturity within the gas window. The Triassic strata of the Central Belt are flanked by the Bentong suture in the west and by the ETZ in the east. Provenance was from highlands in the west and in the east. In the central and southern parts of the Central Belt, sediments are marine to deep-marine turbidites containing common, wide slump intervals. The Triassic strata were deformed into open isoclinal folds, with complex structures in the slump intervals as a result of tectonic overprinting. The thick turbidite and grainflow sandstones are the most likely reservoirs, but porosity is tight in those having tuffaceous admixtures. Near fault zones, the beds are steep to vertical. Therefore, although the sandstone reservoirs are laterally continuous, vertical migration along fractures and faults is probably the norm. No regional seal, except probably sediments associated with the regional unconformity separating Triassic and older rocks from post-Triassic strata, is known. The TOC of Triassic mudstones is higher than that of equivalent Jura-Cretaceous strata. However, the Triassic organic material is over-mature.

Geochemical results suggest that the Triassic strata have better source-rock potential. If Triassic hydrocarbons were still preserved (most likely as gas at the end of the Mesozoic),

Pre-Tertiary Hydrocarbon Potential

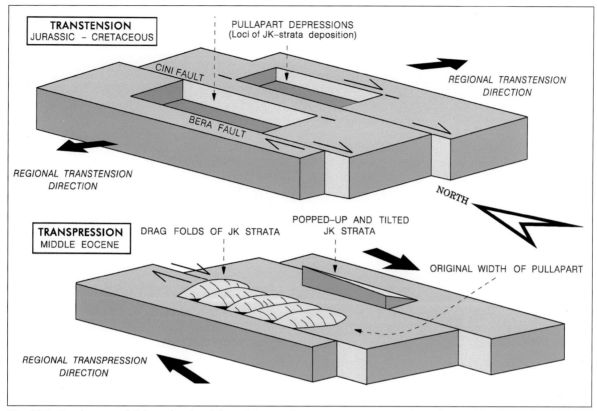

Fig. 26.9a Development of deformed and undeformed Jurassic-Cretaceous strata in Peninsular Malaysia. UPPER. Jurassic-Cretaceous: dextral slip motion on N-S striking faults produces pull aparts between overlapping fault strands. Regional tension is approx. E-W. LOWER. Middle Eocene: the regional stress regime has changed into transpression that re-activates dextral slip along some of the faults. Sedimentary fills of pull aparts are compressed into drag folds striking NW. Where faults remain locked and only accommodate vertical displacements, the sediments of the pull aparts are pushed up to form structural plateaus or cuestas.

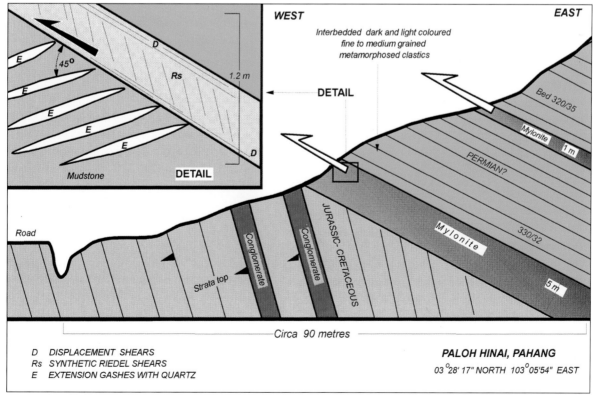

Fig. 26.9b Field sketch to illustrate strong compression on a boundary fault of a JK pull apart depression: reverse fault zone and mylonite bands. ?Permian beds are thrust over near-vertical JK strata. Paloh Hinai, Pahang.

Fig. 26.9c Oblique frontal view of the thrust-fault zone at Paloh Hinai, Pahang. The right edge of the photograph is about 5 metres high.

migration could have been by (1) lateral drainage through Triassic reservoirs with hydrocarbons trapped within those reservoirs, or (2) hydrocarbons could have migrated through fractures and faults into JK reservoirs.

Straits of Melaka

The shallow pre-Tertiary basement in the eastern part of the Straits of Melaka dips gently west to underlie the deep, petroliferous backarc basins of Sumatra. With the exception of its southern tip, the strait resides in the Gondwana continental fragment (Fig. 26.10). A major suture within the Gondwana fragment is the Kerumutan Line, that contains partly 'tectonized melange' and was assigned a Triassic age by Pulunggono and Cameron (1984). The tectonic line may continue onto the peninsula and appears to form the boundary of the Northwest Tectonic Domain (Fig. 26.5). A number of relatively small Tertiary basins were discovered during petroleum exploration in the 1970-1980s. The location of most basins, if not all, appeared to have been controlled by major northerly striking basement faults. North-south regional Jurassic lineaments occur in the Sunda platform that also includes the Straits of Melaka (Tjia, 1997). The small Straits of Melaka basins appear to have been produced by tectonic reactivation of the lineaments during the Oligocene regional extensional stress regime. Evidence for pull apart origin is strong. Liew (1994a) distinguished basin-groupings that are separated by N-S basement highs. The Port Kelang Graben (Fig. 26.11a) and the Johor Graben (Fig. 26.11b) have N-S elongation and are aligned with the Pematang and Lematang troughs of Sumatra, respectively. Smaller faults associated with individual basins commonly strike oblique to that trend. Predominantly NNE-striking normal faults in the pre-Tertiary basement of the east-by-north trending Central Graben, probably resulted from dextral wrench motion parallel to the graben's long axis (Fig. 26.12a). Most of the depressions began to develop in the Oligocene, which is represented by red beds, fanglomerate and lake sediments. Ahmad Munif Koraini (1993) identified Eocene to Oligocene palynomorphs in the basal beds of the Batu Arang basin onshore Peninsular Malaysia. This suggests that regional tensional/extensional stresses may have begun much earlier. Flower structures, *en echelon* faults, and other deformation features in the Tertiary basin-fill, indicate that the basement surface continued to be compressed, possibly until the Late Miocene (Figs. 26.12b, 26.13, 26.14). Farther to the west, in the Central Sumatra Basin, major structural inversion producing traps took place in the Pliocene-Pleistocene (Greenaway and Goh, 1989).

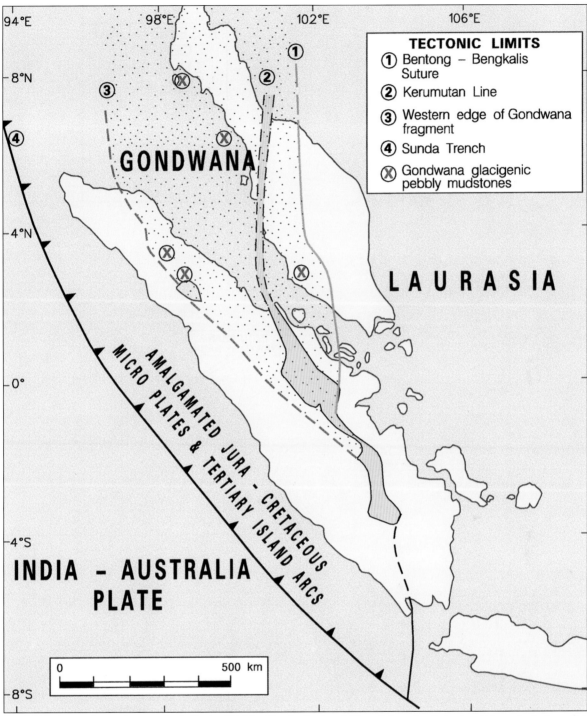

Fig. 26.10 Tectonic limits of terranes in the Straits of Melaka. The distribution of Upper Palaeozoic, glacigenic pebbly mudrock shows that the Kerumutan Line lies within the Gondwana plate fragment. Sources: Cameron et al., 1981; Stauffer and Lee, 1986; Tjia and Anizan Isahak, 1990.

Many of the depressions are half-grabens bounded by faults on the west. Maximum depths of the depressions range between 2700 feet and 13,000 feet (810 m and 3900 m), respectively (Liew, 1994a). A few exploration wells penetrated basement. The Mobil MG-XA well in the Central Graben penetrated 358 feet (109 m) basement consisting of black carbonaceous shale, slate and quartzite overlain by red beds, which Fontaine et al. (1992) correlated with the Carboniferous Singa or Kubang Pasu formations. No hydrocarbon indications, neither source rocks nor seals were found in its Tertiary graben-fill. Dolostone and dolomitic limestone formed the pre-Tertiary basement in wells drilled by Mobil (MSS-XA). Strontium isotopes suggest a Permian age for the bottom dolostone in Sun Malaysia Petroleum Company's wells. $^{87}Sr/^{86}Sr$ dating of calcareous rocks in the lower-most section of the Singa Besar-1 well confirmed a Late Cretaceous or older base below 3 metres of Middle Miocene carbonate. The rich microfossils studied by

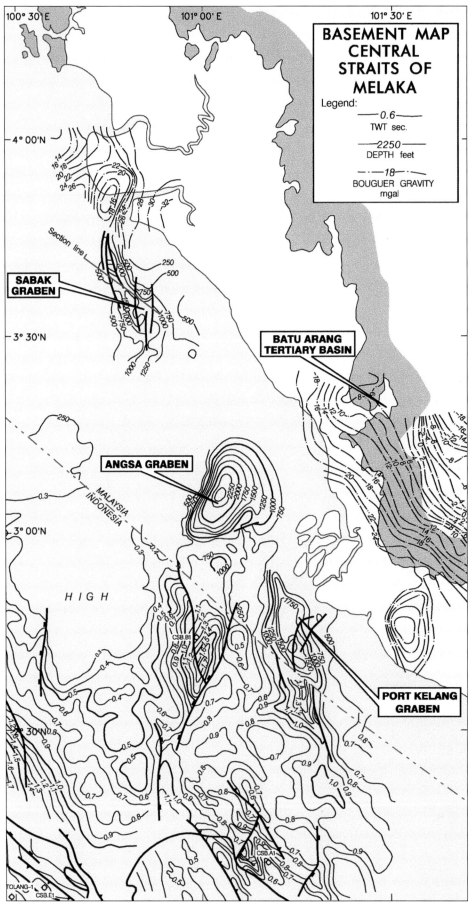

Fig. 26.11a Basement map of the central Straits of Melaka area. Basement depressions are indicated by bouguer gravity anomalies in milligal and by depth contours in feet. Note the N-S zonation of these depressions. Onshore, east of Bukit Jugra the gravity anomaly pattern most probably represents a Tertiary basin buried under Quaternary sediments. The seismic section of Figure 21.13 is indicated..

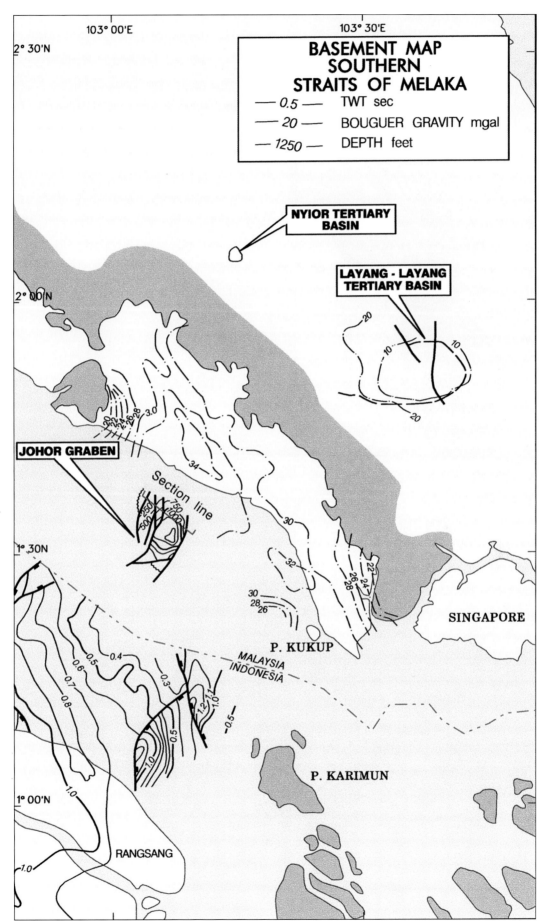

Fig. 26.11b Basement map of the southern Straits of Melaka area. N-S elongation of the offshore depressions is shown. The indicated section line is in Figure 21.14.

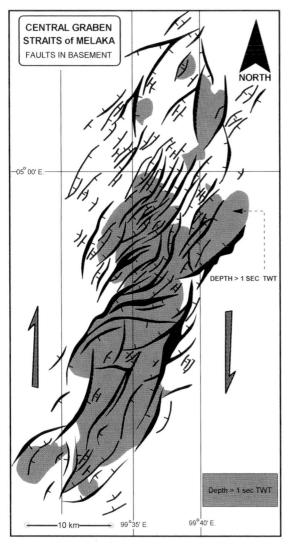

Fig. 26.12a *En echelon* fault pattern in the pre-Tertiary basement of the Central Graben, Straits of Melaka, resulted from dextral slip on a broad, N-S fault zone. The Central Graben is located about 100 km to the SW of Pinang island. Based on unpublished material by Nichol (1990).

Fontaine et al. (1992) indicate a Late Permian age. The strontium-isotopes method returned a 25 Ma or Late Oligocene age for the dolostone to dolomitic limestone basement in Sun's Langgun Timur-1 well.. The bottom limestone in Sun's Dayang-1 well is similar to the Ordovician-Silurian Setul limestone (Fontaine et al. 1992). Only the Singa Besar-1 well tested some gas from fractured carbonate reservoir, the other two wells were dry.

The Port Kelang-1 well penetrated 350 feet (107 m) of grey limestone with rare thin intercalations of shale and siltstone. On account of absent Permian microfossils and the degree of recrystallization, Fontaine et al. (1992) correlated these basement rocks with the older Setul limestone or the Kuala Lumpur/Batu Caves limestone of Ordovician-Silurian age.

Hydrocarbon Potential

The relatively shallow pre-Tertiary basement on the Malaysian side of the Straits of Melaka forms the wide eastern border zone of the prolific petroliferous Central Sumatra Basin and the gas-prone North Sumatra Basin. Although lacustrine shales in the lower part of the Tertiary depressions/ grabens in the Straits are common, and are potential source material, hydrocarbon accumulations have yet to be discovered in the overlying sediments. Similar brown shales in the Pematang Formation (Upper Oligocene-Lower Miocene) of Central Sumatra are known as excellent hydrocarbon source rock. DHIs were reported in exploration reports of Sun Malaysia Petroleum Company.

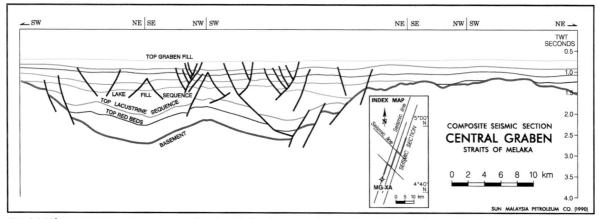

Fig. 26.12b A composite section in a general NNE-SSW direction transecting the Central Graben. Two rootless flower structures represent wrench zones that were active into Late Miocene - Early Pliocene time (approximately equivalent with the Top Graben Fill horizon). Red Beds ~ U. Oligocene, Lacustrine Sequence ~ L. Miocene, Lake Fill Sequence ~ M.-U. Miocene, Top Graben Fill ~ U. Miocene-Recent. Simplified after Nichol (1990).

Pre-Tertiary Hydrocarbon Potential

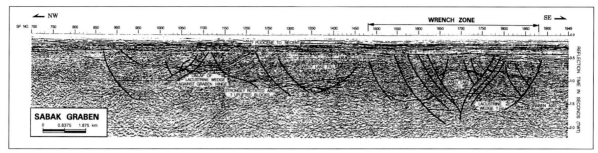

Fig. 26.13 Seismic section obliquely across the Sabak Graben in the Straits of Melaka. After Greenaway and Goh (1989). Sandy Lake Fill ~ M.-U. Miocene, Lacustrine Wedge ~ L. Miocene and older. The line of section is shown on Figure 26.11a.

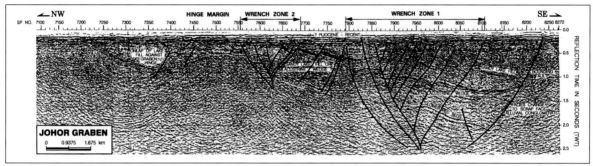

Fig. 26.14 Seismic section across the Johor Graben in the southern Straits of Melaka. After Greenaway and Goh (1989). Sandy Lake Fill ~ M.-U. Miocene, Lacustrine Wedge ~ L. Miocene and older, Lower Red Beds ~ U. Oligocene. The line of section is indicated on Figure 26.11b.

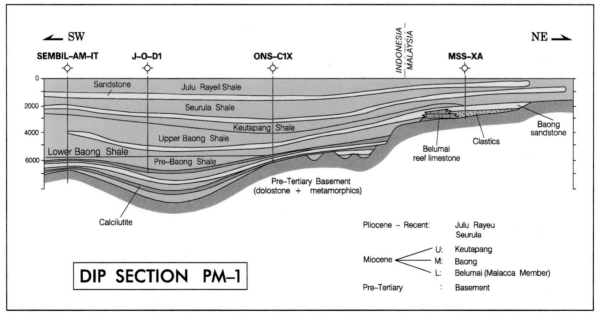

Fig. 26.15 Regional dip section in the northern part of Straits of Melaka. The well MSS-XA is located some 150 km SW of Langkawi island. Laterally continuous sand horizons could act as conduits for fluids originating in the North Sumatra Basin towards fractured basement highs on its flank. After unpublished material by Hatch (1987) and Dyer et al. (1991).

Our knowledge about the type, extent and volume of pre-Tertiary formations in the Straits is still inadequate. As yet there is no evidence of pre-Tertiary hydrocarbon source rocks within the basement on the Malaysian side of the Straits of Melaka. It seems more likely that the role of this basement is that of host to Tertiary hydrocarbons. The dolomitic basement rock drilled so far has good reservoir properties. That of the Singa Besar-1 well flowed 4128 bbl/day water (Nichol, 1990). In the Beruk Northeast field of the Central Sumatra Basin, oil is produced from pre-Tertiary basement rocks (Koning and Darmono, 1984). In the Mekong basin offshore Vietnam, horst blocks of fractured granitoids of Cretaceous age are charged from Oligocene source rock located in the grabens (see Areshev et al,. 1992). In the recent past, only Tertiary carbonate reefs topping basement highs were important exploration targets in Malaysian waters (Fig. 26.15; Hatch, 1987), especially in the northwestern part of the Strait, a few kilometres east of hydrocarbon-bearing features in similar plays in the North Sumatra Basin proper.

Greenaway and Goh (1989) concluded that oil generation in the Central Sumatra troughs: Bengkalis, Pematang and Balam began at various times ranging from Late Pliocene up to the present. Therefore, any migration of these oils towards the Peninsula post-dated the assumed Late Miocene, major structural modification of the basement rocks. In other words, basement highs of the Straits of Melaka were in place before oil migration..

No mature source rock was available in the northern area where the three exploration wells were completed by Sun Malaysia Petroleum Company. Dyer et al. (1991) attributed this to the low temperature gradient of between 1.8 and 2.0 °F/100ft. Long-range migration from the North Sumatra Basin was indicated by oil shows within the Middle Baong sandstone (Middle Miocene) in Dayang-1 and Langgun Timur-1, and probably also by the occurrence of gas in Singa Besar-1. The laterally continuous sandstone horizons are considered to act as conduits (Fig. 26.15).

It is concluded that the pre-Tertiary basement hydrocarbon play of the Straits of Melaka comprises fractured crystalline basement high reservoirs, capped by seals that were at least some 200 metres (a value slightly below the maximum Quaternary sea-level lowstand) below present sea level. Petroleum migrated from the Central or North Sumatra Basin proper.

Basement of Malay and Penyu Basins

The pre-Tertiary basement of the Sunda Shelf, where the Malay and Penyu basins are located, is part of Sundaland, a structural platform that attained relative stability by the end of the Mesozoic (van Bemmelen 1949; Fig. 26.1). It is also widely accepted that Sundaland was an extensive peneplain at the beginning of the Cenozoic. The widespread occurrence of mainly middle-Upper Cretaceous felsic granitoids in the northern part of Sundaland is evidence for strong crustal movements before the platform stabilised. The Malay, Penyu and West Natuna (Indonesia) Tertiary basins were very probably aulacogens associated with an Upper Cretaceous hotspot centred at their triple junction (Tjia, 1994). In the pre-Oligocene this part of Sundaland began to experience regional tension - possibly transtension - that caused the aulacogens to subside at different degrees, the Malay Basin attaining its greatest development and accumulating more than 14 km Tertiary deposits. The regional stress regime had become transpressional near the end of the middle Miocene and caused structural inversion that had involved basement. In the SW end of the Malay Basin reverse faulting resulted in throws exceeding one kilometre at basement level. Compressional stress in the Malay Basin is believed to have continued until today.

Thirty-seven wells were drilled into pre-Tertiary basement, of which two penetrated basement of the Penyu Basin (Fig. 26.16, Table 26.1). Depth of penetration ranged from 1 metre to 534 m (Tok Bidan-1, off the Kelantan coast, Figure 26.17) and 493 m (limestone in Bunga Raya-1, a well near the international boundary in the northeast). Scant fossil evidence was provisionally interpreted as indicative of ?Middle Triassic (Asiah Mohd Salih and Sanatul Salwa Hasan, 1992). Dark coloured, mud-rich limestone from Sotong B-1 was determined by Fontaine et al. (1990) as Upper Triassic. Granitic rock of Belumut-1 has pinkish feldspar. As a rule of thumb, field geologists working in Peninsular Malaysia assign this type of granitoid to the Cretaceous. Pink granitoids also crop out in two small islands, Nyirih and Tokong Burung, upon the Tenggol Arch (Yusoff Johari, 1983).

Hydrocarbon Potential

Wells deeply penetrating pre-Tertiary basement of the Malay Basin indicate that its hydrocarbon potential has not been totally ignored. On the other hand, the high success ratio (average 1 in 4.5 wells) for oil exploration of its Tertiary deposits probably has precluded sustained efforts to target basement reservoirs. The possibility of a pre-Group M (pre-Upper Oligocene) source was proposed by Mohd Jamaal Hoesni and Abolins (1996). An aulacogen or rift origin of the basin may have resulted in restricted depositional environments that favour the preservation of organic matter. Fractured and/or weathered Cretaceous granitic horst blocks of the Mekong basin, in northern Sundaland, may contain up to a billion barrels of oil (Areshev et al., 1992) that was probably sourced from Oligocene shales in adjoining grabens. Dark coloured Permian limestone is believed to be the source of gas produced in the northwestern Khorat Plateau of Thailand.

Along the axial zone of the Malay Basin, the basement is very deep and in the NW portion, crystalline basement is apparently at depths exceeding 14 km. Moreover, above-

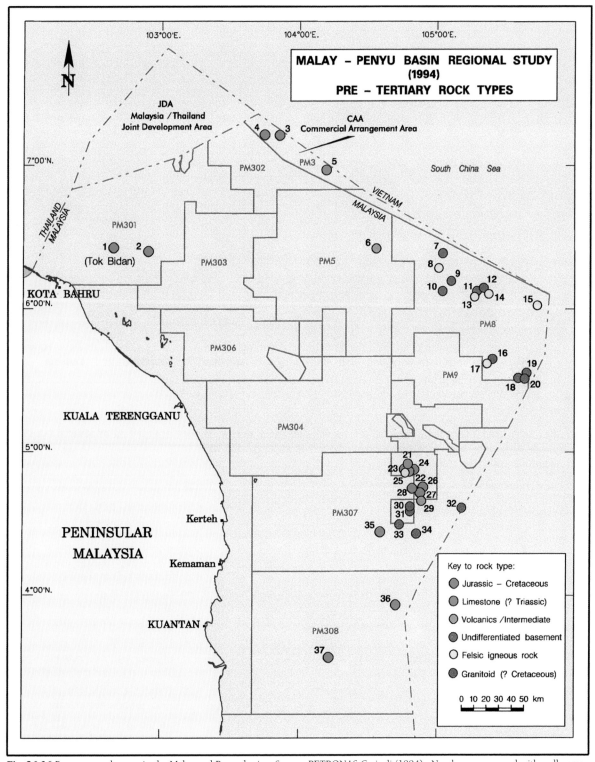

Fig. 26.16 Basement rock types in the Malay and Penyu basins. Source: PETRONAS Carigali (1994). Numbers correspond with well names in Table 26.1.

normal pressures have until now limited production of hydrocarbon to the upper 5 km of the very thick Tertiary section. Exploration targets in basement reservoirs are therefore limited to the basin's margins, and the Tenggol and Belumut arches. In the Malay basin, shallow pre-Tertiary basement underlying proven hydrocarbon accumulations in Tertiary beds occurs in the Sotong field (Triassic limestone), Anding field (JK clastics), along the NE margin (fractured and weathered granitic basement), and the Belumut structure (fractured and weathered granitic rocks). In the Penyu Basin on the SW side of the Tenggol arch, oil was discovered in a Tertiary drape structure over one of the basement horst blocks (well Rhu-1/1A; Texaco 1992). There are ten other basement horsts in the Penyu Basin.

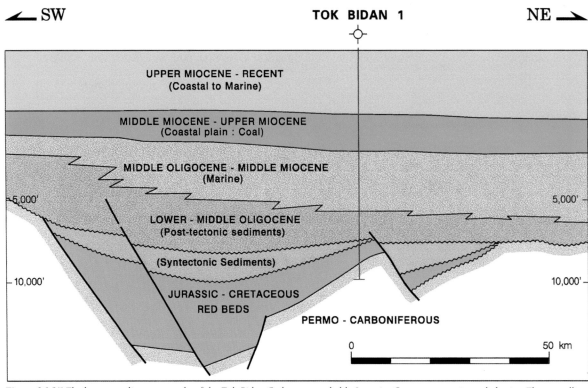

Figure 26.17 The bottom sedimentary rocks of the Tok Bidan Graben are probably Jurassic- Cretaceous continental clastics. The crystalline basement is ?Carbo-Permian. After Liew (1994b). The location of Tok Bidan is indicated on Figure 26.16.

Table 26.1 BOREHOLE RECORDS OF PRE-TERTIARY ROCK TYPES MALAY & PENYU BASINS
Source: PETRONAS Carigali (1994)

Numbered localities are shown on Figure 26.16

1. Tok Bidan-1 (2118 m)
   534 m sandstone, siltstone, shale and metasediments (Jurassic-Cretaceous)
2. Badak-1 (2295 m)
   31 m lithic sandstone and conglomerate (Jurassic-Cretaceous)
3. Bunga Orkid-1 (3671 m)
   174 m continental red beds (Jurassic-Cretaceous)
4. Bunga Pakma-1 (3522 m)
   59 m clastics and metasediments (Jurassic-Cretaceous)
5. Bunga Raya-1 (2905 m)
   493 m limestone *(Tubiphytes, Aulotortus?; ?Triassic)*
6. Larut-1 (2686 m)
   87 m weathered basic volcanics
7. Sumalayang-1 (1438 m)
   13 m undifferentiated
8. North Lukut-2 (1945 m)
   5 m weathered felsic igneous rock
9. Relau-1 (2177 m)
   9 m undifferentiated
10. Penara-1 (2373 m)
    1 m undifferentiated
11. Abu-1 (1910 m)
    1 m undifferentiated
12. Abu Kecil-1 (1935 m)
    8 m weathered granite
13. Abu-2 (1822 m)
    8 m felsic igneous rock
14. Bubu-1 (1609 m)
    28 m weathered felsic basement
15. Lerek-1 (1096 m)
    28 m weathered felsic basement
16. East Raya-1 (2410 m)
    16 m granite
17. South Raya-1 (2261 m)
    1 m felsic basement
18. West Belumut-2 (1455 m)
    1 m granite
19. Belumut-1 (1505 m)
    3 m granite
20. West Belumut-1 (1410 m)
    7 m granite
21. Sotong B2 (2911 m)
    8 m limestone
22. Sotong B1 (2750 m)
    292 m limestone, argillite
23. Sotong B3 (3016 m)
    38 m argillite Jurassic-Cretaceous
24. Sotong B5 (2985 m)
    31 m argillite Jurassic-Cretaceous
25. Sotong B4 (2635 m)
    72 m igneous rock, felsic
26. Anding-1 (2587 m)
27. Anding-2 (2658 m)
    79 m phyllite and red beds (Jurassic-Cretaceous)
28. Anding Barat-1 (3167 m)
    3 m phyllite, dark gray to black (Jurassic-Cretaceous)
29. Feri-1 (2918 m)
    22 m weathered diorite
30. Malong 5G-17.2 (1583 m)
    11 m granite
31. Malong 5G 17.1 (1580 m)
    50 m granite and metamorphics
32. Delah 5H-14.1 (2928 m)
    35 m weathered granite
33. Jelutong 5G-23.1 (1627 m)
    13 m weathered granite
34. Keledang 5G-24.1 (1726 m)
    29 m weathered granite
35. Kempas 5G-22.1 (1799 m)
    157 m Kempas sandstone (Jurassic-Cretaceous)
36. Rhu-1 (2912 m)
    114 m d.gr siltstone + claystone (Jurassic-Cretaceous); felsic volcanic basement
37. Pari-1 (2174 m)
    53 m calcareous siltstone, dark gray - black; many calcite veins

SARAWAK

Stratigraphy

Sarawak occupies most of the northern side of Borneo, the third largest island in the world. Sarawak consists of three major geological domains. (1) Westernmost Sarawak to the west of the Lupar line (by many considered as an Early Eocene suture) is composed mainly of pre-Tertiary crystalline/metamorphosed rocks that in places are overlain by thick Palaeogene-Early Neogene continental deposits. (2) Upper Cretaceous to Lower Eocene imbricated turbidites form a broad fold-thrust belt, also known as the Rajang Accretionary Prism (RAP), and covers the middle zone of highlands in the interior of Sarawak. (3) Another broad zone forms the hill country and part of the mountainous hinterland which mainly consists of folded Neogene, shallow marine to fluviatile-lacustrine sediments. In northeastern Sarawak, this third zone contains large outcrops of Palaeogene rocks that form high-relief topography, such as the Mulu mountain and long and high ridges to its east. Extensive coastal plains of all these three domains are covered by flat-lying Quaternary deposits (Fig. 26.18). Hydrocarbons are being produced from Neogene deposits in the broad frontal belt of the RAP that constitutes the Sarawak basin.

The following pre-Tertiary stratigraphy of Sarawak is based on Liechti et al. (1960) and various annual reports by the Geological Survey of Malaysia, the latest published report being that

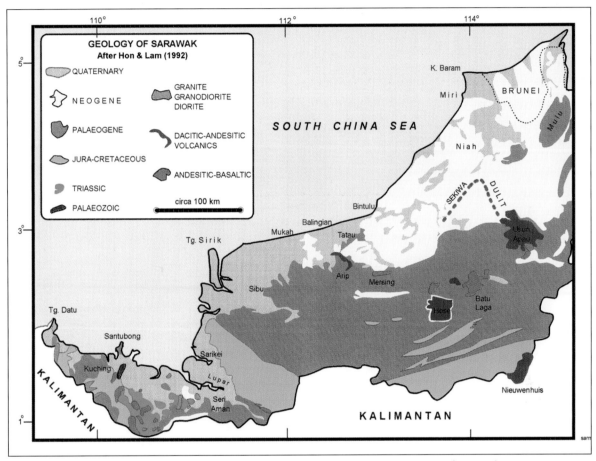

Fig. 26.18 Outline of Sarawak showing crustal growth from the interior towards the South China Sea: from mainly pre-Tertiary western Sarawak to the Neogene sediments of central and northern Sarawak.

of 1995 (Geological Survey of Malaysia, 1995; Fig. 26.19).

Westernmost Sarawak. Pre-Upper Cretaceous rocks only occur in westernmost Sarawak. The oldest known stratigraphic units are marine sedimentary rocks, probably of pre-Upper Carboniferous age: the Kerait Schist and Tuang Formation. The geological age is based on the high metamorphic degree and the presence of possible tentaculitid fossils. An angular unconformity separates these oldest units from overlying fusulinid limestone, chert and some argillite of the Terbat Formation that was deposited in a shallow sea during middle-Late Carboniferous to Early Permian. The limestone is light-grey to black, partly or wholly crystalline, with carbonaceous or bituminous bands in places

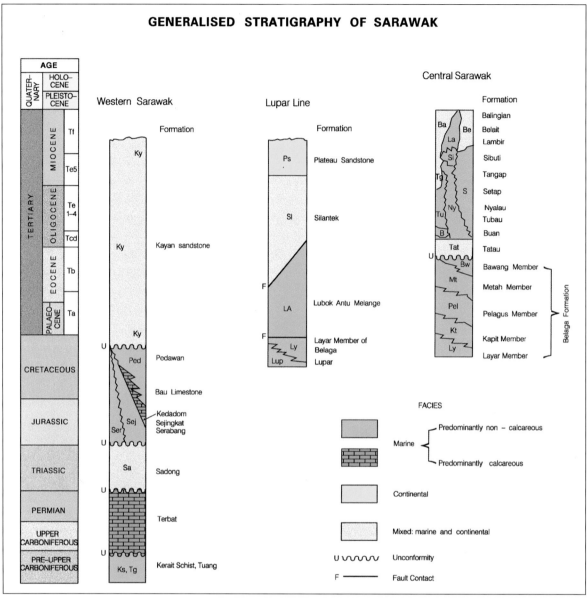

Fig. 26.19 Generalised stratigraphy of Sarawak after Hon and Lam (1992). The various sedimentary environments are highlighted.

(Liechti, 1960, p. 30). Strong folding, uplift and erosion terminated its deposition. In the ?Late Triassic continental to shallow-marine clastics of the Sadong Formation were deposited. This was accompanied by widespread volcanic activity that produced extensive thick successions of andesitic and basaltic lavas and pyroclastics with some trachytic and rhyolitic varieties. The igneous rock association points to an island-arc setting, possibly related to subduction processes. These rocks are included in the Serian Volcanics that also comprise hypabyssal rocks, mainly dioritic stocks and sills.

During the Jurassic and possibly also in the Cretaceous, a trough probably striking WNW - ESE from Tanjung Datu to the Sarawak river delta area north of Kuching accumulated predominantly argillaceous sediments associated with arenites, chert, "tillite-like" bouldery slate, rare lenses of conglomerate and limestone, gabbro, basalt and dolerite. The varied rock association and the bouldery slate may represent an olistostrome interpreted to occur at an oceanic-continental converging plate junction. These rocks were grouped into the Sejingkat and Serabang formations. Their Jurassic age is based on radiolaria. To the south of this probable subduction trench was a Late Jurassic shallow sea where the Bau Limestone and the nearshore clastics of the Kedadom Formation developed. The limestone occurs in lenses reaching 300 m thickness, is pale-coloured, massive, and probably continued developing into the Early Cretaceous. During the same time the Pedawan shales, sandstones, limestones and subordinate cherts and conglomerates were deposited in a deeper sea. The shale is hard, dark grey or blue, and in places rich in plant remains and carbonaceous flakes (Liechti 1960, p. 40). Rapid subsidence and influx of clastics at the end of the Early Cretaceous terminated the Bau Limestone development. These conditions, however, favoured continued deposition of the Pedawan beds that, including some thick layers of dacitic tuff, reached a thickness of more than ·4,600 metres. Sediments of mixed terrestrial and shallow-marine environments in which are dominant thick-bedded sandstone, conglomeratic sandstone, conglomerate, red mudstone interbedded with thick packets of thin sandstone-shale-mudstone intercalations form the Upper Cretaceous to Miocene Kayan Sandstone Formation. In the Bungo basin the 3,000 m thick Kayan Sandstone rests unconformably upon the Pedawan Formation. In the eastern part of West Sarawak, similar sediments, including some significant coal beds known as the Silantek and Plateau Sandstone formations, are Upper Eocene to the Oligocene sequences. Deposition probably continued into the Miocene. Total thickness exceeds 5,000 m. The ages are derived from larger foraminifers for the marine portion, and from pollen for the terrestrial sediments.

Lupar Line. West Sarawak is separated from the rest of the state by the circa 15-km wide, WNW-trending Lupar Line. Associated with the Lupar Line are two stratigraphic units: the Lupar Formation and the Lubuk Antu Melange. The Lupar sediments comprise Upper Cretaceous flysch of probably proximal turbidites that are associated with elongated bodies of basaltic and pillow lavas, and gabbro. The mafic rocks represent oceanic lithosphere that once underlaid the Lupar flysch. The Lubuk Antu rocks comprise sheared tectonic melange comprising radiolarian chert blocks and other fossiliferous clasts that indicate ages from Early Cretaceous to Early Miocene (Tan, 1979). The turbidites are deformed into vertical to steeply SSW-ward overturned isoclinal folds (see Haile et al. 1994) suggesting subduction in the same direction (Fig. 26.20).

Rajang Fold-Thrust Belt. The oldest fossils (mainly diagnostic foraminifers indicating an age range from Late Cretaceous to Eocene) of the Rajang Fold-Thrust Belt (or RAP for Rajang Accretionary Prism) are in the Upper Cretaceous Layar Member of the Belaga Formation. This is a thick succession of possibly distal flysch of slate to phyllitic slate and metagraywacke deposited in bathyal and partly also in outer-neritic environments. Minor constituents are conglomerate, marlstone and limestone. The rocks are strongly deformed. The Belaga Formation is believed to reach great thickness between 10 km and 15 km, even allowing for repetitive sequences through isoclinal folding and thrusting. With decreasing age, the other units of the Belaga Formation are the Kapit, Metah, Pelagus and Bawang members. The Bawang Member is an intensely deformed sequence of dark-coloured, soft, sometimes slaty shale with very subordinate and thin sandstone layers. Slickensided surfaces, cleavage, contorted bedding and crushed foraminifers indicate extremely strong deformation. The Bawang Member is restricted in extent to the Tatau horst. It comprises strongly deformed, usually soft and dark- coloured argillaceous material.

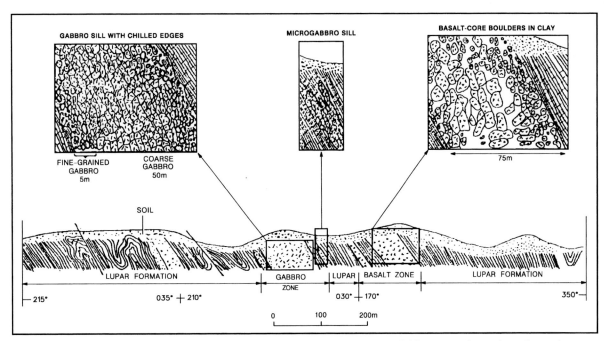

Fig. 26.20 Schematic cross section of part of the Lupar Line (Haile et al. 1994). Beds are probably overturned towards southwest, that is, in the direction of subduction. The location of the section is in Figure 26.18.

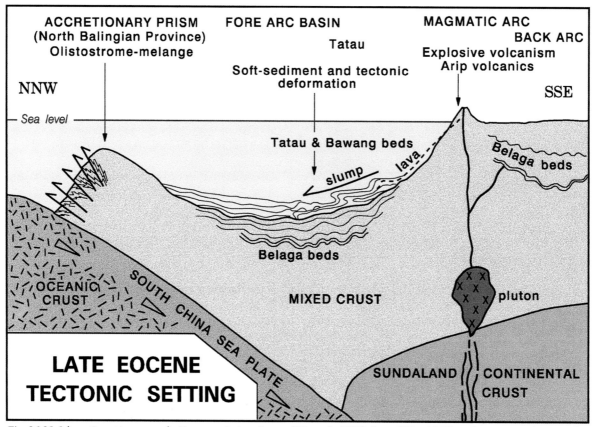

Fig. 26.21 Schematic section across the Tatau area in Late Eocene (Tjia et al., 1987). From SSE to NNW are four tectonic elements of an island arc. Synsedimentary and tectonic deformations affect the Belaga, Bawang and Tatau beds. Tatau is indicated on Figure 26.18.

Tjia et al. (1987) suggested that the Bawang Member and the Tatau Formation (siliciclastics with intercalations of marlstone and limestone of Eocene-Oligocene age) represent slump deposits (olistostrome) at a converging plate junction (Fig. 26.21) The Belaga members tend to young towards north and northeast, which is consistent with accretion in a subducting system where underthrusting is towards south.

The Late Eocene folding was strongest in the rocks in the southwest, that is, in the Lupar zone, and decreased in intensity towards the NNE, where extensive reefs began to develop: the Eocene- Miocene Melinau Limestone and the Oligocene-Miocene Subis Limestone. A distinct unconformity separates the Melinau Limestone from older rocks, but no such contact has been recorded for the younger clastic units and beds of the Rajang Accretionary Prism. The prism is separated from the Neogene zone farther towards the north by the Mersing Line that includes the mafic Bukit Mersing igneous body.

Hydrocarbon Potential

The pre-Tertiary geological development of westernmost Sarawak was most probably similar to that of Peninsular Malaysia, as both belong to Sundaland. However, westernmost Sarawak was also subjected to Miocene intrusions of intermediate to acid character. These intrusions probably seriously downgraded any hydrocarbon potential that may have existed in their immediate surroundings. Away from the intrusions, pre-Tertiary hydrocarbon may have been generated and stored in the Pedawan Formation. Mohd Jamaal Hoesni et al. (1987) found encouraging hydrocarbon characteristics in a carbonaceous shale sample of the Pedawan Formation, which comprise TOC of 10.98%, VRo = 0.88%, Tmax = 453°C corresponding with the main phase of oil generation. The source-potential of ponded sediments of the Rajang Accretionary Prism is unknown, while in northern Sarawak the Melinau Limestone and Kelalan arenaceous flysch may provide repositories for pre-Tertiary and younger oils. E. Honza (Kumamoto University, Japan; personal communication, January 1999) reported an oil seep in the Upper Cretaceous Layar Member of the Belaga Formation near Bukit Batu Tiban (01° 43.5' N., 114° 30.2' E.).

The so called economic basement in Sarawak's offshore may occur at relatively shallow depth. At about 2 km subsea, a well in

Pre-Tertiary Hydrocarbon Potential

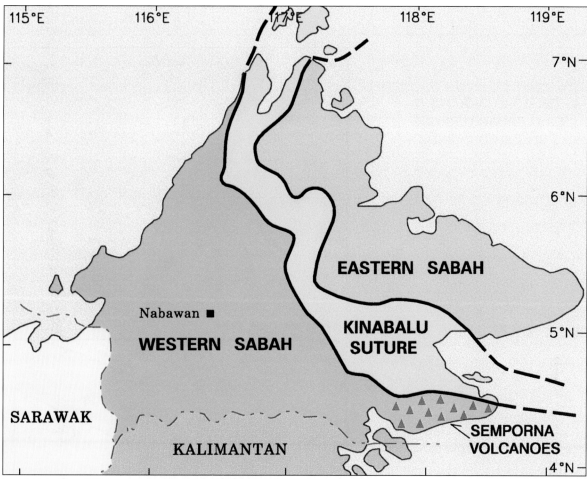

Fig. 26.22a Three-fold division of Sabah: Western Sabah is a Cretaceous-Eocene fold-thrust belt (or accretionary prism); Eastern Sabah is an exotic terrane, probably a microcontinent; the Kinabalu Suture contains Mesozoic ophiolite bodies and is also associated with the Miocene chaotic deposits.

the SW Luconia Province drilled into Paleocene to Lower Eocene strata separated by an unconformity from overlying Cycle II sediments (Mohd. Idrus Ismail, PETRONAS-PMU, personal communication). At Kuala Igan, the Rajang-1 well encountered phyllite and other weakly metamorphosed sedimentary rocks. Therefore, pre-Cycle I hydrocarbon plays in offshore basement highs, such as the Luconia Province, should not be discounted.

Possible pre-Tertiary source rocks in Sarawak are the Jurassic-Cretaceous shales of the Pedawan and Kedadom formations, and shales within the thick, Upper Cretaceous (to Miocene) Kayan Formation. The Kayan beds, however, are mainly arenaceous and could act as reservoirs.

SABAH

Three major geological divisions can be recognised for Sabah: Western Domain, Eastern Domain and the Kinabalu Suture (Figs. 26.1 and 26.22a). The Western Domain is a Cretaceous-Tertiary fold-thrust zone, often referred to as the Crocker Accretionary Prism (CAP) that contains the Upper Cretaceous- Paleocene Sapulut Formation (Fig. 26.23). In the SE corner of the Semporna Peninsula this domain is also marked by the presence of Quaternary volcanoes. The Eastern Domain, as defined in Figure 26.22a, has E-W aligned inliers of Cretaceous Chert-Spilite complex, and Neogene sediments within which diamicts or chaotic deposits are extensive. The third domain, the Kinabalu Suture reaches 80 km width and contains almost all the Cretaceous Chert-Spilite and ?Triassic Crystalline Basement complexes of Sabah.

The oldest pre-Tertiary rocks form the Crystalline Basement complex comprising metamorphic rocks and acid-intermediate igneous rocks (Leong, 1974). Potassium-argon dates of these bodies range from 210 Ma to 87 Ma (Late Triassic to Late Cretaceous). The ?Lower Cretaceous recrystallised Madai-Baturong limestone in the Eastern Domain contains shallow-water foraminifers and algae. Two stratigraphic units straddle the Cretaceous-Tertiary boundary. The Upper Cretaceous to Upper Eocene Sapulut Formation is a fossiliferous, well-bedded, mainly argillaceous

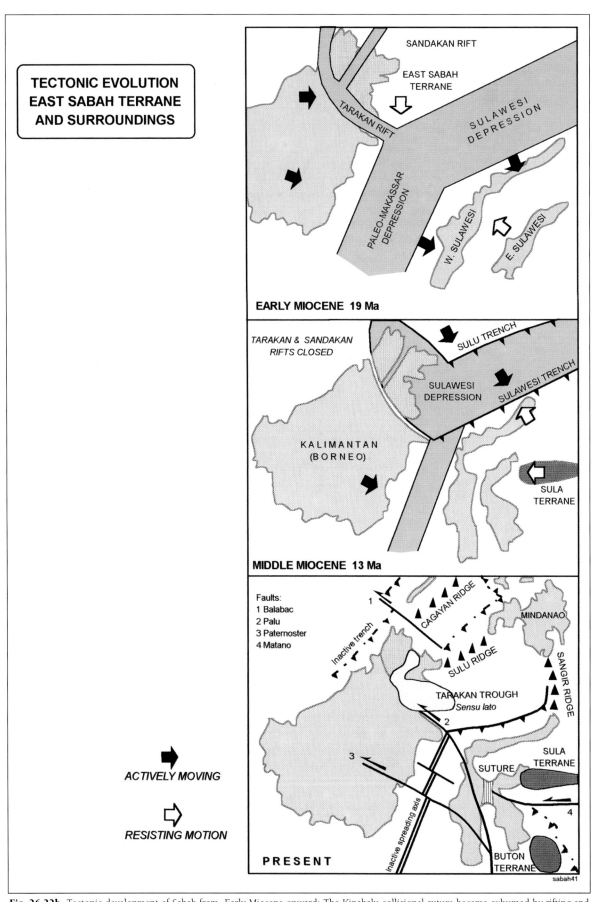

Fig. 26.22b Tectonic development of Sabah from Early Miocene onward: The Kinabalu collisional suture became exhumed by rifting and partially closed for a second time. From Tjia et al. (1990).

Pre-Tertiary Hydrocarbon Potential

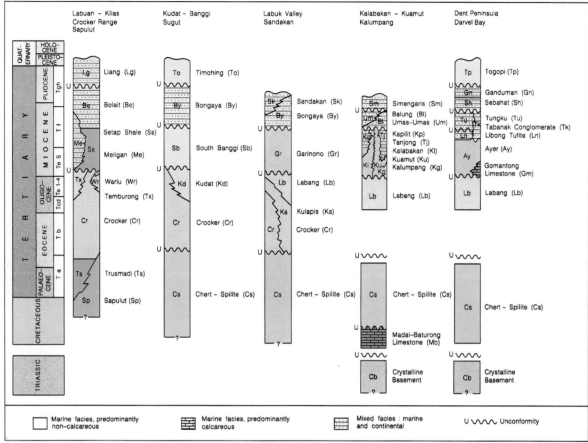

Fig. 26.23 Generalised stratigraphy of Sabah after Lim (1985). Colours indicate facies (compare with legend of Figure 26.19).

succession with minor muddy limestone, limestone, conglomerate, chert, and sandstone. Many of the sandstone layers are thick to massive bedded. Along the road section east of Nabawan, black and red shale sequences are common. Carbonised flakes and shards, ripple marks, and metres-wide slump intervals have also been recorded. Figure 26.24 illustrates the structural style of the Sapulut Formation. The other stratigraphic unit comprises the Cretaceous-Eocene Chert-Spilite complex, that possesses definitive Upper Cretaceous foraminifers.

Hydrocarbon Prospectivity

According to Seah and Mahendran (1986), only the Upper Oligocene Kudat Formation has hydrocarbon source-rock potential among more than 50 outcrop samples representing the Trusmadi, Crocker and Kudat formations in western Sabah and in the Kudat peninsula. They believe that the Upper Eocene-Lower Oligocene Crocker Formation is probably the economic basement (in western Sabah), but they also caution that the geochemical results may not be representative of the true potential, as all the samples studied are partly weathered.

Due to the high degree of induration and/or incipient metamorphism, most Cretaceous stratigraphic units of Sabah have very little hydrocarbon potential. Probably only the Sapulut Formation deserves further attention as to its source potential. The crystalline limestone and rocks of the Crystalline Basement, if suitably fractured and/or weathered, could serve as reservoirs, especially in the vicinity of the Sandakan and NW Tarakan basins.

CONCLUSIONS

Hydrocarbons have yet to be discovered in any of the pre-Tertiary formations of Malaysia. In view of its geological history of multiple deformations, with or without associated magmatic and volcanic activities, the existence of pre-Upper Permian hydrocarbons may be ruled out.

Likely pre-Tertiary source rocks in the Peninsular Malaysia region are the Upper Permian-Triassic limestone, Triassic neritic shales, and Jurassic-Cretaceous lacustrine sediments. Pre-Tertiary reservoir rocks comprise the buried limestone, fluviatile JK sandstone, and fractured/weathered granitic basement highs. These may also act as reservoirs for Tertiary

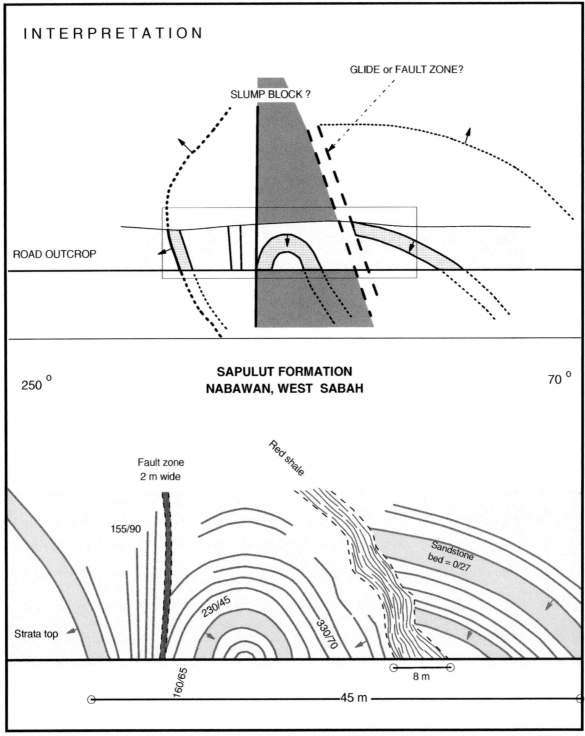

Fig. 26.24 Deformation style of the Sapulut Formation near Nabawan, Sabah (locality indicated on Figure 26.22a). Stratigraphic facing in thick sandstone alternating with thin red shale is indicated by sole markings. Strike and dip values are indicated. The 8-m red shale band contains angular sandstone clasts.

hydrocarbons, especially in the Malay and Penyu basins, and in the Straits of Melaka.

In Sarawak, possible pre-Tertiary source rocks are the Jurassic-Cretaceous shales of the Pedawan and Kedadom formations. If suitably sealed, the sandstones of the Upper Cretaceous-Miocene Kayan Formation, could host hydrocarbons.

In Sabah, the most likely pre-Tertiary source rock could be the black shale of the Sapulut Formation. Its sandstone beds as potential reservoirs need to be studied.

REFERENCES

Ahmad Munif Koraini, 1993. Tertiary palynomorphs from Batu Arang, Malaysia (abstract). Geological Society of Malaysia, Annual Geological Conference 1993, Programme and Abstracts of Papers: 9.

Areshev, E.G., Tran Le Dong, Ngo Thuong San and Shnip, O.A., 1992. Reservoirs in fractured basement on the continental shelf of southern Vietnam. Journal of Petroleum Geology 15 (4), 451-464.

Asiah Mohd Salih and Sanatul Salwa Hasan, 1992. Dating of carbonates found in Bunga Raya-1 well. PETRONAS Research & Scientific Services (internal report).

Cameron, N.R., Clarke, M.G., Aldiss, D.T., Aspden, J.A. and Djunuddin, A., 1981. The geological evolution of northern Sumatra. Proceedings 9th Annual Convention Indonesian Petroleum Association, 149-187.

Clennell, M.B., 1991. The origin and tectonic significance of melanges in eastern Sabah, Malaysia. Journal of Southeast Asian Earth Sciences 6, 407-429.

Dyer, J., Finlay, A., Nichol, M.A. and Wong, D., 1991. Report on the geophysical and geological reevaluation of contract area PM-1, Malacca Straits, Malaysia. Internal report Sun Malaysia Petroleum Company.

Fontaine, H. and Ibrahim Amnan, 1994. The importance of Triassic limestone in the central belt of Peninsular Malaysia. Proceeding of the International Symposium on "Stratigraphic Correlation of Southeast Asia", 15-20 November 1994, Bangkok, Thailand, 195-205.

Fontaine, H., Rodziah Daud and Singh, U., 1990. A Triassic "reefal" limestone in the basement of the Malay Basin, South China Sea: regional implications. Geological Society of Malaysia Bulletin 27, 1-26.

Fontaine, H., Asiah Mohd Salih and Sanatul Salwa Hasan, 1992. Pre-Tertiary sediments found at the bottom of wells drilled in Malacca Straits. CCOP Newsletter 17 (4), 12-17

Foo, K.Y., 1983. The Palaeozoic sedimentary rocks of Peninsular Malaysia - stratigraphy and correlations. Proceedings Workshop Stratigraphic Correlation Thailand and Malaysia. Geological Society of Thailand and Geological Society of Malaysia, Volume 1, 1-19.

Geological Survey of Malaysia, 1995. Annual report 1995. Jabatan Penyiasatan Kajibumi Malaysia, Kuala Lumpur-Ipoh.

Greenaway, P. and Goh, K.K.T., 1989. Preliminary geoscience report contract area block PM-15, Malaysia. Internal report Sun Malaysia Petroleum Company.

Haile, N.S., Lam, S.K. and Banda, R.M., 1994. Relationship of gabbro and pillow lavas in the Lupar Formation, West Sarawak: implications for interpretation of the Lubok Antu melange and the Lupar line. Bulletin of the Geological Society of Malaysia, 36, 1-9.

Hatch, G., 1987. Technical evaluation and project summary for PM-1 block, Straits of Malacca, Malaysia. Internal report Sun Malaysia Petroleum Company.

Hon, V. and Lam, S.K., 1992. Geological Map of Sarawak. 2nd Edition Scale 1:500,000 (2 sheets) (updated after Tan, D.N.K., 1982, First Edition) Geological Survey of Malaysia.

Hutchison, C.S., 1975. Ophiolites in South East Asia. Geological Society of America Bulletin 86, 797-806.

Hutchison, C.S., 1988. Stratigraphic-tectonic model for east Borneo. Bulletin of the Geological Society of Malaysia, 22, 135-151.

Hutchison, C.S., 1990. The Southeast Sulu Sea, a Neogene marginal basin with outcropping extensions in Sabah. Bulletin of the Geological Society of Malaysia, 32, 89-108.

Khoo, H.P., 1983. Mesozoic stratigraphy in Peninsular Malaysia. Proceedings Workshop Stratigraphic Correlation Thailand and Malaysia. Geological Society of Thailand and Geological Society of Malaysia, Volume 1, 370-383.

Koning, T. and Darmono, F.X., 1984. The geology of the Beruk Northeast field, Central Sumatra: oil production from pre-Tertiary basement rocks. Proceedings 13th Annual Convention Indonesian Petroleum Association 1, 385-406.

Leong, K.M., 1974. The Geology and Mineral Resources of the Upper Segama and Darvel Bay area, Sabah, Malaysia. Geological Survey of Malaysia Memoir 4 (revised), 354 p.

Leong, K.M., 1978. The "Sabah blueschist belt" - A preliminary note. Warta Geologi 4 (2), 45-51.

Liechti, P., Roe, F.W. and Haile, N.S., 1960. The Geology of Sarawak, Brunei and the Western Part of Borneo. Geological Survey Department British Territories in Borneo Bulletin 3, Volume I, 360 p.

Liew, K.K., 1994a. Structural patterns within the Tertiary basement of Straits of Malacca. PETRONAS Research & Scientific Services Report PRSS-RP5-94-02 (internal report).

Liew, K.K. 1994b. Structural development at the west-central margin of the Malay basin. Bulletin of the Geological Society of Malaysia 36, 67-80.

Liew, K.K., Mohamad Fauzi Abdul Kadir, Mohd Jamaal Hoesni, Sahalan Abdul Aziz and Tjia, H.D., 1996. Hydrocarbon potential of Mesozoic strata onshore Peninsular Malaysia, results of field and laboratory studies 1995. PETRONAS Research & Scientific Services Report PRSS-RP5-96-10 (internal report).

Lim, P.S., 1985. Geological Map of Sabah. 1:500,000 (3rd Edition). Geological Survey of Malaysia

Mohd Jamaal Hoesni, Abdul Jalil Muhamad, Seah Peng Kiang, E. and Zulkifli Salleh, 1987. Geochemical analyses of outcrop samples from west Sarawak. PETRONAS Laboratory Services Department, internal report MGK/11/87.

Mohd Jamaal Hoesni and Abolins, P., 1996. Occurrence of novel biomarker fingerprints in Malay basin sediments: source implications [abstract]. Warta Geologi 22 (6), 437-438.

Nichol, M.G., 1990. Report on the seismic interpretation of the western margin reconnaissance area, contract area PM-1 Malacca Straits, Malaysia. (Internal report Sun Malaysia Petroleum Company).

PETRONAS Carigali, 1994. Plate 21 [Atlas Malay-Penyu Basin Regional Study]. (Internal report).

Pulunggono, A. and Cameron, N.R., 1984. Sumatra microplates, their characteristics and their role in the evolution of the Central and South Sumatra basins. Proceedings 13th Annual Convention Indonesian Petroleum Association 1, 121-147.

TEXACO, 1992. The significance of the Rhu oil discovery: a preliminary summary, volume 1 Text, PM-14, Malaysia. Internal report Texaco Exploration Penyu, Inc.

Seah, E. and Mahendran, B., 1986. Geochemical analyses and interpretation of the Crocker and Kudat formations, Sabah. PETRONAS Laboratory Services Department, internal report, 9 September 1986.

Shamsudin Jirin and Morley, R., 1994. Lower Cretaceous palynomorphs from the Termus and Mangking formations (Tembeling Group), Peninsular Malaysia: their stratigraphic and climatic significance [abstract]. Warta Geologi 20 (4), 219.

Smiley, C.J., 1970. Later Mesozoic flora from Maran, Pahang. Part 1: geologic considerations. Geological Society of Malaysia Bulletin 3, 77-88.

Stauffer, P.H. and Lee, C.P., 1986. Late Palaeozoic glacial marine facies in Southeast Asia. Bulletin of the Geological Society of Malaysia, 20, 363-397.

Tan, D.N.K., 1979. Lupar valley, West Sarawak, Malaysia. Explanation of sheets 1-111-14, 1-111-15, 1-111-16. Geological Survey of Malaysia, Report 13.

Tan, Y. P., 1988. Stratigrafi-geologi kawasan Parit Sulong, Johor. Universiti Kebangsaan Malaysia, Bangi, MSc thesis, 168 p. (unpublished).

Tjia, H.D., 1986. Geological transport directions in Peninsular Malaysia. Bulletin of the Geological Society of Malaysia, 20, 149-177.

Tjia, H.D, 1988. Accretion tectonics in Sabah: Kinabalu Suture and East Sabah accreted terrane. Bulletin of the Geological Society of Malaysia, 22, 237-251.

Tjia, H.D,. 1989a. Tectonic history of the Bentong-Bengkalis Suture. Geologi Indonesia 12 (1), 89-111.

Tjia, H.D., 1989b. Structural geology of Datai Beds and Macincang Formation, Langkawi. Bulletin of the Geological Society of Malaysia, 23, 85-120.

Tjia, H.D., 1994. Origin and tectonic development of Malay-Penyu-West Natuna basins. Petronas Research & Scientific Services "Technology Day", Kuala Lumpur 21-22 June 1994.

Tjia, H.D., 1997. Meridian-parallel faults and Tertiary basins of Sundaland [abstract]. Geological Society of Malaysia, Petroleum Geology Conference '97, Programme & Abstracts of Papers, 18-20.

Tjia, H.D. and Anizan Isahak, 1990. Permian glacigenic deposits at Salak Tinggi, Selangor. Sains Malaysiana 19 (1), 45-64.

Tjia, H.D., Borhan Sidi and Teoh Chuen Lye, 1987. Superimposed deformations and vergence of Lower Tertiary sediments near Tatau, Sarawak. Bulletin of the Geological Society of Malaysia, 21, 251-271.

Tjia, H.D., Ibrahim Komoo, Lim, P.S. and Tungah Surat, 1990. The Maliau basin, Sabah: Geology and tectonic setting. Bulletin of the Geological Society of Malaysia, 27, 261-292.

Tjia, H.D. and Zaiton Harun, 1985. Regional structures of Peninsular Malaysia. Sains Malaysiana 14 (1), 95-107.

Uyop Said, 1997. Palinologi batuan Jura-Kapur Taman Negara. In Ibrahim Komoo et al., eds. Warisan Geologi Malaysia, Lestari, Universiti Kebangsaan Malaysia, 249-368.

van Bemmelen, R.W., 1949. "The Geology of Indonesia", Volume Ia. Martinus Nijhoff, The Hague.

Yusoff Johari, 1983. A report on the geological reconnaissance on the Tenggol islands. Internal report, PETRONAS.

Part 7

GEOSCIENCE RESEARCH

Chapter 27

Scope For Further Geoscience Research

Mazlan B. Hj. Madon

INTRODUCTION

Since the first oil discovery in Miri in 1910, many oil companies have explored for oil and gas in Malaysia. These companies actively explore in different parts of the sedimentary basins. As multinational and independent oil companies continue to explore in Malaysia, there is further potential for discoveries and development of its petroleum resources. Geoscience knowledge and related technologies have contributed significantly to the success of exploration and development of Malaysian petroleum resources. The exploration and production (E&P) business is dependent on geoscience knowledge that has accumulated over centuries through research by the industry and academia. And as we journey through the new millennium, the task of finding and producing oil will become more and more challenging, both technically and scientifically. More detailed science will be needed to find a barrel of oil than it was two decades ago. There is no doubt that E&P research, in geoscience in particular, has an important role to play in helping the explorationist to find additional oil.

This chapter gives some general ideas on the potential scope for future research in geoscience that are relevant to the E&P industry in the new millennium. What is the future role of geoscience research in helping us to find more petroleum? How can geoscience research contribute to adding to the petroleum reserves in Malaysia? Based on our current understanding of Malaysian basins and hydrocarbon habitats, on what areas of research shall we focus our efforts? These are some of the key issues that will be addressed in this chapter.

E&P RESEARCH FOCUS

The main activity in the E&P business is to increase our petroleum resource base. The explorationist is tasked with finding more oil and gas to add to the discovered reserves, while the petroleum engineer is responsible for maximizing commercial recovery from the discovered reserves through efficient field development and reservoir management. Figure 27.1 schematically illustrates the major components of our petroleum resource, which here means all the discovered and undiscovered petroleum. The total resource at any given time is speculative but finite. Only a portion of this resource has been discovered. Under current economic conditions and using available technologies, only a portion of the discovered resource is expected to be ultimately recoverable, and is here regarded as reserves. Only part of the reserves is being developed and produced. Note that the principal task of the E&P manager is to inflate the first 3 higher-order variables – Resource, Reserves, and Recovery (3Rs) – as shown by the arrows in Figure 27.1. The amount of development and production is generally determined by economics, oil price, and certain regulations and policies, e.g. National Depletion Policy, over which the E&P manager has no control. It is immediately apparent, therefore, that the main contribution from geoscience research would be to help increase the 3Rs. Some of the ways in which this can be achieved are shown in Table 27.1.

Increasing Resource Base

In planning the long-term strategies for exploration and development, it is important to know how much more petroleum is left to be discovered. Since Malaysia has several basins at different stages of exploration maturity, there is a case for doing a proper mass balance study to determine the amount of undiscovered resources. The basins may then be ranked according to the residual reserves to be found. It is also important to find ways to increase the resource base (Table 27.1). Resources may be increased by opening up new exploration acreages or frontier basins,

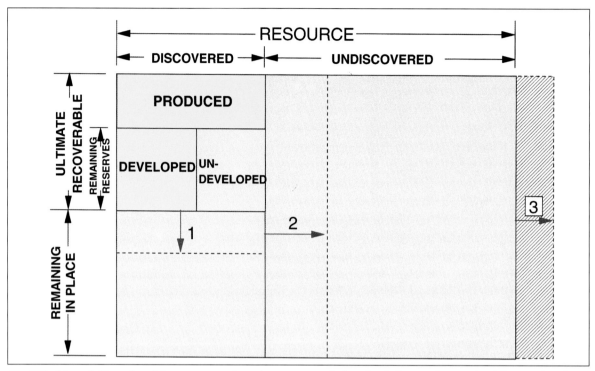

Fig. 27.1. Schematic illustration of the components of petroleum resource, which is classified into undiscovered and discovered. Only part of the latter is recoverable using existing technologies, and only a portion is developed and produced. The red arrows represent the three main potential focus areas in which geoscience research can contribute: 1. Increase recovery. 2. Increase discovery. 3. Increase resource base. (Modified after McKelvey, 1975, reprinted with permission of the American Association of Petroleum Geologists).

	Ways to increase....	How geoscience can contribute....
RESOURCE	• open new basins/acreages for exploration • new play concepts • farm-ins and acquisition of new acreages	• improve understanding of basins and petroleum systems • improve method of estimating undiscovered resources • better handling of geological risks • generate new play concepts
RESERVES	• new discoveries	• explore in poorly explored basins • explore for new accumulations in mature basins • testing new play concepts • revisit old play concepts • testing deeper plays in overpressure zones
	• field growth	• improve reserve estimation • improve reservoir model • better reservoir management • look for bypass oil • better correlation of reservoir units
RECOVERY	• application of new technology • improve existing production technology • small field development	• input into reservoir management • input to formation damage prevention & control • input to reservoir simulation

Table 27.1. Three main focus areas that provide opportunities for further geoscience research.

followed by successful exploration drilling. The introduction of onshore acreage as well as the deep-water PSCs are some of the initiatives in this direction. We can also increase our resource base by generating new play concepts in mature basins or in frontier areas. Geoscience research has a role to play in all these areas. For example, regional studies and analysis of poorly explored basins (e.g. the circular sub-basins of Sabah) would be valuable for understanding the geological risks and for enhancing their prospectivity. Finally, our resource base may be added through farm-in and/or acquisition of exploration blocks overseas. Such efforts would require prior geological understanding of the basin or region of interest before investment decisions are made. Geoscience research towards understanding of petroleum systems and basins worldwide would help us achieve this.

Adding Reserves

The primary focus for the E&P business in the new millennium would be to increase reserves. Reserves addition may be obtained from two main sources: new field discoveries and existing field growth. New discoveries can be either in a completely new area or basin, or between the producing fields and discoveries and deeper reservoir targets in well-explored basins. Field growth is achieved by essentially improving our estimation of reserves in the existing discovered or producing fields. Looking back at the history of exploration in Malaysia during the last decade, half of the new reserves addition have come from field growth (Chapter 3). As new discoveries become increasingly difficult to make, due to smaller size, poor economic viability, and being technically difficult (stratigraphic as opposed to structural plays), a major focus for the new millennium will be to add reserves through field growth, while maintaining a certain level of activity in looking for new discoveries. Clearly, the strategy for research towards finding new accumulations would differ from research on field growth. In a later section, some of the more specific topics for research towards adding reserves will be mentioned.

Optimizing Recovery

Although the efficient recovery of petroleum is ultimately the responsibility of the petroleum/reservoir engineer, geoscience knowledge remains valuable, particularly as input to field development planning and reservoir modelling. The objective is to help optimize the recovery of hydrocarbons without damaging the reservoir. This would include detailed microscopic studies on the porosity characteristics of the reservoir and other factors influencing reservoir behaviour. Detailed petrographic studies on reservoir mineralogy also provide valuable information for mitigating formation damage. On a field scale, detailed studies on reservoir architecture using sequence stratigraphy may improve prediction of reservoir connectivity, which represents essential information for the reservoir engineer.

BASIN-SPECIFIC RESEARCH FOCUS

The focus for research in geoscience would vary from basin to basin depending on the status of our knowledge of the area and the exploration maturity of the basin. The sedimentary basins in Malaysia (Chapter 4) may be classified in terms of commercial success, as follows (see Fig. 27.2):

- *Commercial or productive basins* (referred to herein as P-type basins, Table 27.2): these are well-explored, mature, and productive basins, e.g., Malay, Sarawak, Sabah (including both West and East), and the Baram Delta Province.
- *Noncommercial basins* (N-type): these are basins that have had some exploration activity but have not produced commercial quantities of petroleum, e.g. Penyu Basin, NE Sabah Basin.
- *Unexplored or poorly explored basins* (U-type): these basins are generally poorly understood because of lack of exploration effort, e.g. basins in the Melaka Straits and the SE Sabah Basin.

The disparate tectonic settings and exploration maturities of basins in Malaysia therefore demand a different focus for research in different basins. For example, in mature basins such as the West Baram Delta, the focus could be to add reserves through field growth, perhaps by better reserves estimation, or by looking for bypassed oil in field extensions. In the Malay Basin, besides improving reserve estimation for field growth, future emphasis might be to find stratigraphic traps, and to look for and test alternative or entirely new play concepts, including the possibility of hydrocarbon traps in the pre-Tertiary basement. In the poorly explored basins, on the other hand, it is perhaps more important to focus research efforts that contribute towards enhancing their prospectivity. These would involve increasing our understanding of these sedimentary basins, from tectonic evolution, sedimentation history, to petroleum systems. All these efforts require an in-depth knowledge of sedimentary basins, petroleum systems, and reservoirs.

MANAGING GEOLOGICAL RISKS

Basically, all the research effort towards finding more reserves ultimately relate to identifying,

Scope For Further Geoscience Research

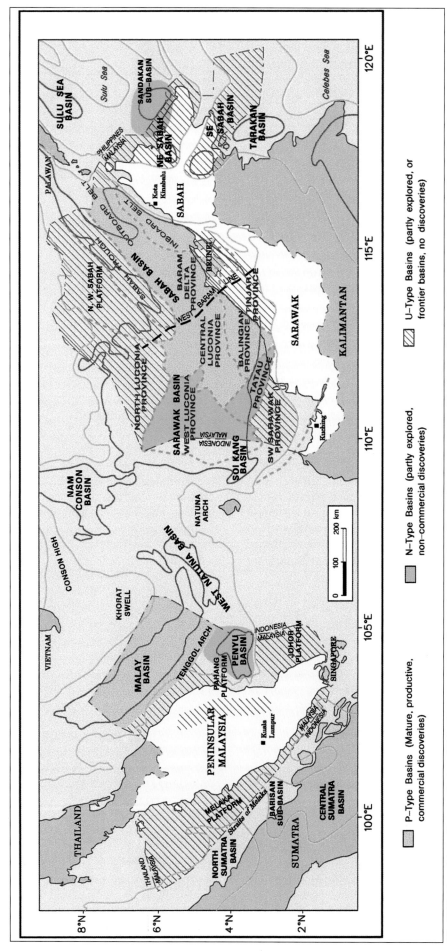

Fig. 27.2. Map of the sedimentary basins of Malaysia, categorised into 3 major basin types based on exploration maturity and the degree of commercial success. See text and Table 27.2 for explanation.

BASIN TYPE	OPERATIONAL FOCUS RESEARCH FOCUS	RESEARCH FOCUS
P-Type (mature, productive, commercial discoveries) e.g. Malay, Central Luconia, West Baram Delta, Sabah	Primary focus: Field growth Secondary focus: New discoveries	• detailed field/reservoir- scale geological studies to improve reservoir model • detailed studies of petroleum systems using available data
N-Type (partly explored, noncommercial discoveries) e.g. Penyu, Tatau Province, NE Sabah	Primary focus: New discoveries Secondary focus: Field growth	• understanding petroleum systems to better predict reservoir, source, seal, and structural timing relationships • basin analysis and modelling
U-type (partly explored or frontier basins, no discoveries) e.g. Melaka Straits, North Luconia, SE Sabah Basin	Primary focus: New discoveries	• regional basin studies • basin analysis and modelling • comparative studies of analogue basins

Table 27.2. Different basin types and their operational and research focuses. See Fig. 27.2 for the location of the different basin types.

quantifying, and reducing the geological risks. The major risks include the presence of source, reservoir, trap, seal, and hydrocarbon charge at the right moment in geological time. Geoscience research has a major role to play in improving our understanding of these risks and in finding ways to manage those risks. In all the 3 basin types (Table 27.2) geological risks are a common theme. Generally, the more information we have on a given basin, the lower the risks are. Thus, the risks are larger in U-type than in the P-type basins. But information is of little value if it is not used and applied properly. How can research add value to geological information? First, proper compilation and archiving of all E&P data are essential. Secondly, the data must then be continuously updated and synthesized into a more comprehensive form. Thirdly, research can then be directed towards better quantification of the geological risks which can be used to guide exploration, e.g. to justify drilling the next well. As basins are continuously being explored, and more data are acquired, it is also important to keep an up-to-date inventory of the risks, as these will change according to the results of recent drilling and because of improved knowledge and understanding of the basins through research activities. It would be a fruitful exercise to regularly conduct post-mortem analyses of drilling results by comparing them with well prognoses/proposals and identify the risks that require further work. For example, if the results of a well indicate that our assumption of the presence of a reservoir at a certain stratigraphic level is wrong, we could then think of ways to improve our prediction in the next well. An interactive and iterative E&P database management system could serve as a powerful tool to identify, quantify, and manage geological risks in exploration.

Cost reduction should be an important factor in guiding future research for the survival of the Malaysian petroleum industry. The importance of cost reduction efforts was emphasized by the official launching in 1996 of the "CORAL Malaysia" initiative by PETRONAS and major production-sharing contractors. CORAL, an acronym for Cost Reduction Alliance, serves as a platform for industry players to interact and network in a concerted effort to reduce the cost of production, development, and drilling. It should be pointed out also, that geological risks need to be managed properly in order to reduce exploration and development costs, which is expected to increase further in the next millennium. The cost of drilling a well may be reduced by using improved drilling technology, but the decision to drill that well in the first place must be based on a thorough understanding of the geological risks involved. Hence, research to improve our understanding of the geological risks in E&P should be given an equal importance as the application of new technologies in order to reduce cost.

SPECIFIC RESEARCH AREAS

Having identified the general focus for geoscience research in the different basins, we can then direct our attention to specific areas of possible research that can help us achieve our goals. For new discoveries in mature basins, alternative or new play concepts have to be generated and tested. In the Malay Basin, for example, future discoveries are more likely to be in subtle stratigraphic traps, particularly on the southwestern flank, and in the deep overpressure play beneath existing fields, as shown by the Bergading Deep-1 discovery. Hence, future research here might be in the area of sequence stratigraphy for better prediction of reservoir distribution in the objective section.

Recent discoveries in overpressured reservoirs indicate the huge potential of this play type. Hence, there is a need for detailed understanding of the distribution of overpressures and its relationship to maturation, expulsion, and migration of hydrocarbons to be able to look for geopressured hydrocarbon accumulations. Although some general studies on overpressure in the Malay Basin have been done (e.g. Singh and Ford, 1982), there are specific aspects of overpressure that need further attention. For example, how does overpressure affect source rock maturation? Does it enhance or retard maturation, and or is there a dependence upon kerogen type? Can we improve the prediction of overpressure ahead of the drill bit, perhaps through improved processing of seismic data or logs? Are there pressure compartments in the basin that result in separate petroleum systems, and can we identify and locate these pressure seals?

On a more geochemical theme, there are uncertainties with regard to the level of thermal maturity at which oil starts to be generated from kerogen. Equivalent vitrinite reflectance values VRo that have been used in thermal modelling, for example, vary from 0.5% to 0.7%. It is uncertain whether these text-book values are applicable to specific basins in Malaysia, especially if there is evidence for suppression of vitrinite reflectance (e.g. Waples et al., 1995).

There are some other possible areas of research in geochemistry, for example, the potential of marine source rocks. Even though marine source rocks have generated the majority of the world's oil reserves, they have been virtually ignored as potential sources in Malaysian basins. Even in basins where marine sequences exist, such as the eastern sides of the Central Luconia and Balingian provinces, little detailed study has been made to determine the possible occurrence of effective marine source rocks. Available source rock data, primarily obtained during post-drilling analyses, have not indicated the presence of promising, on-structure, marine source intervals. Intriguingly, however, geochemical studies have demonstrated some marine contribution in reservoired hydrocarbons in Central Luconia and Balingian. An interesting area of research would therefore be to study the regional distribution of marine influence and the factors controlling the distribution. Are there depositional forces at play that are causing local occurrences of good quality marine source rocks? Can we predict where such forces may be at play and thereby high grade areas of exploration interest?

Some parts of the Sarawak and Sabah basins are relatively poorly explored, with poor success in finding commercial accumulations. In such areas, research may be focused on evaluating the specific geological risks such as the source rock potential or reservoir distribution. Although delta systems in general are proven hydrocarbon-bearing, as in the case of the West Baram Delta, exploration results in other deltaic areas in Malaysian basins such as the West Luconia, East Baram, and Sandakan sub-basin deltas have been rather disappointing. There are only two commercial accumulations in the East Baram Delta (Samarang and Kinabalu), and none in the other deltaic basins. Perhaps more research on deltas, particularly in the West Luconia and Sandakan sub-basins should be carried out in the future.

In the Sabah Basin, as exploration gradually moves into deeper water environments, both present and ancient, it is essential to have a good understanding of deep marine depositional systems, particularly in terms of reservoir architecture and continuity. The turbidite outcrops of onshore Sabah (Crocker and Kudat formations) could be studied in detail as analogues for deep marine rocks offshore.

Carbonate reservoirs are host to about 40% of the total gas reserves in Malaysia, and mostly occur in the Central Luconia Province of the Sarawak Basin (Chapter 15). As natural gas will continue to be an important resource for the country in the new millennium, the development of gas fields will be a major activity in the E&P business. Hence, more research should be done on carbonate reservoirs, for example, in sequence stratigraphy of carbonate systems, carbonate reservoir architecture and diagenesis, and carbonate facies prediction from seismic.

The presence of inert gases impacts drastically on the economic value of a gas accumulation. This is due to the additional costs incurred in the necessary removal of such inert gases. Of most concern in Malaysian basins, particularly the northern Malay Basin and Southwest Luconia is the presence of carbon dioxide. The regional distribution of carbon dioxide, and some of the factors controlling this distribution, are broadly understood. When it comes to specific prospects, however, predicting carbon dioxide levels, or even likely trends in levels, has met with little success. An important area of future research therefore should be to study the physico-chemical, geological, and geochemical factors controlling carbon dioxide abundance at the prospect, and even the reservoir, level. If predictions, ahead of the bit, can be made, then low risk areas can be highlighted. An important aspect of this work would be to establish whether known 'high risk' areas could, in fact, be of less risk, due to the presence of local controls on carbon dioxide abundance.

In less explored areas such as the Penyu Basin, the Tatau Province (Sarawak), and the NE Sabah Basin, new play concepts are needed to spur further exploration activity. Where the structural style is dominated by extensional half-grabens, some research may perhaps be directed towards predicting reservoir-source relationships in the half-graben systems, and looking for stratigraphic traps in the syn-rift sequences. In the NE Sabah Basin many wells have been drilled without much success. The long-held view that the main reservoir objective is the Dent Group (mainly the Sebahat Formation) (e.g. Wong, 1993) clearly needs a re-visit (see Chapter 23). Some wells that were targeted at the Sebahat have found it to be missing, and instead penetrated the pre-Dent formations. We might have overlooked the potential of the pre-Dent (Tanjong Formation equivalent) as a reservoir. Mapping structures in the pre-Dent section might prove useful. Furthermore, the circular outcrops of the pre-Dent section in onshore central Sabah provide an excellent opportunity to study these rocks on the surface. More detailed field work to understand the stratigraphy and evolution of the circular sub-basins of Sabah is needed to properly assess the prospectivity of the NE Sabah Basin. There is also a fundamental problem in correlating between the onshore and offshore formations in this basin (e.g. compare the works of Wong, 1993, and Ismail Che Mat Zin, 1994). A proper correlation scheme has important implications for basin prospectivity. Possible research areas would be in improving the stratigraphic correlation through high-resolution biostratigraphic techniques (Chapter 6), application of sequence stratigraphic techniques to locate potential reservoir and source beds, and integrated geological and basin analysis to determine the petroleum prospectivity of the region.

PETROLEUM SYSTEMS

The generation of new play concepts is possible when there is a good understanding of basins and their hydrocarbon habitats. Geological risk factors such as source, reservoir, trap/seal, and timing of generation/migration in a given basin need to be assessed properly in order to identify further potential. As more and more data are gathered through exploration activities, there is a need to synthesize the data in a systematic way that will help identify data gaps and recognize new plays. Petroleum systems analysis (Magoon and Dow, 1994) is a useful technique that links the source rock to the trap using all available geological/geochemical data to map migration pathways. It also aids the assessment of the potential of complementary prospects. The chapters in this book have not attempted to analyze and describe the petroleum systems rigorously, since this would require more in-depth compilation and integration of all the available data on the Malaysian basins. A more rigorous study of petroleum systems in all the basins should be undertaken in the future to help identify further potential.

FIELD GROWTH

Reserves addition through field growth relies on improving the estimation of reserves. This is achieved through research towards understanding the geometry, continuity, and variability of the reservoirs. In field development, geoscience research are valuable for detailed modelling of the subsurface geology, involving detailed facies analysis and correlation using sequence stratigraphic techniques. It is important to have a wide experience base in dealing with different types of reservoir, from fluvial channel sandstone to deep marine turbidite systems, to carbonate rocks. As more gas fields in the Central Luconia Province will be produced from carbonate reservoirs, there is a need to fully understand carbonate depositional systems with regard to reservoir continuity and behaviour. Some research in deep marine turbidite systems would also help in future development of hydrocarbons in such environments. Geological input to reservoir modelling serves to enhance the reservoir image through advanced seismic reprocessing techniques, better lithology prediction through seismic reprocessing, and the application of 3D seismic in reservoir mapping.

Another potential area for research and development is the interpretation of low-resistivity low-contrast (LRLC) reservoirs. A pilot study on some Malaysian fields have shown proper evaluation of logs in LRLC reservoirs can increase the reserve estimates substantially. This technique needs further investigation for future application to other Malaysian fields.

Technology plays an ever-increasing role in the development of petroleum resources. In field-development, computer-based techniques for imaging the subsurface, including 3D/4D visualization technology (see Slatt et al., 1996), will have a great impact on the way reservoirs are managed. Even in exploration, computer-based applications are increasingly important in modelling the maturation, generation, and migration of hydrocarbons. This will continue to be the case in the near future. Research is needed to refine the techniques available, especially in the improvement of the parameters and variables for input into the basin models.

CONCLUSION

Geoscience research has an increasing role to play in ensuring the future success in E&P in the new millennium. The three main focus areas for research are in enhancing resources, reserves, and recovery. These can be addressed through the broad themes of research: basin analysis, petroleum systems, and field/reservoir studies. A major focus for research, particularly in well-explored mature basins, would be the adding of reserves through field growth. The main thrust would be to improve reserve estimation by better reservoir correlation and improved understanding of reservoir architecture. Application of computer-based technology and the concepts in sequence stratigraphy will be the main thrusts in these areas. Improved techniques of estimating reserves, especially in low-resistivity low-contrast reservoirs, should also be developed further.

Geoscience research aimed towards finding new discoveries will remain important for well-explored basins, where new play concepts are needed to spur exploration activity. In such basins, E&P data need to be systematically integrated and synthesized in terms of petroleum systems. This will help in identifying future plays and guide exploration strategy. Basin analysis and modelling will continue to be the main thrusts in all basins, filling the knowledge gaps in mature basins while providing insights into the petroleum systems in less explored and frontier basins. As field development becomes the main thrust in the future, research in reservoir sedimentology and sequence stratigraphy should be enhanced. Deep marine turbidites and carbonate reservoirs will gain more importance in future developments. Some emphasis to enhance the understanding of these depositional systems would be beneficial. Research on deltaic systems could give a better insight into the prospectivity of the less-explored deltaic provinces such as the West Luconia, East Baram Delta, and the Sandakan sub-basins.

REFERENCES

Ismail Che Mat Zin, 1994. Dent Group and its equivalent in the offshore Kinabatangan area, East Sabah. Bulletin of the Geological Society of Malaysia, 36, 127-143.

Magoon, L.B. and Dow, W.G., 1994. The petroleum system. In: Magoon, L.B. and Dow, W.G., ed., The petroleum system – from source to trap. American Association of Petroleum Geologists, Memoir 60, 3-24.

McKelvey, V.E., 1975. Concept of reserves and resources. American Association of Petroleum Geologists, Studies in Geology 1, 11-14.

Singh, I. and Ford, C.H., 1982. The occurrence, causes and detection of abnormal pressure in the Malay Basin. Offshore South East Asia 82, 9-12 February 1992, Singapore.

Slatt, R.M., Thomasson, M.R., Romig Jr., P.R., Pasternack, E.S., Boulanger, A., Anderson, R.N. and Nelson Jr., H.R., 1996. Visualization technology for the oil and gas industry: today and tomorrow. American Association of Petroleum Geologists Bulletin, 80, 453-459.

Waples, D.W., Mahadir Ramly and Leslie, W., 1995. Implications of vitrinite reflectance suppression for the tectonic and thermal history of the Malay Basin. AAPG International Conference & Exhibition, Kuala Lumpur, Malaysia, August 21-24, 1994. Bulletin of the Geological Society of Malaysia, 37, 269-284.

Wong, R.H.F., 1993. Sequence stratigraphy of the Middle Miocene-Pliocene southern offshore Sandakan Basin. In: Teh, G.H., ed., Proceedings of the Symposium on Tectonic Framework and Energy Resources of the Western Margin of Pacific Basin. Bulletin of the Geological Society of Malaysia, 33, 129-142.

BIODATA OF AUTHORS

(Note : (7) refers to chapter number)

ABDUL JALIL BIN ZAINUL (3)

Abdul Jalil Zainul received his B.Sc. in Chemical Engineering from the Catholic University of America, Washington DC, U.S.A. in 1988. He worked with PETRONAS as a reservoir and production planning engineer from 1989 to 1994. During this time, his involvement was mainly in reservoir management of Peninsular Malaysia fields. From 1995 to 1998, he was in PETRONAS Petroleum Resource Planning Department of Petroleum Management Unit (PMU), where his work area covered national production planning. He has published three papers on Resource/Reserves classifications and Improved Oil Recovery. In late 1998, he joined PETRONAS Carigali (Carigali) as a petroleum economist.

ABDUL JALIL BIN ALI (3)

Abdul Jalil Ali received his B.Sc. in Chemical Engineering from Tulsa University, U.S.A., and Master in Business Administration from La Trobe University, Victoria, Australia. He joined PETRONAS in March 1989 and has been involved in several assignments in project engineering, production technology and production chemistry in Carigali Baram Operations, and also in petroleum economics. Currently, he is in oil resource planning in PMU.

ABOLINS, PETER (8, 14, 15)

Peter Abolins graduated from the University of Newcastle Upon Tyne, U.K., with a B.Sc. (Hons) in Geology (1984) and a M.Sc. in Organic Geochemistry and Organic Petrology (1985). He continued post-graduate studies, in the field of biopolymer applications to EOR, before joining Core Laboratories (Malaysia) Sdn Bhd in 1989 as Geochemistry Laboratory Supervisor. Peter joined PETRONAS Research and Scientific Services Sdn. Bhd. (PRSS) in 1993 where he is presently a senior geochemist. His main areas of interest are reservoir geochemistry and biomarker applications.

AZLINA ANUAR (9, 13, 22, 23, 24)

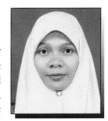

Azlina Anuar obtained her Bachelor in Applied Science (Applied Geology) from the Royal Melbourne Institute of Technology, Australia, in 1987, and her doctorate in Petroleum Geochemistry from Imperial College of Science and Technology, University of London, U.K. in 1994. She joined PRSS (formerly known as Petronas Research Institute, PRI) in 1988, where her tasks entailed palynological work and well log interpretation. Following her Ph.D., she became involved in technical consultancy and services in organic geochemistry, particularly in source rock evaluation and hydrocarbon characterisation. She is currently working on a joint project with Sarawak Shell Berhad (SSB) and Carigali on the SK 307 West Baram Delta block.

CHOW KOK THO (13)

Chow Kok Tho graduated from the University of Malaya, Kuala Lumpur, in 1980 with B.Sc. (Hons) in Geology. He joined Carigali the same year and worked in various capacities as a wellsite geologist and explorationist in Malaysia. Currently, he heads the new business ventures for Nigeria/Chad, Africa Region, responsible for acreage acquisition.

HO WANG KIN (2)

Ho Wang Kin received his B.Sc. (Hons) in Applied Geology from the University of Malaya, Kuala Lumpur, in 1979. He joined PETRONAS upon graduation. During the early part of his career, he had worked in various technical positions from operations geologist to development geologist. He was responsible for overseeing the effective implementation of exploration activities by Production Sharing Contractors. The latter part of his career centred around managerial capacities in attracting oil companies to participate in Malaysia's upstream activities. He had also advised foreign governments on the development and management of their petroleum sector. He is presently attached to Energy Africa as the New Business Development Manager responsible for securing new exploration opportunities within the Gulf of Guinea, West Africa.

LEONG KHEE MENG
(21, 22, 23, 24)

K.M. Leong received his B.Sc. (Hons) in Geology and M.Sc. in Petroleum Geology from Carleton University, Ottawa, Canada and Imperial College of Science and Technology, University of London, U.K. in 1967 and 1977 respectively. He worked as a geologist in the Geological Survey of Malaysia, Sabah from 1967 to 1978 and was involved in regional mapping and mineral investigation. He joined PETRONAS in late 1978 and has held the positions of Geological Manager/Chief Geologist (in Carigali/Carigali-BP), Exploration Manager and E&P Planning Manager. Currently he is attached to the Office of Senior Vice President, E&P Business.

MANSOR BIN AHMAD (8, 10)

Mansor Ahmad graduated from the University of Malaya, Kuala Lumpur, in 1982 with B. Sc. in Geology. He joined Carigali in 1983 working as operations geologist for about 4 years. From 1986 to 1992 he worked as project geologist in Malay and Sabah basins. From 1992 to 1995 he was assigned to Carigali Overseas Office in Vietnam and worked in Vietnam Blocks 1 and 2. From 1995 to 1998 he was attached with Carigali Overseas Office as explorationist for Philippines, China, Pakistan, Bangladesh and CIS. He was transferred to PMU in early 1998, and is currently working on Malay Basin projects as staff geoscientist.

MAZLAN B. HJ. MADON
(1, 4, 5, 6, 8, 9, 10, 12, 14, 17, 18, 19, 22, 27)

Mazlan Madon has been with PRSS for 15 years and is currently senior geologist/project leader in the exploration group. He joined the company in 1984 upon receiving a B.Sc. (Hons.) in Geology from the University of Southhampton (UK). He has a M.Sc. in Geology from the University of Malaya (1992) and D.Phil. in Earth Sciences from the University of Oxford (1996). His main research interests are sedimentology, structural geology, tectonics and basin analysis. His doctoral thesis was on the tectonic evolution of the Malay and Penyu basins. Mazlan has published about 15 papers in local and international journals. His work has included facies analysis, sandstone diagenesis, clay mineralogy, basin modeling and tectonics.

MOHD IDRUS BIN ISMAIL
(16, 25)

Mohd Idrus Ismail received his B.Sc. (Earth Sciences) from Leeds University, U.K. in 1981. He joined as a geologist with the then Exploration Department of PETRONAS and since then worked in various capacities mainly in regional geology. He is currently the staff geoscientist (Regional Exploration) responsible for the Sabah Region in the Petroleum Resource Assessment Department, PMU.

MOHAMMAD JAMAAL
BIN HOESNI (8)

Mohammad Jamaal Hoesni obtained his B.Sc. (Hons) in Geology from the University of Adelaide, Australia, in 1986. Since joining PRSS (formerly Lab Services and PRI) in 1986, he has worked on post-well analytical services and research projects involving organic petrology and source rock characterization. He is currently involved in research projects and technical consultancy services in basin modeling and petroleum system evaluation.

MOHAMMAD YAMIN BIN ALI (15)

Mohammad Yamin Ali received his B.Sc. (Hons) from the Universiti Kebangsaan Malaysia, Selangor in 1983 and M. Phil. in Geology from University of Edinburgh, U.K. in 1989. He started his career in PRSS (previously PRI) as a geologist in 1983 and has worked in various other capacities. During this period, he worked in several basins worldwide particularly on carbonates of Southeast Asia and the Middle East. He was also on job assignments with Sarawak Shell Berhad in 1990 and BP Research in Sunbury in 1992. He received several recognition and outstanding awards. He is currently the E&P Account Manager in PRSS, responsible for exploration and production businesses.

MUNIRAH KHAIRUDDIN (1)

Munirah Khairuddin obtained her B.A. (Hons) in Accounting and Finance from the University of Newcastle Upon Tyne, U.K. in 1998. She joined PETRONAS upon graduation as a Corporate Writer in Group Communications Department, Corporate Affairs. She is currently in the Joint Venture Investment and Audit Department in Carigali and aspires to become a Chartered Accountant.

OTHMAN ALI BIN MAHMUD (20)

Othman Ali Mahmud graduated from the University of Malaya, Kuala Lumpur, in 1986 with B.Sc. (Hons) in Geology. He worked as Mudlogging Geologist in Halliburton Geodata from 1988 to 1990. He joined PETRONAS in 1991 as an operations geologist and later worked in various sections as a development geologist, petroleum resource analyst and explorationist. He is currently a senior geoscientist attached to the Sarawak Section of Petroleum Resource Assessment Department in PMU. His current job involves exploration study and block promotion of the Sarawak area.

RASHIDAH BT ABD. KARIM (6)

Rashidah Karim received her B.Sc. (Hons) from the Universiti Kebangsaan Malaysia, Selangor, in 1984. She worked as a full-time micropalaeontologist and biostratigrapher in PRSS. During this time, she provided technical and consultancy services to Carigali and third parties such as Carigali-Triton, Mobil and Elf-Aquitaine in Malaysian, Middle Eastern and African basins. She is presently the Manager for Human Resources Management in PRSS.

REDZUAN BIN ABU HASSAN (16, 17, 18)

Redzuan Abu Hassan received his B.Sc. (Hons) in Geological Science from Brock University, St. Catherines, Ontario, Canada in 1986. He worked as a mudlogging geologist in Geoservices Eastern from 1988 to 1990. In May 1990, he joined PRSS (previously PRI) as a petroleum geochemist. During this time, he was involved in various petroleum geochemistry technical consultancy and integrated geological projects. His main interests are in reservoir geochemistry and regional geochemical studies.

ROSNAN BIN MISMAN (3)

Rosnan Misman graduated from the University of Arizona, Tucson, U.S.A. in 1989 with B.Sc. in Chemical Engineering. He joined PETRONAS in 1990 as a Reservoir Engineer looking after Baram Delta oil fields and Central Luconia gas fields operated by Sarawak Shell Berhad and Carigali respectively. He was then attached to Sarawak Shell Berhad for six months in 1991 for the formulation of MLNG-Dua Field Development Plan. Since then he has been in the various planning work for oil and gas development and production planning in PMU. In 1999, he joined Carigali as senior petroleum economist.

SALAHUDDIN BIN SALEH (20)

Salahuddin Saleh graduated from the Wichita State University, Wichita, Kansas, U.S.A. in 1986 with a B.Sc. in Geology. He joined PETRONAS and worked in various capacities as an operations geologist, explorationist and petroleum resource analyst. He is currently a senior geoscientist in the Sarawak Team of the Petroleum Resource Assessment Department, PMU, responsible for petroleum resource assessment, technical evaluation and promotion of open exploration blocks.

TJIA, H. D. (7, 26)

H.D. Tjia obtained his Doctorate in mathematics and physics, Institut Teknologi Bandung, Indonesia (1966). Currently he is technical adviser (exploration) in PRSS. Prior to 1993 he taught a wide range of geological subjects at Universiti Kebangsaan Malaysia, Selangor, and at other universities in Malaysia and Indonesia. He has written original textbooks on geomorphology and tectonics and has published articles on structural geology, tectonics, geomorphology, and sea-level behaviour concerning the Southeast Asian region. He has carried out extensive fieldwork in many parts of Malaysia and Indonesia. He is a founding member of Ikatan Ahli Geologi Indonesia (1961), honorary member of Persatuan Geologi Malaysia, and member of the American Geophysical Union.

WONG HIN FATT, ROBERT (6, 11)

Robert Wong graduated from University of Malaya, Kuala Lumpur, in 1981 with a B.Sc. (Hons) in Physics. He started work as a processing geophysicist in Digicon, Singapore, in May 1981. In February 1982, he joined PETRONAS as a trainee geophysicist. He was seconded for attachment training to Esso Production Malaysia Inc. (EPMI) as a seismic interpreter in 1982. Since then he has worked in various capacities as a seismic interpreter, regional explorationist and staff geoscientist. He is currently the Manager/staff geoscientist for the Peninsular Malaysia Section, Petroleum Resource Assessment Department, PMU.

ABDUL HADI ABD. RAHMAN (13)

Abdul Hadi is a lecturer in sedimentology, petroleum geology and well log analysis at the University of Malaya, Kuala Lumpur. He obtained his B.Sc (Hons) in Applied Geology from the University of Malaya, M. Sc. in Petroleum Geology from Imperial College of Science and Technology, University of London, U.K., and Ph.D. in Sedimentology and Petroleum Geology from the University of Reading, U.K. in 1989, 1990 and 1995 respectively. His doctoral thesis was on the reservoir characteristics of Upper Cycle V of Baram Field, offshore Sarawak. He has published several papers on the Baram Delta and Baram Field. His research interests also include the Tertiary basins and tectonic evolution of Peninsular Malaysia.

BAIT, BONIFACE (13)

Boniface Bait graduated from the University of Malaya, Kuala Lumpur in 1980 with a B.Sc. (Hons) in Geology. He joined Sarawak Shell Berhad (SSB) and worked in various capacities as a wellsite petroleum engineer, sedimentologist and development geologist. He also worked in Shell Expro, London as a reservoir geologist from 1987 to 1990. He is currently the Team Leader of the D35 field asset responsible for production optimisation in SSB.

TAN, DENIS N.K. (13)

Denis Tan received his B.Sc. (Hons) and M.Sc. in Geology from the University of Malaya, Kuala Lumpur, in 1972 and 1978 respectively. He worked as a geologist in the Geological Survey of Malaysia in Sarawak from 1972 to 1984. During this time, he was involved in regional mapping and published two reports. In late 1984, he joined Sarawak Shell Berhad where he has worked mainly in regional geological review, exploration geology and operations geology. He is presently responsible for Shell's exploration activities in Peninsular Malaysia.

GLOSSARY

AAPG, American Association of Petroleum Geologists.

ABF, ASEAN Bintulu Fertilizer.

API, American Petroleum Institute.

API gravity, the gravity of oil is calculated according to the formula :

$$\text{API gravity} = \frac{141.5}{\text{specific gravity at } 60°F} - 131.5$$

ASCOPE, ASEAN Council on Petroleum.

ASEAN, Association of Southeast Asian Nations.

AVO, amplitude versus (with) offset; useful for identifying gas reservoirs.

Barrel, one barrel = 159 litres, or 35 imperial gallons, or 42 U.S. gallons.

BDO, Baram Delta Operations.

BG, British Gas.

BOPD, barrels of oil per day.

BOE, barrels of oil equivalent.

BP, British Petroleum.

BSCF, billion standard cubic feet.

BSCF/D, billion standard cubic feet per day.

BSTB, billion stock tank barrel.

CAA, Commercial Arrangement Area.

Carigali, PETRONAS Carigali, the exploration and production arm of PETRONAS.

CBP, Carigali-British Petroleum

CCOP, Committee for Co-ordination of Joint Prospecting for Mineral Resources in Asian Offshore Areas.

CIS, Commonwealth of Independent States.

Condensate, light hydrocarbons that condense when cooled to surface temperature; natural gas liquids.

CONOCO, Continental Oil Company.

CTOC, Carigali-Triton Operating Company.

DHI, direct hydrocarbon indicator on a seismic section.

DST, drill stem test.

EPMI, ESSO Production Malaysia Incorporated.

EPRCO, Exxon Production Research Company.

EUR, Estimated Ultimate Recovery.

GIIP, Gas Initially in-place.

GOC, Gas : oil contact.

GOR, Gas : oil ratio.

IPC, International Petroleum Company.

JDA, Joint Development Area (Malaysia-Thailand).

JAPEX, Japan Petroleum Exploration Company.

LASMO, London and Scottish Marine Oil Company.

Lead, evidence suggesting a possible hydrocarbon trap; percursor to a prospect.

LPG, Liquified Petroleum Gases.

MBDOC, Malaysia Baram Delta Oil Company.

MLNG, Malaysia Liquified Natural Gas.

MMSCF, million standard cubic feet.

MMSCF/D, million standard cubic feet per day.

MMSTB, million stock tank barrel.

MNOC, Multinational Oil Corporation.

NGS, Net gas sand.

NOS, Net oil sand.

RM, Malaysian Ringgit.

OIIP, Oil Initially in-place.

OPEC, Organization of Petroleum Exporting Countries.

OXY, Occidental Oil Company.

Pay, production interval.

PDA, Petroleum Development Act, 1974.

PERTAMINA, Perusahaan Negara Pertambangan Minyak dan Gas Bumi Nasional, the Indonesian state oil company.

PETRONAS, Petroliam Nasional Berhad, Malaysia's National Oil Company.

PGB, PETRONAS Gas Berhad.

PGU, Peninsular Gas Utilization (Project).

PMU, Petroleum Management Unit of PETRONAS.

Play, petroleum traps of a particular genetic type..

PRI, PETRONAS Research Institute.

Prospect, a location where a well could be drilled to locate oil or gas.

PRSS, PETRONAS Research and Scientific Services Sdn. Bhd. (formerly PRI).

PSC(s), Production Sharing Contracts; Production Sharing Contractors.

SEPM, Shell Exploration and Production Malaysia.

Shell, in reference to SSB, SSPC and/or SEPM.

SMDS, Shell Middle Distillate Synthesis.

SSB, Sarawak Shell Berhad.

SSPC, Sabah Shell Petroleum Company Limited.

STATOIL, Norwegian state oil company.

SUN, Sun Oil petroleum company.

TSCF, trillion standard cubic feet.

TWT/TWTT, two-way travel time.

Wildcat, exploration well in a new area.

WMC, Western Mining Corporation of Australia.

VRo, vitrinite reflectance.

Malay Place Names

Bukit, hill.

Gunung or G., mountain or mount.

Kg. (short for 'Kampung'), village.

Kuala, estuary or river-mouth.

Laut, sea.

Teluk, bay.

Tg. (short for 'Tanjung'), cape or headland

Sungai, river.

INDEX

A

Acis Field, 356, 357
Acis Sub-basin, 94, 347, 355 356, 360, 363, 365, 408, 409
Aeromagnetic maps, Dent peninsula of, 552
Aguja Ridge, 103
Anau-Nyalau Fault Zone, 364, 395
Andaman-Sumatra-Java Arc, 71
Anding Field, 623
Angsa Graben, 244
Angsi Field, 175, 184, 186, 264
Apatite fission track analysis (AFTA), 582
Arip-Pelagau Anticline, 395, 401
Arun gas field, 90, 240
Asahan Arch-Kepulauan Arua Nose, 159, 240, 242
Asam Paya Field, 293, 325, 333, 460, 503
Australia, 492
AVO (Amplitude Versus offset) anomalies, 468-470
Axial Malay Fault Zone, 160, 166, 177, 179
Ayer Formation, 483, 551

B

Back-stripping, basin, of, 180, 281
Badong Conglomerate, 150
Bakau Deep Field, 295
Bakau Field, 294, 323
Bako structure, 449, 453
Balabac Line, 504
Balabac Strait Fault, 85, 503
Balai Field, 376, 383, 465
Balambangan Limestone, 485
Balingian Province, 82, 85, 92-4, 275, 287, 346-365, 461-462
Balingian Sub-basin, 94, 356, 365
Balingian Thrust Belt, 356
Balung Formation, 131, 485, 575, 578, 581
Bambazon Field, 601
Bampo Formation, 123, 241
Banda Arc, 492
Bangai-Sula Platform, 490
Baram Field, 52, 287, 294, 312, 322, 327, 460
Baram-A Field, 323, 327
Baram-B Field, 323, 327
Baram-Champion Delta complex, 79, 81, 85, 86, 92, 99-101, 126, 297-317, 348, 371, 396, 429
Barisan fold-thrust belt, 83, 90
Baronia Field, 294, 326, 333
Barton Field, 54, 502, 503, 513, 524, 597
Baong Formation, 123, 237, 241, 246, 247
Baronia Field, 52, 326-327, 460
Basalt, lava flows of, 148
Basement, 89, 103, 129, 158-159, 177, 238-239, 242, 263, 413, 448, 479, 547, 576
    Borneo, of, 276
    drape structure over, 187, 227
    faults in, 186
    gas derived from, 207
    petroleum potential of, 605-632
    Sabah of, 477, 501, 505
Basins,
    classification of, 82-86
    nomenclature of, 79-82
Batu Arang Basin, 90, 243
Batu Arang Beds, 243
Batu Caves limestone, 124, 620
Bau Limestone, 279, 626, 627
Bawang Member, 627
Bayan Field, 53, 345, 351, 360, 408
Bekok field, 51, 175
Bekuyat Limestone, 400
Belait Formation, 400, 409, 485, 512, 513, 528, 529
Belaga Formation, 90, 277, 278, 279, 413, 627
Belumai Formation, 123, 244
Belumut Arch, 623, 623
Bengkalis Trough, 159
Bentara Field, 376, 383, 465
Bentong-Bengkalis suture, 163
Bentong-Raub line, 157, 163
Bentong suture, 607, 609, 614
Bergading Deep Field, 47, 212, 264, 268
Beryl Field, 294
Besar Field, 175, 186
Betty Field, 52, 100, 294, 308, 327-328
Biban Sandstone Member, 400
Bijan Field, 372, 376
Bindu Field, 189
Biostratigraphy, 115-118, 645
Bokor Field, 52, 312, 323, 460
Bonanza Fault, 514, 515

Bongaya Formation, 485, 506, 517, 528, 529, 548, 560
Borneo,
   continental shelf of, 79
   western continental core of, 276
Brown Shale, 90
Bruksah Formation, 123
Brunei, 293, 298, 325
Bukit Arang Basin, 90
Bukit-Mersing Line, 401
Bujang Field, 260
Bunga Kekwa Field, 42, 177, 189, 203, 204, 207, 260
Bunga Orkid Field, 177, 189, 260
Bunga Pakma Field, 177, 260
Bunga Pelaga Field, 376, 465
Bunga Raya Field, 177, 189, 260
Bunyu Formation, 581
Buried hill structure, 444, 453

C

Cagayan Ridge, 551
Cambrian, rocks of, 61, 605
Canada Hill Thrust, 296
Carbon dioxide gas, 206, 211, 365, 423-425, 430, 431, 437-438, 470, 644
Carbon isotope studies, 380, 422, 424, 437, 563
Carbonate buildups, 82, 90, 244, 371, 373, 375, 376, 417, 434, 438, 482, 518, 526, 557, 560
Carbonate depositional model, 384
Carboniferous, rocks of, 617
Carigali, 22, 52,
   stratigraphic scheme of, 120-121
Central Balingian Sub-basin, 409
Central Graben, Melaka Basin, of, 240, 242-243, 384, 616
Central Luconia Province, 82, 92, 95, 104, 126, 275, 286, 320, 345, 371-391, 429, 432, 448, 462-463
Central Sabah Basin, 545, 551, 596
Central Sumatra Basin, 82, 83, 88, 90, 237, 239, 256, 616
Champion Field, 518
Channel,
   amplitude map of, 264
   coherency cube, map of, 265
Cherating structure, 228
Chert, 477, 505
Chert-Spilite Formation, 479, 483, 485, 489, 506, 547, 551, 553, 557, 576, 581, 583, 607, 629, 631
Chuping Limestone, 90, 123
Cimmerides, 64
Circular structures, 104-105, 488, 545, 551, 558, 560-561, 578, 602
Clay, diapirs, 100-101, 324, 325, 326, 433, 487, 448, 492, 508, 518, 519, 524, 550, 552, 559, 578, 609

Coal,
   Balingian Province in, 351-355, 362
   Batu Arang, at, 90, 243
   Bukit Arang, at, 90
   Merit -Pila, at, 277
   Sabah Basin in, 517, 533, 537, 578, 582, 583, 584
   Sarawak, in, 627
   source potential of, 199, 333, 351-355, 382, 407, 452
   Tinjar Province, in, 405
Coherency cube, image of, 265
Collins Field, 597
Compressional anticlines, Malay basin, in, 177, 181, 186
Con Sol Swell, 177
Condensate,
   carbonates, in, 386
   Malaysian, reserves of, 56
   Northeast Sabah Basin, in, 564
   Sabah, reserves in, 597
   Tatau Province, of, 421
Constitution, Malaysia of, 8
Core terrane, 141, 605
Correlation, oils, of, 203, 334, 535
Crocker Fold-Thrust Belt, 63, 67, 81, 85, 99, 276, 278, 280, 345, 374, 399, 415, 480, 489, 501, 503, 504, 505, 508, 509, 600, 625, 627, 628, 629
Crocker Formation, 68, 97, 102, 129, 475, 479, 505, 506, 517, 518, 528, 547, 582, 624
   East, Formation, 480
   North, Formation, 481
   South, Formation, 281
   West, Formation, 480
Cretaceous, Peninsular Malaysian rocks of, 612-613
Crude oil reserves, 39-47
Crust, thinning of, 178
Crystalline Basement, 479B, 483, 489, 547, 576, 582, 629, 631
Cycles,
   carbonate buildups, in, 377-379
   correlation of, 115-133, 126
   North Luconia Province, in, 447-450
   Sarawak shelf, of, 126-131, 399-405
   sedimentary, 100

D

Datai Beds, 606
dLogR, source rock evaluation technique, 196
D 18 field, 94
D 35 field, 53
Danau Formation, 277
Dangerous Ground, 67, 68, 101-102, 281, 475, 479, 504, 505
Dent Graben, 553, 557
Dent Group, 129, 483, 485, 545, 548, 550, 551, 555, 558, 560, 561

Diagenesis, 359, 378-380, 58583
Diapirs, mud/shale, 100-101, 324, 325, 326, 433, 448, 487, 492, 508, 518, 519, 524, 550, 552, 559, 578, 609
Direct Hydrocarbon Indicators (DHIs), 247, 620
Dolomite, 237, 378-380, 385, 617
Dulang Field, 175, 202, 207, 260
Dulit Range, 395, 401
Dungan Fault, 177
Dungan Graben, 177
Duyong Field, 51, 184, 186, 205, 264

E

East Baram Delta, 100, 298, 486, 492, 502, 509, 518-519, 521-524, 599
East Crocker Formation, 90, 97, 277, 477, 479, 489, 501, 547, 549
East Malaya Block, 279
Eastern Neogene Basin, Sabah, of, 483-485
Eastern Tectonic Zone, 609, 614
Electricity, 8
Embalu Group, 277
Emerald Fault Zone, 507
Epidiagenetic porosity evolution, 241
Erb South Field, 502, 524, 597
Erb West Field, 502, 503, 520, 525, 599
Eustatic sea level, changes of, 377, 381, 383
Exploration, history of, 20, 35
Exports, 11
Extrusion hypothesis, 179

F

F6 Field, 36, 47
Facies,
    seismic, 451
    North Luconia Province, of, 453
    West Baram Delta, described, 301-317
Failed rift, 483
Fairley-Baram Field, 294
Fault,
    normal, traps associated with, 187,< 229
    strike-slip, 178, 181
Flat spots, 417
Flower structure, 181, 225, 511, 513, 515, 555, 560
Foraminifera, use in stratigraphy, 115-133
Foreland,
    basin, 280
    loading of, 99
Fracture porosity, 230

G

Gagau Group, 150
Gambang Field, 364
Ganduman Formation, 105, 485, 548, 554
Garinono Formation, 483, 551, 561, 609
Gas,
    biogenic, 205
    Malay Basin, composition of, 205-208
    Malaysian reserves of, 45-56
    non-hydrocarbon, 207-208
    Tatau Province, of, 422-425
Gemas Formation, 147, 613
Geochemistry,
    Malay Basin, study of, 193-200
Geosynclinal theory, 278, 475
Geothermal gradients, 85
    Balingian Province, in, 355
    control on reservoir quality of, 193
    Northeast Sabah Basin, in, 562
    Straits of Melaka, in, 622
    West Baram Delta, in, 326
    West Luconia Province, in, 436, 438
    Tatau Province, in, 421
Glayzer Field, 502, 503, 597
Gomantong Limestone, 547, 557, 560, 561
Gondwana, 603,
    break-up of, 279, 607, 616, 617
    glaciation of, 64, 143
Grabit-Mentebai Syncline, 399
Granites, Malaysian Peninsula, of, 144
Gravity,
    anomalies,
    Malay Basin, in, 178
    Sabah Basin, of, 509
    map, Bukit Arang Basin of, 146
    map, Klang Valley and Kuala Lumpur, 152
    map, Malay Peninsula, 160
    map, Reed Bank, 281
    sliding, 100, 451
    tectonics, 519
Growth faults, 100, 318-328, 433, 434, 438, 448, 516, 518, 519, 520, 550
Gulf of Thailand, 173
Guntong Field, 174, 175, 191, 202, 203, 260, 267

H

Half Graben Province, 415, 420, 424
Heat flow, 210, 226, 421
Helang Field, 47, 287, 294
High Pressure High Temperature Play (HPHT), 460, 465
Hose Mountains, 277
Hot spot, Malay Basin, in, 178, 622

I

Igan-Oya Graben, 286, 414-417
Igan-Oya Line, 416
Inas Field, 260
Indosinian Orogeny, 64
Inversion, structural, 86, 161, 166, 181, 182, 226, 550, 616, 622
Irong Field, 175
Irong Barat Field, 175, 260, 267

J

Jelutong structure, 189
Jerneh Field, 37, 51, 174, 175, 182, 184, 205, 264, 267
Jerudong-Morris Faults, 85, 99-100, 401
Jeuku Limestone, 123, 241
Jintan Field, 47, 287, 372, 381
Johor Graben, 240, 244, 616
Johor Platform, 224
Julu Rayeu Formation, 246
Jurassic, Peninsular Malaysian rocks of, 612-613

K

K Shale, 120, 197-198
Kakus Member, 400
Kalabakan Formation, 104, 483, 575, 576, 578, 609
Kalumpang Formation, 551, 575, 576, 578
Kalutan Group, 487, 506, 537
Kamansu Field, 525
Kantan Field, 382
Kapilit Formation, 104, 489, 577, 585, 587, 600, 602
Kapit Member, 627
Kayan Sandstone Formation, 279, 627, 629, 632
Kebabangan Field, 47, 503, 525
Kedadom Formation, 279, 626, 629, 632
Kelabu Anticline, 399
Kelalan Formation, 278, 628
Keledang structure, 189
Kempas structure, 189
Kenny Hill Formation, 143
Kepong Field, 191
Kerait Schist, 276, 625
Kerum Volcaniclastic, 147, 148, 612
Kerumutan Line, 616, 617
Ketam Field, 37, 502, 503, 512, 518, 524, 597
Keutapang Formation, 124, 246
Khlong Marui Fault, 279
Kinabalu Fault, 487, 488, 513, 578, 607, 630
Kinabalu Field, 47, 53, 503, 518, 520, 522, 599
Kinarut Field, 503, 525
Kinarut-Mangalum Fault, 511, 518
Kodiang Limestone, 143
Kuala Lumpur Limestone, 143, 620
Kuamat Formation, 104, 483, 551, 575, 576, 578
Kubang Pasu Formation, 617
Kuchung Zone, 277, 282
Kudat Formation, 97, 103, 481, 485, 501, 506, 518, 528, 600, 624
Kudat Platform, 487, 502, 506, 515, 526, 596
Kukup Graben, 244
Kuantan Graben, 87, 224, 225, 228
Kulapis Formation, 129, 481, 547, 548, 549, 551

L

L Shales, 197
Labang Formation, 29, 103, 481, 547, 548, 549, 551, 576, 578, 581, 582
Labuan-Paisley Syncline, 513
Lada Hitam Field, 376, 465
Laila Field, 294, 333
Lambir Formation, 299, 324, 400, 403, 409
Langun Timur-1 well, 238
Lanis Conglomerate, 148, 612, 613
Larut Field, 260, 264, 267
Laurasia, 617
Lavas, basaltic, 149
Lawit Field, 51, 205, 264
Layang Field, 47, 294
Lebir Fault, 86, 165, 609
Lembuk Field, 333
Lepar fault zone, 150, 159, 166
Lerek Field, 207
Liang Formation, 485
Libong Tuffite, 560
Lignite,
    Batu Arang, at, 90
Lingan Fan Complex, 507, 525
Lithosphere,
    rigidity of, 281
    stretching of, 180, 226, 509
Lokan Field, 502, 513, 524, 597
Lotong Sandstone Formation, 150
Low Resistivity/Low Contrast Project, 267, 645
Lubok Antu Melange, 277, 627
Luconia Block, 63, 67, 68, 83, 85, 281, 285
Luconia Slope, 97, 451
Lupar Formation, 90, 277, 627
Lupar Line, 67, 86, 277, 279, 429, 625, 627
Lupar-Rajang delta system, 429

M

Ma' Okil Formation, 614
Macincang Formation, 142, 164
Madai-Baturong Limestone, 477, 479, 489, 547, 627
Madalon structure, 518
Mahang Formation, 142
Malawali Sub-basin, 486, 502, 516-517
Malay Basin, 66, 69, 71, 72, 83, 86, 104, 115, 119-122, 153-156, 159-162, 173-212, 2,55, 281
Malay Dome, 159, 165
Malay-Natuna-Lupar Shear Zone, 66
Maliau sub-basin, 104, 578, 580, 583, 584
Malibau sub-basin, 104, 578, 580, 583, 584
Malong Field, 188, 191
Malong-Sotong-Anding Field, 264
Mangking Sandstone Formation, 148, 612
Marine flooding surfaces, 126
Maturation, Malay Basin source rocks, of, 200
Melaka, Straits of, 88, 166
Melaka Straits Basins, 237-247, 255, 265
Melaka Platform, 123,124, 239, 242
Melaka reefs, 123
Melange, 65
Malawali Sub-basin, 102
Mantanani Limestone, 485
Meligan Delta, 99
Meligan Formation, 485, 510, 528, 530
Melinau Limestone Formation, 126, 278, 349, 628
Mentawai Fault, 71
Merchong Graben, 87, 224, 225
Mergui block, 163
Merit-Pila coal field, 277, 407
Merpati Field, 328
Mersing Line, 628
Migration, petroleum, of, 210-211, 363-364
Milankovitch cycles, 377
Minas Formation, 242
Miri Field, 36, 42, 100, 286, 293, 295-297, 318, 333, 371, 396, 399, 457
Miri Formation, 299, 528
Miri Zone, 278, 278, 279
Morris Fault-Jerudong Line, 298, 324, 325, 401, 507, 511
Mud diapirs, see clay diapirs
Mud volcanoes, 551, 552
Multimedia Super-Corridor (MSC), 14
Mulu Formation, 278
Mulu structure, 449, 453

N

Nannofossils, use in stratigraphy, 115-133
Narathiwat High, 177
Natuna Arch, 429
Natuna Platform, 429
Nenering deposits, 90

Nitrogen,
    Tatau Province, in, 423, 425
    West Luconia Province, in, 437
    Sarawak, in, 470
Noring Field, 260
North Bayan Field, 358, 359
North Palawan Block, 102, 281
Northeast Sabah Basin, 79, 102-3, 545-566, 596
Northern Sabah Province, 515, 526
NW (Northwest) Borneo Trend, 101
NW (Northwest) Sabah Margin tectonostratigraphic provinces, 82, 446, 532
NW (Northwest) Sabah Platform, 102, 509, 600
North Luconia Province, 92, 96-97, 430, 443-453
North Sumatra Basin, 83, 88, 122, 237, 239, 256
Northern Tarakan Sub-basin, 573, 578-581, 583
Nosong Field, 520, 521
Nuri Field, 325
Nyalau Formation, 126, 324, 397, 400, 402, 403, 405-406
Nymphe structure, 104

O

Oil,
    reserves, 39-45
    types, Malay Basin, 201-205
Olistostromes, 63, 103, 104, 129, 157, 159, 483, 489, 546, 547, 548, 552, 575, 576, 581, 582, 609, 628
Onshore basins, 265
Outboard Belt, 101-102, 492, 502, 509, 518, 525-526, 596, 599
Ophiolites, 65, 276, 278, 487, 490, 517, 547, 55
Ordovician, rocks of, 620
Overpressure, 189
    Central Luconia Province, in, 381, 382
    Deep play, 643
    Malay Basin, in, 209-210
    Northeast Sabah Basin, in, 562-563
    West Baram Delta, in, 328, 331, 336-337

P

Pad Basin, 553, 554, 556, 558, 561
Padas Field, 518, 521
Pahang Platform, 224
Palawan Continental Terrane, 485
Palaeocurrent analysis,
    Crocker Formation, of, 481
    Payung Formation, of, 148-149
    Tembeling Group, of, 612
Palaeogeography,
    Malay Basin, of, 183
    Sabah, of, 490, 506, 508, 528
    Sarawak Basin, of, 283-285
Palaeomagnetic data, 279

Palas Field, 175, 191, 211, 260
Palawan Block, 68
Palawan Trough, 101
Palynomorphs, use in stratigraphy, 115-133
Pandan-Kakus Sub-basin, 401, 403
Pangaea, 63
Pari Formation, 156, 223
Patok High, 415
Patricia Field, 360, 361
Patricia trend, 349, 356
Pattani Trough, 66, 173, 177
Payung Formation, 614
Pedawan Formation, 279, 627, 628, 629, 632
Pegasus Ridge, 103, 555, 557
Pekan Graben, 87
Pelagus Member, 627
Pematang Group, 124, 238, 242, 247, 620
Pematang-Balam Trough, 159
Penara Field, 203, 207
Penian High, 283, 359, 415
Peninsular Malaysia, 141-166
    petroleum resources of, 253-272
Penyu Basin, 82, 86-88, 104, 115, 162-163, 177, 221-231, 255, 263, 265, 271
    stratigraphic zonation of, 122, 156
Penyu Formation, 156, 163, 223
Permian, Peninsular Malaysian rocks of, 609, 620
Petani Formation, 124, 245
Petrochemicals, 13
Petroleum,
    Development Act, 19
    exports, 29
    resources, 35-57
    state control of, 12
    system, 199
PETRONAS, 7, 19, 22, 28, 56, 57
PETRONAS Twin Towers, 9
Philippine Mobile Belt, 492
Pilong Formation, 156, 163, 223
Pinnacle reefs/buildups, 377
Plate tectonics,
    Malaysia, of, 61-72
    Sabah, of, 491
    theory, of, 278
Plateau Sandstone Formation, 279, 627
Platform reefs/build-ups, 377
Port Kelang Graben, 240, 244, 616
Production Sharing Contracts, (PSCs), 22-27
Pressure profile, 269-270
Pulai Field, 36, 51, 174, 260
Pulai Formation, 119
Pyrolysis, gas chromatography, 201

Q

R

Ramp margin play, 189, 211
Raub-Bentong Suture, 64
Rajang-Crocker Fold-Thrust Belt, 63, 67, 81, 85, 99, 276, 278, 280, 345, 374, 399, 415, 480, 489, 501, 503, 504, 505, 508, 509, 600, 625, 627, 628, 629
Rajang Group, 83, 90, 91, 97, 277, 278, 281, 282, 317, 347, 501, 503, 607
Rajang Sea, 279, 489, 505
Rajang Slope, 96, 429, 432, 446, 448, 451, 452
Rajang-Lupar Delta, 432
Rasau Field, 318
Raya Field, 255
Recovery, enhanced, 641
Red River Fault, 65, 66, 67, 69
Reed Bank Block, 63, 67, 68, 102, 281, 446, 504, 505
Reefs, Miocene, 82, 90, 238, 246, 286, 244, 371, 373, 375, 376
Resak Field, 47, 204, 264
Research, 638-646
Reservoir, quality of, 193
Resources, Malaysia, of, 35-57, 639-641
Rhu Ridge, 229
Rhu wells, 221, 227, 228, 229, 230, 263
Risk, managing, 641-642
Rollover structure, 225, 432, 434, 438
Rumbia Fault, 162, 226
Rumbia Graben, 87, 224, 225

S

Sabah,
    petroleum resources of, 593-602
    stratigraphy of, 631
    tectonic evolution of, 630
Sabah Basin, 66, 79, 85, 86, 97-99, 105, 281, 501-538, 596
    blueschist belt, 488, 513
    geological setting of, 475-492
    palaeogeography of, 506, 508
    sequence stratigraphy of, 132
    stratigraphic zonation of, 127-131
Sabah Trough, 86, 97, 101-102, 276, 298, 475, 509, 596, 600
Sabah Orogeny, 68, 103
Sabah Platform, 596
Sabah Ridges, 101, 508
Sabah Shear, 504
Sabak Graben, 244
Sadong Formation, 626
Samarang Field, 502, 520, 599
Samarang Kecil Field, 502, 503, 518, 521, 522
Sandakan Formation, 105, 130, 485, 548

Sandakan Rift, 483, 549
Sandakan Sub-basin, 79, 103, 488, 545, 552-557, 596, 631
Sapulut Formation, 90, 97, 277, 477, 479, 489, 501, 547, 549, 576, 582, 629, 631, 632
Sarong Limestone Formation, 400, 402
Sarawak
    Basin, 66, 79, 81, 83, 85, 115,
    geology of, 275-287, 371
    petroleum resources of, 457-471
    pre-Tertiary petroleum potential of, 625-629
    stratigraphy of, 125-7
Sarawak Orogeny, 85, 280
Scanning Electron Microscope (SEM) analysis, 380
Seal,
    Malay Basin, of, 199
    Tinjar Province, of, 405
    West Luconia Province, in, 434-435
Sebahat Formation, 105, 485, 548, 554, 556, 558, 560, 561, 562
Sedili Volcanic Series, 279
Sedimentary basins, Malaysia of, 642
Seeps, petroleum, 295, 396, 482, 502, 526, 546, 573, 574, 581-582, 583, 600, 628
Segama Group, 130, 133, 483, 550, 551, 552, 554, 555, 557, 558, 560, 561, 562, 600
Seismic
    facies analysis, 521
    4D, 646
    3D, 600, 601, 646
    sequence stratigraphy, 115, 448
Sejingkat Formation, 626
Seligi Field, 175, 184, 191, 205, 267
Semanggol Formation, 143
Semangkok Field, 174, 175
Semantan Formation, 147, 165, 613
Semporna Fault, 487, 578
Serabang Formation, 626
Serai Field, 372
Seria Field, 318
Serian Volcanic Series, 279, 626
Setap Shale Formation, 126, 299, 318, 323, 337, 397, 400, 402, 405-406, 481, 485, 528
Setul Limestone, 123, 124, 142, 164
Sequence boundary, use in stratigraphy, 119, 126, 182
Sequence stratigraphy, 131-133, 377, 443, 550, 645
Siagut Syncline, 502
Sibumasu, 64, 163
Sibu Zone, 277, 279, 282
Sibuti Formation, 400
Sihapas Group, 124
Sikuatu Member, 481, 526, 528, 529
Silantek Formation, 279, 627
Silurian, rocks of, 620
Simengaris Formation, 131, 485, 578, 601
Singa Formation, 143, 617
Singa Besar, 238, 244, 247
Sirik Fault, 415, 416

Siwa Field, 294, 325
Sinoburmalaya, 64
Sniffer surveys, 238
Song Ma Suture, 65
Sotong Field, 623
    depositional model of, 190
    reservoir quality of, 193
Source rocks,
    Balingian Province, of, 351-356
    Central Luconia Province, of, 382-391
    Fluviodeltaic, 195, 198-200
    Lacustrine, 195
    Malay Basin, of, 193-200
    Northeast Sabah Basin, of, 560-563
    North Luconia Province, of, 452
    Sabah Basin, of, 527-537, 583-587
    Tatau Province, of, 418-421
    Tinjar Province, of, 405-403
    West Baram Delta, of, 328-335
    West Luconia Province, of, 435-437
South Acis Sub-basin, 94
South Banggi Formation, 485, 486, 506, 515, 528
South China Sea Basin, 66, 373, 395, 401, 446, 448, 475-476, 485, 487, 489, 504, 505, 547, 607
Southeast Collins Field, 502, 513, 524, 597
Southeast Sabah Basin, 79, 102, 104-105, 573-587, 596, 597
South Furious Field, 502, 503, 513, 524, 597
South Raya Field, 189
Southwest Emerald Field, 502, 503, 525
Southwest Sarawak Province, 92
Southwest Sulu basin, 545
St Joseph Field, 53, 502, 503, 513, 520, 524, 597
Strait Balabac Fault, 488
Straits of Melaka,
    Basins, 237-247
    stratigraphic correlation of, 122-126
    tectonics of, 616-619
Stratigraphy, Tertiary, correlation of, 115-133
Stretching factor, 227
Strike-slip, faulting, 181, 280, 487
Strontium-isotope stratigraphy, 374
Subis Limestone Formation, 126, 347, 349, 398, 628
Subsidence,
    curves, 281-282, 509
    thermal, 182
Sulu Graben, 553
Sulu Trend, 101, 103, 509, 548
Sulu Sea Basin, 607
Sumatra-Java Trench, 71
Sunda folds, 226
Sunda Shelf, 153, 155, 275, 374, 605, 622
Sundaland, 62, 65, 69, 88, 141, 142, 177, 239, 279, 487, 622
Synthetic Aperture Radar (SAR) surveys, 347, 396, 404, 405, 414, 578, 607
Systems, petroleum, 241, 245, 374, 433, 438, 645

Systems tracts, 127

T

Tabu Field, 174, 191, 203, 211, 260
Tabul Formation, 583
Tajau Sandstone, 481
Tamiang High, 240
Tembungo Field, 502
Tamiang-Yang Besar High, 159
Tampur Formation, 123, 240
Tangap Formation, 400
Tangga Barat Field, 260
Tanjong Rambutan deposits, 90
Tanjong Formation, 103, 104, 129, 130, 133, 483, 489, 548, 549, 550, 552, 554, 555, 558, 559, 560, 561, 562, 583, 585, 586, 587, 600, 602
Tapis Field, 51, 174, 175, 184, 260, 267
Tapis Formation, 120
Tapis Sandstone, 120
Tarakan Basin, 104, 573, 576, 631
Tarakan Rift, 483
Tatau Formation, 399, 628
Tatau Province, 92-93, 275, 286, 413-425
Tatau Horst, 349, 395, 401, 416, 628
Tatau-Mersing Line, 278, 413
Telisa Formation, 242
Telok Field, 260
Tembeling Group, 612-613
Temana Field, 345, 347, 351, 357, 359, 361, 364, 407
Tembeling Group, 147, 148
Temburong Formation, 97, 481, 489, 501, 528
Tembungo Field, 36, 466, 518, 520, 523, 525, 532, 599
Tempadong Limestone, 547, 557
Tenggol Arch, 86, 162, 177, 187, 188, 221, 224, 623,
Tenggol Fault, 177, 188
Terbat Formation, 626
Termus Shale, 612, 614
Terranes, Malaysia, of, 606
Terengganu Formation, 156, 221, 223
Terengganu, Platform, 177
Terengganu Shale, 119
Tertiary, stratigraphic correlation, 115-133
Tethys, 63, 279
Thailand Basin, 69
Thief beds, 382
Three Pagodas, strike-slip fault, 66, 160, 179
Tidung sub-basin, 104, 573, 578, 583
Tiga Papan Field, 502, 503, 525
Tiga Papan Sub-basin, 514
Tiga Papan Unit, 515
Tillite, 626
Timohing Formation, 485, 506, 552
Tinggi Field, 174, 175
Tinjar Fault, 396, 401, 503
Tinjar Province, 85, 92, 94-95, 286, 345, 395-409, 425
Tiong/Kepong Field, 260
Titik Terang Field, 503
Titiwangsa granitoid complex, 609
Togopi Formation, 485, 548, 552, 554, 560, 562
Tok Bidan Graben, 177, 624
Transport system, 10
Transpression, 181, 186, 615
Transtension, 181, 615
Triassic, Peninsular Malaysian rocks of, 609-612
Trusmadi Formation, 90, 97, 277, 477, 479, 489, 501, 547, 631
Tuang Formation, 276, 625
Tubau Fault, 401
Tukau Field, 294, 323, 460
Tukau Formation, 299
Tungku Formation, 554, 560
Turbidites, 147, 277, 280, 326, 337, 375, 391, 434, 438, 4612, 466, 486, 514, 518, 525, 532, 557, 559, 600, 601, 614, 625

U

Ular Field, 202
Ulu Suai Dome, 396, 399
Umas-Umas Formation, 131, 485, 578, 581
Unconformity,
    Base Cycle III, 359
    Base-Pari, 223, 227
    Base-Pliocene, 247
    Cycle 6, 417
    Deep Regional, 68, 127, 448, 485, 492, 505, 513, 521
    Green, 69, 448, 449
    Intra-Cycle 5, 417
    Late Miocene-Pliocene, 71
    Lower Intermediate, 505, 513
    Lower Miocene, 418
    Middle Miocene, 97, 228, 447, 448, 449
    Middle-Upper Miocene, 193
    Shallow Regional, 102, 486, 492, 505, 513, 519, 520
    Top-Pari, 223, 227
    Unconformity 'C', 485, 492
    Upper Intermediate, 505, 507, 513
    Upper Miocene, 241
Usun Apau, 277

V

Vietnam, basement of, 622
Vitrinite reflectance, 210, 229, 230, 245, 247, 331, 420, 436, 584, 613, 628
Volcanics, 477, 483, 487, 489, 492, 546, 548, 550, 551, 552, 557, 560, 576, 578, 581, 582, 613, 626

W

Wallace Formation, 485, 578, 581
Wang Chao strike-slip fault, 66, 69
Wariu Formation, 485
Waxy oils, Central Luconia Province, of, 387-388
West Balingian Line, 94, 347, 356, 415
West Baram Delta, 100, 275, 283, 286, 287, 293-338, 457, 460-461
West Baram Line, 85, 99, 101, 276, 283, 286, 318, 324, 503
West Borneo Basement, 278, 284
West Crocker Formation, 92, 97, 103, 489, 501, 509
West domain, 141
West Luconia
    Delta, 96, 126, 371, 433, 446, 452
    Rim, 96, 433
    Province, 92, 96, 429-438
West Lutong Field, 52, 294, 323, 460
West Natuna Basin, 69, 71, 173, 221, 222
West Sarawak Shelf, 423
West Sulu Basin, 545
Western Hinge Fault zone, 86, 177
Wrench faulting, 161, 281, 348, 349, 396, 400, 401, 416, 492, 503, 509, 515, 518, 548, 555, 560, 583, 613, 614
Woyla Group, 65

X

Y

Yang Besar High, 240
Yinggehai Basin, 179
Yong Field, 255

Y

ZI, 327

Dulang A (DLA)